SURFACTANT SYSTEMS

Their chemistry, pharmacy and biology

D. Attwood
Department of Pharmacy
University of Manchester

A. T. Florence
Department of Pharmacy
University of Strathclyde

LONDON   NEW YORK

CHAPMAN AND HALL

First published 1983 by
Chapman and Hall Ltd
11 New Fetter Lane, London EC4P 4EE
Published in the USA by
Chapman and Hall
733 Third Avenue, New York NY 10017

© 1983 D. Attwood and A. T. Florence

Printed in Great Britain by
J. W. Arrowsmith Ltd, Bristol

ISBN 0 412 14840 4

British Library Cataloguing in Publication Data

Attwood, D.
    Surfactant systems.
    1. Surface active agents
    I. Title   II. Florence, A. T.
    668'.1   TP994

    ISBN 0-412-14840-4

Contents

TP 994
A88
1983
CHEM

Preface

It is now twelve years since *Solubilization by Surface-Active Agents* appeared. Since the publication of that monograph the subject has expanded rapidly as the unique potential of surfactants has become known to a wider circle of scientists. In a recent review Menger (*Accounts of Chemical Research*, **12** (1979) 111) estimated that since 1970 there have been over 2800 publications on micelles and micellization alone. The topic of catalysis in micellar media was in an early stage of development in 1968 but the growth in this subject has given rise to an excellent textbook by Fendler and Fendler. We have felt for some time that a revision of *Solubilization by Surface-Active Agents* was overdue. The book has been out of print for some time. Owing to pressure of other work, Professor P. H. Elworthy and Dr C. B. Macfarlane were unable to undertake the work of revision but while working together on an undergraduate textbook the present authors decided to set to work, realizing both the impossibility of producing a comprehensive textbook and the need to alter the scope of the book.

Micellar solubilization occurs over a relatively small surfactant concentration range; because of this and because the phenomenon is never observed in isolation, we have extended the text to include surface activity, emulsions and suspensions and, as our emphasis is on formulation of medicinal products, to the difficult topic of the toxicology of surface-active agents. It is clearly not possible to produce a text which reviews all of these areas in great detail; what we have attempted to do is to deal with those aspects of the subject matter which are relevant in the formulation and use of surfactants in pharmacy and in formulation of products for human and animal use. The consequences of the inclusion of surfactants into such formulations is considered in some detail. The field is widened to include discussion of related problems of the use of surfactant formulations in agriculture and horticulture. Since 1968 there has been an increased awareness of the potential and limitations of surfactant systems.

Wherever possible we have attempted to give a greater quantitative emphasis to the treatment of topics than was perhaps evident in the original textbook. The book is, however, aimed at the same audience, final year students of pharmacy especially those specializing in pharmaceutics and pharmaceutical technology, postgraduate students of pharmacy, biochemistry, biology, chemistry, and those working in industrial research and development laboratories exploring the value or problems of surfactant systems. As we are both based in Schools of

Pharmacy our emphasis is on surfactant systems containing biologically active molecules, but the problems experienced in pharmaceutical systems are not unique and we hope that our treatment of the subject matter will be of interest to those working in many other spheres of activity. Biochemists, especially those working on membrane structure and function, have long been aware of the selective solubilizing power of surfactants for membrane components; chemists now know that surfactants can modify chemical reactions; pharmaceutical scientists are aware of the formulation potential of surfactants. Here we try to put these threads together and to discuss the newer developments in surfactant technology which will allow even greater progress in the future.

We are grateful to Professor Elworthy and Dr Macfarlane for allowing us to use portions of the original text, although comparatively little of the original remains, and for their encouragement to pursue the rewriting. We would welcome receiving from workers in the field copies of their published work especially if it is felt that we have neglected an area or misinterpreted their viewpoint.

D. Attwood
Manchester

A. T. Florence
Glasgow

1 Surface activity

A surface-active agent is, as the name implies, a compound which will adsorb at an air–water or oil–water interface and at the surface of solids. The structural features of these molecules which are responsible for surface activity are examined in this chapter and some of the factors which influence the extent of adsorption are considered. A study of the adsorption process is of fundamental importance in an understanding of the properties of surface-active compounds since, as we will discuss in later chapters, it is the change in interfacial free energy and surface charge resulting from adsorption which leads to the ability of these compounds to act as emulsifying and suspending agents. A survey of the wide variety of surfactants which are used in pharmaceutical systems is included in the chapter, together with a brief indication of the many and diverse uses of this versatile group of compounds.

1.1 Amphipathic molecules

Surface-active agents are characterized by the possession of both polar and non-polar regions on the same molecule. The polar or hydrophilic region of the molecule may carry a positive or negative charge, giving rise to cationic or anionic surfactants, respectively, or may be composed of a polyoxyethylene chain, as in most of the non-ionic surfactants. The non-polar or hydrophobic portion of the molecule is most commonly a flexible chain hydrocarbon although there is a large number of compounds including many molecules of biological interest, with aromatic hydrophobic groups. The dual nature of a surfactant is typified by sodium dodecyl sulphate (NaDS):

$$\underbrace{CH_3 . CH_2 . CH_2 . CH_2 . CH_2 . CH_2 . CH_2 . CH_2 . CH_2 . CH_2 . CH_2 . CH_2}_{\text{hydrophobic}} . \underbrace{SO_4^- Na^+}_{\text{hydrophilic}}.$$

The existence in the same molecule of two moieties, one of which has affinity for solvent and the other of which is antipathetic to it, is termed amphipathy. This dual nature is responsible for the phenomenon of surface activity, and of micellization and solubilization. As a class these substances, which include soaps and detergents, can be called association colloids, a name indicating their tendency to associate in solution, forming particles of colloidal dimensions. Owing to their tendency to become adsorbed at interfaces, they are often called surface-active agents or colloidal surfactants. Typical examples of the main classes of surfactants are:

(1) *Anionic.* The anion is the surface-active species, e.g.

Potassium laurate	$CH_3(CH_2)_{10}COO^-$	K^+
Sodium dodecyl (lauryl) sulphate	$CH_3(CH_2)_{11}SO_4^-$	Na^+
Hexadecylsulphonic acid	$CH_3(CH_2)_{15}SO_3^-$	H^+
Sodium dioctylsulphosuccinate	$C_8H_{17}OOCCHSO_3^-$	Na^+

$$C_8H_{17}OOCCH_2$$

(2) *Cationic.* The cation of the compound is the surface-active species, e.g.

Hexadecyl(cetyl)trimethylammonium bromide $\qquad CH_3(CH_2)_{15}N^+(CH_3)_3\ Br^-$

Dodecylpyridinium chloride $\qquad \overset{\frown}{\underset{\smile}{\bigcirc}} N^{\pm}C_{12}H_{25} \qquad Cl^-$

Dodecylamine hydrochloride $\qquad CH_3(CH_2)_{11}\overset{+}{N}H_3 \qquad Cl^-$

(3) *Ampholytic.* This type can behave as either an anionic, non-ionic, or cationic species, depending on the pH of the solution. The zwitterionic form of *N*-dodecyl-*N,N*-dimethyl betaine is:

$$C_{12}H_{25}N^+(CH_3)_2CH_2COO^-$$

(4) *Non-ionic.* The water-soluble moiety of this type can contain hydroxyl groups or a polyoxyethylene chain, e.g. polyoxyethylene *p-tert*octylphenyl ether:

$$C_8H_{17}C_6H_4O(CH_2CH_2O)_{10}H$$

Polyoxyethylene monohexadecyl ether:

$$CH_3(CH_2)_{15}(OCH_2CH_2)_{21}OH$$

Fatty acid esters of anhydrous sorbitols which have been treated with ethylene oxide are also used:

$$H(OCH_2CH_2)_nOCH-CH-O(CH_2CH_2O)_nH$$
$$H_2C\underset{O}{\diagdown\diagup}CHCHCH_2O(CH_2CH_2O)_nOCR$$
$$O(CH_2CH_2O)_nH$$

where R is the hydrocarbon part of the fatty acid chain.

(5) *Naturally occurring compounds.* Phosphatides are surface-active agents, e.g. lecithin: dialkylglycerylphosphorylcholine:

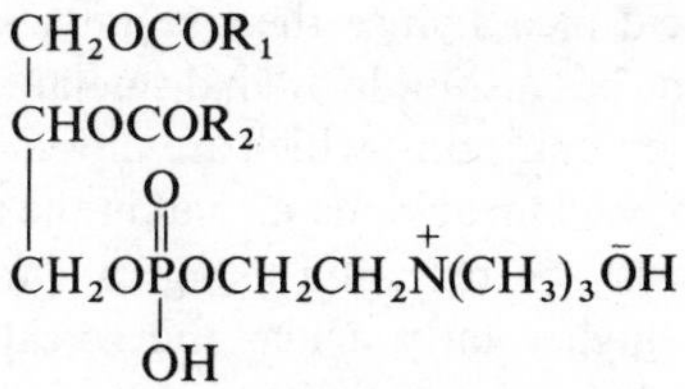

$$CH_2OCOR_1$$
$$|$$
$$CHOCOR_2$$
$$|\quad\quad O$$
$$|\quad\quad \|\quad\quad\quad +$$
$$CH_2OPOCH_2CH_2N(CH_3)_3\bar{O}H$$
$$|$$
$$OH$$

where R_1 and R_2 represent fatty acid residues, generally C_{12}–C_{18}, depending on the source. Lysolecithin, with one fatty acid residue removed, is more water soluble. The lecithins are believed to play some part in the transport of water-insoluble compounds *in vivo*. Cholic acid and desoxycholic acid are the most important of the naturally occurring bile acids, which also behave as association colloids.

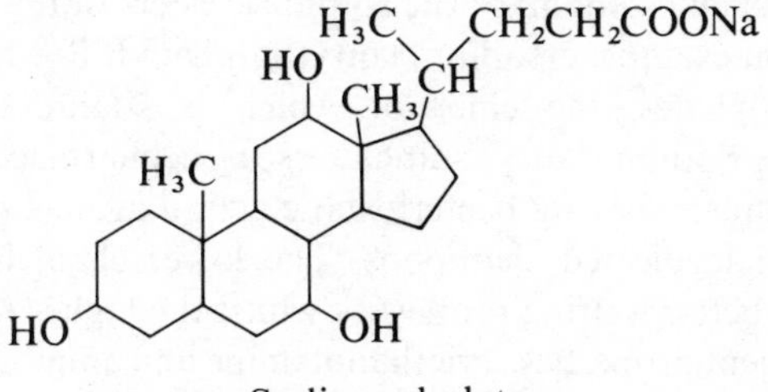

Sodium cholate

The micelles formed by these perhydrocyclopentophenanthrene derivatives have an important biological role.

(6) *Drugs.* A large number of drugs are surface active including the phenothiazine derivatives, e.g. chlorpromazine, diphenylmethane derivatives e.g. diphenhydramine, and tricyclic antidepressants e.g. amitriptyline.

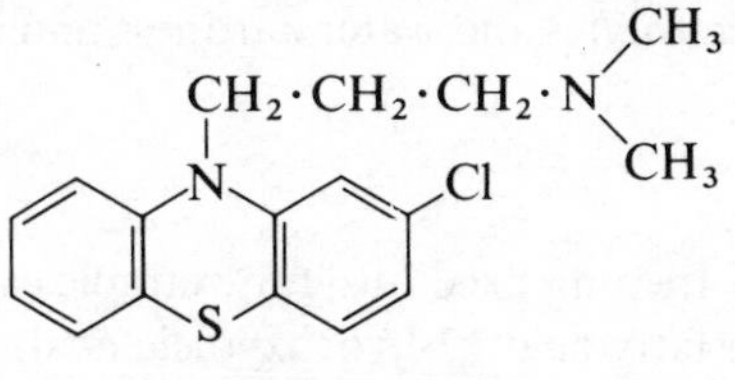

Chlorpromazine

The biological and pharmaceutical consequences of the surface activity and association behaviour of drugs is discussed in Chapter 4.

1.1.1 General properties of some surfactants of pharmaceutical interest

(A) ANIONIC SURFACE-ACTIVE AGENTS

(i) Soaps
The most commonly used soaps are the alkali-metal soaps, RCOOX where X is sodium, potassium or ammonium. The chain length, *R*, of the fatty acid is generally between C_{10} and C_{20}. Lower members of the series possess little surface activity, whilst higher members are not sufficiently soluble in water to be of use.

Soaps are unstable in acid media since the free fatty acid formed under these conditions will tend to be insoluble. Alkali-metal soaps are used in the preparation of oil-in-water emulsions, which are most stable in alkaline solution (above pH 10) and which crack in acid media and in the presence of calcium ions. Water-in-oil emulsions may be prepared using calcium, zinc, magnesium and aluminium salts of the higher fatty acids – the so-called metallic soaps. The combination of amine salts such as triethanolamine, with fatty acids gives the amine soaps. These soaps yield oil-in-water emulsions which are more stable than those prepared with alkali-metal soaps, although they still tend to crack in acid conditions.

(ii) Sulphated fatty alcohols
These are salts (usually sodium) of the sulphuric esters of the higher fatty acids. The most common example is sodium lauryl sulphate B.P. which is a mixture of sodium alkyl sulphates, the chief of which is sodium dodecyl sulphate $C_{12}H_{25}SO_4^- Na^+$. Sodium lauryl sulphate is used pharmaceutically as a pre-operative skin cleanser, having bacteriostatic action against gram-positive bacteria, and also in medicated shampoos. The lower chain length compounds around C_{12} have better wetting properties whilst the higher members, $C_{16}-C_{20}$ have better detergent properties. Triethanolamine and ammonium salts are used in hair shampoos and cosmetics. The sulphated fatty alcohols generally retain their properties over a wide pH range.

(iii) Sulphated polyoxyethylated alcohols
These compounds have the general formula $R(OCH_2CH_2)_xSO_4^- M^+$. The oxyethylene chain length x is usually less than 6. They are similar in properties to the sulphated fatty alcohols but have the advantage of better aqueous solubility, better resistance to electrolytes and water hardness and are generally less irritant to the skin and eyes.

(iv) Sulphated oils
These are prepared by treating fixed oils, for example castor oil, (which contains the triglyceride of the fatty acid 12-hydroxyoleic acid) with sulphuric acid and neutralizing with sodium hydroxide solution. Sulphated castor oil is used pharmaceutically as an emulsifying agent for oil-in-water creams and ointments. It is non-irritant and is used as a cleansing agent when soap is contra-indicated. It is also used in the manufacture of shampoos and deodorant sprays.

(B) CATIONIC SURFACE-ACTIVE AGENTS
The main types of cationic surfactants of pharmaceutical importance have hydrocarbon chains containing between 8 and 18 carbon atoms to which are attached amine, pyridinium or piperidinium groups. There is an important difference between the effect of pH on the properties of the primary, secondary and tertiary amines and on those of the quaternary amines. Whilst the quaternary amines are ionized at all pH values, the other classes of amine are fully charged

only at pH values up to within two pH units of their pK_a which is typically between pH 8 to 10. Since the uncharged forms of the majority of cationic surfactants are insoluble in water, most long chain non-quaternary amines precipitate out of solution at alkaline pH. Piperidinium derivatives may have two pK_a values and hence may exist as a single or doubly charged compound depending on the pH.

Cationic surfactants are important pharmaceutically because of their bactericidal activity against a wide range of gram-positive and some gram-negative organisms. They may be used on the skin especially in the cleansing of wounds. Aqueous solutions are used for cleaning contaminated utensils. Cationic surfactants may be used as emulsifying agents in the formation of oil-in-water creams and lotions into which may be incorporated cationic or non-ionic ingredients. They may also be used as preservatives. Some of the more important cationic surfactants of pharmaceutical importance are:

(i) Cetrimide B.P.
This is a mixture consisting mainly of tetradecyl (approximately 68 %), dodecyl (approximately 22 %) and hexadecyltrimethylammonium bromides (approximately 7 %). Solutions containing 0.1 to 1 per cent of Cetrimide are used for cleansing the skin, wounds and burns, for cleaning contaminated vessels, for storage of sterilized surgical instruments and for cleaning polythene tubing and catheters. Solutions of Cetrimide are also used in shampoos to remove scales in seborrhoea. In the form of Cetrimide emulsifying wax it is used as an emulsifying agent for producing oil-in-water creams.

(ii) Benzalkonium chloride
This is a mixture of alkylbenzyldimethylammonium chlorides of the general formula $[C_6H_5CH_2\overset{+}{N}(CH_3)_2R]Cl^-$ where R represents a mixture of the alkyls from C_8H_{17} to $C_{18}H_{37}$. In dilute solution (1 in 1000 to 1 in 2000) it may be used for the pre-operative disinfection of skin and mucous membranes, for application to burns and wounds and for cleansing polythene and nylon tubing and catheters. Benzalkonium chloride is also used as a preservative for eye-drops of the B.P.C. and U.S.N.F.

(c) NON-IONIC SURFACE-ACTIVE AGENTS
Non-ionic surfactants have the advantage over ionic surfactants in that they are compatible with all other types of surfactant and their properties are generally little affected by pH. However, aqueous solutions of some non-ionic surfactants may become turbid on warming as the cloud point is exceeded.

The amphiphilic nature of non-ionic surfactants may be expressed in terms of the balance between the hydrophobic and hydrophilic portions of the molecule. An empirical scale of HLB (hydrophile–lipophile balance) numbers was devised by Griffin which is useful in the selection of a surfactant mixture for the emulsification of a particular oil (see Chapter 8). Although applied mainly to non-ionic surfactants, the HLB system may also be used for ionic surfactants. For non-ionic surfactants, HLB values range from 0 to 20 on an arbitrary scale. The lower the HLB number, the more lipophilic is the compound and vice versa. HLB values are quoted in Table 1.1 and 1.2 for some commonly used non-ionic surfactants.

Table 1.1 HLB values of sorbitan esters

Chemical name	Commercial name	HLB
Sorbitan monolaurate	Span 20	8.6
Sorbitan monopalmitate	Span 40	6.7
Sorbitan monostearate	Span 60	4.7
Sorbitan tristearate	Span 65	2.1
Sorbitan mono-oleate	Span 80	4.3
Sorbitan tri-oleate	Span 85	1.8

Table 1.2 HLB values of polysorbates

Chemical name	Commercial name	HLB
Polyoxyethylene (20) sorbitan monolaurate	Polysorbate (Tween) 20	16.7
Polyoxyethylene (20) sorbitan monopalmitate	Polysorbate (Tween) 40	15.6
Polyoxyethylene (20) sorbitan monostearate	Polysorbate (Tween) 60	14.9
Polyoxyethylene (20) sorbitan tristearate	Polysorbate (Tween) 65	10.5
Polyoxyethylene (20) sorbitan mono-oleate	Polysorbate (Tween) 80	15.0
Polyoxyethylene (20) sorbitan tri-oleate	Polysorbate (Tween) 85	11.0

(i) · Sorbitan esters

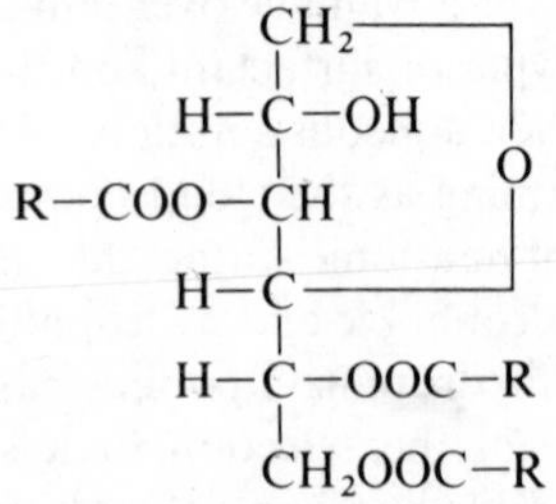

$$
\begin{array}{l}
\quad\quad\ \mathrm{CH_2} \\
\quad\quad\ \ |\quad\quad\quad\ \ \ \rceil \\
\quad\ \ \mathrm{H-C-OH} \\
\quad\quad\ \ |\quad\quad\quad\quad\ \mathrm{O} \\
\mathrm{R-COO-CH} \\
\quad\quad\ \ |\quad\quad\quad\quad\ \rfloor \\
\quad\ \ \mathrm{H-C} \\
\quad\quad\ \ | \\
\ \ \mathrm{H-C-OOC-R} \\
\quad\quad\ \ | \\
\ \ \mathrm{CH_2OOC-R}
\end{array}
$$

where R is H or an alkyl chain

The commercial products are mixtures of the partial esters of sorbitol and its mono- and di-anhydrides with oleic acid. The formula of a representative component is shown above. They are generally insoluble in water and are used as water-in-oil emulsifiers and as wetting agents. The main sorbitan esters are listed in Table 1.1.

(ii) Polysorbates

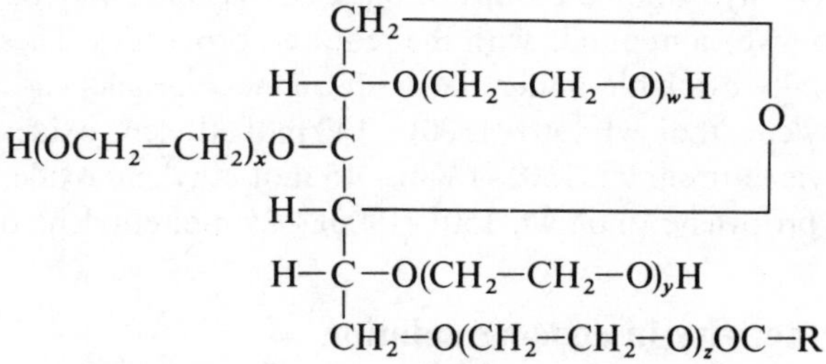

$$CH_2-$$
$$H-C-O(CH_2-CH_2-O)_wH$$
$$H(OCH_2-CH_2)_xO-C \qquad \rangle O$$
$$H-C-$$
$$H-C-O(CH_2-CH_2-O)_yH$$
$$CH_2-O(CH_2-CH_2-O)_zOC-R$$

where $n = x + w + z + 2$ and R is an alkyl chain

Commercial products are complex mixtures of partial esters of sorbitol and its mono- and di-anhydrides condensed with an approximate number of moles of ethylene oxide. The formula of a representative component is shown above. The polysorbates are miscible with water, as reflected in their higher HLB values (see Table 1.2), and are used as emulsifying agents for oil-in-water emulsions.

(iii) Polyoxyethylated glycol monoethers

We may represent the structure of these compounds by the general formula C_xE_y where x and y denote the alkyl and ethylene oxide chain lengths, respectively, e.g. $C_{16}E_7$ represents heptaoxyethylene glycol monohexadecyl ether.

Cetomacrogol 1000 B.P.C. is a water-soluble compound with an alkyl chain length of 15 or 17 and an oxyethylene chain length of between 20 and 24. It is used in the form of cetomacrogol emulsifying wax in the preparation of oil-in-water emulsions and also as a solubilizing agent for volatile oils.

Other polyoxyethylated glycol monoethers are commercially available as the Brij series (Atlas) which includes polyoxyethylene lauryl ethers (Brij 30, 35) cetyl ethers (Brij 52, 56, 58), stearyl ethers (Brij 72, 76, 78) and oleyl ethers (Brij 92, 96, 98).

(iv) Polyoxyethylated alkyl phenols

Polyoxyethylated nonylphenols are available commercially as the Igepal Co series with a wide range of oxyethylene chain lengths from 1.5 to 100. Detergents with low oxyethylene chain lengths are water-insoluble and are water-in-oil emulsifying agents, longer oxyethylene chain length compounds are water-soluble and produce oil-in-water emulsions. Polyoxyethylated *t*-octylphenols are available as the Triton-X series which includes X-114 (E_{7-8}); X-100 (E_{9-10}) and X-102 (E_{12-13}).

(v) Poloxamers

These are polyoxyethylene–polyoxypropylene derivatives with the general formula $HO(CH_2.CH_2.O)_a$ $(CH_2.CH(CH_3)O)_b$ $(CH_2.CH_2.O)_cH$ where $a = c$. The polyoxypropylene hydrophobic groups $—(CH_2CH(CH_3)O)$ and the polyoxyethylene hydrophilic groups of these compounds may be varied over a wide range to give compounds with the required properties. These compounds are commercially available under the trade name *Pluronic*, e.g. *Pluronic* F68: Polyoxypropylene (mol. wt. 1501–1800) + 140 mol ethylene oxide; *Pluronic* L62: polyoxypropylene (mol. wt. 1501–1800) + 15 mol ethylene oxide; and *Pluronic* L64: polyoxypropylene (mol. wt. 1501–1800) + 25 mol ethylene oxide.

1.2 Surface activity in aqueous solution

Two processes have an important influence on the surface activity in aqueous solution. One concerns the effect which a solute has on the structure of water and the other concerns the freedom of motion of the hydrophobic groups.

Many of the current theories of water structure are based on the premise that water is composed of both structured regions, in which the water molecules are hydrogen-bonded together in a tetrahedral arrangement similar to that of ice, and also regions of free, unbound molecules. The continuous reorientation of water molecules resulting in the destruction and reconstruction of ordered regions has led to the apt description of this particular model as the 'flickering cluster' model of water structure [1]. The solution of an amphiphilic molecule in pure water is accompanied by an initial disruption or distortion of the hydrogen bonds as the molecule is accommodated into the highly structured network. In the case of the hydrophobic region of the molecule there is usually no possibility of hydrogen bonding with water molecules to compensate for this bond disruption. Experimental evidence from proton spin relaxation times [2] and elastic neutron scattering suggests that the water molecules in the immediate vicinity of the hydrocarbon chain restructure into an even more ordered arrangement than in pure water. This phenomenon has been termed 'hydrophobic hydration'. The overall effect is of an entropy decrease making the dissolution of hydrocarbon an unfavourable process. The hydrocarbon chain of the amphiphile is brought into solution by virtue of its attachment to a hydrophilic group, either a polar head group as in ionic surfactants, or an oxyethylene chain as in most non-ionic surfactants. Both of these groups are able to form strong hydrogen bonds with the water molecules which more than compensate energetically for the initial disruption process.

The amphiphile in solution may be thought of as being surrounded by a cage of highly structured water. A consequence of this situation is that the internal torsional vibrations of the hydrocarbon chains are restricted in solution as indeed has been demonstrated from investigations of the hydrocarbon proton spin relaxation times [3]. It has been suggested by several authors that it is this process, rather than hydrophobic hydration, which is responsible for the major part of the entropy decrease on the dissolution of a hydrocarbon.

It is clear that the removal of the hydrophobic portions of the molecule from its aqueous environment is an entropically favourable process leading to the disruption of the highly organized water structure and the removal of the mobility constraints on the hydrocarbon chains. For these reasons the amphiphile will tend to accumulate at the air–water or oil–water interface in such an orientation that its hydrophobic portion is extended into the gaseous or oil phases. The molecule is anchored at the interface by the hydrophilic head group which remains in contact with the water. A consequence of the intrusion of the amphiphilic molecules into the surface or interfacial layer is that some of the water molecules are effectively replaced by hydrocarbon or other non-polar groups. Since the forces of intermolecular attraction between water molecules and non-polar groups are less than those existing between two water molecules, the contracting power of the surface is reduced and so, therefore, is the surface tension. In some cases the interfacial tension between two liquids may be reduced to such a low level (10^{-3} mN m^{-1}) that spontaneous emulsification of the two immiscible liquids is observed. These very low interfacial tensions are of relevance in an understanding of the formation and stabilization of emulsions.

1.2.1 Factors affecting surface and interfacial activity

(A) SURFACTANT STRUCTURE

The surface activity of a particular surfactant depends on the balance between its hydrophilic and hydrophobic properties. For the simplest case of a homologous series of surfactants, an increase in the length of the hydrocarbon chain as the series is ascended results in increased surface activity. This relationship between hydrocarbon chain length and surface activity is expressed by Traube's rule which states that 'in dilute aqueous solutions of surfactants belonging to any one homologous series, the molar concentrations required to produce equal lowering of the surface tension of water decrease three-fold for each additional CH$_2$ group in the hydrocarbon chain of the solute'. Traube's rule also applies to the interfacial tension at oil–water interfaces.

Although most investigations of the effect of the length of the hydrocarbon chain on surface activity have been concerned with ionic surfactants, it is also clear from the available data from studies involving polyoxyethylene non-ionic surfactants that these too conform to Traube's rule [4].

Increase in the length of the polyoxyethylene chain of non-ionic surfactants at a constant hydrophobic chain length results in a decreased surface activity. This effect is clearly shown in Fig. 1.1 for a series of non-ionic surfactants with the general formula $CH_3(CH_2)_{15}(OCH_2 \cdot CH_2)_n OH$.

Fluorinated surfactants are currently of interest since they can advantageously replace conventional hydrocarbon surfactants in a number of applications,

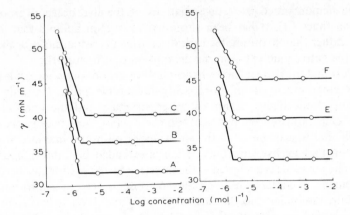

Figure 1.1 Surface tension versus log concentration plots for non-ionic surfactants of the general formula:
$CH_3(CH_2)_{15}(OCH_2CH_2)_nOH$.
  (A) $n = 6$, $C_{16}E_6$; (B) $n = 9$, $C_{16}E_9$; (C) $n = 15$, $C_{16}E_{15}$;
  (D) $n = 7$, $C_{16}E_7$; (E) $n = 12$, $C_{16}E_{12}$; (F) $n = 21$, $C_{16}E_{21}$
From Elworthy and Macfarlane, [5] with permission.

particularly in the preparation of emulsions. Thoai [6] has demonstrated that the partial substitution of some of the hydrogen atoms in the hydrophobic chain of a conventional surfactant, with fluorine atoms generally leads to a greater efficiency in lowering the oil–water interfacial tension. The effect becomes most pronounced when the ratio of fluorine to hydrogen atoms in the chain exceeds 0.50.

(B) EFFECT OF ELECTROLYTE
The addition of an inert electrolyte to solutions of an ionic surfactant results in a pronounced increase in the surface activity (see Fig. 1.2) due to a decrease in the

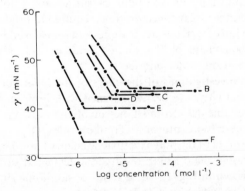

Figure 1.2 Plots of surface tension versus log molar concentration for $C_{16}(OCH_2CH_2)_7OSO_3Na$ in (A) water and (B) 0.005 M; (C) 0.01 M; (D) 0.05 M; (E) 0.10M NaCl solutions, (from [7]); (F) $C_{16}(OCH_2CH_2)_7OH$ in water (from Fig. 1.1).

magnitude of the electric field generated by the charged interface. The presence of the charge at the interface impedes adsorption of the charged surfactant and hence is responsible for the lower surface activity of an ionic surfactant compared with a non-ionic surfactant of equal hydrophobic chain length. Fig. 1.2 shows a very much higher surface activity for $C_{16}E_7$ than for its ionic analogue, $C_{16}E_7SO_4^- Na^+$.

The effect of electrolyte on the surface activity of aqueous solutions of non-ionic surfactants is much less pronounced. Schick [8] has reported a small increase in the saturation adsorption following the addition of sodium chloride to solutions of polyoxyethylene nonylphenols. The slope of the γ versus log concentration plot is also slightly increased in the presence of salt [9] indicating closer packing in the adsorbed monolayers.

(c) EFFECT OF CO-SURFACTANT

In several important applications, ionic surfactants are used in conjunction with a co-surfactant such as a medium chain-length alcohol. Of particular pharmaceutical interest are the water–surfactant–co-surfactant–electrolyte systems used to produce microemulsions. An expression for the interfacial tension of an oil–water interface in the presence of such a surfactant system has been derived by Ruckenstein and Krishnan [10] and used to provide indications about the conditions under which microemulsions form. These authors have demonstrated that under certain conditions the interfacial tension becomes negative producing an unstable interface. As a result, the area of the interface increases until the interfacial tension reaches a small but positive value and a microemulsion forms. The role of the co-surfactant in producing the ultra-low interfacial tension may be visualized in the following way. The amount by which an ionic surfactant is capable of lowering interfacial tension is generally less than the interfacial tension of a typical oil–water interface (approximately $50 \, \text{mN m}^{-1}$). The main limitation on the surface activity is the charge at the interface generated by the surfactant adsorption which acts as a barrier to the further adsorption of the ionic surfactant. However, since the co-surfactant is uncharged its adsorption is not impeded by the electric field and will therefore provide the additional lowering of interfacial tension necessary for microemulsion formation.

Surfactant systems capable of producing ultra-low interfacial tensions are currently attracting attention because of their potential value in increasing the recovery of oil from underground reservoirs [11–15]. The systems used contain co-surfactant, salt, and complex ill-defined mixtures of surfactants known as petroleum sulphonates. There is evidence to suggest that the interfaces present during the recovery processes are not simple liquid–liquid phase boundaries with surfactant monolayers [11, 12, 16, 17] and it has been suggested that the ultra-low tensions are caused by the presence of a surfactant-rich third phase at the interface.

1.3 Adsorption at liquid surfaces

1.3.1 The Gibbs' equation

The Gibbs' equation expresses the equilibrium between the surfactant molecules at the surface or interface and those in the bulk solution. It is a particularly useful equation since it provides a means by which the amount of surfactant adsorbed per unit area of the surface, the 'surface excess', may be calculated. The direct measurement of the surface excess provides almost insuperable experimental problems and hence the Gibbs' equation is widely used as an alternative method of determining this quantity. In the derivation of this equation a definite boundary between the bulk of the solution and the interfacial layer is imagined (see Fig. 1.3). The real system containing the interfacial layer is then compared to this reference system in which it is assumed that the properties of the two bulk phases remain unchanged up to the dividing surface.

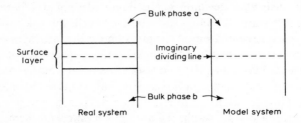

Figure 1.3 Diagrammatic representation of an interface between two bulk phases in the presence of an adsorbed layer.

In its most general form the Gibbs' equation is written

$$d\gamma = -\sum \Gamma_i d\mu_i \qquad (1.1)$$

where $d\gamma$ is the change in surface or interfacial tension of the solvent; Γ_i is the surface excess concentration of the ith component, i.e. it is the excess per unit area of surface of the ith component present in the system over that present in a hypothetical system of the same volume in which the bulk concentrations in the two phases remain constant up to the imaginary dividing surface; $d\mu_i$ is the change in chemical potential of the ith component of the system.

For a two-component system at constant temperature, Equation 1.1 reduces to

$$d\gamma = -\Gamma_1 d\mu_1 - \Gamma_2 d\mu_2 \qquad (1.2)$$

where subscripts 1 and 2 refer to solvent and solute, respectively. A convenient choice of location of the arbitrarily chosen dividing surface is that at which the surface excess concentration of the solvent, Γ_1, is zero. Indeed this is the most realistic position since we are now considering the surface layer of adsorbed solute. Equation 1.2 then becomes

$$d\gamma = -\Gamma_2 d\mu_2 = -\Gamma_2 RT d \ln a_2. \qquad (1.3)$$

For dilute solutions we may substitute concentration, c_2, for activity, a_2, thus

$$\Gamma_2 = -\frac{1}{2.303\,RT}\left(\frac{d\gamma}{d\log c_2}\right) = -\frac{c}{RT}\left(\frac{d\gamma}{dc_2}\right). \tag{1.4}$$

Equations 1.3 and 1.4 are applicable to the adsorption of non-dissociating solutes such as the non-ionic surfactants.

Fig. 1.4 shows typical plots of surface tension against the logarithm of concentration. The surface tension shows an initial gradual decrease with increasing concentration at low concentrations and eventually becomes a linear function of the logarithm of concentration when the concentration is within approximately 20 per cent of the inflection point, which corresponds to the critical micelle concentration, CMC. It is evident from Equation 1.4 that along this linear portion of the plot the surface excess concentration is constant since the slope $d\gamma/d\log c_2$ is constant. Despite this apparent saturation of the surface at a solution concentration well below the CMC, the surface tension continues to decrease until the CMC is reached. An explanation for this apparently contradictory behaviour has been proposed by Schott [18]. The continuing decrease of surface tension during conditions of apparent saturation adsorption has been explained in the following way. While the surface excess concentration of the surfactant reaches a constant value at the onset of saturation, the total surfactant concentration in the surface layer, which consists of the surface excess concentration plus the surfactant concentration present in an equivalent volume of bulk solution, continues to increase slightly as the bulk concentration is increased throughout the saturation region. As the bulk concentration approaches the CMC, the total surface concentration of the surfactant exceeds its surface concentration by small but increasing amounts. Such slight increases in the total surface concentration produce disproportionately large decreases in the surface tension because of the high density of the packing in the saturated monolayer. The effect is similar to that noted when a condensed insoluble monolayer is compressed beyond the point of close packing. The limiting surface

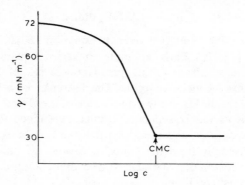

Figure 1.4 Schematic plot of surface or interfacial tension (γ) versus logarithm of the surfactant concentration (c).

area, A, per molecule of the surfactant at the surface or interface may be calculated from

$$A = 1/N_A\Gamma_2 \tag{1.5}$$

where N_A is the Avogadro constant and Γ_2 is the surface excess concentration calculated from the limiting $d\gamma/d\log c$ value using equation 1.4.

For ionic surfactants, note must be taken of the adsorption not only of the surfactant ion and its counterion but also of the ions of any added electrolyte. The simplest ionic system to treat theoretically is that of an ionic surfactant in the presence of excess of a non-surface-active electrolyte with a common counterion. Thermodynamic treatment [19, 20] shows that the Gibbs' equation as expressed by Equation 1.3 should apply. Thus for an anionic surfactant, e.g. sodium dodecyl sulphate (NaDS), in the presence of an excess of sodium chloride, the surface excess of the surface-active dodecyl sulphate anion Γ_{DS^-}, in dilute solution is given by

$$\Gamma_{DS^-} = -\frac{1}{RT}\left(\frac{d\gamma}{d\ln c}\right). \tag{1.6}$$

The validity of Equation 1.6 has been confirmed experimentally by Tajima [20] using a radiotracer method, for aqueous solutions of tritiated NaDS in the presence of excess sodium chloride. Values of Γ_{DS^-} calculated from plots of γ against $\log c$ using Equation 1.6 in the manner described above for non-ionic surfactants, were in excellent agreement with directly measured values.

There has been much dispute over the correct form of the Gibbs' equation for ionic surfactants in the absence of added electrolyte. In such systems, for example NaDS in water, Equation 1.3 is written

$$-d\gamma = RT\Gamma_{Na^+}d\ln a_{Na^+} + RT\Gamma_{DS^-}d\ln a_{DS^-}$$
$$+ RT\Gamma_{H^+}d\ln a_{H^+} + RT\Gamma_{OH^-}d\ln a_{OH^-}. \tag{1.7}$$

If it is assumed that $da_{H^+} = da_{OH^-} = 0$; $\Gamma_{H^+} = \Gamma_{OH^-} = 0$, and $\Gamma_{Na^+} = \Gamma_{DS^-} = \Gamma_2$, Equation 1.7 reduces to

$$d\gamma = -2RT\Gamma_2 d\ln a_2. \tag{1.8}$$

The form of the Gibbs' equation given by Equation 1.8 has been proposed by several workers [21, 22]. Early attempts to verify Equation 1.8 by direct experimental measurement of Γ_2 using spread monolayers [21–23] or radiotracer techniques [24] gave inconclusive results. The discrepancies between experimental and theoretical results led to a re-examination of the Gibbs' equation. Pethica [19], on the basis of the experimental results, questioned the validity of the assumption, $\Gamma_{Na^+} = \Gamma_{DS^-}$, and concluded that the surface excess could not be calculated unequivocally from the Gibbs' equation, the limiting form of the equation varying between that of Equation 1.3 in dilute solution and Equation 1.8 in more concentrated solution.

More recently, Tajima et al. [25] have reported rigorous determinations of the surface excess for aqueous solutions of tritiated NaDS in the absence of added salt, using a radioisotope technique. These workers have demonstrated the validity of Equation 1.8 under these conditions but emphasize that activity rather than concentration must be used. Equation 1.8 was rearranged in the form

$$-\frac{d\gamma}{RT\,d\ln c} = 2\Gamma_2 \left(\frac{1+d\ln f_{\pm}}{d\ln c}\right),\tag{1.9}$$

where $f_{\pm}$ is the mean ion-activity coefficient as calculated from the Debye–Hückel equation.

The form of the Gibbs' equation applicable to anionic surfactants in the presence of less than swamping amounts of added electrolyte has been considered by several workers. On the assumption of negligible adsorption of Cl^- ion, Matijevic and Pethica [26] proposed the following equation

$$d\gamma = -x\,RT\Gamma_2\,d\ln c,\tag{1.10}$$

where $x = 1 + c/(c + c_s)$, and c_s is the concentration of added salt.

The validity of Equation 1.10 was recently established by Tajima [27] who applied the Gibbs' equation from a more general viewpoint with no preliminary assumptions of the adsorbed amount of solute. The surface excess of an anionic surfactant, such as NaDS, is given as

$$\Gamma_{DS^-} = \frac{1}{1+i}\left[I_{Cl^-}(1+bi) - I_{DS^-}\,ai\right],\tag{1.11}$$

where

$$I_{Cl^-} = -\left[\frac{d\gamma}{RT\,d\ln c_{DS^-}}\right]_{NaCl}$$

$$I_{DS^-} = -\left[\frac{d\gamma}{RT\,d\ln c_{Cl^-}}\right]_{NaDS}$$

$$a = c_{DS^-}/c_{Na^+} \text{ and } b = c_{Cl^-}/c_{Na^+}.$$

The activity coefficient in dilute solution, i, is expressed as $i = 1 - A\sqrt{c_{Na^+}}$ where A is the Debye–Hückel constant (1.18 at 25° C). Similar equations are given for the surface excess concentration of Na^+ and Cl^- ions. Equation 1.11 reduces to Equation 1.10 when $\Gamma_{Cl^-} = 0$ and $i = 1$. Application of this theoretical treatment to NaDS solutions containing varying amounts of added NaCl showed that the Γ_{Cl^-} value was indeed zero (or slightly negative) (see Fig. 1.5), thus confirming the validity of the assumptions of the Matijevic and Pethica equation (Equation 1.10). Additional proof of the validity of this equation came from the excellent agreement between predicted values of Γ_{DS^-} and experimental values determined by a radiotracer technique (see Fig. 1.5).

Analogous equations for a system of a cationic surfactant in the presence of varying amounts of added electrolyte were proposed by Ozeki et al. [28]. The surface excess of the surfactant ion Γ_{D^+} was given as

$$\Gamma_{D^+} = \frac{1}{2+\beta(c+c_s)}\left\{\left(1+\frac{c_s}{c+c_s}+\beta c_s\right)\Gamma_2' - \left(\frac{c}{c+c_s}+\beta c\right)\Gamma_3'\right\}\tag{1.12}$$

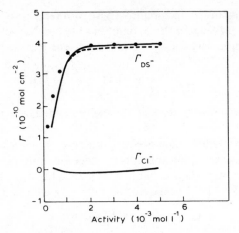

Figure 1.5 Adsorption isotherms of Γ_{DS^-} and Γ_{Cl^-} at the surface of aqueous solutions containing NaDS and 0.01 M NaCl
———: calculated from Equation 1.11
------: calculated from Equation 1.11 with $i = 1$
●: observed values
From Tajima [27] with permission.

where
$$\Gamma'_2 = -\frac{1}{RT}\left(\frac{d\gamma}{d\ln c}\right)_{c_s}$$

$$\Gamma'_3 = -\frac{1}{RT}\left(\frac{d\gamma}{d\ln c_s}\right)_c$$

$$\beta = \frac{d\ln f_\pm{}^2}{d(c + c_s)}.$$

Experimentally, Γ'_2 is determined from the limiting slopes of plots of γ against $\log c$ at a series of constant salt concentrations and Γ'_3 from the limiting slopes of plots of the variation of surface tension with the logarithm of salt concentration at a series of different surfactant concentrations.

The Gibbs' equation has been extended by Ikeda [29] to treat multicomponent aqueous solutions of weak and strong electrolytes. For example, for an aqueous solution of a fatty acid, the combined surface excess of the undissociated acid Γ_{HD} and the dissociated acid Γ_{D^-} is given as

$$\Gamma_{HD} + \Gamma_{D^-} = \left(1 - \frac{\alpha^2 c}{2\alpha c + 2c_{OH^-}}\right)\Gamma_2, \tag{1.13}$$

where α is the degree of dissociation of the weak acid. Similarly for an aqueous solution of sodium laurate,

$$\Gamma_{HD} + \Gamma_{D^-} = \Gamma_2 \left/ \left(2 + \frac{(1-\alpha)^2 c}{(1-\alpha^2)c + 2c_{H^+}}\right)\right. \tag{1.14}$$

Modifications of Equations 1.13 and 1.14 for the presence of NaOH or HCl and NaCl have been proposed.

1.3.2 Equation of state of soluble monolayers

A quantitative prediction of the surface or interfacial tension produced by a given concentration of surfactant can most conveniently be made using an equation of state of the monolayer. In such equations, surface tension lowering is expressed in terms of the surface pressure, π, defined as

$$\pi = \gamma^\circ - \gamma, \tag{1.15}$$

where γ° and γ are the surface tensions of pure solvent and solution, respectively.

For non-ionized monolayers the simplest equation of state is Szyszkowski's equation [30].

$$\pi = RT\Gamma^\infty \ln\left[\frac{c}{a} + 1\right], \tag{1.16}$$

where Γ^∞ is surface excess concentration when the surface is saturated and a is a constant. Equation 1.16 may be derived from a combination of the Gibbs' equation and the Langmuir adsorption isotherm. More complex equations consider the interactions between the adsorbed molecules [31].

The equation of state for ionized monolayers has been discussed by Hachisu [32]. This author has shown by independent derivations using three different approaches that the equation proposed by Davies [33] is applicable in the presence or absence of added electrolyte provided that the Gouy–Chapman electrical double-layer model applies. The Davies equation may be written

$$\pi = \frac{kT}{A - A_0} + \sqrt{\left[\frac{8\varepsilon N_A c (kT)^3}{10^3 \pi e^2}\right]}\left(\cosh\frac{e\psi_0}{2kT} - 1\right), \tag{1.17}$$

where A is the area per ion, A_0 is the co-area, k is the Boltzmann constant, ε is the dielectric constant of the solution, e is the elementary charge and ψ_0 is the surface potential.

In practical applications, surfactants are usually present as mixtures of several surfactants rather than as a single species. Even when dealing with single surfactants the surface properties may easily be affected by minute amounts of highly surface-active contaminants or by other members of the same homologous series present as impurities. Fatty acid soap solutions also fall into this category since they may contain varying ratios of ionized to non-ionized forms of the surfactant depending on the pH. The surface equations of state for single pure surfactants has been extended to cover systems containing two different surfactants by Lucassen-Reynders [34] enabling the surface tension to be predicted as a function of the concentration of the separate constituents. Ionic surfactants are often used in conjunction with medium chain-length alcohols (co-surfactants) as, for example, in microemulsions. Such systems also frequently contain added electrolyte. The equation of state applicable to such systems has recently been proposed by Ruckenstein and Krishnan [10]. Expressions derived by coupling the Gibbs' adsorption equation with a multicomponent adsorption isotherm showed that under certain conditions the interfacial tension would become negative producing instability of the interface. Microemulsion formation would then result as the area of the interface increased. The increase in the interfacial area is associated with accumulation of surfactant and co-surfactant on the interface and continues until the bulk concentrations of surfactant and co-surfactant become sufficiently low for the interfacial tension to reach a small but positive equilibrium value.

1.3.3 Factors affecting the extent of adsorption at the gas–liquid and liquid–liquid interface

(A) SURFACTANT STRUCTURE
The surface excess concentration under conditions of surface saturation, Γ_m, may conveniently be used as a measure of the maximum extent of adsorption of a surfactant. Several factors determine the maximum amount of surfactant which can be adsorbed at an interface.

Rosen [35] has tabulated values of Γ_m for a wide variety of anionic, cationic, non-ionic and zwitterionic surfactants and has discussed the effect of surfactant structure on Γ_m. With hydrocarbon surfactants, the length of the hydrophobic group has little effect except when this exceeds 16 carbon atoms when a significant decrease in Γ_m is noted, possibly due to coiling of the chain. Chain branching has only a small effect on Γ_m, as has introduction of fluorine atoms into the hydrophobic chain. With polyoxyethylene non-ionic surfactants of fixed oxyethylene chain length the value of Γ_m appears to be little influenced by the length of the hydrocarbon chain.

The most pronounced structural influence on Γ_m comes from the nature of the hydrophilic group. With ionic surfactants the cross-sectional area of the hydrated hydrophilic group at the interface determines the extent to which the surfactant is adsorbed. Thus, carboxylates generally have higher values of Γ_m than sulphonates or organic sulphates [36]. Similarly, a decrease in Γ_m is noted with increase in the size of the head group in the quaternary ammonium salts. The presence of two hydrophilic groups in an amphiphilic molecule can cause a considerable increase in the area per molecule at the interface. This effect is particularly pronounced where the charged groups are located at opposite ends of the molecule as in the bolaform electrolytes, since the molecule is now anchored at two places at the surface. Fig. 1.6 shows a schematic representation of the 'wicket-like' conformation proposed for some dicationic bolaform electrolytes with the general structure $R_3\overset{+}{N}-(CH_2)_{12}-\overset{+}{N}R_3$ where R is methyl or *n*-butyl. The areas per molecule of these compounds are approximately twice those of the singly charged analogues, dodecyltrimethyl- and dodecyltributylammonium bromide, respectively [37].

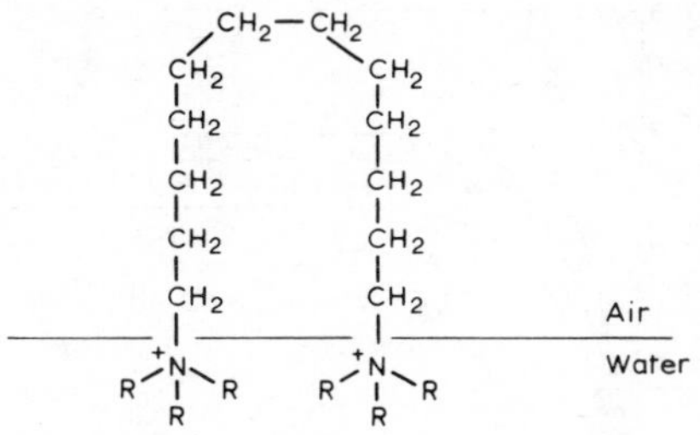

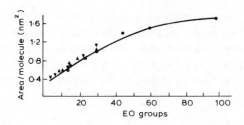

Figure 1.6 Schematic representation of the orientation of $C_{12}Me_6$ and $C_{12}Bu_6$ dibromides at an air–water interface with high surface pressure. From Menger and Wrenn [37] with permission.

The area per molecule of polyoxyethylated non-ionic surfactants at the interface is determined by the length of the hydrophilic oxyethylene chain. In general, for a given hydrophobic chain length the area per molecule increases with increase in the number of oxyethylene groups (see Fig. 1.7).

Figure 1.7 Relation between area per molecule and number of ethylene oxide (EO) groups for *n*-hexadecyl, *n*-dodecyl, and *n*-nonylphenyl ethers of polyoxyethylenes. ● hexadecyl, ▼, dodecyl ▲, nonylphenyl. Adapted from Donbrow [38] with permission.

(B) EFFECT OF ELECTROLYTE

The extent of adsorption of ionic surfactants is greatly affected by addition of electrolyte. The increased adsorption of NaDS at the air–water interface resulting from the addition of increasing amounts of sodium chloride is shown in Fig. 1.8. The electrolyte presumably exerts its effect by decreasing the repulsion between the orientated ionic head groups allowing a closer packing in the surface layer as the ionic strength is increased. For non-ionic surfactants electrolyte addition has only a slight effect on Γ_m. Hsiao *et al.* [9] have, as stated earlier, reported a slight decrease in area per molecule as the ionic strength is increased.

(C) EFFECT OF TEMPERATURE

Relatively few studies of the effect of temperature on the extent of adsorption have been reported. Generally, with ionic surfactants, temperature increase over

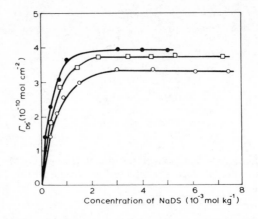

Figure 1.8 Adsorption isotherms of NaDS solutions in the presence of NaCl at 25°C.
—○—: 1.0×10^{-3} M NaCl
—□—: 5.0×10^{-3} M NaCl
—●—: 10.0×10^{-3} M NaCl
From Tajima [27] with permission.

the range 20–85° C appears to cause a slight decrease in Γ_m. A similar lack of any significant effect has been noted with non-ionic surfactants [39, 40] with the exception of a 30 oxyethylene chain length compound reported by Schott [41] in which a gradual increase in Γ_m with increasing temperature was noted. This effect was attributed to the lower hydration of the ethylene oxide groups at the higher temperatures.

1.4 Adsorption at solid surfaces

Adsorption from a dilute aqueous solution onto the walls of a container or on to particulate matter present in suspension may involve specific chemical interaction between adsorbate and adsorbent (chemisorption). The most common interactions of this type include an ion-exchange process in which the counterions of the substrate are replaced by surfactant ions of similar charge; hydrogen bond formation between adsorbate molecule and substrate and an ion-pairing interaction in which the surfactant ions are adsorbed on to oppositely charged sites unoccupied by counterions. Alternatively, the interaction may be less specific as in adsorption through weak van der Waals' forces between the adsorbent and adsorbate molecules. Frequently more than one mechanism may be involved in the adsorption process, for example, the charged groups of the adsorbate may undergo chemical interaction whilst the remainder of the molecule is adsorbed by van der Waals' attraction.

1.4.1 Adsorption isotherms

The experimental determination of the extent of adsorption usually involves shaking a known mass of adsorbent material with a solution of known concentration at a fixed temperature. The concentration of the supernatant solution is then determined by either physical or chemical means and the experiment is continued until no further change in the concentration of the supernatant is observed, that is until equilibrium conditions have been established.

(A) NON-IONIC SURFACTANTS

The Langmuir equation (Equation 1.18) has been widely used in the interpretation of adsorption data of non-ionic surfactants

$$\frac{n_2^s}{n^s} = \frac{bc_2}{1 + bc_2} \qquad (1.18)$$

where n_2^s is the number of moles of solute adsorbed per gram of adsorbent, n^s is the number of moles of adsorption sites per gram, c_2 is the concentration of solute in solution, $b = b'e^{E/RT}/n^s$, b' is a constant and E is the energy of activation for the removal of solute from the adsorbed layer. The constants of the Langmuir equation may conveniently be evaluated by rearrangement of Equation 1.18 into a linear form,

$$\frac{c_2}{n_2^s} = \frac{1}{n^s b} + \frac{c_2}{n^s}. \qquad (1.19)$$

Thus a plot of c_2/n_2^s against c_2 should give a straight line of slope $1/n^s$ and intercept $1/n^s b$.

The Langmuir equation is theoretically valid when the adsorbent is homogeneous, when there are no solute–solute or solute–solvent interactions either in solution or in surface layers and under conditions of monolayer adsorption [42].

Fig. 1.9 shows typical Langmuirian curves for the adsorption of a series of polyoxyethylated non-ionic surfactants on to graphitized carbon black

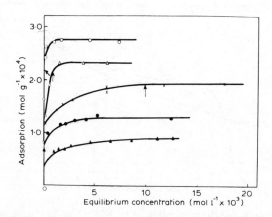

Figure 1.9 Plot of adsorption against equilibrium concentration at 25° C for C_6E_6 (●), C_8E_6 (×), $C_{10}E_6$ (△), $C_{12}E_6$ (○), and hexaoxyethylene glycol (▲). Arrows indicate critical micelle concentrations. From Corkill *et al.* [43] with permission.

(Graphon) [43]. Similar isotherms have been reported by Abe and Kuno [44, 45] for the adsorption of polyoxyethylated nonylphenols onto carbon and calcium carbonate and by Elworthy and Guthrie [46] for the adsorption of a series of polyoxyethylene non-ionics onto the water-insoluble, antifungal antibiotic, griseofulvin. Other authors have examined the adsorption of polyoxyethylene ethers on paraffin wax [47] and silver iodide particles [48, 49]. In general, the plateau region commences at a concentration at or near the CMC.

(B) IONIC SURFACTANTS

The interpretation of adsorption isotherms of ionic surfactants is more complex due to a variety of factors mainly associated with the charge on the surfactant.

Adsorption of ionic surfactants on to non-polar or hydrophobic surfaces generally, but not always, results in Langmuirian (or L-shaped) isotherms as, for example, in the adsorption of sodium dodecyl sulphate and dodecyltrimethyl-ammonium bromide onto Graphon [50]. The influence of any ionic surface groups on the adsorbent on the form of the isotherm was shown by Connor and Ottewill [51] in a study of the adsorption of hexadecyltrimethylammonium bromide on to polystyrene and polystyrene latex particles. Langmuirian iso-therms were observed for the adsorption on to the polystyrene sample without surface charge (see Fig. 1.10) whereas S-shaped isotherms were reported for adsorption of the same surfactant onto the polystyrene latex which had surface carboxyl and hydroxy groups. Similar S-shaped isotherms have been noted, for example, for the adsorption of hexadecylpyridinium bromide on to porous glass [52], sodium dodecyl sulphate, dodecylpyridinium bromide and dodecyl-ammonium chloride [53, 54] and alkylbenzenesulphonates [55] onto aluminium oxide. The S-shaped isotherms are thought to reflect three distinct modes of adsorption. At low concentrations, up to the 'knee', the surfactant adsorbs by ionic interaction with the charge groups of the adsorbent. In the region above

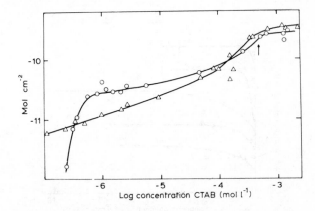

Figure 1.10 Isotherms for the adsorption of hexadecyltrimethylammonium ions, at pH 8.0 in 10^{-3} M KBr solution, on: —△—, polystyrene; —○—, latex-B; ↕, CMC value. From Connor and Ottewill [51] with permission.

the 'knee' adsorption is thought to involve interactions of the hydrophobic chains of the surfactant molecules about to be adsorbed with those already present at the surface. Evidence for the occurrence of lateral interactions was presented by Fuerstenau and co-workers [56–58] in a study of the adsorption of alkyl-ammonium ions onto quartz. It was concluded by these workers that once the adsorbed ions reach a certain critical concentration at the interface (well below the CMC) they begin to associate into two-dimensional patches of ions which Fuerstenau *et al.* termed 'hemi-micelles'. The forces responsible for this association at the surface were assumed to be the same as those operating in the bulk except that coulombic attraction for the surface adsorption sites aided the association. At higher concentrations the isotherm exhibits a plateau, the onset of which often occurs at a concentration at or near to the CMC.

Tamamushi and Tamaki [53] attempted to describe the S-shaped isotherms using a Brunauer–Emmett–Teller (B.E.T) type of equation

$$a = \frac{a_m k c}{(c_m - c)\{1 + (k-1)c/c_m\}},$$ (1.20)

where a is the amount adsorbed per gram of adsorbent, a_m is the amount adsorbed per gram of adsorbent to achieve complete monolayer coverage, c is the equilibrium concentration and c_m is the saturation concentration (CMC). The validity of this approach has been questioned [58] because of a neglect of electrical effects.

An analysis of the adsorption process which occurs at concentrations above the 'knee' of the S-shaped curves and which involves hydrophobic interaction of the hydrocarbon chains of adsorbed surfactant (in the form of hemi-micelles) and those of the adsorbing surfactant, has been given by Wilson and Kennedy [59]. These authors have estimated the binding energy χ_0 of an isolated surfactant ion in the hydrophobically adsorbed layer and the net stabilizing energy, w, associated with van der Waals' attraction and coulombic repulsion between adjacent surfactant ions in this layer. The adsorption isotherm of the surfactant in the second adsorbed layer is given by

$$\frac{c(\theta)}{c'} = \sigma(\theta) = \exp\left(\frac{-\chi_0}{kT}\right)\frac{\theta}{1-\theta}\left(\frac{2-2\theta}{\beta+1-2\theta}\right)^2$$ (1.21)

$$\beta = \{1 - 4\theta(1-\theta)[1 - \exp(-2w/zkT)]\}^{1/2},$$

where θ = fraction of surface sites occupied by surfactant ions

        z = number of nearest neighbours of a surfactant ion in the condensed surface phase,

    $c(\theta)$ = concentration of surfactant ions in the bulk solution, ion cm^{-3}

       c' = $(2\pi mkT/h^2)^{3/2}kT(j^S(T)/j^A(T))$; neglecting the dependence of c' on T and ionic strength

      m = mass of a surfactant ion

       k = Boltzmann's constant

       h = Planck's constant

$j^S(T)$ = partition function for the internal motions of a surfactant ion in solution

$j^A(T)$ = partition function for the internal motions of an adsorbed surfactant ion

$w = w(\text{van der Waals}) + w(\text{Coulomb})$

$2w/z$ = increase in energy when a new pair of surfactant nearest neighbours is formed.

Equation 1.21 has been used to predict the effect of several factors on adsorption in the second layer. Increasing ionic strength increases the shielding of the surfactant ions thereby decreasing their coulombic repulsion energy and permitting the formation of a condensed second layer. An increase in the length of the hydrocarbon chain decreases the concentration of surfactant at which a condensed second layer may form thereby rendering the surface hydrophilic. A 30° C temperature increase results in a roughly three- to fourfold increase in the surfactant concentration at which a densely occupied second layer is formed. The surface area occupied by a surfactant ion in the condensed second layer has a very marked effect on the adsorption isotherms; the smaller the surface area per ion the higher the surfactant concentration required to form a densely occupied second layer.

An interesting feature of the adsorption curves of many ionic surfactants is the occurrence of a distinct maximum in the isotherm at concentrations well in excess of the CMC. Such curves have been reported for the adsorption of sodium dodecyl sulphate and potassium myristate on graphite [60], sodium dodecyl sulphate on cotton and carbon [61, 62] and sodium dodecylbenzene sulphonate on cotton, nickel and lead [63]. The striking decrease in the adsorption of sodium tetradecyl- and dodecyl sulphate on to porous glass (Fig. 1.11) at high concentra-

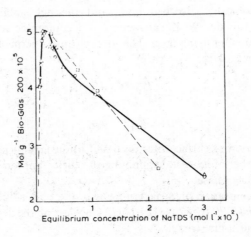

Figure 1.11 Adsorption of sodium tetradecyl sulphate, NaTDS, to Bio-Glas 200 (I) at high concentrations. $\bigcirc$ = 4 days, 30° C; $\triangle$ = 6 days, 30° C; $\square$ = 7 days, 35° C. From Mukerjee and Anavil [52] with permission.

tion is thought [52] to arise because of the exclusion of the highly charged micelles from the similarly charged surface layer. Because of repulsion between the charged double layers there is a volume of the solution adjacent to the surface from which micelles are excluded. Consequently, the concentration of the micelles in the bulk solution is anomalously high and any estimations of the extent of adsorption based on this apparent bulk concentration represents an underestimation of the true extent of adsorption. The effect, thought to be general, was more pronounced with these anionic surfactants than with cationic surfactants which had a higher surface coverage on this adsorbent. The explanation of the adsorption maxima based on the micellar exclusion effect was also thought to be adequate in explaining the gentle decrease in apparent adsorption noted with some solid systems over extended concentration ranges above the CMC but not the more rapid decreases in adsorption reported, for example, by Vold and Phansalkar [62].

(C) DRUGS

A large number of drugs are amphiphilic (see Chapter 4) and are readily adsorbed at solid surfaces. The problems arising from the adsorption of medicaments by adsorbents such as antacids which may be taken simultaneously by a patient [64] has led to extensive studies of the adsorption of drugs on to solid surfaces. The adsorption isotherms observed were generally Langmuirian when the adsorbent was uncharged. For example Langmuirian isotherms were reported for the adsorption of sulphonamides [65], barbiturates [66], phenothiazines [67] and antidepressants [68] on to carbon black. In the limited number of adsorption studies reported in which the adsorbent was charged, departure from Langmuirian behaviour was reported [69, 70].

1.4.2 Factors affecting extent of adsorption on to solid substrates

(A) SURFACTANT STRUCTURE

The effect of the hydrocarbon chain length of a homologous series of ionic surfactants on the adsorption characteristics for a given adsorbent has been the subject of several investigations. In general it has been shown that the longer the chain length the greater is the amount adsorbed at saturation and the lower is the equilibrium concentration at which adsorption attains saturation. This latter effect is due to a decrease in the CMC which, as discussed previously, is coincident in many systems with the concentration at which saturation occurs. Several authors have noted that isotherms for a homologous series are superimposable when plotted against the reduced concentration c/c_m [51–53]. Such a result is considered to indicate an adherence to Traube's rule. A benzene ring is equivalent in its effect on the adsorption characteristics of a molecule to approximately 3 to 3.5 CH_2 groups [55], which is the same contribution as that observed in micelle formation.

The extent of adsorption of non-ionic surfactants decreases with increase in the length of the polyoxyethylene chain [43, 45, 46] and, as seen in Fig. 1.9, increases with increase in the length of the alkyl chains [43].

(B) ELECTROLYTE ADDITION

Addition of neutral electrolyte causes an increase in the adsorption of an ionic surfactant to a similarly charged surface, owing to a reduction in the repulsive electrostatic interaction, and a decrease in adsorption when the surface is of opposite charge. Because of the decrease in CMC of the surfactant with salt addition, the concentration at which the adsorption attains a maximum also decreases. This displacement of the isotherms to lower concentrations is noted for example with the adsorption of dodecylammonium chloride on to aluminium oxide [71] and sodium alkyl sulphates on to carbon black [72].

The adsorption of non-ionic surfactants (polyoxyethylene glycol nonylphenols) is decreased by water-structure breaking anions while structure making anions increase it [73].

(c) pH

A change of pH of the aqueous phase can affect the adsorption process through its effect on the charge on the adsorbent surface and on the degree of ionization of the surfactant. In general, adsorption on to a surface not bearing charged groups increases as the ionization of the drug is suppressed, the extent of adsorption reaching a maximum when the drug is completely unionized. This effect is seen for example in the adsorption of local anaesthetics [74] and alkyl *p*-hydroxybenzoates on to polyethylene [75]. The extent of adsorption of benzocaine on to nylon [76] as a function of pH has been shown to be superimposable on the drug dissociation curve (see Fig. 1.12). The mechanism of adsorption in this system is thought to involve the formation of hydrogen bonds between the drug and the amide group of the nylon which are then stabilized by van der Waals' forces. For surfactants where charge is independent of pH, for example those containing quaternary ammonium groups, the extent of adsorp-

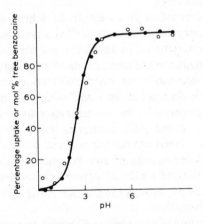

Figure 1.12 pH profile for the adsorption of benzocaine by nylon 6 powder from buffered solutions at 30° and ionic strength 0.5 M (O) and the corresponding drug dissociation curve (●). From Richardson and Meakin [76] with permission.

tion may be determined by the effect of pH on the charge on surface groups on the adsorbent. Thus the adsorption of hexadecyl- and dodecyltrimethylammonium bromide on to a polystyrene latex with surface carboxyl groups was shown to decrease with decrease in pH due to the protonation of these adsorption sites at low pH [51]. Hydrogen and hydroxyl ions are potential-determining ions for silica [57] according to the following scheme

$$-\overset{|}{\underset{|}{Si}}-OH_2^+ \overset{H^+}{\rightleftarrows} -\overset{|}{\underset{|}{Si}}-OH \overset{OH^-}{\rightleftarrows} -\overset{|}{\underset{|}{Si}}-O^- + H_2O.$$

It is not surprising, therefore, that adsorption of charged adsorbates is highly pH dependent. Adsorption of anionic surfactants is enhanced at low pH when the surface is positively charged whilst cationics are adsorbed to a greater extent at high pH when the surface is negatively charged [58, 71].

(D) TEMPERATURE

Temperature increase generally causes a slight decrease in the extent of adsorption of ionic surfactants. The effect is not pronounced and is insignificant compared with the effects of electrolyte and pH. Adsorption of polyoxyethylated non-ionic surfactants on to Graphon has been reported to increase with increase in temperature [43]. This effect has been attributed to a decrease in the hydration of the polyoxyethylene chain. The opposite effect was, however, noted for the adsorption of cetomacrogol on griseofulvin [46].

(E) NATURE OF THE ADSORBENT

Of the many properties of an adsorbent which affect its adsorptive capacity for a given solute the most significant include its state of subdivision, its porosity and the nature of its surface groups. Although an exact prediction of the behaviour of an adsorbent requires a complete characterization of that particular sample, together with a knowledge of its pretreatment and the conditions under which adsorption is to be carried out, it is possible to make generalizations about the adsorption behaviour from a consideration of the structure of the adsorbent. We will consider here only a few of the adsorbents commonly encountered in pharmaceutical systems.

The clays in common usage as pharmaceutical adsorbents include montmorillonite (the major clay component of bentonite), attapulgite and kaolin. The clays occur as either plate-like or fibrous structures [77]. Kaolin is an example of a clay with a plate-like structure, each plate is composed of one sheet consisting of octahedra of alumina and one sheet built from octahedra of silica. In montmorillonite each plate consists of one alumina sheet sandwiched between two silica sheets. This clay has a high cationic-exchange capacity which arises from a process of isomorphous substitution within the layers, for example, from the substitution of tetravalent silicon by trivalent aluminium in the silica sheet and of trivalent aluminium by divalent magnesium in the alumina sheet. Both substitution processes lead to an excess of negative surface charge and consequently a preferential adsorption of cations by this clay. The cation-exchange capacity of kaolin is considerably lower. Attapulgite also has a 2:1 ratio of silica to alumina but crystal growth is limited to one dimension resulting in very thin ribbons interspersed with voids. Attapulgite consequently has a porous nature with a moderately high external surface area. However, the dimensions of the pores are such as to accommodate only small molecules. It has only a limited cation-exchange capacity. The importance of the surface charge on the adsorptive capacity of the clays was demonstrated in a recent study [78] of the adsorption of the antibiotic, tetracycline. In general, clay structures with a high surface charge such as montmorillonite are most effective in binding tetracycline at a pH where it exists in a protonated form. Clays such as kaolin and attapulgite, in contrast, are most effective in binding the zwitterionic form of tetracycline. The presence of impurities within the clays has a significant effect on adsorptive capacity. Thus in the above example, the presence of multivalent, exchangeable cations on the clay surface diminished interaction with the protonated tetracycline, whilst non-clay impurities such as calcite and dolomite increase the interactions of the zwitterionic and anionic forms of tetracycline with the clay.

The adsorption characteristics of activated carbon appear to be associated with a system of capillaries (micropores) and channels (macropores) giving it an extremely high surface area (approx $1000 \, m^2 g^{-1}$). Detailed knowledge of the structural characteristics of activated carbon has been derived from studies of adsorption and from X-ray patterns [79–81]. Although adsorption to carbon is mainly through van der Waals' attraction, evidence has also been presented by Mattson *et al.* [82] for the presence of significant amounts of carbonyl and carboxyl groups which enhance the adsorption of cationic adsorbates.

Prediction of the adsorption behaviour of the plastics encountered in pharmacy is often difficult because of the complex nature of commercial plastics. Several authors have consequently chosen to examine polymers of known composition such as nylon 66 and polyethylene. Nylon 66 (polycaprolactam) is a polyamide with the general formula $-(NH-[CH_2]_x-NH-CO-[CH_2]_y-CO)_2$. Adsorption occurs by van der Waals' attraction and also by weak hydrogen bonding involving the carbonyl oxygen or the amide group [76, 83]. The surface of polyethylene is virtually free of polar groups and the main mechanism of adsorption is by van der Waals' interaction.

1.5 The wettability of solid surfaces

1.5.1 The contact angle

When a drop of liquid is placed on a plane, homogeneous solid surface it assumes a shape which corresponds to a minimum free energy for the system. A representation of the several forces acting on the drop is shown in Fig. 1.13a. The condition for minimum free energy at equilibrium is that given by Young's equation

$$\gamma_{S/A} = \gamma_{S/L} + \gamma_{L/A} \cos\theta \tag{1.22}$$

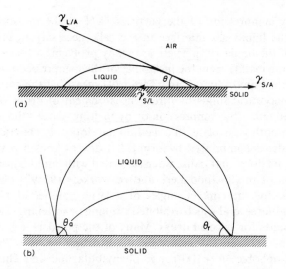

Figure 1.13(a) Equilibrium between forces acting on a drop of liquid on a solid surface. (b) Advancing and receding angles of a drop on a surface.

where $\gamma_{S/A}$, $\gamma_{S/L}$ and $\gamma_{L/A}$ are the surface or interfacial tensions at the solid/air, solid/liquid and liquid/air interfaces, respectively. The angle θ is termed the contact angle. Direct experimental verification of Young's equation is difficult although measurements by Neumann [84] of the temperature coefficients of contact angles give heats of wetting in good agreement with those calculated from the equation. Some of the theoretical objections to Young's equation have been discussed by Pethica [85]. This author draws an interesting analogy between the current doubts concerning the validity of Young's equation and those expressed about the validity of the Gibbs' equation before the refined experimental techniques became available with which it was eventually verified. He comments that 'historical example encourages the continued use of the Young's equation, provided that the thermodynamic arguments on which it is based are well secured, with the expectation that experimental evidence will be forthcoming'.

Real surfaces, however, rarely conform to the rigorous requirements for the validity of Young's equation. In practice it is found that for a given liquid/solid system there are a number of stable contact angles. The most reproducible of these are the advancing angle θ_a and the receding angle θ_r (Fig. 1.13b) so-called since they are often measured by advancing the periphery of a drop over the surface (in the case of θ_a) or pulling it back over the surface (in the measurement of θ_r). The difference between these values is referred to as hysteresis. Methods for determining advancing and receding contact angles have been reviewed by Johnson *et al.* [86].

Where the solid is available as a powder, commonly the case with pharmaceutical compounds, experimental determination of the contact angle is difficult. Early methods involved the observation of the liquid penetration into a cake made by compression of the particles [87] or the measurement of the pressure on the liquid–gas interface in the cake necessary to prevent further penetration of the liquid [88]. There are many problems associated with these methods, arising mainly from the instability of the compressed powder when in contact with the liquid. Zografi and Tam [89] measured the contact angles of several pharmaceutical solids by direct measurement of drops of liquid on a surface formed either by compression or by melting of the solid and resolidification on a smooth glass plate. In a method developed by Heertjes and Kossen [90] the powder is compressed to form a flat disc of known porosity and the height of a liquid drop on this disc when saturated with liquid is measured. Using a modification of this method Lerk and co-workers [91, 92] obtained reproducible results for the contact angles of a large number of pharmaceutical powders including several which exhibited swelling or softening. Table 1.3 shows the contact angles for selected drugs. Many of the powders, e.g. indomethacin, tolbutamide, and nitrofurantoin, are slightly hydrophobic ($\theta = 70$–$90°$) or even strongly hydrophobic ($\theta > 100°$), e.g. phenylbutazone and chloramphenicol palmitate. Formulation of these drugs as suspensions (e.g. chloramphenicol palmitate oral suspension USP) presents wetting problems. Table 1.3 shows that θ can be affected by the crystallographic structure (see chloramphenicol palmitate). An alternative method of determining contact angle was proposed by Studebaker and Snow [93] in which a comparison is made of the times taken for a

Table 1.3 Contact angles of some pharmaceutical powders*

Material	Contact angle, θ (deg)	Material	Contact angle, θ (deg)
Acetylsalicylic acid	74	Lactose	30
Aluminium stearate	120	Magnesium stearate	121
Aminophylline	47	Nitrofurantoin	69
Aminopyrine	60	Pentobarbitone	86
Aminosalicylic acid	57	Phenacetin	78
Ampicillin (anhydrous)	35	Phenobarbitone	70
Ampicillin (trihydrate)	21	Phenylbutazone	109
Amylobarbitone	102	Prednisolone	43
Barbitone	70	Prednisone	63
Boric acid	74	Quinalbarbitone	82
Caffeine	43	Salicylic acid	103
Calcium carbonate	58	Salicylamide	70
Calcium stearate	115	Stearic acid	98
Chloramphenicol	59	Succinylsulphathiazole	64
Chloramphenicol		Sulphacetamide	57
palmitate (α form)	122	Sulphadiazine	71
Chloramphenicol		Sulphamerazine	58
palmitate (β form)	108	Sulphamethazine	48
Diazepam	83	Sulphanilamide	64
Digoxin	49	Sulphathiazole	53
Indomethacin	90	Theophylline	48
Isoniazid	49	Tolbutamide	72

* Selected values from Lerk *et al.* [91, 92].

perfectly wetting liquid (cos $\theta = 1$) and an imperfectly wetting liquid (cos $\theta < 1$) to flow a given distance through a bed of the powder. A modification of this liquid penetration method was recently used by Hansford *et al.* [94] in a study of the wetting characteristics of samples of griseofulvin which were found to be dependent on its method of production, e.g. whether it was prepared by crystallization or milling techniques. Values of contact angle determined by this method differed significantly from those from the modified Heertjes and Kossen method and possible causes of these discrepancies were discussed.

1.5.2 The wetting process

In general terms 'wetting' is taken to mean the process which occurs when a solid–air interface is replaced by a solid–liquid interface. In practice we may distinguish three distinct types of wetting which are involved when a dry particle is placed into a liquid so that it is completely immersed within the liquid [95]. The wetting processes are shown diagrammatically in Fig. 1.14.

(i) In adhesional wetting the solid is brought into contact with the liquid surface which adheres to it. As a result of this process a surface area a cm^2 of the solid–air interface and the same area of the liquid–air interface are lost and replaced by a cm^2 of solid–liquid interface. The change in surface free energy per unit area, $-\Delta G/a$, is given by

$$-\Delta G/a = \gamma_{S/A} + \gamma_{L/A} - \gamma_{S/L}. \tag{1.23}$$

The work, W_a, involved per unit area of surface in this process is given by

$$W_a = \gamma_{S/L} - \gamma_{S/A} - \gamma_{L/A}. \tag{1.24}$$

The work required to revert back to the initial condition is called the work of adhesion, $-W_a$, and for the solid–liquid interface is given by the Dupre equation as

$$W_{S/L} = -W_a = \gamma_{S/A} + \gamma_{L/A} - \gamma_{S/L}. \tag{1.25}$$

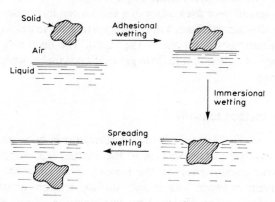

Figure 1.14 Diagrammatic representation of wetting processes.

Since the surface and interfacial tensions of solids cannot be measured directly, it is more usual to define W_a in terms of the contact angle. From the Young equation (Equation 1.22)

$$\gamma_{S/L} - \gamma_{S/A} = -\gamma_{L/A} \cos \theta \qquad (1.26)$$

and substitution into Equation 1.24 yields

$$W_a = -\gamma_{L/A} - \gamma_{L/A} \cos \theta = -\gamma_{L/A} (\cos \theta + 1). \qquad (1.27)$$

(ii) Immersional wetting is the process whereby a unit area of surface is completely immersed in the liquid. In this process a specific area of solid–air interface is converted to solid–liquid interface. The work involved, W_i, per unit area of surface is given by

$$W_i = \gamma_{S/L} - \gamma_{S/A} = -\gamma_{L/A} \cos \theta. \qquad (1.28)$$

(iii) In spreading wetting the liquid spreads over the solid surface displacing the air. As a consequence, a specific area of solid–air interface is lost and replaced by the same area of liquid–air and liquid–solid interface. The work, W_s, involved for a unit area of surface is given by

$$W_s = \gamma_{S/L} + \gamma_{L/A} - \gamma_{S/A} = -\gamma_{L/A} (\cos \theta - 1). \qquad (1.29)$$

The driving force which causes spreading is often described in terms of a spreading coefficient S

$$S = -W_s = \gamma_{S/A} - (\gamma_{S/L} + \gamma_{L/A}) = \gamma_{L/A} (\cos \theta - 1). \qquad (1.30)$$

Equations 1.23 to 1.30 are useful in predicting the conditions whereby each stage of the wetting process will proceed spontaneously. The work involved in adhesional wetting, W_a, is negative for all values of the contact angle less than 180°, i.e. adhesion wetting is always spontaneous. From Equation 1.29 the work involved in immersional wetting, W_i, is negative only when the contact angle is less than 90°. From Equation 1.29 the work involved in spreading wetting is positive for all values of the contact angle except $\theta = 0$, i.e. a liquid will only spontaneously spread over the surface of a solid when its contact angle with the solid is zero. In all other cases work must be done to achieve wetting otherwise the particle may remain at the surface of the liquid. The necessary work to cause submersion may, of course, come from gravitational forces when the particle is sufficiently dense.

1.5.3 Critical surface tension

Zisman and co-workers [96] established a relationship between the advancing contact angle and the surface tension of the wetting liquid for many surfaces with low surface energy, such as the plastics. Fig. 1.15a shows the typical linear relationship between $\cos \theta$ and surface tension obtained for these solids. The surface tension obtained by extrapolation to $\cos \theta = 1$ is known as the critical surface tension, γ_c, of the solid. The γ_c value is a characteristic

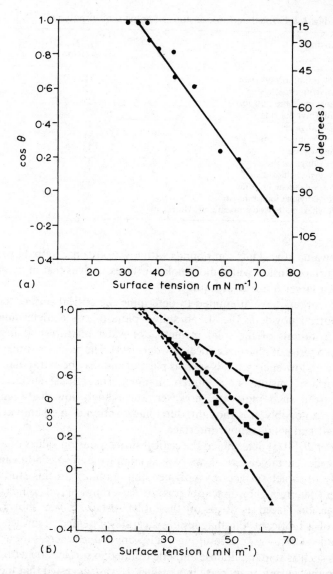

Figure 1.15 Cosine of the contact angle θ versus surface tension for (a) polyethylene and (b) the following pharmaceutical powders: ($\bullet$), acetylsalicylic acid; ($\blacktriangle$), salicylic acid; ($\blacksquare$), phenacetin; ($\blacktriangledown$), paracetamol. From (a) Fox and Zisman [97] and (b) Fell and Efentakis [98].

property of a solid and may be related to the composition of the solid. Table 1.4 gives γ_c values for several solids [99]. In general, it is found that γ_c depends on the nature of exposed groups at the surface and increases in the sequence $-CF_3 < -CF_2H < -CF_2- < -CH_3 < -CH_2-$. The condition for wetting of a solid is that the surface tension of the wetting liquids must not

Table 1.4 Critical surface tensions of some typical polymers [99]

Polymer	γ_c, at 20° C (mN m^{-1})
Polytetrafluoroethylene	18
Polytrifluoroethylene	22
Polyvinylidene fluoride	25
Polyvinyl fluoride	28
Polyethylene	31
Polytrifluorochlorethylene	31
Polystyrene	33
Polyvinyl alcohol	37
Polyvinyl chloride	39
Polyvinylidene chloride	40
Polyethylene terephthalate	43
Polyhexamethylene adipamide (nylon 6, 6)	46

exceed the value of γ_c. Thus compounds with low γ_c are only wetted by liquids of very low surface tension whilst more polar surfaces, such as that of nylon, can be wetted by a larger number of liquids.

Several workers have attempted to determine the critical surface tensions of pharmaceutical powders. Fig. 1.15b shows plots of cos θ (determined using saturated solutions of the solid in methanol–water mixtures) against surface tension for a series of pharmaceutical powders [98]. The plots are similar to those obtained with low-energy solids such as plastics and may be extrapolated to give critical surface tensions in a similar manner. The critical surface tensions obtained from such plots are, however, surprisingly low (between 18 and 24 mN m^{-1}), possibly, it is thought, due to adsorption of methanol at both the solid–liquid and solid–vapour interface.

Harder *et al.* [100] determined the critical surface tension values of a range of acetylsalicylic acid tablets as a means of ascertaining the effect of adjuvants on the wettability of tablet surfaces by polymer film coatings. In this study contact angles were determined by direct observation for a range of test liquids of known surface tension placed as drops on the tablet surface. It was shown that the addition of a lubricant like magnesium stearate presented a surface richer in —CH$_3$ and —CH$_2$— groups resulting in a lower γ_c value, whereas the addition of adjuvants such as starch, cellulose and talc resulted in surfaces richer in =O and —OH groups causing an increase in γ_c values. It was suggested that increased γ_c values would improve the wetting of the tablet by the coating solution and result in an increased bonding force between the tablet surface and the polymer film coating after the solvent has evaporated.

1.6 Modification of the surface properties of solids by adsorbed surfactants

The adsorption of surfactants at the liquid–liquid and solid–liquid interfaces is of practical importance primarily because of the resultant changes in interfacial tension and interfacial charge caused by the adsorbed surfactant film. The role of surfactants in reducing interfacial tension to the very low levels required for the formation of microemulsions has been mentioned in Section 1.2.1. Their use in the formation and stabilization of emulsions is considered in Chapter 7. Adsorption of surfactants on to the surface of insoluble powders added to an aqueous solution not only affects the wetting of the solid particles thus aiding their dispersion, but also the physical stability of the suspension.

The effectiveness of the wetting process may, as discussed in Section 1.5, be expressed in terms of the solid–liquid–vapour contact angle which must be zero for spontaneous wetting of the solid. It is apparent from Equation 1.22 that surfactants can aid wetting by reduction of the interfacial tension at both the liquid–vapour and solid–liquid interfaces. Although surfactants almost invariably reduce $\gamma_{L/A}$, their effect on $\gamma_{S/L}$ depends on the nature of the adsorption process and improved wetting is not an inevitable consequence of surfactant addition. The two factors which must be considered in the prediction of the effect of surfactants on $\gamma_{S/L}$ are the nature of the charges on solid and surfactant ions and the mechanism of adsorption. Where adsorption occurs by van der Waals' attraction, as for example in the adsorption of surfactants on to non-polar adsorbents, the surfactant molecules are orientated with the hydrophilic group towards the aqueous solution, thereby increasing the hydrophilicity of the adsorbent particles and rendering them more wettable by the aqueous solution. In contrast, where the adsorption occurs by charge interaction between the surfactant and adsorbent, for example by an ion-exchange or ion-pairing mechanism, the surfactant ions will be orientated with their hydrophobic groups towards the solution and consequently the surface becomes less readily wetted. The increase of contact angle following the addition of ionic surfactants to a solution in contact with a flat rutile surface [101] is shown in Fig. 1.16. Water

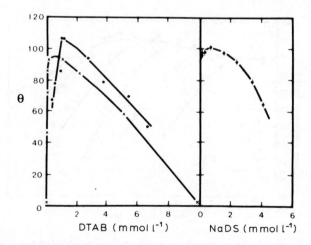

Figure 1.16 Contact angles on rutile as a function of the concentration of NaDS (●) and DTAB (●). On silica (×). From Parfitt and Wharton [101] with permission.

spreads spontaneously over a clean rutile surface but the addition of a small quantity ($< 1 \, \text{mmol} \, \text{l}^{-1}$) surfactant causes the contact angle to increase to 104° for dodecyltrimethylammonium bromide (DTAB) and 101° for NaDS. The high-energy surface of rutile is thus converted to a low-energy surface by the adsorbed layer, the critical surface tension of which may be less than that of the solution thus preventing wetting. Similar curves have been reported [102] for DTAB solutions on flat silica surfaces. At higher concentrations of surfactant above the point at which the surface charge is neutralized, the surfactant ions adsorb by hydrophobic interaction with the adsorbed layer thus exposing their hydrophilic groups to the solution and rendering the surface more readily wetted. The consequent decrease in contact angle is seen in Fig. 1.16.

The presence of an adsorbed surfactant layer on the surface of solid particles dispersed in an aqueous medium can affect the stability of the dispersion in several ways [95]. The adsorption of ionic surfactants on to non-polar surfaces imparts a surface charge to the solid surface which may increase stability through the repulsion of the electrical double layers. One of the ways in which the stability of a dispersion is increased by the adsorption of non-ionic polyoxyethylated surfactants is thought to be associated with the polyoxyethylene chains which extend into the solution. Interaction of the polyoxyethylene chains of the adsorbed layers on neighbouring particles would result in restriction in their movement and hence a decrease of the entropy – a process referred to as entropic stabilization. Stabilization of dispersions by surfactants is discussed in detail in Chapters 8 and 9.

The adsorption of charged surfactants on to adsorbents which possess ionic surface groups of opposite charge may cause flocculation of the suspension. Measurements of the changes in zeta potential of dispersed particles as the concentration of surfactant is increased [57, 103, 104] have indicated a charge

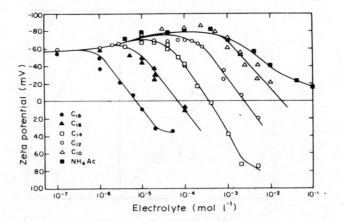

Figure 1.17 Effect of hydrocarbon chain length on the ζ-potential of quartz in solutions of alkylammonium acetates and in solutions of ammonium acetate. From Somasundaran *et al.* [57] with permission.

reversal at concentrations well below the CMC (see Fig. 1.17). Similar effects have been reported for the adsorption of DTAB and NaDS on to rutile [101]. The surfactant concentration at which the zeta potential is zero corresponds to that at which the contact angle reaches its maximum value [57, 101, 105] and the dispersion is the least stable. Adsorption of non-ionic polyoxyethylated surfactants on to charged particles may afford protection against flocculation by electrolytes. Mathai and Ottewill [49] have reported a study of the flocculation of positively and negatively charged silver iodide sols by divalent and trivalent ions in the presence of a series of polyoxyethylated non-ionics. It was shown that as the concentration of non-ionic surfactant was increased, the concentration of the inorganic salt required to produce flocculation also increased. The increase in stability of the sols could be partly accounted for by the reduction of the attraction between the particles due to the presence of the adsorbed layer. Steric hindrance between the adsorbed layers of the particles as they approached also appeared to contribute to stabilization.

References

1. G. NÉMETHY and H. S. SHERAGA (1962) *J. Chem. Phys.* **36**, 3382.
2. F. FRANKS, J. RAVENHILL, P. A. EGELSTAFF and D. I. PAGE (1970) *Proc. Roy. Soc. (London) A* **319**, 189.
3. I. DANIELSSON, B. LINDMAN and L. ODBERG (1969) *Suomen Kemistilehti* **42**, 209.
4. H. LANGE (1966) in *Nonionic Surfactants*, Ch. 14, (ed. M. J. Schick), Marcel Dekker, New York.
5. P. H. ELWORTHY and C. B. MACFARLANE (1962) *J. Pharm. Pharmac.* **14**, 100T.
6. N. THOAI (1977) *J. Colloid Interface Sci.* **62**, 222.
7. D. ATTWOOD (1969) *Kolloid-Z.*, **232**, 788.
8. M. J. SCHICK (1962) *J. Colloid Sci.* **17**, 801.
9. L. HSIAO, H. N. DUNNING and P. B. LORENZ (1956) *J. Phys. Chem.* **60**, 657.
10. E. RUCKENSTEIN and R. KRISHNAN (1980) *J. Colloid Interface Sci.* **76**, 201.
11. R. N. HEALY and R. L. REED (1974) *Soc. Petrol. Eng. J.* **14**, 491.
12. R. N. HEALY, R. L. REED and D. G. STENMARK (1976) *Soc. Petrol. Eng. J.* **16**, 147.
13. J. L. CAYIAS, R. S. SCHECHTER and W. H. WADE (1977) *J. Colloid Interface Sci.* **59**, 31.
14. R. L. CASH, J. L. CAYIAS, G. FOURNIER, D. MACALLISTER, T. SCHANES, R. S. SCHECHTER and W. H. WADE (1977) *J. Colloid Interface Sci.*, **59**, 39.
15. P. M. DUNLAP WILSON and C. F. BRADNER (1977) *J. Colloid Interface Sci.* **60**, 473.
16. C. A. MILLER, R-N. HWAN, W. J. BENTON and T. FORT (1977) *J. Colloid Interface Sci.* **61**, 554.
17. E. I. FRANSES, J. E. PULG, Y. TALMON, W. G. MILLER, L. E. SCRIVEN and H. T. DAVIS (1980), *J. Phys. Chem.* **84**, 1547.
18. H. SCHOTT (1980) *J. Pharm. Sci.* **69**, 852.
19. B. A. PETHICA (1954) *Trans. Faraday Soc.* **50**, 413.
20. K. TAJIMA (1970) *Bull. Chem. Soc. Japan* **43**, 3063.
21. A. P. BRADY (1949) *J. Colloid Sci.* **4**, 417.
22. E. G. COCKBAIN and A. I. MCMULLEN (1951) *Trans. Faraday Soc.* **47**, 322.
23. E. G. COCKBAIN (1954) *Trans. Faraday Soc.* **50**, 874.
24. D. J. SALLEY, A. and J. WEITH, A. A. ARGYLE, and J. K. DIXON (1950) *Proc. Roy. Soc. A* **203**, 42.
25. K. TAJIMA, M. MURAMATSU and T. SASAKI (1970) *Bull. Chem. Soc. Japan* **43**, 1991.
26. E. MATIJEVIC and B. A. PETHICA (1958) *Trans. Faraday Soc.* **54**, 1382, 1390, 1400.

27. K. TAJIMA (1971) *Bull. Chem. Soc. Japan* **44**, 1767.
28. S. OZEKI, M-A. TSUNODA and S. IKEDA (1978) *J. Colloid Interface Sci.* **64**, 28.
29. S. IKEDA (1977) *Bull. Chem. Soc. Japan* **50**, 1403.
30. B. SZYSZKOWSKI (1908) *Z. phys. Chem.* **64**, 385.
31. E. H. LUCASSEN-REYNDERS and M. VAN DER TEMPEL (1964) *Proc. IVth Int. Cong. Surface Active Substances*, Brussels, Vol. II, p. 779.
32. S. HACHISU (1970) *J. Colloid Interface Sci.* **33**, 445.
33. J. T. DAVIES (1951) *Proc. Roy. Soc. Ser A* **A208**, 224.
34. E. H. LUCASSEN-REYNDERS (1972) *J. Colloid Interface Sci.* **41**, 156.
35. M. J. ROSEN (1978) *Surfactants and Interfacial Phenomena*, Ch. 2, Wiley, New York.
36. F. VAN VOORST VADER (1960) *Trans. Faraday Soc.* **56**, 1067.
37. F. M. MENGER and S. WRENN (1974) *J. Phys. Chem.* **78**, 1387.
38. M. DONBROW (1975) *J. Colloid Interfac. Sci.* **53**, 145.
39. J. M. CORKILL, J. F. GOODMAN and S. P. HARROLD (1964) *Trans. Faraday Soc.* **60**, 202.
40. R. A. HUDSON and B. A. PETHICA (1964) *Proc. IVth Int. Congr. Surface Active Substances*, Brussels, Vol. II, p. 631.
41. H. SCHOTT (1969) *J. Pharm. Sci.* **58**, 1521.
42. J. J. BETTS and B. A. PETHICA (1960) *Trans. Faraday Soc.* **56**, 1515.
43. J. M. CORKILL, J. F. GOODMAN and J. R. TATE (1966) *Trans. Faraday Soc.* **62**, 979.
44. H. KUNO and R. ABE (1961) *Kolloid-Z.* **177**, 40.
45. R. ABE and H. KUNO (1962) *Kolloid-Z.* **181**, 70.
46. P. H. ELWORTHY and W. G. GUTHRIE (1970) *J. Pharm. Pharmacol.* **22**, 114S.
47. H. LANGE (1960) *J. Phys. Chem.* **64**, 538.
48. K. G. MATHAI and R. H. OTTEWILL (1962) *Kolloid-Z.* **185**, 55.
49. K. G. MATHAI and R. H. OTTEWILL (1966) *Trans. Faraday Soc.* **62**, 750, 759.
50. F. G. GREENWOOD, G. D. PARFITT, N. H. PICTON and D. G. WHARTON (1968) in *Adsorption from Aqueous Solution* (eds. W. J. Weber and E. Matijevic) American Chemical Society, Washington DC, p. 135.
51. P. CONNOR and R. H. OTTEWILL (1971) *J. Colloid Interface Sci.* **37**, 642.
52. P. MUKERJEE and A. ANAVIL (1975) in *Adsorption at Interfaces*, ACS Symposium Series No. 8. American Chemical Society, Washington, p. 107.
53. B. TAMAMUSHI and K. TAMAKI (1957) *Proc. IInd Intern. Cong. Surface Activity* London, Vol. III. Butterworths, London, p. 449.
54. T. WAKAMATSU and D. W. FUERSTENAU (1968) in *Adsorption from Aqueous Solution*, (eds. W. J. Weber and E. Matijevic) American Chem. Society, Washington DC, p. 161.
55. S. G. DICK, D. W. FUERSTENAU and T. W. HEALY (1971) *J. Colloid Interface Sci.* **37**, 595.
56. D. W. FUERSTENAU (1956) *J. Phys. Chem.* **60**, 981.
57. P. SOMASUNDARAN, T. W. HEALY and D. W. FUERSTENAU (1964) *J. Phys. Chem.* **68**, 3562.
58. P. SOMASUNDARAN and D. W. FUERSTENAU (1966) *J. Phys. Chem.* **70**, 90.
59. D. J. WILSON and R. MOFFATT KENNEDY (1979) *Sepn. Sci. Techn.* **14**, 319.
60. M. L. CORRIN, E. L. LIND, A. ROGINSKY, and W. D. HARKINS (1949) *J. Colloid Sci.* **4**, 485.
61. R. D. VOLD and N. H. SIVARAMAKRISHNAN (1958) *J. Phys. Chem.* **62**, 984.
62. R. D. VOLD and A. K. PHANSALKAR (1955) *Rec. trav. chim.* **74**, 41.
63. A. FAVA and H. EYRING (1956) *J. Phys. Chem.* **60**, 890.
64. A. T. FLORENCE and D. ATTWOOD (1981) *Physicochemical Principles of Pharmacy*, ch. 9, Macmillan, London.
65. H. NOGAMI, T. NAGAI and S. WADA (1970) *Chem. Pharm. Bull.* **18**, 342, 348.
66. H. NOGAMI, T. NAGAI and H. UCHIDA (1969) *Chem. Pharm. Bull.* **17**, 176.
67. H. NOGAMI, T. NAGAI and N. NAMBU (1970) *Chem. Pharm. Bull.* **18**, 1643.
68. H. NOGAMI, S. SAKURAI and T. NAGAI (1975) *Chem. Pharm. Bull.* **23**, 1404.
69. S. EL-MASRY and S. A. H. KHALIL (1974) *J. Pharm. Pharmacol.* **26**, 243.
70. S. A. H. KHALIL and M. IWUAGWA (1978) *J. Pharm. Sci.* **67**, 287.
71. B. TAMAMUSHI and K. TAMAKI (1959) *Trans. Faraday Soc.* **55**, 1007.

72. G. R. F. ROSE, A. S. WEATHERBURN and C. H. BAYLEY (1951) *Textile Research J.* **22**, 797.
73. H. RUPPRECHT (1978) *Progr. Colloid and Polymer Sci.* **65**, 29.
74. G. BAUER and E. ULLMANN (1973) *Arch. Pharmazie* **306**, 86.
75. K. KAKEMI, H. SEZAKI, E. ARAKAWA, K. KIMURA and K. IKEDA (1971) *Chem. Pharm. Bull.* **19**, 2523.
76. N. E. RICHARDSON and B. J. MEAKIN (1974) *J. Pharm. Pharmacol.* **26**, 166.
77. H. VAN OLPHEN (1976) *Progr. Colloid and Polymer Sci.* **61**, 46.
78. J. E. BROWNE, J. R. FELDKAMP, J. L. WHITE and S. L. HEM (1980) *J. Pharm. Sci.* **69**, 816.
79. J. C. ARNELL and W. M. BARSS. (1948) *Can. J. Research* **26A**, 236.
80. P. H. EMMETT (1948) *Chem. Rev.* **43**, 69.
81. W. F. WOLFF (1958) *J. Phys. Chem.* **62**, 829.
82. J. S. MATTSON and H. B. MARK (1969) *J. Colloid Interface Sci.* **31**, 131.
83. A. J. KAPADIA. W. L. GUESS and J. AUTIAN (1964) *J. Pharm. Sci.* **53**, 28.
84. A. W. NEUMANN (1974) *Adv. Colloid Interface Sci.* **4**, 105.
85. B. A. PETHICA (1977) *J. Colloid Interface Sci.* **62**, 568.
86. R. E. JOHNSON and R. H. DETTRE (1969) *Surface and Colloid Science*, (ed. E. Matijevic) Vol. 2, Wiley-Interscience, p. 85.
87. E. W. WASHBURN (1921) *Phys. Rev.* **17**, 273.
88. F. E. BARTELL and H. J. OSTERHOF (1927) *Ind. Eng. Chem.* **19**, 1277.
89. G. ZOGRAFI and S. TAM (1976) *J. Pharm. Sci.* **65**, 1145.
90. P. M. HEERTJES and N. W. F. KOSSEN (1965) *Chem. Eng. Sci.* **20**, 593.
91. C. F. LERK, A. J. M. SCHOONEN and J. T. FELL (1976) *J. Pharm. Sci.* **65**, 843.
92. C. F. LERK, M. LAGAS, J. P. BOELSTRA and P. BROERSMA (1977) *J. Pharm. Sci.* **66**, 1480.
93. M. L. A. STUDEBAKER and C. W. SNOW (1955) *J. Phys. Chem.* **59**, 973.
94. D. T. HANSFORD, D. J. W. GRANT and J. M. NEWTON (1980) *Powder Technol.* **26**, 119.
95. G. D. PARFITT (1973) *Dispersion of Powders in Liquids*, 2nd edn., Applied Science, London, Ch. 1.
96. W. A. ZISMAN (1964) *Adv. Chem.* **43**, 1.
97. H. W. FOX and W. A. ZISMAN (1952) *J. Colloid Sci.* **7**, 428.
98. J. T. FELL and E. EFENTAKIS (1978) *J. Pharm. Pharmacol.* **30**, 538.
99. E. G. SHAFRIN and W. A. ZISMAN (1960) *J. Phys. Chem.* **64**, 519.
100. S. W. HARDER, D. A. ZUCK and J. A. WOOD (1970) *J. Pharm. Sci.* **59**, 1787.
101. G. D. PARFITT and D. G. WHARTON (1972) *J. Colloid Interface Sci.* **38**, 431.
102. G. A. H. ELTON (1957) *Proc. IInd Int. Cong. Surface Activity*, London, vol. III, p. 161.
103. D. P. BENTON and B. D. SPARKS (1966) *Trans. Faraday Soc.* **62**, 3244.
104. J. POWNEY and L. J. WOOD (1941) *Trans. Faraday Soc.* **37**, 220.
105. D. W. FUERSTENAU (1957) *Trans. A.I.M.E.* **208**, 1365.

2 Phase behaviour of surfactants

2.1 Introduction

Although much of the published work on surfactant solutions has been concerned with the properties of dilute aqueous solutions, many surfactants also form liquid crystalline phases at high concentration. The dilute isotropic surfactant solutions represent only a part of the overall picture of surfactant–solvent interaction. In this chapter we examine the structure and properties of the various lyotropic mesophases which may be present in surfactant solutions both in the presence and absence of solubilizate.

2.2 Liquid crystalline phases in binary surfactant systems

It has long been known that structures of the types depicted in Fig. 2.1 may be encountered as the concentration of a surfactant solution is increased. At

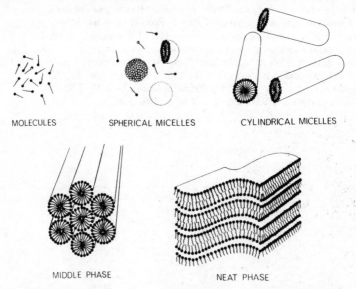

MOLECULES   SPHERICAL MICELLES   CYLINDRICAL MICELLES

MIDDLE PHASE   NEAT PHASE

Figure 2.1 Schematic representation of idealized structures that may be encountered as the concentration of surface-active agent increases in a surface-active agent + water system. From Corkill and Goodman [10] with permission.

concentrations well above the critical micelle concentration there is insufficient water to fill the spaces between the spherical or elongated micelles and more ordered structuring of the solutions occurs. Two main types of mesophase may be identified – the middle phase, M, which is characterized by a hexagonal array of indefinitely long, mutually parallel, rods and the neat phase, G, which has a lamellar structure. In aqueous solutions the cylindrical structures of the middle phase usually have a hydrocarbon core surrounded by an interfacial layer of hydrated polar groups. Such structures are referred to as normal or type 1 mesophases. In anhydrous organic solvents the structure is reversed; the polar groups now form the core and are shielded from the non-aqueous environment by a layer of hydrocarbon chains. These are reverse or type 2 mesophases. The neat phase, which is intermediate between these two types, usually consists of double layers of surfactant molecules with the polar groups protruding into the intervening layers of water molecules. In some cases single layers of surfactant molecules are noted with the polar groups towards opposite interfaces with the intervening layers of water molecules. An additional isotropic region, the viscous isotropic region, may occur in some systems at concentration regions in between the anisotropic middle and neat phases. X-ray diffraction studies have indicated a body-centred cubic lattice and this phase is considered to consist of two interwoven networks of short rod-like elements.

The conditions of temperature and concentration over which each type of mesophase exists may be indicated by a phase diagram. Fig. 2.2 shows typical phase diagrams for a series of polyoxyethylated non-ionic surfactants [1, 2]. The region labelled B represents a single isotropic solution. Although this region remains clear over a very wide concentration range, the aggregates within the solutions exhibit a gradual transition of size and shape with changes in both temperature and concentration (see Chapter 3). Above a critical temperature, the lower consulate temperature or cloud point, the solution becomes turbid and two liquid phases separate out (region A), one of which is rich in surfactant [3, 4]; the other contains a monomeric dispersion with virtually no micelles present. Clouding is a reversible phenomenon, the solution becoming clear again as the temperature is reduced below the critical temperature. For a homologous series of polyoxyethylated non-ionic surfactants the cloud point decreases with increasing length of the hydrocarbon chain (as seen in Fig. 2.2) and decreasing length of the oxyethylene chain [5, 6]. The cloud point may exceed 100° C for surfactants with long ethylene oxide chains. For such compounds it may be measured under pressure or in the presence of an additive which reduces the cloud point to a measurable temperature. The values obtained in the latter case are then extrapolated back to zero additive concentration [3]. The effect of additives on the cloud point is summarized in Section 3.5.4. Definite hydrates with congruent melting points are formed from very concentrated solutions (Fig. 2.2).

The extent of the mesomorphic phase regions which are encountered at high surfactant concentration, increases with increase in the alkyl chain length in a homologous series of polyoxyethylated non-ionic surfactants. This effect is

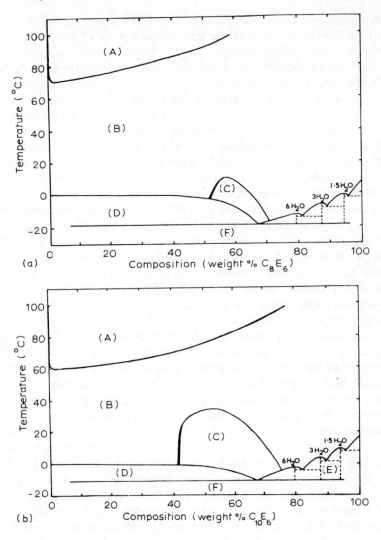

shown in Fig. 2.2 for surfactants with an oxyethylene chain of 6 units. C_8E_6 is virtually homogeneous throughout the whole concentration range, with only a small liquid crystalline region of the middle phase type. The middle phase is of greater extent in $C_{10}E_6$ and both middle and neat phases are present in $C_{12}E_6$. Husson *et al.* [7] have reported the influence of oxyethylene chain length on the phase behaviour of a series of *p*-nonylphenol polyoxyethylene derivatives with the general formula $C_9H_{19}C_6H_4O(CH_2.CH_2O)_nH$. Short chain compounds ($n = 6$ and 8) have neat phase only; both neat and middle phases are noted when $n = 9$, and higher chain length compounds, $n = 10$ to 15, show a middle phase only.

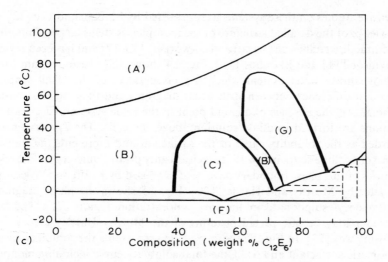

Figure 2.2 Condensed binary phase diagrams for (a) $C_8E_6 + H_2O$ (b) $C_{10}E_6 + H_2O$ and (c) $C_{12}E_6 + H_2O$. A, two isotropic liquids; B, one isotropic liquid; C, middle phase; D, ice + liquid; E, crystals + liquid; F, ice + crystals; G, neat phase. Full lines: experimental boundaries; dotted lines: interpolated boundaries. The estimated extent of the two phase co-existence region between phases B and C and B and G is shown by the thickness of the boundary line. From Corkill *et al.* [1, 10] with permission.

A viscous isotropic region has been detected at concentration regions in between the middle and neat phases in some non-ionic systems [7–9]. Fig. 2.3 shows the occurrence of this phase in the *N,N,N*-trimethylaminododecano-imide–water system which was extensively studied by Clunie *et al.* [9].

For details of the spatial organization within the mesomorphic phases of non-ionic surfactants, the reader is referred to reviews by Corkill and co-workers [9, 10] and Winsor [11].

Much of the published work on the phase behaviour of ionic surfactants has concentrated on the anionic surfactants and in particular the fatty acid soaps.

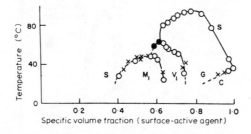

Figure 2.3 Phase diagram for the *N,N,N*-trimethylaminododecanoimide–water system. Phase boundaries were determined from optical ○, density ×, and X-ray diffraction ● measurements. S, isotropic solution; M_1, middle phase; V_1, viscous isotropic phase; G, neat phase; C, crystals. From Winsor [11] with permission.

Extensive studies on these systems were reported by McBain and Lee [12]. Our knowledge of the detailed structure of the mesophases stems largely from the X-ray diffraction studies of Luzzati and co-workers [13–17] and has been reviewed by Winsor [11] and Skoulios [18]. Fig. 2.4 shows the phase diagram for the sodium laurate–water system which is representative of the alkali soaps. An important difference between such phase diagrams and those of the non-ionic surfactants is the absence of a cloud point in the ionic systems. Clear isotropic solutions are formed in dilute solution above curve T_C. The T_C curve may be regarded as the solubility curve of the surfactant and represents the transition from a crystalline soap, below T_C, to a clear isotropic or liquid crystalline region. Theories of this phase boundary have been proposed by Krafft and Wiglow [19] and later by Murray and Hartley [20]. Fig. 2.5 shows the phase diagram of sodium decyl sulphonate in the low concentration region [21]. The phase behaviour may be interpreted in terms of the Murray–Hartley theory in the following way [22]. In Fig. 2.5 the curve $S_1(T)$ represents the solubility curve of monomeric surfactant and $S(T)$, the total solubility curve including monomers and micelles. The two curves are identical at temperatures where the total solubility is less than or equal to the CMC. The temperature at which the monomer solubility equals the CMC is called the Krafft point, which may be regarded as the temperature at which solid hydrated surfactant and micelles are in

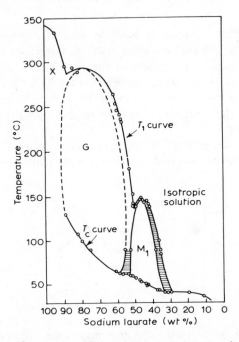

Figure 2.4 Phase diagram for the sodium laurate–water system showing homogeneous fields of each of the different solution phases and isothermal tie lines connecting the phases in heterogeneous equilibria. From Winsor [11] with permission.

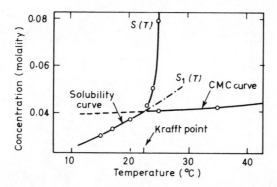

Figure 2.5 Phase diagram close to the Krafft point of sodium decyl sulphonate. From Shinoda and Hutchinson [21] with permission.

equilibrium with monomers. The very pronounced increase in concentration with increase in temperature above the Krafft point may be explained by considering a system at the Krafft point with a surfactant concentration well in excess of the monomer solubility. Under these conditions the system contains dissolved monomers with a concentration equal to the CMC, in equilibrium with solid hydrated surfactant. A slight increase in temperature results in an incremental increase in the solubility of the monomers. However, the hydrated solid surfactant which dissolves at this higher temperature goes into solution as micelles rather than monomer since the total surfactant in solution now exceeds the CMC. The continued attempt by the system to restore the solid–monomer equilibrium at this temperature results in the dissolution of more and more solid until the solid dissolves completely. Hence at a temperature slightly above the Krafft point, the solubility increases dramatically. The temperature at which this occurs is referred to as the critical micelle temperature (CMT). Since both the solubility of the monomer and the CMC can vary with the concentration of any added electrolyte it might be expected that the Krafft point, which is the intersection of the $S_1(T)$ curve and the CMC, would also be dependent on the electrolyte concentration and they may readily be shown to be so by experiment.

Between T_1 and T_C at relatively high surfactant concentration are the neat and middle phase liquid crystalline regions. In addition, X-ray diffraction studies [16] have revealed several other phases intermediate in composition between the neat and middle phases, including the viscous isotropic phase. A comprehensive account of the structure of these and other mesomorphous phases is given by Skoulios [18].

Only a limited number of studies of cationic surfactant–water systems have been reported. Fig. 2.6 shows the phase behaviour for the binary system dodecyltrimethylammonium chloride–water [23]. The interesting feature of this binary system is the appearance of two distinct cubic mesomorphic phases. An extensive range of techniques including optical microscopy, differential thermal analysis, proton magnetic resonance and low-angle X-ray diffraction were used to

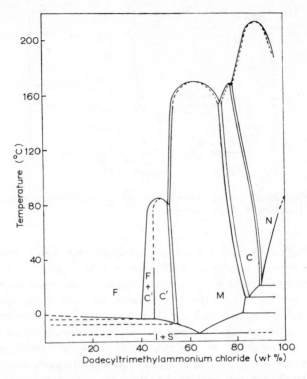

Figure 2.6 Phase diagram for dodecyltrimethylammonium chloride plus water. F = fluid isotropic; C' and C = cubic; M = middle; N = neat; S = solid; I = ice. ———, Experimental boundary; - - - -, interpolated boundary. From Balmbra *et al.* [23] with permission.

examine the structure of these two phases. Both cubic phases exist at room temperature, and are highly viscous and optically isotropic. There are, however, conspicuous qualitative differences in the low-angle X-ray diffraction patterns, which were discussed by Balmbra *et al.* [23]. The formation of the second cubic phase depends on both the length of the alkyl chain and the nature of the counterion. For example, this cubic phase is formed by the decyl, dodecyl, and tetradecyl trimethylammonium chlorides but not by the hexadecyl and octadecyl homologues and also not by any of the corresponding bromides.

2.3 Liquid crystalline phases in ternary surfactant systems

Ternary systems are encountered in pharmacy mainly as emulsifier–oil–water systems in emulsions or as surfactant–solubilizate–water systems in the formulation of solubilized systems. Phase equilibria within ternary systems are most conveniently represented by a triangular phase diagram which delineates the phase regions enabling the formulator to select suitable combinations of the three components to produce stable emulsions or clear isotropic solubilized systems.

To represent a three-component system in one plane, use is made of the geometrical property of an equilateral triangle that the sum of the lengths of perpendiculars from a given point in the triangle to the three sides is equal to the height of the triangle. If one constructs an equilateral triangle, as in Fig. 2.7, and divides each side into 100 parts to correspond to percentage composition, the composition of the system at any point can be obtained by measuring the perpendicular distances to the three sides, e.g. the point X is equivalent to 20 % A, 30 % B, and 50 % C. A further property of a diagram of this type is that a line from the apex to a point on the opposite side represents all possible mixtures which have the same relative amounts of the other two components. This is useful in forecasting what will happen if one dilutes a given system with solvent; line CD represents all possible compositions obtained when a mixture containing originally 70 % A and 30 % B is diluted with component C.

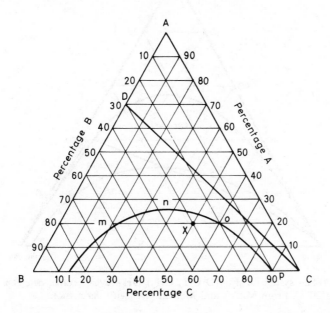

Figure 2.7 A triangular three-component phase diagram.

Simple phase diagrams are produced for ternary systems in which two of the three components are completely miscible. The essential characteristics of such diagrams are shown in Fig. 2.7 in which liquids A and B and A and C are completely miscible whilst B and C are only partially miscible. The region, 2L, under the curve lmnop is an area of immiscibility. Systems with compositions within this region separate into two liquid phases at equilibrium. From experimental determinations it is possible to draw tie lines connecting those systems within the 2L region which separate at equilibrium into two phases with compositions given by the points of intersection of the tie lines and the curve lmnop. In Fig. 2.7 all systems on the tie line, mo, will separate at equilibrium into two phases, one with composition given by point m (20% A, 60% B and 20% C) and the other with composition given by point o (20% A, 20% B and 60% C). Although the compositions of these phases are identical for all systems along mo, the weight ratio of the phase of composition m to that of composition o will be determined by the relative distance of the particular system along the line mo. In a system represented by point X, for example, which is three-quarters of the distance along mo measured from m, the weight ratio of m to o is 1:3. Examples of the simple ternary systems of this type are the C_6E_6–water systems containing solubilized cyclohexane, benzene or octanol studied by Mulley and Metcalf [24] (see Fig. 2.8).

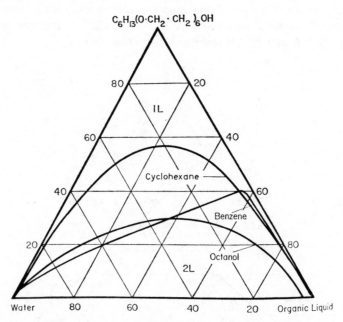

Figure 2.8 The systems $C_6H_{13}(O \cdot CH_2 \cdot CH_2)_6OH$ + cyclohexane, benzene, or octanol + water at 20° C. From Mulley and Metcalf [24] with permission.

Phase diagrams of greater complexity are required to completely describe the phases present in systems in which components are not completely miscible. Although these phase diagrams often differ markedly in appearance, it is possible to identify phase regions which are common to many systems. Fig. 2.9 shows the phase diagram for the potassium caprylate–decanol–water system which has been studied in detail by Ekwall *et al.* [25]. Two isotropic solubilized regions may be identified. Considering water as the solvent, the first, the L_1 phase region, appears where the surfactant–water system incorporates the solubilizate in its micellar structure. The L_1 phase region is a clear isotropic solution in which, at concentrations above the CMC, the solubilizate is distributed between the micelles and the aqueous phase. The second isotropic region, L_2, appears at high

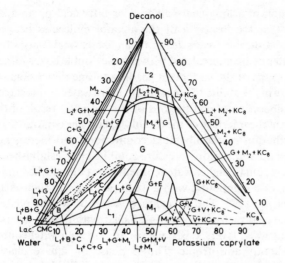

Figure 2.9 Equilibrium diagram for the system potassium caprylate–decanol–water at 20°C.

L_1. Region with homogeneous, isotropic aqueous solutions.

L_2. Region with homogeneous, isotropic decanolic solution.

B. Region with homogeneous, lamellar mesophase (mucous woven type).

G. Region with homogeneous, lamellar mesophase (neat phase type).

C. Region with homogeneous, normal two-dimensional tetragonal mesophase (square phase).

M_1. Region with homogeneous, normal two-dimensional hexagonal mesophase (middle phase).

M_2. Region with homogeneous, reversed two-dimensional hexagonal mesophase (middle phase, reversed).

V. Region with homogeneous, normal cubic mesophase (viscous isotropic).

From Ekwall *et al.* [25] with permission.

concentrations of the water-insoluble component (decanol). In this case the solubilizate–solvent roles are reversed; here water is solubilized and the water-insoluble component is now the solvent. The micelles within this phase are 'reversed', i.e. the polar groups are directed inwards forming a polar environment for the lipophobic solubilizate. The nature and properties of solubilized systems in the L_1 and L_2 regions are discussed in Chapter 5.

The areas B, C, G, M_1 and M_2 of Fig. 2.9 represent homogeneous, liquid crystalline regions, which also may be regarded as solubilized systems. The areas M_1 and M_2 are the normal and reversed middle phases, respectively, the structures of which have been discussed in the previous section. Incorporation of solubilizate into these structures is not thought to alter substantially their basic structures. The liquid crystalline region, G, is the neat phase. This is a lamellar mesophase composed of water layers alternating with (in the case of the potassium caprylate–decanol–water system) double amphiphile layers. The extension of this region towards the water corner of the phase diagram is a typical consequence of the solubilization of a long-chain alkanol or fatty acid by an alkali soap. It is thought [26] that the insertion of the alcohol molecules between surfactant molecules in the double layers leads to an increasingly large fraction of the solubilized water being intercalated between the double layers rather than bound to the polar groups of the surfactant, leading to one-dimensional swelling. The sudden increase in the ability to incorporate more water at this stage is thought to be due to the liberation of counterions of the soap as a result of the decrease in charge density in the interface of the double layers. The conditions are created for Donnan distribution of the liberated counterions and increasing amounts of water are taken up until vapour-pressure equilibrium is established. Extension of the neat phase region to higher water content is not noted when the solubilizate is a hydrocarbon or a compound with a weak polar group such as a methyl ester, nitrile or aldehyde [27]. The B region or mucous woven phase is similar in structure to the neat phase in that it is lamellar with double layers of surfactant with intervening water layers. It is, however, of very high water content, 70 to 90 % and is only slightly anisotropic. The C phase or square phase bears some resemblance to the middle phase in that it consists of indefinitely long, mutually parallel rods but these are arranged in a tetragonal rather than hexagonal array. The region is anisotropic with a relatively high water content (40 to 65 %).

Between the homogeneous liquid crystalline regions are a variety of regions in which two of the single phases discussed above are in equilibrium, and also a number of three-phase triangles. These latter arise where regions of immiscibility touch one another in such a way as to produce an inner triangular phase region. In such regions the compositions of the three phases in equilibrium are fixed at the compositions indicated at the apices of the triangle and only the relative amounts of each phase vary throughout the triangular region. At high concentration of surfactant a solid region exists consisting, in the case of the ternary system under discussion, of solid crystalline potassium caprylate and hydrated potassium caprylate with a fibre structure.

Of the two-phase regions it is generally assumed that only the $(L_1 + L_2)$ region represents a true emulsion system. Dispersion of the two phases produces either an oil-in-water or water-in-oil emulsion stabilized by adsorption of the surfactant at the $(L_1 + L_2)$ interface. However, dispersion of combinations of the L_1 or L_2 phase and a liquid crystalline phase also produce emulsions which have the macroscopic appearance of the conventional $(L_1 + L_2)$ emulsion. There is increasing evidence to suggest that many of the conventional apparently two-liquid phase emulsions actually contain a third phase which usually takes the form of a gel or is liquid crystalline. Indeed, some workers have suggested that the stability and other properties of the emulsion are entirely controlled by this third phase. The phase diagram of the system $C_{10}E_6$–water–dodecane (Fig. 2.10) reported by Ali and Mulley [28] shows an extensive area LBJM in which an aqueous phase containing most of the surfactant is in equilibrium with a gel-like phase, which is thought to have a similar structure to that of the viscous isotropic phase. Emulsions prepared in this region are very much more stable than the normal liquid–liquid emulsions found in the narrow two-liquid region ALNC, an observation which suggested the stabilizing influence of the additional gel phase.

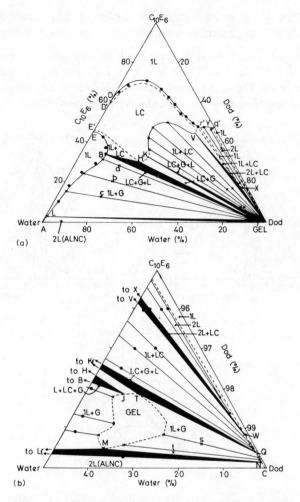

Figure 2.10(a) Phase diagram of the system $C_{10}E_6$–water–dodecane at 25° C. Dod: Dodecane. (b) The dodecane-rich corner of the sytem $C_{10}E_6$–water–dodecane. The terms 1L, 2L, 1L + LC, 2L + LC, 1L + G, 2L + G and 3L represent one-liquid, two-liquid, one-liquid and liquid crystal, two-liquid and liquid crystal, one-liquid and gel, two-liquid and gel and three-liquid phases, respectively. From Ali and Mulley [28] with permission.

The relationship between phase equilibrium and emulsion stability will be considered in greater depth in Chapter 8.

2.4 Factors affecting phase behaviour

2.4.1 Ionic surfactants

Of the many factors which affect the nature of the phase diagram of ionic surfactants, the most influential is probably the nature of the solubilizate. Ekwall and co-workers [27], in an exhaustive survey of this topic, classified the solubilizates into four general categories.

(a) Completely lipophilic substances such as hydrocarbons or chlorinated hydrocarbons are solubilized in the L_1 isotropic region and also in the middle phase. Fig. 2.11a shows that increase in the amount of solubilizate leads to the formation of a viscous isotropic mesophase.

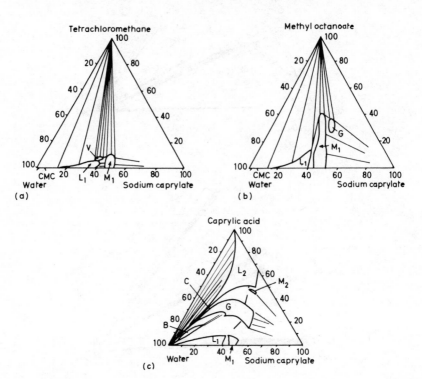

Figure 2.11 Phase diagrams for the ternary systems at $20°$ C: (a) Sodium caprylate–tetrachloromethane–water; (b) Sodium caprylate–methyl octanoate–water; (c) Sodium caprylate–caprylic acid–water.

L_1.　Homogeneous isotropic aqueous solution.
L_2.　Homogeneous isotropic alcoholic solution.
B.　Homogeneous mesomorphous phase displaying lamellar structure (mucous woven type).
C.　Homogeneous mesomorphous phase displaying normal two-dimensional tetragonal structure.
G.　Homogeneous mesomorphous phase displaying lamellar structure (neat phase type).
M_1.　Homogeneous mesomorphous phase displaying normal two-dimensional hexagonal structure.
M_2.　Homogeneous mesomorphous phase displaying reversed two-dimensional hexagonal structure.
V.　Homogeneous viscous isotropic mesomorphous phase displaying cubic structure.
From Ekwall *et al.* [27] with permission.

(b) When amphiphilic compounds with a weak hydrophilic group such as a nitrile, methyl ester or aldehyde group are added there is a considerable increase in the solubilizing capacity of both the L_1 and middle phase regions and a lamellar neat phase appears at high solubilizate concentrations (Fig. 2.11b).

(c) Fig. 2.9 is typical of phase diagrams in which the solubilizate is a monohydric alcohol. This diagram is considerably more complex than the two previous cases in that the lamellar neat phase now extends to very high water contents and additional mesophases (the mucous woven phase B, the square phase C and the reverse middle phase M_2) and an isotropic L_2 phase are apparent.

(d) The solubilization of fatty acid produces similar phase diagrams to those for the solubilization of alcohols except that the L_2 region may extend further towards the water corner (Fig. 2.11c).

The effect of increasing the chain length of the surfactant is shown in Fig. 2.12 for ternary systems consisting of a homologous series of alkyl trimethylammonium bromides with water and hexanoic acid [29]. For simplicity, the ternary liquid crystalline region area is drawn to include both L_1 + liquid crystal and L_2 + liquid crystal. The near independence of the extent of the L_2 and the upper 2L regions on chain length is striking. The increase in solubilizing power with decrease of hydrocarbon chain length of the soap is shown by increase in the extent of the L_1 region.

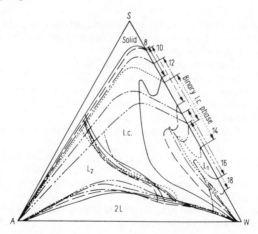

Figure 2.12 Equilibrium diagram for the 8-, 10-, 12-, 14-, 16- and 18-alkyl trimethylammonium bromide soaps with water and hexanoic acid. For simplicity the ternary liquid crystalline area includes both L_1 + lc and L_2 + lc. From Lawrence *et al.* [29] with permission.

A more general indication of the effect of molecular structure on the phase behaviour is shown in Fig. 2.13 which compares the phase diagrams of a variety of both anionic and cationic surfactants [27]. The general similarity of appearance

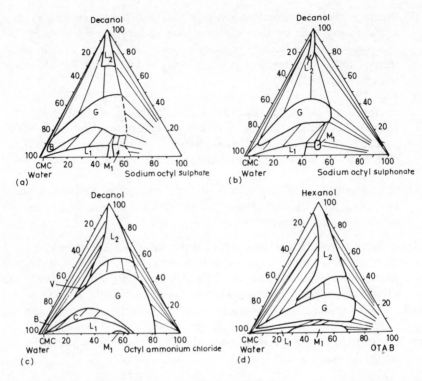

Figure 2.13 Phase diagrams for ternary systems containing an alkyl sulphate, alkyl sulphonate, a primary or a quaternary alkyl ammonium salt, water and an alcohol.

- (a) Sodium octyl sulphate–decanol–water at 20° C
- (b) Sodium octyl sulphonate–decanol–water at 20° C
- (c) Octyl ammonium chloride–decanol–water at 20° C
- (d) Octyl trimethylammonium bromide–hexanol–water at 25° C

L_1. Homogeneous isotropic aqueous solution.
L_2. Homogeneous isotropic alcoholic solution.
B. Homogeneous mesomorphous phase displaying lamellar structure (mucous woven type).
G. Homogeneous mesomorphous phase displaying lamellar structure (neat phase type).
M_1. Homogeneous mesomorphous phase displaying normal two-dimensional hexagonal structure (middle phase type).
V. Homogeneous viscous isotropic mesomorphous phase displaying cubic structure.
From Ekwall *et al.* [27] with permission.

of these phase diagrams is indicative of the considerably smaller effect of molecular structure on phase behaviour as compared with that of solubilizate structure. More major changes in the molecular structure of the surfactant can, however, lead to significant differences in phase behaviour. For example, in the sodium cholate–decanol–water and sodium desoxycholate–decanol–water system the isotropic L regions extend from the alcohol to the water corners and no liquid crystalline phases can be detected [27].

2.4.2 Non-ionic surfactants

The phase behaviour of non-ionic surfactants is much more sensitive to surfactant structure than is that of the ionic surfactants. Friberg and co-workers [30, 31] have demonstrated the effect of changes in hydrophilic character of the surfactant. Comparison of the nonylphenol nonaethylene glycol ether–water–p-xylene and nonylphenol diethylene glycol ether–water–p-xylene systems (Fig. 2.14) shows that reduction of the hydrophilic character leads to an increase in the concentration of the onset of the solubilization of water in the L_2 region; a reduction in the size of the neat phase and the disappearance of the B phase, the micellar solution phase L_1 and the hexagonal M_1 phase. Essentially similar results were noted by Ali and Mulley [28] in a comparison of the $C_{10}E_3$–water–dodecane and $C_{10}E_6$–water–dodecane systems.

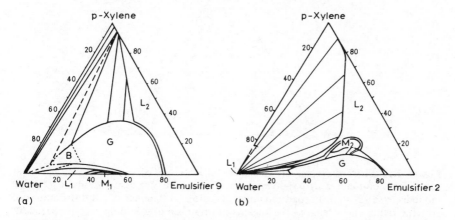

Figure 2.14 Phase diagrams for the three-component systems at 20° C: a, water/nonylphenol nonaethylene glycol ether/p-xylene; b, water/nonylphenol diethylene glycol ether/p-xylene. From Friberg and Mandell [30] with permission.

The influence of the ethylene oxide chain length distribution in a sample of a non-ionic surfactant on its phase behaviour has been reported by Bradley and Lissant [32, 33]. As emphasized by these authors, triangular phase diagrams cannot in a strict sense be applied to a water–hydrocarbon–surfactant systems when the surfactant is made up of a spectrum of closely related compounds. Nevertheless, such mixtures are representative of most commercial non-ionic polyoxyethylene derivatives, only a small proportion of which are produced in a pure homogenous state, and hence such investigations as these are of value from a practical view point. Reproducible results were obtained with these heterogeneous mixtures [32] although the phase diagrams were limited in their use since tie lines could not reasonably be constructed. In the system, Shellflex 131 (a commercial aliphatic solvent)–water–surfactant, three types of surfactant were considered, each of an average composition $C_{10}E_4$. Fig. 2.15 compares the phase behaviour in systems in which the ethylene oxide chain length distribution was (a)

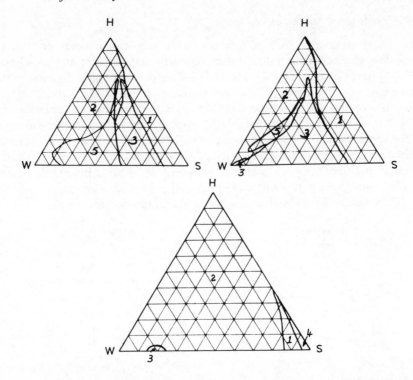

Figure 2.15 Phase diagrams of water (W)–Shellflex-131(H)–surfactant (S) systems at 25° C for non-ionic polyoxyethylene derivatives with average formula $C_{10}E_4$. (a) monodisperse (S-30), (b) broad monomodal (S-75), (c) bimodal (S-68) distributions of ethylene oxide chain length. The regions are numbered according to type: (1) clear, isotropic, oil-rich; (2) two-phase; (3) clear or hazy, anisotropic, water-rich; (4) solid phase; (5) clear water-rich phase. From Bradley and Lissant [33] with permission.

monodisperse (S30), (b) broad but monomodal (S75), and (c) bimodal (S68). Phase diagrams for monodisperse (Fig. 2.15a) and broad monomodal (Fig. 2.15b) distributions are similar in appearance each showing five distinct phase regions. The ternary diagram for the system containing S68 is, however, quite different; most of the volume except in the extreme corners is two-phase, the inverse micellar region (region 1) is much smaller and other regions are absent or much reduced in size. The authors conclude that 'one cannot simulate a particular oxyethylene chain length by an indiscriminate mixture of widely diverse materials'.

Extensive studies on the influence of the nature of the oil phase on the phase behaviour of the Brij 96–oil–water systems have been reported [34, 35]. The considerable influence of the oil phase on the aspect of the phase diagram is apparent from Fig. 2.16 which shows the phase behaviour for nine different oils. In Fig. 2.17a the percentage uptake of water into different isotropic L_2 phases containing 50 per cent Brij 96 is plotted against the dielectric constant ε of the oil

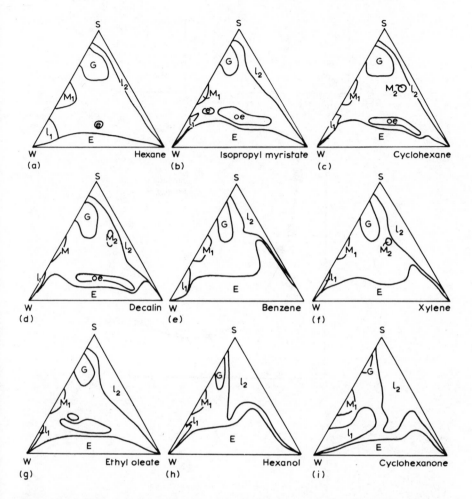

Figure 2.16 Phase diagrams for Brij 96–oil–water systems for the oil as shown. The phases identified include the L_1, L_2, E, M_1, and G phases as noted before and also an isotropic elastic phase (e) and an isotropic elastic phase containing dispersed globules of oil (oe).

phase. The general trend is that as the dielectric constant increases, the total amount solubilized by this oily phase increases. Fig. 2.17b and c show the reverse trend for the uptake of oil in the aqueous neat (G) phase (for $\varepsilon < 8$) and the middle (M_1) phase (for $\varepsilon < 3.4$), although in the latter case the aromatic molecules appear to exhibit anomalous behaviour. Similar conclusions were obtained where vegetable oils were considered [34]; Fig. 2.18 shows phase diagrams for the sweet almond, maize and castor oil–Brij 96–water systems.

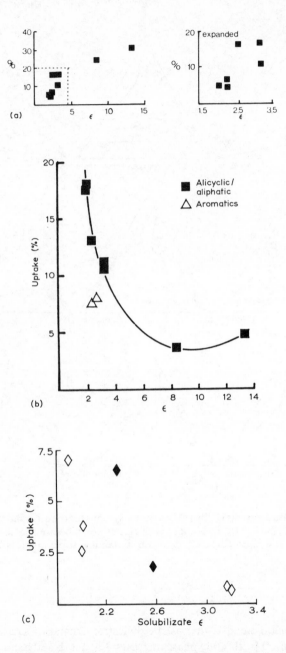

Figure 2.17(a) Water uptake in the isotropic oil phase, L_2, as a function of the solvent dielectric constant, ε. (b) Uptake of solubilisate molecules in the aqueous neat (G) phase of Brij 96 as a function of the dielectric constant of ■, aliphatic/alicyclic molecules, and △, aromatic molecules. (c) Uptake in the middle (M_1) phase at a concentration of surfactant of 50% as a function of the dielectric constant, ε, of the solubilisate. ◇, Aliphatics/alicyclics; ◆, aromatics. From Lo *et al.* [35] with permission.

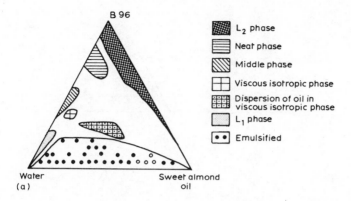

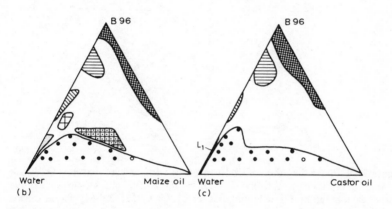

Figure 2.18 Phase diagrams for Brij 96–water–vegetable oil ternary systems, for the following oils: (a) sweet almond oil; (b) maize oil; (c) castor oil. From Kabbani *et al.* [34] with permission.

The influence of the hydrophile–lipophile balance (HLB) of the non-ionic surfactant on the phase behaviour has recently been studied [34–37] with the aim of better understanding the mechanisms of emulsion and suspension stabilization. Ternary phase diagrams for dodecane, water and mixtures of Brij 92, and Brij 96 (polyoxyethylated oleyl alcohol derivatives with oxyethylene chain lengths of 2 and 10, respectively), with a range of HLB values, are shown in Fig. 2.19. The areas of the L_2 and the inverse middle phase M_2 (an interesting feature of these systems not usually observed in polyoxyethylene non-ionic systems) increase as the HLB increases, reaching a maximum at HLB 8. Maximum water uptake in the L_2 phase in 25 %, 40 % and 50 % surfactant solutions in oil as a function of HLB is

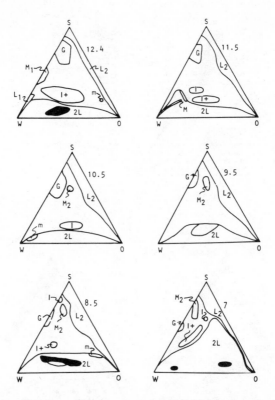

Figure 2.19 Phase diagrams of dodecane–water–Brij 92/96 mixtures. O, W, S = 100 %
oil, water, surfactant, respectively. HLB values of the surfactant mixtures are shown to the
right of each diagram. Phase boundaries for L_1, L_2, G, M_1 and M_2 phases shown. Other
symbols represent: 2L, two-phase oil–water system (emulsions of varying stability); M,
microemulsion; I, isotropic elastic; I +, isotropic elastic + disperse oil phase. Filled black
regions: stable emulsion zones. From Lo *et al.* [37] with permission.

given in Fig. 2.20. Maximal solubilization in dodecane occurs at an HLB of 8
where the molar ratio of polyoxyethylene (2) oleyl ether to polyoxyethylene (10)
oleyl ether is approximately 3:1. At surfactant levels below 25% there is
negligible uptake of water by polyoxyethylene (2) oleyl ether in dodecane. The
concentration dependence of the optimal HLB values for solubilization, apparent
in Fig. 2.20 diminishes the value of HLB as an index for the prediction of
solubilization capabilities of mixed systems. The middle phase, M_1, disappears on
addition of more than 20% Brij 92 to the surfactant mixture. The neat phase, G,
which because of its more linear structure, one would expect to be more
dependent on surfactant size and shape, diminishes in size on addition of the
short-chain surfactant and indeed does not form when Brij 92 is present as the
sole surfactant. Nevertheless, addition of Brij 92 to the system allows formation
of the neat phase at lower surfactant concentrations than in solutions of Brij 96.
The capacity for solubilization of dodecane, however, is greater the higher

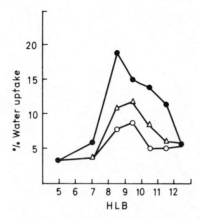

Figure 2.20 Water uptake into non-aqueous L_2 dodecane phase as a function of the calculated HLB of Brij 92/96 mixtures. The dodecane results obtained at ○ 25%, △ 40% and ● 50% w/w surfactant levels. From Lo *et al.* [37] with permission.

the concentration of Brij 96 in the mixture, i.e. Brij 92 depresses solubilization in the neat phase.

In contrast to the small effects which temperature change has on the phase behaviour of ionic surfactants [38] there is a very pronounced change in the appearance of phase diagrams of oil–water–non-ionic surfactant systems with increase in temperature. Changes induced by temperature in the relative positions and extent of isotropic and liquid crystal phases present in the ascorbic acid–water–polysorbate 80 system have been recorded by Nixon and Chawla [39] (Fig. 2.21). Temperature increase decreases the width of the liquid crystal band; the most pronounced effect occurring between temperatures of 25 and 30° C where the polysorbate concentration at which liquid crystals first appear $(L_1 + LC)$ is increased from about 35 to 36% to 44% polysorbate in the presence of ascorbic acid.

A more comprehensive study of the effect of temperature on phase behaviour has been reported by Friberg and Lapczynska [40] for the hexadecane–water–tetraethylene glycol dodecyl ether system (Fig. 2.22). At 18.5° C the isotropic liquid region extended from the water corner in a sector which ended at a water percentage of 40. A temperature increase to 20° C caused thinning of the sector which also reached further toward the hexadecane corner. The connection with the water corner was broken after a further increase in temperature to 25° C and the area was reduced in extent and transferred towards the water–hexadecane axis. As the temperature was increased further the region again shifted parallel to the water–hydrocarbon axis towards the water–hexadecane axis until at 35° C a coalescence with the L_2 area took place, giving rise to a narrow salient towards the water corner. Similar temperature-induced changes were noted by these workers when decane or hexane constituted the hydrocarbon. When the third component was benzene or dodecyl benzene the phase diagram at temperatures as low as 10

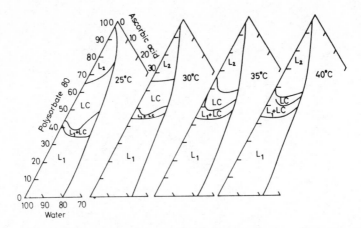

Figure 2.21 Solubility of ascorbic acid in polysorbate 80–water systems: effect of temperature on phases present. L_1 and L_2, isotropic phases: LC, liquid crystal.

Boundary of liquid crystal regions (%w/w polysorbate 80)

Temperature	Absence of ascorbic acid			5% w/w ascorbic acid			10% w/w ascorbic acid			Solubility limit of ascorbic acid		
(°C)	a	b	c	a	b	c	a	b	c	a	b	c
25	41	44	64	36	39	67	34.5	40	—	34	49	76
30	44	50	66	44	46	65	44	48	66	47	51	66.5
35	44	57	66	43	48	59	43	48.5	58	48	54	59
40	44	—	59	43	—	54	43	50	55	48	55	56

(a) Lower limit of liquid crystal + L_1; (b) lower limit of liquid crystal only; (c) upper limit of liquid crystal + L_2.
From Nixon and Chawla [39] with permission.

to 15° C resembled those of the other alkanes at 30° C, i.e showed coalescence with the L_2 region. These pronounced changes in the isotropic regions are accompanied by only very slight alterations in the extent of the liquid crystalline regions of the phase diagram. At a temperature of 25° C in the hexadecane system the solutions require a minimum of surfactant in order to solubilize large amounts of both water and hydrocarbon. This temperature corresponds to the phase inversion temperature as defined by Shinoda [41] which is the narrow temperature range in which solubilization behaviour changes from oil-in-water to water-in-oil with increasing temperature. Detailed examination of the phase behaviour of ternary oil-water-surfactant systems in this manner provides valuable information of use in formulation of stable solubilized systems. Fig. 2.22, for example, demonstrates the optimal surfactant concentration required in order that the temperature range for solubilization may be the largest possible. The temperature sensitivity is clearly displayed by the maximal stability range of 6° C at a surfactant concentration of 13%.

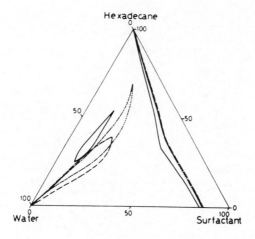

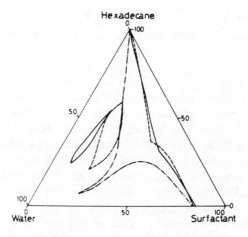

Figure 2.22 Isotropic liquid solution regions for the system water, hexadecane and tetraethyleneglycol dodecyl ether

	Temperature (°C)
—·—·—·—·—·—·—·—.	18.5
- - - - - - - - - - - -	20
———————————	25
— — — — — — — —	30
—··—··—··—··—··—··—	35

From Friberg and Lapczynska [40] with permission.

2.5 Quaternary phase systems

A four-component system at constant temperature and pressure may be represented graphically using a four-sided, three-dimensional space model. The most convenient model is that of a regular tetrahedron (Fig. 2.23). Each side of the tetrahedron is an equilateral triangle representing a ternary system, the corners of

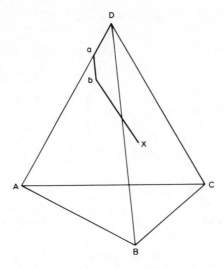

Figure 2.23 A tetrahedral four-component phase diagram.

the tetrahedron represent pure components, and the edges represent binary mixtures. The method of plotting data may be illustrated by the following example in which a mixture X containing 20% A, 10% B, 40% C and 30% D is to be plotted. In Fig. 2.23 the percentage of A has been measured along the line DA towards A and is represented by point a. The percentage of B (ab) is then measured along a line starting from point a and moving towards B in a direction parallel to the edge DB. The percentage of C (bX) is then measured from the point b along a line parallel to DC which goes into the interior of the tetrahedron. From the properties of a regular tetrahedron the percentage of D is given by [100 − (A + B + C)]. In Fig. 2.23 the percentage of D is the vertical distance from X to the base of the tetrahedron.

Relatively few four-component systems have been studied in any detail. Fig. 2.24a shows diagrammatically the way in which the quaternary phase diagram for the system phosphated nonyl phenolethoxylate (PNE)–phosphated fatty alcohol ethoxylate (PFE)–*n*-hexane–water (Fig. 2.24b) is built up [42].

$$CH_3(CH_2)_8 \bigcirc (OCH_2CH_2)_{8.5}O \cdot P(OH)_2 \atop \downarrow \atop O$$

PNE

$$CH_3(CH_2)_8(OCH_2CH_2)_3O \cdot P(OH)_2 \atop \downarrow \atop O$$

PFE

Fig. 2.24b has been simplified to show the relationship between the two liquid crystalline phases. The main points of interest in this system are the formation of liquid crystalline material immediately water is added to the anhydrous mixture

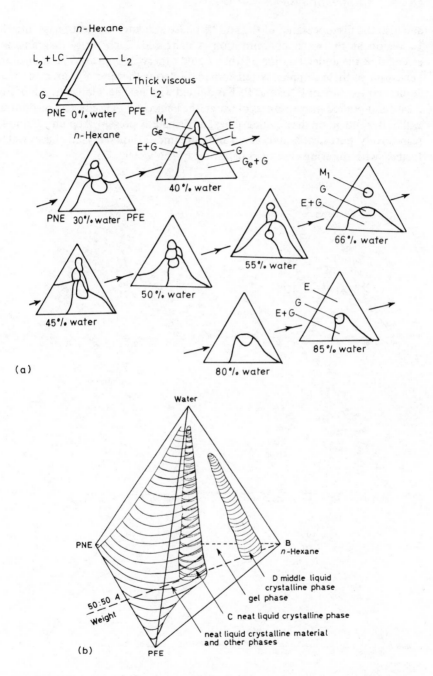

Figure 2.24(a) Diagrammatic representation of the effect of progressively adding water to a mixture of PNE, PFE and *n*-hexane. L, liquid; G_e, gel; E, emulsion; M_1, middle phase liquid crystal; G, neat phase liquid crystal. (b) Diagrammatic quaternary phase diagram for the system PNE–PFE–water–*n*-hexane at 25° C. From Groves *et al.* [42] with permission.

and also the disappearance of the middle phase and later the neat phase into the L_1 region as the water concentration is increased. Such phase diagrams are essential to the understanding of what would otherwise appear to be anomalous behaviour of these complex systems when diluted with water. For example, if an equimolar mixture of PNE and PFE is diluted with hexane (along line AB of Fig. 2.24b) and poured into an excess of water, the behaviour of the system on dilution will be dependent on the amount of hexane originally present. At points C and D, respectively, pure neat liquid crystal and pure middle liquid crystal phases will be formed with differing ease of penetration of the water.

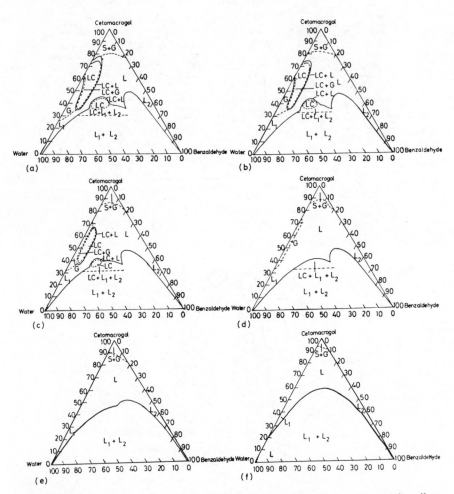

Figure 2.25 Phase equilibria in systems containing cetomacrogol–propyl gallate–benzaldehyde–water at 25° C. Ratio w/w cetomacrogol : propyl gallate; A 29 : 1, B 19 : 1, C 9 : 1, D 6 : 1, E 4 : 1, F 2 : 1. LC = liquid crystal, G = isotropic gel phase, S = solid. From Nixon *et al.* [43] with permission.

The influence of gallate antioxidants on the phase behaviour of the ceto-macrogol 1000–water–benzaldehyde system has been reported by Nixon *et al.* [43]. Because of the phenolic nature of the gallates it is likely that bonded complexes with oxyethylene groups of the cetomacrogol will be formed and indeed evidence for this was obtained by differential scanning calorimetry. As a consequence, the system was treated as a ternary system, the surfactant being considered as a cetomacrogol–gallate mixture. Fig. 2.25 shows the effect on the phase diagram of a gradual increase of the proportion of propyl gallate. The most dramatic change is the disappearance of the liquid crystalline phases as the proportion of propyl gallate in the system is increased. Minor changes occur in other regions of the phase diagram; the size of the $(L_1 + L_2)$ region representing unstable emulsions of the conjugate liquids shows little variation up to a cetomacrogol:propyl gallate ratio of 6:1 and thereafter the boundary moves to higher surfactant concentrations, the L_1 phase increases with increasing proportion of propyl gallate to a maximum at a cetomacrogol:gallate ratio of 6:1, the width of this region then decreases rapidly, the width of the L_2 region is little affected by increases in gallate concentration. Quaternary phase diagrams are particularly valuable in pharmacy for the representation of the phase behaviour of emulsions which often contain a stabilizer (or co-surfactant) in addition to the oil, water and emulsifier. The properties, such as stability, emulsion type and viscosity of these four-component systems may be strongly influenced by variations in the relative proportions of the emulsifying agent and stabilizer. Marland and Mulley [44] have reported details of the phase behaviour of the C_8E_6–water–dodecane–octanol (stabilizer) system. Data were presented using a projection method described by Woodman [45] which shows the four ternary faces of the tetrahedral model (see Fig. 2.26) as if the regular tetrahedron were peeled open and laid flat in a single plane. At 25° C this system comprises two pairs of immiscible liquids, (octanol–water, dodecane–water), the remaining four binary systems being completely miscible at this temperature. An unusual feature of this system is the presence of a three-liquid (3L) region which occurs near the ternary face which contains the two pairs of partial miscibility. The position of this 3L region is clearly shown in Fig. 2.27 which is a photograph of a three-dimensional model of the system. It is clear from this study that the surfactant, although miscible with water, octanol and dodecane, is distributed completely differently in the two pairs, water–octanol and water–dodecane, and that this will have important effects on the nature of the emulsions formed within the two regions. The presence of three-liquid phases in some regions also means that the dispersions produced will be more complex than in two-phase emulsions.

A similar three-liquid region was observed by Ali and Mulley [28] in a quaternary system containing dodecane, water and two non-ionic surfactants, $C_{10}E_6$ and $C_{10}E_3$. The compositions involved were near the face of the tetrahedron representing the four-component system and containing about equal quantities of water and dodecane and 1 to 2.5% $C_{10}E_6$. Addition of 3 to 7.2% $C_{10}E_3$ to this two-phase, three-component system in the face of the tetrahedron, produced the system with three liquid phases in equilibrium. The lower liquid

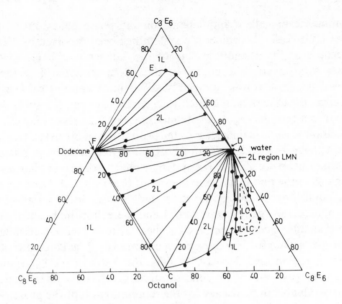

Figure 2.26 Phase diagram of the four ternary faces of the tetrahedral model representing the system C_8E_6–water–octanol–dodecane at 25° C plotted by Woodman's method. From Marland and Mulley [44] with permission.

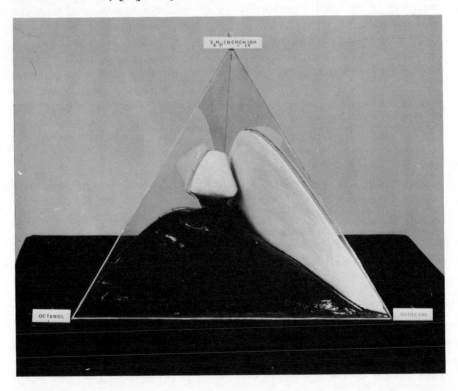

phase consisted essentially of water with small traces of emulsifier; the middle phase contains the two emulsifying agents together with a substantial amount of water and dodecane and the upper phase was essentially dodecane with small amounts of the two emulsifiers. Multiple-drop formation readily occurred in the 3L region and the authors have suggested that the presence of three-liquid phases could be responsible for some of the reported cases of multiple emulsion globules. [46, 47].

Quaternary systems which have recently received a great deal of attention are the so-called 'microemulsion' systems. These homogeneous transparent systems, first described by Hoar and Schulman [48], contain high proportions of oil and yet are of low viscosity and, in contrast to the typical emulsion systems, are thermodynamically stable. The phase behaviour of these four-component systems is commonly represented in the form of ternary phase diagrams of oil–water–total emulsifier (co-surfactant + surfactant). However, the usefulness of such diagrams is limited by the extreme sensitivity of the phase behaviour to the co-surfactant/surfactant ratio and it is more informative to use quaternary phase diagrams. Fig. 2.28 shows the areas of water-in-oil microemulsion in a four-component system comprising p-xylene–water–sodium dodecyl sulphate and pentanol [49]. The quaternary phase diagram shows clearly the relationship of these areas to the isotropic liquid L_2 phase in the sodium dodecyl

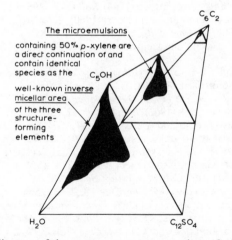

Figure 2.28 Phase diagram of the quaternary system: p-xylene (C_6C_2)–sodium dodecyl sulphate ($C_{12}SO_4$)–pentanol (C_5OH)–H_2O. From Rance and Friberg [49] with permission.

Figure 2.27 Tetrahedral model of the system at 25°C looking towards the C_8E_6–octanol–dodecane face. The 3L region may just be seen as a line between the black and the white 2L regions. The small light coloured volume extending from the octanol–water–C_8E_6 face is a liquid plus liquid crystal region. From Marland and Mulley [44] with permission.

sulphate–water–pentanol system which forms the base triangle. No microemulsion is, of course, present in this three-component system because of the absence of hydrocarbon. The continuation between the microemulsion regions at higher hydrocarbon content and the L_2 regions in the absence of hydrocarbon suggests that the microemulsions are better thought of as solubilized systems in which the 'droplets' are swollen micelles containing solubilized oil. Similar diagrams were presented to explain the relationship between oil-in-water microemulsions and the isotropic micellar L_1 region although here the link is less direct. The conception of the colloidal nature of microemulsions [40, 49–54] contrasts with the opposing view of these systems held by Schulman and co-workers [55–58] and later by Prince [59, 60] in which the role of a negative interfacial tension between the oil and aqueous surfactant phase has been stressed (see Chapter 8).

References

1. J. S. CLUNIE, J. M. CORKILL, J. F. GOODMAN, P. C. SYMONS and J. R. TATE (1967) *Trans. Faraday Soc.* **63**, 2839.
2. J. S. CLUNIE, J. M. CORKILL and P. C. SYMONS (1969) *Trans. Faraday Soc.* **65**, 287.
3. W. N. MACLAY (1956) *J. Colloid Sci.* **11**, 272.
4. T. NAKAGAWA and K. TORI (1960) *Kolloid-Z.* **168**, 132.
5. L. RAPHAEL (1954) *Proc. Ist Intern. Congr. Surface Activity*, Vol. 1, Paris, p. 36.
6. P. H. ELWORTHY and C. MCDONALD (1964) *Kolloid-Z.* **195**, 16.
7. F. HUSSON, H. MUSTACCHI and V. LUZZATI (1960) *Acta Cryst.* **13**, 668.
8. E. S. LUTTON (1966) *J. Amer. Oil Chemists Soc.* **43**, 28.
9. J. S. CLUNIE, J. M. CORKILL and J. F. GOODMAN (1965) *Proc. Roy. Soc. (London)* **A 285**, 520.
10. J. M. CORKILL and J. F. GOODMAN (1969) *Adv. Colloid Interface Sci.* **2**, 297.
11. P. A. WINSOR (1968) *Chem. Rev.* **68**, 1.
12. J. W. MCBAIN and W. W. LEE (1943) *Oil and Soap* p. 17.
13. V. LUZZATI and F. HUSSON (1962) *J. Cell Biol.* **12**, 207.
14. V. LUZZATI, H. MUSTACCHI and A. E. SKOULIOS (1957) *Nature* **180**, 600.
15. V. LUZZATI, H. MUSTACCHI and A. E. SKOULIOS (1958) *Discussions Faraday Soc.* **25**, 43.
16. V. LUZZATI, H. MUSTACCHI, A. SKOULIOS and F. HUSSON (1960) *Acta Cryst.* **13**, 660.
17. V. LUZZATI and F. REISS-HUSSON (1966) *Nature* **210**, 1351.
18. A. SKOULIOS (1967) *Adv. Colloid Interface Sci.* **1**, 79.
19. F. KRAFFT and H. WIGLOW (1895) *Berichte* **28**, 2543, 2566.
20. R. C. MURRAY and G. S. HARTLEY (1935) *Trans. Faraday Soc.* **31**, 183.
21. K. SHINODA and E. HUTCHINSON (1962) *J. Phys. Chem.* **66**, 577.
22. N. A. MAZER, G. B. BENEDEK and M. C. CAREY (1976) *J. Phys. Chem.* **80**, 1075.
23. R. R. BALMBRA, J. S. CLUNIE and J. F. GOODMAN (1969) *Nature* **222**, 1159.
24. B. A. MULLEY and A. D. METCALF (1964) *J. Colloid Sci.* **19**, 501.
25. P. EKWALL, L. MANDELL and K. FONTELL (1969) *J. Colloid Interface Sci.* **31**, 508.
26. P. EKWALL, I. DANIELLSON and P. STENIUS (1973) in *MTP International Review of Science, Surface Chemistry and Colloids* (ed. M. Kerker), Series 1 Vol. 7, ch. 4, Butterworths, London.
27. P. EKWALL, L. MANDELL and K. FONTELL (1969) *Molec. Crystals* **8**, 157.
28. A. A. ALI and B. A. MULLEY (1978) *J. Pharm. Pharmac.* **30**, 205.
29. A. S. C. LAWRENCE, B. BOFFEY, A. BINGHAM and K. TALBOT (1964) *Proc. IVth Int. Cong. Surface Activity*, Vol. 2, p. 673,
30. S. FRIBERG and L. MANDELL (1970) *J. Amer. Oil Chem. Soc.* **47**, 149.

31. S. FRIBERG, L. MANDELL and K. FONTELL (1969) *Acta chem. Scand.* **23**, 1055.
32. K. J. LISSANT and G. M. BRADLEY (1979) in *Solution chemistry of surfactants*, (ed. K. L. Mittal), Vol. II, Plenum, New York, p. 919.
33. G. M. BRADLEY and K. J. LISSANT (1980) *J. Phys. Chem.* **84**, 1567.
34. B. KABBANI, F. PUISIEUX, J-P. TREGUIER, M. SEILLER and A. T. FLORENCE (1977), *Proc. 1st Int. Congr. Pharm. Tech.*, Paris.
35. I. LO. A. T. FLORENCE, J-P. TREGUIER, M. SEILLER and F. PUISIEUX (1977) *J. Colloid Interface Sci.* **59**, 319.
36. J-P. TREGUIER, I. LO, M. SEILLER and F. PUISIEUX (1975) *Pharm. Acta. Helv.* **50**, 421.
37. I. LO, F. MADSEN, A. T. FLORENCE, J-P. TREGUIER, M. SEILLER and F. PUISIEUX in *Micellization, Solubilization and Microemulsions*, (ed. K. L. Mittal), Vol. 1, Plenum, New York, p. 455.
38. R. M. A. MUSTAFA, M. A. HASSAN and H. T. FIKRAT (1979) *Canad. J. Pharm. Sci.* **14**, 43.
39. J. R. NIXON and B. P. S. CHAWLA (1969) *J. Pharm. Pharmac.* **21**, 79.
40. S. FRIBERG and I. LAPCZYNSKA (1975) *Progr. Colloid Polymer Sci.* **56**, 16.
41. K. SHINODA (1967) *J. Colloid Interface Sci.* **24**, 14.
42. M. J. GROVES, R. M. A. MUSTAFA and J. E. CARLESS (1974) *J. Pharm. Pharmac.* **26**, 616.
43. J. R. NIXON, R. S. UL HAQUE and J. E. CARLESS (1971) *J. Pharm. Pharmac.* **23**, 1.
44. J. S. MARLAND and B. A. MULLEY (1971) *J. Pharm. Pharmac.* **23**, 561.
45. R. M. WOODMAN (1964) *Emulsion Technology*, 2nd edn. Brooklyn Chemical Publishing Co. New York, p. 169.
46. W. CLAYTON (1943) *The Theory of Emulsions and their Technical Treatment*, 4th edn. Churchill, London, p. 252.
47. S. S. DAVIS (1976) *J. Clin. Pharm.* **1**, 11.
48. T. P. HOAR and J. H. SCHULMAN (1943) *Nature* **152**, 102.
49. D. G. RANCE and S. FRIBERG (1977) *J. Colloid Interface Sci.* **60**, 207.
50. A. W. ADAMSON (1969) *J. Colloid Interface Sci.* **29**, 261.
51. G. H. GILLBERG, S. LEHTINEN and S. FRIBERG (1970) *J. Colloid Interface Sci.* **33**, 40.
52. K. SHINODA and H. KUNIEDA (1973) *J. Colloid Interface Sci.* **42**, 381.
53. S. I. AHMED, K. SHINODA and S. FRIBERG (1974) *J. Colloid Interface Sci.* **47**, 32.
54. K. SHINODA and S. FRIBERG (1975) *Adv. Colloid Interface Sci.* **4**, 281.
55. J. H. SCHULMAN and D. P. RILEY (1948) *J. Colloid Sci.* **3**, 383.
56. J. E. C. BOWCOTT and J. H. SCHULMAN (1955) *Z. Electrochem.* **59**, 283.
57. W. STOCKENIUS, J. H. SCHULMAN and L. PRINCE (1960) *Kolloid-Z.* **169**, 170.
58. J. H. SCHULMAN and J. B. MONTAGUE (1961) *Ann. NY Acad. Sci.* **92**, 3661.
59. L. M. PRINCE (1967) *J. Colloid Interface Sci.* **23**, 165.
60. L. M. PRINCE (1975) *J. Colloid Interface Sci.* **52**, 182.

3 Micellization

3.1 Introduction

In Chapter 1 the characteristic property of amphiphiles in accumulating at air–water or oil–water interfaces was discussed in terms of the entropy gain caused by disruption of organized water structure and the removal of mobility constraints as the hydrophobic region was removed from an aqueous environment. An alternative to the crowding of the interface as the surfactant concentration is increased is provided by the formation of small aggregates or micelles in the bulk of the solution. The hydrophobic moieties compose the core of the micelle, being shielded from the surrounding solvent by the shell of ionic head groups. There is much experimental evidence to suggest that the mobility of hydrocarbon chains in the micellar interior resembles that in a liquid hydrocarbon. Hence the process of micellization does indeed represent a method by which the hydrocarbon chains may regain their mobility.

The concentration at which micelles first appear in solution is termed the critical micelle concentration (CMC). Experimentally the CMC is determined from the inflection point of plots of some physical property of the solution as a function of concentration. A wide variety of techniques involving the measurement of such physical properties as the surface tension, conductivity, light scattering intensity and osmotic pressure have been used in the determination of the CMC. The intrinsic reliability of these techniques has been reviewed [1]. It is important to note that the change in physical properties at the CMC occurs over a narrow concentration range rather than at a precise point. The magnitude of this concentration range may be dependent on the physical property which is measured. Similarly there are systematic differences between the various techniques for determining the CMC [2] depending, for example, on whether the technique is sensitive to changes in monomer concentration (as in surface tension techniques) or micelle concentration (as in light scattering). Such differences become important in cases where a precise CMC value is required. Some of the factors which affect both the CMC and the micellar size will be discussed in this chapter.

One of the most important consequences of micellization from a pharmaceutical point of view is that micelles are capable of solubilizing drugs of limited water solubility (see Chapter 6). Considerable interest centres on the exact location of solubilizates within the micellar structure, as discussed in Chapter 5.

To aid in an understanding of this problem the various micro-environments within the micelle will be treated in some detail.

3.2 Micellar structure

3.2.1 General structure of ionic and non-ionic micelles

The generalized structure of the cross-section of a typical ionic micelle containing n molecules is shown in Fig. 3.1. It may be thought of as having a liquid core formed by the n associated hydrocarbon chains with the fully ionized head groups projecting out into the water. Immediately surrounding the core is the Stern layer which contains not only the ionic head groups but also $(1 - \alpha)n$ counterions, where the degree of ionization α is about 0.2 to 0.5. The Stern layer constitutes the inner part of the electrical double layer surrounding the micelle the outer, more diffuse layer, which contains the remaining αn counterions is termed the Gouy–Chapman layer. The average location of the interface between the core and the water has been reported [3] to be 0.08 ± 0.04 nm above the α carbon atom of the associated alkyl chains, although as discussed below there may be appreciable water penetration into the micellar core. The outermost boundary of the Stern layer corresponds with the hydrodynamic shear surface of the micelle. The core and the Stern layer form the kinetic micelle with a total charge $\alpha n e$ where e is the elementary charge. The surface potential of the kinetic micelle is the electrophoretic zeta potential.

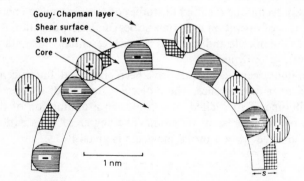

Figure 3.1 Partial cross-section of sodium dodecyl sulphate micelle. Cross-hatched area in Stern layer is available to the centres of sodium ions. From Stigter [3].

In polyoxyethylated non-ionic micelles the core is surrounded by a layer composed of the polyoxyethylene chains to which solvent molecules may be hydrogen bonded. This region of the micelle is often termed the palisade layer.

Recent years have seen a renewed interest in the investigation of the nature of the micellar regions as more sophisticated experimental techniques have become available. An understanding of the detailed nature of the micellar micro-environments is important in understanding the ability of the micelles to act as solubilizing agents and for the prediction of the location of the solubilizate molecules within the micellar entity.

3.2.2 Nature of the micellar core

Early deductions from observations on the macroscopic properties of micelles suggested an essentially liquid-like hydrocarbon core. Thus, for example, there was observed to be a similarity between the heat capacities [4] and compressibilities [5] of the micelles and those of the bulk hydrocarbon of which the core was composed. Mukerjee [6], however, drew attention to the irregular variation of the CMC with chain length in a homologous series of sodium alkyl sulphates which suggested a partial structuring of the core.

Fluorescent probes, which have for some considerable time been used in investigations of the nature of biological macromolecules and membrane systems, have recently been applied to micellar systems. Early applications of fluorescent probes in investigations of micellar systems employed probes which were highly fluorescent in non-polar regions and virtually non-fluorescent in water. Unfortunately, these probes which exhibited such a sensitivity to the polarity of the medium tended to be large and ionic. A typical probe was 8-anilino-1-naphthalene sulphonate. The use of such probes has been criticized [7, 8] on the grounds of uncertainty of the location of the probe and the possibility of perturbation of the micro-environment. Later workers utilized aromatic hydrocarbons as probes which would be less likely to perturb the micellar properties. The fluorescent technique is suited to studies of micellar micro-environments in that it is capable of operation over time scales of shorter duration than those of the lifetime of the micelle and can thus provide essentially instantaneous rather than time-average information.

Two main fluorescent techniques have been employed: fluorescent depolarization and excimer fluorescence. The molecular motion of a fluorescent probe within the lifetime of its excited state results in a diminution of the extent of polarization of the fluorescent radiation. The degree of polarization, P, of the fluorescence emitted from a probe molecule is given by

$$P = \frac{I_\parallel - I_\perp}{I_\parallel + I_\perp}, \tag{3.1}$$

where $I_\parallel$ and $I_\perp$ are the fluorescence intensities observed through a polarizer orientated parallel and perpendicular to the plane of polarization of the excitation beam. When the probe is located in the micellar core its motion is restricted to an extent which is proportional to the viscosity of its environment. The relationship between the observed P and the microviscosity, η, of the medium is given by the Perrin equation [9]

$$\left(\frac{1}{P} - \frac{1}{3}\right) \bigg/ \left(\frac{1}{P_0} + \frac{1}{3}\right) = 1 + (\tau kT/V_0\eta), \tag{3.2}$$

where P_0 is the degree of polarization in extremely viscous or rigid media, τ is the lifetime of the fluorescence, V_0 is the effective volume of the fluorescent molecule and k is the Boltzmann constant. Measurements using fluorescent probes have, in general, supported the notion of a liquid-like nature of the core. For example, Shinitzsky *et al.* [9] using two aromatic probes, perylene and methylanthracene, concluded that the interiors of micelles of several long chain cationic surfactants were of a liquid nature but less fluid than hydrocarbons of similar chain length, presumably because of anchoring of the chains at the micellar surface. A similar conclusion was reached by Rehfeld for micellar solutions of sodium phenyl-undecanoate [10].

Some molecules, e.g., pyrene, have a sufficiently long lifetime in their excited state, P^*, to interact with molecules in the ground state P to form excimers P_2^*. The process of excimer formation may be represented as follows:

$$P \to P^* \tag{3.3}$$

$$P + P^* \rightleftarrows P_2^*. \tag{3.4}$$

The forward reaction of Equation 3.4 is diffusion controlled and consequently its rate will vary inversely with the viscosity of the medium. The ratio I_E/I_M is commonly used as a measure of the ease of excimer formation, I_E and I_M being the excimer and monomer emission, respectively. Excimer formation in micellar systems requires at least two probe molecules per micelle for the reaction of Equation 3.4 to occur within the micelles. The ratio I_E/I_M is thus dependent on the distribution of probe molecules among the micelles which is assumed to follow a Poisson distribution. At the commonly used probe/surfactant molar ratio of 0.01, Zachariasse [13] calculates from Poisson statistics that 27% of sodium dodecyl sulphate (NaDS) micelles are more than singly occupied. There is difficulty in the interpretation of the fluorescence data since excimer emission occurs alongside the partly quenched monomer fluorescence in doubly or higher occupied micelles, whereas singly occupied micelles show only unquenched monomer fluorescence. This situation leads to uncertainty in the calculated microviscosity and may explain the anomalous value of 150 cP proposed by Pownall and Smith [11] for the microviscosity of the micellar core of hexadecyltrimethylammonium bromide.

A way of avoiding these uncertainties is to use probes in which the chromophores are linked together by short methylene chains, O linkages or N atoms. Such probes may form excimers intramolecularly by a rearrangement or bending of the molecule.

$$P \text{\small\bfseries}\text{P} \to P^*\text{\small\bfseries}\text{P} \to \overset{\frown}{P^* \quad P} \tag{3.5}$$

The deformation of the molecule in this manner is dependent on the solvent viscosity and hence solubilized probes give a measure of the microviscosity of their micellar environment. Providing the probes are used at low enough concentrations to avoid any intermolecular excimer formation arising from double occupancy, any problems arising from uncertainty in the distribution are avoided. A probe of this type, 1,3-di-α-naphthylpropane (I) was used by Turro *et*

(I)

al. [12] in a study of the fluidity of the micellar core of hexadecyltrimethyl ammonium bromide; a microviscosity of 39 cP was obtained in good agreement with values from fluorescence depolarization measurements (15–35 cP). Zachariasse [13] calculated a microviscosity of 19 cP for the micellar core of NaDS at 20° C using dipyrenylpropane. A much lower value (4 cP) was, however, obtained using the smaller probe 1,3-diphenylpropane indicating the difficulties still associated with the use of probe molecules to determine micellar characteristics.

An alternative spectroscopic technique which has been employed in the investigation of micellar structure involves the measurement of electron spin resonance spectra from free radicals, notably nitroxides, incorporated in probes within the micelle. Restricted motion of the probe (the resonance time is 10^{-8} s) results in hyperfine splitting in the e.s.r. spectrum which is resolvable into transition moments parallel and perpendicular to the applied magnetic field. Waggoner *et al.* [14] concluded from the observed lack of hyperfine splitting of the spectrum of nitroxide probes incorporated into sodium dodecylsulphate, that the probe was in an essentially liquid environment. However, a broadening of the spectral lines was noted indicating a fairly high local viscosity. In contrast, Povich *et al.* [15] reported that the environment of a nitroxide free-radical solubilized in hexadecyltrimethylammonium bromide was relatively rigid being similar to that of hexadecane at very low temperatures ($-22°$ C).

A disadvantage of both fluorescence and e.s.r. spectroscopic techniques which may, in part, be responsible for the conflicting estimations of the micellar core viscosity is the necessity for the use of probes. Not only is there a possibility of disruption of the environment adjacent to the probe but also in some cases a lack of knowledge of the exact location of the probe. A spectroscopic method which avoids this criticism involves the determination of [13]carbon spin-lattice relaxation times (T_1) of the micellar systems. In general, such measurements have shown an increase in the segmental motion of the hydrocarbon chain on moving from the polar head group to the terminal CH_3 group. Segmental motion in micelles of *n*-octyl trimethylammonium bromide, for example, has been shown [16] to cause a monotonic increase in methylene spin lattice relaxation times from 0.9 s at the polar head to 2.9 s at the tail. Similarly, Menger and Jerkunica [17] have interpreted T_1 values for a series of ω-phenylalkanoic acids in terms of the microviscosity of the micellar core. Anisotropic motion within the micelles of these acids was shown to depend on the depth of penetration of the phenyl group into the core.

The conformational state of the hydrocarbon chain in the micellar core has been studied using Raman spectroscopy [18–20] and n.m.r. techniques [21, 22]. Laser Raman spectra of several cationic and anionic surfactants were compared in the crystalline and micellar states. The hydrocarbon chains existed in an all-trans conformation in the crystalline state whilst the presence of gauche isomers in the micellar state was indicative of the liquid-like motion of the terminal carbon atoms of the chains in the micelle interior. From an analysis of the chemical shifts in p.m.r. spectra of *n*-acyl sarcosinates it has been suggested that an increase in the percentage of all-trans form of this surfactant occurs as concentration is increased. A similar conclusion was drawn from ^{13}C n.m.r. chemical shifts during the micellization of *n*-nonylammonium bromides. Laser Raman [18] and fluorescence depolarization [23] measurements have indicated a considerable increase in the microviscosity of the core as the micelles become more asymmetric following electrolyte addition.

The extent of water penetration into the micellar core has been the subject of much controversy. In a series of papers, Muller and co-workers [24–27] examined ionic and non-ionic surfactants having $CF_3(CH_2)_n$ groups in place of the usual hydrocarbon chain. From an analysis of the fluorine magnetic resonance chemical shifts for these surfactants, it was concluded that the average environment of the *terminal* CF_3 group, was at least partially aqueous in nature. A similar conclusion was drawn from measurements of ^{19}F spin-lattice relaxation times of micelles of heptafluorobutyric acid [28]. However, an alternative explanation of the chemical shifts in partially fluorinated surfactants has been proposed. Because of the pronounced non-ideality of interactions between fluorocarbons and hydrocarbons it has been suggested [29] that the trifluoromethyl group, rather than sampling the core, is in fact concentrated at the core's surface. An alternative method of investigation which avoids this problem of non-ideality was used by Menger *et al.* [30] and was based on the observation that the ^{13}C chemical shifts of carbonyls are solvent sensitive. The ^{13}C chemical shifts of carbonyl groups either inserted into micelles in the form of solubilized probes (e.g. octanal in hexadecyltrimethylammonium bromide) or incorporated into the surfactant molecule (e.g. 8-ketodecyltrimethylammonium bromide) were compared with chemical shifts in a range of solvents and it was suggested that water penetration into the micelles reaches at least the first seven carbon atoms of the chain. These studies have, however, been criticized [31] on the grounds that there was neither an interpretation of the molecular cause of the chemical shift nor an independent investigation of the distribution of the carbonyl group in the micelle. Proton n.m.r. [32] and spin-lattice relaxation studies [33] on polyoxyethylated non-ionic surfactants failed to detect any significant water penetration of the micelle interior.

Menger and Boyer [34] have recently presented optical rotary dispersion data which indicate appreciable penetration of water into the micellar core. A large change in the sign of the Cotton effect is induced on transferrance of (+)trans-2-chloro-5-methylcyclohexanone from heptane to water. This is ascribed to a diaxial–diequatorial equilibrium which responds to the nature of its environment, lying further to the right in water than in heptane.

in heptane   in water

The Cotton effect of the probe when incorporated in micelles of both hexadecyltrimethylammonium bromide and sodium dodecyl sulphate more closely resembles that of water than heptane, suggesting an aqueous environment for the solubilized probe.

Clearly more evidence is required before the classical picture of the Hartley micelle is abandoned although, as pointed out by Menger and Boyer, 'proof' in the usual sense of the word is precluded by the assumptions and limitations necessary in the interpretation of data from experiments designed to define the structure of transient aggregates in solution.

3.2.3 The Stern layer of ionic micelles

The electrical potential, ψ, at the interface between the micellar core and the surrounding water may be estimated by the Gouy–Chapman theory of the electrical double layer. In the classical theory, a uniform continuous interfacial surface charge is assumed, which is neutralized by a diffuse ionic layer of charges in the aqueous solution. In a detailed model of the Stern layer proposed by Stigter [35–37], this theory is refined to allow for the size and high concentration of the charge carriers at the micelle surface.

The experimental measurement of the electrical potential, ψ, has been reported by several workers using a variety of probes. A probe of molecular size which is known to be solubilized at the core–water interface and which does not itself disturb the system, is required for this purpose. In 1940 Hartley and Roe [38] suggested the use of pH indicators to determine the hydrogen ion concentration at the micellar surface, and this suggestion was later implemented by Mukerjee and Banerjee [39] using bromophenol blue and bromocresol green as indicators. The apparent pK of the indicators when solubilized in ionic micelles differed from that in the bulk and this shift was related to ψ assuming a Boltzmann distribution of the hydrogen ion concentration at the surface. One of the problems with this method is that the equilibrium of an indicator at the micellar surface will also be affected by the lower dielectric constant, ε, at the surface compared with that of the bulk solution. A method by which the observable shifts of apparent pK may be separated explicitly into a component due to electrical potential and one caused by a change in polarity has been proposed by Fernandez and Fromherz [40]. These workers attributed pK shifts for a non-ionic micelle (Triton X-100) to a reduced polarity at the micelle surface for which a dielectric constant of approximately 32 was estimated. For charged micelles, a similar dielectric constant at the micellar surface was responsible for part of the pK shift. By assuming that the remaining part was a measure of the electric potential, values of ψ of $-134\,\text{mV}$ and $+148\,\text{mV}$ were calculated for sodium dodecyl sulphate and cetyltrimethylammonium bromide, respectively. There is general agreement that the dielectric constant of the Stern layer is intermediate between the value of 79 for water and 2 for hydrocarbons. Kosower [41] showed that the charge transfer absorption spectra exhibited by micelles of alkyl pyridinium iodide were very sensitive to the polarity of the surrounding medium. Comparison of the micellar spectra and those in a series of solvents showed an effective dielectric constant of the interface of about 36. Kalyanasundaran and Thomas [42] have reported ε values ranging between 15 and 50 for several ionic and non-ionic surfactants based on measurements of fluorescence from the probe, pyrene-3-carboxaldehyde. Studies on non-ionic and zwitterionic surfactants [43] yielded values of between 36 and 46.

Estimates of the extent of binding of counterions to the kinetic micelle have come from a variety of techniques. Light scattering has been widely used to determine the effective thermodynamic degree of dissociation of the micelles expressed as $\alpha = p/n$ (where p is the effective charge per micelle of aggregation number n). The values quoted have usually been in the range $\alpha = 0.2$ to 0.3, i.e. the Stern layer retained about 70 to 80% of the counterions. More recently, ion selective electrodes have been employed to measure directly the activity of the counterions in micellar solutions [44–49]. Table 3.1 shows the range of values observed for cationic surfactants using a variety of techniques. From this Table it is seen that α decreases with increasing chain length. This reduction in α possibly arises because of increased interchain attraction leading to closer packing of the ionic head groups in the micelle with the consequence that more counterions are bound in order to reduce the increased ionic repulsion. An increase in α with increasing size of counterion and decreasing concentration of added electrolyte has been noted [50] for the anionic surfactant, sodium dodecanoate.

Table 3.1 Degree of micellar dissociation, α, in solutions of n-decyl, n-dodecyl, n-tetradecyl and n-hexadecyl trimethylammonium bromides (adapted from [49])

α values				Technique	Reference
C_{10}TAB	C_{12}TAB	C_{14}TAB	C_{16}TAB		
0.27	0.24	0.13	0.12	Ion-selective electrodes	[49]
0.224	0.193	0.131	0.091*	Light scattering	[51]
—	0.256	—	—	Ion-selective electrodes	[45]
—	0.22	0.15	—	Potentiometric method	[52]
—	—	0.249	—	Osmometry	[53]
—	—	0.321	—	Conductimetry	[53]
—	—	—	0.11	Osmometry	[54]

* In the presence of 0.013M KBr.

3.2.4 The polyoxyethylene layer of non-ionic surfactants

[13]C relaxation data for micelles of a series of alkylpolyoxyethylene glycols and *p-tert*-octylphenyl polyoxyethylene ethers are consistent with maximal restriction of segmental motion at the hydrophobic–hydrophilic interface, i.e. not only is there increasing mobility of the alkyl chains towards the centre of the micelle but the oxyethylene chain also exhibits a similar gradient of increased mobility from the interface towards the terminal unit [32, 55, 56].

The conformation of the oxyethylene chain has been discussed using 'zig-zag' and 'meander' models [57]. It is thought that polyoxyethylene chains with more than 10 ethylene oxide groups are predominantly in the form of expanded helical coils (meander model) rather than in fully extended (zig-zag) conformations. Recent laser Raman scattering studies have indicated a dihedral helical conformation of the polyoxyethylene chain in a series of alkylpolyoxyethylene glycol monoethers and alkylphenoxypolyethylene oxyethanols (Igepals and Tritons) [18].

The essentially open structure of the chain leads to physical entrapping of large amounts of water in addition to the hydration of the ester linkage by hydrogen bond formation which accounts for a maximum of two water molecules per ethylene oxide (EO) group [58]. Evidence for physically bound water in the kinetic micelle has come from hydrodynamic studies [59–61], studies of proton chemical shifts [32, 62] and proton relaxation times [24–27]. The extent of hydration is generally thought to increase with increasing ethylene oxide chain length [61] although increased crowding of the ethylene oxide groups in compounds of very long polyoxyethylene chain length has been suggested to result in dehydration. For example the micelles of $C_{16}E_{21}$ hold six water molecules per ethylene oxide unit [63] compared with only 2.35 molecules per ethylene oxide unit for $C_{12}E_{28}$ [64], and only 0.03 molecules per ethylene oxide unit for $C_{32}E_{41}$ and $C_{35}E_{40}$ calculated from the data of Arnarson and Elworthy [65].

The latter authors have suggested that part of the polyoxyethylene chain of these long chain nonionic surfactants may intrude into the hydrocarbon core of the micelles.

3.3 Micellar shape

Since micelles are dynamic structures with a liquid core it is probably unrealistic to regard them as rigid structures with a precise shape. It is, however, instructive to consider an average micellar shape.

The experimental determination of an unequivocal shape for small micelles such as those formed by ionic surfactants in the absence of added electrolyte and close to the CMC has not yet proved feasible. Although for the purposes of interpretation of experimental data it is usual to assume micellar sphericity, several authors [66–69] have shown from geometrical considerations that most of the common surfactants with a single unbranched hydrocarbon chain cannot form truly spherical micelles. Because no holes may exist in the centre of the micelle, the micellar radius is always limited by the maximum possible extension of the hydrocarbon chain. It is thus possible using experimental density values for the hydrocarbons, to establish the maximum possible aggregation number for a given hydrocarbon chain length. (see Table 3.2) Fig. 3.2 shows that the experimental aggregation numbers of micelles of surfactants which have a single normal alkyl chain as their hydrophobic moiety are generally larger than the maximum *n* values consistent with spherical micelles.

Table 3.2 Maximum values for the aggregation number of micelles of *n*-alkyl surfactants consistent with spherical shape [67]

No. of C atoms in hydrocarbon chain	Extended chain length (nm)	$d_{25}^{°}/_{25}^{°}$*	Max. aggregation number n_{max}
10	1.405	0.791	39.2
12	1.657	0.802	54.3
14	1.908	0.811	72.0
16	2.160	0.818	92.2
18	2.412	0.823	114.9

* Specific gravity of the hydrocarbon moiety.

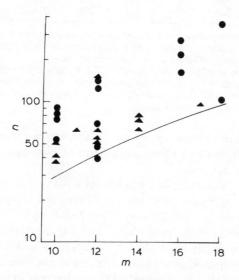

Figure 3.2 Aggregation numbers, *n*, of surfactants versus the number, *m*, of carbon atoms in the normal alkyl moiety. Key: ▲, ionic surfactants; and ●, non-ionic surfactants. Points above the curve refer to non-spherical micelles. From Schott [66] with permission.

Although there is little disagreement on the lack of sphericity, there remains some dispute as to the exact shape of these small micelles. Tanford [69] has proposed that the distortion of the micellar shape into an ellipsoid of revolution is the simplest way of incorporating a larger number of molecules into the micelles. Although the minor semiaxis of such an ellipsoid would not exceed the maximum extension of the hydrocarbon chain no such limitations would exist for the major semiaxis which could now increase to provide the additional volume. Only small values of the axial ratio a_0/b_0 (see Table 3.3) were necessary to allow a sufficient increase in the permitted aggregation number such that most hydrocarbon chain surfactants may be accounted for.

Table 3.3 Micelle aggregation number for globular micelles*[†][69]

	m_C				
	6	10	12	15	20
Ellipsoids, $b_0 = l_{max}$					
Prolate					
$a_0/b_0 = 1.25$	21	50	70	105	178
$a_0/b_0 = 1.5$	25	60	84	126	214
$a_0/b_0 = 1.75$	29	70	97	146	250
$a_0/b_0 = 2.0$	33	80	111	167	285
Oblate					
$a_0/b_0 = 1.25$	26	63	87	131	223
$a_0/b_0 = 1.5$	38	90	125	188	321
$a_0/b_0 = 1.75$	51	123	171	256	437
$a_0/b_0 = 2.0$	67	160	223	335	570

* m_C represents the number of carbon atoms in that portion of the alkyl chain which is incorporated in the hydrophobic core. This will generally be less than the total length of the hydrocarbon chain.
† The tabulated figures are the number of hydrocarbon chains per micelle. This is equal to the micelle aggregation number n for amphiphiles with a single hydrocarbon chain. It would be equal to $2n$ for amphiphiles with two chains, each of m_C carbon atoms.

There is some controversy concerning the shape of the larger micelles formed by non-ionic surfactants and ionic surfactants at high concentration or in the presence of added electrolyte.

The Triton X-100 micelle is considered by many workers to be spherical [70, 71]. From geometrical considerations, Robson and Dennis [72] have shown, however, that a spherical micelle would be possible only if several oxyethylene chains were embedded in the hydrophobic core (Fig. 3.3a). These authors consider that an oblate (Fig. 3.3b) rather than prolate (Fig. 3.3c) micelle would be most consistent with intrinsic viscosity measurements and volume calculations. Small-angle X-ray scattering measurements [73], conductivity and viscosity measurements [74] were also more consistent with an oblate ellipsoid of revolution rather than a prolate equivalent.

A disc-like (oblate) shape has been assigned to the micelles of two *n*-alkylpolyoxyethylene glycol monoethers, $C_{12}E_8$ and Lubrol WX ($C_{17}E_{16}$) from sedimentation and viscosity data [75].

From theoretical considerations, Tanford has maintained that, except when head group repulsion is very strong, oblate ellipsoids are thermodynamically

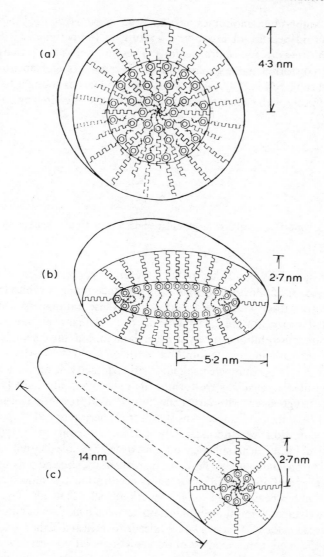

Figure 3.3 Schematic diagrams of one-half of the micelle of Triton X-100 based on geometrical calculations by Robson and Dennis [72] for (a) spherical, (b) oblate, and (c) prolate micelle models. The spherical model necessitates intrusion of the oxyethylene chains into the micellar core.

favoured over prolate ellipsoids for most micelles. Non-ionic surfactants and ionic surfactants in the presence of electrolyte would be included in this category. For ionic surfactants in the absence of added salt a transition to prolate ellipsoids with increase in solution concentration has been proposed, in agreement with experimental findings. A detailed examination of the geometric constraints on the packing of amphiphilic molecules has led Israelachvili *et al.* [76] to reject the oblate spheroid because of excessive curvature of the peripheral regions and excessive thickness of the central regions. These authors have proposed as an alternative, a distorted oblate spheroid which bears some resemblance in shape to a red blood cell (see Fig. 3.4). Further micellar growth was thought to lead to a cylindrical micelle with hemispherical ends, a model also favoured by Leibner and Jacobus [77].

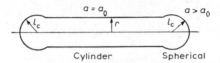

Figure 3.4 Cylindrical micelle with globular ends. From Israelachvili *et al.* [76] with permission.

Micellar shape may be affected by such factors as concentration, temperature and the presence of added electrolyte. It is not uncommon for micelle shape to undergo a transition with increase in one of these parameters from a near spheroidal shape to a more asymmetric form. ^{14}N n.m.r. relaxation measurements on aqueous solutions of cetyltrimethylammonium bromide (CTAB) have shown an increase in correlation time with concentration which has been interpreted in terms of an increasing micellar asymmetry; it was concluded that flexible rod-like aggregates were formed at higher concentrations [78]. These findings are in agreement with earlier small-angle X-ray scattering measurements on CTAB [79] which indicate a transition from spherical to rod-shaped micelles at a concentration of approximately 0.15 M. In contrast, cetyltrimethylammonium chloride forms spherical micelles over the whole concentration range from the CMC to the solubility limit. An increase in the size and polydispersity of NaDS micelles with increase in surfactant concentration in solutions of high salt concentration has been noted by several workers [80–82]. Ikeda *et al.* [82] have taken these changes to imply an equilibrium between two types of micelle; a small micelle formed at the CMC and a larger micelle formed at a higher concentration. A detailed thermodynamic analysis of the growth of NaDS micelles with increase in surfactant and electrolyte concentration has been proposed by Missel *et al.* [83].

The changes which occur on addition of electrolyte to aqueous solutions of sodium dodecyl sulphate have been the subject of much recent investigation. Mazer and co-workers [80, 81] using quasi-elastic light scattering techniques noted that at a fixed NaDS concentration and temperature, an increase of concentration of added sodium chloride over the range 0 to 0.6 M resulted in a dramatic increase in micellar weight and a change in shape from roughly spheroidal aggregates to a polydisperse distribution of spherocylindrical aggregates. The presence of rod-like micelles in aqueous solutions of NaDS in 0.6 M NaCl at temperature below 40° C was later confirmed by Young *et al.* [84]. Some limitations of these studies were pointed out by Hayashi and Ikeda [85] who investigated the effect of electrolyte on NaDS micelles using classical light-scattering methods. A transition from spherical to rod-like micelles was noted when the concentration of added sodium chloride was increased beyond 0.45 M and the concentration of the NaDS was well in excess of the CMC. In an extension of this study to include the effects of other sodium halides on the micellar shape of NaDS [82] it was concluded that the sphere–rod transition is not influenced by the halide ion species of the added salt, rather, it is caused by the electrostatic effect of counterion binding on the micelle.

A change in shape of the micelles of dodecyldimethylammonium chloride from spheroidal to rod-like aggregates when the added NaCl concentration exceeded 0.8 M has been reported [86] and the flexibility of the rod-like aggregates has been examined [87].

3.4 Polydispersity of micelle size

Although the polydispersity of micelle size is an important property of a micellar system it has so far, except for a few isolated cases, eluded direct measurement. The reason for this is purely a question of the limitations of experimental technique.

Polydispersity of size is conveniently expressed as the ratio of the weight-average n_w, to the number-average n_n degree of association where n_n and n_w are defined as

$$n_n = \sum_{n=2}^{\infty} X_n \bigg/ \sum_{n=2}^{\infty} X_n/n \tag{3.6}$$

$$n_w = \sum_{n=2}^{\infty} nX_n \bigg/ \sum_{n=2}^{\infty} X_n \tag{3.7}$$

where X_n is the mole fraction of amphiphile contained in micelles of aggregation number, n. A ratio of n_w/n_n of unity, of course, represents a completely monosized system. Even small deviations of this ratio from unity can represent an appreciable spread of micellar sizes and thus a high degree of accuracy is demanded of the measurements of the two average degrees of association. The determination of n_w is relatively straightforward using light-scattering or ultracentrifugation techniques. Even so, the errors in the absolute values of weight-average micellar weight may approach $\pm 10\%$ particularly with the former technique which necessitates calibration against a primary standard. It is, however, the determination of number-average micellar weight which presents the most problems. Vapour-pressure osmometry is not of sufficient sensitivity for determinations on the large micelles of non-ionic surfactants and the uncertainty in the counterion concentration in micellar systems of ionic surfactants renders this colligative technique of little value. Membrane osmometry, which is an absolute technique capable of a precision of about $\pm 5\%$, may be used for aggregates of sufficient size to be retained by the membrane. (Usually a micellar weight exceeding about 6 to 8×10^3 is required). This limitation generally restricts its use mainly to non-ionic surfactants, although successful measurements on ionic systems have been reported [88–91]. Measurements on small micelles generally require correction for leakage through the membrane and such corrections increase the error in the micellar weight value. Attwood *et al.* [90] have reported n_w/n_n ratios for the non-ionic surfactants, $C_{12}E_6$ and $C_{16}E_9$ of $1.00 (\pm 0.13)$ and $1.10 (\pm 0.14)$, respectively, indicating a lack of any appreciable polydispersity in these systems. Similarly the number-average micellar weight of cetomacrogol was within 5% of the weight-average value from light-scattering techniques [92]. A more promising experimental approach to the determination of polydispersity may lie in the technique of quasi-elastic light-scattering spectroscopy. This technique utilizes the temporal fluctuations in the intensity of the scattered light which arise from Brownian motion, in a determination of the transitional diffusion coefficients of the micellar species. Not only is the mean diffusion coefficient determined, but also its variance which is related to the polydispersity of the system. The technique is, however, extremely sensitive to slight contamination of the solutions and its meaningful application is limited to a study of large aggregates. Mazer *et al.* [80] have reported a significant polydispersity of micellar sizes in sodium dodecyl sulphate solutions containing a high concentration of added electrolyte. The estimated range of micellar species extended to $\pm 70\%$ of the weight-averaged mean value.

From an analysis of the variation of weight-average micellar weight with concentration for several non-ionic and ionic surfactants using a multiple equilibrium model of micellization, Mukerjee [93] concluded that for small micelles, the size distributions are quite narrow, for example the n_w/n_n ratio for sodium dodecyl sulphate in 0.1M NaCl was calculated to be < 1.03. In contrast, the polydispersity of sizes in solutions containing larger aggregates, e.g. the polyoxyethylated non-ionic surfactants, was shown to be appreciable with n_w/n_n approaching a value of 2. Similarly, Corkill *et al.* [94] also using a multiple equilibrium micellar model, have demonstrated polydispersity of micellar size in the micelles of the zwitterionic surfactant $C_8H_{17}N^+(CH_3)_2(CH_2)_3SO_3^-$ (see Fig. 3.5).

Recent developments of the theory of kinetics of micellar equilibrium by Aniansson *et al.* [95] have demonstrated the potential of those chemical relaxation techniques, such as ultrasonic absorption, which are capable of following the kinetics of monomer exchange between micellar species and the bulk solution, in the determination of micellar size distribution curves. Further details of this and other theories of the kinetics of micelle formation are given in Section 3.7. Table 3.4 shows that the ratio of the half-width σ of the size distribution curve (assumed to be Gaussian) to the aggregation number n, decreases with increase in the length of the hydrophobic tail in a homologous series of ionic surfactants indicating that the micelles became more monodisperse with increasing hydrophobic chain length. A broadening of the size distribution curve of a given surfactant with increase in temperature was also noted.

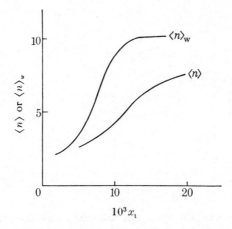

Figure 3.5 Aggregation numbers for $C_8H_{17}N^+(CH_3)_2(CH_2)_3SO_3^-$, $\langle n \rangle_w$ and $\langle n \rangle$ against total concentration (expressed as mole fraction). From Corkill *et al.* [94] with permission.

Table 3.4 Values of the aggregation numbers n and distribution widths, σ, for sodium alkyl sulphates at 25° C [95]

Surfactant	n	σ	σ/n
NaC_6SO_4	17	6	0.353
NaC_7SO_4	22	10	0.454
NaC_8SO_4	27	—	—
NaC_9SO_4	33	—	—
$NaC_{10}SO_4$	41	—	—
$NaC_{11}SO_4$	52	—	—
$NaC_{12}SO_4$	64	13	0.203
$NaC_{14}SO_4$	80	16.5	0.206
$NaC_{16}SO_4$	100	11	0.110

Size distribution curves have also been calculated by Tanford from a theoretical treatment of micelle formation [96, 97].

3.5 Factors affecting the CMC and micellar size

3.5.1 Nature of the hydrophobic group

The large majority of amphiphiles, whether ionic or non-ionic have hydrophobic regions composed of hydrocarbon chains. For ionic amphiphiles, increase in the number of carbon atoms in unbranched hydrocarbon chains leads to a decrease in the CMC. Fig. 3.6 shows a linear relationship between log CMC and the number

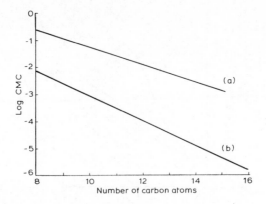

Figure 3.6 Variation of critical micelle concentration with hydrocarbon chain length for: (a) sodium alkyl sulphates and alkyltrimethylammonium bromides; (b) hexaoxyethylene monoalkyl ethers.

of carbon atoms in the chain, m, for compounds with the same head group. The dependence of CMC on m may be expressed by the empirical equation

$$\log CMC = A - Bm, \tag{3.8}$$

where A and B are constants for a homologous series. The theoretical basis of Equation 3.8 is considered in Section 3.6. As a general rule for ionic surfactants, the CMC is halved when the length of the straight hydrocarbon chain is increased by one methylene group. For chains of greater length than 16 carbon atoms this relationship no longer holds and further increase in chain length often has no appreciable effect on the CMC, possibly due to the coiling of the long chains in solution [2]. An even more pronounced decrease in CMC with increase of hydrocarbon length is noted with non-ionic surfactants, the addition of one methylene group causing the CMC to decrease to approximately one-third of its original value [98]. For branched hydrocarbon chains the effect on the CMC of an increase in the number of carbon atoms in the branched segments of the chain is not as great as that following a similar increase when the carbon atoms are in a straight chain.

As might be expected, the increased hydrophobicity conferred by an increase of chain length also causes an increase in micellar size. In many cases a linear relationship has been established between log (micellar weight) and the hydrocarbon chain length. Arnarson and Elworthy [65] have demonstrated a linear relationship between the aggregation numbers of polyoxyethylene non-ionic surfactants and the number of carbon atoms in the hydrocarbon chain (see Fig. 3.7) for surfactants with similar ratios of (number of ethylene oxide units)/(number of carbon atoms in hydrocarbon chain).

The effect on micellar properties of the introduction of a fluorine atom into the hydrocarbon chain has been investigated. Muller *et al.* [24–27] reported that substitution of the CF_3 group for the terminal CH_3 group of the surfactants

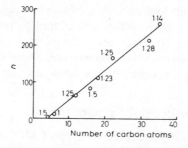

Figure 3.7 Micellar aggregation number, *n*, against the number of carbon atoms in the hydrocarbon chain. The numbers are the ratio number of ethylene oxide units: number of carbon atoms in the hydrocarbon chain. C_{22}, interpolated from Arnarson and Elworthy, unpublished data. C_{18}, Elworthy and Patel, unpublished. C_{16}, Elworthy [99] C_{12}, interpolated from Becher [100] C_4 and C_6, Elworthy and Florence [101]. From Arnarson and Elworthy [65] with permission.

hydrocarbon chain roughly doubled the CMC. This effect was confirmed by Gerry *et al.* [102] who also noted corresponding decreases in aggregation number. Such results are surprising since fluorocarbons are more hydrophobic than hydrocarbons and consequently the introduction of a F atom might be expected to promote micellization rather than impede it. It has been suggested [29] as mentioned earlier that because of mutual phobicity of hydrocarbons and fluorocarbons, the CF_3 groups tend to occupy positions at the micellar surface rather than in the core thus reducing the tendency of the partially fluorinated compounds to micellize. If this is true then the effect of replacement of all of the protons of the chain with F atoms, as in the perfluoroalkane carboxylic acids which results in an increase in hydrophobicity [103] is expected.

Some surfactants possessing two hydrocarbon chains attached to a single head group, e.g. the dialkyldimethylammonium chlorides, form lamellar structures when dispersed in aqueous solution. When such turbid solutions are subjected to ultrasonic irradiation, optically clear solutions are formed in which the surfactant is dispersed in the form of closed vesicles [104] similar in structure to the liposomes formed by phospholipids. The use of these totally synthetic bilayer vesicles as model membranes and as potential drug delivery systems is currently under investigation.

A phenyl group is roughly equivalent in its effect on the CMC to three and a half methylene groups, when introduced into a straight hydrocarbon chain. Substituents on a phenyl ring or other aromatic ring systems may profoundly effect the micellar properties. In general such groups as —Cl, —Br, —I, —F, and —CH_3 increase the hydrophobicity as evidenced by a decrease in CMC and increase in aggregation number.

Perhaps the most dramatic effect of the hydrophobic group on the association characteristics is to be found in the small group of amphiphiles with aromatic hydrophobic regions. An appreciable number of drug molecules are based on a diphenylmethane moiety and associate in solution in a typically micellar manner.

Conversion of this flexible hydrophobic moiety into a rigid planar structure can, in some cases, lead to a non-micellar association pattern in which the mean aggregate size increases continuously with increase in concentration. Such systems have no CMC and exhibit a considerable degree of polydispersity in the aggregate size. This topic is dealt with in detail in Chapter 4.

3.5.2 Nature of the hydrophilic group

There is a pronounced difference between the CMCs of ionic and non-ionic surfactants with identical hydrophobic moieties. The lower CMCs of the non-ionic surfactants are a consequence of the lack of electrical work necessary in forming the micelles.

A detailed study of the effect of the nature of the polar group of ionic surfactants on the micellar properties has been reported by Anacker and co-workers. These authors concluded that an important factor controlling the micellar size was the mean distance of closest approach of a counterion to the charge centre of the surfactant [105]. Thus, for example, decylammonium bromide forms very much larger micelles than decyltrimethylammonium bromide because the Br⁻ counterions are able to approach more closely the charged nitrogen atom of decylammonium thus effectively shielding the repulsive electrical forces and allowing larger micelles to form. Solvent interaction may also be an influential factor [106]. Hydrogen bonding between the oxygen atom of decylmorpholinium bromide and water is thought to be responsible for the smaller micellar size of this compound as compared to decylpiperidinium bromide which does not interact with the solvent in this way. The replacement of an ethyl group associated with the polar head by an ethanol group in a series of cationic surfactants which included decylethyl-, decyldiethyl- and decyltriethyl-ammonium bromides caused an increase in aggregation number [107]. The cause of this effect was attributed to changes in the effective dielectric constant produced when the polar head structure is changed, intermolecular hydrogen bonding or hydrogen bonding between the polar head hydroxyethanol group and water. The effect of changing the charge-bearing atom in the polar head has been reported [108]. Replacement of nitrogen in decyltrimethylammonium bromide by phosphorus or arsenic increased the aggregation number by at least 20% with a corresponding 35% decrease in CMC. In most cationic surfactants the charge is localized on a single N atom which is typically located at the end of the hydrocarbon chain. However, when the polar group is a piperazinium or pyridinium ring the mode of delocalization of the charge around the ring may affect the aggregation number. An increase in the positive charge on that particular atom of the ring to which the hydrocarbon chain is attached leads to a decrease in aggregation number [109].

As might be expected, the more ionized groups present in the surfactants, the higher the CMC due to the increase in electrical work required to form the micelle (Table 3.5).

The position of the ionic group also affects the micellar properties. Evans [113]

Table 3.5 Effect of number of ionized groups on CMC

Compound	$CH_3(CH_2)_{11}COOK$	$CH_3(CH_2)_9CH(COOK)_2$	$CH_3(CH_2)_7CH(COOK)CH(COOK)_2$
CMC (mol l^{-1})	0.0125	0.13	0.28
Ref.	[110]	[111]	[112]

demonstrated an increase in the CMC of a series of sodium alkyl sulphates as the sulphate group was moved from the terminal position to a medial position along the chain.

For the polyoxyethylated ether type of non-ionic surfactant, increase of the length of the polyoxyethylene chain caused an increase in the CMC and a decrease in the micelle size (Table 3.6). Increasing the polyoxyethylene chain length makes the monomer more hydrophilic and the CMC increases. The same effect may be partially responsible for the decrease in micellar size with the increased chain length, but other factors, including possible geometric considerations of the packing of the monomer into the micelles may be involved.

Table 3.6 CMCs and micelle weights of hexadecyl polyoxyethylene ethers, $CH_3(CH_2)_{15}(OCH_2CH_2)_xOH$ [61, 114]

	x					
	6	7	9	12	15	21
10^6 CMC (mol l^{-1})	1.7	1.7	2.1	2.3	3.1	3.9
10^{-5} micellar weight	12.3	3.27	1.4	1.17	—	0.82
n	2430	590	220	150	—	70

3.5.3 Nature of the counterion

The counterions associated with an ionic amphiphile may have a pronounced effect on micellar properties. In extreme cases, for example with amphiphilic drugs such as mepyramine and brompheniramine maleate containing pyridine rings, a proton transfer interaction may occur between maleate counterions and the nitrogen of the pyridine ring [115]. A consequence of this interaction is a very polydisperse micellar solution with no clear CMC [116] (see Chapter 4).

In conventional ionic surfactants a change in counterion to one of greater polarizability or valence leads to a decrease in CMC and corresponding increase in aggregation number. The size of the counterion is also a determining factor, an increase of CMC is noted with increase in hydrated radius. Counterion effects on the CMC of two typical surfactants, are shown in Table 3.7 [2].

3.5.4 Effect of additives

Numerous studies have been reported of the effects of added electrolyte on the micellar properties of ionic surfactants. Values of CMC in the presence of

Table 3.7 Variation of the CMC with different counterions [2]

Lauryl sulphates	CMC in water at 25° $(\text{mol}\,1^{-1} \times 10^3)$	Dodecyltrimethyl-ammonium salts	CMC at 31–32° in 0.500 M sodium salt[†] $(\text{mol}\,1^{-1} \times 10^3)$
Li^+	8.92*	IO_3^-	5.1
Na^+	8.32*	CHO_2^-	6.0
K^+	7.17 (32°)*	BrO_3^-	3.3
Cs^+	6.09*	F^-	8.4
$(CH_3)_4N^+$	5.52*	Cl^-	3.8
$(C_2H_5)_4N^+$	3.85*	NO_3^-	0.8
$(n\text{-}C_3H_7)_4N^+$	2.24*	Br^-	1.9
$n\text{-}C_4H_9(CH_3)_3N^+$	2.38*		
$n\text{-}C_6H_{13}(CH_3)_3N^+$	1.25*		

* From specific conductance versus concentration plots.
† From light-scattering data. The solutions were made up by dissolving dodecyltrimethylammonium bromide in 0.500 M NaX, where X is IO_3^-, CHO_2^-, etc.

electrolyte are available in the reference text by Mukerjee and Mysels [1]. Addition of electrolyte causes a reduction in the thickness of the ionic atmosphere surrounding the polar head groups and a consequent decreased repulsion between them. These effects are manifest as a reduction in CMC and an increase in aggregation number. An empirical equation relating the CMC to the electrolyte concentration has been used [117]

$$\log \text{CMC} = -a \log C_c + b \tag{3.9}$$

where a and b are constants for a particular ionic group and C_c denotes the total counterion concentration. The theoretical basis of Equation 3.9 is considered in Section 3.6.

As an illustration of the effect of electrolyte on micellar size of ionic surfactants, the sodium dodecyl sulphate system will be considered; a system which has been subjected to recent intensive investigation. Kratohvil [118] has critically examined the literature values for the aggregation number of NaDS in the presence of added sodium chloride from a variety of techniques including quasi-elastic light scattering [119], classical light scattering [120], sedimentation equilibrium [121] and a technique based on the quenching of luminescence [122]. A wide variation in values was noted; the most rigorous studies were considered to be those of Huisman [120] and Doughty [121] and their values were recommended as standard values for NaDS (see Table 3.8). As discussed earlier, the shape of the micelles of NaDS undergoes a transition from sphere to rod at a concentration of approximately 0.45 M NaCl at 25° C. A corresponding dramatic increase of aggregation number (see Fig. 3.8) has been reported by several workers at a similar electrolyte concentration from light scattering [82, 85], membrane osmometry [91] and a technique involving the use of pyrene excimer formation [124].

The effect of urea addition on the CMC is of interest in view of the disruptive

Table 3.8 Aggregation numbers for sodium dodecyl sulphate in the presence of added sodium chloride at 25° C [118]

Concentration NaCl*	Aggregation number	
	[120]	[121]
0	58	—
0.1	91	91
0.2	104†	105
0.3	116	118

* Concentration of NaCl in molarity in [120] and molality in [121].
† Interpolated values.

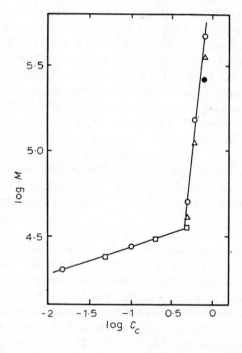

Figure 3.8 Logarithmic relation of apparent micelle molecular weight with ionic strength: (○) 25° C; (△) 30° C; (●) 35° C; (□) data of Emerson and Holtzer [123]. From Hayashi and Ikeda [85] with permission.

effect which this compound has on the structure of water. Table 3.9 shows the effects of urea on the CMC of dodecylpyridinium iodide [125]. The increase in CMC with increase in urea concentration confirms the role of water structure in micelle formation (see Chapter 1). However, as pointed out by Mukerjee and Ray [125], the effects on the CMC are relatively small compared to hydrophobic

Table 3.9 CMC data on dodecylpyridinium iodide (from [125])

	Urea concentration (mol l^{-1})				
Medium	0	0.96	3.4	5.9	8.0
Water, 25° C	0.00526	—	0.00934	0.0136	—
Water, 45° C	0.00670	—	0.0118	0.0171	0.0213

effects; for example, a decrease in the hydrocarbon chain length by only two methylene groups increases the CMC by a factor of 4.

A detailed study of the effect of the inorganic additives on solutions of non-ionic surfactants has been reported by Schott and co-workers [126–128]. Inorganic electrolytes may have appreciable effects on the cloud point of the non-ionic surfactant, i.e. the temperature at which phase separation occurs (see Section 3.5.5). Schott has identified two categories of additive which cause considerable increases in cloud point: (a) urea and salts with anions known to break the structure of water such as iodides, thiocyanates and perchlorates, and (b) salts with cations capable of forming complexes with model ethers such as dioxane. In the latter case the resulting complexation increases the solubility of the surfactant molecules above that in water, so increasing the cloud point. This phenomenon is referred to as salting-in. Electrolytes in the second category include strong acids and salts of lithium and polyvalent cations such as lead, cadmium, magnesium, nickel, aluminium and calcium. Lowering of the cloud point (i.e. salting out) was noted for relatively few electrolytes, being restricted to electrolytes with non-complexing cations (sodium, potassium, ammonium, cesium and rubidium) in association with anions such as nitrate with lyotropic numbers below 11.7.

A lowering of the CMC of polyoxyethylated non-ionic surfactants following the addition of electrolyte has been noted by many workers [127–131]. The magnitude of the lowering is a lot smaller than electrolyte effects on ionic surfactants. As seen from Table 3.10 the most effective electrolytes in causing CMC lowering are the nitrates of sodium and potassium. These two cations were also the most effective in lowering the cloud point. However, even electrolytes such as lithium, calcium, nickel, lead and aluminium nitrates which increase the cloud point are capable of lowering the CMC. The CMC-increasing effect of urea shown in this table has also been reported for polyoxyethylated lauryl and cetyl alcohols [132, 133].

From a study of the charge transfer interaction between micelles of dodecyl heptaoxyethylene glycol monoether ($C_{12}E_7$) and the strong electron acceptor 7,7,8,8-tetracyanoquinodimethane (TCNQ) it was concluded [134] that the compactness of the $C_{12}E_7$ micelles increased with addition of LiCl, NaCl, KCl, KBr and K_2SO_4 and decreased with addition of KNO_3, KSCN and urea.

The addition of lower alcohols to ionic surfactants causes a decrease in the CMC which becomes more pronounced with increase in hydrophobicity of

Table 3.10 Effect of additives on the CMC of a polyoxyethylated oleyl alcohol* (Brij 96) [127, 128]

Additive	CMC, % (w/w), at additive molalities of			
	0.50	1.0	2.0	3.0
Sodium nitrate	0.0025	0.0025	0.0015	0.0015
Potassium nitrate	0.003	0.0015	0.0015	
Lithium nitrate	0.002	0.0015	0.001	0.0006
Hydrochloric acid	0.003	0.004	0.003	0.003
Sulphuric acid	0.005	0.004	0.003	0.003
Magnesium nitrate	0.009	0.006	0.007	0.010
Calcium nitrate	0.0065	0.004	0.005	0.004
Aluminium nitrate	0.0035	0.003	0.005	
Lead nitrate	0.0045	0.003		
Nickel nitrate	0.004	0.002	0.0025	
Urea	0.003	0.0095	0.010	0.010
Cadmium nitrate	0.0025	0.0045	0.0055	0.005
	0.10	0.25		
Cadmium nitrate	0.0025	0.003		

* The CMC with no additive was 0.0055–0.0060 % (w/w).

the added alcohol. A linear relationship between the CMC decreasing power and the number of carbon atoms in the alcohol molecule was established for a series of potassium soaps [134]. Several authors have discussed the effect of alcohol in terms of the standard free energy of transfer of the alcohol from the aqueous to the micellar phases, ΔG_p° [135–137]. The main factor which causes a decrease in CMC is likely to be the reduction of the free energy of the micelle due to the diluted surface charge density on the micelle.

The literature studies on the effect of *n*-alcohols on the aggregation number, *n*, of ionic surfactants have recently been discussed by Backlund *et al.* [138]. The data suggest that water soluble alcohols (methanol to butanol) are predominantly dissolved in the water phase and may increase or decrease the aggregation number depending on the alcohol concentration. Moderately soluble alcohols (pentanol, hexanol) are distributed between the aqueous and micellar phases and at low concentrations may increase *n*. Sparingly soluble alcohols such as heptanol and octanol are almost entirely solubilized in the micelles and thereby increase *n*.

In contrast to the extensive study of the effect of added alcohol to ionic micelles, only a few studies [139, 140] have been reported for non-ionic systems. Whereas methanol and ethanol cause a CMC increase, the higher alcohols butanol and pentanol, cause a decrease in this property. Propanol exhibits an intermediate effect, low concentrations causing a decrease in CMC, higher concentrations ($> 1 \, \text{mol} \, \text{l}^{-1}$) causing an increase. The CMC-increasing effect of the lower alcohols has been attributed to a weakening of the hydrophobic bonding. The CMC-decreasing effects are thought to be a consequence of the penetration of the alcohols into the palisade layer of the micelle, forming a mixed micelle.

3.5.5 Effect of temperature

Fig. 3.9 shows the variation of the CMC with temperature for an ionized surfactant (sodium dodecyl sulphate) and a non-ionic surfactant ($C_{10}E_5$). The minimum in the curve for ionic surfactants occurs typically between 20 and 30° C. Isolated examples exist of minima in CMC–temperature curves for non-ionic surfactants, e.g. minima were noted at approximately 50° C in a series of octylphenoxyethoxyethanols with oxyethylene chain lengths of between 6 and 10 [141]. The general failure to detect minima in curves for non-ionics could conceivably be a consequence of a lack of data at sufficiently high temperatures. In many cases such measurements would not be feasible due to phase separation at elevated temperatures. The decrease in the CMC of ionic surfactants with temperature increase at lower temperatures is possibly due to dehydration of the monomers, whilst further temperature increase causes disruption of the structured water around the hydrophobic groups which opposes micellization.

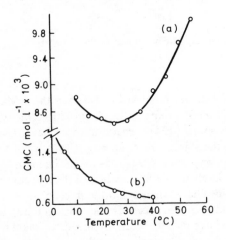

Figure 3.9 Variation of CMC with temperature for: (a) sodium dodecyl sulphate; (b) $CH_3(CH_2)_9(OCH_2CH_2)_5OH$. (After Goddard and Benson [142]).

Whereas a decrease of the micelle size of ionic surfactants with increase in temperature has been reported [143], micelles of many polyoxyethylene non-ionic surfactants increase rapidly in size with rising temperature [59, 144–149]. In some systems the rapid increase in micelle size is noted only above a characteristic transition temperature and is accompanied by an increase in micelle asymmetry. From viscosity data at elevated temperatures, it has been suggested that an increase in the extension of the polyoxyethylene chains occurs as the temperature is increased, resulting in an increase in the amount of water physically trapped by the micelles [59].

The most drastic effect of temperature on non-ionic surfactants is the effect on solubility. Non-ionic surfactants form isotropic solutions below a lower Krafft point and an upper cloud point. Heating a clear solution to above the cloud point causes a reversible phase separation: a phase rich in surfactant separates out of solution leaving an aqueous phase containing surfactant monomers.

3.5.6 Effect of pressure

The effect of pressure on the CMC of a series of alkyltrimethylammonium bromides and on sodium dodecyl sulphate has been mainly studied by conductivity techniques [150–156]. An increase in CMC with pressure increase was reported up to pressures of about 150 MPa, followed by a CMC decrease at higher pressures. Such behaviour has been rationalized in terms of a solidification of the micellar interior [150], a pressure-induced increase in the dielectric constant of water [151] and other aspects related to water structure [155, 156]. More recent studies using differential absorbance measurements [157] have, however, shown a monotonic increase in the CMC of NaDS with pressure up to 550 MPa and it has been suggested that the reversal of behaviour reported by earlier workers may be an artefact arising from the conductivity technique. The differential absorbance measurements involved the use of an optical probe (naphthalene) which was solubilized within the NaDS micelles. This method has been criticized on the grounds that the CMCs determined by these workers were those for the formation of mixed NaDS-naphthalene micelles and thus related to the effect of pressure on the solubilization process [158]. Further evidence for a maximum in the CMC–pressure plot was presented by Nishikido *et al.* [158] using the optical method of Rodriguez and Offen, and measuring the absorbance from the charge transfer band formed on micellization of dodecylpyridinium bromide. Fig. 3.10 compares CMC values from this method and those previously determined by electroconductivity methods.

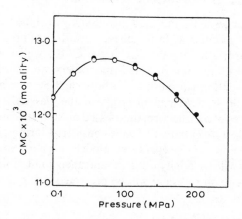

Figure 3.10 The CMC of dodecylpyridinium bromide as a function of pressure at 303 K: (○) optical method; (●) electroconductivity method. From Nishikido *et al.* [158] with permission.

The effect of pressure on the micellar size in aqueous solutions of NaDS and $C_{12}E_6$ has been examined by Nishikido and co-workers [159]. Fig. 3.11a shows a minimum in the aggregation number–pressure curve of NaDS which corresponds to the maximum in the CMC–pressure curve (Fig. 3.10). The results suggest that the dissociation of micelles into monomers is caused by compression up to about 100 MPa while the association of monomers into micelles is promoted at higher pressures. The aggregation number of $C_{12}E_6$ decreased with increasing pressure (Fig. 3.11b) the rate of decrease being most marked at lower pressures. This observation was discussed in terms of the effect of pressure on hydrophobic and hydrogen bonding.

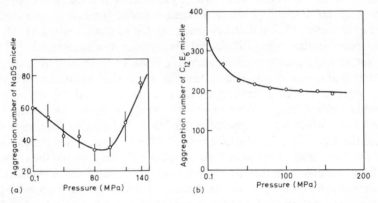

Figure 3.11 Aggregation number as a function of pressure of micelles of (a) sodium dodecyl sulphate at $30°C$, (b) $C_{12}E_6$ at $25°C$. From Nishikido *et al.* [159] with permission.

3.6 Thermodynamics of micelle formation

Two main approaches to the thermodynamic analysis of the micellization process have gained wide acceptance. In the phase separation approach the micelles are considered to form a separate phase at the CMC, whilst in the mass-action approach micelles and unassociated monomers are considered to be in association–dissociation equilibrium. In both of these treatments the micellization phenomenon is described in terms of the classical system of thermodynamics. Theories of micelle formation based on statistical mechanics have also been proposed [160–162] but will not be considered further. The application of the mass-action and phase-separation models to both ionic and non-ionic micellar systems will be briefly outlined and their limitations discussed. More recent developments in this field will be presented.

3.6.1 Phase-separation model

In this approach the micelle, and in the case of ionic micelles this includes the counterions, is treated as a separate phase. It is clear from previous sections that micelles do not constitute a 'phase' according to the true definition of this concept since they are not homogeneous and uniform throughout. Similarly there are problems associated with the application of the phase rule when considering the micelles as a separate phase [163].

(A) APPLICATION OF THE PHASE-SEPARATION MODELS TO
NON-IONIC SURFACTANTS

To calculate the thermodynamic parameters for the micellization process we require to define the standard states. The hypothetical standard state for the surfactant in the aqueous phase is taken to be the solvated monomer at unit mole fraction with the properties of the infinitely dilute solution. For the surfactant in the micellar state, the micellar state itself is considered to be the standard state.

If μ_s and μ_m are the chemical potentials per mole of the unassociated surfactant in the aqueous phase and associated surfactant in the micellar phase, respectively, then since these two phases are in equilibrium

$$\mu_s = \mu_m. \tag{3.10}$$

For a non-ionized amphiphile,

$$\mu_s = \mu_s^{\ominus} + RT \ln a_s. \tag{3.11}$$

If it is assumed that the concentration of free monomers is low then the activity of surfactant monomer, a_s, may be replaced by the mole fraction of monomers x_s and Equation 3.11 becomes

$$\mu_s = \mu_s^{\ominus} + RT \ln x_s \tag{3.12}$$

where $\mu_s^{\ominus}$ is the chemical potential of the standard state.

Since the micellar material is in its standard state

$$\mu_m = \mu_m^{\ominus}. \tag{3.13}$$

If $\Delta G_m^{\ominus}$ is the standard free energy change for the transfer of one mole of amphiphile from solution to micellar phase, then

$$\Delta G_m^{\ominus} = \mu_m^{\ominus} - \mu_s^{\ominus}$$
$$= \mu_m - \mu_s + RT \ln x_s$$
$$= RT \ln x_s. \tag{3.14}$$

Assuming that the concentration of free surfactant in the presence of micelle is constant and equal to the CMC value, x_{CMC}, then

$$\Delta G_m^{\ominus} = RT \ln x_{CMC}. \tag{3.15}$$

x_{CMC} is the CMC expressed as a mole fraction and is defined by

$$x_{CMC} = \frac{n_s}{n_s + n_{H_2O}}. \tag{3.16}$$

Since the number of moles of free surfactant, n_s, is small compared to the number of moles of water, n_{H_2O}, Equation 3.16 may be approximated to

$$x_{CMC} = n_s / n_{H_2O}. \tag{3.17}$$

Substituting Equation 3.17 into Equation 3.15 and converting to decadic logarithms

$$\Delta G_m^{\ominus} = 2.303 RT (\log CMC - \log w) \qquad (3.18)$$

where $w = $ mol dm^{-3} water (55.40 mol dm^{-3} at 20° C).

Application of the Gibbs–Helmholtz equation to Equation 3.15 yields

$$\frac{\partial}{\partial T}\left(\frac{\Delta G_m^{\ominus}}{T}\right)_P = -R\left(\frac{\partial \ln x_{CMC}}{\partial T}\right)_P = \frac{-\Delta H_m^{\ominus}}{T^2}. \qquad (3.19)$$

Hence the standard free energy of micellization per mole of monomer $\Delta H_m^{\ominus}$, is

$$\Delta H_m^{\ominus} = -RT^2\left(\frac{\partial \ln x_{CMC}}{\partial T}\right)_P = R\left(\frac{\partial \ln x_{CMC}}{\partial (1/T)}\right)_P. \qquad (3.20)$$

Finally the standard entropy of micellization per mole of monomer, $\Delta S_m^{\ominus}$, may be obtained from

$$\Delta S_m^{\ominus} = (\Delta H_m^{\ominus} - \Delta G_m^{\ominus})/T. \qquad (3.21)$$

(B) APPLICATION OF THE PHASE-SEPARATION MODEL TO IONIC SURFACTANTS

In the calculation of $\Delta G_m^{\ominus}$ it is necessary to consider not only the transfer of surfactant molecules from the aqueous to the micellar phase but also the transfer of $(1 - \alpha)$ moles of counterion from its standard state to the micelle. Equation 3.14 is thus written

$$\Delta G_m^{\ominus} = RT \ln x_s + (1 - \alpha)RT \ln x_x \qquad (3.22)$$

where x_s and x_x are the mole fractions of surfactant ion and counterion respectively.

The analogous equations to Equations 3.15 and 3.18 for an ionic surfactant in the absence of added electrolyte are

$$\Delta G_m^{\ominus} = (2 - \alpha)RT \ln x_{CMC} \qquad (3.23)$$

and

$$\Delta G_m^{\ominus} = (2 - \alpha)2.303 RT (\log CMC - \log w). \qquad (3.24)$$

It is often assumed that the micellar phase is composed of the charged aggregate together with an *equivalent* number of counterions, and Equations 3.23 and 3.24 are approximated to

$$\Delta G_m^{\ominus} = 2RT \ln x_{CMC} \qquad (3.25)$$

and

$$\Delta G_m^{\ominus} = 4.606 RT (\log CMC - \log w). \qquad (3.26)$$

The analogous equation to Equation 3.20 for ionic surfactants is

$$\Delta H_m^{\ominus} = -2RT^2\left(\frac{\partial \ln x_{CMC}}{\partial T}\right)_P. \qquad (3.27)$$

One of the main criticisms of the phase-separation model is that it predicts that the activity of the monomers above the CMC remains constant. Dialysis [164], surface tension [165], and emf measurements [48, 49, 166, 167], however, indicate a decrease in monomer activity above the CMC of ionic surfactants.

3.6.2 Mass-action model

Micelles and unassociated surfactant ions are assumed to be in association–dissociation equilibrium and the law of mass action is applied. The mass-action approach was originally applied mainly to ionic surfactants [168–170]. Its application to non-ionic surfactants has been discussed by Corkill *et al.* [171].

(A) APPLICATION OF MASS-ACTION MODEL TO NON-IONIC SURFACTANTS
Micelles, M, are considered to be formed by a single step reaction from n monomers, D, according to

$$nD \rightleftharpoons M. \tag{3.28}$$

The equilibrium constant for micelle formation, K_m, is given by

$$K_m = \frac{a_m}{(a_s)^n}. \tag{3.29}$$

Assuming ideality, we may write as an approximation

$$K_m = \frac{x_m}{(x_s)^n}. \tag{3.30}$$

Corkill *et al.* [171] have shown that at the CMC the free energy of micellization, $\Delta G_m^\ominus$, is given by

$$\Delta G_m^\ominus = RT\left[\left(1 - \frac{1}{n}\right)\ln x_{CMC} + f(n)\right] \tag{3.31}$$

where

$$f(n) = \frac{1}{n}\left[\ln n^2\left(\frac{2n-1}{n-2}\right) + (n-1)\ln\frac{n(2n-1)}{2(n^2-1)}\right]. \tag{3.32}$$

If n is large, Equation 3.31 reduces to

$$\Delta G_m^\ominus = RT \ln x_{CMC}. \tag{3.33}$$

Applying the Gibbs–Helmholtz equation and assuming the aggregation number, n, to be large and independent of temperature

$$\Delta H_m^\ominus = -RT^2\left(\frac{\partial \ln x_{CMC}}{\partial T}\right)_P = R\left(\frac{\partial \ln x_{CMC}}{\partial(1/T)}\right)_P. \tag{3.34}$$

However, as discussed in Section 3.5.5, the aggregation numbers of many non-ionic surfactants vary with temperature and in some cases a concentration dependence of n has also been reported. In such cases Equation 3.34 is not applicable.

(B) APPLICATION OF MASS-ACTION MODEL TO IONIC SURFACTANTS

The ionic micelle, M^{+p}, is considered to be formed by the association of n surfactant ions, D^+, and $(n-p)$ firmly bound counterions, X^-.

$$nD^+ + (n-p)X^- \rightleftharpoons M^{+p}. \tag{3.35}$$

The equilibrium constant for micelle formation assuming ideality, is thus,

$$K_m = \frac{x_m}{(x_s)^n (x_x)^{n-p}} \tag{3.36}$$

where x_x is the mole fraction of counterion.

The standard free energy of micellization per mole of monomeric surfactant is given by

$$\Delta G_m^{\ominus} = \frac{-RT}{n} \ln K_m = \frac{-RT}{n} \ln \frac{x_m}{(x_s)^n (x_x)^{n-p}}. \tag{3.37}$$

When the aggregation number, n, is large, and when data in the region of the CMC are considered, Equation 3.37 may be reduced to

$$\Delta G_m^{\ominus} = \left(2 - \frac{p}{n}\right) RT \ln x_{CMC}. \tag{3.38}$$

Equation 3.38 is of the same form as Equation 3.23 from the phase-separation model, since $\alpha = p/n$. The two equations differ slightly because of differences in the way in which the mole fractions are calculated. In the phase-separation model the total number of moles present at the CMC is equal to the sum of the moles of water and surfactant (Equation 3.16) whereas the total number of moles in the mass-action model is equal to the sum of the moles of water, surfactant ions, micelles, and free counterions.

The standard enthalpy of micellization (per mole of monomer) is given by

$$\Delta H_m^{\ominus} = -(2-\alpha) RT^2 \left(\frac{\partial \ln x_{CMC}}{\partial T}\right)_p$$
$$= (2-\alpha)R[(\partial \ln x_{CMC})/(\partial(1/T)]_p. \tag{3.39}$$

The mass-action model is a more realistic model than the phase-separation model in describing the variation of monomer concentration with total concentration above the CMC. Using an arbitrary value of $K_m = 1$ the concentrations of the various species have been calculated for aggregation numbers of 10 and 100 (Fig. 3.12). The dotted lines show how the concentrations vary according to the phase-separation approach and the bold ones according to the mass-action law. At lower aggregation numbers there is a notable increase of the monomer concentration above the CMC from the mass-action treatment, and none from the phase-separation one. At an aggregation number of 100 there is little difference between the monomer concentrations according to either approach, which is reasonable since for $n = \infty$ the mass-action and phase-separation models are equivalent.

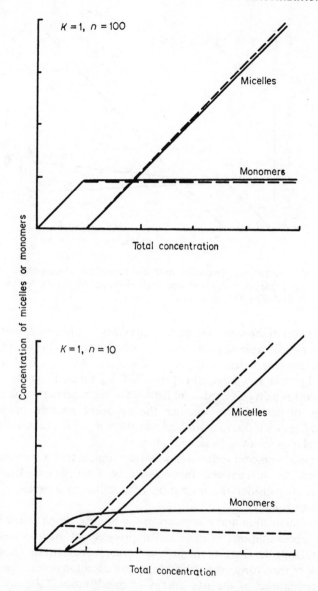

Figure 3.12 Concentrations of micellar or monomeric species against total concentration (arbitrary units), calculated from mass action (full lines) and from phase separation theory (broken lines), for micellar aggregation numbers of 10 and 100.

For an ionized surfactant with $n = 100$, $p = 15$ and $K_m = 1$, the concentrations of the different species vary with total concentration of surfactant as shown in Fig. 3.13. Although the surfactant ion concentration decreases above the CMC, the activity of the monomers, $(x_s, x_x)^{\frac{1}{2}}$, still increases. The predicted decrease in monomer concentration is in agreement with experiment, as discussed previously.

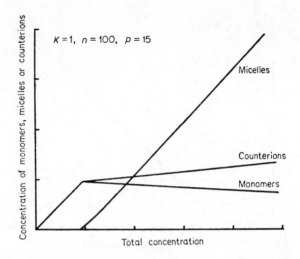

Figure 3.13 Concentrations of micellar, monomeric, and counterion species against total concentration (arbitrary units), calculated from mass action theory for an aggregation number of 100, and with 85% of counterions bound.

The sharpness of the inflection point at the CMC in plots of the concentration dependence of a suitable physical property is dependent on the relative values of the mass-action parameters, n, K_m and α. For low n and K_m a gradual change in the physical property in the region of the CMC is predicted. Simulation of light-scattering curves using selected combinations of the mass-action parameters is a useful way of determining whether the apparent absence of a CMC in experimental curves is due to a combination of low n and K_m values or arises from a non-micellar mode of association [116].

The mass-action model suffers a serious limitation in that it considers only one species of micelle, i.e. it assumes monodispersity of micelle size. A more realistic approach is to consider the formation of micelles by a series of successive equilibria (see below).

Both the mass-action and phase-separation models, despite their limitations, are useful representations of the micellar process and may be used to derive equations relating the CMC to the various factors that determine it. Some insight into the role of the hydrocarbon chain in the micellization process may be gained from determinations of the free energy of micellization, $\Delta G_m^{\ominus}$. A convenient method of determining $\Delta G_m^{\ominus}$ of ionic surfactants is from measurements of the effect of electrolyte on the CMC [172].

Rearrangement of Equation 3.37 gives

$$\log x_s = -\left(1 - \frac{p}{n}\right)\log x_x + \frac{\Delta G_m^{\ominus}}{2.303RT} + \frac{1}{n}\log x_m. \tag{3.40}$$

Assuming that the monomeric surfactant concentration x_s, in the presence of micelles may be equated to the CMC, it is seen that Equation 3.40 is of the same form as the empirical equation (Equation 3.9) which describes the experimentally observed decreases of CMC with added electrolyte. The slope of plots of log CMC against the total counterion concentration (added electrolyte plus CMC) may thus be equated with $-(1-\alpha)$ and the intercept at $\log x_x = 0$ with $(\Delta G_m^{\ominus}/2.303RT) + (1/n \log x_m)$. It is often assumed for the purpose of calculation that 2 per cent of the total surfactant concentration is in micellar form at the CMC. Conceptually, $\Delta G_m^{\ominus}$ may be imagined to be divided into an electrical component $F_{el}^{\ominus}$ arising from the ionic head groups and a hydrocarbon contribution $F_{hc}^{\ominus}$.

$$\Delta G_m^{\ominus} = F_{hc}^{\ominus} + F_{el}^{\ominus}. \tag{3.41}$$

$F_{el}^{\ominus}$ may be estimated from experimental measurements of the zeta potential [173] using numerical solutions of the Poisson–Boltzmann equations [174]. $F_{el}^{\ominus}$ is, of course, positive and its contribution to the total $\Delta G_m^{\ominus}$ value is generally small (see Table 3.11).

Table 3.11 Free energies of formation of micelles of antihistamine drugs at 303 K*

	$\Delta G_m^{\ominus}$ (kJ mol^{-1})	$F_{el}^{\ominus}$ (kJ mol^{-1})	$F_{hc}^{\ominus}$ (kJ mol^{-1})
Diphenhydramine	−28.49	1.63	−30.12
Bromodiphenhydramine	−30.68	1.24	−31.92
Chlorcyclizine	−32.64	1.55	−34.19
Diphenylpyraline	−29.53	0.92	−30.45

* From Attwood and Udeala [175].

The $F_{hc}^{\ominus}$ value obtained from Equation 3.41 may be divided into free energy contributions from the component —CH$_2$— groups, $\Delta G_{CH_2}^{\ominus}$, and the terminal —CH$_3$ group, $\Delta G_{CH_3}^{\ominus}$. Measurements of the solubility of alkanes in water indicate that the free energy contribution of the terminal —CH$_3$ group is not dependent on the chain length. Thus for a homologous series, $\Delta G_{CH_3}^{\ominus}$ may be regarded as a constant term and may be written as $\Delta G_{CH_2}^{\ominus} + k$ where k is constant. Assuming that the free energy contribution of the hydrophilic group is also constant and represented by F_h we may write for non-ionic surfactants from Equation 3.18:

$$\log \text{CMC} = \frac{\Delta G_m^{\ominus}}{2.303RT} + \log w \tag{3.42}$$

$$\log \text{CMC} = \left[\frac{k + F_h}{2.303RT}\right] + \log w + m\left[\frac{\Delta G_{CH_2}^{\ominus}}{2.303RT}\right], \tag{3.43}$$

where m is the total number of carbon atoms in the hydrocarbon chain. Equation 3.43 is of the same form as the empirical equation (Equation 3.8) representing the dependence of the CMC on the hydrocarbon chain length. An analogous equation may be derived for ionic surfactants using Equation 3.24 if it is assumed that the fraction of counterions bound to the micelle is independent of the hydrocarbon chain length in the homologous series. $\Delta G^{\ominus}_{CH_2}$ values of -2.93 kJ (mol CH$_2$)$^{-1}$ have been quoted [176] for a series of polyoxyethylene non-ionic surfactants and little dependence on the structure of the hydrophilic group has been noted. For a wide variety of ionic surfactants including both cationic and anionic, $\Delta G^{\ominus}_{CH_2}$ is approximately -2.72 kJ mol^{-1} [177].

The enthalpy of micellization, $\Delta H^{\ominus}_m$ has been measured directly by calorimetry[178–182] and also, more usually, estimated from the temperature dependence of the CMC using Equation 3.34 or 3.39. This latter method is not considered to be very precise and is based on the assumption that the size and shape of the micelles do not change with temperature, which is often not the case with non-ionic surfactants. With ionic surfactants there is the added complication of possible variation of α with temperature. Several workers have attempted to divide $\Delta H^{\ominus}_m$ into separate contributions from the hydrocarbon chain and the polar head group by studying the dependence of the thermodynamic quantities upon the alkyl chain length [178, 181]. Fig. 3.14 shows the variation of the thermodynamic parameters for a series of the non-ionic surfactants, n-alkyl methyl sulphoxides. The incremental change in $\Delta H^{\ominus}_m$ per CH$_2$ group in these systems is approximately 40% of the total incremental change in $\Delta G^{\ominus}_m$. For a series of alkyl hexaoxyethylene glycol monoethers, the corresponding value is

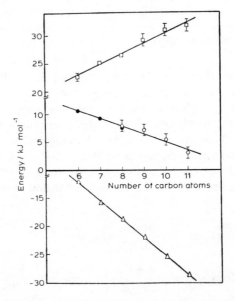

Figure 3.14 Thermodynamic quantities of micellization for n-alkyl methyl sulphoxides at 296.7 K plotted against number of carbon atoms in the alkyl chain. $\bigcirc$, heat of micellization (ΔH_m) from temperature dependence of the CMC; $\bullet$, heat measured calorimetrically; $\triangle$, standard free energy of micellization ($\Delta G^{\ominus}_m$); $\square$, $T\Delta S^{\ominus}_m$. From Clint and Walker [181] with permission.

approximately 50%. From these results the change in free energy of micellization per CH_2 group would appear to be partly enthalpic and partly entropic with possibly a slightly greater emphasis on the entropic term.

3.6.3 Application of other thermodynamic models to the micellization process

The thermodynamics of small systems developed by Hill [183] has been applied to non-ionized, non-interacting surfactant systems by Hall and Pethica [184]. In this approach the aggregation number is treated as a thermodynamic variable, thereby enabling variations in the thermodynamic functions of micelle formation with the mean aggregation number $\bar{n}$ to be examined. The thermodynamic functions of micellization assuming solution ideality are

$$\Delta G_m^{\ominus} = RT[\ln x_s - (\ln x_m/\bar{n})] \tag{3.44}$$

$$\Delta H_m^{\ominus} = -RT^2\left[\left(\frac{d\ln x_s}{dT}\right)_p - \frac{1}{\bar{n}}\left(\frac{d\ln x_m}{dT}\right)_p\right] \tag{3.45}$$

$$\Delta S_m^{\ominus} = -RT\left(\frac{d\ln x_s}{dT}\right)_p + \frac{RT}{\bar{n}}\left(\frac{d\ln x_m}{dT}\right)_p$$
$$- R\ln x_s + (R/\bar{n})\ln x_m. \tag{3.46}$$

The authors concluded that for systems where $\bar{n}$ is large and changes little with temperature, Equations 3.44 to 3.46 may be approximated to the corresponding equations from the mass-action or phase-separation models. A more detailed treatment of multicomponent micelles has been developed by Hall [185].

An essentially equivalent approach to that of small-systems thermodynamics has been formulated by Corkill and co-workers and applied to systems of non-ionic surfactants [94, 176]. As with the small-systems approach, this multiple-equilibrium model considers equilibria between all micellar species present in solution rather than a single micellar species, as was considered by the mass-action theory. The intrinsic properties of the individual micellar species are then removed from the relationships by a suitable averaging procedure. The standard free energy and enthalpy of micellization are given by equations of similar form to Equations 3.44 and 3.45 and are shown to approximate satisfactorily to the appropriate mass-action equations for systems in which the mean aggregation number exceeds 20.

Application of the small-systems/multiple-equilibrium models to solutions of ionic surfactants is less satisfactory because of the failure of these models to deal with interactions between micelles or to give a satisfactory description of the role of counterions in micellization. Alternative, more rigorous, thermodynamic formalisms for describing systems of interacting aggregates have been developed by Hall [184–188]. This theoretical approach has led to precise expressions for the effect of temperature, pressure and electrolyte concentration on the CMC which allow for solution non-ideality. Although these expressions are of a similar form to those from the mass-action treatment and it is shown that in many cases the consideration of non-ideality probably makes little significant difference, this theoretical treatment is more rigorous and clearly defines the micellar degree of association, α, in terms of the negative adsorption of co-ions and surfactant monomer by the micelle.

An interesting model of micelle formation based on geometrical considerations of micelle shape has been proposed by Tanford [97]. Equations are presented which relate the micelle size and CMC to a size-dependent free energy of micellization. The calculations are based on the assumption of an ellipsoidal shape. The hydrophobic component of the free energy is estimated in terms of the area of contact between the hydrophobic core and the solvent. The hydrophilic component of ΔG_m^*, i.e. the free energy of repulsion between the head groups, is assumed to be inversely proportional to the surface area per head group. This approach has been further developed by Ruckenstein and Nagarajan [189] and used in the prediction of the properties of sodium octanoate micellar solutions [190].

3.7 Kinetics of micelle formation

The rates at which micelles form and break up and monomers enter and leave the micelles are very rapid and it is only comparatively recently that techniques have become available which are capable of following these processes. There is now a general agreement between workers [191–196] that the relaxation spectra of micellar solutions are characterized by at least two relaxation times differing by a factor of 10^2 to 10^3. The slow process, which has a relaxation time in the millisecond range has been attributed to the micellization–dissolution equilibrium, i.e. the equilibrium between the complete dissociation of the micelle into n monomers and its complete reformation from n monomers. This process is amenable to study using temperature jump [197–203], pressure jump [201, 202, 204, 205] and stopped flow [199, 206–215] techniques. The fast process which has been detected using ultrasonic absorption [209–212], n.m.r. [213–215] and e.s.r. [216–218] techniques has a relaxation time which is typically less than 10 μs. This process has been identified with the exchange of monomers between the micellar species and the bulk solution. An additional relaxation process has been detected from ultrasonic absorption studies on concentrated micellar solutions and has been related to the change of micellar shape [219].

Many different kinetic models have been described which allow the evaluation of an expression for the various relaxation times. Sams *et al.* [220] have proposed a 'two-state' model which considers a monomeric state and an associated state consisting of all species larger than the monomer unit. This model describes only the fast process and makes the assumption that the rate constants for association and dissociation of the monomer from the micelle are independent of micellar size. The association process is regarded as a collision between a small particle and a large sphere. The rate of monomer association was considered to be proportional to the concentration of monomers, the concentration of micelles and the aggregation number (which determines the cross-sectional area of the micelle). The relaxation time, τ_1, was related to the overall concentration, c, by

$$\tau_1^{-1} = k_f c - k_b, \tag{3.47}$$

where k_f and k_b are the forward and reverse rate constants, respectively. The linear relationship predicted by Equation 3.47 was experimentally verified for a large number of surfactants.

Further development of the two-state model considered the manner in which monomer units are packed into a micelle [221]. Incorporation of a monomer into an existing micelle was assumed to occur only when the monomer collided with a part of the surface not already covered by a head group. The rate constants, k_f' and k_b' of this modified model are related to the rate constants of equation 3.47 by

$$k_f = k_f' a_0 (1 - \alpha)/\alpha \tag{3.48}$$
$$k_b = k_b' a_0, \tag{3.49}$$

where the term $(1 - \alpha)/\alpha$ is considered as a packing factor describing the distribution of monomer units on the micelle surface.

Further extensions of the model to account for relaxation spectra of mixed micellar solutions have been reported [222].

A more detailed kinetic model which accounts for both the fast and slow relaxation processes has been proposed by Aniansson and co-workers [95, 194–196, 223]. These authors consider that the association and dissociation of the micelles proceed by a stepwise process involving the entry and departure of one monomer at a time from the micelle. There is thus a series of equilibria

$$A_1 + A_{n-1} \underset{k_n^-}{\overset{k_n^+}{\rightleftharpoons}} A_n, \quad n = 2, 3, \ldots, \tag{3.50}$$

where A_n denotes an aggregate containing n monomers, and k_n^+ and k_n^- are the forward and reverse rate constants for a given step. The reattainment of equilibrium following perturbation of the system by a sudden change in the temperature or pressure or passage of ultrasound involves change in the micelle size distribution. If the aggregates are assumed to occur in substantial amounts in two regions (one around the mean aggregation number and the other in the region of monomers, dimers and trimers) then in the attainment of equilibrium the fast process represents a change in the position of the peak of the distribution curve from state 1 to state 2 as one or a few monomers dissociate from or associate to existing micelles. The slow process involves the change in the total number of micelles which occurs when a few micelles dissociate completely to (or are formed from) monomers. The corresponding change on the distribution curve is from state 2 to state 3. There is no change in the aggregation number. These two processes are shown diagrammatically in Fig. 3.15. The slow process has been likened to a flow from one block to another through the region between monomers and micelles. Assuming the aggregation number to be a continuous variable and applying a treatment analogous to that of heat conduction,

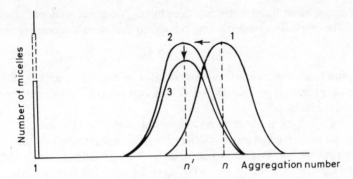

Figure 3.15 Changes of the distribution curve of micelle size. From Lang *et al.* [224].

Aniansson and co-workers proposed the following expression for the fast process

$$\tau_1^{-1} = \frac{k^-}{\sigma^2} + \frac{k^-}{n}\,a(1+c_0) \tag{3.51}$$

where σ is the distribution width of the distribution curve of micellar sizes (assumed to be Gaussian), k^- is the stepwise rate constant which is assumed to be independent of n in the micellar region, $a = (A_{tot}-\bar{A}_1)/\bar{A}_1$, and A_{tot} and $\bar{A}_1$ are the total surfactant concentration and mean monomer concentration, respectively. c_0 is a measure of the average deviation from equilibrium and is usually less than 1%. Assuming c_0 can be neglected and equating $\bar{A}_1$ with the CMC gives

$$\tau^{-1} = \frac{k^-}{\sigma^2} + \frac{k^-}{n}(A_{tot}-\text{CMC})/\text{CMC}. \tag{3.52}$$

Equation 3.52 thus predicts a linear relationship between τ^{-1} and total concentration, in agreement with experiment. Since n may be readily determined by light scattering, Equation 3.52 affords a method of calculation of not only the stepwise rate constants but also the width of the size distribution curve.

The expression derived by Aniansson and co-workers for the relaxation time of the slow process may be simplified to

$$\tau_2^{-1} \simeq \frac{n^2}{\bar{A}_1 R}\left[1+\frac{\sigma^2}{n}\,a\right]^{-1}, \tag{3.53}$$

where R is a term which may be visualized as the resistance to flow through the critical region and is given by

$$R = \sum_{s=s_1+1}^{s_2} 1/(k_s^- \bar{A}_s),$$

where s is the aggregation number of some particular aggregate and $\bar{A}_s$ is the equilibrium concentration of aggregates of order s.

The dependence of τ_2^{-1} upon ionic strength, concentration and temperature have all been interpreted in terms of their effect upon R.

Although many other kinetic treatments have been proposed [192], that of Aniansson and co-workers is possibly the most comprehensive. Current developments of this theory have dealt with the dynamics and extent of *partial* motions of the monomers out into the aqueous environment and back again into the micelle [225]. A comprehensive list of kinetic studies on surfactants has been given by Muller [226].

3.8 Non-micellar association

The distribution of aggregates sizes in a truly micellar system is bimodal with one peak in the region of monomers and possibly including some dimers and trimers, the other peak being in the region of the mean aggregation number (see Fig. 3.16). The half-width of this second micellar region is, of course, a measure of the polydispersity of the system. As suggested before, micellization is essentially a multistep process involving a series of equilibria thus

$$A_1 + A_1 \underset{}{\overset{K_2}{\rightleftharpoons}} A_2$$

$$A_1 + A_2 \underset{}{\overset{K_3}{\rightleftharpoons}} A_3$$

$$A_1 + A_3 \underset{}{\overset{K_4}{\rightleftharpoons}} A_4$$

$$. \quad . \quad . \quad .$$

$$A_1 + A_{n-1} \underset{}{\overset{K_n}{\rightleftharpoons}} A_n \tag{3.54}$$

which, in theory, could result in a wide range of micellar sizes in the solution. Each of the steps represented in Equation 3.54 will be associated with a particular equilibrium constant K_n. The fact that, within the limitation of experimental

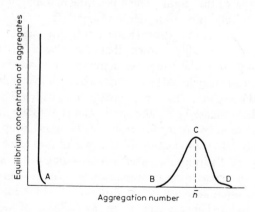

Figure 3.16 Typical distribution curve for a micellar system with mean aggregation number, $\bar{n}$.

technique, it would appear that most micellar systems are reasonably monodisperse suggests a specific relationship between K_n and n which we will examine now in more detail.

The overall equilibrium constant, β_n, for the self-association reaction

$$nA_1 \rightleftharpoons A_n \tag{3.55}$$

in which multimer A_n is formed from n monomers is expressed as

$$\beta_n = [A_n]/[A_1]^n. \tag{3.56}$$

β_n is the product of all the stepwise association constants, K_2, K_3, etc. up to K_n

$$\text{i.e. } \beta_n = K_2 K_3 \ldots K_n = \prod_{2}^{n} K_n. \tag{3.57}$$

Equations 3.55 to 3.57 are not directly applicable to ionic surfactants since no account has been taken of the counterions associated with the micelles or the micellar charge. The equilibrium constant, β_n, for ionic surfactants will incorporate charge effects and, to simplify the treatment, these are assumed to remain constant with increase in aggregate size. This assumption is most likely to be valid for systems containing added electrolyte [93]. Whereas the magnitude of β_n gives an indication of the tendency for association, it is the way in which K_n varies with aggregate size which defines the association pattern.

The particular dependence of K_n on n which gives rise to a micellar type of association has been discussed by Mukerjee [227–229]. During the early stages of micellar growth, i.e. at point B (Fig. 3.16) K_n must increase with n, a concept referred to as co-operativity. However, it is clear that K_n cannot continue to increase, otherwise larger and larger aggregates would be formed, rather than aggregates of a reasonably uniform size. K_n must therefore eventually decrease with n (anti- or negative co-operativity) and the K_n versus n plot shows a maximum at the mean aggregation number. This self-association pattern may be rationalized in terms of the factors which promote and inhibit micellar growth. The impetus for micelle formation derives mainly from the entropic changes accompanying the transfer of hydrocarbon from an aqueous environment to the micelle interior. It is readily shown that the efficiency of shielding the hydrophobic group from the water, as a monomer is added to the micelle, increases with increase in aggregate size. On the basis of hydrophobic interaction alone there would be expected to be an increase of K_n with n. The factor which is responsible for the decrease in K_n at point C and thus limits micelle size arises from the progressive increase in the density of the head groups at the micellar surface as the micellar size increases. In the case of non-ionic surfactants this increased density results in an increased self-crowding at the surface and with ionic surfactants, an increased change repulsion between head groups. It has been shown that only a slight decrease (of the order of 2%) in the free energy profile (the variation of the free energy change on the addition of monomer with the aggregation number) is sufficient to produce micelles with narrow size distributions (Fig. 3.17).

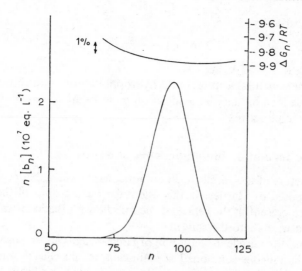

Figure 3.17 Variation in the concentration of monomers existing as micelles, $n[b_n]$, as a function of the number of monomers in the micelle n, for an assumed free energy profile, ΔG_n versus n (upper curve). From Mukerjee [227] with permission.

In the typical micellar system the K_n values in the intermediate region A–B between the two peaks in the size distribution curve (Fig. 3.16) are thought to be of insignificant magnitude. There is, however, a limited number of amphiphilic systems in which this is not so. The simplest of these non-micellar association schemes is that of continuous open-ended self-association. Here there are no geometrical or charge factors which limit association and it is often possible to analyse data from techniques such as light-scattering and vapour-pressure osmometry in terms of association models which assume simple relationships between successive K_n values.

Ghosh and Mukerjee [230] have described the self-association of the cationic dye, methylene blue, using a stepwise association model in which all the equilibrium constants are assumed to be of equal value

i.e.
$$K_2 = K_3 = K_4 = K_n = K. \tag{3.58}$$

A similar scheme has been applied to the association of nucleotides [231] and some narcotic drugs [232]. The association of the narcotic drug, pethidine, has been described in terms of a co-operative association model in which K_n is related to a generalized equilibrium constant, K, by

$$K_n = K(n-1)/n, \qquad n \geqslant 2, \tag{3.59}$$

i.e. $K_2 = K/2$, $K_3 = 2K/3$, $K_4 = 3K/4$, etc.

The association of the antihypertensive drug, pavatrine, on the other hand, follows an antico-operative association scheme in the absence of added electrolyte [233] in which association to form aggregates of greater size than the dimer is described by

$$K_n = K(n-1)/(n-2), \qquad n \geqslant 3. \tag{3.60}$$

i.e. $K_3 = 2K$, $K_4 = 3/2\,K$, etc.

Some of the structural features of the hydrophobic group which are thought to be responsible for non-micellar association patterns are discussed in Chapter 4 and other examples given.

3.9 Micelle formation in non-aqueous solvents

The usual picture of the micelle in a non-aqueous solvent is that of the 'inverted micelle' proposed by Hartley. In this, the polar head groups of the surfactant monomer are present in the centre of the micelle with the hydrocarbon chains extending outwards into the solvent.

The reasons for micelle formation in organic solvents are somewhat different from those in aqueous solution. The main cause of micellization is the energy change due to dipole–dipole interactions between the polar head groups of the surfactant molecules. In certain cases hydrogen bond formation between head groups may also occur. Opposing micelle formation is the possible loss of translational, vibrational and rotational freedom of monomers when in the micelle.

Estimates of micellar size have shown that, in general, aggregation numbers are small in organic media, often not exceeding five monomers per micelle. It has been suggested by Kertes and Gutmann in a comprehensive review of this topic [234] that for such small aggregates, a spherical micellar shape would not be able to provide sufficient shielding of polar regions. A lamellar micellar model was proposed as an alternative, comprising double layers of orientated molecules placed end-to-end and tail-to-tail with sheets of solvent molecules between the surfactant layers (Fig. 3.18). Weak van der Waals' forces acting between the hydrocarbon chains of parallel layers and strong dipole–dipole bonds between the polar heads hold together the lamellar structure. Lamellar aggregates have been reported for metal carboxylates in hydrocarbons [235–238], alkylbenzene and dialkylnaphthalene sulphonates [239, 240], sulphosuccinates [239] and lecithin in benzene [241–243].

There is some dispute over the type of association in non-aqueous systems. Ruckenstein and Nagarajan [244], from a consideration of the physical factors controlling aggregation, have argued against the existence of a CMC in these systems. The study of aggregation in organic solvents presents many more problems than are encountered with aqueous solutions. Many of these problems arise because of the very much smaller aggregates present in organic solvents. Determination of a CMC presents a particular problem, as association often commences at very low concentration. As the solutions have little surface activity and the aggregates are not ionized, it is not possible to use two of the most commonly used methods in aqueous systems, i.e. surface tension and conductivity. The refractive index increments of the aggregates are very low and as a

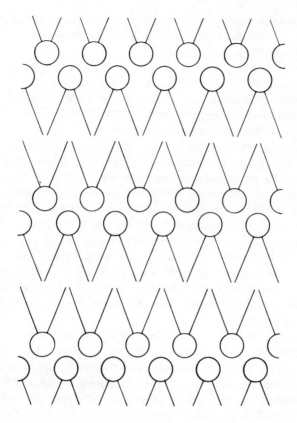

Figure 3.18 Idealized diagram of lamellar structure of normal secondary alkylammonium salts. From Kertes and Gutmann [234] with permission.

consequence the light-scattering intensity is also extremely low so that this technique also is not sufficiently sensitive to determine a CMC. Solubilization methods for CMC determination have been widely used particularly in early work but these are open to criticism [234]. A recent technique which has been applied in a study of these systems is that of positron annihilation. This method is based on the observation that the reactions, as well as the formation, of the positronium atom (the bound state of an electron and a positron) are greatly dependent on the environment in which these interactions occur [245]. By this technique, abrupt CMCs were noted for Aerosol OT (sodium 2-ethylhexyl sulphosuccinate) and dodecylammonium propionate in benzene and cyclohexane [246, 247] which agreed well with 'CMC' values determined by light-scattering and vapour-pressure depression [248], dye absorption techniques [249], dielectric increment measurements [250] and ^{1}H nmr [251, 252].

In other systems it has not been possible to detect a CMC even though the techniques used were of sufficient sensitivity to detect aggregation. Thus Lo *et al.*

[253] have reported an absence of a CMC for dodecylammonium propionate in benzene and cyclohexane. In such systems the association has been described in terms of a stepwise association model [234, 253–255].

In an attempt to rationalize the self-association behaviour of ionic surfactants in apolar solvents, Muller [256] has suggested that when the solvent has a relatively high dielectric constant or the sum of the radii of the ionic head groups is large, association is best represented by stepwise association. When these two parameters are both small, compact clusters are preferred and the association is more analogous to micellization in aqueous solution. Although this categorization correctly predicts the association patterns of dodecylammonium propionate in benzene or cyclohexane and Aerosol OT in benzene, it was not found to apply to solutions of alkylammonium carboxylates in cyclohexane, all of which exhibited apparent CMCs as detected by positron annihilation techniques [257]. Change in the counterion associated with Aerosol OT in benzene from Na^+ to alkylammonium$^+$ caused a corresponding change in the association pattern from a micellar type to a continuous association pattern [258]. Similar results were reported by Eicke *et al.* [250] for di-2-ethylhexyl sulphosuccinate with different counterions in benzene. The possibility of such differences in association being attributable to small amounts of water present as an impurity in the systems has been considered [258].

A discussion of the factors which affect the size of the aggregates in non-polar solvents is often confused by unreliable data. In many cases, light-scattering techniques have yielded micellar weights which were very much higher than those from vapour-pressure osmometry. In some instances such discrepancies have been attributed to the presence of impurities [259] or small traces of water. Attwood and McDonald [260] have shown that the high micellar weights reported [261, 262] for the non-ionic surfactant, sorbitan monostearate (Span 60), may be a consequence of fluorescence which was appreciable at excitation wavelengths of 436 nm. Correction of the scattered light intensity for the contribution of fluorescence considerably reduced the estimates of the extent of association.

A decrease in aggregation number with increase in the number of carbon atoms in a homologous series has been noted for zinc [263], magnesium and copper carboxylates [264] and sodium dialkyl sulphosuccinates [265]. Several authors have reported on the effect of the polar group on the extent of aggregation. For example, the aggregation numbers of tridodecylammonium [266] and dialkyldimethylammonium [265] salts increase in the order $Cl^- < NO_3^- < Br^- < ClO_4^- < HSO_4^-$. For non-ionic surfactants the length of both the lipophilic and hydrophilic groups determine the degree of association. The nature of the solvent and, in particular, its dielectric constant, solubility parameter and ability to solvate the surfactant may have a very pronounced effect on the aggregate size [242]. In general, the CMC values increase with increasing solvent polarity and decrease with increasing dielectric constant. Solvents may be classified into (a) those in which solvophobic interactions are possible, (b) those in which inverted micelles are formed and (c) those in which micellization does not occur [267, 268]. The first group includes such solvents as glycol and 2-aminoethanol which have two or more hydrogen bonding centres, and are structured in a similar way as water. Micelles formed in such solvents are thought to be of the same type as in aqueous solution. The second group of solvents includes the typical hydrocarbon solvents such as benzene and cyclohexane. The solvents of the third group which do not support micelle formation have a single hydrogen bonding centre and include methanol, ethanol and dimethylformamide.

References

1. P. MUKERJEE and K. J. MYSELS *Critical micelle concentrations of aqueous surfactant systems*, NSRDS–NBS 36. US Gov. Printing Office, Washington, DC (1971).
2. P. MUKERJEE (1967) *Adv. Colloid Interfac. Sci.* **1**, 241.
3. D. STIGTER (1974) *J. Phys. Chem.* **78**, 2480.
4. E. D. GODDARD, C. A. J. HOEVE and G. C. BENSON (1957) *J. Phys. Chem.* **61**, 593.
5. K. SHIGEHARA (1965) *Bull. Chem. Soc. Japan* **38**, 1700.
6. P. MUKERJEE (1970) *Kolloid-Z.-u-Z. Polymere* **236**, 76.
7. G. A. DAVIS (1972) *J. Amer. Chem. Soc.* **94**, 5089.
8. R. R. HAUTALA, N. E. SCHORE and N. J. TURRO (1973) *J. Amer. Chem. Soc.* **95**, 5508.
9. M. SHINITZKY, A-C. DIANOUX, C. GITLER and G. WEBER (1971) *Biochem.* **10**, 2106.
10. S. J. REHFELD (1970) *J. Colloid Interfac. Sci.* **34**, 518.
11. H. J. POWNALL and L. C. SMITH (1973) *J. Amer. Chem. Soc.* **95**, 3136.
12. N. J. TURRO, M. AIKAWA and A. YEKTA (1979) *J. Amer. Chem. Soc.* **101**, 772.
13. K. ZACHARIASSE (1978) *Chem. Phys. Lett.* **57**, 429.
14. A. S. WAGGONER, O. H. GRIFFITH and C. R. CHRISTENSEN (1967) *Proc. Nat. Acad. Sci.* **57**, 1198.
15. M. J. POVICH, J. A. MANN and A. KAWAMOTO (1972) *J. Colloid Interfac. Sci.* **41**, 145.
16. E. WILLIAMS, B. SEARS, A. ALLERHAND and E. H. CORDES (1973) *J. Amer. Chem. Soc.* **95**, 4871.
17. F. M. MENGER and J. M. JERKUNICA (1978) *J. Amer. Chem. Soc.* **100**, 688.
18. K. KALYANASUNDARAM and J. K. THOMAS (1976) *J. Phys. Chem.* **80**, 1462.
19. J. B. ROSENHOLM, P. STENIUS and I. DANIELSSON (1976) *J. Colloid Interfac. Sci.* **57**, 551.
20. H. OKABAYASHI, M. OKUYAMA and T. KITAGAWA (1974) *Bull. Chem. Soc. Japan* **47**, 1075.
20a H. OKABAYASHI, M. OKUYAMA and T. KITAGAWA (1975) *Bull. Chem. Soc. Japan* **48**, 2264.
21. B-O. PERSSON, T. DRAKENBERG and B. LINDMAN (1976) *J. Phys. Chem.* **80**, 2124.
22. H. TAKAHASHI, Y. NAKAYAMA, H. HORI, K. KIHARA, H. OKABAYASHI and M. OKUYAMA (1976) *J. Colloid Interfac. Sci.* **54**, 102.
23. K. KALYANASUNDARAM, M. GRATZEL and J. K. THOMAS (1975) *J. Amer. Chem. Soc.* **97**, 3915.
24. N. MULLER and R. H. BIRKHAHN (1967) *J. Phys. Chem.* **71**, 957.
25. N. MULLER and R. H. BIRKHAHN (1968) *J. Phys. Chem.* **72**, 583.
26. N. MULLER and T. W. JOHNSON (1969) *J. Phys. Chem.* **73**, 2042.
27. N. MULLER and H. SIMSOHN (1971) *J. Phys. Chem.* **75**, 952.
28. U. HENRIKSSON and L. ODBERG (1974) *J. Colloid Interfac. Sci.* **46**, 212.
29. P. MUKERJEE and K. J. MYSELS (1975) in *Colloidal Dispersions and Micellar Behaviour*, (ed. K. L. Mittal), ACS symposium series No. 9 *Amer. Chem. Soc.* Washington, DC, pp. 239–252.

30. F. M. MENGER, J. M. JERKUNICA and J. C. JOHNSTON (1978) *J. Am. Chem. Soc.* **100**, 4676.
31. H. WENNERSTROM and B. LINDMAN (1979) *J. Phys. Chem.* **83**, 2931.
32. F. PODO, A. RAY and G. NÉMETHY (1973) *J. Am. Chem. Soc.* **95**, 6164.
33. C. J. CLEMETT (1970) *J. Chem. Soc. A.* 2251.
34. F. M. MENGER and B. J. BOYER (1980) *J. Am. Chem. Soc.* **102**, 5938.
35. D. STIGTER (1967) *J. Colloid Interface Sci.* **23**, 379.
36. D. STIGTER (1974) *J. Colloid Interface Sci.* **47**, 473.
37. D. STIGTER (1975) *J. Phys. Chem.* **79**, 1008, 1015.
38. G. S. HARTLEY and J. W. ROE (1940) *Trans. Faraday Soc.* **36**, 101.
39. P. MUKERJEE and K. BANERJEE (1964) *J. Phys. Chem.* **68**, 3567.
40. M. S. FERNANDEZ and P. FROMHERZ (1977) *J. Phys. Chem.* **81**, 1755.
41. E. M. KOSOWER (1958) *J. Amer. Chem. Soc.* **80**, 3253.
42. K. KALYANASUNDARAM and J. K. THOMAS (1977) *J. Phys. Chem.* **81**, 2176.
43. P. MUKERJEE, J. R. CARDINAL and N. R. DESAI (1977) in *Micellization, Solubilization and Microemulsions*, Vol. 1, (ed. K. L. Mittal), Plenum, New York, pp. 241–61.
44. C. BOTRÉ, V. L. CRESCENZI and A. MELE (1959) *J. Phys. Chem.* **63**, 650.
45. T. INGRAM and M. N. JONES (1969) *Trans. Faraday Soc.* **65**, 297.
46. L. SHEDLOVSKY, C. W. JAKOB and M. B. EPSTEIN (1963) *J. Phys. Chem.* **67**, 2075.
47. J. T. PEARSON and C. J. HUMPHREYS (1970) *J. Pharm. Pharmacol.* **22**, 126S.
48. S. G. CUTLER, P. MEARES and D. G. HALL (1978) *J. Chem. Soc. Farad I* **74**, 1758.
49. T. SASKI, M. HATTORI, J. SASAKI and K. NUKINA (1975) *Bull. Chem. Soc. Japan* **48**, 1397.
50. T. S. BRUN, H. HØILAND and E. VIKINGSTAD (1978) *J. Colloid Interfac. Sci.* **63**, 590.
51. K. J. MYSELS (1955) *J. Colloid Sci.* **10**, 507.
52. J. F. PADDAY (1967) *J. Phys. Chem.* **71**, 3488.
53. A. CUSHMAN, A. P. BRADY and J. W. MCBAIN (1948) *J. Colloid Sci.* **3**, 425.
54. M. J. VOLD (1950) *J. Colloid Sci.* **5**, 506.
55. A. A. RIBEIRO and E. A. DENNIS (1976) *J. Phys. Chem.* **80**, 1746.
56. A. A. RIBEIRO and E. A. DENNIS (1977) *J. Phys. Chem.* **81**, 957.
57. M. ROSCH (1967) in *Nonionic surfactants*, (ed. M. J. Schick), Arnold, London, pp. 753–773.
58. H. SCHOTT (1966) *J. Chem. Eng. Data* **11**, 417.
59. D. ATTWOOD (1968) *J. Phys. Chem.* **72**, 339.
60. D. I. D. EL EINI, B. W. BARRY and C. T. RHODES (1976) *J. Colloid Interfac. Sci.* **54**, 348.
61. P. H. ELWORTHY and C. B. MACFARLANE (1962) *J. Chem. Soc.* 537.
62. J. M. CORKILL, J. F. GOODMAN and J. WYER (1969) *Trans. Faraday Soc.* **65**, 9.
63. H. SCHOTT (1967) *J. Colloid Interfac. Sci.* **24**, 193.
64. H. SCHOTT (1969) *J. Pharm. Sci.* **58**, 1521.
65. T. ARNARSON and P. H. ELWORTHY (1981) *J. Pharm. Pharmacol.* **33**, 141.
66. H. SCHOTT (1971) *J. Pharm. Sci.* **60**, 1596.
67. H. SCHOTT (1973) *J. Pharm. Sci.* **62**, 162.
68. A. D. ABBOT and H. V. TARTAR (1955) *J. Phys. Chem.* **59**, 1195.
69. C. TANFORD (1972) *J. Phys. Chem.* **76**, 3020.
70. L. M. KUSHNER and W. D. HUBBARD (1954) *J. Phys. Chem.* **58**, 1163.
71. M. CORTI and J. DEGIORGIO (1975) *Opt. Commun.* **14**, 358.
72. R. J. ROBSON and E. A. DENNIS (1977) *J. Phys. Chem.* **81**, 1075.
73. H. H. PARADIES (1980) *J. Phys. Chem.* **84**, 599.
74. A. B. MANDAL, S. RAY, A. M. BISWAS and S. P. MOULIK (1980) *J. Phys. Chem.* **84**, 856.
75. C. TANFORD, Y. NOZAKI and M. F. RHODE (1977) *J. Phys. Chem.* **81**, 1555.
76. J. N. ISRAELACHVILI, D. J. MITCHELL and B. W. NINHAM (1976) *J. Chem. Soc. Farad. II* **72**, 1525.
77. J. E. LEIBNER and J. JACOBUS (1977) *J. Phys. Chem.* **81**, 130.

78. U. HENRIKSSON, L. ODBERG, J. C. ERIKSON and L. WESTMAN (1977) *J. Phys. Chem.* **81**, 76.
79. F. REISS-HUSSON and V. LUZZATI (1964) *J. Phys. Chem.* **68**, 3504.
80. N. A. MAZER, G. B. BENEDEK and M. C. CAREY (1976) *J. Phys. Chem.* **80**, 1075.
81. N. A. MAZER, M. C. CAREY and G. B. BENEDEK (1977) in *Micellization, Solubilization and Microemulsions*, Vol. 1, (ed. K. L. Mittal), Plenum, New York, N.Y., p. 359.
82. S. IKEDA, S. HAYASHI and T. IMAE (1981) *J. Phys. Chem.* **85**, 106.
83. P. J. MISSEL, N. A. MAZER, G. B. BENEDEK, C. Y. YOUNG and M. C. CAREY (1980) *J. Phys. Chem.* **84**, 1044.
84. C. Y. YOUNG, P. J. MISSEL, N. A. MAZER, G. B. BENEDEK and M. C. CAREY (1978) *J. Phys. Chem.* **82**, 1375.
85. S. HAYASHI and S. IKEDA (1980) *J. Phys. Chem.* **84**, 744.
86. S. IKEDA, S. OZEKI and M. TSUNODA (1980) *J. Colloid Interfac. Sci.* **73**, 27.
87. S. OZEKI and S. IKEDA (1980) *J. Colloid Interfac. Sci.* **77**, 219.
88. H. COLL (1970) *J. Colloid Interfac. Sci.* **74**, 520.
89. K. S. BIRDI (1972) *Kolloid-Z.-u-Z. Polymere* **250**, 731.
90. D. ATTWOOD, P. H. ELWORTHY and S. B. KAYNE (1970) *J. Phys. Chem.* **74**, 3529.
91. K. S. BIRDI, S. V. DALSAGER and S. BACKLUND (1980) *J. Chem. Soc. Farad. I* **76**, 2035.
92. D. ATTWOOD, P. H. ELWORTHY and S. B. KAYNE (1969) *J. Pharm. Pharmacol.* **21**, 619.
93. P. MUKERJEE (1972) *J. Phys. Chem.* **76**, 565.
94. J. M. CORKILL, J. F. GOODMAN, T. WALKER and J. WYER (1969) *Proc. Roy. Soc. A* **312**, 243.
95. E. A. G. ANIANSSON, S. N. WALL, M. ALMGREN, H. HOFFMANN, I. KIELMANN, W. ULBRICHT, R. ZANA, J. LANG and C. TONDRE (1976) *J. Phys. Chem.* **80**, 905.
96. C. TANFORD (1974) *Proc. Nat. Acad. Sci. USA* **71**, 1811.
97. C. TANFORD (1974) *J. Phys. Chem.* **78**, 2469.
98. K. SHINODA, T. YAMAGUCHI and R. HORI (1961) *Bull. Chem. Soc. Japan* **34**, 237.
99. P. H. ELWORTHY (1960) *J. Pharm. Pharmacol.* **12**, 260T.
100. P. BECHER (1961) *J. Colloid Sci.* **16**, 49.
101. P. H. ELWORTHY and A. T. FLORENCE (1965) *Kolloid-Z.* **204**, 105.
102. H. E. GERRY, P. T. JACOBS and E. W. ANACKER (1977) *J. Colloid Interfac. Sci.* **62**, 556.
103. H. KUNIEDA and K. SHINODA (1976) *J. Phys. Chem.* **80**, 2468.
104. T. KUNITAKE and Y. OKAHATA (1977) *J. Amer. Chem. Soc.* **99**, 3860.
105. R. D. GEER, E. H. EYLAR and E. W. ANACKER (1971) *J. Phys. Chem.* **75**, 369.
106. E. W. ANACKER and R. D. GEER (1971) *J. Colloid Interfac. Sci.* **35**, 441.
107. P. T. JACOBS and E. W. ANACKER (1976) *J. Colloid Interface Sci.* **56**, 255.
108. E. W. ANACKER, H. E. GERRY, P. T. JACOBS and I. PETRARIU (1977) *J. Colloid Interface Sci.* **60**, 514.
109. P. T. JACOBS and E. W. ANACKER (1973) *J. Colloid Interface Sci.* **44**, 505.
110. H. B. KLEVENS (1948) *J. Phys. Colloid Chem.* **52**, 130.
111. K. SHINODA (1955) *J. Phys. Colloid Chem.* **59**, 432.
112. K. SHINODA (1956) *J. Phys. Colloid Chem.* **60**, 1439.
113. H. C. EVANS (1956) *J. Phys. Colloid Chem.* **60**, 576.
114. P. H. ELWORTHY and C. B. MACFARLANE (1963) *J. Chem. Soc.* 907.
115. J. GETTINS, R. GREENWOOD, J. E. RASSING and E. WYN-JONES (1976) *J. Chem. Soc. Chem. Comm.* **24**, 1030.
116. D. ATTWOOD and O. K. UDEALA (1975) *J. Phys. Chem.* **79**, 889.
117. M. L. CORRIN and W. D. HARKINS (1974) *J. Am. Chem. Soc.* **69**, 684.
118. J. P. KRATOHVIL (1980) *J. Colloid Interfac. Sci.* **75**, 271.
119. A. RHODE and E. SACKMANN (1979) *J. Colloid. Interfac. Sci.* **70**, 494.
120. H. F. HUISMAN (1964) *Proc. Kon. Ned. Akad. Wetensch.* **B67**, 367, 376, 388, 407.
121. D. A. DOUGHTY (1979) *J. Phys. Chem.* **83**, 2621.
122. N. J. TURRO and A. YEKTA (1978) *J. Amer. Chem. Soc.* **100**, 5951.

123. M. F. EMERSON and A. HOLTZER (1967) *J. Phys. Chem.* **71**, 1898.
124. P. LLANOS and R. ZANA (1980) *J. Phys. Chem.* **84**, 3339.
125. P. MUKERJEE and A. RAY (1962) *J. Phys. Chem.* **67**, 190.
126. H. SCHOTT (1973) *J. Colloid Interfac. Sci.* **43**, 150.
127. H. SCHOTT and S. K. HAN (1975) *J. Pharm. Sci.* **64**, 658.
128. H. SCHOTT and S. K. HAN (1976) *J. Pharm. Sci.* **65**, 975.
129. L. HSIAO, H. N. DUNNING and P. B. LORENZ (1956) *J. Phys. Chem.* **60**, 657.
130. P. BECHER (1962) *J. Colloid Sci.* **17**, 325.
131. P. BECHER (1963) *J. Colloid* **18**, 196.
132. M. J. SCHICK (1964) *J. Phys. Chem.* **68**, 3585.
133. M. J. SCHICK and A. H. GILBERT (1965) *J. Colloid Sci.* **20**, 464.
134. K. SHINODA (1953) *Bull. Chem. Soc. Japan* **26**, 101.
135. S. H. HERZFELD, M. L. CORRIN and W. D. HARKINS (1950) *J. Phys. Colloid Chem.* **54**, 271.
136. K. SHIRAHAMA and T. KASHIWABARA (1971) *J. Colloid Interfac. Sci.* **36**, 65.
137. K. HAYASE and S. HAYONO (1978) *J. Colloid Interfac. Sci.* **63**, 446.
138. S. BACKLUND, K. RUNDT, K. S. BIRDI and S. DALAGER (1981) *J. Colloid Interfac. Sci.* **79**, 578.
139. F. A. GREEN (1972) *J. Colloid Interfac. Sci.* **41**, 124.
140. N. NISHIKIDO, Y. MOROI, H. UEHARA and R. MATUURA (1974) *Bull. Chem. Soc. Japan* **47**, 2634.
141. J. SWARBRICK and J. DARAWALA (1969) *J. Phys. Chem.* **73**, 2627.
142. E. D. GODDARD AND G. C. BENSON (1957) *Canad. J. Chem.* **35**, 986.
143. K. KUMIYAMA, H. INOUE and T. NAKAGAWA (1962) *Kolloid-Z-u-Z. Polymere* **183**, 68.
144. R. R. BALMBRA, J. S. CLUNIE, J. M. CORKILL and J. F. GOODMAN (1962) *Trans. Faraday Soc.* **58**, 1661.
145. R. R. BALMBRA, J. S. CLUNIE, J. M. CORKILL and J. F. GOODMAN (1964) *Trans. Faraday Soc.* **60**, 979.
146. K. KURIYAMA (1962) *Kolloid-Z.* **181**, 144.
147. P. H. ELWORTHY and C. McDONALD (1965) *Kolloid-Z.* **195**, 16.
148. R. H. OTTEWILL, C. C. STORER and T. WALKER (1967) *Trans. Faraday Soc.* **63**, 2796.
149. P. H. ELWORTHY and A. T. FLORENCE (1965) *Kolloid-Z.* **204**, 105.
150. S. D. HAMANN (1962) *J. Phys. Chem.* **66**, 1359.
151. R. F. TUDDENHAM and A. E. ALEXANDER (1962) *J. Phys. Chem.* **66**, 1839.
152. J. OSUGI, M. SATO and N. IFUKU (1965) *Rev. Phys. Chem. Japan* **35**, 32.
153. J. OSUGI, M. SATO and N. IFUKU (1968) *Rev. Phys. Chem. Japan* **38**, 58.
154. M. TANAKA, S. KANESHINA, K. SHIN-NO, T. OKAJIMA and T. TOMIDA (1974) *J. Colloid Interfac. Sci.* **46**, 132.
155. S. KANESHINA, M. TANAKA, T. TOMIDA and R. MATUURA (1974) *J. Colloid Interfac. Sci.* **48**, 450.
156. M. TANAKA, S. KANESHINA, S. KURAMOTO and R. MATUURA (1975) *Bull. Chem. Soc. Japan* **48**, 432.
157. S. RODRIGUEZ and H. OFFEN (1977) *J. Phys. Chem.* **81**, 47.
158. N. NISHIKIDO, N. YOSHIMURA and M. TANAKA (1980) **84**, 558.
159. N. NISHIKIDO, M. SHINOZAKA, G. SUGIHARA and M. TANAKA (1980) *J. Colloid Interfac. Sci.* **74**, 474.
160. C. A. T. HOEVE and G. C. BENSON (1957) *J. Colloid Interfac. Sci.* **61**, 1149.
161. R. H. ARANOW (1963) *J. Colloid Interfac. Sci.* **67**, 556.
162. A. WULF (1978) *J. Colloid Interfac. Sci.* **82**, 804.
163. B. A. PETHICA (1960) *Proc. IIIrd Inter. Cong. Surface Activity, Cologne,* Vol. 1, p. 212.
164. ABU-HAMDIYYAH and K. J. MYSELS (1967) *J. Phys. Chem.* **71**, 418.
165. P. H. ELWORTHY and K. J. MYSELS (1966) *J. Colloid Interfac. Sci.* **21**, 331.
166. K. KAIBARA, T. NAKAHARA, I. SATAKE and R. MATUURA (1970) *Mem. Fac. Sci. Kyushu Univ. Ser. C.* 71.

167. C. BOTRÉ, D. G. HALL and R. W. SCOWEN (1972) *Kolloid-Z.* **250**, 900.
168. R. C. MURRAY and G. S. HARTLEY (1935) *Trans. Faraday Soc.* **31**, 185.
169. M. J. VOLD (1950) *J. Colloid Sci.* **5**, 506.
170. J. N. PHILLIPS (1955) *Trans. Faraday Soc.* **51**, 561.
171. J. M. CORKILL, J. F. GOODMAN and S. P. HARROLD (1964) *Trans. Faraday Soc.* **60**, 202.
172. E. W. ANACKER (1970) in *Cationic Surfactants*, (ed. Jungermann), Marcel Dekker, New York.
173. J. Th. G. OVERBEEK and D. STIGTER (1956) *Rec. Trav. Chim.* **75**, 1263.
174. A. L. LOEB and J. Th. G. OVERBEEK (1966) *J. Colloid Interfac. Sci.* **22**, 78.
175. D. ATTWOOD and O. K. UDEALA (1974) *J. Pharm. Pharmacol.* **26**, 854.
176. J. M. CORKILL and J. F. GOODMAN (1969) *Adv. Colloid Interfac. Sci.* **2**, 297.
177. P. MOLYNEUX, C. T. RHODES and J. SWARBRICK (1965) *Trans. Faraday Soc.* **61**, 1043.
178. J. M. CORKILL, J. F. GOODMAN and J. R. TATE (1964) *Trans. Faraday Soc.* **60**, 996.
179. M. N. JONES, G. PILCHER, L. ESPADA and H. A. SKINNER (1969) *J. Chem. Thermodynam.* **1**, 381.
180. L. ESPADA, M. N. JONES and G. PILCHER (1970) *J. Chem. Thermodynam.* **2**, 333.
181. J. H. CLINT and T. WALKER (1975) *J. Chem. Soc. Farad. I.* **71**, 1946.
182. J. L. WOODHEAD, J. A. LEWIS, G. N. MALCOLM and I. D. WATSON (1981) *J. Colloid Interfac. Sci.* **79**, 454.
183. T. L. HILL (1963/1964) *Thermodynamics of Small Systems*, Vols. 1, 2, Benjamin: New York.
184. D. G. HALL and B. A. PETHICA (1967) in *Nonionic Surfactants* (ed. M. J. Schick) Arnold, London: Ch. 16.
185. D. G. HALL (1970) *Trans. Faraday Soc.* **66**, 1351, 1359.
186. D. G. HALL (1972) *J. Chem. Soc. Farad. II* **68**, 1439.
187. D. G. HALL (1972) *J. Chem. Soc. Farad. II* **68**, 25.
188. D. G. HALL (1977) *J. Chem. Soc. Farad. II* **73**, 897.
189. E. RUCKENSTEIN and R. NAGARAJAN (1975) *J. Chem. Soc. Farad. II* **79**, 2622.
190. F. ERIKSSON, J. C. ERIKSSON and PER STENIUS (1979) in *Solution Chemistry of Surfactants*, Vol. 1, Plenum: New York.
191. E. WYN-JONES (ed.) (1975) *Chemical and Biological Applications of Relaxation Spectrometry* Reidel: Dordrecht, pp. 133–264.
192. T. NAKAGAWA (1974) *Colloid Polymer Sci.* **252**, 56.
193. N. MULLER (1972) *J. Phys. Chem.* **76**, 3017.
194. E. A. G. ANIANSSON and S. N. WALL (1974) *J. Phys. Chem.* **78**, 1024.
195. E. A. G. ANIANSSON and S. N. WALL (1975) *J. Phys. Chem.* **79**, 857.
196. S. N. WALL and E. A. G. ANIANSSON (1980) *J. Phys. Chem.* **84**, 727.
197. B. C. BENNION and E. M. EYRING (1970) *J. Colloid Interfac. Sci.* **32**, 286.
198. B. C. BENNION, L. K. J. TONG and E. M. EYRING (1969) *J. Phys. Chem.* **73**, 3288.
199. C. TONDRÉ and R. ZANA (1978) *J. Colloid Interfac. Sci.* **66**, 544.
200. T. INOUE, R. TASHINO, Y. SHIBUYA and R. SHINOZAWA (1978) *J. Phys. Chem.* **82**, 2037.
201. T. INOUE, R. TASHINO, Y. SHIBUYA and R. SHINOZAWA (1980) *J. Colloid Interfac. Sci.* **73**, 105.
202. J. LANG and E. M. EYRING (1972) *J. Polym. Sci. Part A-2* **10**, 89.
203. G. C. KRESCHECK, E. HAMORI, G. DAVENPORT and H. A. SCHERAGA (1966) *J. Amer. Chem. Soc.* **88**, 246.
204. P. F. MIJNLIEFF and R. DITMARSCH (1965) *Nature* **208**, 889.
205. K. TAKEDA and T. YASUNAGA (1972) *J. Colloid Interfac. Sci.* **40**, 127.
206. T. YASUNAGA, K. TAKEDA, and S. HARADA (1973) *J. Colloid Interfac. Sci.* **42**, 457.
207. J. LANG, J. J. AUBORN and E. M. EYRING (1972) *J. Colloid Interfac. Sci.* **41**, 404.
208. M. J. JAYCOCK and R. H. OTTEWILL (1967) in *Chem. Phys. Appl. Surface Active Subst.*, Proc. IVth Int. Congr., (ed. J. Th. G Overbeek) Vol. 2, Gordon and Breach: New York, p. 545.
209. E. GRABER, J. LANG and R. ZANA (1970) *Kolloid-Z.* **238**, 470.

210. J. RASSING, P. J. SAMS and E. WYN-JONES (1973) *J. Chem. Soc.* **69**, 180.
211. T. YASINAGA, S. FUJII and M. MIURA (1969) *J. Colloid Interfac. Sci.* **30**, 399.
212. R. ZANA and J. LANG (1968) *C. R. Acad. Sci. Ser. C* **266**, 893.
213. T. NAKAGAWA, H. INOUE, H. JIZOMOTO and K. HORIUCHI (1969) *Kolloid-Z.* **229**, 159.
214. T. NAKAGAWA and K. TORI (1964) *Kolloid-Z.* **194**, 143.
215. N. MULLER and F. PLATKO (1971) *J. Phys. Chem.* **75**, 547.
216. K. FOX (1971) *Trans. Faraday Soc.* **67**, 2802.
217. J. OAKES (1972) *J. Chem. Soc. Farad. Trans. II* **68**, 1464.
218. T. NAKAGAWA and H. JIZOMOTO (1972) *Kolloid-Z.* **250**, 591.
219. J. RASSING and E. WYN-JONES (1973) *Chem. Phys. Lett.* **21**, 93.
220. P. J. SAMS, E. WYN-JONES and J. RASSING (1972) *Chem. Phys. Lett.* **13**, 233.
221. P. J. SAMS, J. E. RASSING and E. WYN-JONES (1974) *J. Chem. Soc. Farad. II* **70**, 1247.
222. P. J. SAMS, J. E. RASSING and E. WYN-JONES (1975) *Adv. Mol. Relax. Processes* **6**, 255.
223. E. A. G. ANIANSSON (1978) *Ber. Bunsenges Phys. Chem.* **82**, 981.
224. J. LANG, C. TONDRE, R. ZANA, R. BAUER, R. HOFFMANN and W. ULBRICHT (1975) *J. Phys. Chem.* **79**, 276.
225. E. A. G. ANIANSSON (1978) *J. Phys. Chem.* **82**, 2805.
226. N. MULLER (1979) in *Solution Chemistry of Surfactants*, (ed. K. L. Mittal) Vol. 1 Plenum, New York (1979).
227. P. MUKERJEE (1974) *J. Pharm. Sci.* **63**, 972.
228. P. MUKERJEE (1975) in *Physical Chemistry: Enriching topics from Colloid Surface Science.* IUPAC, eds. H. V. M. Olphen and K. J. Mysels, Theorex, La Jolla, California, p. 135.
229. P. MUKERJEE (1978) *Ber. Bunsenges Phys. Chem.* **82**, 931.
230. A. K. GHOSH and P. MUKERJEE (1970) *J. Amer. Chem. Soc.* **92**, 6403, 6408, 6413.
231. P. O. P. TS'O, I. S. MELVIN and A. C. OLSON (1963) *J. Amer. Chem. Soc.* **85**, 1239.
232. D. ATTWOOD and J. A. TOLLEY (1980) *J. Pharm. Pharmac.* **32**, 761.
233. D. ATTWOOD, S. P. AGARWAL and R. D. WAIGH (1980) *J. Chem. Soc. Farad. I* **76**, 2187.
234. A. S. KERTES and H. GUTMANN, in *Surface and Colloid Science*, ed. E. Matijevic, Vol. 8, J. Wiley, New York.
235. T. M. DOSCHER and R. D. VOLD (1948) *J. Phys. Chem.* **52**, 97.
236. S. S. MARSDEN, K. J. MYSELS and G. H. SMITH (1947) *J. Colloid Sci.* **2**, 265.
237. W. PHILIPPOFF and J. W. MCBAIN (1949) *Nature (London)* **164**, 885.
238. R. D. VOLD and M. J. VOLD (1948) *J. Phys. Chem.* **52**, 1424.
239. M. B. MATHEWS and E. HISCHHORN (1953) *J. Colloid Sci.* **8**, 86.
240. M. VAN DER WAARDEN (1950) *J. Colloid Sci.* **5**, 448.
241. P. H. ELWORTHY and D. S. MCINTOSH (1964) *J. Phys. Chem.* **68**, 3448.
242. I. BLEI and R. E. LEE (1963) *J. Phys. Chem.* **67**, 2085.
243. A. J. FRYER and S. KAUFMAN (1969) *J. Colloid Interfac. Sci.* **29**, 444.
244. E. RUCKENSTEIN and R. NAGARAJAN (1980) *J. Phys. Chem.* **84**, 1349.
245. J. A. MERRIGAN, S. J. TAO and J. H. GREEN (1972) in *Physical Methods of Chemistry*, Vol. 1, Part 111 (eds. D. A. Weissberger & B. W. Rossiter) Wiley, New York.
246. Y-C. JEAN and H. J. ACHE (1978) *J. Amer. Chem. Soc.* **100**, 6320.
247. Y-C. JEAN and H. J. ACHE (1978) *J. Amer. Chem. Soc.* **100**, 984.
248. K. KON-NO and A. KITAHARA (1965) *Kogyo Kagaku Zasshi* **68**, 2058.
249. S. MUTO and K. MEGURO (1973) *Bull Chem. Soc. Japan* **46**, 1316.
250. H. F. EICKE and H. CHRISTEN (1978) *Helv. Chim. Acta.* **61**, 2258.
251. J. H. FENDLER, E. J. FENDLER, R. T. MEDARY and O. A. EL SEOUD (1973) *J. Chem. Soc. Farad. II* **69**, 280.
252. J. H. FENDLER, E. J. FENDLER, R. T. MEDARY and O. A. EL SEOUD (1973) *J. Phys. Chem.* **77**, 1432.
253. F. Y-F. LO, B. M. ESCOTT, E. J. FENDLER, E. T. ADAMS, R. D. LARSEN and P. W. SMITH (1975) *J. Phys. Chem.* **79**, 2609.

254. S. GOLDMAN and G. C. B. CAVE (1971) *Can. J. Chem.* **49**, 1716, 1726, 4096.
255. P. S. SHEIH and J. H. FENDLER (1977) *J. Chem. Soc. Farad. 1* **73**, 1480.
256. N. J. MULLER (1978) *J. Colloid Interfac. Sci.* **63**, 383.
257. L. A. FUCUGAUCHI, B. DJERMOUNI, E. D. HANDEL and H. J. ACHE (1979) *J. Amer. Chem. Soc.* **101**, 2841.
258. B. DJERMOUNI and H. J. ACHE (1979) *J. Phys. Chem.* **83**, 2476.
259. P. DEBYE and H. COLL (1962) *J. Colloid Sci.* **17**, 220.
260. D. ATTWOOD and C. MCDONALD (1974) *Colloid and Polymer Sci.* **252**, 138.
261. A. F. SIRIANNI, J. M. G. COWIE and I. E. PIDDINGTON (1962) *Can. J. Chem.* **40**, 957.
262. P. BECHER and N. K. CLIFTON (1959) *J. Colloid Sci.* **14**, 519.
263. S. M. NELSON and R. C. PINK (1952) *J. Chem. Soc.* 1744.
264. N. PILPEL (1964) *Nature (London)* **204**, 378.
265. K. KON-NO and A. KITAHARA (1971) *J. Colloid Interfac. Sci.* **35**, 636.
266. G. MARKOVITS and A. S. KERTES (1967) in *Solvent Extraction Chemistry* (eds. D. Dyrssen, J. O. Liljenzin and J. Rydberg). North-Holland: Amsterdam, p. 390.
267. A. RAY (1969) *J. Amer. Chem. Soc.* **91**, 6511.
268. A. RAY (1971) *Nature* **231**, 313.

4 Surface activity and colloidal properties of drugs and naturally occurring substances

4.1 Colloidal properties of drugs

4.1.1 Introduction

A large number of drugs have been found to exhibit typical colloidal behaviour in aqueous solution in that they accumulate at interfaces, depressing the surface tension, and form aggregates in solution at sufficiently high concentrations. The biological and pharmaceutical implications of this behaviour have been reviewed [1–3] and are discussed later in this chapter. The first part of the chapter will be concerned primarily with the solution properties of the colloidal drugs, with particular emphasis on their mode of association.

Drugs represent an interesting variety of amphiphilic structures ranging at one extreme from the cationic quaternary ammonium germicides, which are easily recognized as typical surfactants, to more complex aromatic or heterocyclic molecules such as the phenanthrene narcotic analgesics. It is important to recognize that micellization is merely one pattern of association, albeit an important one, which amphiphilic molecules may exhibit in solution. Typical surfactants, as discussed in Chapter 3, have hydrocarbon groups which can intertwine during the micellization process to form approximately spheroidal aggregates. Replacement of this flexible hydrophobic moiety with a rigid aromatic or heterocyclic ring system can have very pronounced effects on the way in which molecules are disposed within the aggregates to such an extent that the process of association may no longer be regarded as micellization. A well known illustration of this effect is the association of the cationic dyes and the purine and pyrimidine bases of nucleotides which associate by a stacking process. This self-association process is generally continuous, that is, there is no equivalent to a CMC and there is a wide range of aggregate sizes in solution. The side chains of such molecules are generally small with respect to the hydrophobic ring system and the self-association is controlled by hydrophobic interactions, charge repulsion playing an insignificant role in the aggregation process. In between these two extremes lie many of the drug molecules. Although the hydrophobic groups of most drugs are aromatic they may resemble typical surfactants in that these groups have a high degree of flexibility. This is true, for example, of the quite extensive group of drugs which are derivatives of diphenylmethane. On the other hand, those drugs with rigid aromatic ring systems, for example the pheno-thiazines, differ from the cationic dyes in that their charge is generally localized at a terminal group of a relatively long side chain rather than delocalized in the ring system, as is common with dye molecules. Drugs thus provide an opportunity to investigate those factors which are responsible for the type of association which is exhibited by a particular amphiphilic molecule in solution. It is this aspect of the studies on colloidal properties of drugs, rather than any pharmacological consequences of the colloidal behaviour, which will be emphasized in this first section of the chapter.

4.1.2 Antihistamines

(A) ASSOCIATION CHARACTERISTICS

(i) Diphenylmethane derivatives

The structures of some diphenylmethane antihistamines which have been studied in detail are shown in Table 4.1. Light-scattering measurements on these compounds [4] invariably produce plots showing distinct inflections which for conventional surfactants would be identified with the CMC (see Fig. 4.1). Aggregation numbers are, however, much lower (usually around 9 to 12 monomers per micelle in the absence of added electrolyte) than those of flexible chain surfactants and in view of this it is necessary to examine critically the evidence for micellization as opposed to continuous association. A compu-

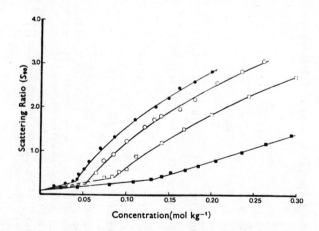

Figure 4.1 Variation of the scattering ratio, S_{90}, with concentration for aqueous solutions of ●, chlorcyclizine hydrochloride; ○, bromodiphenhydramine hydrochloride; □, diphenylpyraline hydrochloride; ■, diphenhydramine hydrochloride. (—) calculated from mass-action theory (Equation 3.36). From Attwood and Udeala [5] with permission.

Table 4.1 Micellar properties of some diphenylmethane antihistamines in water and 0.9% NaCl

Drug	R_1	R_2	Additive	CMC (mol l^{-1})	Aggregation number[‡]
Diphenhydramine HCl	—OCH$_2$CH$_2$N(CH$_3$)$_2$	H	—	0.140* 0.127[†]	3
			NaCl	0.077§ 0.085[‡]	6
Bromodiphenhydramine HCl	—OCH$_2$CH$_2$N(CH$_3$)$_2$	p-Br	—	0.042* 0.048[†] 0.054[¶] 0.052∥	11
			NaCl	0.033§ 0.020[‡]	29
Chlorcyclizine HCl	—N⟋＼N—CH$_3$	p-Cl	—	0.040[†],[¶]	9
			NaCl	0.012[‡]	24
Diphenylpyraline HCl	—N⟋＼O (N—CH$_3$)	H	—	0.090[†] 0.087[¶]	9
			NaCl	0.044[‡]	14
Medrylamine maleate	—OCH$_2$CH$_2$N(CH$_3$)$_2$	p-OCH$_3$	NaCl	0.012§	—
Phenyltoloxamine citrate	$R_2 = o$-OCH$_2$CH$_2$N(CH$_3$)$_2$	R_1 = H	NaCl	0.013§	—

* Mean value at 20°C from several techniques including cryoscopy, spectrophotometry, surface tension, potentiometry and conductivity, Thoma and Siemer [7]. [†] Mean value at 30°C from several techniques including conductivity, surface tension and light scattering, Attwood and Udeala [4], [8], [16]. [‡] Value at 30°C by light scattering [8]. § Value at 30°C by surface tension [7]. [¶] Value at 25°C by ultrasonic relaxation, Causon et al. [10]. ∥ Value at 25°C by vapour pressure osmometry, Farhadieh et al. [11].

tational procedure has been devised by which the mass-action theory, in the form of Equation 3.36, may be applied to the light-scattering data in an attempt to simulate the experimental variation of light-scattering intensity with solution concentration [5]. The lines through the experimental scattering points in Fig. 4.1 have been computed in this manner and show clearly that the scattering behaviour of these compounds may be described by the mass-action equation of micellization. The possession of a CMC may be viewed as a criterion of a micellar type of association. However, as discussed by Mukerjee [6] it is all too easy to imagine inflection points in experimental curves. It is preferable to examine the system using a variety of techniques and to compare the CMC values obtained from inflections of the data. Table 4.1 shows such a comparison for several antihistamine drugs. Taking into account inherent differences in CMC values which arise between various techniques, the agreement between CMC values is further evidence that the process of aggregation is indeed micellar.

The addition of electrolyte to solutions of these diphenylmethane anti-histamines produces an increase in aggregation number and decrease in CMC which is typical of the behaviour of surfactants. Plots of log CMC as a function of log counterion concentration X^- (Fig. 4.2) are linear [8] as predicted from Equation 3.40 which is derived from the mass-action theory. Values of the degree of ionization, α, derived from the slopes of such plots are in agreement with those from light scattering and furthermore, ΔG_h° values determined from the intercepts of such plots are in reasonable agreement with expected values derived from a consideration of the free energy change associated with the transference of two phenyl rings from an aqueous to a non-aqueous environment.

The micellar charge and hydration of the diphenylmethane antihistamines have been examined in detail [9].

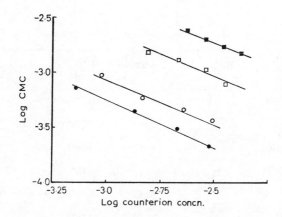

Figure 4.2 Log CMC against counterion concentration for ○, bromodiphenhydramine hydrochloride; ●, chlorcyclizine hydrochloride; □, diphenylpyraline hydrochloride; and ■, diphenhydramine hydrochloride. Concentrations are expressed as mole fractions. From Attwood and Udeala [8] with permission.

(ii) Pyridine derivatives

Fig. 4.3 and Table 4.2 show examples of pyridine derivatives with antihistaminic activity. Evidence for the association of mepyramine maleate and tripelennamine hydrochloride was first presented by Farhadieh et al. [11] from conductivity and vapour-pressure osmometry data. It was suggested that small aggregates are present at high concentrations although no CMC could be detected. In later studies by Attwood and Udeala [12, 13] an interesting distinction was noted in the association characteristics of tripelennamine and thenyldiamine hydrochlorides, which are micellar, (see Table 4.2) and mepyramine, pheniramine, chlorpheniramine and brompheniramine maleates, the light-scattering behaviour of which indicates a continuous association process with no apparent CMC. Fig. 4.4 shows a typical curve for the latter group of drugs and the best fit which

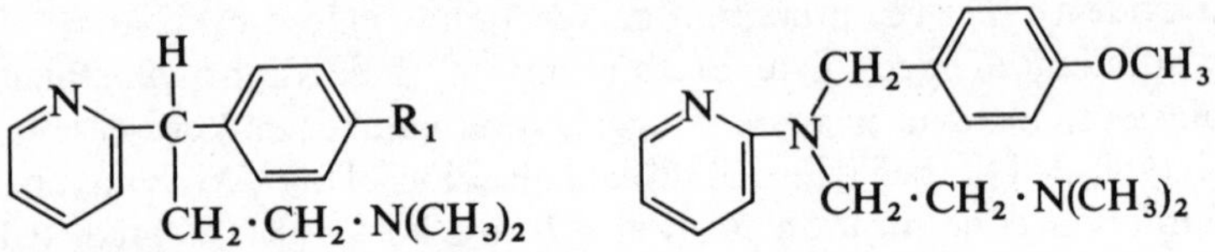

Pheniramine R₁ = H
Chlorpheniramine R₁ = Cl
Brompheniramine R₁ = Br

Mepyramine

Figure 4.3 Structures of some antihistaminic drugs containing a pyridine ring.

Table 4.2 Micellar properties in water and 0.9% NaCl of pyridine derivatives with antihistamine activity

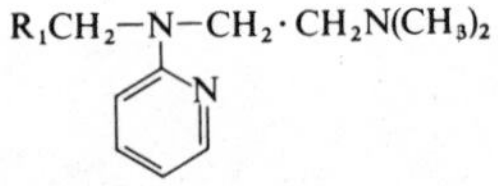

$$R_1CH_2-N-CH_2\cdot CH_2N(CH_3)_2$$

Drug	R_1	Additive	CMC (mol l⁻¹)	Aggregation number‡
Tripelennamine HCl	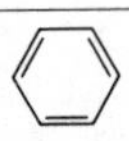	—	0.2* 0.130†	—
		NaCl	0.078† 0.11‡ 0.12*	3
Chloropyramine HCl	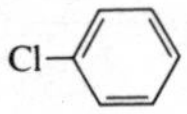 Cl–	NaCl	0.035†	—
Thenyldiamine HCl	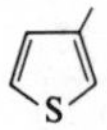	—	0.18*	—
		NaCl	0.185† 0.130‡ 0.10*	3
Methapyrilene HCl	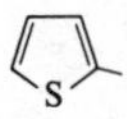	NaCl	0.181†	—

* At 30° C by surface tension, Attwood and Udeala [16]. † At 20° C by various techniques, Thoma and Siemer [7]. ‡ From Attwood and Udeala [13].

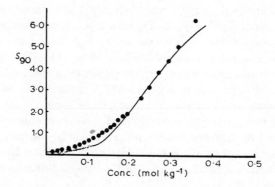

Figure 4.4 Simulation of the concentration dependence of the scattering ratio, S_{90}, for mepyramine maleate, using mass-action equations. The continuous line represents values calculated using Equation 3.36 with $n = 10$, $K_m = 10^{42}$ and $\alpha = 0.2$. (●) = experimental data. From Attwood and Udeala [5] with permission.

could be obtained from application of the mass-action equation to the light-scattering data. In order to reproduce the appreciable scatter at higher concentrations, it is necessary to assume aggregation numbers $n \geqslant 10$ and high equilibrium constants, K_m. Such combinations invariably produce significant inflections in the light-scattering plots which were not present in experimental data. Similarly, plots of molar conductivity against (concentration)$^{\frac{1}{2}}$, although showing appreciable deviation from the Onsager plots, did not exhibit any significant inflections which could be identified with a CMC. In later studies [14] it was shown that the association of mepyramine maleate could be described by a stepwise association model in which association constants

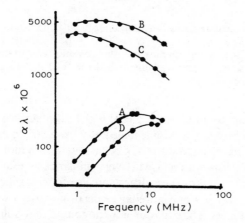

Figure 4.5 Plots of relaxation spectra of (A) 0.1 M tripelennamine hydrochloride, (B) 0.1 M chlorpheniramine maleate, pH 4.9, (C) 0.1 M pyridine + 0.05 M sodium maleate, pH 5.0, and (D) 0.1 M chlorpheniramine maleate, pH 1.8. From Gettins *et al.* [15] with permission.

increased sequentially with aggregation number. The cause of the non-micellar association is clearly attributable to the maleate counterion since change of the counterion associated with tripelennamine to maleate induces a non-micellar association pattern in this previously micellar drug [13]. Ultrasonic relaxation studies on the pyridine derivatives [15] have indicated a single relaxation of weak amplitude for all compounds associated with Cl^- counterions. In contrast, those compounds associated with maleate counterions show an intense relaxation associated with more than one relaxation time. It was established that this intense relaxation arose from a proton transfer process involving the maleate counterions and the pyridine ring of the drug molecule. It could be eliminated at a pH of 1.8 when the spectrum was similar to that of the hydrochloride salts (see Fig. 4.5).

(B) SURFACE ACTIVITY

The surface activity at the air–water interface of a range of antihistamines has been reported independently by Thomas and Siemer [7] and Attwood and Udeala [16] with similar results. Fig. 4.6 shows the decrease in CMC and increase of surface activity resulting from the presence of chloro- or bromo-substituents on the phenyl rings when compared with the unsubstituted analogues.

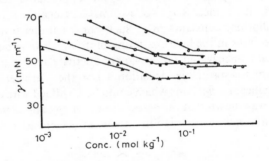

Figure 4.6 Surface tension, γ, as a function of log molal concentration, m, showing effect of changes in nature of hydrophobic group for -○- pheniramine and -●- brompheniramine maleates; -□- diphenhydramine and -■- bromodiphenhydramine hydrochlorides; -△- cyclizine and -▲- chlorcyclizine hydrochlorides in H_2O at 303 K. From Attwood and Udeala [16] with permission.

It is interesting to note from Fig. 4.6 that the surface tension plots of pheniramine and brompheniramine maleates show clearly defined inflections, in marked contrast to the light-scattering behaviour, which indicates a continuous increase of micellar size with no CMC. Very similar behaviour was also noted by Thomas and Siemer [7] for these drugs and also for chlorpheniramine maleate, another drug for which a non-micellar association pattern was indicated from light-scattering studies. The situation is analogous to that noted for several non-ionic detergents. Light-scattering studies [17] on aqueous solutions of hepta-oxyethylene glycol monohexadecyl ether ($C_{16}E_7$) for example, have shown a pronounced concentration dependence of micellar size at low concentrations, with scattering curves of similar appearance to those obtained for the anti-histamine maleates having no CMC. Surface tension graphs [18] for $C_{16}E_7$ (see Fig. 1.1), however, show an abrupt change of slope at a well-defined CMC. Later studies on other drugs with non-micellar association patterns showed that the apparent CMC detected by surface tension techniques arose because of the very limited change of monomer concentration with total solution concentration at high concentrations. This effect will be discussed below (p. 136).

The penetration of lecithin monolayers by a series of antihistamines correlated with their surface activity at the air–solution interface [19].

4.1.3 Antiacetylcholine drugs

(A) ASSOCIATION CHARACTERISTICS

Many of antiacetylcholine drugs have structures based on the diphenylmethane moiety (see Table 4.3) and, as expected, their association is micellar in nature [20]. The effect on the micellar properties of modification of the chemical structure of the hydrophobic and hydrophilic portions of the molecule may be assessed from the data presented in Table 4.3. In this respect the drugs investigated here present an interesting series of model surfactants. The most pronounced effect on the micellar properties results from the —OH or —$CONH_2$ groups in the R_3 position. As expected, these groups confer a greater hydrophilicity, as evidenced by the higher CMCs and lower aggregation numbers of compounds with these substituents. The magnitude of the effect may be assessed by considering pairs of compounds with otherwise similar structures, for example, piperidolate compared with pipenzolate and adiphenine with lachesine. A comparison of the properties of lachesine, poldine, pipenzolate, clidinium and benzilonium illustrates the well-known effect on the hydrophobicity of increasing the number of CH_2 groups in the polar chain (R_4). There is a decrease in CMC (increased hydrophobicity) with increase in the number of CH_2 groups, in the order lachesine (6CH_2) > poldine (7CH_2) > pipenzolate, clidinium and benzilonium (8CH_2). It is of interest to compare orphenadrine and chlorphenoxamine with the antihistamine, diphenhydramine ($R_1 = R_2 = R_3 = H$) which has a CMC of $0.132 \, \text{mol} \, l^{-1}$ and forms trimers in aqueous solution. These three compounds have identical R_4 substituents and a comparison of their properties illustrates the influence of substituents on the diphenylmethane nucleus. The gradual increase in aggregation number, n, and decrease in CMC in the order diphen-hydramine – orphenadrine – chlorphenoxamine, illustrates the increasing hydrophobicity resulting from the ring-substituted methyl group of orphenadrine and the substitution of Cl and CH_3 in the R_1 and R_2 positions, respectively, in chlorphenoxamine.

Departure of the molecular structure from the basic diphenylmethane structure so far considered produces interesting effects on the aggregation characteristics. All of the drugs included in Table 4.4 retain sufficient flexibility to form micellar aggregates (see Fig. 4.7 for structures). The effect on micellar properties of modifications of the chemical structure of the hydrophobic groups

Table 4.3 Micellar properties of anticholinergic drugs in water

Compound	R_1	R_2	R_3	R_4	CMC (mol l^{-1})	Aggregation number
Adiphenine HCl	H	H	H	$COO(CH_2)_2 \cdot N(C_2H_5)_2$	0.082*	10
Piperidolate HCl	H	H	H	*(structure)*	0.082*	12
Benztropine mesylate	H	H	H	*(structure)*	0.041*	7
Orphenadrine HCl	H	CH_3	H	$O(CH_2)_2 \cdot N(CH_3)_2$	0.096* 0.100†	7
Chlorphenoxamine HCl	Cl	H	CH_3	$O(CH_2)_2 \cdot N(CH_3)_2$	0.045* 0.039†	13
Lachesine Cl	H	H	OH	$COO(CH_2)_2 \cdot N(CH_3)_2C_2H_5$	0.204*	2
Poldine methylsulphate	H	H	OH	*(structure)*	0.150*	3

Pipenzolate Br	H	H	OH	(structure)	0.105*	3
Clidinium Br	H	H	OH	(structure)	0.123*	4
Benzilonium Br	H	H	OH	(structure)	0.143*	4
Ambutonium Br	H	H	$CONH_2$	$(CH_2)_2{}^+N(CH_3)_2C_2H_5$	0.136*	2

* At 30° C by light scattering, Attwood [20]. † At 20° C by various techniques, Thoma and Siemer [7].

Table 4.4 Micellar properties of some antiacetylcholine drugs in water

Drug	CMC (mol l^{-1})	Aggregation number*
Oxyphenonium bromide	0.108*	10
Tricyclamol chloride	0.150*	6
Dicyclomine HCl	0.050* 0.049[†]	10
Glycopyrronium bromide	0.189*	5
Penthienate bromide	0.220*[†]	6

* At 30°C by light-scattering, Attwood [21].
[†] At 25°C by ultrasonic relaxation, Causon *et al.* [10].

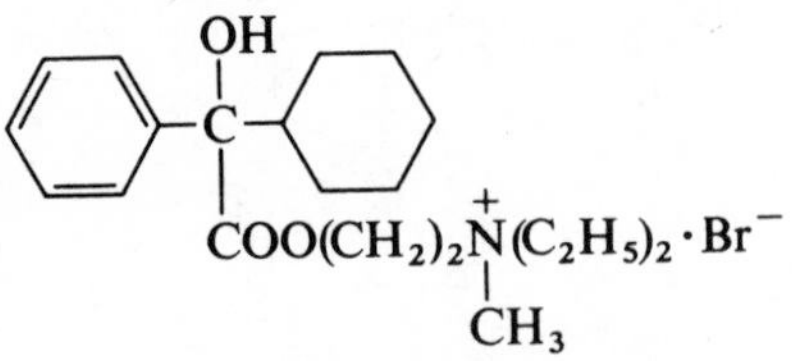

Oxyphenonium bromide

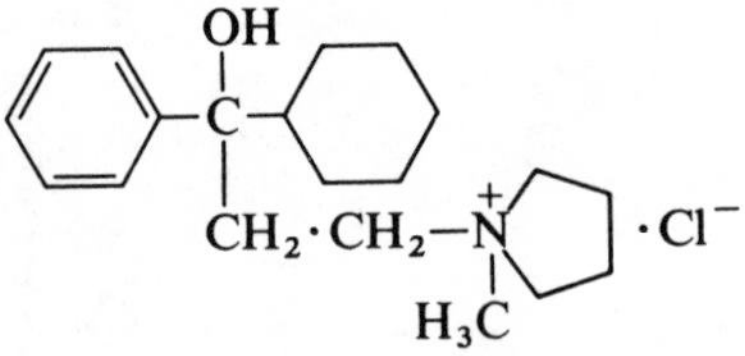

Tricyclamol chloride

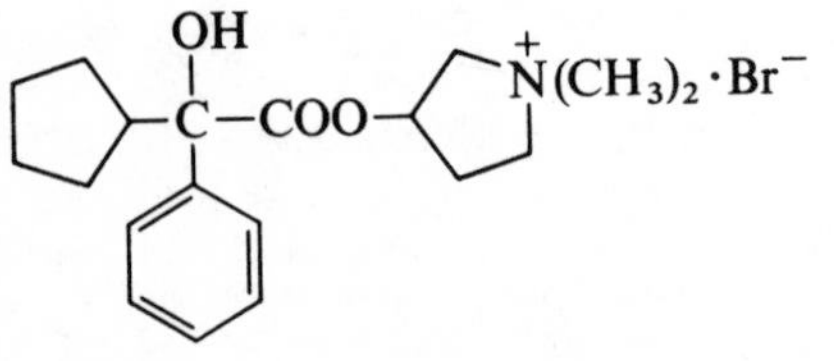

Glycopyrronium bromide

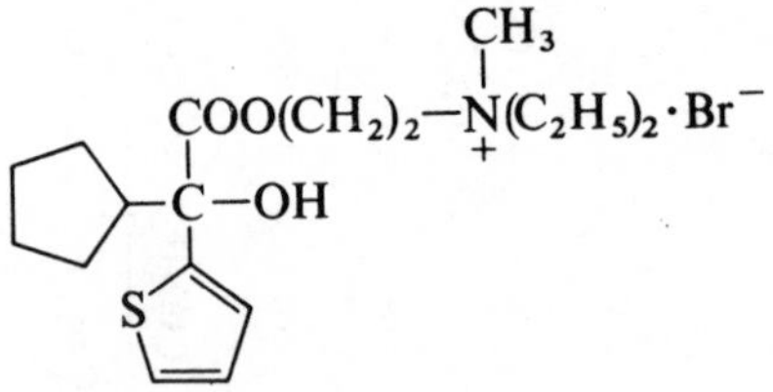

Penthienate bromide

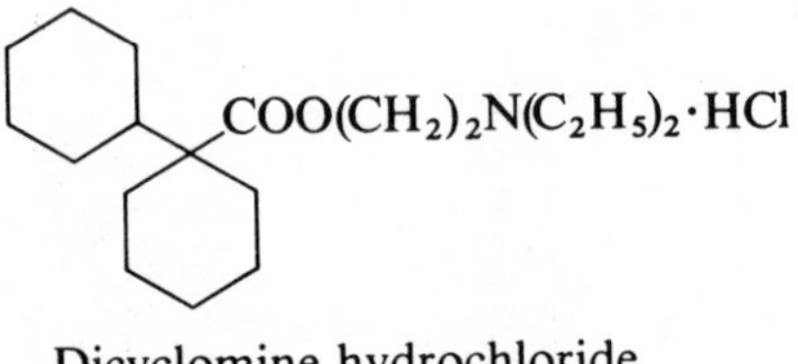

Dicyclomine hydrochloride

Figure 4.7 Structures of the antiacetylcholine drugs included in Table 4.4.

may be assessed by comparisons with those drugs of Table 4.3 with similar hydrophilic chains. Comparison of oxyphenonium bromide with lachesine chloride shows that replacement of one of the phenyl rings of the diphenylmethane moiety with a cyclohexane ring considerably increases the hydrophobicity, as evidenced by the lower CMC and higher aggregation number. Conversely, the replacement of a phenyl ring with a cyclopentane ring has little apparent effect on hydrophobicity; the diphenylmethane derivative, benzilonium bromide, has similar properties to glycopyrronium bromide.

When the two phenyl rings of the diphenylmethane moiety of adiphenine are linked together in the form of a rigid fluorene group as in pavatrine hydrochloride (I), the association pattern no longer conforms to that of a monodisperse micellar system [22]. The concentration dependence of the light-scattering intensity of pavatrine exhibits no such inflection as that detected in the curve of adiphenine (Fig. 4.8) and cannot be simulated by the mass-action theory.

$$COOCH_2 \cdot CH_2 \cdot N(C_2H_5)_2$$

(I)

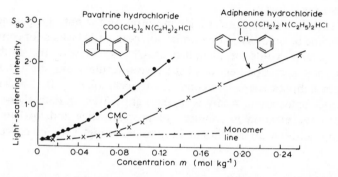

Figure 4.8 Comparison of light-scattering curves for adiphenine hydrochloride and its rigid analogue pavatrine hydrochloride showing the influence of the rigidity of the hydrophobic group on the aggregation characteristics. From Attwood *et al.* [22] with permission.

In describing the association of pavatrine it is necessary to assume a stepwise association model (see Section 3.8) in which the dimerization constant is treated as an independent variable and higher association constants, K_n, decrease with aggregation number, n, according to the relationship $K_n = K(n-1)/(n-2)$. Addition of sodium chloride causes not only an increase in the magnitude of the association constants but also a change in the pattern of association. In the presence of $0.1 \, \text{mol} \, l^{-1}$ NaCl the association is also best represented by a two-parameter model but with K_n values for $n \geqslant 3$ of equal magnitude. At higher electrolyte concentration (0.2 and $0.5 \, \text{mol} \, l^{-1}$) the association follows a one-parameter model in which association constants increase with increasing n according to $K_n = K(n-1)/n$.

The results obtained for pavatrine clearly indicate the influence of the structure of the hydrophobic group on the association pattern, a rigid ring structure promoting non-micellar association. This conclusion is further substantiated by the association behaviour of the two antiacetylcholine drugs, propantheline (II) and methantheline bromide (III) both of which have rigid tricyclic ring systems.

COO(CH$_2$)$_2$$\overset{+}{N}$·[CH(CH$_3$)$_2$]$_2$
$\underset{CH_3}{|}$

(II)

COO(CH$_2$)$_2$$\overset{+}{N}$·(C$_2H_5$)$_2$
$\underset{CH_3}{|}$

(III)

The light-scattering plots of both of these drugs indicate a non-micellar association pattern. Analysis of the data reveals that the association can best be described by a stepwise association model similar to that of pavatrine in high electrolyte concentration with association constants increasing with aggregation number according to the relationship, $K_n = K(n-1)/n$.

(B) SURFACE ACTIVITY

Plots of surface tension against log concentration for both propantheline and methantheline bromide show clear inflection points [23] which for conventional micellar systems would be identified as CMCs. As with the antihistamine maleates, there is an apparent conflict between results from this technique and those from light-scattering and conductivity techniques both of which suggest non-micellar association. Analysis of the light-scattering data shows that the monomer concentration in solutions of propantheline and methantheline in water and electrolyte approaches limiting values at high solution concentrations

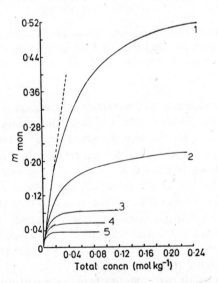

Figure 4.9 Variation of monomer concentration m_{mon}, (mol kg^{-1}) with solution concentration calculated from light-scattering data assuming stepwise association. 100% monomers (– – –). 1, methantheline bromide in H$_2$O; 2, propantheline bromide in H$_2$O, 3, 0.15 mol kg^{-1} NaBr, 4, 0.30 mol kg^{-1} NaBr, and 5, 0.45 mol kg^{-1} NaBr. From Attwood [23] with permission.

(Fig. 4.9) and furthermore these limiting monomer concentrations are in good agreement with the apparent CMC values from surface tension plots (Table 4.5). The inflections in the surface tension plots are thus a consequence of a limiting monomer concentration in the solutions rather than a micellar type of association.

Table 4.5 Limiting monomer concentrations for non-micellar antiacetylcholine drugs [21]

Drug	Additive	Limiting monomer concentrations (mol l^{-1} × 10^3) from	
		Surface tension*	Light scattering†
Propantheline Br	—	13.5	~ 22
	0.15 M NaBr	8.2	8.3
	0.30 M NaBr	5.1	5.7
	0.45 M NaBr	4.6	3.7
Methantheline Br	—	44	52

* Inflection of plot of γ versus log concentration.
† Calculated assuming stepwise association.

4.1.4 Tricyclic antidepressants

The micellar properties of a series of tricyclic antidepressants were reported by Attwood and Gibson [24] and later by Thoma and Albert [25] and are summarized in Tables 4.6 and 4.7. Agreement between CMC values from a wide range of techniques and also the form of the light-scattering plots clearly indicate the micellar nature of the association process. This is perhaps surprising in view of the only limited flexibility of the tricyclic ring systems [26] and suggests other requirements apart from a rigid hydrophobic group for a non-micellar association pattern. Possible additional factors are discussed on p. 161.

The antidepressant drugs provide an interesting series of compounds with which to illustrate the effect of substituents on solution properties. In the following examples, pairs of compounds have been selected in which each member of the pair has an identical side chain (R_1 group) so that any changes in hydrophobicity may be attributed entirely to the structural changes of the hydrophobic moiety. As all of these drugs are hydrochloride salts, the effect of the counterion is also eliminated. A comparison of imipramine with clomipramine shows the increased hydrophobicity (as evidenced by a decrease in the CMC) which is conferred by a —Cl substituent on one of the phenyl rings. This is a well-known effect and is noted, for example, in a comparison of the phenothiazine drugs, promazine and chlorpromazine (see p. 140) and the antihistamines, diphenhydramine and bromodiphenhydramine (see p. 126). A comparison of amitriptyline, doxepin and dothiepin demonstrates the effect on solution properties of a heteroatom in position 10 of the dibenzocycloheptadiene ring

Table 4.6 Micellar properties of iminodibenzyl and dibenzocycloheptadiene antidepressant drugs in water [24] (at 30° C) and 0.9 % NaCl [25] (at 20° C)

Iminodibenzyl derivatives

Drug	R_1	R_2	Additive	CMC $(mol\,l^{-1} \times 10^2)$	Aggregation number
Imipramine HCl	$(CH_2)_3N(CH_3)_2$	H	—	4.7	8
			NaCl	2.1	—
Clomipramine HCl	$(CH_2)_3N(CH_3)_2$	Cl	—	2.2	6
			NaCl	0.5	—
Desipramine HCl	$(CH_2)_3NHCH_3$	H	—	4.9	7
			NaCl	1.7	—
Trimipramine mesylate	$CH_2.CH(CH_3)CH_2.N(CH_3)_2$	H	NaCl	1.3	—

Dibenzocycloheptadiene derivatives

Drug	R_1	Additive	CMC $(mol\,l^{-1} \times 10^2)$	Aggregation number
Amitriptyline HCl	$= CH(CH_2)_2N(CH_3)_2$	—	3.6	7
		NaCl	1.4	—
Nortriptyline HCl	$= CH(CH_2)_2NHCH_3$	—	2.3	4 ·
		NaCl	0.8	—
Noxiptiline HCl	$= NO(CH_2)_2N(CH_3)_2$	NaCl	2.9	—
Butriptiline HCl	$-CH_2.CH(CH_3)CH_2N(CH_3)_2$	—	4.2	9

system. Tables 4.6 and 4.7 show that the hydrophobicity increases according to $O < CH_2 < S$.

A comparison of the surface activities of $10^{-3}\,mol\,l^{-1}$ solutions of the antidepressants in 0.9 % NaCl and acetate buffer [25] leads to similar conclusions concerning the influence of substituents on the hydrophobicity. For example, a $10^{-3}\,mol\,l^{-1}$ solution of clomipramine hydrochloride has a surface pressure of $12.9\,mN\,m^{-1}$ in 0.9 % NaCl which is approximately twice that of a $10^{-3}\,mol\,l^{-1}$ solution of imipramine hydrochloride due to the hydrophobic nature of the —Cl substituent on the tricyclic ring of clomipramine.

Other workers have reported the surface activity of selected tricyclic antidepressants. Vilallonga *et al.* [27] report the surface activity of imipramine in 0.1 M HCl; Kitler and Lamy [28] have determined the interfacial activity of several antidepressants at pH 2.00 in an aqueous buffer/cyclohexane system; and Nambu *et al.* [29] have measured the surface activity of $10^{-3}\,mol\,l^{-1}$ solutions of a range of antidepressants in buffer solutions at pH 6.

Table 4.7 Micellar properties of some antidepressant drugs in water [25] (at 30°C) and 0.9% NaCl [15] (at 20°C)

Drug	Structure	Additive	CMC (mol l^{-1} × 10^2)	Aggregation number
Doxepin HCl		—	6.0	7
		NaCl	2.2	—
Dothiepin HCl		—	2.9	10
Iprindole HCl		—	3.1	19
Protryptiline HCl		—	4.1	9
		NaCl	1.3	—
Melitracene HCl		NaCl	7.9	—
Dimetacrine tartrate		NaCl	7.9	—
Opipramol* di HCl		—	—	5
		NaCl	27.5	—

* Micellar properties are pH dependent (see [14]).

4.1.5 Phenothiazine and thioxanthene tranquillizers

A detailed study of the solution properties of the phenothiazine and thioxanthene drugs has clearly indicated that the association process is micellar [30]. Light-scattering graphs show clear inflection points [14] and CMC values from this and other techniques are in good agreement. Tables 4.8 and 4.9 summarize the micellar properties of a large number of these drugs. Taking the CMC and aggregation number as indications of the hydrophobicity of the molecules some interesting conclusions may be drawn from comparisons of molecules of similar structures with identical counterions. The effect on the hydrophobicity of substituents in the 2 position of the phenothiazine nucleus may be determined

Table 4.8 Micellar properties of phenothiazine tranquillizers in water and 0.9% NaCl

Drug	R_1	R_2	Additive	CMC $(mol\,l^{-1} \times 10^3)$	Aggregation number
Promazine HCl	$(CH_2)_3N(CH_3)_2$	H	—	36.5* 30.5†	11*
			NaCl	16.2§	37§
Chlorpromazine HCl	$(CH_2)_3N(CH_3)_2$	Cl	—	18.9* 18.0† 18.8‡	11*
			NaCl	5.9‖ 5.3§ 4.6† 5.6‡	64§ 40¶
Chlorproethazine HCl	$(CH_2)_3N(C_2H_5)_2$	Cl	NaCl	3.3‖	39¶ —
Methoxypromazine maleate	$(CH_2)_3N(CH_3)_2$	OCH_3	NaCl	4.7‖	
Methpromazine HCl	$(CH_2)_3N(CH_3)_2$	CH_3	—	—	46§
Promethazine HCl	$CH_2 \cdot CH(CH_3)N(CH_3)_2$	H	—	43.9* 43.5† 44.0‡	9*
			NaCl	16.7† 24.8‡ 22.7§	25§
Dimethothiazine mesylate	$CH_2 \cdot CH(CH_3) \cdot N(CH_3)_2$	$SO_2N(CH_3)_2$	NaCl	7.6‖	—
Methiomeprazine HCl	$CH_2 \cdot CH(CH_3)N(CH_3)_2$	SCH_3	NaCl	2.9‖	48¶
Trimeprazine maleate	$CH_2 \cdot CH(CH_3)CH_2N(CH_3)_2$	H	NaCl	13.0‖	31¶

Compound	Side chain	Substituent	Added salt		
Methiotrimeprazine HCl	$CH_2 \cdot CH(CH_3)CH_2N(CH_3)_2$	OCH_3	NaCl	9.1‖	23¶
Diethazine HCl	$CH_2 \cdot CH_2N(C_2H_5)_2$	H	NaCl	17.0‖	26§
Thioridazine HCl	CH_3 / $CH_2 \cdot CH_2$ (N-methylpiperidine side chain)	SCH_3	—	5.9* 6.5† 5.8‡	8*

Structure: phenothiazine ring bearing R_2 and the side chain $CH_2 \cdot CH_2 \cdot CH_2 - N$(piperazine)$N - R_1$

| | | | NaCl | 1.8* | 57* 67¶ |

Compound	R_2	R_1	Added salt		
Perazine dimalonate**	CH_3	H	NaCl	14.0‖	30¶
Trifluoperazine di HCl**	CH_3	CF_3	NaCl	9.8‖	59¶
Thioperazine dimethane sulphonate**	CH_3	$SO_2N(CH_3)_2$	NaCl	8.0‖	6¶
Thiethylperazine dimaleate**	CH_3	$SCH_2 \cdot CH_3$	NaCl	0.5‖	—
Perphenazine di HCl**	CH_2CH_2OH	Cl	NaCl	5.2‖	17¶
Fluphenazine di HCl**	CH_2CH_2OH	CF_3	NaCl	7.4‖	36¶
Thiopropazate di HCl**	$(CH_2)_2OOC.CH_3$	Cl	NaCl	6.0‖	43¶

* At 34°C, Attwood et al. [30]. †,‡ At 20°C and 34°C, respectively, Florence and Parfitt [31]. § At 23°C Scholtan [34]; ¶,‖ At 20°C, Thoma and Arning [32, 33], respectively.
** Micellar properties are pH dependent (see [14]).

Table 4.9 Micellar properties of thioxanthene tranquillizers in water and 0.9% NaCl at 20°C

Drug	R_1	R_2	Additive	CMC (mol l^{-1} × 10^3)	Aggregation number
Flupenthixol di HCl*	—N N—(CH$_2$)$_2$OH	CF$_3$	—	8.5† 6.0‡	19†
			NaCl	1.3§ 2.0†	44†
Clopenthixol di HCl*	—N N—(CH$_2$)$_2$OH	Cl	NaCl	1.24§	—
Thiothixene di HCl*	—N N—CH$_3$	SO$_2$N(CH$_3$)$_2$	NaCl	1.78§	—
Chlorprothixene HCl	—N(CH$_3$)$_2$	Cl	NaCl	1.99§	—

* Micellar properties are pH dependent (see [14]).
† At 30° C, Attwood, unpublished data.
‡ At 18° C, Enever et al. [35].
§ At 20° C, Thoma and Albert [25].

from comparisons of promazine and chlorpromazine, promethazine and methiomeprazine, perazine and thiethylperazine which demonstrate the hydrophobicity of the Cl, SCH_3 and SCH_2-CH_3 substituents, respectively. A branched side chain is less hydrophobic than a straight chain with an identical number of C atoms. Thus promethazine has a higher CMC and lower aggregation number than promazine. It is interesting to compare the phenothiazines with the tricyclic antidepressants. A direct comparison of compounds with identical side chains is possible in the case of the pairs, imipramine and promazine and also clomipramine and chlorpromazine. In both cases, the S atom of the phenothiazines confers a greater hydrophobicity than the two CH_2 groups which replace it in the antidepressant drugs. Consequently, the antidepressants have, in general, higher CMCs and slightly lower aggregation numbers than the phenothiazine drugs.

The effect of external factors such as temperature, pH and electrolyte on the solution properties of the phenothiazines is typical of that of micellar compounds. Addition of electrolyte causes an increase in micellar weight and linear decrease in log CMC with increasing log of the counterion concentration [33]. The micellar weights of thioridazine and methiomeprazine decrease over the temperature range 22 to 40° C [32]. The magnitude of the effect is not large, for example the aggregation number of thioridazine in 0.45 % NaCl decreases from 35 at 22° C to 24 at 40° C. A decrease in micellar weight from 39 800 to 20 000 was reported by Scholtan [34] over the same temperature range for chlorpromazine in 2.5 % NaCl. A study of the effect of pH on the micellar properties of several drugs including some phenothiazines was reported by Attwood and Natarajan [14]. One of the problems associated with such studies is the marked effect which electrolyte, added in the form of the buffer components, has on micellar properties. When care is taken to reduce such effects to a minimum it can be demonstrated that pH has no significant effect on the CMC or aggregation number of chlorpromazine over the pH range 2.0 to 5.5. At pH values approaching the pK_a, a marked increase in n and eventual precipitation of the drugs might be expected as the percentage of non-ionized insoluble base increases. The situation is complicated by the reduction of pK_a which is normally associated with micelle formation. The micellar properties of drugs such as trifluoperazine and thiopropazate which have a piperazine ring in the side chain exhibit marked sensitivity to pH. Table 4.10 shows the changes in CMC, aggregation number and ionization, α, of trifluoperazine dihydrochloride ($pK_{a_1} = 3.8$ $pK_{a_2} = 8.4$). It is interesting to note that because of the lowering of both pK_{a_1} and pK_{a_2} the micellar characteristics do not reach constant values representative of those for complete ionization of both N atoms or one N atom until the pH is at least 3 pH units below the stated pK_{a_1} and pK_{a_2} values, respectively. As expected, a higher aggregation number and lower CMC are noted when drugs containing piperazine moieties are present as singly charged compounds (see Table 4.11).

The surface and interfacial properties of the phenothiazine tranquillizers have been examined in detail in a series of papers by Zografi and co-workers [36–39]. Several points of general interest are reported in these studies. In agreement with

Table 4.10 Effect of pH on micellar properties of trifluoperazine dihydrochloride at ionic strength = 0.37

pH	CMC (mol kg^{-1})	Aggregation number	Degree of ionization α
0.5	0.020	17	0.46
1.0	0.019	17	0.46
2.0	0.016	24	0.41
2.9	0.007	35	0.42
3.4	0.005	40	0.23
4.3	0.003	48	0.20
5.0	0.003	49	0.18
5.5	0.002	82	0.20

From Attwood and Natarajan [14].

Table 4.11 Effect of pH on micellar properties of drugs containing a piperazine moiety. Ionic strength = 0.37

Compound	pH	CMC (mol kg^{-1})	Aggregation number	Degree of ionization, α
Clopenthixol di HCl	1.0	0.005	36	0.46
	4.8	0.0005	87	0.38
Flupenthixol di HCl	1.0	0.005	41	0.44
	4.8	0.001	62	0.28
Opipramol di HCl	1.0	0.034	8	0.49
	5.5	0.011	23	0.36
Thiopropazate di HCl	1.0	0.016	23	0.44
	4.8	0.003	47	0.33

From Attwood and Natarajan [14].

the effect of pH on the micellar properties of chlorpromazine discussed above, it was noted [38] that in order to obtain pH independence of the surface properties the pH should be 5.0 or less; the presence of non-protonated drug at higher pH caused a dramatic increase of surface pressure, π. Anionic buffer ingredients appeared to have an effect on the surface activity. Phthalate, citrate and succinate buffers tended to increase surface activity whilst acetate buffer had the opposite effect.

An earlier study [37] considered the effect of a large number of inorganic and organic ions on surface activity. Marked inhibitory effects were noted in the presence of organic cations, e.g. the tetra-alkylammonium ions; the greater the chain length, the greater the inhibition. The inhibition, which was also noted in the presence of sodium methanesulphonate, was attributed to an effect of these ions on water structure. On the other hand, bromide, iodide, propanesulphonate, benzenesulphonate and naphthalenesulphonate ions all increased surface activity to a greater extent than that normally associated with electrolyte addition, and some type of interaction of the electrolyte with the phenothiazine molecules was implied.

The effect on surface activity of substituents on the phenothiazine molecule was examined by Zografi and Munshi [38] and later by Thoma and Arning [33]. Substitution in the 2-position on the phenothiazine ring enhances the surface activity in the order $CF_3 > SCH_3 > Cl > H > OCH_3$. Changing the position of the chloro-group of chlorpromazine significantly influences the surface activity, the order being 3-chloro > 2-chloro > 1-chloro [38]. Comparison of ethyl, propyl and butyl derivatives of chlorpromazine shows the well-known effect of increasing alkyl chain length on surface activity. An increase of one $—CH_2—$ group gives about a two-fold increase in activity. Comparison of the isomers, promazine and promethazine shows a lower surface activity for the branched chain compound (promethazine) compared with its straight chain analogue in agreement with conclusions from a consideration of their micellar properties. Because of the high polarity of the sulphoxide group, oxidation of the ring S atom drastically reduces the surface activity of chlorpromazine, promethazine and trifluoperazine.

Vilallonga *et al.* [27] suggested from a study of the surface tension of several phenothiazines in 0.1 M HCl that the most probable arrangement of the molecules at the air–water interface is with the phenothiazine nucleus lying flat and the side chains directed into the aqueous solution.

Several authors have studied the interaction of the phenothiazines with various lipid membranes [40–43], and in some cases correlations with biological activities have been reported. Whilst some workers have interpreted the results of monolayer studies in terms of penetration of the monolayer by the drug molecules [40], an alternative view has been taken by Sears and Brandes [43] who conclude that the phenothiazines act immediately below the monolayer.

4.1.6 Analgesics

It is convenient in considering the colloidal properties of the analgesics to separate those drugs based on the diphenylmethane or similar moiety, such as methadone (IV) and dextropropoxyphene (V) since, as expected, their properties conform to those of typical surfactants. The association of dextropropoxyphene has been examined by several workers. Conine [44] demonstrated the ability of this drug to solubilize sparingly soluble organic acids. Thakkar *et al.* [45] using

$CH_3CH_2CO \cdot C \cdot CH_2 \cdot CH(CH_3)N(CH_3)_2$

$CH_3 \cdot CH_2 \cdot COO \cdot C—CH(CH_3)N(CH_3)_2$
$\qquad\qquad\qquad\quad CH_2$

(IV)　　　　　　　　　　　　　　(V)

n.m.r. techniques have shown evidence of hydrophobic interactions in aqueous solutions at high concentration. CMC values of between 0.10 and 0.12 $mol l^{-1}$ have been reported [46, 47] from a variety of techniques and an aggregation number of 7 has been determined by light-scattering methods. Methadone hydrochloride forms only trimers in aqueous solution [47] at a critical concentration of 0.1 $mol l^{-1}$.

The solution properties of the narcotic analgesics based on phenanthrene, such as morphine and codeine, and those based on piperidine, such as pethidine (see Fig. 4.10 for details of structures), have been examined by Attwood and Tolley [48]. Although results from optical rotary dispersion and conductivity techniques have been interpreted to suggest association of salts of morphine, codeine and hydromorphine [49] in the absence of added electrolyte, no evidence for any significant association could be detected for morphine sulphate or codeine phosphate from a variety of other techniques. Only in the presence of 0.5 $mol l^{-1}$ electrolyte, was association noted for ethylmorphine hydrochloride, oxycodone hydrochloride, codeine phosphate and pethidine hydrochloride. The association is non-micellar and can be described by stepwise association models. For pethidine hydrochloride the most appropriate model is that in which association constants K_n increase sequentially with aggregation number, n, according to the relation $K_n = K(n-1)/n$. The association of the remaining drugs is best described by an association model in which the stepwise association constants are

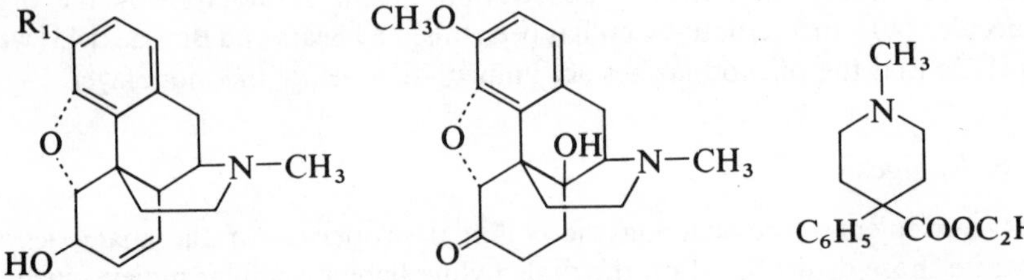

Morphine R_1 = OH
Codeine R_1 = OCH_3
Ethylmorphine R_1 = OC_2H_5

Oxycodone

Pethidine

Figure 4.10 Structures of some narcotic analgesics.

Table 4.12 Association models of analgesics in presence of 0.5 $mol l^{-1}$ electrolyte

	Association model	$K(l mol^{-1})$
Codeine phosphate	$K_n = K$	1.26
Ethylmorphine HCl	$K_n = K$	3.08
Oxycodone HCl	$K_n = K$	2.99
Pethidine HCl	$K_n = K(n-1)/n$	3.58

From Attwood and Tolley [48].

independent of the aggregation number. There is a pronounced difference between the order of magnitude of the association constants for the analgesics (see Table 4.12) and those of other non-micellar drugs, e.g. pavatrine and propantheline for which the K_n values are a factor of approximately 10^2 greater. This difference is attributable to the bulky, non-planar nature of the phenanthrene analgesics which inhibits the formation of large aggregates in solution. These analgesics more closely resemble the nucleotides [50] both in their mode of association and the order of magnitude of the association constants.

4.1.7 Antibacterials

A large number of antibacterials are quaternary ammonium or pyridinium cationic surfactants. The micellar properties of such compounds have been adequately covered in other texts on surface active agents. Although in many cases a correlation between surface activity and antibacterial action has been established it should be noted that surface activity *per se* is not a prerequisite for antibacterial action since the non-ionic surfactants are generally inactive. Several cationic dyes, for example acriflavine, proflavine and methylene blue, are used as bactericides. The colloidal properties of dyes have been reviewed by Duff and Giles [51]. The association is generally of a limited nature and in many cases is restricted to a monomer–dimer equilibrium. Spectrophotometric methods have been widely used in the interpretation of such equilibria, the association constant for dimerization being determined from the spectral shifts to higher wave numbers (metachromasy) which accompany increase in solution concentration. The association of crystal violet [52, 53], and methylene blue (VI) [54–56] have

(VI)

been determined in this manner. The validity of the spectral method for the determination of association constants for larger multimers has been questioned by Mukerjee and Ghosh [57]. These authors have used an 'iso-extraction' method based on the partition of the dye salt between an organic and an aqueous phase to determine the monomer concentration in solutions of methylene blue over a wide concentration range [57, 58]. The extent of self-association, which at very low concentration involved only dimers, increased rapidly to involve dimers, trimers, tetramers and higher multimers as the concentration was increased. The pattern of association was shown to conform to a two-parameter model in which the higher association constants $K_n (n \geqslant 3)$ decreased in a mild sequence with increasing aggregation number, n, according to the relationship $K_n = K(n-1)/(n-2)$. The kinetics of the monomer–dimer equilibrium in the concentration range 10^{-5} to 10^{-3} mol l^{-1} where the extent of dimerization is thought to be limited to dimerization, has been examined by the temperature-jump method [59].

The surface activities of a series of amino acridines including aminacrine, proflavine, mepacrine and acriflavine have been related to their chemical structure [60]. The results suggest that marked surface active properties are unnecessary for good antiseptic properties in this series.

The association characteristics of antibacterial agents, dequalinium acetate (VII) and chlorhexidine acetate (VIII) are of interest since these compounds have structures resembling those of the bolaform electrolytes, that is, they are symmetrical molecules with two charge centres separated at a relatively large distance.

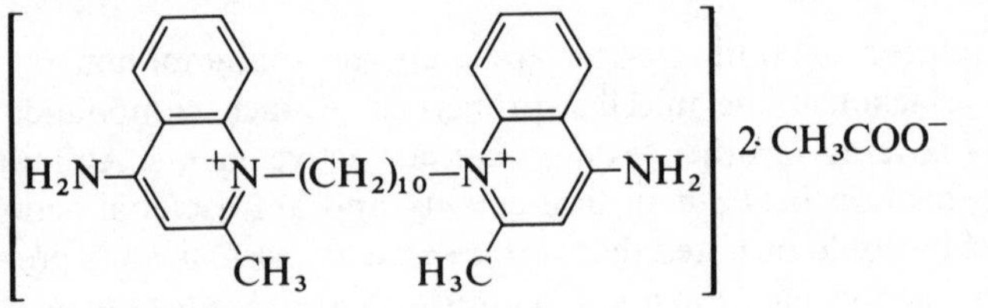

(VII)

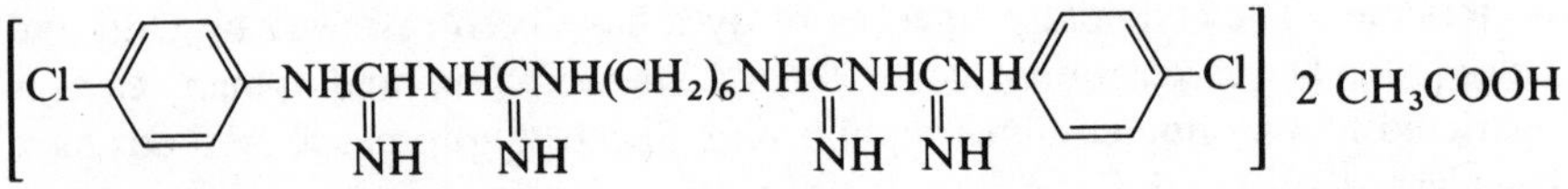

(VIII)

Dequalinium acetate forms micelles with an aggregation number of 16, at a critical concentration of $4 \times 10^{-3} \, mol \, l^{-1}$ [61]. The association of chlorhexidine salts has been examined by several workers with conflicting results. Heard and Ashworth [62] reported evidence from a variety of techniques for the association of both acetate and gluconate salts of this compound. Perrin and Witzke [63] determined a CMC for chlorhexidine gluconate from measurements of optical rotary dispersion techniques. Later workers [61] were, however, unable to detect any significant association of chlorhexidine acetate.

4.1.8 Antibiotics

The study of the colloidal properties of many of the antibiotics is complicated by difficulties of purification and the results of some of the earlier work must be regarded critically. This is particularly so of the measurements of surface properties which are susceptible to error arising from the presence of low concentrations of surface active impurities. The effect of impurities on the bulk properties is not so serious, although the release of even small amounts of solubilized water-insoluble material as the solutions are diluted below the CMC can be a problem in such techniques as light scattering.

The formation of aggregates by Penicillin G (sodium benzyl-penicillin) in aqueous solution was first observed as early as 1947 by Hauser *et al.* [64]. Light-scattering measurements by Hocking [65] indicated that the aggregates must be small (with molecular weights not greater than 3000) although no actual value of aggregate size was given. The CMC is relatively high; values in the region of $0.25 \, \text{mol} \, l^{-1}$ ($130\,000$ units ml^{-1} or 8.26% w/v) were determined from conductance and surface tension techniques [66] and later confirmed from a study of the n.m.r. chemical shift of the aromatic protons [67]. Other penicillin and streptomycin salts have been reported to form micelles [64, 68, 69].

There have been conflicting opinions concerning the surface activity of the penicillins. Hauser and Marlowe [69] reported that the sodium and potassium salts of Penicillin G were highly surface active whilst Kumler and Alpen [70] found only a slight surface activity. Careful purification of the sodium salt removed highly surface-active impurities in some samples [71] giving batches with little surface activity, except at pH values below 4.1.

Several of the polyene antifungal antibiotics have been shown to aggregate. Amphotericin B (IX) forms micellar aggregates in an aqueous dimethylsul-

(IX)

phoxide (DMSO)–cetyldimethyl ammonium chloride system [72]. In a solution of $100 \, \mu\text{g} \, ml^{-1}$ of the antibiotic with 1% DMSO and $50 \, \mu\text{g} \, ml^{-1}$ of the cetyldimethylbenzylammonium chloride the micellar weight is 4×10^5; alteration of any of the ratios causes the micelles to grow larger than 10^6. Diumycin and prasinomycin are closely related mixtures of antibiotics that aggregate in aqueous solution giving micelles of approximately 16 sub units [73]. Another antifungal antibiotic, saramycetin, aggregates above concentrations of $5 \, \text{mg} \, ml^{-1}$ to form micelles with a molecular weight of $55\,000$ in $1 \, \text{M}$ acetate buffer [74]. The penetration of lipid monolayers by amphotericin B, nystatin, filipin, pimaricin and etruscomycin has been related to their ability to produce membrane damage [75].

The self-association of the cytotoxic antibiotics, daunorubicin (X) and doxorubicin (XI) has been studied by Barthelemy-Clavey et al. [76] and Eksborg [77]. The spectral and distribution studies of Eksborg gave constants for the formation of dimers and tetramers of the order of $10^{4.5}$ and 10^{12}, respectively. It was suggested from the magnitude of these constants that self-association must be taken into account in the determination of distribution ratios of these antibiotics and also in the determinations of their binding to DNA.

(X) Daunorubicin, R = C(=O)CH₃

(XI) Doxorubicin, R = C(=O)CH₂OH

The surface properties of the cyclic decapeptide antibiotics have been reported by Few and co-workers [78, 79]. The surface tension at the air–water interface of tyrocidine A is greater than that of gramicidin SA, whilst polymyxin E causes only a slight decrease in the surface tension of the water [78]. The surface properties of the polymyxins A, B, C, D and E when spread as monomolecular films have been related to their nephrotoxic action by Few and Schulman [79].

The colloidal properties of the fungal antibiotic, sodium fusidate (XII), have been examined by Carey and Small [80] and later by Richard [81]. These workers have suggested that primary micelles, possibly pentamers, are formed through hydrophobic interactions of the steroid-like structure, followed at increasing ionic strength (> 0.15 M NaCl) by the formation of secondary micelles through hydrogen bond-type interactions of the hydrophilic groups. A similar type of association was suggested by Small [82] for dihydroxy bile salt micelles. However, in more recent studies, the association of the dihydroxy bile salts has been visualized by some workers as an association of hydrogen-bonded dimers [83–85] and by others as a continuous association process [86–88] (see Section 4.4.1).

There is a close resemblance between the structure of sodium fusidate (XII) and the bile salt, sodium cholate (XIII). One surface of the steroidal nucleus of each carries all of the hydroxyl groups and is consequently more hydrophilic than the other. In view of this similarity in structure and the differing opinions currently held on the association pattern of the bile salts, a re-examination of the association characteristics of sodium fusidate might prove of interest.

Actinomycin D forms relatively stable aggregates in aqueous solution. Initial discrepancies regarding the stoichiometry of the aggregates [89–91] were resolved by equilibrium centrifugation studies of Crothers *et al.* [92] which showed conclusively that the dimeric form predominates at concentrations > 10⁻⁴ mol l⁻¹ with no detectable formation of higher aggregates even at concentrations approaching saturation. The conformation of the dimer has been examined in a detailed 220 MHz p.m.r. study [93] the results of which have established that the dimer is formed by a vertical stacking interaction of the actinocyl chromophore groups with one chromophore inverted with respect to the second (see Fig. 4.11).

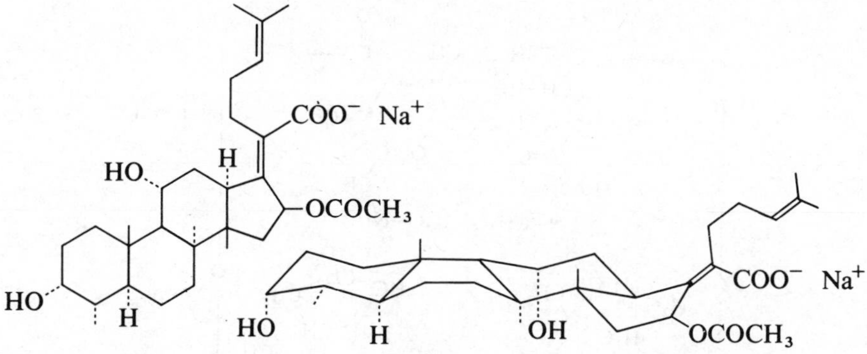

(XII)

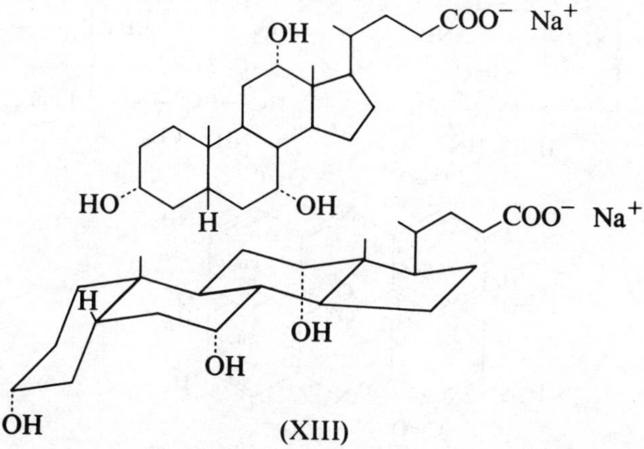

(XIII)

Examination of the solution properties of the 2-palmitate ester of clindamycin (XIV) [94, 95] has indicated association of this antibiotic in aqueous solution. Determination of the aggregation number and the CMC is, however, complicated by the release of solubilized free base originating from hydrolysis, as the solutions are diluted to below the CMC. Lincomycin-2-palmitate does not appear to form aggregates [95].

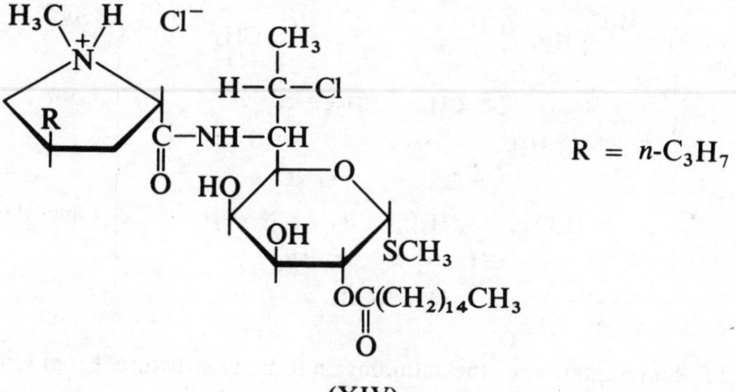

$R = n\text{-}C_3H_7$

(XIV)

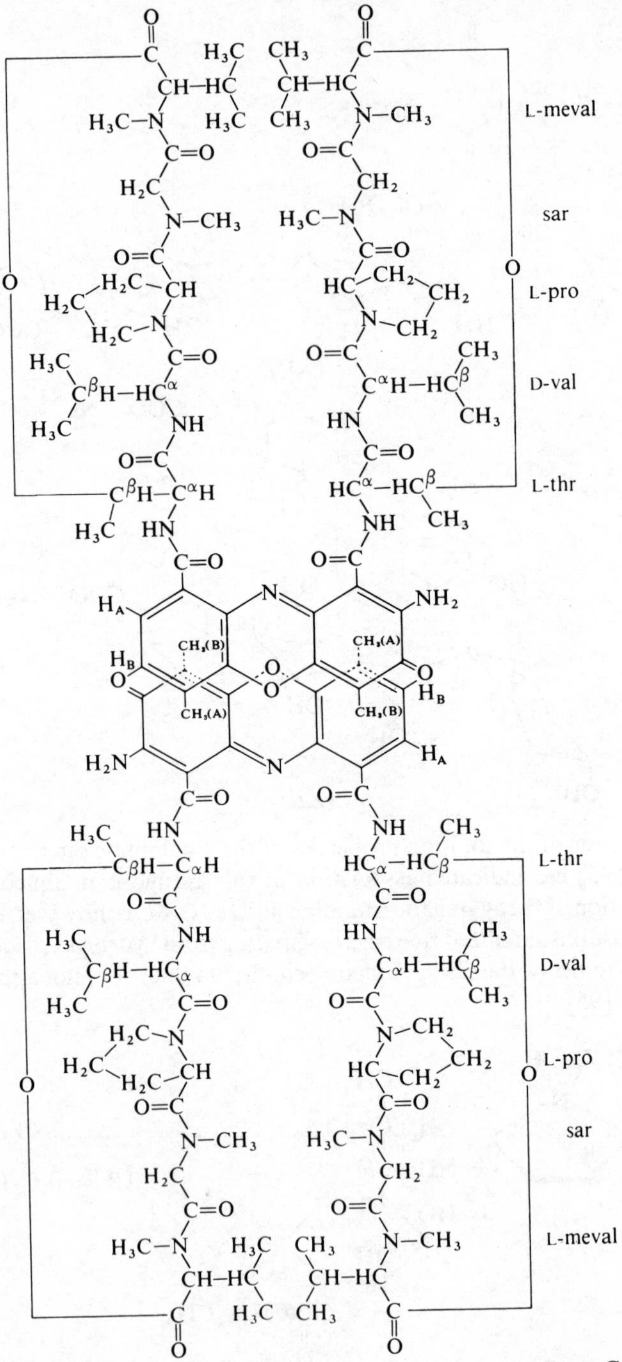

Figure 4.11 Representation of the actinomycin D dimer structure. From Crothers *et al.*
[92].

4.1.9 Local anaesthetics

The colloidal properties of local anaesthetics have been reported by several workers [11, 96–100]. Jaenicke [96] studied the association of tetracaine, stadacaine and its higher homologues, and cinchocaine. The association process appears to be micellar and aggregation numbers, determined by ultracentrifugation, are given in Table 4.13. Johnson and Ludlum [97] also reported the micellar properties of cinchocaine hydrochloride from light-scattering determinations, in good agreement with the previous work. However, the light-scattering plots for tetracaine and also procaine ($R_1 = NH_2$) hydrochloride both in water and added electrolyte were not of a typically micellar form and were interpreted as a multiple equilibrium process. A similar conclusion was reached by Farhadieh *et al.* [11] who also failed to detect CMCs for these two drugs from vapour-pressure and conductivity techniques although association was indicated. Rohman *et al.* [101] examined the colloidal properties of the surface anaesthetics similar in structure to stadacaine but having the R_1 groups at positions 2 and 4 on the benzene ring.

The surface activity of local anaesthetics has been measured by several workers [102, 103] in attempts to relate surface properties to local anaesthetic action (see Section 4.2.4). The penetration of monolayers formed from lipids extracted from

Table 4.13 Micellar properties of some local anaesthetics in water and 0.9% NaCl

$$R_1 - \langle \bigcirc \rangle - CO \cdot O \cdot CH_2 \cdot CH_2 \cdot N(R_2)_2$$

Drug	R_1	R_2	Additive	CMC*,† $(mol\,l^{-1} \times 10^2)$	Aggregation* number
Tetracaine HCl	C_4H_9NH-	CH_3	—	13.3	—
			NaCl	7.0	7
Stadacaine HCl	C_4H_9O-	C_2H_5	—	7.6	—
			NaCl	3.6	17
—	$C_6H_{13}O-$	C_2H_5	—	1.7	—
			NaCl	0.8	48
—	$C_8H_{17}O-$	C_2H_5	—	0.4	—
			NaCl	0.2	86

$$CO \cdot NH \cdot CH_2 \cdot CH_2 \cdot N(C_2H_5)_2$$

Cinchocaine	C_4H_9O-		—	6.6† 6.1‡,§ 6.0¶	15‡
			NaCl	3.2*,†	25*
—	$C_5H_{10}O-$		—	3.0†	—
			NaCl	0.8†	—
—	$C_6H_{13}O-$		—	1.4†	—
			NaCl	0.2†	—

* 20°C, Jaenicke [96]. † 20°C, Eckert *et al.* [99]. ‡ Johnson and Ludlum [97] (room temperature). § 25°C, Farhadieh *et al.* [11]. ¶ Hammarlund and Pedersen-Bjergaard [100].

nerve tissue, by cocaine, tropocaine, tetracaine and cinchocaine, correlated well with the blocking potency of these drugs. Hersh [104] observed that the minimum blocking concentrations of some eight drugs with local anaesthetic activity all lowered the surface tension of a monolayer of L-α-dipalmitoyl lecithin by approximately the same amount.

The properties of monolayers of the cyclodepsipeptide, valinomycin (Fig. 4.12) have been the subject of recent investigations. Ries and Swift [105] have determined a molecular area of 3.70 nm² from extrapolations of pressure–area isotherms for monolayers of valinomycin from which they infer a horizontal orientation of molecules within the monolayer. Mixtures of valinomycin and cholesterol (which has a vertical orientation in monolayers) have been investigated in view of their similarity to biological membranes and the possibility of their use as models for naturally occurring membranes. Monolayers of valinomycin are also of interest because of the ability of valinomycin to stimulate the transport of K^+ ions across mitochondrial and red blood cell membranes [106]. Several workers have studied the interaction of electrolytes with valinomycin monolayers [106–108]. These studies have shown a specific interaction of

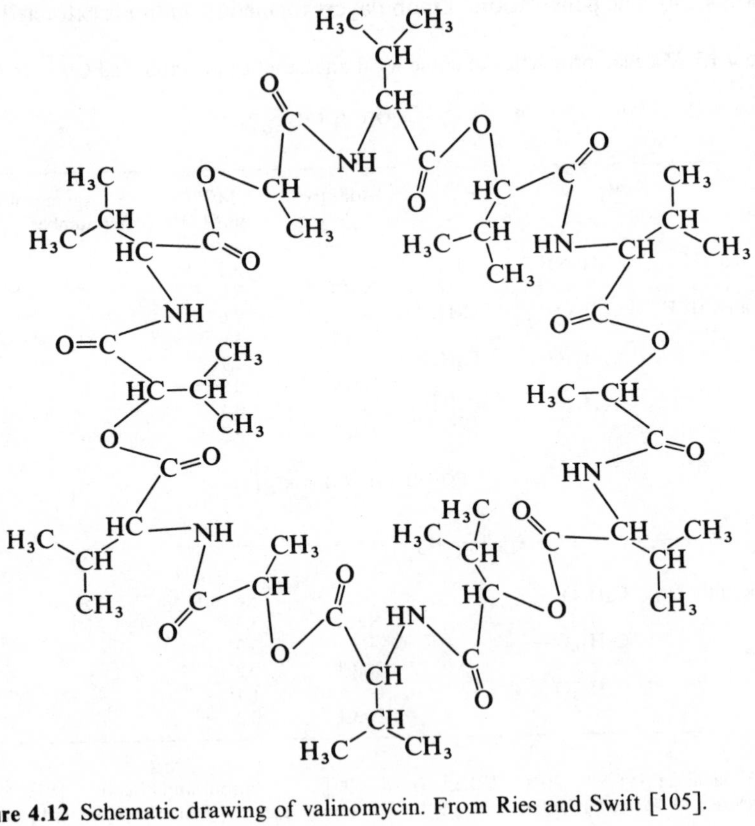

Figure 4.12 Schematic drawing of valinomycin. From Ries and Swift [105].

the monolayers with K$^+$ in preference to other cations and also demonstrated the role of the anion in the formation of the K$^+$–valinomycin complex in the monolayers.

4.1.10 Antihypertensives

Of the many antihypertensive drugs with β-adrenoceptor blocking action whose solution properties have been examined only propranolol hydrochloride (see Fig. 4.13) shows any significant association in the absence of electrolyte. Elliott *et al.* [110] report CMCs and aggregation numbers ranging from 0.108 mol l^{-1} and 13 in water to 0.069 mol l^{-1} and 36 in the presence of 0.2 mol l^{-1} KCl. Oxprenolol and acebutolol hydrochlorides form small micelles ($n = 4$ and 3, respectively) in 0.5 mol l^{-1} NaCl, whilst sotalol and metoprolol are thought to undergo monomer–dimer equilibria under these conditions [111]. Many of the β-adrenoceptor blocking agents exhibit a range of pharmacological effects which are independent of their β-blocking activity and which arise as a result of modification of the cell membrane. These effects, collectively referred to as the membrane stabilizing activity, include non-specific cardiac depression, depres-

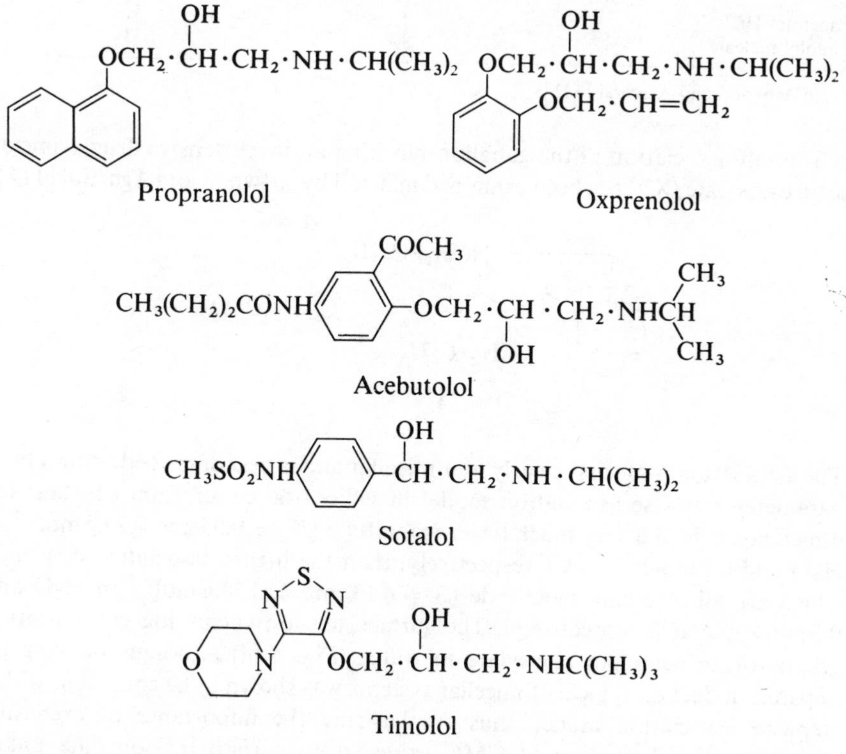

Figure 4.13 Structures of some antihypertensive drugs.

sion of myocardial conduction velocity and local anaesthetic activity. The surface activity of several β-adrenoceptor blocking agents has been examined and attempts made to relate this to local anaesthetic action of the drugs with limited success [111–113]. Table 4.14 compares the surface activity of several β-adrenoceptor blocking agents [111]. Only a limited correlation of the surface activity with the local anaesthetic potency of these drugs was obtained (see also Section 4.2.4).

Table 4.14 Surface activity of β-blocking agents in water and 0.5 mol kg^{-1} NaCl

Compound	Additive	CMC mol kg^{-1}	concn (mol kg^{-1} × 10^3) for π = 10 mN m^{-1}	area per molecule m^2 × 10^{20}
Propranolol HCl	—	0.095	11	62
Oxprenolol HCl	—		12	54
	NaCl	0.170	9	54
Acebutolol HCl	—	0.170	20	51
	NaCl	0.070	11	51
Sotalol HCl	—		30	42
	NaCl	0.180	25	35
Metoprolol tartrate	—		5	57
	NaCl	0.140	7	44
Labetolol HCl	—		3	51
Timolol maleate	—		8	33

From Attwood and Agarwal [111].

The self-association of the ganglion-blocking antihypertensive drug, trimetaphan camsylate (XV) has been examined in detail by Attwood and Agarwal [114].

$$\text{(XV)}$$

The association pattern is clearly non-micellar and is best described using a two-parameter stepwise association model in which the equilibrium constant for dimerization is of a very much lower magnitude ($K_2 = 0.05$ and 0.2 kg mol^{-1} in H_2O and 0.5 mol l^{-1} NaCl, respectively) than the higher association constants which are all of equal magnitude ($K = 6.13$ and 12.85 kg mol^{-1} in H_2O and 0.5 mol l^{-1} NaCl, respectively). The surface tension versus log concentration graph for trimetaphan camsylate in water (Fig. 4.14) although showing an apparent inflection typical of micellar systems was shown to be consistent with a stepwise association model, thus emphasizing the importance of exercising caution in the derivation of CMC values from surface tension data unless confirmatory evidence of micellar behaviour is available.

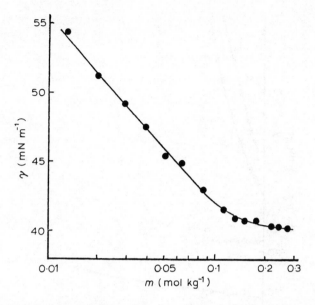

Figure 4.14 Variation of surface tension, γ, with log molal concentration, m, for trimetaphan camphor-sulphonate in water ●, experimental; (———), surface tension predicted using n_w values from light scattering. From Attwood and Agarwal [114] with permission.

4.1.11 Xanthine derivatives

Evidence for the self-association of ethyltheobromine and a number of derivatives of theophylline (**XVI**) came from a study by Guttman and Higuchi

(XVI) R = H
(XVII) R = CH₃

[115] of the partitioning of these compounds between water and an organic solvent. Fig. 4.15 shows clearly an appreciable increase in the distribution coefficient of some of the compounds towards water, which was attributed to the formation of multimolecular species in the aqueous phase. Although this figure shows no significant change in the partition coefficient of theophylline, later studies [116, 117] established that self-association did occur albeit to a lesser extent than with the theophylline derivatives. Kirschbaum [116] using ultracentrifugation techniques presented evidence for the existence of dimers, trimers and tetramers and Thakkar *et al.* [117] showed from n.m.r. spectra that the self-association involved hydrophobic interaction. Theophylline was also shown by

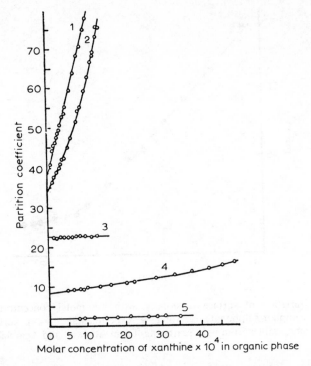

Figure 4.15 A plot of the partition coefficients of a number of xanthines between water and an organic solvent at 30° C. The organic solvent used for all of the studies except theophylline was isooctane. In the case of theophylline, chloroform–isooctane (90:10) was used. 1, ethyltheobromine; 2, 7-ethyltheophylline; 3, theophylline; 4, 7-propyltheophylline; 5, butyltheophylline. From Guttman and Higuchi [115] with permission.

Ng [118] to self-associate by hydrogen bonding in non-aqueous (deuterochloroform) solutions.

The self-association of caffeine (XVII) is well established from osmometric [50, 119] and n.m.r. [120] techniques.

The evidence from n.m.r. studies suggests that the association of the xanthine derivatives is similar to that for the structurally analogous purines [121] involving a vertical stacking of the monomers.

4.1.12 Xanthones and thioxanthones

The colloidal properties of some 23 derivatives of xanthone (XVIII) and thioxanthone (XIX) were investigated by Scholtan [122]. Compounds differed not only in the nature of the side chain at R but also had varying substituents on the tricyclic ring. The aggregation numbers of the xanthones in 0.2 % NaCl range from 3 to 9 with the notable exception of Miracil B (R = $NH(CH_2)_2 N(C_2H_5)_2$ with 6–Cl substituent) which has an aggregation number of 970. As might be

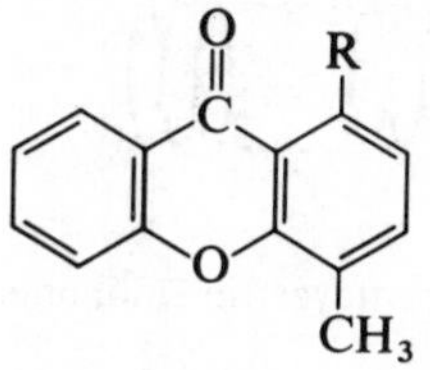

(XVIII) 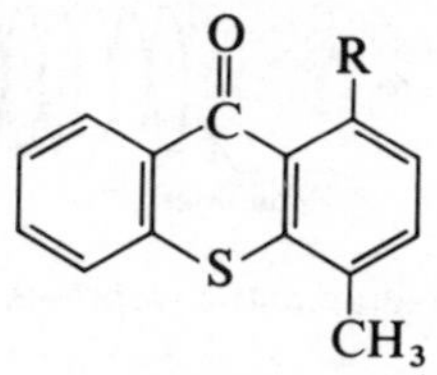   (XIX)

expected the thioxanthone derivatives generally have larger aggregation numbers; Lucanthone ($R = NH(CH_2)_2 \cdot N(C_2H_5)_2$) forms micelles in 0.2% NaCl composed of 54 monomers. A relationship between the micellar weight and activity of several xanthones in the treatment of schistosomiasis has been demonstrated by Scholtan and Gonnert [123].

4.1.13 Miscellaneous drugs

Many studies of the surface properties of hormones have appeared in the literature. These studies have been concerned with interaction of hormones with monomolecular films [124–128] and the properties of monolayers of the hormones themselves [129]. Very few studies of the association of hormones have, however, been reported, the notable exception being that of insulin. Light-scattering investigations by Doty and co-workers [130] showed that below pH 2.2 the association could be described in terms of a monomer–dimer equilibrium. At higher pH, tetramers and probably trimers exist in solution. The data were fitted to a stepwise association model [131] using an analytical technique developed by Steiner [132].

The solution properties of the linear octapeptide, angiotensin II and also of pentagastrin (XX) have been examined by Attwood *et al.* [133]. Evidence for the

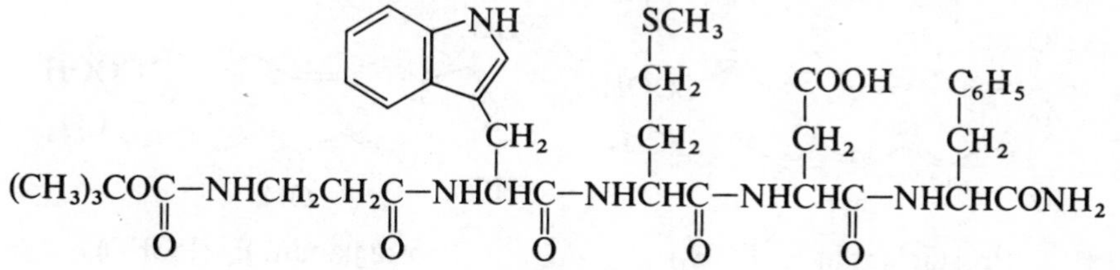

(XX)

aggregation of pentagastrin at solution concentrations in excess of about $0.1\ \mathrm{g\,dl^{-1}}$ came from capillary viscometry, light-scattering, dialysis and absorbance measurements. An axial ratio of about 19 for the aggregated species as derived from viscosity data led to the suggestion that the association of pentagastrin might proceed by linear association (see Fig. 4.16) in a manner similar to that suggested by Minton [134] for haemoglobin. In 1 $\mathrm{mol\,l^{-1}}$ urea, aggregation did not proceed beyond dimers. Angiotensin showed none of the

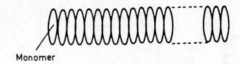

Monomer

Figure 4.16 Diagrammatic representation of pentagastrin monomers in a linear aggregate.

tendencies to aggregate that are found with pentagastrin and at most the molecules form only dimers.

The formation of micelles in aqueous solutions of the tromethamine salt of the biologically active lipid, prostaglandin $F_{2\alpha}$, was reported by Roseman and Yalkowsky [135]. Surface tension, titration and solubility measurements indicated a CMC of between 0.018 and 0.033 mol l^{-1} depending on the pH and electrolyte concentration. The degree of flexibility of the hydrophobic chains of this compound is probably sufficient for a co-operative association in the form of true micelles. The surface activities of four naturally occurring prostaglandins PGE_2, $PGF_{2\alpha}$, PGA_1 and PGB_1 (see Fig. 4.17) have been reported by Sims and Holder [136]. All four compounds were observed to bring about increased instability of stearic acid monolayers spread on subphases containing the prostaglandins. PGE_2 did not readily penetrate the monolayer but appeared capable of associating with the polar groups of the film bringing about instability. Other prostaglandins appeared to penetrate the film giving rise to increased surface pressure. The order of surface activities observed for the prostaglandins was $PGA_1 > PGB_1 > PGF_{2\alpha} > PGE_2$, as might be expected from a consideration of the relative hydrophobicities of the structures.

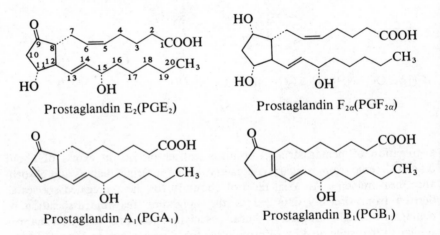

Prostaglandin E₂(PGE₂)   Prostaglandin F₂α(PGF₂α)

Prostaglandin A₁(PGA₁)   Prostaglandin B₁(PGB₁)

Figure 4.17 Structure of prostaglandins.

4.1.14 Structural features of drug molecules which influence association behaviour

In the preceeding survey of the literature on the colloidal properties of drugs, emphasis has been placed on the type of association shown by compounds of widely differing structure. Several points are clear from these studies. There is little doubt that the flexibility of the hydrophobic moiety is an important property which influences the mode of association. Micellar association is exhibited by the diphenylmethane derivative, adiphenine but not when the two phenyl rings are linked together in the rigid fluorene ring structure of pavatrine. It is perhaps surprising in view of this that the phenothiazine tranquillizers and the tricyclic antidepressants which possess only very limited flexibility are able to form micelles. A possible indication of additional structural features which are responsible for a non-micellar association mode comes from an examination of the structures of those drugs with simple counterions such as propantheline, methantheline and pavatrine which undergo a stepwise association. These drugs, unlike any of the other tricyclic micellar compounds, have an ester linkage directly attached to the hydrophobic ring. This group may be sufficiently hydrophilic to cause the micellar surface to be located close to the tricyclic ring system. In the other tricyclic compounds such as the phenothiazines, the tricyclic rings are directly attached to a hydrocarbon chain and the micellar surface is located at the terminal charge group of this chain. This hydrocarbon chain must be considered as part of the hydrophobic moiety in these compounds and its flexibility is such as to facilitate micelle formation. A recent comparison [137] of the association patterns of isothipendyl hydrochloride (XXI) and promethazine hydrochloride (XXII) further substantiates this hypothesis.

$CH_2 \cdot CH(CH_3) \cdot N(CH_3)_2$

$CH_2 \cdot CH(CH_3) \cdot N(CH_3)_2$

(XXI)

(XXII)

Isothipendyl exhibits a non-micellar association mode in contrast to the typically micellar association pattern shown by promethazine. This difference in association behaviour must arise from the presence of the ring N and it is suggested that this pyridine-like N atom is sufficiently hydrophilic to act in an analogous manner to the ester groups of propantheline, methantheline and pavatrine. These drugs with a hydrophilic group or atom either within or directly attached to the rigid tricyclic ring system may be compared to compounds such as the cationic dyes, the xanthine derivatives and the phenanthrene narcotic analgesics which do not possess long hydrocarbon side chains and which also associate in a non-micellar mode.

The comparison between the maleate and hydrochloride salts of those antihistamines which contain pyridine rings highlights the role which can be played by the counterion in determining the association pattern in conditions where proton transfer between counterion and drug ion can occur.

4.2 Some biological consequences of drug surface activity

As most of the drugs that we have so far considered form micelles at concentrations which they do not attain *in vivo*, it is most likely that it is their surface-active characteristics which are more important biologically, although the propensity of the molecules to form associations by hydrophobic bonding will manifest itself. Surface-active drugs will tend, therefore, to bind hydrophobically to proteins and to other biological macromolecules and will tend also to associate with other amphipathic substances such as dyes and other drug substances, with bile salts and of course with receptors. Accumulation of drug molecules in certain sites in the body to such an extent that they reach micellar concentrations is possible in certain instances; the biological consequences of micelle formation will be discussed later. In considering the possible biological implications of surface activity we will concentrate first on the phenothiazines on which there is an extensive literature, although the surfactant properties are not always recognized or discussed by the workers concerned.

4.2.1 Phenothiazines

In their extensive review of the biochemical and biophysical actions of the phenothiazines, Guth and Spirtes [138] select the effect on membrane permeability as the central cause of the diversity of the activities of this remarkable class of compounds. They write

'much of the seemingly confusing evidence in the literature may be explained by postulating as the basic action of phenothiazines an alteration in membrane function and wherever membranes exist, this alteration may occur and produce the measured result. On the other hand, the phenothiazines do produce changes in membraneless systems. For example certain purified enzymes are inhibited by chlorpromazine. The enzyme inhibition and the alterations in membrane permeability could have a common cause at the molecular level.'

The fundamental link is their surface activity, and it is thus possible to link the biological activities of the phenothiazines with other groups of drugs which bear little structural similarity but which are themselves amphipathic. Guth and Spirtes (whose literature review ended in August 1963) survey the effects of phenothiazines, Guth and Spirtes [138] select the effect on membrane permeability and transport of solutes across membranes in whole organs and tissues, and erythrocytes, the effect on permeability of storage organelles and other subcellular particles. The binding of phenothiazines to biological materials is also considered and reference made to the data of Berti and Ferrari [139] who found a rough correlation between the binding of 16 phenothiazines by brain tissue and therapeutic activity. Such correlations now abound; the problem is, noting the ubiquity of interfaces and possibilities for non-specific binding, in divining the relevance of a particular interaction to the biological activity of the drug. The concentration of phenothiazines in the eye and the consequences of its presence there are also discussed. In view of the discussion in Chapter 7 on the biological effects of surfactants it is noteworthy that the phenothiazines also display biphasic activity, their action at low concentrations often being reversed at higher concentrations. Table 4.15, from Guth and Spirtes' review, shows examples of this and the wide range of activities affected by the presence of phenothiazines.

Table 4.15 Reversal of effects with high and low doses of phenothiazines

System	High dose*	Low dose*	Reference
Behavioural effects methamphetamine	−	+	[140]
Growth axolotl larvae	−	+	[141]
Cytochrome c oxidase	−	+	[142]
Mitochondrial respiration and oxidative phosphorylation		+	[143]
	−		[144] and others
Release catecholamines from rabbit adrenal granules	+	−	[145]
^{14}C-acetate incorporation into fatty acid of liver	+	−	[146]
^{32}P incorporation into brain phospholipid	−		[147] and others
^{32}P incorporation into brain phospholipid		+	[148]
^{14}C-mevalonate incorporation into brain lipids	−	+	[149]
^{14}C-glucose incorporation into brain lipids	−	+	[149]
Glutamine synthesis *in vitro*	−	+	[150]
Dinitrophenol-stimulated ATPase of liver	−	+	[151]
Growth of vitamin B$_6$-deficient rats	−, 0	+	[152]

* + = enhancement; − = inhibition; 0 = no effect.

In considering more recent literature one can discern continuing interest in the wide spectrum of properties of the phenothiazines. Some recent papers which lend weight to the belief that these compounds act by altering the conformation and activity of enzymes and by altering membrane permeability and function include the following: the antibacterial action of chlorpromazine (CPZ) and R-factor inhibiting activity have been explored by Molnar et al. [153]; the bactericidal action against *E. coli* and *Staph. aureus* is shown in Fig. 4.18; several phenothiazines have been included in the drugs found to prevent the emergence of resistance in bacteria [154] as do other amphipathic drugs such as chloroquine and acridine orange; synergistic effects of CPZ and perphenazine (PPZ) with several chemotherapeutic agents have been reported [155]. Interactions were tested using *E. coli* and *Ps. aeruginosa* and a filter paper strip agar diffusion method with β-lactam antibiotics and nalidixic acid derivatives. There is the potential here for interaction of drug with the components of the diffusion medium and for cation–anion interactions to confuse the observed effects; nevertheless both phenothiazines exhibited some synergy, perhaps due to their inhibition of ATPase and DNAase as well as specific binding to DNA. CPZ and

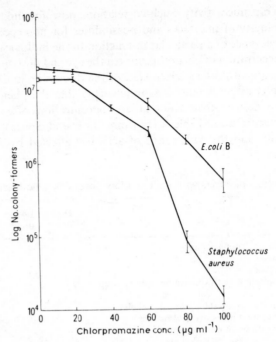

Figure 4.18 Bactericidal effect of chlorpromazine. From Molnar *et al.* [153] with permission.

promazine, among other surface active drugs such as propranolol and haloperidol exhibit a lytic action on retroviruses *in vitro* [156]. An example of an observed effect which cannot have any biological relevance as far as pharmacological activity in man is concerned, is the observed effects of CPZ on growth of barley seedlings caused by the production of abnormal nuclear membranes preventing normal mitotic activity and reduction in growth [157].

Great activity has centred around the effects of phenothiazines on erythrocyte membrane stabilization. Human erythrocytes are protected or stabilized against hypotonic and mechanical haemolysis in the presence of low concentrations of phenothiazines; high concentrations cause lysis [158–160]. The ability of phenothiazines to preserve stored blood was recognized in 1950 by Halpern *et al.* [161] and further studied by Freeman and Spirtes [162] who obtained results which agreed with earlier data [163] suggesting both stabilization and haemolysis in stored blood, depending on the concentration of the phenothiazine. The stabilizing activity which is long-lasting correlates approximately with the clinical potency [158]. The stabilizing and lytic effects of some phenothiazines and tranquillizers is shown in Fig. 4.19. The extent of adsorption of the phenothiazine onto the erythrocyte membrane can be correlated with the haemolytic effects. Seeman and Weinstein have examined this closely; some of their results are shown in Fig. 4.20. Erythrocyte stabilization is associated with the accumulation of around 10^8 molecules per cell; the volume of these would be of the order of 1 to

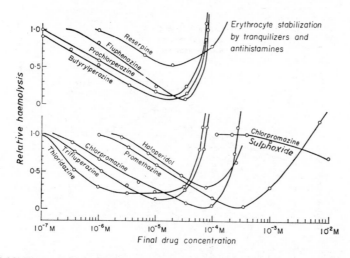

Figure 4.19 The stabilizing and lytic effects of various tranquilizers and antihistamines on human erythrocytes. Stabilization of the erythrocytes against hypotonic haemolysis is caused by butyrylperazine dimaleate, prochlorperazine ethane disulphonate, fluphenazine diHCl, reserpine phosphate, thioridazine HCl, trifluoperazine diHCl, chlorpromazine HCl, promethazine HCl, and haloperidol. High concentrations of all the compounds caused direct lysis. Stabilization by chlorpromazine sulphoxide occurred at concentrations higher than equiosmolar concentrations of NaCl or sucrose. All the compounds were dissolved in aqueous solution except haloperidol which was added to the final aqueous solution from concentrated stock ethanolic solutions. The final concentration of the erythrocytes was 2.4×10^7 cells ml^{-1}. A relative haemolysis of 1.0 indicates an absolute degree of haemolysis of around 40%. From Seeman and Weinstein [158] with permission.

2×10^7 nm^3; if these molecules occupy space within the cell membrane which has a volume of around 10^8 nm^3 there should be an expansion of the membrane of between 10 and 2%, which has been experimentally observed [159]. Cell wall rigidity is increased [164]. Electron spin resonance studies suggest some change in protein conformation in the membrane as phenothiazines penetrate to accessible sites [165]. The drug SKF 525-A, which is surface active [166] has similar effects on erythrocytes as CPZ [167]. Mohandas and Feo [168] have compared cationic and anionic phenothiazines and their interaction with erythrocytes, in particular the morphological changes they induce. The normally encountered cationic derivatives cause a stomatocytic morphology while the anionics cause an echinocytic change. It has been proposed that anionic drugs penetrate mainly into the exterior layer of bilayers and expand this relative to the inner layer at the cytoplasmic side, whereas permeable cationics react in an opposite manner causing the membrane to swell or expand on the cytoplasmic side, causing the morphological changes observed. Cationic phenothiazines inhibit the sickling of cells, although the mechanism is as yet unclear [169], but it is likely that membrane stabilization or changes in radius or curvature are involved.

Several tricyclic dyestuffs have been found to prevent the spontaneous haemolysis of pathologically fragile erythrocytes and to inhibit the haemolysis of

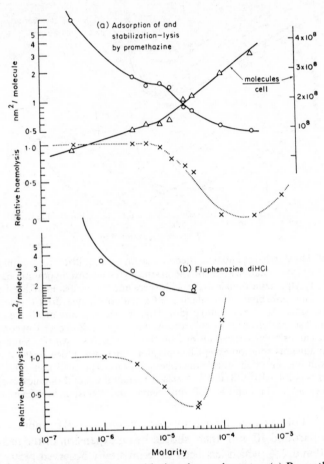

Figure 4.20 The adsorption of phenothiazines by erythrocytes. (a) Promethazine HCl. (b) Fluphenazine diHCl. The phenothiazine stabilization of the erythrocytes against hypotonic haemolysis is indicated by the dashed lines. The amount of phenothiazine adsorbed by the erythrocytes concomitant with stabilization is shown by the line with the triangles (molecules/cell). Dividing the number of molecules/cell by the average erythrocyte area gives the amount of membrane area associated with one stabilizing molecule. It is seen that at maximal stabilization there is about 0.6 nm²/molecule for promethazine and 1.70 to 1.80 nm²/molecule for fluphenazine diHCl. The adsorption of further molecules is associated with direct lysis of the cells. From Seeman and Weinstein [158] with permission.

normal erythrocytes induced by phenothiazines [170]. Inhibition of chlorpromazine-induced haemolysis by fluorescein derivatives is depicted in Fig. 4.21. The structures of the compounds are shown below the figure. Although no evidence of CPZ-inhibitor interactions were noted spectrophotometrically and chromatographically, there seems to be the possibility of anion–cation interactions here which would affect the behaviour of the CPZ; Baur [170], suggests that because of the structural similarities between the CPZ and the inhibitors some

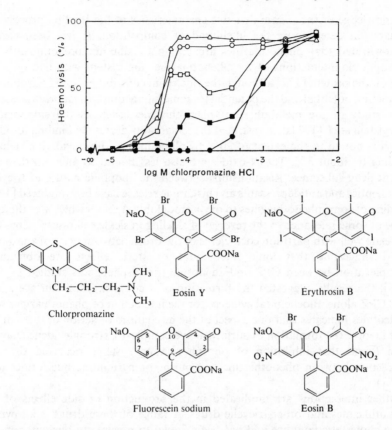

Figure 4.21 Inhibition of chlorpromazine-induced haemolysis by fluorescein derivatives (5×10^{-4} M). ○, no inhibitor; △, 2′,4′,5′,7′-tetrachlorofluorescein; □, 2′,4′,5′,7′-tetrabromofluorescein (eosin Y); ■, 2′,4′,5′,7′-tetraiodofluorescein (erythrosin B); and ●, 4′,5′-dibromo-2′,7′-dinitrofluorescein (eosin B). From Baur [170] with permission.

competitive interaction at the cell surface is involved. Certainly binding of the antihistaminic cyproheptadine (XXIII) to blood platelets is inhibited by phenothiazines and tricyclic antidepressants, presumably by displacement and competitive mechanisms [171].

(XXIII) Cyproheptadine

Knowledge of the thermodynamic characteristics of the binding process are valuable in determining the likelihood of competition. It has been clearly demonstrated that phenothiazines are bound to albumin predominantly by hydrophobic interactions of the phenothiazine ring system with hydrophobic sites on the protein [172] although the side chain contributes both hydrophobic interaction potential and the potential for ionic interaction with appropriate sites. In a study of the metabolites of phenothiazines, such as the sulphoxides, Krieglstein *et al.* [173] demonstrated the remarkably decreased binding for CPZ which is bound to the extent of 70% by albumin. 1% sulphoxidation reduces binding to about 3%. The *N*-oxides are also less bound to albumin than the parent drug substance. Relationships between the lipophilic nature of tricyclic neuroleptics and antidepressants and histamine release have been adduced [174]. Binding of some phenothiazines to oxyhaemoglobin A and S which was thought to be of some relevance to the reversal of sickling in sickle cell disease, does not correlate well with partition coefficients of the drugs between octanol and water [175] suggesting that ionic interactions or steric effects are obtruding. Complexation between CPZ and adenosine triphosphate was studied by Blei [176]; the results suggested the formation of a complex, more surface active than CPZ alone. Biochemical evidence for the interaction of phenothiazines with nucleotides indicates that one aspect of the mechanism of action of these drugs may involve the formation of surface-active complexes, strong evidence having been found that inhibition of phosphorylating systems occurred through interaction between phenothiazine and purine or pyrimidine nucleotides (e.g. [177]).

Other interactions are implicated in the production of side effects of the phenothiazines and other tricyclic drugs. A range of these drugs is known to cause cholestatic jaundice and has been shown to precipitate bile components and protein components *in vitro* [178]. The interaction involves ionic bonding between carboxyl groups on protein and glycoprotein components of bile and the amine groups of the drugs in solution. The suggestion is that precipitation of bile components by tricyclic drugs may occur within the bile canaliculi causing obstruction and playing some role in the development of cholestatic jaundice. Chlorpromazine and its metabolites are concentrated in the liver, secreted into bile and undergo limited enterohepatic circulation [179]. Carey and colleagues [179] point out that at physiological pH one would expect bile salts, which are anionic detergents, to interact with molecules such as chlorpromazine hydrochloride, which are cationic detergents and under certain conditions precipitate. 1:1 complexes were found between CPZ and bile salts, through ionic interaction supported by secondary hydrophobic bonds. In excess bile salt or drug, the complex is solubilized. Electrostatic interactions between phenothiazines and other anionic biological materials include those between CPZ and organic polyphosphate [180], chondroitin sulphate [181], and DNA [182], all of which lead to precipitation. A detailed thermodynamic study [183] of the interaction of phenothiazines with model anionic surfactants, the alkyl sulphates, showed that the entropy of interaction was positive and increased for a given alkyl sulphate with increasing hydrophobicity of the phenothiazine confirming the importance of the hydrophobic interaction [183]. Mequitazine has been found to interact with both the hydrophobic interior and polar head groups of lipids in egg phosphatidylcholine liposomes [184]. The incorporation of the drug into the bilayers decreases the enthalpy and entropy of transition from the gel to liquid crystalline state, ΔH_{trans} and ΔS_{trans}, until at the CMC of the drug the bilayer is disrupted with the formation of mixed phospholipid–phenothiazine micelles following extraction of phospholipid from the bilayers. The presence of cholesterol in the bilayer inhibits the interaction between CPZ and liposomes, it being suggested that this perhaps indicates that the CPZ occupies the same space around the phospholipid molecules as the cholesterol [185]. Nevertheless, high concentrations of neuroleptic drugs interact with biological membranes which contain cholesterol. They can, for example, fluidize pre-synaptic and vesicle membranes and cause a spontaneous release of neurotransmitters [186].

Uptake of paraquat by rat lung slices was found to be inhibited by chlorpromazine and efflux of paraquat from the slices was enhanced by the drug *in vitro*. However, *in vivo* chlorpromazine potentiated the toxicity of paraquat rather than reducing it as expected [187]. The metabolism of CPZ *in vivo* may account for the unexpected findings, although the decreased surface activity of some of the metabolites might be expected to give rise to a decreased membrane interaction. The increased toxicity was ascribed to reduced urinary excretion of paraquat and its increased pulmonary levels in the presence of CPZ, by mechanisms as yet not understood.

Phenothiazines and tricyclic antidepressants inhibit red cell agglutinations, possibly by interacting with the neuraminic acid groups at the red cell membrane surface [188]. The *N,N*-dimethyl side chain is reckoned to be important as *N,N*-diethyl substituted compounds such as procaine do not inhibit the agglutination reaction. Tait suggests that the interaction may be of relevance clinically in respect of ABO incompatibilities between mother and foetus and also in laboratory serum antibody testing [188]. Chlorpromazine decreases the electrophoretic mobility of platelets suggesting interaction with the neuraminic acid moieties responsible for the initial negative charge, but interpretation of these results is complicated by the fact that addition of CPZ to the platelet suspension causes release of 5-HT [189].

Differences in membrane damage caused by tricyclic compounds may be related to their surface activities and extent of adsorption on to cell membranes [190, 192]. Although in experiments on enzyme leakage from isolated hepatocytes the concentrations of tricyclic used was much higher than therapeutic plasma levels, this is perhaps reasonable as it has been found that the liver/plasma ratio of tricyclic antidepressants is of the order of 20–100:1 [192]. The order of toxicity of tricyclic compounds estimated by the extent of leakage of cytoplasmic and lysosomal enzymes from isolated hepatocytes, was in the order CPZ > amitryptiline > imipramine [191], which is the ranking of these compounds according to their surface activity and cell uptake.

4.2.2 Hepatotoxicity and surface activity of other agents

Erythromycin estolate, which is a complex salt of the propionate of erythromycin (XXIV), and the detergent moiety lauryl sulphate is the only derivative of this antibiotic to produce hepatotoxicity in man [193, 194]. An investigational drug erythromycin cetyl sulphate was withdrawn from trials in man when the incidence of hepatotoxic effects reached 10 to 15%. Consequently, Dujovne has examined the role of the detergent anions in these preparations on their toxic actions [193, 194].

(XXIV) Structural formula of erythromycin.

Chemical structure of the erythromycins

	R	R^1
Erythromycin base	H	
Erythromycin propionate	$CH_3CH_2CO_2$	
Erythromycin estolate	$CH_3CH_2CO_2$	$C_{12}H_{26}OSO_3$
Erythromycin cetyl sulfate	H	$C_{16}H_{34}OSO_3$

Surface-pressure measurements of solutions of various erythromycins and cetyl and lauryl sulphate are shown in Fig. 4.22; erythromycin propionate is surface active and results showed that when such a drug is combined in solution with a surfactant there is competition for adsorption at the cell surface. The cytotoxic effects of the erythromycin cetyl sulphate resided in the detergent molecule while the estolate toxicity was derived from both the propionate and the lauryl sulphate. In Table 4.16, Dujovne's results on the interactions of these species in Chang cell cultures are shown, enzyme leakage and drug uptake being measured. Morphological changes in hepatocytes exposed to erythromycin base and estolate have been compared, the latter producing marked cytological changes [195]. An endogenous surfactant, chenodeoxycholic acid, has been observed to induce hepatotoxic reactions in patients receiving the compound for treatment of gallstones [196]. DOSS, alone or combined with oxiphenisatin, has also been implicated with hepatotoxicity [197, 198]. The review by Dujovne [199] on the problem of drug-induced hepatotoxicity should be read for a more comprehensive account of this complex problem. Of particular interest is the

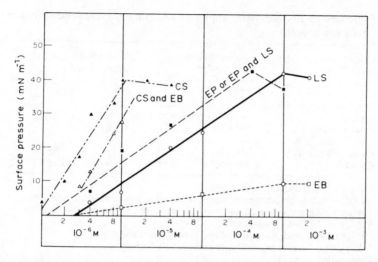

Figure 4.22 Surface pressure of solutions of various concentrations of erythromycin drugs and their components. Abbreviations: CS, cetyl sulphate; EB, erythromycin base; EP, erythromycin propionate; and LS, lauryl sulphate. From Dujovne [193, 194] with permission.

question why, if a physical mechanism of toxicity is involved, the incidence of toxic symptoms is idiosyncratic.

In a recent clinical study [200] of intravenous fusidic acid, 34 % of the patients (38 patients) developed jaundice. The mechanism of the induction of jaundice in these patients is unknown, but the structural resemblance of fusidic acid to the steroids, which can also induce jaundice, and its surface activity and resemblance to the bile salts may well be important factors.

Table 4.16(a) Erythromycin base and cetyl sulphate interactions in Chang cell cultures

Drugs* $(8 \times 10^{-5}$ M)	Enzyme leakage (units ml^{-1})		Drug uptake (nmol/10^6 cells)	
	LDH[†]	BG[‡]	EB	CS
Control	25 ± 5	3 ± 1		
Erythromycin base (EB)	27 ± 4	3 ± 1	3.7 ± 0.1	
Cetyl sulfate (CS)	465 ± 67[§]	25 ± 4[§]		14 ± 1
EB + CS	350 ± 46	16 ± 3‖	3.6 ± 0.1	10 ± 0.7‖

* Four to six cultures in each group.
[†] LDH = lactate dehydrogenase leaked from cells into surrounding medium.
[‡] BG = beta glucuronidase leaked from cells into surrounding medium.
[§] $P < 0.01$ difference from control and each of the single drugs.
‖ $P < 0.01$ difference from control and each of the single drugs.

Table 4.16(b) Erythromycin propionate and lauryl sulphate interactions in Chang cell cultures

Drugs* (8×10^{-5} M)	Enzyme leakage (units ml^{-1})		Drug uptake (nmol/10^6 cells)	
	LDH[†]	BG[‡]	EP	LS
Control	19 ± 3	4 ± 0.5		
Erythromycin propionate (EP)	47 ± 3[§]	18 ± 1[§]	8 ± 0.1	
Lauryl sulphate (LS)	46 ± 2[§]	12 ± 1[§]		7 ± 0.2
EP + LS	300 ± 30[ǀ]	30 ± 2[ǀ]	7 ± 0.4	12 ± 1.2[¶]

* Four to six cultures in each group.
† LDH = lactate dehydrogenase leaked from cells into surrounding medium.
‡ BG = beta glucuronidase activity leaked from cells into surrounding medium.
§ $P < 0.01$ difference from control.
ǀ $P < 0.01$ difference from control and each of the single drugs.
¶ $P < 0.01$ difference from cultures exposed to LS singly.
From Salhab *et al.* [191].

4.2.3 Surface activity and drug-induced lipidosis

A large number of drugs of diverse pharmacology produce intralysosomal accumulation of phospholipids which often are observable as multilamellar objects within the cell. In several reports attention has been drawn to the fact that many of the drugs which are implicated in phospholipidosis induction are amphipathic compounds [201, 202]. Comprehensive reviews of the topic have been published by Blohm [203] and Lüllmann and colleagues [201], the latter postulating that interactions between the surfactant drug molecules and phospholipids render the phospholipid resistant to degradation by lysosomal enzymes resulting in their accumulation in cells. To account for the accumulation of basic amphipathic drugs within lysosomes, De Duve *et al.* [204] took account of the markedly lower interior pH of the lysosomes which would result in significant ionization of the drugs with pK_as around 8. It was calculated that a maximal distribution ratio of 1 : 1000 between plasma and lysosomal compartments would be achieved by this mechanism alone, and for every rise in pK_a by one unit it has been estimated that the time taken to reach equilibrium would rise by a factor of 10 [204]. The complexation with anionic lipids would retard the efflux of the cationic drugs from the cells. The functional significance of phospholipidosis is uncertain, but the striking feature has been the observation of the physicochemical similarity of the causative agents which have included chlorphentermine, fenfluramine, triparanol, chloroquine, chlorcyclizine, iprindole, amitriptyline, chlorpromazine and thioridazine [201].

4.2.4 Surface activity and local anaesthesia

While synthetic cationic and anionic surface-active agents generally are not anaesthetic compounds, some non-ionic surfactants are, and one, alkyl polyoxyethylene ether is used as an endoanaesthetic; many local anaesthetics have significant surface activity. It is thus tempting to relate the two. Indeed the first correlations of surface activity with local anaesthetic potency were made in 1926 with the 4-aminobenzoic acid alkyl esters. However, the overall relationship between surface activity and anaesthetic activity is not very convincing [205], although there can be discerned relationships within homologous series [103]. This is not to say that surface activity is unimportant in anaesthetic action but that other factors obtrude. These other factors include the partitioning of drug into the nerve membrane, a factor which will depend on pK_a, and the spatial distribution of hydrophobic and cationic groups which must be important for the appropriate disruption of nerve membrane function. Nevertheless there appear many instances in the literature testimony to the similarity between local anaesthetic activity and the activity of other amphipathic drugs which suggest that modes of incorporation into membranes are similar and somehow important for the activity of the drug molecules. Local anaesthetics and amphipathic amines are effective, for example, in stimulating ciliary reversal in *Paramecium*; the interaction of drugs with this organism bears a striking similarity to interactions with erythrocytes. The relative potency for ciliary reversal for several local anaesthetic compounds correlated well with relative anaesthetic potency [206] (see Table 4.17).

Table 4.17 Comparison between the ability of various local anaesthetics to stimulate ciliary reversal in *Paramecium aurelia* and their local anaesthetic potency

Drug	Relative potency* for ciliary reversal	Relative potency for local anaesthesia
Procaine	1 (3)	1
Benzocaine‡	1 (3)	1
Lidocaine	6 (0.5)	4
Tetracaine	40 (0.075)	36
Dibucaine	46 (0.065)	53

* Values in parentheses are the ED (mM) for stimulation of ciliary reversal by the various anaesthetics.
‡ Cells responded very weakly to benzocaine.
From Clark and Hughes [206].

In erythrocyte membranes the asymmetric distribution of membrane phospholipids is believed to produce an asymmetry in the polar regions of the two faces of the bilayer; the negative inner face is more likely to hold and attract a cationic amine. Differentiation between the effects of cationic and anionic phenothiazines was referred to above. Browning and Nelson's [205] findings on ciliary reversal in *Paramecium* parallel the effects of the same drugs on erythrocyte membranes. Local anaesthetics, they believe, affect *Paramecium* through perturbation of one half of the lipid bilayer opening the Ca^{2+} 'gate'.

Some β-adrenoceptor blocking agents have significant local anaesthetic or membrane activity, while others have none. The surface activity of a series of β-blocking agents has been examined [112, 114]. The surface activity of these drugs, however, is apparently unrelated to their *in vitro* myocardial β-adrenergic blocking potency, although there is a limited correlation between local anaesthetic activity and surface activity.

Clark and Hughes [206] have studied the blocking action of some quaternary derivatives of 2-(2,6-xylyloxy)ethylamine on adrenergic nerves and made measurements of surface tension and local anaesthetic potency (187). One example of the sort of correlations that are achieved is shown in Fig. 4.23 in which results for 5 series of compounds are shown.

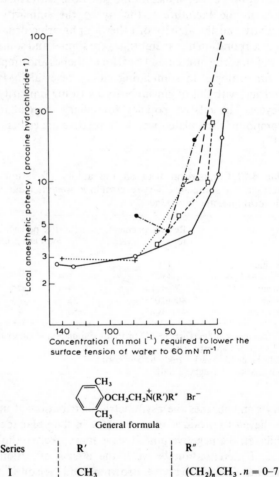

Series	R′	R″
I	CH_3	$(CH_2)_n CH_3 . n = 0\text{–}7$
II	C_2H_5	$(CH_2)_n CH_3 . n = 0\text{–}3$
III	$n\text{–}C_3H_7$	$(CH_2)_n CH_3 . n = 0\text{–}3$
IV	$CH_3 . C_2H_5 .$	$CH_2 = CHCH_2$
V	$n\text{–}C_3H_7$ or $n\text{–}C_4H_9$	H_2C

4.2.5 Antihaemolytic potency and surface activity

Non-specific membrane effects of eight therapeutic β-adrenoceptor blocking drugs have been studied by Wiethold *et al.* [207]. Data from many sources suggest that these effects are due to conformational changes in the cell membranes of the affected system. Wiethold and co-workers found that the compounds protected erythrocytes against osmotic haemolysis (as seen in Fig. 4.24).

Use of the anionic fluorophore, l-anilino-8-naphthalene sulphonate (ANS) to detect membrane conformational changes has shown that the order of fluorescence increase which results on addition of drug to erythrocyte ghost suspensions, corresponds with the antihaemolytic potency. This is clearly seen by comparing Figs. 4.25 and 4.24. A linear correlation between per cent increase of ANS binding sites and the relative negative isotropic activity of these compounds has been found.

The great variety of surface-active agents which at low concentrations protect erythrocytes against hypotonic haemolysis is such that no significance can be drawn from the fact that a drug molecule has this action; it is a consequence only of surface activity. This surface activity may or may not have a biological significance, but already we have seen in this short chapter that the commonality of biochemical effects induced by surface-active agents gives to this property a greater significance than liposolubility or partition coefficient. Yet it cannot be denied that other structural factors will sometimes superimpose themselves to make correlations difficult to deduce. Seeman has examined a range of compounds for their effect on human erythrocytes, and in the process has highlighted the action of specific haemolysins such as digitonin, saponin and filipin [160] (see Fig. 4.26). Low concentrations of the polyene antibiotics, pimaricin and ascosin, produce a transient stabilization but high concentrations elicit rapid haemolysis. There is considerable evidence to show that polyene antibiotics induce permeability changes in sensitive organisms which result in the leakage of essential components [208]. Irradiated filipin has neither haemolytic activity nor antifungal activity. Early work was carried out with an impure filipin sample, but later work confirmed three major pentaene components which all induced erythrocyte haemolysis and were all growth inhibitory to a variety of fungi. Filipin is too toxic for clinical use; amphotericin B, however, is useful in human medicine [209]. The difference between the two antibiotics has been ascribed to their differential affinities for cholesterol and ergosterol. The preferential toxicity

---

Figure 4.23 Variation of local anaesthetic potency with 'surface activity' of compounds. Local anaesthetic potency (ordinate) is expressed as mean molar potency (relative to procaine hydrochloride = 1) as determined in guinea-pig intradermal weal test. 'Surface activity' (abscissa) is expressed as concentration (mmol l^{-1}) required to reduce the surface tension of water to $60 \, \text{mN m}^{-1}$. The five series of compound are shown as follows: O——O *N-n*-alkyl-*N,N*-dimethyl series; + · · · + *N-n*-alkyl-*N,N*-diethyl series; □---□ *N-n*-alkyl-*N,N*-di-*n*-propyl series; ● · — · ● *N*-allyl-*N,N*-di-*n*-alkyl series; △——△ *N*-benzyl-*N,N*-di-*n*-alkyl series. From Clark and Hughes [206] with permission.

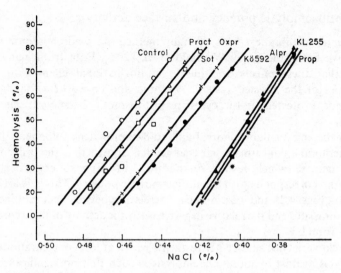

Figure 4.24 The stabilization of human erythrocytes against hypotonic haemolysis by equimolar concentrations of therapeutically used β-receptor blocking drugs (3×10^{-3} M). For details see text. From Wiethold *et al.* [207] with permission.

Non-proprietary name	Abbreviation	Chemical structure
Alprenolol	Alpr	
KL 255	KL	
Oxprenolol	Oxpr	
Pindolol	Pind	
Practolol	Pract	
Propranolol	Prop	
Sotalol	Sot	

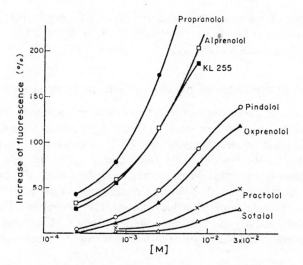

Figure 4.25 Effect of beta-adrenergic blocking drugs (therapeutically used compounds) on ANS fluorescence in human erythrocyte ghosts. Experimental conditions: titration with drugs, ANS 10^{-5} M, ghosts 2 mg protein/ml buffer pH 7.0. From Wiethold *et al.* [207] with permission.

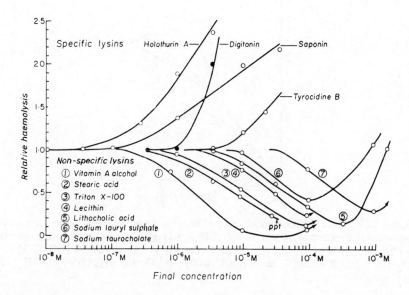

Figure 4.26 A heterogeneous group of compounds, cationic, anionic, and non-ionic (1 to 7, above), in low concentrations protect or stabilize human erythrocytes against hypotonic haemolysis; high concentrations of these compounds lead to complete haemolysis (arrowheads, above). The steroid glycoside haemolysins, digitonin, holothurin A, and saponin do not in low concentrations cause stabilization against hypotonic haemolysis. All compounds above were dissolved in hypotonic NaCl (66 to 69 mM, pH 7) except ▶

177

of amphotericin B for yeast cells is due to its avidity for ergosterol present in the fungal membrane. Filipin, on the other hand, is more avidly attracted to cholesterol of mammalian cell walls.

4.3 Biological relevance of micelle formation by drug molecules

As we have seen in the first part of this chapter a large number of drug molecules are amphipathic and associate, usually at fairly high non-physiological concentrations to form aggregates of varying size. While accumulation of drug molecules in certain sites in the body is likely and while this may possibly, as with the biogenic amines and porphyrins lead to aggregate formation, it is probably only in pharmaceutical systems that micellar concentrations are reached. The nonionic polyether surfactants have local anaesthetic properties. The change in thermodynamic activity of these compounds at their CMC, shown in Fig. 4.27 obviously complicates their behaviour when an homologous series is under study.

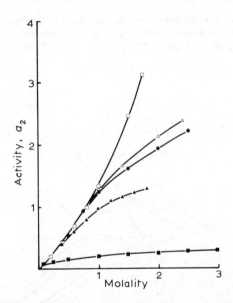

Figure 4.27 Solute thermodynamic activity of alkyl polyoxyethylene ethers in aqueous solutions as a function of molality. $\square$ $CH_3CH_2(OCH_2CH_2)_6OH$ – does not form micelles. $\triangle$ $CH_3(CH_2)_3(OCH_2CH_2)_6OH$ at $20°C$ and $\blacktriangle$ at $30°C$ and $\bullet$ $(CH_3)_2CH.CH_2$-$(OCH_2CH_2)_6OH$ at $20°C$; $\blacksquare$ $(CH_3CH_2)_2CH.CH_2(OCH_2CH_2)_6OH$ at $20°C$. From Florence [3].

lithocholic acid, stearic acid, and lecithin, which were added from concentrated ethanolic solutions; ppt indicates that stearic acid precipitated before the cells were added. A relative haemolysis of 1.0 indicates an absolute haemolysis of about 45 %. The cell density was 2.4 $\times 10^7$ cells ml^{-1}. From Seeman [160] with permission.

It is in homologous series of anaesthetics (and many other drugs) that a decrease in biological activity is seen at higher hydrocarbon chain lengths. While some have ascribed this decreased activity to limitation of aqueous solubility it may also arise from micelle formation, as Fig. 4.27 shows.

It is probably best to separate two distinct events that occur on micelle formation: the change in the properties of the *monomer* and the change in the properties of the *solution*. Both have biological significance. To this can be added the importance of the presence of a micellar 'phase' in relation to solubilization of active molecules or membrane components and to the direct involvement of the micellar phase in catalysis of reactions.

When amphipathic molecules aggregate, several biologically significant changes occur in the system of which they constitute a part: the concentration of monomeric species may increase only slowly or may decrease with increase in total concentration and the transport and colligative properties of the system are changed. If the amphipathic molecules have an intrinsic biological activity, then the formation of aggregates might well lead directly to an alteration in biological activity. One can envisage the aggregation of naturally occurring molecules leading directly to a change in biological activity due to decreased transport rates or decreased ability to pass through biological barriers or the ability of the aggregated species to interact which other biological species may change and there may be a physical alteration in the environment caused by micelle formation. To neglect the possibility of aggregation of biological molecules is to neglect what is undoubtedly in some cases the mechanism of controlling the thermodynamic activity and hence biological activity of the compound *in vivo*.

There is in medicine a growing need for systems which control the release of drugs, systems such as polymer (Silastic[R]) implants and microcapsules. Use of

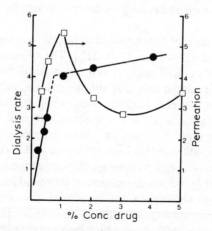

Figure 4.28 Rates of dialysis of chlorpromazine, a micelle forming drug with a CMC ~ 0.7% in water, through cellophane membranes in aqueous solutions above and below its critical micelle concentration. This may be compared with the rate of permeation of the drug through Silastic[R] membranes (both in arbitrary units). From Florence [3].

concentrated solutions of surface-active drugs or of solubilized systems leads to a degree of control like that exercised in storage granules. Fig. 4.28 shows the variation in penetration rate of chlorpromazine from aqueous solutions above and below its CMC across cellophane and polymethylsiloxane membranes. In the former a simple filtration of monomers occurs. The movement across the lipophilic silicone membrane, which probably resembles more a biological barrier, results from diffusion of the drug into the polymer; the changed partition characteristic matters here as the permeability coefficient, P (a measure of drug transfer from bulk solution to the solution on the other side of the membrane), is the product of the drug partition coefficient (K_p) and the drug diffusion coefficient (D_m) within the membrane, provided that diffusion through the aqueous diffusion layers is not rate- limiting. Under steady state conditions, P can be calculated from

$$P = \frac{(\mathrm{d}q/\mathrm{d}t)l}{AC} \qquad (4.1)$$

where $\mathrm{d}q/\mathrm{d}t$ is the rate of permeation at steady state, l and A are the membrane thickness and area, respectively, and C is the concentration of unionized drug in the bulk fluid. Thus

$$\frac{\mathrm{d}q}{\mathrm{d}t} = (\text{constant}) (K_p D_m)C \qquad (4.2)$$

D_m will be unaltered by micelle formation but both K_p and C are altered, K_p decreasing on aggregation and the effective concentration, C, being affected both directly by micellization and by an alteration in the ionization of the total system following on aggregation. This would qualitatively explain the results with silicone membranes and micellar drug. As biological membranes are more akin to lipophilic polymer membranes than cellophane membranes these measurements have some relevance in understanding the biological activity of micellar solutions of drug molecules.

Theoretical predictions of the transport of micelle-forming solutes have been made by Amidon and co-workers [210, 211]. The objective was to assess the importance of the diffusion of micellar and monomeric components of micelle-forming solutes through an aqueous diffusion layer in series with a semi-permeable membrane. The major assumptions used in the derivation were (i) that either a stagnant diffusion layer or a convective diffusion boundary layer is in series with the membrane, (ii) that steady state is achieved, (iii) the CMC is a constant, (iv) rapid equilibrium between monomer and micelles occurs, and that (v) only monomeric species diffuse through the semipermeable membrane. Below the CMC (Regime I, Fig. 4.29a) only monomer is present and flux is proportional to bulk concentration. Above the CMC micelles diffuse into the diffusion layer, and at moderate concentrations (Regime II) the concentration in the diffusion layer can drop below the CMC. At very high bulk concentrations (Regime III) the solute concentration remains above the CMC everywhere within the diffusion layer and the flux becomes constant. The results of theoretical predictions may be seen in Fig. 4.29b.

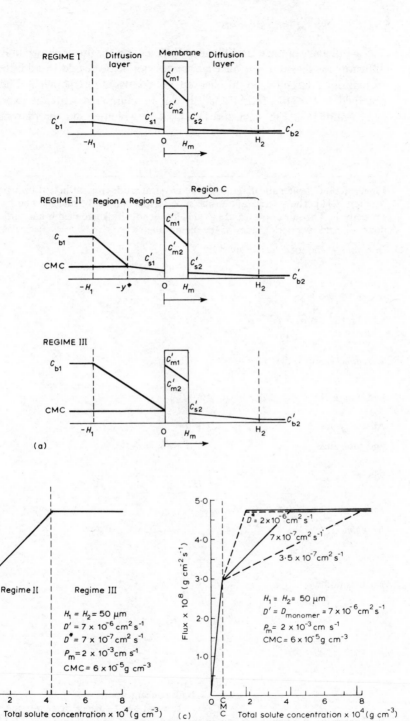

Figure 4.29 (caption on p. 182)

The influence of micellar diffusion coefficient on the flux is shown in Fig. 4.29c; diffusion coefficient of the micellar species can be decreased by addition of salt, formation of large mixed micelles, etc., as discussed in Chapter 3. The profiles obtained in this theoretical analysis can be compared with the experimental results shown in Fig. 4.28 and in Fig. 4.30 which shows the dialysis rate of

Figure 4.29(a) Schematic diagram of the physical model for surfactant transport. From Amidon [211]. The asterisk and prime superscripts denote the micellar and free species, respectively. The subscripts b and s used below denote bulk and membrane surface for the donor, 1, and receiver, 2, sides of the membrane.

In regime I: the total flux is given by

$$J_T = P_{\text{eff}}(C_{b_1} - C_{b_2})$$

where $P_{\text{eff}} = \left(\dfrac{1}{\dfrac{1}{P_1} + \dfrac{1}{P_m} + \dfrac{1}{P_2}} \right)$ $P_i = \dfrac{D_i}{H_i}$ $P_m = \dfrac{KD_m}{H_m}$.

In regime II, region A, $J_A = \dfrac{D}{H_A}(C_{b_1} - CMC)$

and region B, $J_B = \dfrac{D}{y^x}(CMC - C_{s'})$.

At steady state $J_A = J_B = J_m = J_2$

and total flux $J_T = P_{\text{eff}}(C_{b_1} - C_{b_2})$

where $P_{\text{eff}} = \left(\dfrac{1}{\dfrac{H_A}{D^x} + \dfrac{y^x}{D'} + \dfrac{1}{P_m} + \dfrac{1}{P_2}} \right)$

In regime III $J_T = \left(\dfrac{1}{\dfrac{1}{P_m} + \dfrac{1}{P_2}} \right)(CMC - C_{b_2})$

Membrane flux $J_m = \dfrac{KD_m}{H_m}(C_{s_1} - C_{s_2})$

$$J_1 = \dfrac{D_1'}{H_1}(C_{b_1} - C_{s_1})$$

$$J_2 = \dfrac{D_2}{H_2}(C_{s_1} - C_{b_2})$$

$$J_1 = J_m = J_2$$
C_{b_1} = bulk concentration.
C_s = concentration at surface of membrane.

(b) Theoretical predictions of flux in the three regimes in (a). From [210, 211].
(c) Theoretical predictions of the influence of micelle diffusion coefficient on solute flux through a membrane as depicted in (a).

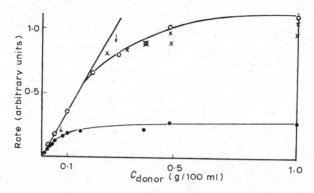

Figure 4.30 Rate of dialysis of thioridazine across cellophane membranes as a function of drug concentration above and below the CMC in water (○) and 0.154 M NaCl (●) at room temperature. × = calculated rates of transport assuming exclusive passage of monomer across the membrane, based on NMR data. Arrows denote CMC values. From Attwood *et al.* [30] with permission.

thioridazine across cellophane membranes above and below the CMC in water and in 0.154M NaCl. Calculated rates of dialysis assuming simple transport only of monomeric species are shown for the case of pure water and are seen to underestimate flux at the higher concentrations.

Amidon *et al.*'s predictions show that the flux of solute is increased by 40 % over the situation in which diffusion of the micelles across the diffusion layer is neglected, although obviously the importance of regime II is diminished if the diffusion layer is not a rate-limiting step in transport. In biological systems the influence of the surfactant species on the permeability of the membrane complicates the picture; the concentration dependence of this effect referred to many times in this book, reinforces the difficulty in making predictions of the flux of surfactants across biological membranes.

The biological activity of micelle-forming solutes is potentially changed above the CMC. The evidence for this is, understandably, difficult to interpret; it comes either from studies on the activity of homologous series of surface-active agents, the lower members of which having high CMCs and the higher members of the series having CMCs in the range of concentrations of the experiment, or it comes from study of the activity–concentration profiles of a single compound below and above the CMC. While it is easy to demonstrate plateau effects *in vitro, in vivo* the multifarious actions of surface-active drugs make a clear definition of effect unlikely, but evidence has, however, been adduced in local anaesthetics for a maximal effect at the CMC [101] both in duration and potency. In some series of compounds, for example the alkylbenzyldimethylammonium chlorides (ABDACs) the CMC has been taken to be the limiting solubility of the monomeric species and has been considered as a factor limiting biological activity of the hydrophobic members of the series [214]. In Fig. 4.31, from Tomlinson *et al.* [212], the hypothesis that colloidal association interferes with the antibacterial

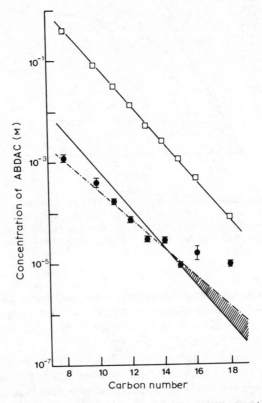

Figure 4.31 Illustration of hypothesis that colloidal association interferes with the measurement of the antibacterial activity of ABDAC's in a MIC test: (□) CMC measured in deionized distilled water; (●) MIC values in CDM and the standard deviations; (—) limiting solubility of the monomeric form of the quaternaries in CDM (i.e. critical micelle concentration in CDM); (—·—·—) extrapolated line of the C_8–C_{14} activity data of the ABDAC's. Shaded portion represents the amount of monomer required to be released from the micellar state before an effective equilibrium concentration is reached. From [212].

activity of the ABDACs in a minimum inhibitory concentration (MIC) test is examined. The influence of the chemically defined medium (CDM) for bacterial growth on the CMC values is shown. The MIC values obtained are shown for the compounds as a function of alkyl chain length. Results for C_{16} and C_{18} are higher than might be expected by extrapolation of the other results. The limiting solubility line is shown intersecting with the extrapolated activity prediction at about C_{14-15}. Table 4.18 clearly shows the manner in which the MIC/CMC ratio alters for the last members of the series. Others have shown an alteration in activity of antimicrobials in micellar systems (e.g. [213, 214]).

Table 4.18 Thermodynamic activities of ABDAC against *Pseudomonas aeruginosa* calculated from MIC, sterilization kinetics, and CMC data

Alkyl chain length	$\dfrac{\text{MIC}}{\text{CMC*}}$	$\dfrac{\text{MIC}}{\text{CMC}^\dagger}$	$\dfrac{\text{30 min}^\ddagger}{\text{CMC}}$	$\dfrac{\text{2 h}^\ddagger}{\text{CMC}}$
8	0.0040	0.3800	0.0011	0.0003
10	0.0068	0.6400	0.0012	0.0007
11	0.0068	0.7171	0.0014	0.0007
12	0.0057	0.5763	0.0019	0.0011
13	0.0057	0.5489	0.0022	0.0016
14	0.0126	1.1431	0.0032	0.0018
15	0.0098	0.8800	0.0047	0.0021
16	0.0330	3.4090	0.0055	0.0014
18	0.1220	11.9650	0.0289	0.0050

* CMC values measured in de-ionized distilled water.
† CMC values obtained from extrapolation of the hexadecylbenzyldimethylammonium chloride measurements in chemically defined media.
‡ 30 min and 2 h refer to concentrations necessary to reduce colony count to 10 % in these time intervals.
From Tomlinson *et al.* [212].

4.4 Naturally occurring micelle formers: the bile salts, phospholipids and related systems

We begin this section with a discussion of the bile salts as biological detergents, concentrating first on their physico-chemical properties and some of their interactions with other solutes and with membranes. There then follows a brief look at the phospholipids and a consideration of ternary bile salt–phospholipid–water systems and quaternary cholesterol–bile salt–phospholipid–water systems which play an important role in various biological processes. The chapter ends with a survey of some miscellaneous amphipathic compounds of biological interest.

4.4.1 Bile salts

Bile salts are biological detergents synthesized in the liver; they form small aggregates because of the bulky nature of the monomer. The structures of the commonly occurring bile acids, with the ring numbering system illustrated, are shown on p. 186.

The chemistry of the bile acids is that expected from alicyclic compounds possessing hydroxyl and carboxyl groups. It is the positioning of the hydrophilic groups in relation to the hydrophobic steroidal nucleus that gives to the bile salts their surface activity and determines the ability to aggregate. Fig. 4.32 shows the possible orientation of cholic acid at the air–water interface, the hydrophilic groups being oriented towards the aqueous phase. The steroid portion of the molecule is shaped like a 'saucer' as the A ring is *cis* with respect to the B ring.

Lithocholic acid
3α-Hydroxy-5β-cholan-
24-oic acid

Deoxycholic acid
3α,12α-Dihydroxy-5β-cholan-
24-oic acid

Chenodeoxycholic acid
3α,7α-Dihydroxy-5β-cholan-
24-oic acid

Hyodeoxycholic acid
3α,6α-Dihydroxy-5β-cholan-
24-oic acid

Cholic acid
3α,7α,12α-Trihydroxy-5β-cholan-
24-oic acid

Hyocholic acid
3α,5α,7α-Trihydroxy-5β-cholan-
24-oic acid

Small [82] has proposed that small or primary aggregates with up to 10 monomers form above the CMC by hydrophobic interactions between the non-polar side of the monomers, and that when conditions favour the formation of

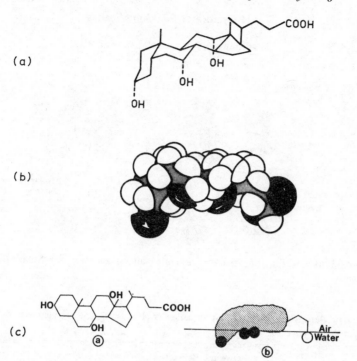

(a)

(b)

(c)

Figure 4.32 (a) Structural formula of cholic acid showing the *cis* position of the A ring. (b) Courtauld space filling model of cholic acid. (c) Diagrammatic representation of the orientation of cholic acid at the air–water interface, the hydroxyl groups being represented by the filled circles and the carboxylic acid group by the open circle. From Barry and Gray [215] and Oakenfull and Fisher [216].

larger aggregates, secondary micelles form by hydrogen bonding between the primary micelles. The structure of primary and secondary aggregates is shown diagrammatically in Fig. 4.33. Only the di- and trihydroxy bile salts were thought to form the primary and secondary structures; trihydroxy bile salts do not form secondary aggregates, although there is evidence that in the three types there are concentration limits at which the properties of the micelle, including its solubilizing power, change. The cholate and deoxycholate groups show approximately the same changes. Table 4.19 illustrates the phenomenon with alterations in the saturation capacity of bile salt micelles at 40°C for *p*-xylene between these concentration limits.

Oakenfull and Fisher [83, 85] have stressed the role of hydrogen bonding rather than hydrophobic bonding in the association of bile salts. They consider that the first stage is the formation of co-operatively hydrogen-bonded dimers and it is these rather than the larger primary units suggested by Small that are the basic building blocks for the formation of larger units. This viewpoint has, however, been criticised by Zana [86] who regards the association as a continuous process with hydrophobic interaction as the main driving force. The

Structure of Bile Salt Micelles

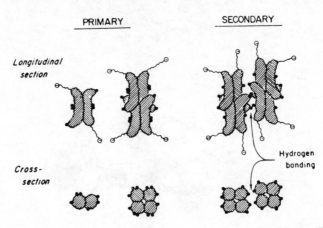

Figure 4.33 Model for the structure of bile salt micelles. From Small [82] with permission.

Table 4.19 Saturation capacity of micelles of bile salts for *p*-xylene at 40°C (mol xylene/mol bile salt)

	Between concentration limits 1 and 2	Between concentration limits 2 and 3	Above limit 3	Concentration limits (M)		
				1	2	3
Sodium cholate	0.085	0.23	0.35	0.014	0.048	0.10
Sodium taurocholate	0.091	0.22	0.32	0.013	0.053	0.12
Sodium glycocholate	0.12	0.25	0.35	0.014	0.050	0.10
Sodium deoxycholate	0.12	0.25	0.36	0.0048	0.010	0.048
Sodium taurodeoxycholate	0.12	0.28	0.40	0.005	0.010	0.045

From [217].

concept of a continuous association has been adopted by other workers to explain data obtained for the association of sodium cholate from a variety of techniques including ultrasonic absorption [87] and self-diffusion [88].

The CMCs of the bile salts are strongly influenced by their structure, the trihydroxy cholanoic acids having higher CMCs (3 to 8 mM) than the less hydrophilic dihydroxy derivatives with CMC values in the range 2 to 4 mM. Conjugated bile salts have higher CMCs than the free bile salts. As might be expected, the pH of solutions of these carboxylic acid salts has an influence on micelle formation. At sufficiently low pH, bile acids which are sparingly soluble will be precipitated from solution, initially being incorporated or solubilized in the existing micelles. The pH at which precipitation occurs, on saturation of the micellar system, is generally about one pH unit higher than the pK_a of the bile acid. Lowering of the pH of sodium cholate and deoxycholate produces different effects [218]. Lowering the pH from 9 to 7 does not affect aggregation of the former, but in the case of deoxycholate the CMC falls drastically and the aggregation number increases significantly. pH–solubility relations of chenodeoxycholic acid and ursodeoxycholic acid have been studied to determine whether or not this had any physiological significance [217]. The solubilities of the undissociated species of several bile acids are shown in Table 4.20.

Table 4.20 Absolute aqueous solubilities and capillary melting points of undissociated bile acids

Bile acid (H$_2$O, pH 2.4)	Temperature (°C)	Value (μM)	Melting point (uncorrected)
Cholic	20	428	201–202
	37	460	
Chenodeoxycholic	20	229	166–168
	37	256	
Deoxycholic	20	111	176–178
	37	114	
Ursodeoxycholic	20	51	201–203
	37	53	
Lithocholic	20	1	187–189
	37	1	

From Igimi and Carey [219].

There is some correlation between the number of hydroxyl groups and the aqueous solubility, but in the dihydroxy acids a 12α-hydroxy group is not as effective in promoting solubility as a 7α-hydroxyl, and a 7β-hydroxyl is only about half as effective as a 12α group [219]. Despite comparable pK_a values ursodeoxycholic acid and its glycine conjugate precipitated at pH values of 8 to 8.1 and 6.5 to 7.4 whereas the chenodeoxycholic acid and its glycine conjugate precipitated at pH values of 7 to 7.1 and 4.8 to 5.0, respectively. Igimi and Carey relate these differences to the low solubility of ursodeoxycholic acid in water (53 μM) and in ursodeoxycholate micelles, in which the saturation ratio of anion:acid is 90–400:1. Chenodeoxycholic acid, on the other hand, is more soluble in water (256 μM) and in micelles the equivalent anion:acid ratio is 5–25:1. The results suggest a physico-chemical basis for the differences in behaviour of these two compounds *in vivo*: the solubility and absorption of the ursodeoxycholic acid from the duodenum–jejunum may be limited; it will precipitate from the colon at pH values below 8. The precipitation of the ursodeoxycholate conjugate falls within the physiological range and it is possible, Igami and Carey point out, that this bile acid may thus 'short-circuit' the enterohepatic circulation and may precipitate from the bile or gut contents as crystals. These facts may explain why it rarely induces diarrhoea, and why serum and biliary lithocholate levels remain normal following ingestion.

Hypersecretion of acid leads to precipitation of bile salts in the intestine and malabsorption syndromes. Neomycin and kanamycin precipitate bile salts and among other adverse effects also lead to malabsorption [220].

A review by Heaton and Morris [221] recalls the development of ideas about bile salts and 'bitter humour'. The enterohepatic circulation of the bile salts was an early discovery; the site of absorption of the bile salts in the intestine was a later topic of study. Absorption by the jejunum of sodium taurocholate is negligible both below and above the CMC and it was absorbed mostly in the ileum [222]. It has been established that bile salts are absorbed by an active transport process, absorption being related to the number of hydroxyl groups on the molecule.

The bile salts have a well-known role in fat absorption processes. It is in fact generally agreed that one of their principal physiological functions is aiding fat absorption [223]. They render fat absorption more efficient although in the total absence of bile, about 70 % dietary fat can be absorbed. Hofmann and Borgström [224] have shown that mixed micelles of bile salts, fatty acids and monoglycerides can act as vehicles of fat transport. The role of the bile salts in lipolysis is not an obligatory one as bile salts can be replaced by non-surfactant molecules or synthetic surfactants [227]. Below their CMCs bile salts prevent the inactivation of lipase by preventing unfolding at the fat–water interface. An inhibitory effect of bile salts above their CMCs has been reported recently but only in the absence of the amphipathic peptide 'co-lipase'. Naturally in the intestinal contents the bile is above the CMC and co-lipase is a necessary factor [225, 226]. At the CMC Borgström and Erlandson [225] have suggested that a monolayer of bile salts prevents the lipase reaching the substrate. As it is surface active, co-lipase can penetrate this bile salt barrier.

Bile salts above their CMC displace lipase from hydrophobic interfaces, and this parallels the loss of lipase activity [228, 229]. Anionic detergents of the acyltaurine, decanoyl taurine and dodeconylsarcosyltaurine type also inhibit lipase when present above their CMCs, a process reversed by co-lipase. These detergents behave like bile salts and it may be of interest that analogues have been isolated from the gastric contents of the crab [230]. A number of steroidal antibiotics, 3-acetylfusidic acid, cephalosporin P_1 and helvolic acid have recently been found by Carey *et al.* [231] to have micellar properties very similar to the bile salts and it has been suggested that they may serve as model compounds for detergent replacement in bile salt deficiency syndromes.

The biliary route is a major route of drug elimination, but the importance of bile salt micelles in drug transport, absorption and metabolism is not as well understood as their role in fat absorption. Whether binding to micelles has importance in hepatic transport and biliary excretion of organic anions has been discussed by Vonk *et al.* [232]. Most authors agree that there is not a simple relationship between stimulation of bile flow and stimulation of biliary output of organic anions, but taurocholate (TC), which promotes formation of biliary micelles, stimulates biliary excretion of certain compounds more than agents like theophylline, dehydrocholate and hydrocortisone which induce biliary excretion without micelle formation. Several suggestions implicating micelles have been put forward [231] [233] including facilitation of transport from liver to bile by direct effect on canicular membranes, stimulation of micelle formation inside the liver cells, binding of the drug anions to micelles and subsequent exocytosis of these aggregates into bile canaliculae and binding of anions to micelles in the canaliculae with consequent reduction of free drug concentration leading to decreased transfer of drug from bile back into the liver.

Taurocholate is able to form micelles; infusion of TC has been shown [232] to increase output of phospholipid and cholesterol with which it forms mixed micelles. Dehydrocholate has a low tendency to micelle formation and has little influence on the biliary excretion of cholesterol and phospholipid. However, Hardison and Apter [233] have found that while micelle formation is an important factor in the biliary excretion of lipids, it cannot alone explain the results they have obtained. Dehydrocholate, a synthetic agent, is a potent choleretic presumably because its osmotic activity in bile is not diminished by micelle formation and it may therefore exert an osmotic force in biliary canaliculae approximately equal to its concentration. As a result, bile flow is increased more by this substance than by micelle-forming bile salts. Some metabolites of dehydrocholate are, however, capable of micelle formation [234, 235].

The bile salt concentration is 100 to 300 mM in human bile, and about one-tenth of this in human intestinal content and as the CMCs are in the region of 2 to 3 mM both bile and intestinal fluids contain bile salt micelles [236]. The ability of bile salt solutions to solubilize insoluble drugs such as griseofulvin and hexoestrol [237] suggests that the bile salts may be involved in the solubilization of drugs prior to absorption. Bates *et al.* [238] have found that physiological concentrations (0.04 M) of sodium cholate and sodium deoxycholate enhance the rate of solution of hexoestrol and griseofulvin over their rate of solution in water, as shown in Table 4.21. This effect strengthens the view of their supposed action; one notes that the oxidized bile salt dehydrocholate does not micellize and does not enhance fat absorption [236].

Table 4.21 Relative dissolution rates for griseofulvin and hexo-estrol at 37°C in sodium deoxycholate and sodium cholate solutions [238]

Dissolution medium	Griseofulvin time (min)			Hexoestrol time (min)		
	2	5	10	2	5	10
Water	1.0	1.0	1.0	1.0	1.0	1.0
Sodium deoxycholate	7.5	6.6	6.0	14.6	18.6	24.0
Sodium cholate	7.2	6.1	5.5	20.3	30.4	36.0

The enhanced absorption of medicinals on administration with deoxycholic acid may be due to reduction in interfacial tension or micelle formation. The inefficient absorption of reserpine promoted an investigation into its absorption in combination with deoxycholic acid [239]. Deoxycholic acid was found to increase the rapidity of absorption of reserpine and to increase its potency. The rates of dissolution of a deoxycholate–reserpine co-precipitate and recrystallized reserpine of the same particle size, have been found to be identical, ruling out the likelihood that the bile salts increase the bulk solubility in these systems, although the differences in solution rate can be quite marked when different bile acids are used (Fig. 4.34) [240]. Comparison of the effects of bile salts on the dissolution rate and solubility of indomethacin and phenylbutazone led Miyazaki *et al.* [241] to suggest that bile salts increased the solution rate of indomethacin by micellar solubilization, but that of phenylbutazone by wetting effects. Their results are shown in Fig. 4.35 and can be compared with the results obtained for the co-precipitates although only two of the bile salts are common to both studies.

The administration of quinine and other *Cinchona* alkaloids in combination with bile acids has been claimed to enhance their parasiticidal action [242, 243]. Quinine, taken orally, is considered to be absorbed mainly from the intestine. A considerable quantity of bile salts is required to maintain a colloidal solution of quinine; one might argue that an efficient supply of bile salts was therefore a

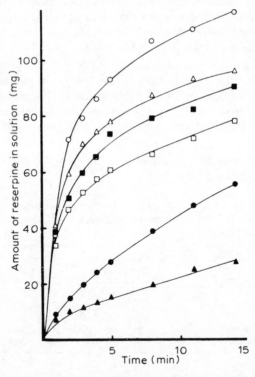

Figure 4.34 *In vitro* dissolution rates of various cholanic acid derivative–reserpine co-precipitates in ethyl acetate at 37° C. ○, lithocholic acid; △, cholic acid; ■, deoxycholic acid; □, 5β-cholanic acid; ▲, 3,12,24-trihydroxycholane; ●, precipitated reserpine only. From Stoll *et al.* [240] with permission.

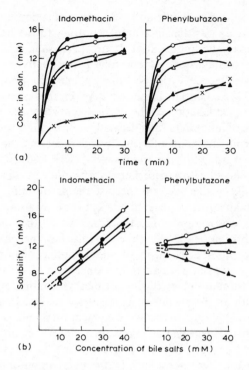

Figure 4.35(a) Effects of bile salts on the dissolution behaviour of indomethacin and phenylbutazone in pH 7.3 phosphate buffer at 37° C.

—×—: control.

—○—: sodium desoxycholate.

—●—: sodium cholate.

—△—: sodium glycocholate.

—▲—: sodium taurocholate.

The concentration of bile salts was 40 mM.

(b) Effect of bile salts concentration on the solubilities of indomethacin and phenylbutazone in pH 7.3 phosphate buffer at 37° C. The solubility of phenylbutazone is reduced in taurocholate, yet dissolution is enhanced.

—○—: sodium desoxycholate.

—●—: sodium cholate.

—△—: sodium glycocholate.

—▲—: sodium taurocholate.

From Miyazaki *et al.* [241] with permission.

prerequisite of quinine absorption [244]. It is interesting to note in this context that recurrent attacks of malaria are sometimes found to be accompanied by hepatic disturbances which may prejudice the normal flow of bile [245].

Bile salts may also influence drug absorption either by affecting membrane permeability or by altering normal gastric emptying rates, the latter having been explored by Feldman *et al.* [246, 247]. Sodium taurocholate increases the absorption of sulphaguanidine from the stomach, jejunum and ileum, the effect being most pronounced in the ileum [248]; the absorption of imipramine was, however, inhibited, as was that of quinine, contrary to the findings referred to above in relation to the antimalarial. It has been suggested that absorption increases are due to increases in membrane permeability induced by calcium depletion and interference with the bonding between phospholipids in the membrane [249]. Mixed micelle formation between the imipramine and the bile salt is suggested in explanation of the decreased absorption of this compound [248]. The uptake of imipramine and desmethylimipramine into bile micelles has been discussed from the point of view of understanding the metabolism and biliary excretion of lipophilic drug molecules [250]. Reductions in absorption have also been observed with a diethylaminobenzamide [251] and isopropamine iodide [252]. While *in vitro* bile salts increase the partitioning of the latter drug into octanol from an aqueous phase at bile salt concentrations below the CMC, the reduction in absorption suggests, along with other data adduced for decamethonium bromide [253], that the bile salts have no physiological role to play in the absorption mechanism of water-soluble protonated quaternary ammonium compounds in normal circumstances.

Mixed micelle formation between alkyltrimethylammonium salts and sodium cholate occurs with the production of aggregates which are small, approximately spherical, and only slightly charged, with aggregation numbers of 19 to 32 [215]. Micelle formation and coacervation in mixtures of alkyltrimethylammonium salts and di- and trihydroxy bile salts have been studied by Barry and Gray [254]. Coacervation occurs when approximately equimolar amounts of the cationic and anionic species are present in the system; micelles in the sodium cholate systems were small while those in sodium deoxycholate mixtures grew large, an interaction that no doubt would have an effect on bioavailability as it would affect not only the thermodynamic activity of the cation but also that of the bile salt, such that its effect on membrane permeability might be inhibited. The effect of bile salts on membranes has been the subject of some attention in the literature. Oesophageal mucosal permeability is increased by bile salts and this has been suggested to be the cause of oesophagitis and ulceration associated with reflux of gastric contents into the oesophagus [255]. The normal defence mechanism against the ravages of the hydrogen ion seem ineffective in the presence of bile salts. They can increase the passive permeability of the intestinal epithelium and gastric mucosa [256], leading to inhibition of net sodium and water absorption; in this regard the bile salts seem to behave no differently from other ionic surfactants. Sodium glycocholate induces lysis of erythrocytes at low concentrations when the erythrocytes have a high phosphatidylcholine/low sphingomyelin content, as in rat and guinea-pig, and at higher concentrations when, as in sheep, there is a low phosphatidylcholine/high sphingomyelin ratio [257, 258]. At sub-lytic levels, the membranes lose only phospholipid but in membranes with a higher phospholipid content both phospholipid and vesicles were lost (Table 4.22).

Solubilization studies by the same group [259] suggest that dihydroxy bile salts are potentially more damaging to membranes than trihydroxy bile salts in that

Table 4.22 Relationship between sphingomyelin/phosphatidylcholine contents of erythrocyte membranes and the effects of glycocholate

Erythrocytes	Sphingomyelin / (Sphingomyelin + phosphatidylcholine)	Concentration of bile salt for approx. 10% lysis (mM)	Distribution of material in pre-lytic supernatant
Rat and guinea-pig	Low (0.2–0.25)	10	Microvesicles > soluble
Human and pig	Intermediate (0.45–0.55)	25–35	Microvesicles ≃ soluble
Sheep and ox	High (1.0)	40–50	Microvesicles < soluble

From Coleman and Billington [257].

they are able to cause the release of not only phospholipid but proteins normally held in the lipid bilayer. Several of the morphological and metabolic changes in the liver which are the result of cholestasis according to Coleman and Holdsworth [259] are better correlated with the accumulation of dihydroxy bile salts. Conjugated and unconjugated bile salts increase the rate of swelling of *Pseudomonas aeruginosa* but have no effect on enzyme induction [260]. Swelling is the result of changes in cell permeability with the entry of water into the cell from a salt medium. The results with this organism corroborate the views on the difference between the trihydroxy and the dihydroxy derivatives. Of the trihydroxy group of compounds, only one increased swelling of *Ps. aeruginosa*. Two series of dihydroxycholanic acids compounds were quite effective in enhancing swelling, compounds of the chenodeoxycholate series being most effective.

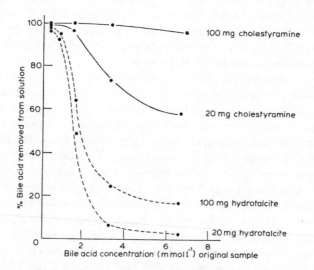

Figure 4.36 Percentage bile acid binding from duodenal fluid by cholestyramine resin and hydrotalcite. From Llewellyn *et al.* [262] with permission.

In vitro inhibition of platelet aggregation by bile salts has been attributed to the detergent action of the salts on the platelet membrane [261]. Interaction of bile salts with other components of the gastro-intestinal tract will affect their performance. It has been suggested that for the treatment of disorders involving the gastric reflux of bile acids it might be desirable to use a substance which would bind the bile salts without affecting the enterohepatic cycle. The relative binding capacity of antacids and cholestyramines has been measured [262]. Hydrotalcite, while effective in lowering bile acid levels from solutions, does not significantly affect enterohepatic circulation but lowers bile acid concentration in gastric fluids and not in the duodenum. Cholestyramine lowers both gastric and duodenal levels. Fig. 4.36 shows the results of *in vitro* binding experiments. Hydrotalcite was more effective in removing dihydroxy bile salts than it was in adsorbing trihydroxy derivatives [263].

4.4.2 Ternary and quaternary systems of lecithin, cholesterol, and bile salts

In order to gain some understanding of the biological problems involving lecithin, cholesterol, and bile salts, model systems have been investigated. The problems involved include those of fat absorption, interactions with proteins and biological membranes. Bile salt–cholesterol mixed micelles were found to be slightly larger than pure bile salt micelles at low steroid concentrations, and at higher concentrations the size became still greater. Sitosterol, a plant steroid, has a lower solubility in the micellar phase than cholesterol in these systems. In the presence of sitosterol, however, the solubilization of cholesterol was increased, yet cholesterol reduced the solubilization of sitosterol. Cholesterol esters have a lower micellar solubility than the free steroid, the logarithm of the percentage distribution to the micelle varying inversely with carbon-chain length of the ester from methyl to amyl [264]. Sitosterol inhibits cholesterol absorption *in vivo*, but these observations do not clarify the matter.

Ekwall and Baltcheffsky [265] have discussed the formation of cholesterol mesomorphous phases in the presence of protein–surfactant complexes. In some cases when cholesterol is added to these solutions a mesomorphous phase forms, e.g. in serum albumin–sodium dodecyl sulphate systems, but this does not occur in serum albumin–sodium taurocholate solutions [266]. Cholesterol solubility in bile salt solutions is increased by the addition of lecithin [236]. The bile salt micelle is said to be 'swollen' by the lecithin until the micellar structure breaks down and lamellar aggregates form in solution; the solution is anisotropic. Bile salt–cholesterol–lecithin systems have been studied in detail by Small and co-workers [267–269]. The system sodium cholate–lecithin–water studied by these workers gives three paracrystalline phases I, II, and III shown in Fig. 4.37. Phase I is equivalent to a 'neat-soap' phase, phase II is isotropic and is probably made up of dodecahedrally shaped lecithin micelles and bile salts. Phase III is of 'middle soap' form. The isotropic micellar solution is represented by phase IV. The addition of cholesterol in increasing quantities reduces the extent of the isotropic

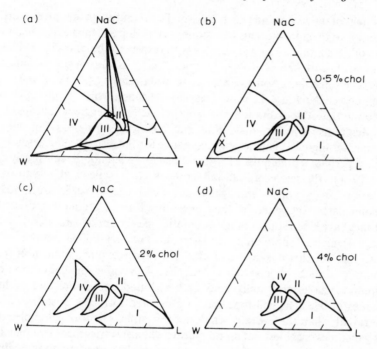

Figure 4.37 Phase diagrams of sodium cholate (NaC), water (W), lecithin (L). Systems (a) with no added cholesterol; (b) with 0.5 % cholesterol; (c) with 2 % cholesterol; and (d) with 4 % cholesterol. In the system without cholesterol the three paracrystalline phases are labelled I, II, and III. Phase I is analogous to 'neat soap' of aqueous soap systems. Phase II is a 'cubic phase', and phase III is analogous to the 'middle soap' phase of common soap systems. Phase IV is isotropic micellar system. X in diagram (b) represents the composition of normal human gall-bladder bile. From Small *et al.* [267, 268].

micellar region. Gall-bladder bile [270] whose composition varies, especially in disease states, must have a composition in phase IV if it is to remain micellar. The phase diagrams serve to predict the extent of alteration in composition which would result in the formation of cholesterol crystals or paracrystalline phases. Where the cholesterol is in association with the lecithin this is due to 'a mutual solution of the paraffin parts . . . a consequence of the orientation of the two species induced by their contact with water' [269] and not to a normal molecular association.

4.4.3 Solubilization of cholesterol in bile acid–phospholipid systems

The ternary diagrams in Fig. 4.37 indicate the complexity of the bile salt–lecithin systems especially in the presence of cholesterol which limits the extent of the isotropic micellar phase (IV) in particular. Intense interest in these systems has been generated in the search for the cause of cholelithiasis. Of great importance in understanding gallstone formation is how cholesterol, which is a major component of the stones and which is normally in solution in bile, precipitates out of solution to form the nucleus of a stone. According to Holzbach *et al.*, normal human bile is commonly supersaturated with respect to cholesterol [271] but bile from patients with cholesterol cholelithiasis has a greater degree of supersaturation. There has been some debate as to whether the zone IV in the *in vitro* systems depicted in Fig. 4.37 has an *in vivo* significance. Holzbach and co-workers have found that it does, but that the zone is much smaller than indicated by the phase diagrams. A male–female difference has not been detected in the phase diagrams, yet there is a female preponderance for the disease; supersaturation of the bile is perhaps a prerequisite for stone formation but not the sole precipitating factor [271]. Physico-chemical influences such as bile acid structure, pH, counterions, bile pigments and agents which could act as 'seeds' are all likely to play some part in the genesis of the stone. Once formed, attempts to dissolve the gallstones have led to studies of the dissolution behaviour of cholesterol itself and formed elements containing cholesterol. The crystalline nature of the cholesterol – whether it is in its hydrated or anhydrous form – influences the rate of solution in model bile systems [272]. Fig. 4.38 shows the dissolution of co-precipitated cholesterol, anhydrous and hydrated cholesterol and of human gallstones in bile. The gallstone was in powdered form and is seen to dissolve more rapidly than the hydrated cholesterol sample. The form of cholesterol in most gallstones is cholesterol monohydrate; extensive treatment by Higuchi and co-workers has demonstrated that dissolution of the stones and monohydrate in bile salt–lecithin systems is controlled largely by an interfacial barrier [273–276]. The possibility that this arises from an adsorbed phospholipid layer which interacts with cholesterol molecules to form a condensed layer has been considered [277]. Certainly cholesterol dissolution rates in bile acid solutions are about 2 to 20 times slower depending on the degree of agitation than diffusion-controlled rates would predict [274].

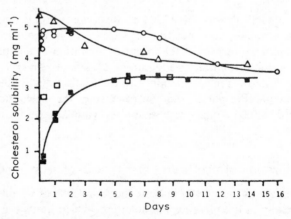

Figure 4.38 Dissolution behavior of coprecipitated (△), anhydrous (○), and hydrated (■) cholesterol and of human gallstone (□) in a model bile system. From Mufson *et al.* [272] with permission.

A commercial emulsifying agent, glyceryl monooctanoate (monooctanoin), has been found to be an excellent solvent for cholesterol [278]. *In vitro* it dissolved mixed cholesterol gallstones more than twice as fast as did sodium cholate solutions which have been used as infusions for dissolution of retained cholesterol bile duct stones. Monooctanoin has also been tried by T-tube infusion in an attempt to effect direct dissolution of stones. Cholesterol solubility and dissolution kinetics are enhanced (Fig. 4.39). Monooctanoin infusions were well tolerated and some or all stones in 10 out of 12 patients were removed by dissolution by biliary tract infusion of monooctanoin over periods ranging from 4 to 21 days.

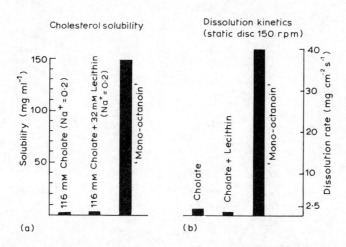

Figure 4.39(a) Solubility (mg ml^{-1}) of cholesterol in dilute sodium cholate, this solution with added lecithin and in mono-octanoin. (b) Dissolution rate (mg cm^{-2} s^{-1}) of cholesterol in these solutions under conditions of rapid stirring (150 rpm) using a static disc dissolution kinetics apparatus. From Thistle *et al.* [278] with permission.

Neomycin precipitates bile acids and fatty acids from micellar solutions *in vitro* [279] and promotes faecal excretion of bile acids in man [280], facts supporting the hypothesis that this is due to the ability of the drug to interact with bile and fatty acids during the micellar phase of lipid absorption [281]. Precipitation is more complete with taurochenodeoxycholate micelles than with taurocholate micelles and of the derivatives of neomycin studied dimethylaminopropyl neomycin was the most active, followed in order by neomycin, dodeca-*N*-methyl neomycin hexamethochloride and *N*-methylated neomycin [281]. Hexa-*N*-acetyl neomycin failed to precipitate any of the micellar components. The same order of activity was found when the compounds were tested for their hypocholesterolaemic effect in newborn chicks [281].

The importance of mixed micelle formation in the solubilization and biliary excretion of lipids is well established but little is known about the role of mixed micelles in the excretion of other solutes by the biliary route. A variety of drugs

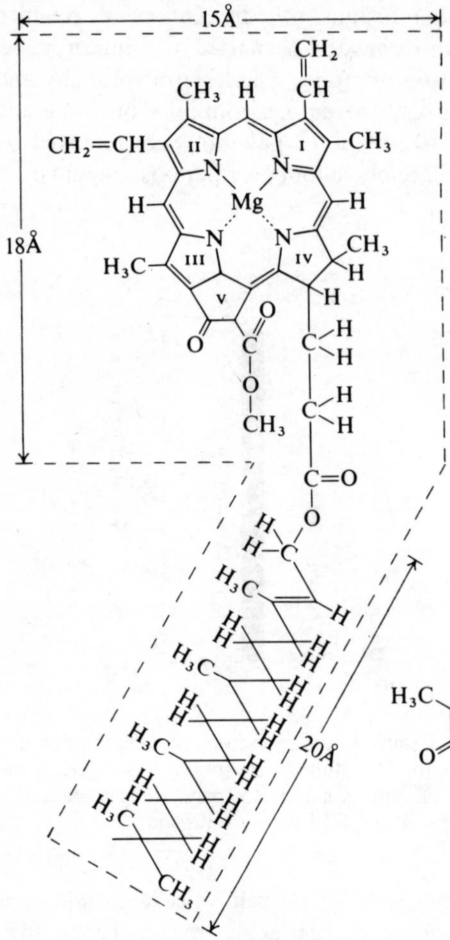

Chlorophyll A

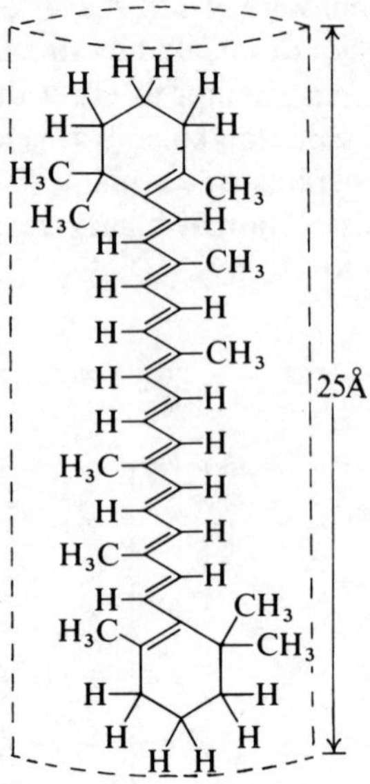

β-Carotene

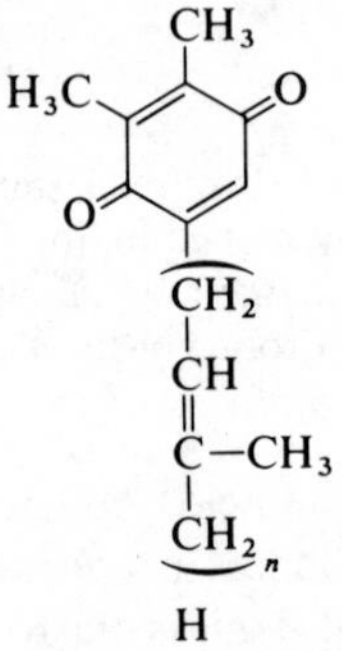

Plastoquinone

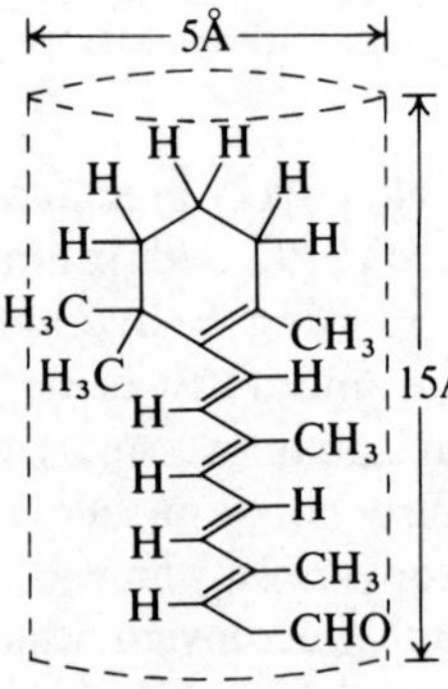

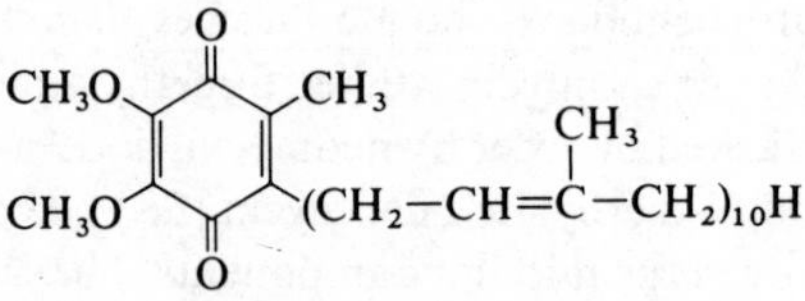

Ubiquinone C_0Q_{10}

Retinene
(retinaldehyde)

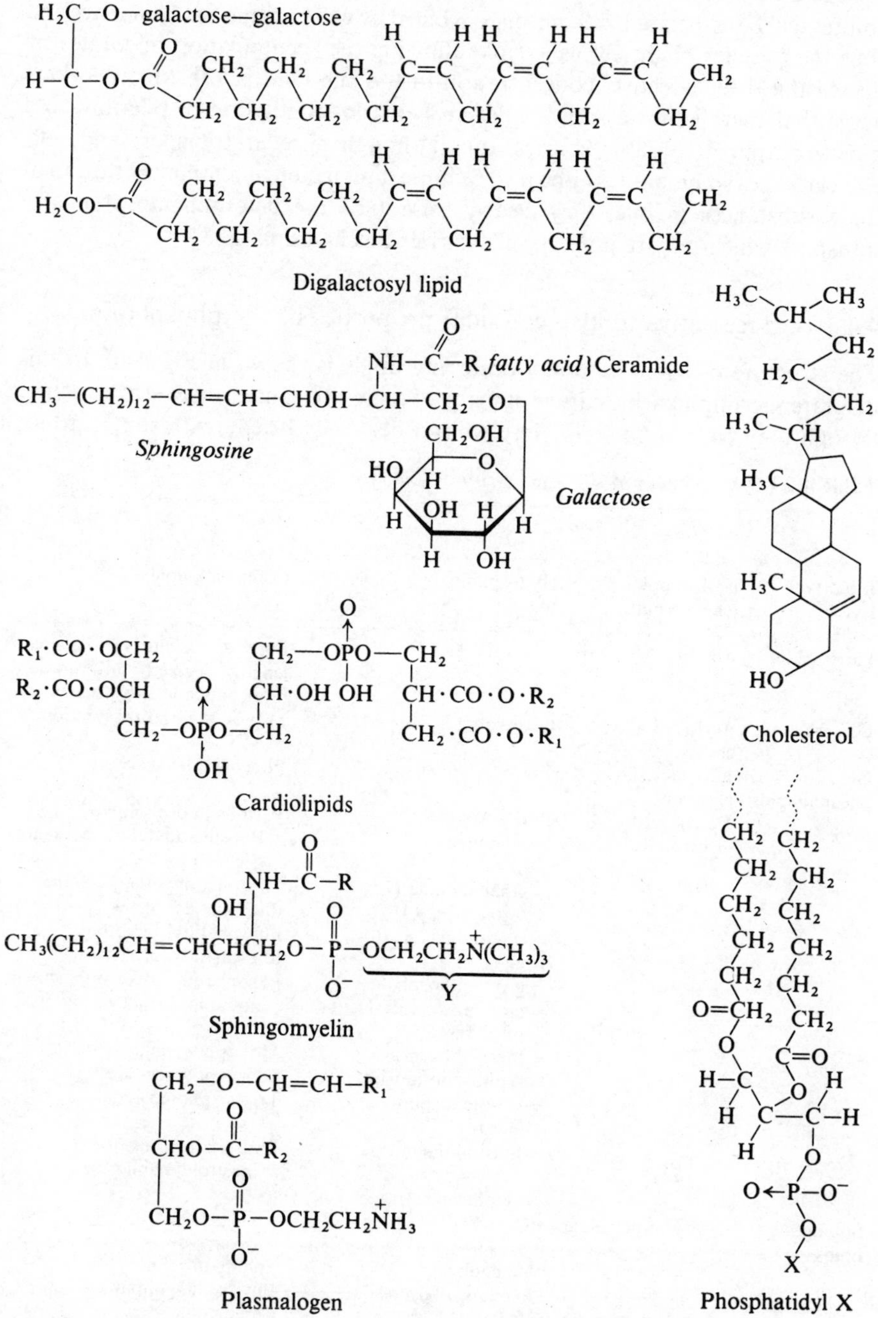

Figure 4.40 Structure of some naturally occurring lipids (from ref 284).

are excreted primarily by the liver. It has been suggested that as many of these solutes will be associated with micelles in bile they will be osmotically inactive and thus the micelles might act as a 'sink' allowing the accumulation of solutes by decreasing their 'effective' concentration in the bile fluid [282]. Most workers agree that there is not a simple relation between stimulation of bile flow and biliary output of organic solutes, thus although micellar transport and self-association of some solutes appear to be important factors in the concentration of these substances in bile, they clearly cannot be the determinants of biliary transport which may require specific carrier mechanisms [283].

4.4.4 Surface-active lipids – colloidal properties of the phospholipids

The structure of some naturally occurring lipids is shown in Fig. 4.40. In this diagram the amphipathic nature of many of these substances, diverse in structure though they be, can be seen. Here we can deal only briefly with some surface

Table 4.23 A classification scheme for lipids [284]

| Lipid type | Substituents | | Composite lipid |
	Hydrophobic (chain length)	Hydrophilic (pK)	
Fatty acid	Alkyl-(8–22 even)	COO$^-$ (4–5)	Fatty acids with various chain lengths, degree of unsaturation, branching and substituents.
Glyceride	Acyl-(1–3 alkyl chains)	Glycerol (12)	Mono-, di-, and triglycerides.
Glycero-phospholipids	1,2-Diacyl-L-glycerol	PO$_4$ (1–2, 6–8)	Phosphatidic acid
	1,2-Diacyl-L-glycerol	PO$_4$ + X X = choline	Various phospholipids Phosphatidyl choline (isoelectric 3–10)
		= ethanolamine (7.5)	Phosphatidyl ethanolamine
		= serine (2.2, 9.1)	Phosphatidyl serine
		= threonine (2.6, 10.4)	Phosphatidyl threonine
		= glycerol	Phosphatidyl glycerol
		= glycerophosphate	Phosphatidyl glycerophosphate
		= o-amino acid derivative of glycerol	o-amino acid ester of glycerol
		= inositol (*myo*-)	Monophosphoinositide
		= 4-phospho-inositol	Diphosphoinositide
		= 4,5-diphospho-inositol	Triphosphoinositide
		= inositoldimannose	Phosphatidyl (*myo*) inositoldimannoside
		= sulphosugar	
Sphingo-lipids	N-acylsphingosine	PO$_4$ + Y	
		Y = choline	Sphingomyelin
		= inositolglycoside	Phytosphingolipid
	N-acylsphingosine	Glucose	Cerebroside
		Oligosaccharide containing neuraminic acid	Gangliosides

chemical and colloidal properties of some of these compounds, in particular the phospholipids, which are not only widely found in biological membranes but are used as emulsifiers especially for intravenous fat emulsions, and as a key ingredient of microspherical drug delivery systems or liposomes. A classification scheme for lipids may be found in Table 4.23 which categorizes fatty acids, glycerides, glycerophospholipids and sphingolipids.

Fig. 4.41 gives the structure of some common phospholipids. Phosphatidyl-choline (PC) (lecithin) is an abundant phospholipid present in the soluble lipoproteins of plasma and in the lipoprotein of cellular and intracellular membranes. Like other glycerophospholipids it forms coarse, turbid dispersions of large aggregates in aqueous solution which on ultrasonic irradiation break down to smaller size. The initial large and highly asymmetric particles in the coarse lecithin dispersions in water are broken down during irradiation to produce almost spherical aggregates with a unit weight of around 2×10^6 [285] representing some 2740 monomers with a degree of hydration of approximately 0.16 g water per g lecithin. The effect of sodium chloride and calcium chloride on the shape of irradiated lecithin aggregates has been investigated [286]. Calcium chloride has no effect but sodium chloride increases the size and asymmetry of the aggregates as does choline chloride. The sodium chloride also affects the progress of the size reduction during ultrasonic irradiation as shown in Fig. 4.42.

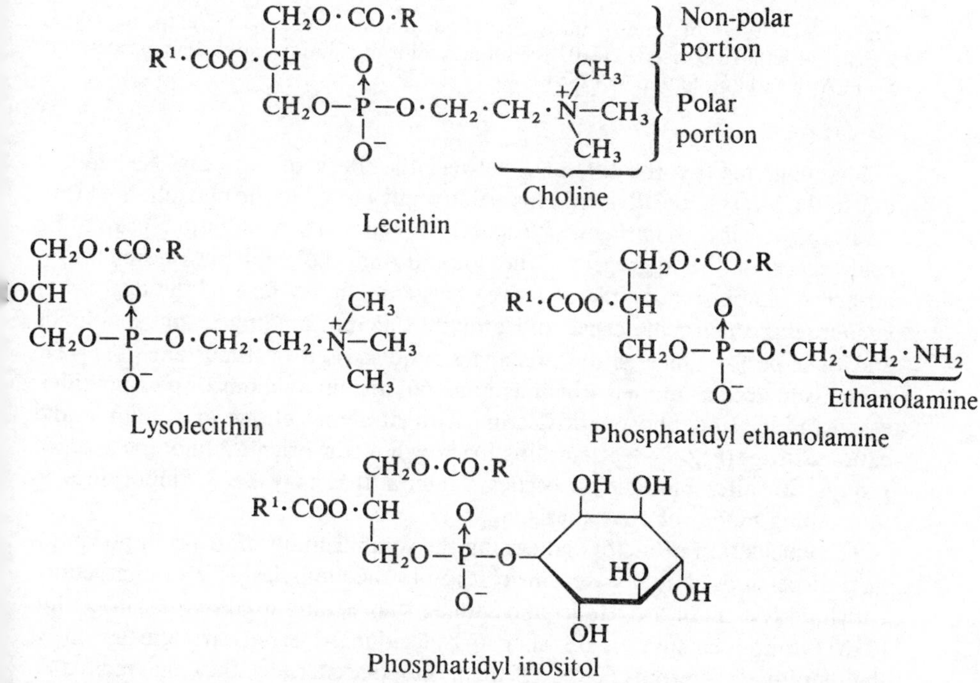

Figure 4.41 Structure of phosphatidylcholine (lecithin), lysolecithin, phosphatidyl-ethanolamine and phosphatidyl inositol.

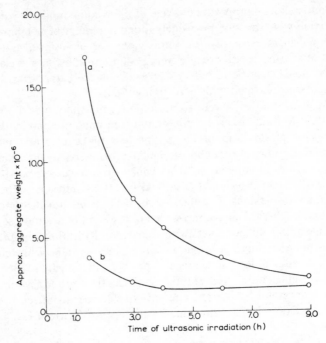

Figure 4.42 Effect of time of ultrasonic irradiation on the weight-average aggregate weight of lecithin dispersed in (a) 0.01 M sodium chloride solution and (b) deionized water. From Attwood and Saunders [285].

It is suggested that the difference between the effects of Ca^{2+} and Na^+ salts is due to the fact that the divalent ions are strongly bound to the phospholipid thus creating a positive charge which stabilizes the units in solution preventing coalescence of the aggregates. The sodium and choline being more weakly adsorbed simply reduce the negative charge on the surface of the aggregates encouraging their coalescence and growth. Calcium binding to phospholipids and their performance as ion-exchangers is discussed by Blaustein [287]. He presents molecular models which demonstrate the interactions of phosphatidyl-ethanolamine and phosphatidylserine with divalent cations and anionic and cationic drugs (Fig. 4.43) attempting to show how the orientation of the charged groups can alter binding properties, a factor that may be of importance in unravelling modes of anaesthesia.

The elucidation of factors governing the solubilization of drugs in phospho-lipid dispersions should give some clues as to the biological role of interactions with lipid systems *in vivo*. Hoyes and Saunders have solubilized steroids in lecithin [288]. Monopolar steroids are taken up by lecithin dispersions to a greater extent than multipolar steroids [289]. Of the monopolar steroids, those having a 5-en structure were not taken up to the same extent as the 4-en-3-one compounds or those with a saturated nucleus; the effect of the chain length of the ester group in

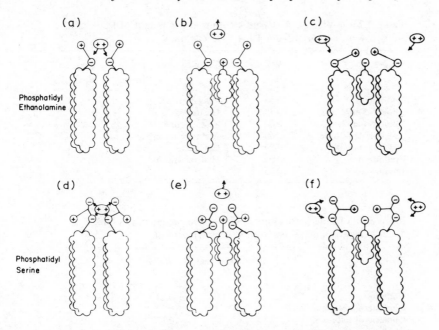

Figure 4.43 Molecular models demonstrating interactions of phospholipids, divalent cations and anionic and cationic drugs. From Blaustein [287] who explains, "As shown in (a) and (d) we might expect that 'at rest', most polar groups would be tied up by divalent cations. The proportion of polar groups bound in this manner would be a function of the divalent cation concentration and the binding constant for the particular cation involved. During depolarization, the change in the electric field across the polar heads might be expected to change the orientation of unbound polar heads, thereby changing their ion exchange properties.

We would expect cationic drugs such as procaine to compete with the divalent ions for binding to the phospholipid polar heads. The lipid-soluble properties of these drug molecules would help them to compete successfully, since the lipid-soluble moiety would likely be inserted between the fatty acid chains of the phospholipids, as in (b) and (e).

The lipid-soluble characteristics of anionic drugs would also favour their insertion between the phospholipid fatty acid chains (c and f). In this case, however, the increase in net negative charge in the region of the phospholipid polar groups, would be expected to result in increased divalent cation binding. Changing the electric field across the polar heads, by depolarization, should result in the release of divalent cations and some of the anionic or cationic drug molecules. The lipid-soluble nature of the drug molecules would, however, keep most of these molecules within the lipid layer. Non-polar molecules such as the alcohols, diethyl ether and fluothane might also be expected to dissolve in the lipid portion of the membrane."

cholesterol esters on uptake into 4 % lecithin sols was also measured [289]. Some of these results are given in Table 4.24. As the number of carbon atoms in these esters increases there is a marked reduction in solubilization until C_{12} when the amount solubilized increases significantly. In Chapter 10 we discuss the significance of the C_{12} hydrocarbon chain in interaction with biological membranes. It is evident from these solubilization results that the dodecyl chain

Table 4.24(a) Weight of steroid solubilized per weight of lecithin. All sols were prepared from 4% (w/w) lecithin.

Steroid	Weight (g)	Steroid	Weight (g)
Cholesterol	0.53	Progesterone	0.08
Cholest-5-en-3-one	0.50	Oestrone	0.05
Cholest-4-en-3-one	0.60	Testosterone	0.08
5α-Cholestan-3-one	0.60	Methyltestosterone	0.075
5α-Cholestan-3β-ol	0.59	Cortisone	0.08
Prednisolone	0.01	Hydrocortisone	0.08
Prednisone	0.05		

(b) Weight of cholesterol esters solubilized per weight of lecithin. All sols were prepared from 4% (w/w) lecithin.

Steroid	C atoms in acid radical	Weight (g)
Cholesterol	0	0.53
Cholesterol acetate	2	0.20
Cholesterol butyrate	4	0.12
Cholesterol caproate	6	0.08
Cholesterol octanoate	8	0.07
Cholesterol decanoate	10	0.09
Cholesterol laurate	12	0.34
Cholesterol myristate	14	0.36
Cholesterol palmitate	16	0.32
Cholesterol stearate	18	0.39

From Kellaway and Saunders [289].

has a special significance, Kellaway and Saunders suggesting that at this length and above the alkyl chain is sufficiently long to be able to fold back on itself and readily pack in the lecithin micelles. In mixed lecithin–lysolecithin sols the amount of progesterone solubilized increased with the weight of lysolecithin incorporated in the micelle.

The light-scattering molecular weight of lysolecithin (lysophosphatidyl choline, LPC) which forms spherical micelles (without ultrasonic treatment) was found to be about 9×10^4 indicating 180 monomers per micelle. At molar ratios of lysolecithin to lecithin of 2:1 a maximum limiting viscosity number of 900 is obtained indicating a high asymmetry; these mixed aggregates are thought to be helical with micellar weights in the region of 1.5×10^6 [290]. Intermediate data are shown in Table 4.25. Solubilization by lysolecithin of some local anaesthetics has been studied by Hunt and Saunders [291]; solubilization occurs in the hydocarbon interior and increases, therefore, in the order ethyl, *n*-propyl and *n*-butyl ester. A high concentration of LPC has been found in the duodenum which may facilitate micellar absorption of fatty acids from the gut [292], and perhaps influence the absorption of poorly soluble drugs from the intestine. LPC

Table 4.25 The limiting viscosity of ultrasonically irradiated PC/LPC sols in water at 25° C. The limiting viscosity (η) was obtained from the expression

$$\lim_{c \to 0} \frac{\eta/\eta_0 - 1}{c} = [\eta]$$

where c = concentration of solute in g/100 ml; η = solution viscosity, η_0 = pure solvent viscosity.

Solute (weight fraction)		Mol ratio	Limiting viscosity no. (η)
PC		100 : 0	3.8
PC/LPC	4 : 1	72.4 : 27.6	6.6
PC/LPC	3 : 2	49.5 : 50.5	18.8
PC/LPC	1 : 1	39.6 : 60.4	317
PC/LPC	2 : 3	30.4 : 69.6	520
PC/LPC	3 : 7	21.9 : 78.1	188
PC/LPC	1 : 4	14.1 : 89.9	7.5
LPC		0 : 100	4.2

From Martin *et al.* [293].

increases the solubilization of fatty acids by bile-salts. Sodium deoxy-cholate–phospholipid sols do not show linear solubilizing power for progesterone [293]. A maximum occurs at a bile salt–phospholipid ratio of 1 : 4 and introduction of sodium deoxycholate into LPC/phospholipid (lecithin) mixtures results in decreased progesterone solubility.

4.4.5 Liposomes

In the last few years an explosion of papers on liposomes has occurred. Liposomes are smectic mesophases of phospholipids organized into bilayers which assume a multilamellar or unilamellar structure. The multilamellar species are heterogeneous aggregates, most commonly prepared by dispersal of a thin film of phospholipid and cholesterol into water. Sonication of the first formed multilamellar units can give rise to the unilamellar liposomes. The net charge of the liposome can be varied by incorporation say of a long chain amine, generally stearylamine, (to give positively charged vesicles) or dicetyl phosphate (giving negatively charged species). Both lipid-soluble and water-soluble drugs can be entrapped in liposomes, the latter being intercalated in the aqueous layers, while liposoluble drugs are solubilized in the hydrocarbon interiors of the lipid bilayers. The use of liposomes as drug carriers is the subject of a recent review by Fendler and Romero [294], a paper to be consulted for further details. Since liposomes can encapsulate drugs, proteins and enzymes the systems can be administered intravenously, orally or intramuscularly in order to decrease toxicity, increase specificity of uptake of drug and in some instances to control release.

We can consider here just a few aspects of their behaviour as aggregates rather than deal with the many facets of their use. In order that liposomes can be used effectively as drug delivery systems it is essential that we appreciate that the liposomes interact with components of serum, and with amphipathic substances in particular. These interactions can lead to their breakdown. Surfactants such as Triton X 100 and lysolecithin can intercalate into the bilayer and increase the permeability of the vesicles to any entrapped compounds [295–297]. The lytic action of a series of sodium alkyl sulphates on egg phosphatidylcholine liposomes increases with increasing alkyl chain length of the surfactant [298]. Electron spin resonance studies of the interaction of liposomal structures with Triton X 100 have shown that the effect of the Triton depended on the particular phospholipid used in the preparation. In PC vesicles Triton at low concentrations fluidizes the membrane without disruption but when a surfactant:PC ratio of 0.4 is reached the vesicle is solubilized [299]; the presence of cholesterol oddly did not have an effect on the sensitivity of the membrane system to the surfactant. The sensitivity of dimyristyl PC liposomes is greater than that of PC liposomes as seen in Fig. 4.44, showing the result of addition of surfactant to liposomal dispersions. The presence of sphingomyelin in PC liposomes increases the sensitivity to Triton X 100 [300]; the higher the mole fraction of the sphingomyelin (XXV) the less Triton required to cause complete solubilization.

In simple sphingomyelin–Triton X 100 systems, the sphingomyelin is dispersed as bilayers below a surfactant concentration of 0.2 mM. Above this concentration mixed micelles formed [301, 302]. Studies of the transfer of cholesterol solubilized in sodium dodecyl sulphate to PC liposomes has shown that 90 % of

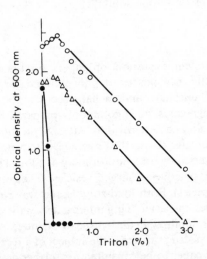

Figure 4.44 Effects of Triton X-100 on the turbidity of dimyristoylphosphatidylcholine liposomes and egg phosphatidylcholine liposomes.
The liposomes were prepared from 3 m phosphatidylcholine. ○, egg phosphatidylcholine–33 mol % of cholesterol; △, egg phosphatidylcholine; ●, dimyristoylphosphatidylcholine. From Anzai *et al.* [299] with permission.

$$CH_3 \cdot (CH_2)_{12}CH=CH \cdot \underset{\underset{HO}{|}}{CH} \cdot \underset{\underset{NH}{|}}{CH} \cdot CH_2O - \underset{\underset{O^-}{\overset{\overset{O}{\uparrow}}{P}}}{P} - O \cdot CH_2 \cdot CH_2 \cdot \underset{\underset{CH_3}{\diagdown}}{\overset{\overset{CH_3}{\diagup}}{\overset{+}{N}}} - CH_3$$

$$\left. \underset{\underset{R}{|}}{\overset{\overset{CO}{|}}{}} \right\} \text{FATTY ACID}$$

CHOLINE

(XXV) Sphingomyelin

the cholesterol is slowly transferred from the micellar state into the liposome where it has considerably greater rotational freedom [303]. Recently there have been attempts to 'solubilize' atherosclerotic plaques which contain cholesterol and cholesterol esters using a variety of detergents including polyunsaturated PC [304]. It is likely that such a transfer process will be slow *in vivo*.

Liposomes, like micelles, may provide a special medium for reactions to occur between molecules intercalated in the lipid bilayers or between molecules trapped in the vesicle and free solute molecules. Rate enhancements can be induced by altering the micro-environment of trapped polar reactants: osmotic shrinkage of the vesicle can result in a change in the viscosity of the bilayer because of alteration in packing, which can also lead to changes in the hydration of the solutes. Kano and Fendler [305] have probed the effects of osmotic contraction of liposomes using charged and uncharged dipalmitoyl-D,L-α-phosphatidylcholine vesicles and fluorescent probes. Indications of the microviscosities of the surroundings of these probes can be gained by measurement of fluorescence depolarization. The variety of solubilization sites and how these are altered is shown in Fig. 4.45. There is a substantial increase in microviscosities on osmotic shrinkage, the reported viscosities in the interior of the vesicles being several times greater in the presence of an osmotic gradient than in its absence. Kano and Fendler have made some calculations of the molecular state assuming the mean radius of the vesicles to be 15 nm and the bilayer thickness to be 3.8 nm [306]. The volume of the aqueous core can be worked out as 6×10^{-18} cm^3. Each vesicle contains, therefore, 2×10^5 molecules of water and assuming that each phospholipid molecule is hydrated by 23 molecules of water [307], the number of water molecules which hydrate the interior of the vesicle containing 1720 molecules of phospholipid is estimated to be 4×10^4, i.e. the ratio of free water to bound water is 4 : 1. On application of an osmotic gradient using 1M NaCl, the number of water molecules reduces to 2×10^4 leaving no free water (see diagrammatic representation in Fig. 4.45 whose legend contains an explanation of the results found in Table 4.26.)

Recently the formation of disc-like aggregates by single-chain phosphocholine molecules was reported [308]. These aggregates are different from normal lecithin vesicles as might be expected; the aggregate weight of 1.5×10^6 and 3.8×10^6 for the compounds, respectively, indicate large aggregates which are shown in electron micrographs not to be simple micellar species. Hargreaves and Deamer [309] have made liposomes from dilute aqueous solutions of C_8–C_{18} single-chain amphiphiles. While some single-chain fatty acids and lysophospho-

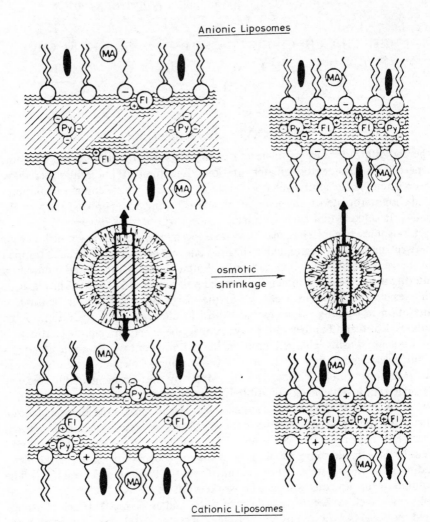

Anionic Liposomes

osmotic
shrinkage

Cationic Liposomes

Figure 4.45 An oversimplified representation of the proposed solubilization sites of methyl anthracene MA, pyranine, PY, and acriflavine, Fl, prior and subsequent to osmotically shrinking cationic and anionic single compartment dipalmitoyl-D,L-α-phosphatidylcholine liposomes. Bound and free water are indicated by waved and straight lines. The charged long chain additives are shown to be interspersed with the phospholipid and cholesterol (⬤). "The viscosity 'reported' by pyranine for anionic and that by acriflavine for cationic single compartment liposomes, ~ 1.0 cP, indicate the aqueous environments of these probes. Increased viscosities following osmotic shrinkages have been rationalized in terms of changing the nature of the liposome entrapped water. Following the release of free water, some bound water is also released as the result of osmotic shrinkage. The determined shrinkage rates support this postulate. The viscosity of the environment of pyranine in cationic, 9.6 ± 0.3 cP, and that of acriflavine in anionic single compartment liposomes, 74 ± 5 cP, indicate electrostatic attractions of the probes to the charged liposome surface. Osmotic shrinkage results in lowering the viscosity of the environments of the probes presumably because the more concentrated sodium chloride

Table 4.26 Microviscosities of the environments of methyl anthracene, pyranine and acriflavine and their fluorescence lifetimes in liposomes, at 25.0° C

Liposome*	$\Delta(NaCl)_{0-i}$, M†	MA‡ η(cP)	τ(ns)	Pyranine‡ η(cP)	τ(ns)	Acriflavine‡ η(cP)	τ(ns)
S_0O	0	420 ± 30	3.6 ± 0.2			19 ± 4	
S_0O	1.0	450 ± 30	3.7 ± 0.2				
M_0O	0	410 ± 30	3.9 ± 0.2				
M_0O	1.0	430 ± 30	3.9 ± 0.2				
SO	0	900 ± 30	4.4 ± 0.1	5.4 ± 0.3		21 ± 5	4.3 ± 0.2
SO	1.0	1100 ± 50	4.7 ± 0.1	9.2 ± 0.4		5.9 ± 0.4	3.9 ± 0.3
S −	0	950 ± 60	4.0 ± 0.2	1.4 ± 0.1		74 ± 5	
S −	1.0	1200 ± 50	4.7 ± 0.1	5.1 ± 0.3		6.3 ± 0.4	
M −	0	850 ± 50	3.7 ± 0.2	1.1 ± 0.2		22 ± 4	4.4 ± 0.2
M −	1.0	840 ± 50	3.8 ± 0.2	1.9 ± 0.1		27 ± 3	4.0 ± 0.3
S +	0	1100 ± 50	4.4 ± 0.1	9.6 ± 0.3		1.1 ± 0.1	4.6 ± 0.1
S +	1.0	1300 ± 50	4.5 ± 0.1	5.1 ± 0.2		2.0 ± 0.2	3.9 ± 0.3
M +	0	1000 ± 50	4.2 ± 0.1	8.5 ± 0.3		1.1 ± 0.1	4.5 ± 0.2
M +	1.0	1000 ± 50	4.3 ± 0.1	4.4 ± 0.2		1.4 ± 0.1	4.2 ± 0.2

* S_0, M_0, S and M stand for single-compartment liposomes in the absence of cholesterol, multicompartment liposomes in the absence of cholesterol, single and multicompartment liposomes in the presence of cholesterol; 0, − and + indicate the charges (neutral, anionic, cationic) of the liposomes.
† Osmotic gradient across the liposomes (i.e., [NaCl] outside–inside).
‡ Concentration of probes = 5.0×10^{-5} M; fluorescence polarizations were determined 1.0 h subsequent to the preparation of the vesicles or subsequent to providing the osmotic shock. From Kano and Fendler [305].

Two phosphocholine compounds [308]

lipids are found in biomembranes they have generally been considered to be incapable of forming stable bilayers. However, there have been reports of the formation of lipid vesicles from *cis*-Δ^9-octadecanoic acid in dilute dispersions [310]. Also penetration of water into single-chain alkyl sulphates in admixture

replaces them from their sites. The high viscosities reported by MA, ~ 1000 cP, suggest the intercalation of this probe in the phospholipid bilayers. Osmotic shrinkage does not alter the environment of MA" [305].

with long-chain alcohols can give rise to vesicle-like formations under the microscope. Hargreaves and Deamer [309] refer to the formation of uni-, oligo- and multilamellar vesicles in the size range 1 to 100 μm in such systems. Apart from the fatty acid vesicles, which they prepared by titrating alkaline solutions with acid or by mixing equimolar quantities of fatty alcohols, sodium dodecyl sulphate–dodecanol mixtures also formed the basis of the vesicles. Addition of fatty alcohols or indeed hydrophobic derivatives of non-ionic surfactants to micellar systems of more hydrophilic homologues can lead to the formation of very asymmetric micelles which at some point must begin to assume the properties of stable bilayers [311]. This is clearly an area for intensive study.

An attempt has been made to distinguish the boundaries between micelles, microemulsion particles and vesicles in Table 4.27 from Fendler [312]. The main distinguishing feature of vesicular structures is their permanence on a micellar timescale as established by studies of monomer kinetics and the kinetics of solubilizate exchange with the environment as well as of their stability to dilution by water.

Table 4.27 Comparison of micelles, microemulsions and vesicles

	Micelles	Microemulsions	Vesicles
Weight averaged molecular weight	2000–6000	10^5–10^6	$> 10^7$
Diameter (nm)	3–6	5–100	30–500
No. of solubilizate molecules per aggregate	Few	Large	Large
Kinetic stability (leaving rate of monomers), s	$\sim 10^{-5}$	$\sim 10^{-5}$	$>$ sec
Solubilizate residence time, s	10^3–10^5	10^3–10^5	$>$ sec
Dilution by water	Destroyed	Altered	Remain stable

From Fendler [312]

4.4.6 Lipoprotein aggregates

None of the plasma lipids are sufficiently polar to circulate as separate entities in solution; rather they depend on interactions with protein and are consequently referred to by the generic name 'lipoprotein'. The soluble lipoprotein aggregates bear some resemblance to micellar aggregates and vesicles although there is still much uncertainty about their structure. The most important groups of lipo- protein are the high-density or α-lipoproteins (HDL), the low-density or β- lipoproteins (LDL), the very low-density (VLDL) and chylomicra. VLDL and chylomicra appear to be spherical; a lipid bilayer structure proposed for the LDL aggregate is shown in Fig. 4.46. The components include proteins and phos- pholipids, the apolar cholesteryl esters and triglycerides and free cholesterol. All lipoprotein models have some common features and are all based on the assumption that, as in micelles, the polar constituents are oriented outwards providing a hydrophilic surface to the lipoidal particle; non-polar constituents present themselves in or near the core of the particles. Some workers (e.g. [314]) have evidence that HDL aggregates have outer and inner phospholipid layers.

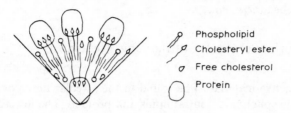

Phospholipid
Cholesteryl ester
Free cholesterol
Protein

Figure 4.46 Lipid bilayer model of LDL structure. From Lewis [313]. Lewis describes the model in this way: 'A protein network is envisaged, with icosahedral symmetry; it is suggested that there are 60 such units, existing as trimers. Instead of the conventional view of a protein-coated molecule with a lipid core, it was proposed that the lipids are organized into a spherical bilayer. The two lipid layers are mirror images, the non-polar regions of their main constituents, phospholipid and cholesteryl ester, oriented towards each other. At the outer surface of the outer layer and the inner surface of the inner layer are situated the polar groups of the phospholipids and the cholesterol side chains; these regions of the major constituents are thus adjacent to the protein units on the surface and to a presumptive protein component at the centre of the particle.'

Lewis [313] claims that there is no cogent reason for regarding VLDL and chylomicra as distinct species. The upper limit of VLDL and the lower limit of chylomicra diameters is frequently given as 80 nm but there will be the same problem in assigning a name to particles at the border as there is in distinguishing microemulsions from swollen micelles.

The mode of interaction of these lipoproteins with surfactants, with liposomes and with drugs may well be of some import in deciding the fate of these species *in vivo*. Very little is known about such interactions. NaDS protects both VLDL and LDL against the denaturing effect of organic solvents and allows the separation in the laboratory of high yields of VLDL and LDL apoprotein. Non-ionic detergents (e.g. Triton) or cationic surfactants (benzalkonium chloride) failed to exhibit this protection [315]. Scanu had earlier [316] proposed that in HDL there are a number of protein units held together by lipid bridges; it is possible that such substructures exist in VLDL and LDL also. One would expect that on solubilization of lipid by surfactant the proteins would form insoluble aggregates unless the surfactant adsorbed to prevent this association occurring. The difference between the detergents studied must lie in their different abilities to solubilize lipid and to associate with the protein of the particles. As sustained administration of some non-ionic detergents induces hyperlipaemia observed by increase in the quantity of circulating lipoprotein, an understanding of the surfactant–lipoprotein interaction is critical. It has been suggested, for example, that the formation of a more soluble surfactant–lipoprotein complex might lead to the trapping of lipids in the plasma preventing their exchange with tissue lipids [316]. There is, however, conflicting evidence from experimental studies. Kellner *et al.* [318] have shown that Triton A25 may initially protect cholesterol fed rabbits from atherosclerosis, but Scanu *et al.* [319] has found the non-ionic surfactant to be atherogenic.

4.4.7 The surface chemistry of the lung, eye and ear

(A) LUNG

The surface-active material to be found in the alveolar lining of the lung is a mixture of phospholipids, neutral lipids and proteins. The lowering of surface tension by the lung surfactant system and the surface elasticity of the surface layers assists alveolar expansion and contraction. Deficiency of lung surfactant in the newborn leads to a respiratory distress syndrome, and has turned attention to the possibility of the instillation of artificial surfactant in these cases (see, for example, [320]). The predominant phospholipid of natural lung surfactant is dipalmitoylphosphatidylcholine (DPPC) which has a melting temperature of 43° C and thus cannot be used alone as an artificial substitute. Mixed surfactant systems are more likely to have the desired surface characteristics. Unsaturated phosphatidylglycerol (PG), which is another natural component, probably lowers the transition temperature to body temperature. It is postulated that during inspiration the PG and DPPC molecules adsorb at the alveolar interface allowing rapid expansion. During expiration the molecules are squeezed out of the interface raising the surface tension thus providing some resistance to compression and preventing collapse. It has recently been postulated that natural surfactant is secreted in a nearly dry form [321] and that hydration of the polar head groups provides the driving force for spreading at the air–liquid interface. Particles of dry surfactant and aqueous suspensions of surfactant deposited in the tracheal fluid of premature rabbits improve the pressure–volume characteristics of the lungs [321, 322]. Bangham *et al.* [323] have recently developed a dry surfactant system containing DPPC and unsaturated PG in a 7:3 ratio. Emulsified phospholipid systems have also been administered to newborn rabbits but these were inferior to natural surfactant [324].

Does the surfactant layer have any relevance to the absorption of drugs administered by the respiratory route? There has been a suggestion that the alveolar surfactants are involved in the solubilization of inhaled gases such as halothane [325]. The hypothesis is that on compression of the lung, surfactant molecules are ejected into the subphase, and form micelles capable of solubilizing gas molecules; on the release of pressure the micelles deaggregate as the surfactant returns to the interface, releasing the anaesthetic to be absorbed by the blood stream. Oxygen solubilization by lung surfactant has been found to be aided by micelle formation [326, 327]. Aerosols containing detergents have been used for some time to improve the hydration and wetting of adhesive sputum and thus facilitate expulsion of the material; the possibility arises that excessive deposition in the alveoli of the surfactants used, might lead to an interference with the natural activity of the lung surfactant [328]. The value of aerosols containing detergents has not, however, been established.

(B) EAR

The major function of the Eustachian tube is to equalize pressure between the middle ear and the external atmosphere; the tube is normally closed but frequently opens during swallowing or sneezing. The surface tension of mucus is thought to be important, Bauer [329] having postulated that a surfactant similar to that in the alveolar lining might be active in the mucosa of the Eustachian tube, middle ear mucosa and the mastoid air cells, affecting formation and evacuation of middle ear exudates [330]. It is reasonable to postulate that deficiency of such surfactant can lead to poor Eustachian tube function; removal of surfactant by saline washing leads to a substantial increase in pressure required to open the tube [330].

(C) EYE

The surface chemistry of the eye is probably better understood than that of the ear. The cornea is covered with a thin, fluid film the so-called tear film, which is believed to consist of an aqueous phase, approximately 10^{-3} cm thick, with an adsorbed lipid and mucin layer at the air–water interface and an adsorbed mucin layer on the corneal side. The latter renders the cornea hydrophilic and enables the tear film to spread. In dry eye syndrome local areas of dewetting occur due to increased contact angle. Dry eye is sometimes precipitated by drug therapy, and there is a search for adequate artificial tear fluids. The surface chemistry of tear film components has been discussed by Holly [331]. Adsorption of cationic surfactants present in eye drops as preservatives can lead to the production of a hydrophobic surface due to electrostatic adsorption of the cations with the hydrocarbon chains oriented towards the tear film. Such a process can itself result in dewetting and thus cationics should be excluded from artificial tear fluids.

Within the eye the photoreceptors – the retinal rods and cones – are highly ordered lamellar structures which have a quasi-crystalline nature [332] which is probably a device for increasing the efficiency of light capture.

4.4.8 Aggregation of naturally occurring molecules

The critical concentration for the association of many naturally occurring compounds is high and often is greater than circulating levels of the compound. This does not diminish the significance of association, for many hormones and other substances are concentrated in cells or special organelles. For example, the surface-active steroidal antibiotic, fusidic acid, is concentrated approximately fifty times inside *E. coli* in solutions of 2×10^{-6} g ml^{-1} [333] and adenosine triphosphate and biogenic amines attain concentrations of the order of 20% in storage granules [334, 335]. While it has been recognized that the quaternary structure of many proteins is crucial for their activity and function, it has also been recognized that many protein molecules aggregate [336]. Frieden [337] has suggested that the reversible association of the enzyme bovine liver glutamate dehydrogenase (GDH-ase) may serve a critical function in the control of certain metabolic pathways *in vivo*. The early observation that effector molecules known to modify GDH-ase activity also dramatically altered the state of aggregation led to the hypothesis that association was probably linked to a metabolic control mechanism, a proposal which is realistic as GDH-ase reaches concentrations in liver cell mitochondria (several milligrams per litre) above the concentrations at which Eisenberg [338] found it to aggregate *in vitro*. Because of aggregation, the enzyme undergoes an allosteric transition at a lower level of activation than the monomeric form. The tendency for aggregation of the active form is much greater than for the inactive form. As active polymers are formed, the concentration of active monomer is depleted since the equilibrium between active and inactive forms is maintained. This pulls some inactive monomers into the active form [339]. There seems to be little reason then that aggregation of other active molecules will not induce biological changes. In the case of the enzyme, it is the change in conformation and masking of reactive groups on association that causes change in activity.

4.4.9 Membranes and micellization

One of the regulatory structures in living organisms is the cell membrane. The simple bimolecular leaflet structure for the lipid portion of the membrane probably still is considered to form the main bulk of the mammalian membrane, but the concept of a more flexible and mobile structure has been suggested by Lucy and others [340]. While the current views on membrane morphology are constantly shifting and recent texts should be consulted, it is salutory to consider work which has discussed the micellar nature of membrane lipids. Much of this is speculative but illustrates one approach to the problem.

Association of the hydrophobic tails of membrane lipids leads, as depicted in Fig. 4.47a, to what Watkins [341] has described as polar discontinuities. Transition to the micellar state is considered to be essential to allow cell fusion to occur as the biomolecular leaflet is thermodynamically a stable system which would resist coalescence with similar structures. External influences can, however, induce phospholipid aggregation and can thus alter the permeability of cell membranes to water-soluble and oil-soluble species. Calcium ions, for example, induce inverse micelle formation in phospholipid systems [342]. Other metal ions also result in this transformation (Fig. 4.47b); addition of adenosine triphosphate (ATP) removes the metal and leads to a reversion to the normal micellar pattern. Maas and Coleman [343] have postulated that such transitions may have significance in nerve membrane operation. Metal–ATP–phospholipid complexes

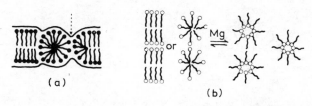

(a)

(b)

Figure 4.47(a) A representation of the association of the lipids in a biological membrane and the formation of 'polar discontinuities' caused by the juxtaposition of polar head groups, after Watkins [340]. (b) Inverse micelle formation in membrane in bilayer or micellar state induced by Magnesium ion, after Maas and Coleman [343].

can be seen as forming lamellar sheets; on removal of ATP, for example by ATP-ase, the metal–phospholipid complex will assume a quite different orientation. Replacement of ATP from some metabolic source will then cause a reversion to the original configuration. Changes in the solution bathing cells lead to shifts in membrane properties.

The glycoprotein components of cell membranes are amphipathic and it has been suggested that increases in cation, protein or polycation content in the aqueous phase cause surface aggregation of these glycoprotein molecules [344, 345]. Cations bind to the surface negative groups and reduce surface negative charge. This allows aggregation of mobile glycoprotein units to occur and membrane resistance decreases immediately on aggregation [346]. The change in the composition of the surface layer has been thought to provide a trigger for pinocytosis (Fig. 4.48a to d). The increase in surface pressure resulting from aggregation would be relieved by the expansion and folding of the membrane. Elworthy's suggestion that changes in local dielectric constant can cause micellar changes in membranes (based on work on the effect of dielectric constant on the transition of micellar states of lecithins) predates the globular micelle membrane model [347].

Calcium ions are vital to the functioning of cell membranes and to many biological processes. The surface micelles that Bray has postulated [348] have analogues in simpler surfactant systems. At around physiological pH for example, monolayers of stearic acid molecules associate in the presence of calcium chloride into surface micelles approximately 6 nm in diameter (Fig. 4.48c). The formation of surface micelles in membranes may be closely related to the structural changes and entropy decrease reported to occur during membrane excitation [349].

Molecules added to biological systems may find their way into membranes and associate in the lipid system. Several compounds of fungal or bacterial origin (such as alamethacin, which forms micelles in aqueous systems [350]) readily incorporate into planar lipid bilayers and form channels by self-association, channels which allow the translocation of ions and other hydrophilic species through the otherwise hydrophobic membranes. Based on calculations made of the rates of association possible for channel formation by lateral diffusion of

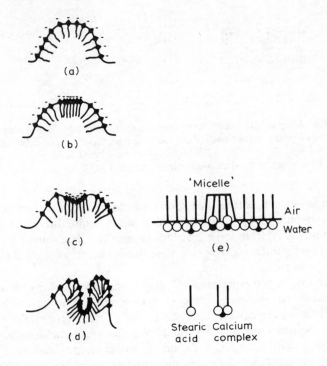

Figure 4.48(a) to (d) Aggregation of amphipathic glycoprotein as a trigger for pinocytosis after Gingell [345]. (a) Normal state of cell surface, (b) salt or protein in medium aggregates glycoproteins, (c) filaments contract and change ionic milieu, (d) channel formation. (e) Surface 'micelle' formation by stearic acid in the presence of calcium ion from Neumann [349] by permission of Academic Press.

random molecules of alamethacin located in the membrane surface, Mueller [351] has concluded that the molecules are probably already aggregated in the local structures at the membrane surface.

The stability of flat arrays of globular micelles may require interactions with protein or mucoprotein molecules because of the proximity and concentration of charged head groups: on the other hand, the ability of phospholipid–cholesterol dispersions to form neat and middle phases in which there is a similar juxtaposition of head groups suggests that this might not be a problem. Labilization and stabilization of membranes may be connected respectively with induction and reduction of the micellar state. On increasing the micellar fraction, the permeability to water will rise dramatically and lead to eventual osmotic damage to the membrane system. The volume of the membrane must change as the micellar phase is much less economical in the use of space. Surface-active drugs such as chlorpromazine, SKF 525 A (Proadifen) [166] and local anaesthetics and surfactant molecules may well induce micelle or mixed micelle formation in membranes. Chlorpromazine hydrochloride has been shown to efficiently solubilize pure membrane lipids [352]. Certainly the penetration theory of their action does not lead to an easy explanation of the ability of these drugs to increase membrane transport whereas penetration and induction of a micellar phase might.

4.4.10 Storage and release of natural substances in the body

Hormones and transmitter substances are frequently stored in the body in relatively high concentrations awaiting signals for their release. There are three possible explanations of the ability of high concentrations of organic electrolytes to be stored within an organelle enclosed by a lipid membrane: the free drug concentration is reduced to negligible proportions (i) by complexation with a macromolecule, or (ii) by self-association or mixed micelle formation, or (iii) the permeability of the membrane is only increased when release is required. The association explanations have been suggested for the storage of biogenic amines in storage vessels within various sites *in vivo* [353]. One of the striking observations about norepinephrine (NE) storage vessels is that they contain considerable quantities of ATP nearly always in a stoichiometric ratio to NE of 1:4. An important means of storing amines in the adrenal and neuronal vesicles may, therefore, involve interactions between NE and ATP, possibly facilitated by divalent cations (Fig. 4.49). The micellar aggregates postulated by Pletscher *et al.* [355] are large and non-diffusable. Berneis [356] has found that the size of the aggregates increases with increasing concentration of the solutes and is dependent on calcium ion concentrations.

Micelle formation can also explain the osmotic stability of the organelles despite their high levels of ATP (20% w/v) and amines such as 5-hydroxy-

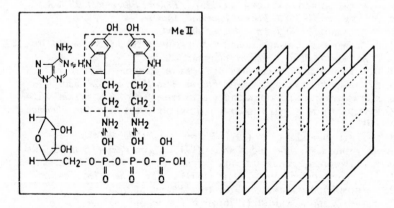

Figure 4.49 Hypothetical structure of aggregates formed by 5-hydroxytryptamine and ATP. Reproduced with permission from Pletscher *et al.* [354].

tryptamine (25%). The rapidity of release of the biogenic amines is of the order of 1 ms, judged by synaptic delay times and much of this time must be accounted for by the time required for molecules to traverse the neuronal gap which is 100 nm wide. Rates of micellar breakdown of ~ 0.1 ms have been observed for surfactant micelles (see Chapter 3) while rates of exocytosis (the alternative mode of extracting the amines from inside the storage vesicles) are too long – of the order of 100 to 200 ms.

References

1. A. T. FLORENCE (1968) *Adv. Colloid Interface Sci.* **2**, 115.
2. A. FELMEISTER (1972) *J. Pharm. Sci.* **61**, 151.
3. A. T. FLORENCE (1977) in *Micellization, Solubilization and Microemulsions* (ed. K. L. Mittal) Vol. 1, Plenum Press, New York p. 55.
4. D. ATTWOOD (1972) *J. Pharm. Pharmac.* **24**, 751.
5. D. ATTWOOD and O. K. UDEALA (1975) *J. Phys. Chem.* **79**, 889.
6. P. MUKERJEE (1974) *J. Pharm. Sci.* **63**, 972.
7. K. THOMA and E. SIEMER (1976) *Pharm. Acta Helv.* **51**, 50.
8. D. ATTWOOD and O. K. UDEALA (1975) *J. Pharm. Pharmac.* **27**, 395.
9. D. ATTWOOD and O. K. UDEALA (1974) *J. Pharm. Pharmac.* **26**, 854.
10. D. CAUSON, J. GETTINS, J. GORMALLY, R. GREENWOOD, R. NATARAJAN and E. WYN-JONES (1981) *J.C.S. Farad. II* **77**, 143.
11. B. FARHADIEH, N. A. HALL and E. R. HAMMARLUND (1967) *J. Pharm. Sci.* **56**, 18.
12. D. ATTWOOD and O. K. UDEALA (1975) *J. Phys. Chem.* **79**, 889.
13. D. ATTWOOD and O. K. UDEALA (1976) *J. Pharm. Sci.* **65**, 1053.
14. D. ATTWOOD and R. NATARAJAN (1981) *J. Pharm. Pharmacol.* **33**, 136.
15. J. GETTINS, R. GREENWOOD, J. E. RASSING and E. WYN-JONES (1976) *J.C.S. Chem. Comm.* **24**, 1030.
16. D. ATTWOOD and O. K. UDEALA (1975) *J. Pharm. Pharmac.* **27**, 754.
17. D. ATTWOOD (1968) *J. Phys. Chem.* **72**, 339.
18. P. H. ELWORTHY and C. B. MACFARLANE (1962) *J. Pharm. Pharmac.* **14**, 100T.
19. D. ATTWOOD and O. K. UDEALA (1975) *J. Pharm. Pharmacol.* **27**, 806.
20. D. ATTWOOD (1976) *J. Pharm. Pharmac.* **28**, 407.
21. D. ATTWOOD (1976) *J. Phys. Chem.* **80**, 1984.
22. D. ATTWOOD, S. P. AGARWAL and R. D. WAIGH (1980) *J.C.S. Farad 1* **76**, 2187.
23. D. ATTWOOD (1976) *J. Pharm. Pharmac.* **28**, 762.
24. D. ATTWOOD and J. GIBSON (1978) *J. Pharm. Pharmac.* **30**, 176.
25. K. THOMA and K. ALBERT (1979) *Pharm. Acta Helv.* **54**, 324, 330.
26. R. J. ABRAHAM, L. J. KRICKA and A. LEDWITH (1975) *J.C.S. Perkin II* 1648.
27. F. VILALLONGA, E. FRIED and J. A. IZQUIERDO (1961) *Arch. Int. Pharmacodyn.* **130**, 260.
28. M. E. KITLER and P. LAMY (1971) *Pharm. Acta. Helv.* **46**, 483.
29. N. NAMBU, S. SAKURAI and T. NAGAI (1975) *Chem. Pharm. Bull.* **23**, 1404.
30. D. ATTWOOD, A. T. FLORENCE and J. M. N. GILLAN (1974) *J. Pharm. Sci.* **63**, 988.
31. A. T. FLORENCE and R. T. PARFITT (1971) *J. Phys. Chem.* **75**, 3554.
32. K. THOMA and M. ARNING (1976) *Arch. Pharm.* **309**, 851.
33. K. THOMA and M. ARNING (1976) *Arch. Pharm.* **309**, 837.
34. W. SCHOLTAN (1955) *Kolloid-Z* **142**, 84.
35. R. P. ENEVER, A. LI WAN PO and E. SHOTTON (1976) *J. Pharm. Pharmacol.* **28**, 32P.

36. G. ZOGRAFI and I. ZARENDA (1966) *Biochem. Pharmacol.* **15**, 591.
37. R. M. PATEL and G. ZOGRAFI (1966) *J. Pharm. Sci.* **55**, 1345.
38. G. ZOGRAFI and M. V. MUNSHI (1970) *J. Pharm. Sci.* **59**, 819.
39. G. ZOGRAFI, D. E. AUSLANDER and P. L. LYTELL (1964) *J. Pharm. Sci.* **53**, 573.
40. G. ZOGRAFI and D. E. AUSLANDER (1965) *J. Pharm. Sci.* **54**, 1313.
41. L. L. M. VAN DEENEN and R. A. DEMEL (1965) *Biochim. Biophys. Acta.* **94**, 314.
42. A. FELMEISTER and R. SCHAUBMAN (1968/69) *J. Pharm. Sci.* **57**, 178; **58**, 64, 1232.
43. D. F. SEARS and K. K. BRANDES (1969) *Agents Actions* **1**, 28.
44. J. W. CONINE (1965) *J. Pharm. Sci.* **54**, 1580.
45. A. L. THAKKAR, W. L. WILHAM and P. V. DEMARCO (1970) *J. Pharm. Sci.* **59**, 281.
46. J. H. PERRIN, W. L. WILHAM and A. L. THAKKAR (1972) *J. Pharm. Pharmac.* **24**, 258.
47. D. ATTWOOD and J. A. TOLLEY (1980) *J. Pharm. Pharmacol.* **32**, 533.
48. D. ATTWOOD and J. A. TOLLEY (1980) *J. Pharm. Pharmacol.* **32**, 761.
49. J. H. PERRIN and A. ISHAG (1971) *J. Pharm. Pharmacol.* **23**, 770.
50. P. O. P. TS'O, I. S. MELVIN and A. C. OLSON (1963) *J. Amer. Chem. Soc.* **85**, 1289.
51. D. G. DUFF and C. H. GILES (1975) in *Water: A comprehensive treatise*, (ed. F. Franks) Vol. 4, Plenum, New York, Ch. 3.
52. M. SCHUBERT and A. LEVINE (1955) *J. Amer. Chem. Soc.* **77**, 418.
53. W. H. J. STORK, G. J. M. LIPPITS and M. MANDEL (1972) *J. Phys. Chem.* **76**, 1772.
54. E. RABINOWITCH and L. EPSTEIN (1941) *J. Amer. Chem. Soc.* **63**, 69.
55. K. BERGMAN and C. T. O'KONSKI (1963) *J. Phys. Chem.* **67**, 2169.
56. E. BASWELL (1968) *J. Phys. Chem.* **72**, 2477.
57. P. MUKERJEE and A. K. GHOSH (1970) *J. Amer. Chem. Soc.* **92**, 6403.
58. A. K. GHOSH and P. MUKERJEE (1970) *J. Amer. Chem. Soc.* **92**, 6408, 6413.
59. W. SPENCER and J. R. SUTTER (1979) *J. Phys. Chem.* **83**, 1573.
60. A. ALBERT, R. J. GOLDACRE and E. HEYMANN (1943) *J. Chem. Soc.* 651.
61. D. ATTWOOD and R. NATARAJAN (1980) *J. Pharm. Pharmacol.* **32**, 460.
62. D. D. HEARD and R. W. ASHWORTH (1980) *J. Pharm. Pharmacol.* **20**, 505.
63. J. H. PERRIN and E. WITZKE (1971) *J. Pharm. Pharmacol.* **23**, 76.
64. E. A. HAUSER, R. G. PHILLIPS and J. W. PHILLIPS (1947) *Science* **106**, 616.
65. C. S. HOCKING (1951) *Nature* **168**, 423.
66. J. W. MCBAIN, H. HUFF and A. P. BRADY (1949) *J. Amer. Chem. Soc.* **71**, 373.
67. A. L. THAKKAR and W. L. WILHAM (1971) *J. Chem. Soc. Chem. Commun.*, 320.
68. E. A. HAUSER (1948) *Kolloid-Z* **111**, 103.
69. E. A. HAUSER and G. J. MARLOWE (1950) *J. Phys. Chem.* **54**, 1077.
70. W. D. KUMLER and E. L. ALPEN (1948) *Science* **107**, 567.
71. A. V. FEW and J. H. SCHULMAN (1953) *Biochim. Biophys. Acta.* **10**, 302.
72. T. KIRSCHBAUM and S. G. KAHN (1967) *J. Pharm. Sci.* **56**, 278.
73. T. KIRSCHBAUM, W. A. SLUSARCHYK and F. L. WEISENBORN (1970) *J. Pharm. Sci.* **59**, 749.
74. T. KIRSCHBAUM and A. ASZALOS (1967) *J. Pharm. Sci.* **56**, 410.
75. R. A. DEMEL, F. J. C. CROMBAG, L. L. M. VAN DEENEN and S. C. KINSKY (1968) *Biochim. Biophys. Acta* **150**, 1.
76. V. BARTHELEMY-CLAVEY, J. C. MAURIZOT, J.-L. DIMICOLI and P. SICARD (1974) *FEBS Lett.* **46**, 5.
77. S. EKSBORG (1978) *J. Pharm. Sci.* **67**, 782.
78. A. V. FEW (1957) *Second International Congress of Surface Activity* Vol. 4, Butterworths, London, p. 288.
79. A. V. FEW and J. H. SCHULMAN (1953) *Biochem. J.* **54**, 171.
80. M. C. CAREY and D. M. SMALL (1971) *J. Lipid Research* **12**, 604.
81. A. J. RICHARD (1975) *J. Pharm. Sci.* **64**, 873.
82. D. M. SMALL (1968) *Advan. Chem. Ser.* **84**, 31.
83. D. G. OAKENFULL and L. R. FISHER (1977) *J. Phys. Chem.* **81**, 1838.
84. D. G. OAKENFULL and L. R. FISHER (1978) *J. Phys. Chem.* **82**, 2443.

85. L. R. FISHER and D. G. OAKENFULL (1980) *J. Phys. Chem.* **84**, 936.
86. R. ZANA (1978) *J. Phys. Chem.* **82**, 2440.
87. A. DJAVANBAKHT, K. M. KALE and R. ZANA (1977) *J. Coll. Interfac. Sci.* **59**, 139.
88. B. LINDMAN, N. KAMENKA, H. FABRE, J. ULMIUS and T. WIELOCH (1980) *J. Coll. Interfac. Sci.* **73**, 556.
89. M. GELLERT, C. E. SMITH, D. NEVILLE and G. FELSENFELD (1965) *J. Mol. Biol.* **11**, 445.
90. W. MÜLLER and I. EMME (1965) *Z. Naturforsch B* **20**, 834.
91. H. BERG (1965) *J. Electroanal. Chem.* **10**, 371.
92. D. M. CROTHERS, S. L. SABOL, D. I. RATNER and W. MÜLLER (1968) *Biochemistry* **5**, 1817.
93. N. S. ANGERMAN, T. A. VICTOR, C. L. BELL and S. S. DANYLUK (1972) *Biochemistry* **11**, 2402.
94. T. I. ABBOT, D. P. BENTON and C. A. HAMPSON (1977) *J. Pharm. Pharmacol.* **29**, 529.
95. E. L. ROWE (1979) *J. Pharm. Sci.* **68**, 1292.
96. R. JAENICKE (1966) *Kolloid-Z* (1969) **212**, 36.
97. E. M. JOHNSON and D. B. LUDLUM (1969) *Biochem. Pharmacol.* **18**, 2675.
98. K. THOMA and C. D. HERZFELDT (1977) *Congr. Int. Technol. Pharm. 1st* **1**, 157.
99. TH. ECKERT, E. KILB and H. HOFFMANN (1964) *Arch. Pharm.* **297**, 31.
99a TH. ECKERT, R. JAENICKE and E. WACHTEL (1966) *Arzneimittel-Forsch.*, **16**, 1140.
100. E. R. HAMMARLUND and K. PEDERSEN-BJERGAARD (1958) *J. Amer. Pharm. Assoc. Sci. Ed.* **47**, 107.
101. C. ROHMANN, TH. ECKERT and G. HEIL (1959) *Arch. Pharm.* **292**, 255.
102. A. SEKERA and C. URBA (1960) *J. Amer. Pharm. Assoc.* **49**, 394.
103. J. C. SKOU (1954) *Acta Pharmacol. Toxicol.* **10**, 280, 317, 325.
104. L. HERSH (1967) *Mol. Pharmacol.* **3**, 581.
105. H. E. RIES and H. S. SWIFT (1978) *J. Colloid Interfac. Sci.* **64**, 111.
106. D. A. HAYDON and S. B. HLADKY (1972) *Quart. Rev. Biophys.* **5**, 187.
107. G. COLACICCO, E. E. GORDON and G. BERCHENKO (1968) *Biophys. J.* **8**, 22.
108. G. KEMP and C. WENNER (1972) *Biochim. Biophys. Acta* **282**, 1.
109. G. COLACICCO and E. E. GORDON (1978) *J. Coll. Interfac. Sci.* **63**, 76.
110. D. N. ELLIOTT, P. H. ELWORTHY and D. ATTWOOD (1973) *J. Pharm. Pharmacol.* **25**, 188P.
111. D. ATTWOOD and S. P. AGARWAL (1979) *J. Pharm. Pharmacol.* **31**, 392.
112. D. HELLENBRECHT, B. LEMMER, G. WIETHOLD and H. GROBECKER (1973) *Naunyn-Schmiedeberg's Arch Pharmacol.* **277**, 211.
113. J. V. LEVY (1968) *J. Pharm. Pharmacol.* **20**, 813.
114. D. ATTWOOD and S. P. AGARWAL (1980) *J. Chem. Soc. Farad. 1* **76**, 570.
115. D. GUTTMAN and T. HIGUCHI (1957) *J. Amer. Pharm. Assoc. Sci. Ed.* **46**, 4.
116. J. KIRSCHBAUM (1973) *J. Pharm. Sci.* **62**, 168.
117. A. THAKKAR, L. G. TENSMEYER and W. L. WILHAM (1971) *J. Pharm. Sci.* **60**, 1267.
118. S. NG (1971) *Mol. Pharmacol.* **7**, 177.
119. F. M. GOYAN and H. N. BORAZAN (1968) *J. Pharm. Sci.* **57**, 861.
120. A. L. THAKKAR, L. G. TENSMEYER, R. B. HERMANN and W. L. WILHAM (1970) *J. Chem. Soc. Chem. Commun.* 524.
121. P. O. P. TS'O (1968) in *Molecular Associations in Biology* (ed. B. Pullman) Academic Press, New York, p. 39.
122. W. SCHOLTAN (1960) *Kolloid-Z* **170**, 19.
123. W. SCHOLTAN and R. GONNERT (1956) *Med. Chem.* **5**, 314.
124. N. L. GERSHFELD and E. HEFTMANN (1963) *Experientia* **19**, 2.
125. J. L. TAYLOR and D. A. HAYDON (1965) *Biochim. Biophys. Acta* **94**, 488.
126. C. Y. C. PAK and N. L. GERSHELD (1967) *Nature* **214**, 818.
127. R. S. SNART and N. N. SANYAL (1968) *Biochem. J.* **108**, 369.
128. K. S. BIRDI (1976) *J. Colloid Interfac. Sci.* **57**, 228.
129. D. A. CADENHEAD and M. C. PHILLIPS (1967) *J. Colloid Interface Sci.*, **24**, 491.
130. P. DOTY, M. GELLERT and B. RABINOVITCH (1952) *J. Amer. Chem. Soc.* **74**, 2065.

131. P. DOTY and G. E. MYERS (1953) *Discuss. Farad. Soc.* **13**, 51.
132. R. F. STEINER (1952) *Arch. Biochem. Biophys.* **39**, 333.
133. D. ATTWOOD, A. T. FLORENCE, R. GREIG and G. A. SMAIL (1974) *J. Pharm. Pharmacol.* **26**, 847.
134. A. P. MINTON (1973) *J. Mol. Biol.* **75**, 559.
135. T. J. ROSEMAN and S. H. YALKOWSKY (1973) *J. Pharm. Sci.* **62**, 1680.
136. J. B. SIMS and S. L. HOLDER (1974) *J. Pharm. Sci.* **63**, 1540.
137. D. ATTWOOD (1982) *J. Chem. Soc. Farad. 1* **78**, 2011.
138. P. S. GUTH and M. A. SPIRTES (1964) *Int. Rev. Neurobiol.* **7**, 231.
139. T. BERTI and M. FERRARI (1959) *Atti. 1st Veneto Sci. Lettere Arti* **117**, 173.
140. L. STEIN (1962) in *Psychosomatic Medicine* (eds. Nodine and Moyer) Lea and Febiger, Philadelphia, Pensylvania.
141. Z. EYAL and H. EYAL-GILADI (1963) *Exptl. Cell Res.* **29**, 394.
142. M. KUROKAMA, H. NARUSE, M. KATO and T. YAKI (1957) *Folia Psychiat. Neurol. Japan.* **10**, 354.
143. W. LUEHRS, G. BACIGALUPO, B. KADENBACH and E. HEISE (1959) *Experientia* **15**, 376.
144. G. M. ALLENBY and H. B. COLLIER (1952) *Can. J. Med. Sci.* **30**, 549.
145. H. WEIL-MALHERBE and H. S. POSNER (1963) *J. Pharmacol. Exptl. Therap.* **140**, 93.
146. J. A. CHRISTENSEN and A. W. WASE (1960) *Federation Proc.* **19**, 228.
147. A. W. WASE, J. CHRISTENSEN and E. POLLEY (1956) *Federation Proc.* **15**, 496.
148. J. A. CHRISTENSEN and A. W. WASE (1959) *Federation Proc.* **18**, 204.
149. E. GROSSI, P. PAOLETTI and R. PAOLETTI (1960) *J. Neurochem.* **6**, 73.
150. M. MESSER (1958) *Australian J. Exptl. Biol. Med. Sci.* **36**, 65.
151. H. LÖW (1959) *Biochim. Biophys. Acta* **32**, 11.
152. J. K. MATHUES, S. M. GREENBERG, J. F. HERNDON, E. T. PARMELEE and E. J. VANLOON (1959) *Proc. Soc. Exptl. Biol. Med.* **102**, 594.
153. J. MOLNAR, J. KIRALY and Y. MANDI (1975) *Experientia* **31**, 444.
154. C. S. HELLER and M. G. SERAG (1966) *Appl. Microbiol.* **14**, 879.
155. S. YAMABE (1978) *Chemotherapy* **24**, 81.
156. V. WUNDERLICH and G. SYDOW (1980) *Europ. J. Cancer Chemother.* **16**, 1127.
157. J. D. ADAMS (1975) *Pharmacology* **13**, 137.
158. P. SEEMAN and J. WEINSTEIN (1966) *Biochem. Pharmacol.* **15**, 1737.
159. P. SEEMAN (1966) *Biochem. Pharmacol.* **15**, 1753.
160. P. SEEMAN (1966) *Biochem. Pharmacol.* **15**, 1767.
161. N. HALPERN, B. DREYFUS and BOURDAN (1950) *Presse Med.* **58**, 1151.
162. A. R. FREEMAN and M. A. SPIRTES (1962) *Biochem. Pharmacol.* **11**, 161.
163. H. CHAPLIN, H. CRAWFORD, M. CUTBUSH and P. O. MOLLISON (1952) *J. Clin. Path.* **5**, 91.
164. W. O. KWANT and J. VAN STEVENINCK (1968) *J. Clin. Path.* **17**, 2215.
165. D. E. HOLMES and L. H. PIETTE (1970) *J. Pharmacol. Exp. Ther.* **173**, 78.
166. A. T. FLORENCE (1970) *J. Pharm. Pharmacol.* **22**, 1.
167. I. P. LEE, H. I. YAMAMURA and R. L. DIXON (1968) *Biochem. Pharmacol.* **17**, 1671.
168. N. MOHANDAS and C. FEO (1975) *Blood Cells* **1**, 375.
169. R. A. LEWIS and F. N. GYANG (1965) *Arch. Int. Pharmacodyn.*, **153**, 158.
170. E. W. BAUR (1973) *Biochem. Pharmacol.* **22**, 1509.
171. Z. N. GANT (1973) *J. Pharmacol. Exp. Therap.* **185**, 171.
172. J. KRIEGLSTEIN and G. KUSHONSKY (1968) *Arzneim. Forsch.* **18**, 287.
173. J. KRIEGLSTEIN, F. LIER and J. MICHAELIS (1972) *N. S. Arch. Pharmacol.* **272**, 121.
174. M. FRISK-HOLMBERG and E. VAN DER KLEIGN (1972) *Europ. J. Pharmacol.* **18**, 139.
175. M. WIND, A. BERLINER and A. STERN (1973) *Res. Comm. Chem. Path. Pharmacol.* **5**, 759.
176. I. BLEI (1965) *Arch. Biochem. Biophys.* **109**, 321.
177. M. J. CARVER (1963) *Biochem. Pharmacol.* **12**, 19.
178. A. E. CLARKE, V. M. MARITZ and M. A. DENBOROUGH (1972) *Chem. Biol. Interactions* **5**, 265.

224 · *Surfactant systems*

179. M. C. CAREY, P. C. HIROM and D. M. SMALL (1976) *Biochem. J.* **153**, 519.
180. P. HELE (1964) *Biochem. Pharmacol.* **13**, 1261.
181. A. F. HARRIS, A. SCINFER and B. W. VOLK (1960) *Proc. Soc. Exp. Biol. Med.* **104**, 542.
182. P. KANTARESIA and P. MARFEY (1975) *Physiol. Chem. Phys.* **7**, 53.
183. E. TOMLINSON, S. S. DAVIS and G. I. MUKHAYER (1979) in *Solution Chemistry of Surfactants*, (ed. K. L. Mittal) Vol. 2, Plenum, New York.
184. M. AHMED, J. HADGRAFT, J. S. BURTON and I. W. KELLAWAY (1980) *Chem. Phys. Lipids* **27**, 251.
185. R. A. SCHWENDENER and H. G. WEDER (1978) *Biochem. Pharmacol.*, **27**, 2721.
186. P. SEEMAN (1977) *Biochem. Pharmacol.* **26**, 1741.
187. Z. H. SIDDICK, R. DREW and T. E. GRAM (1979) *Toxicol. Appl. Pharmacol.* **50**, 443.
188. B. TAIT (1970) *J. Pharm. Pharmacol.* **22**, 738.
189. Y. NOMURA and H. TAKAGI (1974) *Japan J. Pharmacol.* **24**, 205.
190. H. YASUHARA, H. MATSUO, K. SAKAMOTO and I. UEDA (1980) *Japan J. Pharmacol.* **30**, 397.
191. A. S. SALHAB, H. YASUHARA and C. A. DUJOVNE (1979) *Biochem. Pharmacol.* **28**, 1713.
192. C. O. ABERNATHY, L. LUKACS and H. J. ZIMMERMAN (1975) *Biochem. Pharmacol.* **24**, 347.
193. C. A. DUJOVNE (1978) *Biochem. Pharmacol.* **27**, 1925.
194. C. A. DUJOVNE et al. (1972) *J. Lab. Clin. Med.* **79**, 832.
195. C. A. DUJOVNE and A. S. SALHAB (1980) *Pharmacology* **20**, 285.
196. J. L. THISTLE and A. F. HOFFMAN (1973) *New Engl. J. Med.* **289**, 655.
197. G. B. GOLDSTEIN, K. C. LAM and S. P. MISTILIS (1973) *Digestive Dis.* **18**, 177.
198. K. G. TOLMAN, S. HAMMER and J. SAMELLA (1976) *Amer. Intern. Med.* **84**, 290.
199. C. A. DUJOVNE (1977) *Pharmacol. Res. Comm.* **9**, 1.
200. M. W. HUMBLE, S. J. EYKYN and I. PHILLIPS (1980) *Br. Med. J.* **2**, 1495.
201. H. LULLMANN, R. LULLMANN-RAUCH and O. WASSERMAN (1975) *CRC Crit. Res., Toxicol.* **4**, 185.
202. M. R. PARWARESCH, G. H. REIL and K. U. SEILER (1973) *Res. Exp. Med.* **161**, 272.
203. T. R. BLOHM (1979) *Pharmacol. Rev.* **30**, 593.
204. C. DE DUVE, T. DE BARSY, B. POOLE, A. TROUET, P. TULKENS and F. VAN HOOF (1974) *Biochem. Pharmacol.* **23**, 2495.
205. J. L. BROWNING and D. L. NELSON (1976) *Proc. Nat. Acad. Sci. USA* **73**, 452.
206. E. R. CLARK and I. E. HUGHES (1966) *Br. J. Pharmac. Chemother.* **28**, 105.
207. G. WIETHOLD, D. HELLENBRECHT, B. LEMMER and D. PALM (1973) *Biochem. Pharmacol.* **22**, 1437.
208. S. C. KINSKY, R. A. DEMEL and L. L. M. VAN DEENAN (1967) *Biochim. Biophys. Acta* **135**, 835.
209. J. KOTLER-BRAJTBURG, H. D. PRICE, G. MEDOFF, D. SCHLESSINGER and G. S. KOBAYASHI (1974) *Antimicrob. Ag. Chemotherap.* **5**, 377.
210. G. E. AMIDON, W. I. HIGUCHI, N. F. H. HO and J. D. GODDARD (1979) abstracts, *27th American Pharmaceutical Association APhSci meeting*, Kansas City, Nov. 1979.
211. G. E. AMIDON (1980) PhD Thesis, University of Michigan.
212. E. TOMLINSON, M. R. W. BROWN and S. S. DAVIS (1977) *J. Med. Chem.* **20**, 1277.
213. S. ROSS, C. E. KWARTLER and J. H. BAILEY (1953) *J. Colloid Sci.* **8**, 385.
214. P. KOELZER and J. BUECHI (1971) *Arzneim Forsch.* **21**, 1721.
215. B. W. BARRY and G. M. T. GRAY (1975) *J. Colloid Interface Sci.* **52**, 314.
216. D. G. OAKENFULL and L. R. FISHER (1977) *J. Phys. Chem.* **81**, 1838.
217. P. EKWALL, K. FONTELL and A. STEN (1957) *Proceedings 2nd International Congress on Surface Activity*, Butterworths, London, Vol. 1, 357.
218. W. T. BEHER (1976) *Bile Acids*, Karger, Basel.
219. H. IGIMI and M. C. CAREY (1980) *J. Lipid Res.* **21**, 72.
220. W. W. FALOOM, I. C. PAES, D. WOOLFOLK, H. NANKIN et al. (1966) *Ann. N.Y. Acad. Sci.* **132**, 879.

221. K. W. HEATON and J. S. MORRIS (1971) *J. Roy. Coll. Physicians, London* **6**, 83.
222. T. KIMURA, H. SEZAKI and K. KAKEMI (1972) *Chem. Pharm. Bull.* **20**, 1656.
223. H. BROCKERHOFF and R. G. JENSEN (1974) in *Lipolytic Enzymes* Academic Press, New York, 1974.
224. A. F. HOFMANN and B. BORGSTRÖM (1964) *J. Clin. Invest.* **43**, 247.
225. B. BORGSTRÖM and C. ERLANDSON (1973) *Eur. J. Biochem.* **37**, 60.
226. M. F. MAYLIE, M. CHARLES, M. ASTIER and P. DESNEULLE (1973) *Biochem. Biophys. Res. Comm.* **52**, 291.
227. B. BORGSTRÖM (1975) *J. Lipid Res.* **16**, 411.
228. B. BORGSTRÖM and J. DONNER (1975) *J. Lipid Res.* **16**, 287.
229. R. LESTER, M. C. CAREY, J. M. LITTLE, L. A. COOPERSTEIN and S. R. DOWD (1975) *Science* **189**, 1098.
230. A. VAN DEN OORD *et al.* (1965) *J. Biol. Chem.* **240**, 2242.
231. M. C. CAREY, J-C. MONTET and D. M. SMALL (1975) *Biochemistry* **14**, 4896.
232. R. J. VONK, P. JEKEL and D. K. F. MEIJER (1975) *Naunyn Schmiedeberg's Arch. Pharmacol.* **290**, 375.
233. W. G. HARDISON and J. T. APTER (1972) *Amer. J. Physiol.* **222**, 61.
234. G. E. GIBSON and E. L. FORKER (1974) *Gastroenterology* **66**, 1046.
235. W. G. HARDISON (1971) *J. Lab. Clin. Med.* **77**, 811.
236. A. F. HOFFMAN (1965) *Gastroenterology* **48**, 484.
237. T. R. BATES, M. GIBALDI and J. L. KANIG (1966) *J. Pharm. Sci.* **55**, 191.
238. T. R. BATES, M. GIBALDI and J. L. KANIG (1966) *Nature* **210**, 1331.
239. M. N. MALONE, H. I. HOCHMAN and K. A. NIEFORTH (1966) *J. Pharm. Sci.* **55**, 972.
240. R. G. STOLL, T. R. BATES, K. A. NIEFORTH and J. SWARBRICK (1969) *J. Pharm. Sci.* **58**, 1457.
241. S. MIYAZAKI, H. INOYE, T. YAMAHIRA and T. NADAI (1979) *Chem. Pharm. Bull.* **27**, 2468.
242. British Patent 378, 935, (1931).
243. Swiss Patents 126 502: 130 091/3 (1948).
244. K. P. BASU, S. MUKHERJEE and R. P. BANERJEE (1947) *J. Amer. Pharm. Assoc.* **36**, 266.
245. (1945) *J. Amer. Med. Assoc.* **128**, 495.
246. S. FELDMAN and M. GIBALDI (1968) *Gastroenterology* **54**, 918.
247. S. FELDMAN, R. J. WYNN and M. GIBALDI (1968) *J. Pharm. Sci.* **57**, 1493.
248. T. KIMURA, H. SEZAKI and K. KAKEMI (1972) *Chem. Pharm. Bull.* **20**, 1656.
249. K. KAKEMI, H. SEZAKI, R. KONISHI, T. KIMURA and A. OKITA (1970) *Chem. Pharm. Bull.* **18**, 103.
250. M. H. BICKEL and R. MUNDER (1970) *Biochem. Pharmacol.* **19**, 2437.
251. K. KAKEMI, H. SEZAKI, R. KONISHI, T. KIMURA and M. MURAKAMI (1970) *Chem. Pharm. Bull.* **18**, 275.
252. T. S. GAGINELLA, P. BASS, J. H. PERRIN and J. J. VALLNER (1973) *J. Pharm. Sci.* **62**, 1121.
253. T. S. GAGINELLA, J. H. PERRIN, J. J. VALLNER and P. BASS (1974) *J. Pharm. Sci.* **63**, 790.
254. B. W. BARRY and G. M. T. GRAY (1975) *J. Colloid Interface Sci.* **52**, 327.
255. S. SAFAIE-SHIRAZI, L. DEN BESTEN and W. L. ZIKE (1975) *Gastroenterology* **68**, 728.
256. R. B. SUND (1975) *Acta pharmacol et toxicol.* **37**, 297.
257. R. COLEMAN and D. BILLINGTON (1979) *Biochem. Soc. Trans.* **7**, 948.
258. R. COLEMAN and P. LOWE (1980) *Biochem. Soc. Trans.* **8**, 126.
259. R. COLEMAN and G. HOLDSWORTH (1975) *Biochem. Soc. Trans.* **3**, 747.
260. F. BERNHEIM and L. LACK (1967) *J. Med. Chem.* **10**, 1096.
261. G. BASLE, R. BEKE and F. BARBIER (1980) *Thrombosis & Haemostasis* **44**, 62.
262. A. F. LLEWELLYN, G. H. TOMKIN and G. M. MURPHY (1977) *Pharm. Acta Helv.* **52**, 1.
263. D. MENDELSOHN and L. MENDELSOHN (1975) *South African Med. J.* **49**, 1011.
264. E. B. FELDMANN and B. BORGSTRÖM (1966) *Biochim. Biophys. Acta.* **125**, 136.
265. P. EKWALL and H. BALTCHEFFSKY (1961) *Acta. Chem. Scand.* **15**, 1198.

266. P. EKWALL, H. BALTCHEFFSKY and L. MANDELL (1961) *Acta. Chem. Scand.* **15**, 1195.
267. D. M. SMALL, M. BOURGES and D. G. DERVICHIAN (1966) *Nature* **211**, 86.
268. D. M. SMALL and M. BOURGES (1966) *Mol. Crystals* **1**, 541.
269. M. BOURGES, D. M. SMALL and D. G. DERVICHIAN (1967) *Biochim. Biophys. Acta.* **137**, 157; **144**, 189.
270. B. ISAKSSON (1953–4) *Acta. Soc. Med. Upsal.* **59**, 277.
271. R. HOLZBACH, M. MARSH, M. OLSZEWSKI and K. HOLAN (1973) *J. Clin. Invert.* **52**, 1467.
272. D. MUFSON, K. TRIJANOND, J. E. ZAREMBO and L. J. RAVIN (1974) *J. Pharm. Sci.* **63**, 327.
273. W. I. HIGUCHI, S. PRAKONGPAN and F. YOUNG (1973) *J. Pharm. Sci.* **62**, 945.
274. S. PRAKONGPAN, W. I. HIGUCHI, K. H. KWAN and A. M. MOLOKHIA (1976) *J. Pharm. Sci.* **65**, 685.
275. K. H. KWAN, W. I. HIGUCHI, A. M. MOLOKHIA and A. F. HOFMANN (1977) *J. Pharm. Sci.* **66**, 1094.
276. A. M. MOLOKHIA, A. F. HOFMANN, W. I. HIGUCHI, M. TUCHINDA, K. FELD, S. PRAKONGPAN and R. G. DANZINGER (1977) *J. Pharm. Sci.* **66**, 1101.
277. A. HOELGAARD and S. FROKJAES (1980) *J. Pharm. Sci.* **69**, 413.
278. J. L. THISTLE, G. L. CARLSON, A. F. HOFMANN, N. F. LARUSSO, R. L. MACCARTY, G. L. FLYNN, W. I. HIGUCHI and V. K. BABAYAN (1980) *Gastroenterology* **78**, 1016.
279. J. F. VAN DEN BOSCH and P. J. CLAES (1967) in *Progress in Biochemical Pharmacology*, (eds. D. Kritchvesky, R. Paoletti and D. Steinberg) Vol. 2, p. 97, Karger, Basel, 1967.
280. R. C. POWELL, W. J. NUNES, R. S. HARDING and J. B. VACCA (1962) *Am. J. Clin. Nutr.* **11**, 156.
281. H. EYSSEN, P. CLAES, W. WUYTS, H. VANDERHAEGHE and P. DE SOMER (1975) *Biochem. Pharmacol.* **24**, 1593.
282. B. F. SCHARSCHMIDT and R. SCHMID (1978) *J. Clin. Invert.* **62**, 1122.
283. R. J. VORK, P. JEKEL and D. K. F. MEIJER (1975) *N. S. Arch. Pharmacol.* **290**, 375.
284. M. K. JAIN (1972) *The Bimolecular Lipid Membrane*, Van Nostrand Reinhold, New York.
285. D. ATTWOOD and L. SAUNDERS (1965) *Biochim. Biophys. Acta.* **98**, 344.
286. D. ATTWOOD and L. SAUNDERS (1966) *Biochim. Biophys. Acta.* **116**, 108.
287. M. P. BLAUSTEIN (1967) *Biochim. Biophys. Acta.* **135**, 653.
288. S. D. HOYES and L. SAUNDERS (1966) *Biochim. Biophys. Acta.* **116**, 184.
289. I. W. KELLAWAY and L. SAUNDERS (1967) *Biochim. Biophys. Acta.* **144**, 145.
290. L. SAUNDERS (1966) *Biochim. Biophys. Acta.* **125**, 70.
291. M. J. HUNT and L. SAUNDERS (1975) *J. Pharm. Pharmacol.* **27**, 119.
292. A. K. LOUGH and A. SMITH (1976) *Br. J. Nutr.* **35**, 89.
293. G. P. MARTIN, I. W. KELLAWAY and C. MARRIOTT (1978) *Chem. Phys. Lipids* **22**, 227.
294. J. H. FENDLER and A. ROMERO (1977) *Life Sci.* **20**, 1109.
295. K. INOUE (1974) *Biochim. Biophys. Acta.* **339**, 390.
296. K. INOUE and T. KITAGAWA (1976) *Biochim. Biophys. Acta.* **426**, 1.
297. T. KITAGAWA, K. INOUE and S. NOJIMA (1976) *J. Biochemistry (Tokyo).* **79**, 1123.
298. B. YU. ZASLAVSKY, A. A. BOROVSKAYA, A. K. LAVRINENKO, A. YU. LISICHKIN, Y. A. DAVIDOVICH and S. V. ROGOZHIN (1980) *Chem. Phys. Lipids* **26**, 49.
299. K. ANZAI, H. UTSUMI, K. INOUE, S. NOJIMA and T. KWAN (1980) *Chem. Pharm. Bull.* **28**, 1762.
300. R. HERTZ and Y. BARENHOLZ (1977) *J. Colloid Interface Sci.* **60**, 188.
301. S. YEDGAR, Y. BARENHOLZ and G. V. COOPER (1974) *Biochem. Biophys. Acta.* **363**, 98.
302. S. YEDGAR, R. HERZ and S. GATT (1975) *Chem. Phys. Lipids* **13**, 404.
303. C. F. SCHMIDT, J. K. CHUN, A. V. BROCCOLI and R. P. TAYLOR (1978) *Chem. Phys. Lipids* **22**, 125.
304. W. W. STAFFORD and C. E. DAY (1975) *Artery* **1**, 106.
305. K. KANO and J. H. FENDLER (1979) *Chem. Phys. Lipids* **23**, 189.
306. S. M. JOHNSON (1973) *Biochim. Biophys. Acta.* **307**, 27.
307. E. G. FINER and A. DARKE (1974) *Chem. Phys. Lipids.* **12**, 1.

308. Y. OKAHATA, H. IHARA, M. SHIMOMURA, S. TAWAKI and T. KUMIKATE (1980) *Chem. Lett.* 1169.

309. W. R. HARGREAVES and D. W. DEAMER (1978) *Biochemistry* **17**, 3759.

310. J. M. GEBIEKI and M. HICKS (1973) *Nature* **243**, 232.

311. A. A. AL-SADEN, A. T. FLORENCE, T. L. WHATELEY, F. PUISIEUX and C. VAUTION (1981) *J. Colloid Interface Sci.* in press.

312. J. H. FENDLER (1980) *J. Phys. Chem.* **84**, 1485.

313. B. LEWIS (1976) *The Hyperlipidaemias*, Blackwell Scientific, Oxford.

314. G. ASSMAN, H. B. BREWER and D. S. FREDERIKSON (1974) *Proc. Nat. Acad. Sci.* **71**, 1534.

315. J. L. GRANDA and A. SCANN (1966) *Biochemistry* **5**, 3301.

316. A. SCANU (1965) *Proc. Nat. Acad. Sci.* **54**, 1699.

317. A. SCANU (1965) *Adv. Lipid Res.* **3**, 63.

318. A. KELLNER, J. W. CORRELL and A. T. LADD (1951) *J. Expt. Med.* **93**, 373.

319. A. SCANU and I. H. PAGE (1961) *J. Lipid Res.* **2**, 161.

320. B. ROBERTSON (1980) *Lung* **158**, 57.

321. C. J. MORLEY, A. D. BANGHAM, P. JOHNSON, G. D. THORBURN and G. JENKINS (1978) *Nature* **271**, 162.

322. G. ENHORNING and B. ROBERTSON (1972) *Paediatrics* **50**, 58.

323. A. D. BANGHAM, C. J. MORLEY and M. C. PHILLIPS (1979) *Biochim. Biophys. Acta.* **573**, 552.

324. G. GROSSMANN, I. LARSSON, R. NILSSON, B. ROBERTSON, L. RYDHAG and P. STENIUS (1979) *Pathol. Res. Pract.* **165**, 100.

325. B. ECANOW, R. C. BALAGOT and V. SANTABIES (1967) *Nature* **215**, 1400.

326. W. F. STANASZEK, B. ECANOW and R. S. LEVINSON (1976) *J. Pharm. Sci.* **65**, 142.

327. R. S. LEVINSON, S. BUYUKYAYLACI and L. V. ALLEN (1980) *J. Pharm. Sci.* **69**, 62.

328. A. D. BASTON (1974) *Am. Rev. Resp. Dis.* **110**, 104.

329. F. BAUER (1970) *Paed. Clin North Am.* **3**, 67.

330. P. N. RAPPORT, D. J. LIM and H. S. WEISS (1975) *Arch. Otolaryngology* **101**, 305.

331. F. J. HOLLY (1974) *J. Colloid Interface Sci.* **49**, 221.

332. J. J. WOLKEN (1962) *J. Theoret. Biol.* **3**, 192.

333. P. M. BENNETT and O. MAALE (1974) *J. Mol. Biol.* **90**, 541.

334. R. J. BALDESSARINI (1975) in *Handbook of Psychopharmacology*, Vol. 3, (eds. S. D. Iverson and S. H. Snyder) Plenum Press, New York.

335. H. G. WEDER and U. W. WIEGARD (1973) *FEBS Lett.* **38**, 64.

336. I. M. KLOTZ (1967) *Science* **155**, 697.

337. C. FRIEDEN (1963) *J. Mol. Biol.* **238**, 3286.

338. H. EISENBERG (1970) *Acc. Chem. Res.* **4**, 379.

339. R. J. COHEN, J. A. JEDZINIAK and G. B. BENEDEK (1975) *Proc. Roy. Soc. Lond.* **A345**, 73.

340. J. A. LUCY (1964) *J. Theoret. Biol.* **7**, 360.

341. J. C. WATKINS (1965) *J. Theoret. Biol.* **9**, 37.

342. M. WOLMAN and H. WEINER (1963) *Nature* **200**, 486.

343. J. W. MAAS and R. W. COLEMAN (1965) *Nature* **208**, 41.

344. D. GINGELL (1973) *J. Theoret. Biol.* **38**, 677.

345. D. GINGELL (1976) in *Mammalian Cell Membranes*, Vol. 1, (eds. G. A. Jamieson and D. M. Robinson) Butterworths, London.

346. P. W. BRANDT and A. R. FREEMAN (1967) *Science* **155**, 582.

347. P. H. ELWORTHY and D. S. MCINTOSH (1964) *Kolloid-Z.* **195**, 27.

348. D. BRAY (1973) *Nature* **244**, 93.

349. R. D. NEUMAN (1975) *J. Colloid Interface Sci.* **53**, 161.

350. A. I. MCMULLEN and J. A. STIRRUP (1971) *Biochim. Biophys. Acta.* **241**, 807.

351. P. MUELLER (1975) *Ann N.Y. Acad. Sci.* **264**, 247.

352. M. C. COREY, P. HIROM and D. M. SMALL (1976) *Biochem. J.* **153**, 519.

353. R. A. O'BRIEN, M. DA PRADA and A. PLETSCHER (1972) *Life Sci.* **11**, 749.

354. A. PLETSCHER (1972) *Biochem. Psychopharmacol.* **2**, 205.
355. A. PLETSCHER, M. DA PRADA, H. STEFFEN, B. LÜTOLD and K. H. BERNEIS (1973) *Brain Res.* **62**, 317.
356. K. H. BERNEIS, U. GOETZ, M. DA PRADA and A. PLETSCHER (1973) *Naunyn-Schmiedeberg's Arch. Pharmacol.* **277**, 291.

5 Solubilization

5.1 Introduction

The increased solubility in a surfactant solution of an organic substance, insoluble or sparingly soluble in water, is a phenomenon which has been applied empirically for a very long time; it is only in the last century that an attempt at a rather less empirical approach to solubilization has been made.

Solubilization, as defined by McBain and Hutchinson [1], is 'a particular mode of bringing into solution substances that are otherwise insoluble in a given medium, involving the previous presence of a colloidal solution whose particles take up and incorporate within or upon themselves the otherwise insoluble material'. Taken literally, such a definition has the disadvantages that it is only really applicable to solutions where micelles or colloidally sized particles exist; it does not take into account any changes occurring in the structure of the colloidal particle when the third component is incorporated, or differences in the mechanism of incorporation.

Modern usage of the word solubilization appears to have given it a more general meaning, and would perhaps warrant the very broad definition of 'the preparation of a thermodynamically stable isotropic solution of a substance normally insoluble or very slightly soluble in a given solvent by the introduction of an additional amphiphilic component or components'. Such a description encompasses dilute and concentrated solutions, taking into account whether the solubilizates are polar or not, and includes hydrotropy and co-micellization, but it does not infer a similar mechanism of incorporation; simply defining the final solution as isotropic. It is not synonymous with 'dissolve', as appears from certain publications.

As discussed in Chapter 2, liquid crystalline regions such as the neat and middle phases formed in concentrated surfactant solutions are capable of incorporation of solubilizates. The ternary systems so formed represent a type of solubilized system although such systems, being anisotropic, are not strictly in accordance with the definition of a solubilized system proposed above. Relatively few studies of these systems have been reported and we shall in this chapter concentrate exclusively on solubilization within the micellar, L_1, and the reverse micellar, L_2, isotropic regions.

Early work in the field of solubilization has been adequately reviewed by

Klevens [2] and by McBain and Hutchinson [1], and for a fuller account the reader is referred to these authors. Much of this early work was on aqueous solutions of bile salts and soaps. In 1892 Engler and Diekhoff [3] showed that for the range of soaps studied, both ascending a homologous series and addition of the corresponding acid to that of the soap increased the solubilizing power, whereas addition of electrolyte decreased it.

It was not until the early 1930s, with the work of Lester-Smith [4–6] and others [7, 8], that an attempt was made to rationalize the position and correlate the increased solubility with colloidal properties. Verzar [9] came to the conclusion that the solvent action of bile salts might be due to the formation of a ring of the bile-salt molecules around the molecules being solubilized, the ionized groups of the former pointing outwards.

Lawrence [10–12] extended this work, suggesting that the increased solubility of a solubilizate in a surfactant solution was due to some form of attachment of the solubilizate to the exterior of the micelle or solution in it. He also pointed out a difference in behaviour when polar and non-polar molecules were solubilized, and the increased solubility of the soap itself with the addition of a polar solubilizate. Hartley [13] made the further contribution that solubilization occurred only above the CMC; above this, the amount of substance solubilized increased with soap concentration. This work in turn evolved a widely used technique for determining the CMC, but great care must be taken in evaluating results from this technique, as it involves the use of ternary systems to investigate a binary one.

5.2 Experimental methods of studying solubilization

5.2.1 Determination of the maximum additive concentration

The study of solubilized systems must obviously start with the determination of the concentration of solubilizate which can be incorporated into a given system with the maintenance of a single isotropic solution. This saturation concentration of solubilizate for a given concentration of surfactant is termed the maximum additive concentration (MAC).

Basically the methods of measuring this concentration are the same as those for the determination of the solubility of any compound in a given solvent with the additional difficulties due to the presence of a colloidal solution. As these solubilization properties are temperature sensitive, it is important that all these methods should be subject to adequate temperature control; a factor often missing in much experimental work.

Apparatus for the study of solubilization from the vapour phase has been described by McBain and O'Connor [14], simply measuring the volume of the gas which dissolves in the pure solvent and the solution, respectively. Determination of the MAC of liquids or solids in surfactant solutions involves the observation of a second phase.

If the refractive indices of the solubilizing solution and the solubilizate are sufficiently different, saturation is detected by the appearance of supra-colloidal aggregates with a concomitant increase in the opacity. A series of vials containing known weights or volumes of surfactant solution are prepared and varying weights or volumes of solubilizate added. The vials are sealed and then agitated at the specified temperature until equilibrium is reached. Times required to attain equilibrium varying from hours [15], up to 3 months for solid polycyclics [16], have been reported. The surfactant concentration has also been shown to influence the time required for equilibration [15]. In the simplest form the maximum concentration of solubilizate forming a clear solution can be determined by visual inspection, the experiment being repeated over a narrower range of concentrations to obtain a more precise value. Alternatively, one can plot the optical density versus solubilizate concentration; the appearance of solubilizate droplets or an anisotropic phase being accompanied by a sharp increase in the slope of the plot. The light scattered at right angles to an incident beam passed through the system either in an ultramicroscope or in a light-scattering photometer possibly provides one of the most sensitive means of detecting an increased turbidity. Microscopical examination with polarized light will reveal a departure from a single isotropic solution.

Detection of the appearance of a second phase gives a direct measure of the MAC, but methods based on this are not always suitable. Analytical determination of the amount solubilized is then required, spectrophotometry being very useful in certain cases [16–19].

An excess of solubilizate is shaken up with the surfactant solution until equilibrium is attained and the two phases separated. High-speed centrifugation has been used, particularly for liquid solubilizates or if one phase is liquid crystalline [20, 21]. Filtration through filters of 0.4–1.5 μm pore size has been used when the solubilizate is solid [22, 23]. Supersaturation is a potential source of trouble, especially as surfactants tend to inhibit crystallization, but can be minimized by adequate temperature control.

Carless and Nixon [24] have reported that methyl linoleate added directly to dilute cetomacrogol solutions formed an emulsion with no signs of solubilization, even after several months, whereas mixing of the methyl linoleate and cetomacrogol before addition of the water produced a solubilized system.

There is always the doubt, particularly in the case of solids, whether one is getting conventional solubilization in the micelles or a breaking off of colloidally sized particles of the solubilizate, which are stabilized in suspension by the adsorbed surfactant. Such particles would be small enough to pass through the commonly used filters, but would be too fine to be recognized as another phase. This effect has been shown with a dye solution [25].

Winsor [26] has described a method where the surfactant–solvent solution was titrated with solubilizate until a second phase was visually observed. Concomitant measurements of electrical resistance showed a break in the plot of resistance versus solubilizate concentration at the point where separation occurred. Titration of a concentrated surfactant–solubilizate solution with water until the appearance of turbidity has also been employed [27, 28]. As the endpoints of such titrations are taken as 'turbidity which remained for 2 minutes', these methods do not allow for sufficient time to ensure that equilibrium has been reached; suggested to be important by other workers [15, 16, 29], but with various modifications, this method has been widely used. Attempts to reduce equilibration time by addition of the solubilizate as a molecular dispersion in a volatile solvent suffer from the disadvantages that removal of this solvent gives rise to potential loss of solubilizate and water, and the very nature of the surfactant solution may render complete removal of this solvent very difficult. Melting and mixing of the solubilizate and surfactant before addition to the water does appear to reduce the equilibrium time.

Harkins *et al.* [30] have described a method based on measurements of the density of the solution as a function of solubilizate added and, for concentrated solutions [31], a technique using X-ray diffraction patterns.

5.2.2 Presentation of solubility data

When appraising results of solubilization studies one must always take into account the concentration units used. Units varying from grams solubilizate per 100 ml surfactant solution to moles solubilizate per gram micelle interior appear in the literature.

Furthermore, the means of obtaining the published maximum additive concentration, over and above variations from experimental techniques, can lead to misinterpretation of apparently comparable data. Results obtained for a given surfactant concentration need not necessarily correspond to those at a second surfactant concentration, as can be shown by the change in slope of a differential solubility curve with increasing concentration of surfactant (see Fig. 5.1).

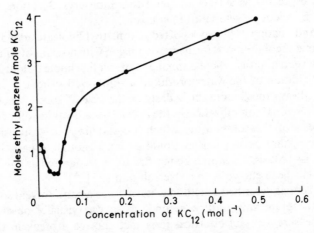

Figure 5.1 The variation in the maximum additive concentration of solubilizate (ethylbenzene) with concentration of surfactant (potassium dodecanoate, KC_{12}). From Heller and Klevens [25] with permission.

Data from solubilization measurements are best expressed either as solubility curves or phase diagrams.

(1) *Solubility curves.* Two types of solubility curves can be used: either a straightforward plot of concentration of solubilizate dissolved versus concentration of surfactant or a differential solubility curve of concentration of solubilizate dissolved/concentration of detergent versus concentration of detergent.

(2) *Phase diagrams.* Solubilized systems contain at least three components, solubilizate, solvent, and surfactant. The amount of information one can obtain from a solubility curve is limited, but the system can be completely described at a given temperature and pressure by a ternary or quaternary phase diagram (see Chapter 2).

5.2.3 Determination of micelle–water distribution equilibria

(A) EXPERIMENTAL METHODS

In solubilized systems in which the solubilizate has a significant water solubility it is of interest to know not only the distribution ratio of solubilizate between the micelles and water under saturation conditions but also at varying degrees of saturation of the system with solubilizate. Such information cannot, of course, be obtained using the solubility methods discussed in Section 5.2.1. A dialysis technique has been described by Patel and Kostenbauder [32] and with various modifications has become a widely used technique [33–41]. In principle, the surfactant solution is separated from an aqueous solution of the solubilizate by a membrane permeable to solubilizate but not to micelles. In a typical dialysis cell the membrane is clamped between two Perspex half cells of approximately 150 cm³ capacity (see Fig. 5.2). Provision is made for stirring and pH control if

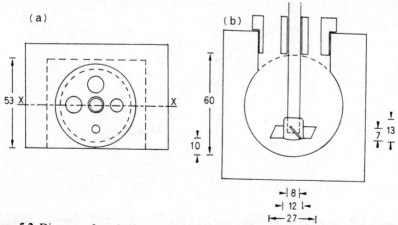

Figure 5.2 Diagram of one half cell. (a) Top view with the collar through which the stirrer shaft, pH electrodes and titrant delivery tube pass and (b) a section through X − X. Only the hydrodynamically significant dimensions are given (mm). From Withington and Collett [41].

required. Temperature control is achieved by partial immersion in a water bath. After equilibration, samples are removed from either side of the membrane and the solubilizate concentration determined using a suitable assay procedure e.g. ultraviolet absorption. The success of this method depends to a large extent on the selection of a suitable dialysis membrane. Nylon membranes have been reported to swell and to bind phenolic compounds [34], rubber membranes may vary in thickness [37] and methylcellulose membranes may be attacked in certain systems [42].

Donbrow and co-workers [43–48] and Evans [49] have used potentiometric titration for solubilization studies on ionizing solubilizates such as acids, amines and phenol. The method utilizes the displacements of the pH of titration curves of such compounds which occur in the presence of surfactant (see Fig. 5.3) to calculate the amount of solubilizate which has been removed from solution. It is applicable only to those solubilizates in which only the unionized form undergoes solubilization and if the surfactant does not influence the pK_a or ionic strength. Azaz and Donbrow [50] have re-examined the original method and have proposed a modification in which the pH difference is measured between paired solutions identical in degree of neutralization and concentration of the weak electrolyte except that one of the pair contains the surfactant in the required concentration. If the concentration of unionized and ionized acid originally present in the aqueous blank are C_a and C_s, respectively, and ΔC_a represents the amount of unionized acid solubilized in the micellar phase of the solution containing the surfactant, it may be shown that the pH difference between the

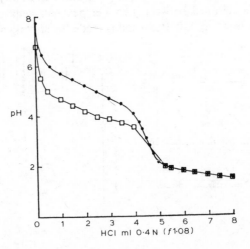

Figure 5.3 Titration of sodium benzoate (77.8 mM) with hydrochloric acid (f 1.08) (25 ml solution containing 1.95×10^{-3} mol sodium benzoate).
●—● 20% w/v cetomacrogol present.
□—□ No cetomacrogol present, precipitation occurred at about 2.5 ml.
From Donbrow and Rhodes [45] with permission.

surfactant and aqueous solution is given by

$$\Delta pH = -\log\left[\frac{C_a - \Delta C_a}{C_s}\right] + \log\left[\frac{C_a}{C_s}\right].$$ (5.1)

Hence

$$\Delta pH = -\log(1 - \Delta C_a/C_a).$$ (5.2)

Equation 5.2 is applicable at any degree of neutralization provided the paired solutions are at the same degree of neutralization with respect to the total unionized acid present. From Equation 5.2, the ratio of solubilized to free acid is

$$\frac{\Delta C_a}{(C_a - \Delta C_a)} = 10^{\Delta pH} - 1.$$ (5.3)

Hence the apparent distribution constant, P_{app}, for known volumes of the micellar and aqueous phases, V_m and V_w, is given by

$$P_{app} = V_m/V_w(10^{\Delta pH} - 1).$$ (5.4)

The equivalent equation for the solubilization of weak bases is

$$P_{app} = V_m/V_w(10^{-\Delta pH} - 1).$$ (5.5)

A molecular sieve technique has been described by Ashworth and Heard [51] and Donbrow and co-workers [42] in which the free solubilization concentration is estimated by adsorption on to a dextran gel.

An ultrafiltration technique, first described by Hutchinson and Schaffer [52] has been used by several workers [38, 53–56]. In a typical experiment a portion of the equilibrated surfactant solution containing a known amount of solubilizate and surfactant is passed through a membrane which is impermeable to micelles but which allows free passage of solubilizate molecules. Filtrand and filtrate are then analysed to determine the composition of the micellar and free phases. As with the equilibrium dialysis technique the selection of a suitable membrane is essential for the success of the technique.

Gel filtration techniques were first applied to solubilized systems by Herries *et al.* [57] and Borgström [58]. A development of this technique [59–64] involves tail analysis of the elution curves to obtain information not only about the elution behaviour of the solubilizates but also of the micelles themselves. Fig. 5.4 shows a typical elution curve of methylparaben solubilized in solutions of dodecyl-hexaoxyethylene glycol monoether, $C_{12}E_6$. The heights of the plateaux $(S)_t$ and $(D)_t$ correspond to the concentrations of $C_{12}E_6$ and methylparaben, respectively, in the original sample. $(D)_m$ and $(D)_f$ are the concentrations of methylparaben solubilized in the micellar phase and of free methyl paraben in the aqueous phase, respectively. The low plateaux of the elution curve of $C_{12}E_6$ corresponds to the CMC in the presence of methylparaben. The micellar concentration of $C_{12}E_6$, $(S)_m$, is obtained by subtraction of the CMC from $(S)_t$.

Several authors have used a fluorescence quenching method to measure the distribution of solubilizate between micelles and the bulk solution. The method is

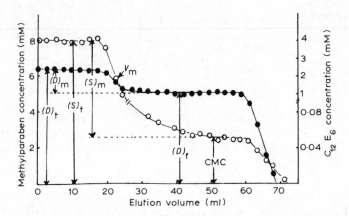

Figure 5.4 Tail analysis of the elution curves of 6.36 mM methylparaben in 4.0 mM $C_{12}E_6$ on Sephadex G-200. The sample was eluted with water on a gel column pre-equilibrated with 160 ml sample solution. (●) methylparaben in eluate; (○) $C_{12}E_6$ in eluate. From Goto *et al.* [64] with permission.

based on the observation that ionic quenching agents localized in the aqueous phase fail to quench the fluorescence of aromatic solubilizates incorporated in the micellar phase when the ion and the micelle surface are like-charged. For example, the fluorescence of anthracene solubilized in sodium dodecylsulphate, NaDS, is not quenched by iodide ions in solution [66], the fluorescence of naphthalene incorporated in NaDS is not quenched by bromide ions [67] and the fluorescence of anthracene in micelles of cetyltrimethylammonium bromide, CTAB, is not quenched by pyridinium chloride [66]. Selective quenching was used to determine the distribution between water and NaDS micelles of the fluorescence probe 4-(1-pyrene) butyric acid [65], the distribution between water and CTAB micelles of a series of alcohols of general formula $HO(CH_2)_nC_6H_5 (n = 0-3)$ [68], and the distribution constants of a series of arenes including, 1-bromo-naphthalene, 1-methylnaphthalene, biphenyl, anthracene, 4-bromo-*p*-terphenyl, benzene, toluene, *p*-xylene and pyrene in micelles of several alkyltrimethylammonium bromides and sodium alkyl sulphates [69].

A micellar isolation method using an analytical ultracentrifuge and so avoiding the inherent errors associated with the use of membranes in the dialysis and ultrafiltration techniques, has been proposed by Park and Rippie [70]. In principle the solubilized systems are centrifuged under selected conditions such that about 40 % separation of micelles is achieved. By assuming that the apparent partition coefficient is independent of surfactant concentration, equations are presented whereby the distribution of the solubilizate can be calculated from analysis of the upper and lower portions of the contents of the centrifuge tube.

(B) DATA TREATMENT

In treating data from partition experiments, most workers have utilized the phase separation model of micellization, i.e. they have treated the micelles as a separate phase and considered the distribution of solubilizate between this phase and the non-micellar fluid. The distribution coefficient, P_m, is then simply defined as [1]

$$P_m = C_3^m / C_3^a \tag{5.6}$$

where C_3^m is the moles of solubilizate per mole of micellar surfactant and C_3^a is the moles of free solubilizate per mole of water. However Equation 5.6 does not include the volumes of the aqueous or micellar phases and the values of P_m cannot, therefore, be compared with classical oil–water partition coefficients. An estimate of micellar volume can be made from the partial molar values of the surfactant and P_m may be written [71]:

$$P_m = \frac{D_b / V}{D_f / (1 - V)} \tag{5.7}$$

where D_b and D_f, are the amount of solute in the micellar and aqueous phases, respectively, V is the volume of the micellar phase and $(1 - V)$ is the volume fraction of the aqueous phase. The volume of the micellar phase, however, is somewhat arbitrary in that V could be the volume of either the hydrocarbon core, the entire micelle or the entire micelle including bound water.

The degree of solubilization of weakly ionizing substrates, for example the substituted benzoic acids, is pH dependent and the solubilization data may be treated by a method proposed by Collett and Koo [73]. If the concentration of the solubilizate in the micellar and aqueous phases is denoted by C_m and C_a, respectively, and superscripts o and − used to denote unionized and ionized species, then the ratio R of the solubility in the surfactant to that in the aqueous phase is

$$R = \left[\frac{C_m^- + C_m^o}{C_a^- + C_a^o} \right] + 1. \tag{5.8}$$

A plot of R against volume fraction, ϕ, of surfactant has a slope, S, given by

$$S = \left[\frac{C_m^- + C_m^o}{C_a^- + C_a^o} \right] \frac{1}{\phi}. \tag{5.9}$$

Values of S at different pH values can be plotted against the fraction, f_i, of ionized solubilizate present; the value of S at $f_i = 1$ is the partition coefficient of the ionized species between surfactant micelles and water.

An alternative method of expressing solubilization data, favoured by many workers [32, 33, 51, 72] uses the linear equation

$$D_t / D_f = 1 + k[M]. \tag{5.10}$$

D_t is the total solute concentration and $[M]$ is the surfactant concentration. k is a measure of the 'binding' capacity of the surfactant and is the slope of plots of D_t / D_f against $[M]$. It has been shown [40] that both the partition and simple binding approaches to solubilization, represented by Equations 5.7 and 5.10, respectively, are equivalent and fit of data to either equation does not permit any assumption to be made about the mechanism of the interaction.

Analogy between the solubilization phenomenon in which solutes are 'bound' to surfactant micelles and the binding of solutes to macromolecules such as protein led Garrett [74] to suggest the application of the classical binding equations in the treatment of solubilization data.

A convenient equation, derived from a mass-action treatment of the binding is

$$r = \frac{nK[D_f]}{1 + K[D_f]} \tag{5.11}$$

where r is the molar ratio of bound solute to total surfactant, i.e. $r = [D_b]/[M]$, n is the maximum number of independent binding sites on the surfactant and K is an intrinsic dissociation constant for the binding of solute molecules to one of the sites. Equation 5.11 has the same form as the Langmuir adsorption isotherm. Linear plots may be produced using a rearranged form of Equation 5.11 (see Fig. 5.5):

$$\frac{1}{r} = \frac{1}{n} + \frac{1}{nK[D_f]}. \tag{5.12}$$

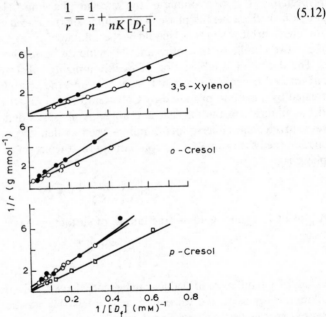

Figure 5.5 Langmuir reciprocal plots of some phenols in cetomacrogol at 25°C: (—●—●) in water, (—O—O) in 0.1N NaCl, (—□—□) in N NaCl. From Azaz and Donbrow [48] with permission.

An alternative rearrangement of Equation 5.11 is known as the Scatchard equation [75]:

$$\frac{r}{[D_f]} = nK - rK. \tag{5.13}$$

Curvature of plots of $r/[D_f]$ against r (see Fig. 5.6) is often taken as evidence of the existence of more than one type of binding site and the binding constants are

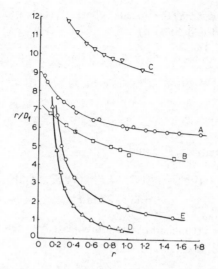

Figure 5.6 Scatchard plot for the interaction of preservatives with cetomacrogol solutions: A, benzoic acid $(r/D_f \times 10^{-1})$. B, *p*-hydroxybenzoic acid $(r/D_f \times 10^{-1})$. C, methyl *p*-hydroxybenzoate $(r/D_f \times 10^{-1})$. D, propyl *p*-hydroxybenzoate $(r/D_f \times 10^{-3})$. E, chloroxylenol $(r/D_f \times 10^{-3})$. From Donbrow and Rhodes [43, 44] with permission.

estimated from the slope of the curve in the region of interest. In some systems, two binding sites have been assumed and the binding parameters evaluated using an expanded version of Equation 5.11:

$$r = \frac{n_1 K_1 [D_f]}{1 + K_1 [D_f]} + \frac{n_2 K_2 [D_f]}{1 + K_2 [D_f]}. \tag{5.14}$$

Analysis of the data allows $n_1 \, n_2 \, K_1$ and K_2 to be estimated. The success of the Langmuir approach in describing solubilization data over a wide concentration range for a variety of systems has led several authors to suggest that solubilization may be described in terms of fixed binding sites within the micelle as indeed is implicit in the Langmuir treatment. Such a concept is, however, difficult to reconcile with the picture of an essentially fluid micelle (Section 3.2).

An alternative treatment of the micelle–water distribution equilibrium in terms of the thermodynamics of small systems (see Section 3.6) has been proposed by Mukerjee [76, 77]. The chemical potential of a solubilizate molecule (component 3) in the aqueous phase, μ_3^a, may be written as

$$\mu_3^a = \mu_3^{\ominus a}(T, P) + kT \ln x_3^a \gamma_3^a \tag{5.15}$$

where superscript a denotes the aqueous phase and $\ominus$ denotes the standard state. x_3^a is the mole fraction of the solubilizate and γ_3^a, its activity coefficient. k is the Boltzmann constant. For the solubilizate in the micellar phase we may write

$$\mu_3^m = \mu_3^{\ominus m}(T, P, \varepsilon_m) + kT \ln x_3^m \gamma_3^m, \tag{5.16}$$

where superscript m denotes the micellar phase. ε_m is the subdivision potential of the small system and is defined by

$$\varepsilon_m = -kT \ln x_m, \tag{5.17}$$

where x_m is the mole fraction of the micelle.

At equilibrium

$$\mu_3^m = \mu_3^a \tag{5.18}$$

and hence from Equations 5.15 and 5.16

$$(\mu_3^{*m} - \mu_3^{*a}) = -kT \ln (x_3^m \gamma_3^m / x_3^a \gamma_3^a)$$
$$= -kT \ln P_3. \tag{5.19}$$

P_3 is the distribution coefficient of the solubilizate between the micelles and the surrounding solution

$$P_3 = \frac{x_3^m \gamma_3^m}{x_3^a \gamma_3^a}. \tag{5.20}$$

In most real systems, particularly those involving ionic micelles and/or solubilizates, interactions between the solubilizate molecules and the micellar surfactant molecules are likely to be severe. In cases of solubilization of uncharged solubilizates by non-ionic micelles where the quantity of free solubilizate is small, γ_3^a may be considered to be unity and Equation 5.20 rewritten as

$$P'_m = \frac{xf}{c} \tag{5.21}$$

or

$$P''_m = \frac{x\gamma}{c} \tag{5.22}$$

where x is the mole fraction of solubilizate in the micelle ($= x_3^m$), c is the molar concentration of the solubilizate in the aqueous phase (proportional to x_3^a), f and γ are the activity coefficients of the solubilized component defined with respect to standard states corresponding to the pure component and infinitely dilute solution, respectively. The value of f is unity for the case of ideal mixing of components within the micelle. Mukerjee [76] has applied the regular solution theory for the calculation of f in simple cases of non-ideality of mixing. He has shown that experimental data giving the typical Langmuir type plots when plotted as x against c, are linear within experimental error when plotted as the product xf against c in accordance with the distribution law. Thus a consideration of non-ideality of mixing of components within the micelle can account for the curvature of plots of x against c without necessitating a Langmuir treatment and its accompanying implications regarding the nature of the binding to micellar sites.

5.2.4 Determination of location of solubilizate

The site of incorporation of the solubilizate is believed to be closely related to its chemical nature. In aqueous systems it is generally accepted that non-polar solubilizates, e.g. aliphatic hydrocarbons, are dissolved in the hydrocarbon core of the micelle (Fig. 5.7a). Semi-polar and polar solubilizates, e.g. fatty acids and alkanols, may be oriented radially in the micelle with the polar group either buried (deep penetration) or near the micellar surface (short penetration). Adsorption on the micellar surface has been postulated for some solubilizates, e.g. dimethylphthalate (Fig. 5.7d). It has also been suggested that certain solubilizates e.g. griseofulvin [78] and chloroxylenol [79] may be incorporated in the polyoxyethylene exterior of the micelles of non-ionic surfactants of the poly-oxyethylene type (Fig. 5.7e). The polyoxyethylene chain is thought to be arranged in an expanding spiral with the narrower end at the surface of the hydrocarbon core (see Chapter 3). As a consequence there is likely to be crowding of the chains in the region closest to the core and hence little space for hydrating water molecules. This part of the polyoxyethylene mantle adjacent to the hydrocarbon region is most likely to be almost pure polyoxyethylene (rather than polyoxyethylene–water) in nature and a preferred site for solubilizates with a high solubility in polyoxyethylene glycol.

Although in many cases a particular location is preferred, the lifetime of a solubilizate within the micelle is long enough for a rapid interchange between different locations [80]. Mukerjee [81] has discussed the solubilization of benzoic acid derivatives by a series of polyoxyethylene surfactants (the Myrj class of non-ionic surfactants) in terms of an equilibrium distribution between two loci – the polyoxyethylene shell and the hydrocarbon core. The amount of the solubilizate solubilized in the core and polyoxyethylene region was assumed to be proportional to the number of equivalents of the alkyl chain moiety (stearate), C_R, and the number of equivalents of oxyethylene groups, C_{EO}, respectively. The

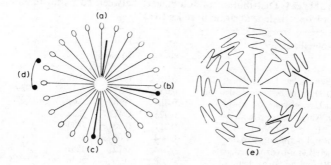

Figure 5.7 Possible sites of incorporation of solubilizate in a micelle: (a) in the hydrocarbon core; (b) short penetration of the palisade layer; (c) deep penetration of the palisade layer; (d) adsorption on the surface of the micelle; (e) in the polyoxyethylene shell of the micelle of a non-ionic detergent (a to d are adapted from the results of Riegelman *et al.* [93]).

total amount solubilized, S', is then

$$S' = aC_{EO} + bC_R, \qquad (5.23)$$

where a and b are proportionality constants. Rearrangement of Equation 5.23 gives

$$\frac{S'}{C_{EO}} = a + b\frac{C_R}{C_{EO}}. \qquad (5.24)$$

Hence a plot of S'/C_{EO} against the inverse of the EO/R mole ratio for the surfactant should be linear with an intercept, a, representing solubilization in the polyoxyethylene palisade layer and a slope, b, representing solubilization in the core (Fig. 5.8). The ratio of the amount solubilized in the polyoxyethylene shell to that in the core, the Y values of Table 5.1, shows good qualitative

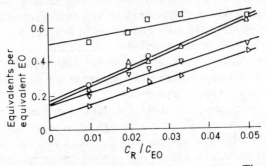

Figure 5.8 Micellar solubilization in polyoxyethylene stearates. The number of equivalents solubilized per equivalent of oxyethylene group is plotted as the ordinate against the stearate–ethylene oxide mole ratio for the surfactants. $\square$, *p*-hydroxybenzoic acid; $\bigcirc$, benzoic acid; $\triangle$, *o*-hydroxybenzoic acid; $\triangledown$, methyl *p*-hydroxybenzoate; and $\triangleright$, propyl *p*-hydroxybenzoate. From Mukerjee [81] with permission.

Table 5.1 Distribution of solubilizate between core and mantle of polyoxyethylene stearate micelles [81]

Solubilizate	a* eq./eq.	b† eq./eq.	Y‡	a'§
Benzoic acid	0.016_9	0.98_2	0.69	—
p-Hydroxybenzoic acid	0.050_8	0.37_7	5.4	0.029
o-Hydroxybenzoic acid	0.015_1	0.98_5	0.61	0.0070
p-Aminobenzoic acid	0.025_9	0.47_4	2.2	—
Ethyl *p*-aminobenzoate	0.011_9	0.56_5	0.84	—
Butyl *p*-aminobenzoate	0.004_2	1.01_9	0.16	—
Methyl *p*-hydroxybenzoate	0.014_6	0.72_4	0.81	0.0073
Propyl *p*-hydroxybenzoate	0.006_9	0.76_3	0.36	0.0013_5

* Amount solubilized in the mantle per equivalent of ethylene oxide. † Amount solubilized in the core per equivalent of stearyl group. ‡ Ratio of the amount in the mantle and the amount in the core calculated for stearyl (EO)$_{10}$ micelles. § Binding in polyethylene glycol 4000 in equivalents per equivalent of oxyethylene group.

correlation with the structure of the solubilizate. Thus benzoic acid which has a polar group at only one end is located mainly in the polyoxyethylene region (low Y value). p-hydroxy and p-amino benzoic acids, on the other hand, have polar groups on both ends of the molecule and consequently have lower solubility in the core and a greater solubility in the polyoxyethylene region giving higher Y values than benzoic acid. Esterification of either the p-hydroxy or p-amino benzoic acids makes the compounds more hydrophobic and this is accompanied by a decrease in Y. Other workers [82, 83] have described the distribution of solubilizate within non-ionic micelles using Equation 5.24.

(A) X-RAY DIFFRACTION

X-ray-diffraction techniques have been used extensively in the investigation of the position of the solubilizate, and although the method of interpretation of the data obtained has been the subject of varying opinion [84–86], the information obtained does show an appropriate qualitative differentiation between various types of solubilizate.

Interpretation of X-ray diffraction has been mainly based on adaptations of Bragg's relationship:

$$n\lambda = 2d \sin \theta \qquad (5.25)$$

where d is the distance between two parallel planes; n is an integer; θ is the angle of incidence to the plane of a beam of X-rays of wavelength λ.

In addition to diffraction caused by the solvent, three diffraction bands have been attributed to the surfactant:

(1) The 'S' or short spacing band; generally giving a repeating distance of 0.4–0.5 nm. It is thought to correspond to the thickness of the hydrocarbon chain.

(2) The 'M' or micelle thickness band; this band varies with the surfactant chain length and gives a value slightly less than twice the extended length of the hydrocarbon chain.

(3) The 'I' or long spacing band; this is greater than either the 'M' or 'S' bands, but unlike them, it is sensitive to surfactant concentration.

Both 'M' and 'I' bands show an increase in length of spacing with the addition of apolar solubilizates, but show little difference or a slight increase with the addition of polar solubilizates [87].

X-ray diffraction results were originally thought to be confirmation of the laminar structure of micelles, but it has been pointed out [88] that to account for all the water in the system studied, the laminae would have to be separated by a greater depth of water than that calculated to be between the ionic faces. Furthermore, the increase in long spacing obtained on solubilizing an apolar substance was greater than could be accounted for as additional volume inside the micelles. On these grounds, and drawing a comparison between the solubilized systems and the emulsions studied by Schulman and Riley [89], Hartley [86] argued that the long spacing should be considered as a regular distance, d, between geometrically similar groups of diffracting atoms. Spherical micelles fitted the experimental data most satisfactorily.

On the assumption of spherical micelles, their radius could be obtained from the long spacing using:

$$d = \left(\frac{8\pi}{3\sqrt{2}}\right)^{\frac{1}{2}} \phi^{-\frac{1}{3}} r \qquad (5.26)$$

where r is the radius of a sphere occupying a fraction, ϕ, of the total volume.

A different small-angle X-ray scattering technique involving the measurement of intensities on an absolute scale was applied to micellar systems by Reiss-Husson and Luzzati [90, 91] and later used in a study of solubilization by Svens and Rosenholm [92]. Curves of scattering intensity $I(S)$ as a function of $2 \sin \theta / \lambda$ for many surfactant solutions containing micelles show a weak and diffuse maximum in the small-angle region. Fig. 5.9 shows the effect which different additives have on the intensity and position of this maximum. Such changes have been interpreted in terms of changes in the radii and electron densities of the core and polar regions of the micelle and hence may often give a good indication of the location of the solubilizate molecule within the micelle. Table 5.2 shows the changes which occur in these parameters following the solubilization of N,N'-dimethylaniline, decanol and p-xylene by the micelles of sodium octanoate. The only changes which occur when dimethylaniline and also low concentrations of decanol are solubilized are small increases in the radius of the polar regions suggesting that these weakly polar solubilizates are built into the polar layer. The considerable increase in the intensity maximum and its shift to lower angles (Fig. 5.9) as the concentration of added decanol increases is due to a swelling of the micelle (increase of the radii of both core and polar region).

Table 5.2 Effect of solubilizates on the radii of the core and polar regions of the micelles of sodium octanoate NaC_8 ($1.2 \, mol \, l^{-1}$) as calculated from X-ray scattering data (adapted from [92])

Solubilizate	Concentration mmol solubilizate/ mol NaC_8	R_{pol}* (nm)	R_{par}[†] (nm)
N,N'-dimethyl aniline	0.933	2.75	0.60
	1.323	2.75	0.60
	1.756	2.80	0.60
Decanol	0.927	2.80	0.65
	1.385	2.95	0.65
	1.762	3.00	0.70
	4.998	3.55	0.80
	7.154	3.80	0.90
	9.708	4.05	1.00
p-xylene	0.296	2.65	0.60
	0.705	2.70	0.65
	1.026	2.75	0.60
	1.333	2.75	0.75

* $R_{pol} = R_{par} +$ thickness of polar layer.
[†] $R_{par} =$ radius of paraffin chain region.

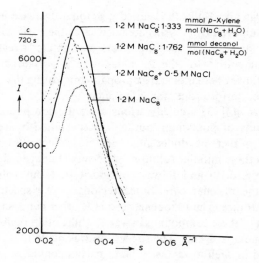

Figure 5.9 The influence of different additives on the scattering curve $I(s)$ of a 1.2 M sodium octanoate solution. From Svens and Rosenholm [92] with permission.

p-Xylene is absorbed to a slight degree in the polar layer (there is a slight increase in the total radius of the core and polar layer) while the larger part is solubilized in the micellar core.

(B) ABSORPTION SPECTROMETRY
The amount of vibrational fine structure shown in the ultraviolet absorption spectrum of a compound in solution is a function of the interaction between solvent and solute. Moreover, the extent of interaction between the solvent and solute increases with increasing solvent polarity, thereby decreasing the fine structure. As the micelle is characterized by regions of different polarity, ultraviolet spectra have been used as a means of obtaining information on the environment of the solubilizate in the micelle [18, 79, 93–101] Bjaastad and Hall [102] utilized an empirical method of determining the polarity of the micro-environment of the solubilizate molecule based on the so-called Z-value method of Kosower [103, 104].

The Z value was defined by Kosower as the energy of electronic transition corresponding to the charge-transfer absorption band of 1-ethyl-4-carbo-methoxy pyridinium iodide, a substance which has an ultraviolet spectrum remarkably sensitive to solvent polarity. Application of this method involves the determination of the spectral characteristics of the solubilizate in a number of solvents of known Z values and the establishment of a relationship between Z values and the energy of transition which is readily obtained from the wavelength of maximum absorbance. From this relationship the Z value of the micellar micro-environment may be determined from the wavelength of maximum absorbance of the solubilizate when solubilized within the micelle. Utilizing this method, Thakkar and Hall [105] noted a gradual decrease in the Z value of testosterone solubilized by polysorbate micelles as the surfactant concentration was increased, indicating that the testosterone encountered a region of decreasing polarity. The magnitude of the Z values in 2% polysorbate solutions was comparable to those of ethanol and isopropanol indicating that the environment of the solubilized testosterone was quite polar.

Riegelmann *et al.* [93] studied various aromatic compounds solubilized in aqueous solutions of potassium laurate, dodecylamine hydrochloride, and a polyoxyethylene ether of dodecanol (Brij 35). The ultraviolet spectra of ethylbenzene in these micellar solutions were very similar to those in non-polar solvents (see Fig. 5.10) and it was concluded that this solubilizate resided completely in the micellar core. Some regions of the spectra of solubilized naphthalene, anthracene and azobenzene, on the other hand, showed similarities with the spectra of these compounds in water, whilst other regions resembled the spectra in non-polar solvents. The suggestion was made that these compounds were solubilized in such a way as to be in partial contact with both the polar micellar surface and the non-polar micellar core, i.e. a position of 'deep penetration'. By similar reasoning it was concluded that *o*-nitroaniline was located at a position of 'short penetration' whereas dimethylphthalate, whose spectrum closely resembled the spectrum in water, was thought to be adsorbed on the micellar surface.

In an earlier paper, Mulley and Metcalf [79] suggested that chloroxylenol solubilized in micelles of cetomacrogol formed hydrogen bonds between the phenolic hydroxyl groups and the oxygen atoms of the polyoxyethylene chains. Donbrow and Rhodes [94] suggested from an examination of the solubilization of benzoic acid in cetomacrogol that the position of the benzoic acid molecules in

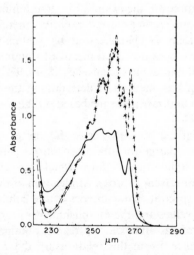

Figure 5.10 The ultraviolet spectrum of ethyl benzene in water (——), octane (– – –) and 0.28 M potassium laurate (-x-x-x). From Riegelman *et al.* [93] with permission.

the micelle is at the junction of the hydrocarbon core and the polyoxyethylene chains with the benzene ring enclosed in the former and the carboxylic acid group protruding outwards: such a position would still allow for hydrogen bonding between the carboxyl group and the innermost ether oxygen. These deductions were substantiated by n.m.r. spectral studies.

Rehfeld [95, 96] has used differential spectroscopy to establish the site of solubilization of benzene in micelles of sodium dodecyl sulphate saturated with this solubilizate. Comparison of the spectra with those of benzene dissolved in hexane, water, methanol, 1-dodecanol and n-dodecane led to the conclusion that the solubilized molecules were situated within the hydrocarbon core at different distances from the polar surface of the micelle. A similar location was indicated for benzene molecules solubilized within micelles of cetyltrimethylammonium bromide (CTAB). These conclusions regarding the solubilization site of benzene in CTAB are, however, at variance with the n.m.r. results of Eriksson and Gillberg [106] who reported that benzene, when present at low concentrations, resides mainly at the micellar surface and Gordon et al. [107] who also found rather polar environments for the solubilized benzene molecules. The reason for these discrepancies has been attributed by Mukerjee and Cardinal [100, 101] to differences in the amount of benzene in the solubilized systems. These workers determined the average polarities of the micro-environments of benzene, several of its derivatives, Triton X-100 and naphthalene when solubilized in micelles at both low and high solubilizate to surfactant ratios. The ultraviolet spectral studies were based on the observation that some spectral parameters for the aromatic molecules, obtained as ratios of absorbances at two selected wavelengths, are highly sensitive to the polarity of some reference solvents. A calibration curve relating this absorbance ratio to dielectric constant was prepared using a range of reference solvents of suitable dielectric constant and used to determine the effective dielectric constant, ε_{eff}, of the aromatic molecule when solubilized in several micellar systems. Table 5.3 shows that the ε_{eff} of solubilized benzene is comparable to that of Triton X-100 which is known to be solubilized exclusively at the micellar–water interface. Progressive alkyl substitution on the benzene leads to a concomitant decrease in polarity suggesting a move from the surface to the hydrocarbon core of the micelle as the hydrophobicity of the solubilizate is increased. The results were rationalized by consideration of the interfacial activity of benzene and its alkyl substituted derivatives. Because of the extremely high surface to volume ratio of the micelles, even the mild interfacial activity of benzene was sufficient to cause preferential adsorption at the surface. The less pronounced interfacial activity of the benzene derivatives was in accordance with the increased tendency for location in the core rather than the surface. Mukerjee and Cardinal found evidence that as the mole fraction of solubilizate was increased to values approaching 0.7, the solubilization in the hydrocarbon core became important. These findings may resolve the controversy between those workers who, using low mole fractions, reported solubilization of benzene at the surface, and other workers using high mole fractions who concluded that solubilization occurred in the micellar core.

Table 5.3 Effective polarities of the micro-environments of solubilized species [77]

Species	Micelles	ε_{eff}	Interfacial tension (mN m^{-1})
Benzene	NaDS*	49	34.1
	CTAC†	49	
p-Xylene	NaDS	40	37.8
Ethylbenzene	NaDS	41	37.7
n-Butylbenzene	NaDS	36	41.4
Hexamethylbenzene	NaDS	23	
p-Di-t-butylbenzene	NaDS	16	
β-Phenethyl alcohol	NaDS	47	
Triton X-100	NaDS	42	
	NaMS‡	42	
	CTAC	43	
	Brij-35§	39	
Naphthalene	NaDS		

* Sodium dodecyl sulphate. † Cetyl trimethylammonium chloride. ‡ Sodium myristyl sulphate. § Non-ionic surfactant with average composition of $C_{12}H_{25}(OCH_2CH_2)_{23}OH$.

(c) NUCLEAR MAGNETIC RESONANCE

NMR was first applied to solubilized systems by Eriksson [108] to study the solubilization of benzene and bromobenzene in cetyltrimethylammonium bromide (CTAB) solutions.

Eriksson and Gillberg [106] later expanded this work to the solubilization of cyclohexane, isopropylbenzene, benzene, *N,N*-dimethylaniline, and nitrobenzene in 0.1729 M CTAB.

Extrapolation of solubilizate peak positions, measured over a range of concentrations down to zero solubilizate concentration, suggested that, with the exception of cyclohexane, the first molecules of solubilizate added were adsorbed at the micelle–water interface. Values for isopropylbenzene showed that the molecule was oriented with the aromatic ring towards the water and the isopropyl group towards the centre of the micelle.

Rapid shifts of the N–CH$_3$, CH$_2$, and especially the αCH$_2$ hydrogen resonance lines to higher fields on the further addition of benzene and of *N,N*-dimethylaniline indicated the predominating solubilization mechanism to be adsorption close to the αCH$_2$ groups; the magnitude of the αCH$_2$ shifts suggested this was a combination of the effect of the aromatic ring and removal of water from this region. The change in the slope of the shift curves at concentration ratios of approximately 1.0 mole benzene, and 0.7 mole *N,N*-dimethylaniline, respectively, per mole CTAB was interpreted as saturation of the adsorption near αCH$_2$ sites and a transition to solubilization in the centre of the micelle.

CH$_2$ and αCH$_2$ hydrogen signals shifted rapidly to higher fields, but the NCH$_3$ resonances were hardly affected by further addition of nitrobenzene; suggesting orientation of the nitro group in the interface and the aromatic ring near the

αCH_2. Transition to solution in the centre of the micelle appeared to occur at about 0.85 mole nitrobenzene per mole CTAB.

Data on isopropylbenzene pointed to a gathering of the solubilizate molecules at the centre of the micelles subsequent to the initial interfacial adsorption at very low solubilizate content.

Jacobs *et al.* [109] have used ^1H n.m.r. in an investigation of the solubilization of phenol by micelles of sodium dodecyl sulphate (NaDS). Fig. 5.11 shows the observed upfield shift of the aromatic protons of phenol as the concentration of NaDS is increased. The results are indicative of a change to a more hydrophobic environment. It is interesting to note that the meta and para protons experience a greater shift than the ortho protons implying a smaller environmental change for the latter. It was concluded that the phenol was solubilized in such an orientation that the hydroxyl group was closest to the polar micellar surface.

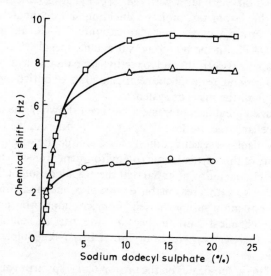

Figure 5.11 Upfield chemical shifts of the aromatic protons of 2 % phenol (relative to 2 % phenol in water) as a function of sodium dodecyl sulphate concentration: □, meta protons; △, para protons; ○, ortho protons. From Jacobs *et al.* [109] with permission.

The solubilization of benzene and nitrobenzene in the zwitterionic surfactant 3-(dimethyldodecylammonium)propane 1-sulphonate (DDAPS) has been reported by Fendler *et al.* [110]. Upfield shifts of the proton magnetic resonance frequencies of benzene and nitrobenzene protons and also those of the terminal CH_3, the dodecyl CH_2, the $N(CH_3)_2$ and the CH_2SO_3 protons of the surfactant showed a linear dependence on the surfactant concentration expressed by

$$v = v_0 + a[X] \tag{5.27}$$

where v and v_0 are the observed shift and the extrapolated shift at zero concentration of X, respectively. X denotes either the solubilizate or the surfactant. Similar linear relationships have been noted for the solubilization of these two solubilizates in micelles of CTAB [106], and sodium decanoate, cetamacrogol, tetramethylammonium bromide and tetra-*n*-butylammonium bromide [107]. From a comparison of *a* values, Fendler and Fendler [111] conclude that benzene is probably solubilized near the surface of CTAB and DDAPS micelles whereas it is closer to the interior in NaDS micelles.

The site of solubilization of acetophenone and benzophenone in micellar CTAB, hexadecylpyridinium chloride, NaDS, DDAPS, and polyoxyethylene (15) nonylphenol (Igepal CO-730) has been investigated by Fendler *et al.* [97] using ^1H n.m.r. and ultraviolet spectroscopic techniques. The concentration dependence of chemical shifts of solubilizate and surfactant protons was linear, i.e. Equation 5.27 was obeyed in all systems, within given concentration ranges. The results indicated that both solubilizates have average locations between the micellar core and the Stern layer with the carbonyl group orientated toward the surface and the aromatic moiety shielding approximately one-half of the methylene protons in the micellar cationic CTAB, anionic NaDS and zwitterionic DDAPS. In the non-ionic surfactant, Igepal CO-730, the solubilizates were located between the polyoxyethylene palisade layer and the hydrocarbon core. In general, benzophenone appeared to be buried more deeply than acetophenone in all the micellar systems.

In recent years a great deal of interest has been shown in the use of probes to investigate micellar structure (see Chapter 3). Probes suitable for such investigations usually display a selective affinity for a unique site within the micelle and reflect the nature of this particular micro-environment. For such investigations to be meaningful it is, of course, essential that the exact location of the probe within the micelle is known with a reasonable degree of confidence. Grätzel *et al.* [112] report an investigation using pulsed Fourier transform proton magnetic resonance and chemical shift analysis of the site of solubilization of the photoactive probes, pyrene, pyrene butyrate (PBA) pyrene sulphate (PSA) within CTAB micelles.

The effect of the solubilizates on the proton n.m.r. spectrum of CTAB is shown in Fig. 5.12. The most prominent changes occur in the methylene resonances following solubilization. Pyrene dramatically affects the CH_2 line by resolving it into two well-defined peaks. Less pronounced splitting is caused by PBA and only broadening toward higher fields by PSA. Splitting of the methylene peak which has also been noted in CTAB solutions containing solubilized benzoic and toluic acids [113, 114] is thought to arise when the aromatic ring of the solubilizate is located in close proximity to the inner methylene groups which are responsible for this peak. The absence of splitting in the case of PSA was then taken to imply that only inefficient contact exists between the inner methylene groups and the pyrene residue, i.e. that PSA is located in the surface region of the micelle. In contrast, the strong splitting which is noted when pyrene is solubilized is indicative of frequent contact of the solubilizate with the inner methylene groups, i.e. a solubilization site in the micellar core. Finally the splitting produced by PBA indicates location of the pyrene ring in an intermediate position: more deeply

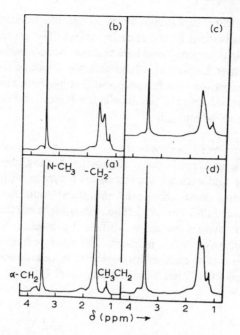

Figure 5.12 Proton nmr spectra of CTAB in solubilizate-free (a) solution and in the presence of pyrene (b), PSA (c), and PBA (d): CTAB = 0.1 M; solubilizates = 0.01 M. From Grätzel *et al.* [112] with permission.

buried than PSA but with less access to inner CH_2 groups than pyrene. This interpretation of the n.m.r. data has been questioned by Ulmius *et al.* [115] who noted a similar peak splitting of the main methylene peak following solubilization of 1-methylnaphthalene in CTAB micelles. These authors have argued that the marked upfield shift of the $N^+(CH_3)_3$ signal and the downfield shift of the ω–CH_3 group which is noted with pyrene and 1-methylnaphthalene solubiliz-ation is indicative of contact between the $N^+(CH_3)_3$ group and the aromatic ring rather than the ω–CH_3 group and the ring as originally suggested. Ulmius *et al.* conclude that both pyrene and 1-methylnaphthalene are solubilized towards the polar part of CTAB. Support for this conclusion came from ^{13}C n.m.r. spectra which showed that 1-methylnaphthalene affected appreciably only the shifts and widths of signals corresponding to groups at or close to the polar end of the CTAB molecule. Further indications of a solubilization site for pyrene close to the micellar surface of CTAB came from the fluorescence depolarization measurements of Dorrance and Hunter [116] which are discussed below. In contrast with CTAB micelles, solubilization of pyrene and 1-methylnaphthalene in NaDS micelles did not produce any splitting of the methylene signal [117].

A novel approach to the problem of determination of solubilization site has been reported by Fox *et al.* [118]. The solubilization of *p*-xylene in NaDS micelles was studied by a n.m.r. technique utilizing the broadening of the *p*-xylene lines (both methyl and aromatic protons) caused by paramagnetic ions such as Mn^{2+}, Gd^{3+} and Cu^{2+} adsorbed at the micelle surface. The distance of separation between the adsorbed paramagnetic ions and the solubilized p-xylene molecules, i.e. the depth of penetration of the solubilizate was obtained from measurements of the spin–spin relaxation rate. It was found that p-xylene cannot be considered to be solubilized at the centre of the micelle but instead is distributed uniformly throughout the micellar interior.

(D) FLUORESCENCE DEPOLARIZATION

Dorrance *et al.* have presented a series of papers reporting results of observations of both monomer and excimer fluorescence of pyrene within micelles of the cationic surfactants, decyl-, dodecyl-, tetradecyl- and hexadecyl trimethyl-ammonium bromide [116, 119–121]. The concept is presented of a shell within which the planar pyrene molecules diffuse by essentially two-dimensional slippage through the hydrocarbon chains. The measurement of the temperature dependence of the rate of excimer formation in micelles containing two pyrene molecules revealed [121] that this mobility shell did not occupy the whole

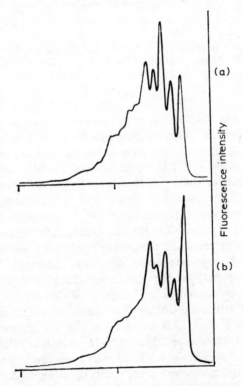

Figure 5.13 Monomer fluorescence spectrum of pyrene in (a) hexadecane and in (b) CTAB showing the Ham effect on the vibronic band intensities. From Dorrance and Hunter [116] with permission.

micellar volume and that mobility through the micellar centre was not allowed. The suggested shell was thought to be located close to the double layer. Further information on the location of the solubilizate molecule was obtained by utilization of the so-called 'Ham effect' [116]. The vibronic structure of the absorption and emission spectra of high symmetry aromatic molecules such as benzene, naphthalene and pyrene show interesting features. Two sets of vibronic bands are noted in pyrene (Fig. 5.13) corresponding to totally symmetric and non-totally symmetric vibrations. When placed in a solvent containing highly polarizable groups the totally symmetric part of the spectrum increases in intensity. This is the Ham effect. The ratio of the height of one of the bands in this spectral region to the total area under the monomer fluorescence spectrum is directly correlated with the dielectric constant of the environment around the pyrene molecule [122]. From comparisons of the values of this ratio for pyrene solubilized in the cationic micelles with values obtained for solvents of known dielectric constant it was concluded that the dielectric constant of the micro-environment of pyrene ranged from 4 in hexadecyl- to 6 in decyl-trimethyl-ammonium bromide, i.e. the pyrene molecules diffused in a shell fairly close in its outer boundary to the double layer.

(E) ELECTRON SPIN RESONANCE

The location of free radical solubilizates within the micelle has been determined from electron spin resonance (e.s.r.) spectra. As discussed in Chapter 3, these solubilizates can provide information on the nature of the various micro-environments of the micelle. Several workers have used nitroxide spin probes, the spectra of which are characterized by three sharp lines produced by nitrogen hyperfine interaction (Fig. 5.14). The distance between the resonance lines is determined by the hyperfine coupling constant, a_N, and this provides a sensitive

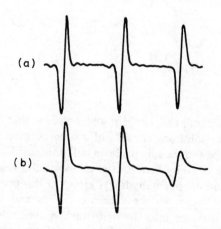

Figure 5.14 X-band spectrum of nitroxide I at 23° C in (a) water and (b) water containing 5 % NaDS. The distance between the outermost lines of (a) is 32.3 G. From Waggoner [80] with permission.

measure of the environment of the probe. Waggoner *et al.* [80] utilized the stable free radicals, the 2,4-dinitrophenyl hydrazone of 2,2,6,6-tetramethyl-4-piperidone nitrogen oxide (I) and 2,2,4,4-tetramethyl-1,2,3,4-tetrahydro-γ-carboline-3-oxyl (II).

(I)   (II)

The hyperfine coupling constants of both (I) and (II) within the micelles of NaDS were compared with those of the probes in water and in dodecane. The coupling constants of the solubilized probes lay between the extreme values in these solvents, being closer to the values observed in water. Similar conclusions were drawn from an examination of ultraviolet spectra arising from the chromophores in these two probes. The results were interpreted to indicate a relatively polar environment for the probes.

The solubilization of two further nitroxide probes, (III) and (IV) by NaDS was investigated by Waggoner *et al.* [123] and these too were thought to be located in a relatively polar environment.

(III)   (IV)

Later studies by Oakes [124, 125] showed, however, that the e.s.r. spectrum of (I) in NaDS micellar solutions consists of a superposition of spectra from the solubilized spin probe and the spin probe in bulk solution. Further measurements of the solubilization of (I) by NaDS, after allowance for this partitioning effect, suggested that the dinitrophenylhydrazyl group of this probe is in the micellar surface, probably adjacent to the sulphate groups and the nitroxide group penetrates a small distance into the hydrocarbon core. The nitroxide (V) was shown to be located in the micelle surface and (VI) was probably incorporated into the NaDS micelle in the same way as the surfactant molecule, with its nitroxide in the surface.

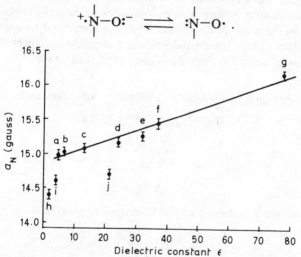

(V)   (VI)

The relationship between the hyperfine coupling constant a_N and the dielectric constant ε of the solvent which was used by the above workers in describing the micro-environment of the solubilizate was examined in more detail by Yoshioka [126]. Although Fig. 5.15 shows a linearity between a_N and ε for probes in solvents possessing a hydroxyl group, the a_N values in these solvents were, however, larger than those obtained by the assumption that a_N was determined only by the polarity of the solvent. The explanation proposed for this discrepancy is as follows. The N–O group of probes is in equilibrium with the following two structures,

$$\overset{+}{\cdot}\text{N}-\text{O:}^- \rightleftharpoons \text{:N}-\text{O}\cdot \ .$$

Figure 5.15 Relationship between a_N and dielectric constant. a, chloroform; b, dodecanol; c, hexanol; d, ethanol; e, methanol; f, ethylene glycol; g, water; h, dodecane; i, ethylene glycol dimethyl ether; j, acetone. From Yoshioka [126] with permission.

A hydroxyl group must stabilize a polarized structure through the hydrogen bond and the density of the unpaired electron increases, resulting in the increase in a_N

$$\overset{+}{\cdot}\text{N}-\text{O:}^- \cdots\cdots \text{H}-\text{O}- \ .$$

Thus although acetone is more polar than chloroform, it has a lower a_N value because unlike chloroform it has no active hydrogen with which to form H bonds.

Table 5.4 a_N in water, dodecane, and NaDS micelle [126]

Probe	Water	Dodecane	NaDS micelle	Deviation from water (%)*
VII	15.93	14.10	15.29	35
VII + NaOH			15.29	35
VIII	15.93	14.10	15.25	37
VIII + NaOH			15.43	27
IX		14.20	15.43	
X	16.25	14.33	15.79	24
XI	16.16	14.38	15.75	23

* This is the value of $[a_N \text{ (water)} - a_N \text{ (micelle)}]/[a_N \text{ (water)} - a_N \text{ (dodecane)}] \times 100$.

Yoshioka studied the location of several spin probes in NaDS micelles (see Table 5.4). The values in the last column indicate how a_N in the micelle deviates from that in water. In general the probes are located in a reasonably polar environment, i.e. close to the micelle surface. The fact that the deviation of probes (VII) and (VIII) from water was larger than that of probes (X) and (XI) is consistent with the greater hydrophobicity of these probes. Ionization of the carboxylic acid group of probe (VIII) by the presence of NaOH results in a solubilization site closer to the surface than that of the non-ionized analogue, as might be expected, although a similar change of solubilization site is not noted with probe (VII).

2-(3-carboxypropyl)-4,4-dimethyl-2-tridecyl-3-oxazolidinyloxyl

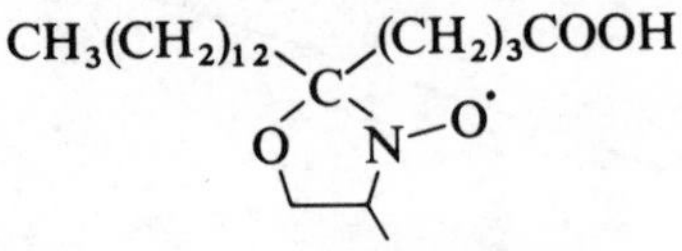

(VII)

2-(14-carboxytetradecyl)-2-ethyl-4,4-dimethyl-3-oxazolidinyloxyl

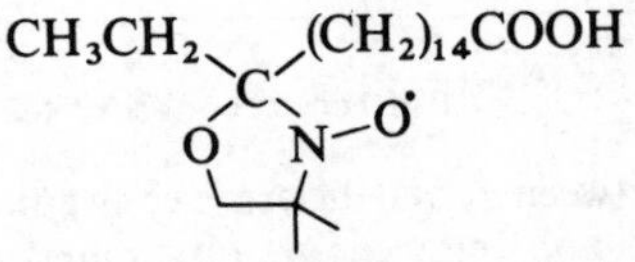

(VIII)

17β-hydroxy-4',4'-dimethylspiro[5α-androstan-3,2-oxazolidin]-3'-yloxyl

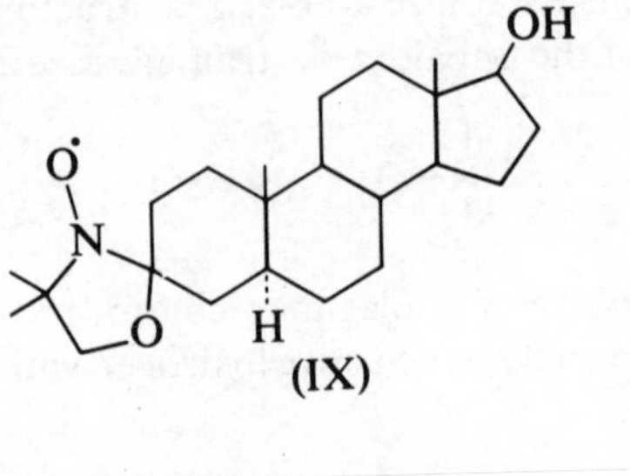

(IX)

2,4-dinitrophenylhydrazone of 2,2,6,6-tetramethyl-4-piperidone *N*-oxide

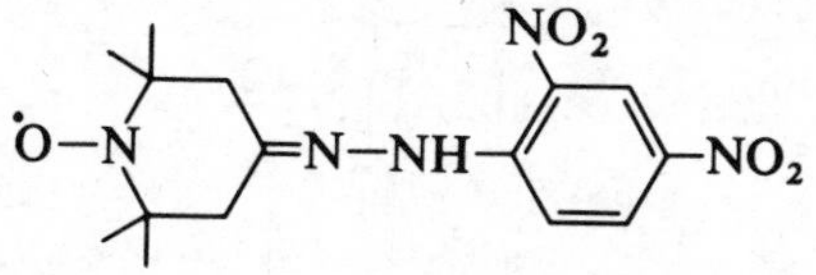

(X)

2,2,6,6-tetramethyl-4-piperidone *N*-oxide

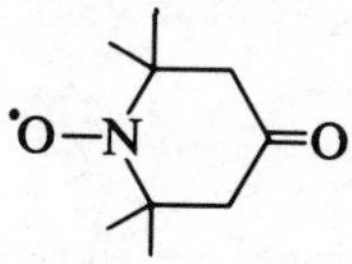

(XI)

5.3 Mobility of solubilizate molecules

Recent studies indicate that like the surfactant monomers themselves, the solubilized molecules are not rigidly fixed in the micelle but have a freedom of motion which is dependent to some extent on the solubilization site. In fact, as discussed in Chapter 3, solubilized probes are used as indicators of the fluidity of their micellar micro-environment. In particular, a decrease in the polarization of the fluorescent radiation from fluorescent probes located in the micellar core is indicative of the molecular motion of these probes. Similarly, tumbling of solubilized nitroxide probes proceeds at a more rapid rate than can be accounted for by simple rotation of the micelle itself indicating that the solubilized probe undergoes dynamic motion within the micelle. Such motion produces a characteristic hyperfine pattern on e.s.r. spectra.

Not only does the solubilizate have freedom of movement, albeit restricted within the micelle but it is also in constant dynamic equilibrium with the bulk aqueous phase. Only a limited number of studies have been reported of the determination of the rate constants for entry and exit of solubilizate molecules.

One of the earliest of such determinations [127] attempted to estimate the residence time of a benzene molecule in NaDS micelle by analysing the n.m.r. signal of NaDS solutions saturated with benzene. The lifetime of the solubilized molecule within the micelle was, however, too short to be determined by this technique, but was thought to be not longer that 10^{-4} s. Electron spin resonance has been used by Nakagawa and Jizomoto [128] in the determination of the lifetime of the stable radical *t*-butyl-(1,1-dimethylpentyl)-nitroxide when solubilized to saturation in a series of NaDS solutions of increasing concentration. The characteristic triplet signal in the e.s.r. spectra of nitroxide spin probes in water or low concentrations of NaDS (Fig. 5.14) is transformed in concentrated NaDS solution to a broad singlet (Fig. 5.16). The spectral change was interpreted

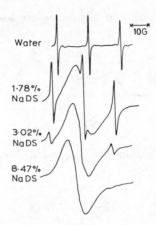

Figure 5.16 Change of signal shape as a function of NaDS concentration. From Nakagawa and Jizomoto [128] with permission.

in terms of the continual interchange of radical between micelles and the bulk phase using a computer simulation technique. An average residence time of 3.3 $\times 10^{-6}$s was obtained at all concentrations of NaDS above the CMC. Similar experiments were later carried out with sodium octyl, decyl and tetradecyl sulphates [129] from which it was concluded that an increase in alkyl chain length increases the residence time of the solubilizate molecule.

Almgren *et al.* [69] have studied the kinetics of solubilization of a phosphorescent probe, 1-bromonaphthalene in micelles of NaDS using a phosphorescence quenching method. The method is restricted to probes which are far more soluble in the micellar phase than in water. A water-soluble quencher is added to the system, which is capable of quenching the excited probe in the aqueous phase only, i.e. there is no quenching of the probe in the micelle. Increase in the concentration of the quencher causes a decrease in the observed phosphorescence lifetime until the exit rate of probe from the micelles becomes the controlling step to the quenching. At this point a further increase in the quencher concentration does not change the observed phosphorescence lifetime. This effect is shown in Fig. 5.17. A kinetic treatment of the limiting values from such plots yielded an exit rate constant of $2.5 \times 10^4 \text{s}^{-1}$ and a re-entry rate constant of $5\text{--}8 \times 10^9 \text{ mol}^{-1}\text{l s}^{-1}$.

Ultrasonic relaxation studies of the kinetics of the exchange of butanol, pentanol and hexanol between the aqueous phase and micelles of CTAB have been reported [130, 131]. The ultrasonic relaxation associated with the alcohol exchange process was well separated from that associated with the exchange process involving the CTAB monomers and could be treated independently. A simple collision-type kinetic model was used to describe the alcohol–micelle exchange process in such a way that rate constants could be evaluated. Table 5.5 shows that the rate constant associated with the dissociation of alcohol monomers from the micelles, k_{-1}, increases as the hydrocarbon chain length of

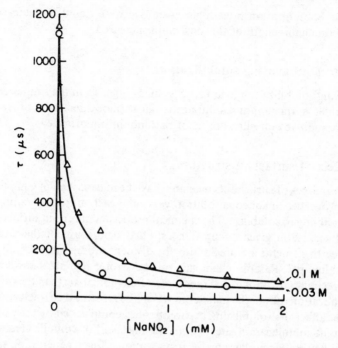

Figure 5.17 Phosphorescence lifetime of 1-bromonaphthalene as a function of [NaDS] and [NaNO$_2$]. From Almgren *et al.* [69] with permission.

Table 5.5 Values of k_1, k_{-1} and $\Delta V°$ obtained from the relaxation data. Values of K obtained from the solubility studies are also tabulated for comparison [131]

Alcohol	K (dm^3 mol^{-1})	k_1 (s^{-1} mol^{-1} dm^3 × 10^{-6})	k_{-1} (s^{-1} × 10^{-6})	k_1/k_{-1} (dm^3 mol^{-1})	$\Delta V°$ (m^3 mol^{-1} × 10^6)
Propan-1-ol	0.5*	42	81	0.5	5.6
Butan-1-ol	1.0	49	58	0.9	5.5
Pentan-1-ol	3.7	76	26	2.9	5.4
Hexan-1-ol	10.2	47	9	5.2	4.5

* 'Trial and error' value.

the alcohols decreases. The equilibrium constants, K, calculated from the ratio of the rate constants for association and dissociation of alcohol monomers i.e. $K = k_1/k_{-1}$ showed reasonable agreement with corresponding value determined from solubility studies.

A study by stopped-flow techniques of the rate of solubilization of positively charged acridine (and related) dyes by the micelles of several sodium *n*-alkyl sulphates was reported by Robinson and co-workers [132]. The solubilization process was relatively slow ($k = 47.6$ s^{-1} for acridine orange in NaDS) due largely to charge effects and the rate was very dependent on the geometrical shape of the dye. Solubilization rates for a given dye were found to decrease as the hydrocarbon chain-length of the surfactant increased.

5.4 Factors influencing solubilization

In discussing solubilization in aqueous systems one has, in the simplest systems, two variables: surfactant and solubilizate. How these affect each other and the maximum additive concentration is of paramount importance.

5.4.1 Effect of surfactant structure

The various types of surfactants available have been described in Chapter 1. Just as their properties in aqueous solution vary with each individual compound so does their ability to solubilize. The chemical properties of certain surfactants may obviate their use in given circumstances. Fatty acid soaps of the monovalent alkaline-earth elements are precipitated by divalent cations and are only suitable at an alkaline pH. Paraffin chain salts of the strong acid sulphates are not so readily precipitated; cationic detergents are often more toxic than the other types.

Generalizations about the manner in which the structural characteristics of the surfactant affects its solubilizing capabilities are complicated by the existence of the different solubilization sites within the micelle. For solubilizates which are located either in the micellar core or in a position of deep penetration within the micelle it might be expected that the solubilizing capacity should show a pronounced dependence on the alkyl chain length of the surfactant. Table 5.6 shows this effect quite clearly for the solubilization of ethylbenzene in a series of potassium fatty acid soaps. Although the calculations are based on the assumption of constancy of monomer concentration above the CMC and as such are necessarily approximate, they do show an increase in the amount of ethylbenzene solubilized per mole of micellar surfactant as the homologous series is ascended. A similar increase in solubilizing capacity with increasing alkyl chain length has been reported for non-ionic surfactants where the solubilizate is located in the hydrophobic core rather than the oxyethylene region. Table 5.7 shows an increase of solubilizing capacity of a series of polysorbates for selected barbiturates as the alkyl chain length is increased from C_{12} (polysorbate 20) to C_{18} (polysorbate 80) [133]. A similar effect was reported [134] for the solubilization of barbiturates in polyoxyethylene monoalkyl ethers of varying hydrophobic chain length. The same trend was also reported for the solubilization of salicylic acid [82] and testosterone [106] in polysorbate micelles. Arnarson and Elworthy [135], however, showed that increasing the hydrocarbon chain length in non-ionic surfactants from C_{16} to C_{22}, although producing larger micelles, did not result in a corresponding increase in solubilization. Thus the micellar size of a polyoxyethylene behenyl ether, $C_{22}E_{21}$, was 2.5 times as large as $C_{16}E_{20}$ and yet five out of eight solubilizates tested were solubilized to a greater extent in $C_{16}E_{20}$ than in $C_{22}E_{21}$. This effect was confirmed in a later paper by these workers [136] who showed that the polyoxyethylene monoethers of

Table 5.6 Solubilization of ethylbenzene in soap solutions (25° C) (values corrected for water solubility) (from Klevens [2])

Mol surfactant per 1000 g solution	Mol solubilizate per 1000 g solution	Mol solubilizate per mol surfactant	Mol solubilizate per mol micellar surfactant
		Potassium octanoate	
0	0.0016		
0.300	0.0012	0.004	
0.480	0.012	0.025	0.141
0.662	0.033	0.048	0.124
0.827	0.066	0.080	0.152
		Potassium decanoate	
0.10	0.0014	0.014	
0.232	0.027	0.116	0.197
0.435	0.067	0.154	0.197
0.500	0.087	0.174	0.214
0.717	0.145	0.202	0.233
		Potassium dodecanoate	
0.042	0.007	0.166	0.411
0.195	0.062	0.318	0.364
0.500	0.212	0.424	0.446
0.603	0.273	0.452	0.472
0.860	0.435	0.506	0.522
		Potassium tetradecanoate	
0.096	0.054	0.563	0.600
0.242	0.176	0.728	0.745
0.500	0.427	0.855	0.866
0.566	0.492	0.872	0.888
		Potassium hexadecanoate	
0.070	0.074	1.06	1.09
0.154	0.176	1.14	1.15
0.228	0.302	1.32	1.33
0.292	0.430	1.47	1.48

dotriacontanol ($C_{32}E_{41}$) and 4,9-dimethyl tritiacontanol ($C_{35}E_{40}$) both had lower solubilizing capacities than that of cetomacrogol. Intrusion of part of the polyoxyethylene chain into the hydrocarbon core was suggested as a possible explanation of the decrease in solubilization with increased hydrocarbon chain length in these compounds.

Kolthoff and Stricks [137] showed an almost linear increase in the solubilizing capacity for dimethylaminobenzene, (DMAB) within the series, potassium dodecanoate to octadecanoate. Introduction of a double bond (potassium oleate), however, produced a decrease in solubilizing power as compared to a saturated compound of similar length. A branched chain compound (di-amylsulphosuccinate) solubilized less DMAB than did potassium tetradecanoate, but more than potassium decanoate (same number of carbon atoms in the longer chain), whereas dodecylamine hydrochloride showed a greater solubilizing power than potassium dodecanoate.

Several workers have reported the effect on the solubilizing capacity of changes

Table 5.7 Solubilizing capacity of polysorbates for the barbiturates at 30° C [133]

Drug	Surfactant	Solubility	
		mg drug/g surfactant	mol drug/mol surfactant $\times 10^2$
Barbital	Polysorbate 20	30.0	19.9
	Polysorbate 40	33.0	23.0
	Polysorbate 60	35.3	25.1
	Polysorbate 80	35.0	24.6
Diallylbarbituric acid	Polysorbate 20	24.0	14.4
	Polysorbate 40	27.0	16.4
	Polysorbate 60	28.0	17.3
	Polysorbate 80	28.0	17.4
Butethal	Polysorbate 20	100.0	57.5
	Polysorbate 40	—	—
	Polysorbate 60	—	—
	Polysorbate 80	115.0	71.1
Cyclobarbital	Polysorbate 20	52.4	27.2
	Polysorbate 40	58.0	31.6
	Polysorbate 60	61.0	33.8
	Polysorbate 80	61.0	34.0
Phenobarbital	Polysorbate 20	55.1	29.1
	Polysorbate 40	61.0	33.7
	Polysorbate 60	63.0	35.5
	Polysorbate 80	66.0	37.2
Amobarbital	Polysorbate 20	32.0	17.2
	Polysorbate 40	38.0	21.7
	Polysorbate 60	—	—
	Polysorbate 80	40.0	22.9
Secobarbital	Polysorbate 20	111.0	57.0
	Polysorbate 40	—	—
	Polysorbate 60	—	—
	Polysorbate 80	144.0	78.8

of the counterion associated with ionic surfactants. Varying the anion in cetylpyridinium surfactants [13] showed that the order effective in solubilizing *trans*azobenzene was $(Br^- > Cl^- > SO_4^{2-} > OCOCH_3^-)$. At 25° C, McBain and Green [138] found that sodium dodecanoate solubilized more of the water-insoluble dye orange OT than did potassium dodecanoate, but Kolthoff and Stricks [137] could find no difference in the solubilizing capacities of the two surfactants at 30° C. The latter workers also reported equal solubilization of DMAB by the sodium and potassium salts of *n*-decanoic, *n*-dodecanoic and *n*-tetradecanoic acids. Similarly, Rigg and Liu [139] found no difference in the solubilizing power of sodium and potassium laurates to both orange OT and to DMAB.

The effect of changes in polar head structure of cationic surfactants on their ability to solubilize the water-insoluble dye orange OT have been examined by Jacobs and Anacker [140, 141]. Dye solubilization efficiency increases with aggregation number for surfactants containing polar heads of similar size and shape and identical hydrophobic moieties and which form spherical or nearly spherical micelles. An increased efficiency is also noted when the polar head acquires the ability to serve as a site for solubilization. Thus the aggregation numbers of decyltriethyl-, decyltripropyl- and decyltributylammonium bromide in 0.5 molal NaBr are approximately the same (37, 35 and 39, respectively) whilst their dye solubilization efficiencies increase with increasing length of the alkyl chain attached to the nitrogen of the head group.

A few workers have attempted a comparison of the solubilizing power of ionic and non-ionic surfactants for solubilizates which are located in the micellar interior [2, 142, 143]. In general, the solubilizing capacity for surfactants with the same hydrocarbon chain length increases in the order anionics < cationics < non-ionics. This effect has been attributed to a corresponding increase in the area per head group in this series leading to 'looser' micelles with less dense hydrocarbon cores which can accommodate more solubilizate [143]. Non-ionic surfactants have the additional advantage in that, by virtue of their lower CMCs, they are effective in much lower concentration than the ionic surfactants.

The effect of an increase in the ethylene oxide chain length of a poly-oxyethylated non-ionic surfactant on its solubilizing capacity is often complex and is dependent on the location of the solubilizate. Several workers [83, 144–146] have noted that the solubilizing power expressed as moles of solubilizate per mole of surfactant increases gradually as the polyoxyethylene chain length increases. When calculated as moles of solubilizate per equivalent of ethylene oxide in the surfactant, however, the solubilizing capacity showed a decrease with increasing ethylene oxide chain length. The reason for the increase in solubilization efficiency with increase in the polyoxyethylene chain length when expressed on a mole to mole basis has been explained by Barry and El Eini [145] in terms of the changes in micellar size which accompany the increase in ethylene oxide chain length. Table 5.8 shows the decrease in aggregation number with increase in hydrophobic chain length for a series of hexadecyloxyethylene glycol monoethers. Although the number of steroid molecules solubilized per micelle decreases with increase in ethylene oxide chain length the total amount solubilized per mole of surfactant in fact increases because of the increasing number of micelles as the ethylene oxide chain length is increased.

Table 5.8 Micellar solubilization parameters for steroids in *n*-alkylpolyoxyethylene surfactants at 25° C. From Barry and El Eini [145].

Surfactant	Partial specific volume (ml g^{-1})	Aggre-gation number	Micelles mol^{-1} × 10^{-21}	Steroid molecules per micelle			
				Hydro-cortisone	Dexa-methasone	Testo-sterone	Proges-terone
$C_{16}E_{17}$	0.9376	99	6.08	9.1	6.7	6.0	5.6
$C_{16}E_{32}$	0.9171	56	10.8	7.6	5.3	4.6	4.3
$C_{16}E_{44}$	0.8972	39	15.4	5.8	4.2	3.6	3.3
$C_{16}E_{63}$	0.8751	25	24.1	4.0	3.3	2.4	2.3

5.4.2 Effect of solubilizate structure

Many factors can effect the amount of a given substance which can be solubilized. Polarity and polarizability, chain length and chain branching, molecular size, shape, and structure have all been shown to have various effects.

Latent heat of fusion has been put forward as an explanation of why the solubilization of a crystalline solid is less than that of the same compound existing in a super-cooled liquid [147]. Klevens [2, 16] suggested that this effect would also partially account for the greater time required to dissolve solid polycyclics than to dissolve liquid hydrocarbons.

Molar volume has been studied widely [2, 128, 148, 149] but no simple relationship has been shown between molar volume and the amount of solubilizate dissolved. Stearns *et al.* [148], studying hexane, heptane, and octane, and benzene, toluene, ethylbenzene, propylbenzene, and butylbenzene concluded that there was inverse proportionality between the volume of hydrocarbon solubilized and molar volume. The slope of the plots of ml hydrocarbon dissolved per 100 g solution against molar volume of hydrocarbon are different for the aliphatic and aromatic series. Klevens [16], with polycyclic compounds in sodium laurate, found linear relationships between the log volume solubilized and molar volume, the slope of plots for linear polycyclics varying from that for the non-linear polycyclics. Schwuger [149] reported that the amount of naphthalene, anthracene, pyrene, perylene and dibenzanthracene solubilized by micelles of dodecylpentaglycol ether was inversely related to the molecular size of these solubilizates.

Although obviously having certain effects, molecular weight as such has little correlation with the amount solubilized; more than three times the weight of methylisobutylketone (molecular weight 100) than heptane (molecular weight 100) is dissolved in 0.1 M dodecylamine hydrochloride.

The most common classification of solubilizates has been by polarity; polar solubilizates forming one group, apolar solubilizates the other. As might be expected, such a sharp division as polar and apolar is always accompanied by intermediate compounds which are difficult to classify categorically, having properties of both groups. An unsaturated ring such as benzene does not have a specific polar group as in octanol and octanoic acid, but its readily polarized electrons lead to its having properties more akin to a polar solubilizate than an apolar one. Cyclohexane, on the other hand, behaves as an apolar solubilizate [150, 151].

Tables 5.9 and 5.10, selected from the results of Nakagawa and Tori [152] and of McBain and Richards [153], respectively, give an indication of the MACs of various solubilizates in a range of surfactants.

Table 5.11 shows the MACs of polycyclic hydrocarbons in 0.5 M potassium laurate.

No simple relationship exists between any single property of a solubilizate and its MAC in a given surfactant. Generalizations correlating MACs and their physical and/or chemical structure are very limited and depend largely on the

Table 5.9 Maximum additive concentrations of long-chain alkyl compounds in 1% aqueous solutions of $C_{10}H_{21}O(CH_2CH_2O)_{10}CH_3$ at 27° C (from Nakagawa and Tori [152])

Solubilizate	MAC $(g \, l^{-1})$
n-Octane	0.90
n-Decane	0.39
n-Dodecane	0.16
n-Octanol	3.12
n-Decanol	2.38
n-Dodecanol	2.07
n-Decyl chloride	0.45
n-Decylamine	3.78
Capric Acid	2.30

Table 5.10 Maximum additive concentrations of a range of solubilizates in $0.1 N$ solutions of dodecylamine hydrochloride ($C_{12}HCl$), sodium oleate (NaC_{18}), and potassium laurate (KC_{12}), respectively at 25° C (from McBain and Richards [153]

Solubilizate	MAC (moles solubilizate per mole surfactant)		
	$C_{12}HCl$	NaC_{18}	KC_{12}
n-Hexane	0.75	0.46	0.18
n-Heptane	0.54	0.34	0.12
n-Octane	0.29	0.18	0.08
n-Decane	0.13	0.052	0.03
n-Dodecane	0.063	0.009	0.005
2,3-Dimethylpentane	0.62	0.35	0.11
3,3-Dimethylpentane	0.55	0.31	0.10
Diisobutylene	0.43	0.38	0.10
Methylcyclopentane	0.40	0.26	0.032
Cyclohexane	0.87	0.56	0.23
1,2,4-Trimethylcyclohexane	0.019	0.012	0.012
Benzene	0.65	0.76	0.29
Toluene	0.49	0.51	0.13
Ethylbenzene	0.38	0.40	0.20
p-Xylene	0.34	0.36	0.20
Methylisobutylketone	1.78	1.82	1.20
Octylamine	0.13	0.07	0.07
n-Octanol	0.18	0.59	0.29
2-Ethylhexanol	0.36	0.47	0.064

systems being studied; comparison between the MACs of octane and octanol, and cyclohexane and benzene in dodecylamine hydrochloride and in the other surfactants shown (Tables 5.9 and 5.10) illustrates this. Furthermore, for apolar hydrocarbons such as cyclohexane and octane the MACs in the respective surfactants are dodecylamine hydrochloride > sodium oleate > potassium laurate, whereas for polar compounds such as octanol or methylisobutylketone the order is sodium oleate > dodecylamine hydrochloride > potassium laurate.

Table 5.11 Maximum additive concentrations of polycyclic hydro-carbons in 0.5M potassium laurate at 25° C (from Klevens [16])

	Moles solubilizate ($l^{-1} \times 10^{-3}$)
Benzene	391
Butylbenzene	112
Naphthalene	33.3
Phenanthrene	6.65
Anthracene	0.85
Chrysene	0.627
1,2-Benzanthracene	0.635
Triphenylene	0.336
Naphthacene	0.100
1,2,5,6-Dibenzanthracene	0.086

Klevens [2] pointed out that in most cases ascending a homologous series, i.e. increasing the length of the alkyl chain either in a straight-chain compound or substituted on a benzene ring, decreases the solubility in a surfactant solution, and that unsaturated compounds are more soluble than their saturated counterparts, and cyclization results in enhanced solubility, but branching of the chain has little effect. Unfortunately, these generalizations apply only to the simplest solubilizates; the presence of a second ring may reduce solubility, e.g. naphthalene is less soluble than *n*-butylbenzene or *n*-decane.

Size and steric factors make the orientation of the larger polycyclics in the micelle more formidable than for a simple straight-chain compound. From results showing that the solubility of the steroid hormones was much greater than those of the polycyclic hydrocarbons in surfactant solutions, Ekwall *et al.* [154] concluded that the latter, like the smaller apolar solubilizates, were situated in the hydrocarbon parts of the micelles, whereas the hormones were situated in the palisade layers of the micelles.

Table 5.12 shows that the solubilities of oestrone and oestradiol in sodium lauryl sulphate solutions are between those of the polycyclic aromatic hydro-carbons and the Δ^4-3-ketosteroids. As the main difference between the oestrogen and the other steroid hormones is that the former possess an aromatic ring, and

Table 5.12 Maximum additive concentrations in sodium lauryl sulphate for polycyclic aromatic hydrocarbons and steroid hormones [154]

Solubilizate	Moles solubilizate per mole surfactant
20-Methylcholanthrene	0.001 12
3:4-Benzpyrene	0.001 38
Oestrone	0.013 8
Oestradiol	0.015
Progesterone	0.18
Testosterone	0.21
Desoxycorticosterone	0.47

methylcholanthrene and benzpyrene contain four and five aromatic rings respectively, it would appear that the presence of the aromatic structures is the cause of the decreased solubility. The various structural characteristics of the steroid molecule which affect its solubilization are discussed in detail in Chapter 6. Only selected examples are included here.

In a study of the solubilization of selected steroids by lysophosphatidylcholine, Saunders and co-workers [155, 156] showed that monopolar steroids such as cholesterol are much more readily solubilized than dipolar steroids such as progesterone, prednisolone, testosterone, oestrone, hydrocortisone and prednisone.

The importance of polarity was further illustrated by an examination of the solubilization characteristics of substituted steroids. 17α substitution of an ethinyl group in testosterone decreased the solubilization in lysophosphatidyl-choline five-fold whereas 17α-ethinyl oestradiol was solubilized to a five-fold greater extent than oestradiol [157]. This surprising result was discussed in terms of the relative polarity of the two terminal rings of the two steroids using molecular orbital calculations of the total dipole moments of the molecules. Table 5.13 relates the algebraic sum of the total dipoles and also the sum of the dipole moments in the direction of the D ring moment to both the solubilization and the aqueous solubility of the parent and substituted steroids. In both cases it is seen that the effects of introducing the ethinyl group on the net polarity of the molecule parallels the effects of this group on both solubility in water and solubilization. Thus the ethinyl group decreases the net dipole moment of testosterone and so reduces its solubilization whereas the ethinyl group increases the net moment of oestradiol and so enhances its solubilization.

Several authors have noted a relationship between the lipophilicity of the solubilizate, expressed as its partition coefficient between octanol and water, $P_{octanol}$, and its saturation distribution between micelles and the aqueous phase. Rank-order correlations between the amount of substituted barbituric acid solubilized by polyoxyethylene stearates and their $P_{octanol}$ values have been reported by Ismail *et al.* [133]. Similarly the partition coefficients of several steroids between water and micelles of long-chain polyoxyethylene non-ionic surfactants have been correlated with their ether/water partition coefficients [145]. Fig. 5.18 shows a linear relationship between the partition coefficient of a

Table 5.13 Coupled dipoles for the steroid molecules.

Steroid	Algebraic sum	Projection sum	Solubilisation at 20° (moles/mole of lysophosphatidylcholine)	Solubility at 20° (μmol l^{-1})
Testosterone	−1.50	−1.37	0.113	84.6
Ethinyl testosterone	−1.04	−0.98	0.0235	1.92
Oestradiol	0.12	1.26	0.0636	16.52
Ethinyl oestradiol	0.48	1.80	0.262	34.41

From Gale and Saunders [157].

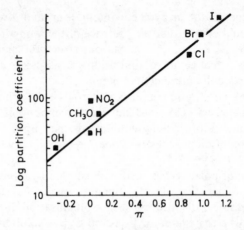

Figure 5.18 Micellar–aqueous partition coefficients of substituted benzoic acids and π values of the functional groups. From Collett and Koo [73] with permission.

series of unionized substituted benzoic acid derivatives between micelles of polysorbate 20 and water and the value of π defined by $\pi = \log P_x - \log P_H$ (where P_x is the octanol/H_2O partition coefficient of the benzoic acid derivative and P_H is the octanol/H_2O partition coefficient of benzoic acid itself) [73]. It can be seen from this figure that an increased lipophilicity of the benzoic acid derivative results in an increased tendency for solubilization. Tomida *et al.* [158] extended these measurements to cover some 34 benzoic acid derivatives including ortho, meta- and para-substituted compounds and also dicarboxylic acid derivatives. Three parallel lines were required to adequately represent $\log K$ versus $\log P_{octanol}$ data. Although most derivatives could be represented by one of these lines, the additional lines were required for the nitro and cyano substituents and also the dicarboxylic acid derivatives, indicating the solubilization of these derivatives in differing regions of the micelle. Similarly the solubilization of several steroids by polyoxyethylene dodecyl ether micelles [159] could be represented by two lines when plotted as $\log K$ against $\log P_{octanol}$; one line representing steroids possessing a fluorine atom, e.g. triamcinolone, betamethasone and dexamethasone, the other line representing those steroids without this substituent, e.g. hydrocortisone, corticosterone, progesterone, testosterone and prednisolone. It was suggested that the fluorine-containing steroids were solubilized in a more hydrophilic region of the micelle than the remaining steroids.

The importance of a consideration of both solubilizate and solubilizer structure in the selection of a particular surfactant for the solubilization of a given solute is illustrated by studies of the solubilization of para-substituted acetanilides by a series of structurally related poloxamer (polyoxyethylene–polypropylene) ABA block co-polymers (Pluronic surfactants) [83]. Fig. 5.19 shows the relation between the π value for the *p*-substituent group of the acetanilides and the slope K of plots of the amount of drug solubilized per

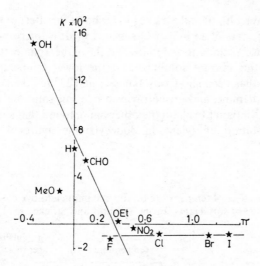

Figure 5.19 Slope (K) for plots of mol of *p*-substituted acetanilide solubilized mol^{-1} of poloxamer against the percentage oxyethylene in the poloxamer molecule as a function of the π value of the substituent group on the acetanilide molecule. From Collett and Tobin [83] with permission.

mole of poloxamer against the percentage of ethylene oxide in the poloxamer. The poloxamers are not simple surfactants; some aspects of their behaviour are discussed in Chapter 6. For the less hydrophobic drugs, K decreases linearly with π but beyond a π value of around 0.4 the amount solubilized was low and reasonably constant. These results indicate that the solubilizates may be divided into two types which differ in their site of solubilization. Analysis of the data by the partition method proposed by Mukerjee indicated that the less hydrophobic derivatives are solubilized primarily in the oxyethylene mantle whilst the hydrophobic derivatives are located in the micellar core. On a more fundamental level the solubilization site of particular solute is determined by the solubility parameter of the solute and the solubility parameter of the surfactant itself [160]. Selection of the appropriate surfactant to obtain maximum solubilization of a particular solubilizate involves the correct matching of these solubility parameters.

5.4.3 The effect of temperature

Temperature has an effect on the extent of micellar solubilization which is dependent on the structure of the solubilizate and of the surfactant. In most cases the amount of solubilization increases with temperature. This effect has been considered to be due to:

(*a*) changes in the aqueous solubility properties of the solubilizate;
(*b*) changes in the properties of the micelles.

Table 5.14 from Kolthoff and Stricks [137] shows the effect of temperature on the solubilizing powers of a number of surfactants. It is interesting to note that the relative increase in amounts solubilized at the higher temperature is inversely proportional to the amount solubilized at the lower temperature. A similar pattern of results has been shown by Bates *et al.* [22] for the solubilization of hexoestrol, glutethamine, and griseofulvin in bile salt solutions (Table 5.15).

Kaminski and McBain [161], on the other hand, found that slight warming of the saturated solution of xylene in dodecylamine hydrochloride produced turbidity.

Table 5.14 The effect of temperature on the maximum additive concentration of DMAB in a range of surfactants. From Kolthoff and Stricks [137].

Surfactant	Temperature (° C)	g solubilizate per mole surfactant
Sodium or potassium caprate	30	0.64
	50	1.19
Sodium or potassium laurate	30	1.50
	50	2.43
Sodium or potassium myristate	30	2.71
	50	4.15
Dodecylamine hydrochloride	30	4.32
	50	5.63

Table 5.15 The effect of temperature on the maximum additive concentrations of griseofulvin, hexoestrol and glutethimide in bile salts. From Bates *et al.* [22].

Solubilizate	Surfactant	$MAC \times 10^3$ (moles solubilizate per mole surfactant)		
		27° C	37° C	45° C
Griseofulvin	None	4.59×10^{-4}	7.14×10^{-4}	10.2×10^{-4}
	Sodium cholate	5.36	6.18	6.80
	Sodium desoxycholate	4.68	6.18	7.54
	Sodium taurocholate	3.77	4.90	6.15
	Sodium glycocholate	3.85	5.13	5.29
Hexoestrol	None	4.66×10^{-4}	6.66×10^{-4}	9.32×10^{-4}
	Sodium cholate	187	195	197
	Sodium desoxycholate	164	167	179
	Sodium taurocholate	220	225	223
	Sodium glycocholate	221	231	251
		27°	32°	37°
Glutethimide	None	7.13×10^{-2}	8.32×10^{-2}	9.94×10^{-2}
	Sodium cholate	59.8	96.2	104
	Sodium desoxycholate	103	119	163
	Sodium taurocholate	61.2	100	108
	Sodium glycocholate	54.3	92.0	71.8

Light-scattering results for a non-ionic detergent containing solubilized *n*-decane or *n*-decanol at a constant ratio showed that the micellar weight increased as the temperature was increased, this being especially marked near the cloud point. The comparable values for the surfactant/solvent system are given (see Tables 5.16 and 5.17). Surfactant solutions saturated with *n*-decane have also been studied by Kuriyama [162] (see Table 5.18). As with the ionic detergents previously mentioned, the maximum additive concentration increased with temperature. Micellar weights measured at various temperatures show that both the number of surfactant monomers and the number of solubilizate molecules per micelle increased with temperature.

In contrast, Barry and El Eini [145] have reported a decrease in the amount of steroid solubilized in aqueous solutions of long-chain polyoxyethylene non-ionic surfactants with increase in temperature. The essential difference between these

Table 5.16 The effect of temperature on the micellar weight of methoxydodecaoxyethylene decyl ether. From Kuriyama [162].

Temperature	Micellar weight $(\times 10^{-4})$	Number of surfactant molecules per micelle
9.7	3.29	47
29.0	3.71	53
50.7	4.55	65
58.5	5.15	73
69.7	7.09	101
73.4	9.26	131
75.0	11.6	165

Table 5.17 The effect of temperature on the micellar constitution of methoxy-dodecaoxyethylene decyl ether containing a definite amount of solubilizate. From Kuriyama [162].

Temperature	Micellar weight $(\times 10^{-4})$	Number of surfactant molecules per micelle	Number of solubilizate molecules per micelle
(A) 1.86% *n*-Decane in methoxydodecaoxyethylene decyl ether			
9.6	4.67	65	5.9
30.0	4.83	67	6.1
50.0	5.10	71	6.5
60.0	6.12	85	7.8
66.6	6.90	96	8.8
69.0	7.87	110	10.1
(B) 9.17% *n*-Decanol in methoxydodecaoxyethylene decyl ether			
10.0	5.62	73	30
29.9	6.42	83	33
43.4	8.45	110	44
49.7	10.75	140	57
55.4	14.3	186	76
61.4	13.1	404	163

Table 5.18 The effect of temperature on the micellar constitution of methoxydodecaoxyethylene decyl ether solutions saturated with *n*-decane From Kuriyama [163].

Temperature (°C)	Maximum additive concentration decane (weight %)	Micellar weight (×10⁻⁴)	Number of surfactant molecules per micelle	Number of solubilizate molecules per micelle
10.0	2.6	5.67	78	10
30.0	4.4	6.27	85	18
50.0	8.7	7.41	97	41
60.0	13.5	10.2	127	84

studies and those of Kuriyama is that the amount of steroid solubilized was only small (representing about 3% of the micellar weight) and measurements were made at temperatures below the cloud point, in a region where temperature change was thought to have only a minimal effect on micellar properties. The decrease in the steroid micelle/water partition ratio P_m with temperature increase in Table 5.19 was attributed to a concomitant increase in the water solubility of the steroids as shown in Table 5.20.

Table 5.19 Solubilization of steroids by *n*-alkyl-polyoxyethylene surfactants [145].

Surfactant	°C	Hydrocortisone P_m	Dexamethasone P_m	Testosterone P_m	Progesterone P_m
$C_{16}E_{17}$	10	177	471	1030	2970
	20	120	346	819	2580
	25	110	314	786	2160
	30	99	289	676	2000
	40	83	224	534	1840
	50	55	185	385	1640
$C_{16}E_{32}$	10	149	414	824	2370
	20	105	295	688	2220
	25	101	273	661	1790
	30	85	261	586	1730
	40	71	203	463	1640
	50	46	170	362	1490
$C_{16}E_{44}$	10	135	364	718	2170
	20	94	274	592	1870
	25	86	244	570	1550
	30	74	239	489	1410
	40	63	185	426	1390
	50	42	159	321	1270
$C_{16}E_{63}$	10	118	304	573	1670
	20	79	215	475	1420
	25	68	199	452	1250
	30	60	188	405	1200
	40	48	151	355	1120
	50	33	123	268	1070

Table 5.20 Aqueous solubilities of steroids at various temperatures [145].

Steroid	Solubility $mol\, l^{-1} \times 10^4$ at					
	10° C	20° C	25° C	30° C	40° C	50° C
Hydrocortisone	4.78	7.43	8.82	10.34	12.65	15.19
Dexamethasone	0.82	1.58	2.27	2.52	3.56	4.60
Testosterone	0.56	0.79	0.81	1.06	1.40	2.10
Progesterone	0.17	0.22	0.28	0.36	0.38	0.49

The extent of solubilization of benzoic acid by several *n*-hexadecylpolyoxy-ethylene surfactants of varying ethylene oxide chain length showed an apparent increase with increasing temperature [163]. However, benzoic acid has an appreciable water solubility which also increases with temperature and this is, of course, a contributing factor towards the overall increase in the amount of benzoic acid taken into solution. Allowance for this effect was made by expressing the solubilization data in terms of the micelle/water distribution coefficient, P_m. Table 5.21 shows a minimum in P_m at about 300 K. The decrease in P_m below this minimum is possibly due to the increase in aqueous solubility of benzoic acid; at higher temperatures this effect would be counteracted by a rapid increase in micellar size as the cloud point is approached which would cause an increased uptake into the micelles.

Table 5.21 Micelle/water distribution coefficient, P_m, for the solubilization of benzoic acid by *n*-alkyl-polyoxyethylene surfactants as a function of temperature [163].

Surfactant formula	P_m				
	291.0 K	298.0 K	304.0 K	310.0 K	318.0 K
$C_{16}E_{16}$	59.51	50.07	43.75	44.11	. . .*
$C_{16}E_{30}$	47.80	45.55	35.42	38.23	38.66
$C_{16}E_{40}$	37.07	32.72	28.76	29.90	37.06
$C_{16}E_{96}$	31.22	27.43	25.43	27.46	32.25

* No P_m value determined at this temperature for $C_{16}E_{16}$ because the cloud point temperature was exceeded.

5.4.4 Effects of added electrolyte

Addition of electrolyte to micellar systems has been discussed in Chapter 3. For ionic micelles the effect of electrolyte addition is to cause an increase in micellar size and a decrease in the CMC. Although such changes in micellar properties are well established their effects on the solubilizing capacity is often not predictable. Clearly the displacement of the CMC to lower concentrations should result in an increased solubilization in this concentration region because of the increased amount of micellar material. This effect has been reported by many workers. Stearns *et al.* [148] found that at concentrations just above the CMC for potassium laurate in water the amount of nitrodiphenylamine solubilized increased with increase in salt concentration. Kolthoff and Graydon [164] studying solubilization in dodecylammonium chloride solutions found an increase in the solubility of orange OT, DMAB, *trans*azobenzene and naphthalene with increasing salt concentration. McBain *et al.* [165] showed an increase in solubilization of orange OT in bile salt solutions on adding sodium chloride and sodium sulphate.

The situation is, however, more unpredictable at surfactant concentrations well in excess of the CMC when other effects of electrolyte on micellar properties predominate. The complexity is demonstrated by a detailed study by Ekwall *et al.* [166] of the effect of sodium chloride on the solubilization of decanol by sodium octanoate. At low octanoate concentrations (up to about 0.3 mol l^{-1}) the decanol solubility, as expected, begins to increase only when the CMC is exceeded and the increase proceeds up to the highest chloride additions studied (Fig. 5.20). However, at octanoate concentrations > 0.3 mol l^{-1} the first chloride additions caused a more or less pronounced increase of decanol solubility, but this increase is changed to a decrease at further increase of the salt concentration; the maximum of the solubility curve occurs at lower sodium chloride when the octanoate content of the system is increased. At octanoate concentrations above about 0.7 mol l^{-1} the addition of salt caused an immediate decrease in the solubilizing capacity of the system. This complex effect of added salt was

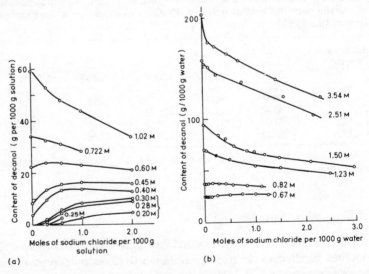

Figure 5.20 The effect of sodium chloride addition on decanol solubility in aqueous solutions of sodium octanoate at 20° C in (a) the concentration range of 0.2 to 1 mol octanoate per 1000 g decanol-free solution and (b) the concentration range of 0.67 to 3.5 mol octanoate per 1000 g water. From Ekwall *et al.* [166] with permission.

explained in terms of the influence of the Na^+ ions not only on the location of the CMC but also on the properties of the mixed octanoate–decanoate micelles. The predominant effect of the Na^+ ions at low octanoate concentration was the lowering of the CMC and the concomitant increase in decanol solubility was due to the increased amount of micellar material. The decreased solubility at higher octanoate concentrations was thought to be a consequence of an increased binding of counterions at the micellar surface which decreased the stability of the micellar solution and resulted in mesophase formation at lower decanol content than in salt-free solutions.

The location of the solubilizate molecule within the micelle may be an important factor in determining the effect of electrolyte addition on the solubilizing capacity of a particular surfactant. In cases where the solubilizate is located within the core or deep within the palisade layer it is readily appreciated that solubilization may be increased due to the increase in micellar volume following electrolyte addition. However, the reduction in repulsion between the head groups when electrolyte is added leads to a closer packing of surfactant molecules in the palisade layer and might be expected to result in a decrease in the solubilization of polar compounds which are located in this region. A decrease in the distribution coefficient of several alkylparabens between micelles of sodium dodecyl sulphate and water with increase in added NaCl has been noted [167]. These solubilizates are thought to penetrate only slightly into the palisade region of the micelle and the decreased solubilization was attributed not only to an increased crowding of the head groups in salt-containing solutions arising from a reduction in charge repulsion as suggested above, but also from a decrease in surface area due to a change from a spherical to an ellipsoidal shape in electrolyte.

Where electrolyte addition to solutions of non-ionic surfactants leads to an increase in micellar weight, the solubilization capacity for hydrocarbons is enhanced [168, 169] (Fig. 5.21). The order of increase in solubilization follows a similar trend to that for depression of the cloud point. The effect of electrolyte on

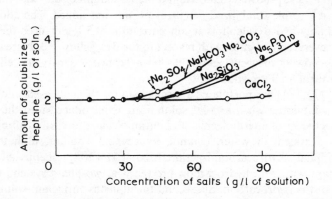

Figure 5.21 The effect of added electrolytes on the solubilization of heptane in 1% aqueous solution i-$R_9C_6H_4O(CH_2CH_2O)_{9.2}H$ at 25° C. From Saito and Shinoda [168] with permission.

Table 5.22 Apparent partition coefficients of sodium 2-naphthalenesulphonate between polysorbate 80 micelles and water*.

Sulphonate ($\%$ (w/w))	Polysorbate 80 ($\%$ (w/w))	Potassium chloride (mol l^{-1})	P_m
0.2	1	0	4.1
0.2	1	0.1	8.8
0.2	1	0.2	13.9
0.2	2	0	4.6
0.2	2	0.1	8.5
0.2	2	0.2	13.4
0.2	4	0	4.6
0.2	4	0.1	8.6
0.2	4	0.2	13.5
0.02	2	0	4.3
0.02	2	0.1	8.8
0.02	2	0.2	13.1
0.002	2	0	4.4
0.002	2	0.1	8.7
0.002	2	0.2	12.8

* All solutions were adjusted to pH 7.
From Park and Rippie [70].

the solubilization of polar solubilizate by non-ionic surfactants follows no clear pattern [169]. Table 5.22 shows an increased partition coefficient of sodium 2-naphthalene sulphonate between polysorbate 80 micelles and water [69].

5.4.5 Effects of addition of non-electrolytes

Non-electrolyte additives can also have a profound effect on the solubilizing properties of surfactant solutions.

Early workers examined the effect of phenol on various solubilizates [3, 170]. Weicherz [171, 172] showed that certain toluene–sodium oleate–water mixtures only formed a stable solubilized system if phenol was added. The addition of mono- and poly-hydroxyalcohols at concentrations of 5 % enhanced the solvent powers of several solubilizates with respect to the dye, yellow AB. Increasing the solubilizing powers of a surfactant has been termed a synergistic effect and decreasing them an antigistic effect.

Winsor [173, 174] has postulated a mechanism for this increased solubility of an apolar solubilizate when a polar solubilizate is introduced into the system, based on his theory of intermolecular and intramolecular forces and affinities. He considered a system of water, octanol, undecane-3, sodium sulphate and a saturated aliphatic hydrocarbon fraction. Mixing 5 ml 20 % aqueous solution of the surfactant with 5 ml hydrocarbon produced a two-phase system; the first composed of the hydrocarbon, the second the aqueous surfactant solution, the micelles of which, according to Winsor, had not solubilized the hydrocarbon to any appreciable extent. Octanol added to such a system penetrated the micelles, orienting with its hydroxyl groups between the sulphate groups of the surfactant

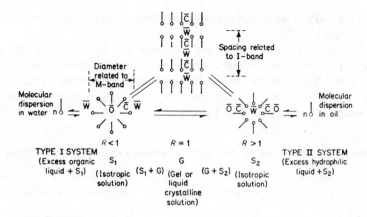

Figure 5.22 Intermicellar equilibrium and associated phase changes shown by certain series of amphiphilic solutions. Suggested correlation with S, M, and I X-ray-diffraction bands.

$\overline{W}$ = Hydrophilic section of micelle
$\overline{O}$ = Amphiphilic section of micelle
$\overline{C}$ = Lipophilic section of micelle
R = Ratio of dispersing tendencies on lipophilic and hydrophilic faces of $\overline{C}$ respectively.
(The S-band arises from an as far as possible random arrangement of the hydrocarbon chains in the $\overline{C}$ and $\overline{O}$ sections all types of micelle.) From Winsor [174] with permission.

and its alkyl tail into the centre of the micelle. This 'dilution' of the highly polar sulphate groups with the less polar hydroxyl groups reduced the interaction between the surface of the micelles and the solvent, allowing a reduction in its curvature and a concomitant uptake of hydrocarbon. Addition of sufficient octanol produced a one-phase solubilized system (see Fig. 5.22).

Further addition of octanol produced a situation where the stability of the solubilizate-containing micelles (at this concentration in a laminar form) was sufficient to allow their separation as a distinct liquid crystalline phase. Continuing addition of octanol produced first a completely gelled liquid crystalline system, followed by the separation of an isotropic liquid in which water was the solubilizate.

McBain and Green [175] showed that the solubility of Orange OT in potassium laurate and potassium oleate increased with the addition of benzene, toluene, and hexane at concentrations equivalent to 75 % of their respective maximum additive concentrations for the particular surfactant–solvent systems under examination. They further showed by comparison of the solubilities of Orange OT in the surfactant + additive solutions and in the respective additives that the calculated solubilities did not agree with the solubilities measured. From these results they concluded that the solubilizing properties of the mixed micelles were not an additive function of the components.

Ethyl alcohol (5 %), on the other hand, decreased the MAC of Orange OT in the two soaps, potassium laurate and potassium oleate. Ethyl alcohol has been shown to affect the CMC of surfactants [176], and in larger concentration even to inhibit micellization [177, 178].

Kolthoff and Graydon [164] found that amyl alcohol in low concentrations produced a reduction but at higher concentrations an increase in the solubilization of DMAB and Orange OT, but had no such complex effect on the solubility of *trans*-azobenzene and naphthalene, simply causing a monotonic increase. Their results are summarized in Table 5.23.

Table 5.23 Summary of observations on systems containing co-solubilized materials [164].

Solubilizate	Co-solubilized additive	Effect of additive on MAC
DMAB Orange OT	} *n*-Amyl alcohol	Decreased, passed through minimum and then increased
trans-Azobenzene Naphthalene	} *n*-Amyl alcohol	Increased continuously
DMAB Orange OT *trans*-Azobenzene	Pentane Octane Di-isoamyl	Increased continuously
trans-Azobenzene Naphthalene	} *n*-Octyl alcohol	Increased continuously
DMAB Orange OT *m*-Dinitrobenzene α-Naphthol	} *n*-Octyl alcohol	Small variations

The synergistic effect on the solubilization of *n*-heptane and *n*-alkanols has been shown to increase with the alcohol chain length over the range octanol to dodecanol [179] and changing the polar group on an octyl radical showed the synergistic powers in the order octyl mercaptan > octylamine > octanol [180]. In both cases this is in opposite order to the maximum additive concentrations of the additives in the surfactant–solvent systems.

5.4.6 Solubilization of mixtures of solubilizates

The investigation of the effect of a second solubilizate on the solubilizing capacity of a surfactant for a particular compound has received little attention, despite the fact that the solubilization of mixtures of solubilizates often occurs in pharmaceutical preparations. Crooks and Brown [181] have reported the solubilization of several pairs of preservatives by the non-ionic surfactant, cetomacrogol. Their results indicate that the solubilities of preservatives solubilized as mixtures may differ substantially from those determined for the compounds individually. No clear pattern of behaviour has emerged. Fig. 5.23a shows that increasing concentrations of benzoic acid produce an apparently linear increase in the amount of methyl parabens solubilized by cetomacrogol whilst with dichloroxylenol the solubility initially decreases to a minimum and then rises again.

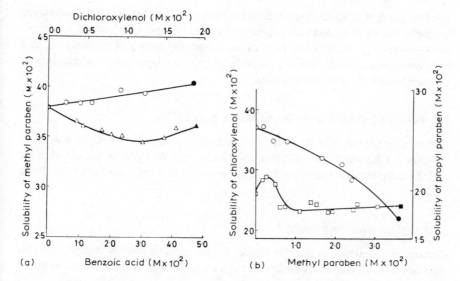

Figure 5.23 (a) The solubility of methyl paraben in 0.019 M cetomacrogol solutions at 25° C as a function of △—△ dichloroxylenol and O—O benzoic acid concentration. Solid points indicate solutions saturated with respect to both methyl paraben and co-solute. (b) The influence of varying concentration of methyl paraben on the solubility of □—□ propyl paraben and O—O chloroxylenol in solutions containing 0.019 M cetomacrogol at 25° C. Shaded points indicate systems saturated with respect to both solute and co-solute. From Crooks and Brown [181] with permission.

The influence of increasing concentrations of methyl paraben on the solubilization of propyl paraben and chloroxylenol is shown in Fig. 5.23b. At low concentrations, the methyl ester produces a sharp increase in solubility of the propyl ester which reaches a maximum then declines. The solubility of chloroxylenol is dramatically reduced by the addition of methyl paraben. Cetomacrogol solutions saturated with methyl paraben will dissolve only 61 % of the chloroxylenol that can be solubilized in solutions free from the methyl ester. Several possible explanations of the differences in effect have been proposed. Where compounds are solubilized within similar micellar regions it is likely that the solubilizates will compete with each other for the solubilization site leading to a diminished solubility of each. In addition there is the possibility of a co-solubilization effect where one solubilizate causes structural alterations in the micelle so enchancing its capacity for another. The simultaneous operation of two such mutually antagonistic processes would explain the occurrence of maxima and minima in the solubility plots.

The simultaneous solubilization of some steroids in aqueous polysorbate 40, tetradecyltrimethylammonium bromide and sodium dodecyl sulphate has been studied [182, 183]. The results are discussed in detail in Chapter 6 and in general support the view that in systems where the solubilizates are located in different sites, they are solubilized independently of each other. For example, the solubilization of testosterone which is thought to be located in the palisade layer is unaffected by the simultaneous solubilization of oestradiol which is accommodated in the micellar core. In contrast, co-solubilization of testosterone and ethinyloestradiol (which is also located in the palisade layer) causes alterations in the solubilities of both solubilizates.

5.5 Effect of solubilizate on micellar properties

The influence of solubilizate on the critical micelle concentration was discussed in Chapter 3. Changes in two other properties of micellar systems, the cloud point and the micellar size, following the incorporation of solubilizate are considered here.

5.5.1 Effect on cloud point

Certain non-ionic detergents, whether or not solubilizate is present, exhibit separation into two phases on heating at a critical temperature referred to as the cloud point (see p. 41).

The effect of solubilizates on the cloud point of Triton X 100 is shown in Fig. 5.24. Alkanes such as cetane or dodecane raise the cloud point whereas dodecanol, benzene and phenol depress it. Several other workers have reported a decrease of cloud point of this surfactant following the addition of phenols

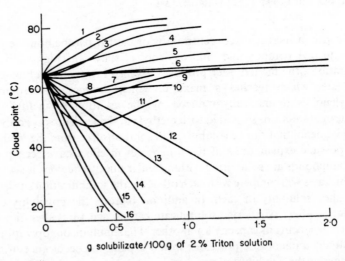

Figure 5.24 Effect of added solubilizates on the cloud point of 2% Triton X-100 solutions. Solubilizates: (1) cetane; (2) dodecane; (3) decane; (4) tetradecene-1; (5) *n*-tetradecyl mercaptan; (6) acetone; (7) citric acid; (8) *n*-octene; (9) hexane; (10) 2-ethylhexene; (11) cyclohexane; (12) aniline; (13) butyl acetate; (14) ethylene dichloride; (15) phenol and oleic acid; (16) *n*-dodecanol and nitrobenzene; (17) benzene. From McLay [184] with permission.

[184–186]. Phenolic solutes have been reported to decrease the cloud point of cetomacrogol [187]. A systematic study of the effect of phenols, cresols and xylenols on the cloud point of cetomacrogol has been carried out by Donbrow and Azaz [188] (see Chapter 6). All of these solubilizates caused a pronounced depression of the cloud point to room temperature, the effect of a phenol being inversely related to its hydrophobicity.

There are striking resemblances between the behaviour of non-ionic surfactants at the cloud point and at maximum additive concentrations for any one solubilizate–surfactant system. Nakagawa and Tori [152] examined the effect of the various solubilizates on the cloud point of methoxydecaoxyethylene decyl ether. Addition of *n*-alkanes such as *n*-decane give solubilization and cloud-point curves (Fig. 5.25) characterized by a slight negative slope of the cloud-point curve BC, leading to a virtually horizontal line CD and a rapid increase in solubilization near the cloud point as shown by the line AC. Fatty alcohols, such as *n*-octanol to *n*-dodecanol, and fatty acids, such as *n*-decanoic acid, give a plot (Fig. 5.26) with a rather larger negative slope in curve BC, followed by a maximum in CD. By use of added dyes and analytical procedures the composition of the

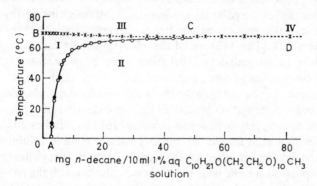

Figure 5.25 Solubilization and cloud point curves 1 % $C_{10}H_{21}O(CH_2CH_2O)_{10}CH_3$ solutions in the presence of *n*-decane. After Nagakawa and Tori [152].

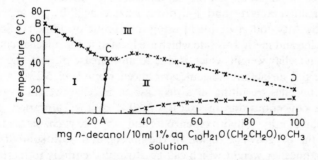

Figure 5.26 Solubilization and cloud-point curves of 1 % $C_{10}H_{21}O(CH_2CH_2O)_{10}CH_3$ solutions in the presence of added *n*-decanol. After Nagakawa and Tori [152].

various sections of the diagram (numerically designated I–IV was worked out thus:

I   consists of an isotropic 'solubilized' system;
II   two phases, the first as in I, the second a phase rich in solubilizate;
III   two phases, the first an aqueous solution containing mono-molecularly dispersed surfactant, the second rich in surfactant with a small amount of water and the solubilizate dissolved in it;
IV   three phases, the first and second as in III, the third phase rich in solubilizate.

5.5.2 Effect on micellar size

The effect of solubilizate on micellar size has been examined in only a few systems. In a light-scattering study of solubilization by hexadecyltrimethylammonium bromide Hyde and Robb [150] showed that the incorporation of increasing amounts of the non-polar molecules, decane, octane and cyclohexane, causes a pronounced increase in the micellar molecular weight. This is due to increases in the numbers of solubilizate and surfactant molecules in each micelle. However, the solubilization of the polar molecule, octanol, although increasing the micellar weight causes a decrease in the number of surfactant molecules in each micelle.

Nakagawa et al. [189, 190] found the solubilization of decane and decanol by three methoxypolyoxyethylene decyl ethers to result in increases in the micellar weight of the micelles of these non-ionic surfactants. Each weight increase is a consequence of increases in the amount of solubilizate and surfactant per micelle. Viscosity and sedimentation studies of the solubilization of 1,2,4-trichloroben-zene and toluene by cetylpyridinium chloride [151] have indicated an increase in micellar weight and in micellar asymmetry with increase in solubilizate concentration up to a maximum, after which further solubilizate promotes the formation of a more spherical micelle which exists in equilibrium with the rod-like micelles produced initially. In contrast, the solubilization of methyl cyclohexane by the same surfactant results in only a small regular increase in micellar weight and viscosity.

The effect on micellar size of solubilization of decane, ethyl p-hydroxy-benzoate, methyl anisate and p-hydroxybenzoic acid by cetomacrogol was examined by Attwood et al. [191] using membrane osmometry. Solubilizates such as decane and methyl anisate which are located in the micellar core cause an increase in micellar weight which is due to an increase in both the number of solubilizate and surfactant molecules per micellar unit (Fig. 5.27a). A restructur-ing of the micelles, resulting in a decrease in micellar weight, was noted at solubilizate concentrations in excess of 80% of the saturation value. The solubilization of ethyl p-hydroxybenzoate and p-hydroxybenzoic acid is thought to involve the oxyethylene region of the micelle and both solubilizates cause increases in micellar weight which can be attributed entirely to increases in the number of solubilizate molecules per micelle; the number of surfactant molecules per micelle remaining constant (Fig. 5.27b).

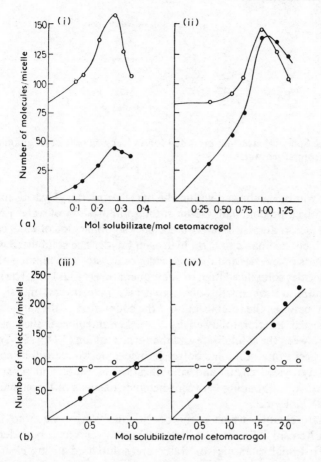

Figure 5.27 (a) The number of molecules per micelle of ○ cetomacrogol and ● solubilizate as a function of the molar ratio of solubilizate in the mixture, for the solubilization of decane (i) and methyl anisate (ii). (b) The number of molecules per micelle of ○ cetomacrogol and ● solubilizate as a function of the molar ratio of solubilizate in the mixture, for the solubilization of ethyl *p*-hydroxybenzoate (iii) and *p*-hydroxybenzoic acid (iv). From Attwood *et al.* [191] with permission.

5.6 Solubilization in non-aqueous solvents

Most of the work using non-aqueous solvents has been concerned with the solubilization of water and small polar molecules and has examined the effect of surfactant structure and the nature of the solvent phase on the solubilizing capacity.

One of the structures proposed for an ionic reverse micelle containing solubilized water is shown in Fig. 5.28 (see Chapter 3 for alternative models). Since micelles in non-aqueous solvents have their polar groups directed inwards and their hydrophobic groups in contact with the solvent, water or small polar

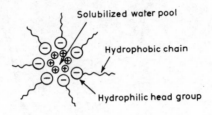

Figure 5.28 Spherical structure proposed for a reverse micelle of an anionic surfactant containing solubilized water.

molecules will be solubilized within the micellar interior. Hydrogen bonding is thought to be a predominant factor in the solubilization of water [2] although additional factors are also of importance since solubilization of water can occur in systems which are unable to form hydrogen bonds. The solubilized water is not uniform in its properties and there is evidence for stronger binding of the initial water molecules solubilized than of subsequent ones [192, 193]. The mechanism for solubilization of small polar molecules involves an initial ion-dipole interaction between the solubilizate and the counterion of the surfactant present in the micellar interior followed by a weaker interaction such as hydrogen bonding between the solubilizate and the surfactant ion [194, 195]. The core of a reversed micelle of a non-ionic polyoxyethylene derivative will be composed of the polyoxyethylene chains and solubilization of water and polar molecules involves hydrogen bonding between the ether oxygens of these chains and the solubilizate molecules.

The phase changes which occur on the addition of water to the anionic surfactant Aerosol OT in various hydrocarbon solvents have been demonstrated [196, 197]. Initial quantities of water are solubilized giving clear solutions. Continued addition of water gives rise to a turbid region followed by a blue translucent region and a subsequent turbid region (Fig. 5.29). A characteristic feature of the phase diagrams is the narrow temperature range terminating at point A in which a high water content is solubilized. Similar phase changes have been noted for the solubilization of water by dodecylammonium propionate in carbon tetrachloride [198] and polyoxyethylated non-ionics in cyclohexane [199] and the composition of the various regions has been discussed [197, 198]. The first turbid region is considered to be a dispersion of water containing a small amount of the surfactant in a water-saturated non-aqueous solution, i.e. a w/o emulsion. The open-circle line of Fig. 5.29 is thus the solubilization curve of water. The second turbid region was considered to be a dispersion of the surfactant containing water and solvent in the solvent. Hence the half-filled circles represent the solubility curve of the surfactant. The colourless and blue translucent regions are the solubilization regions for water, the extent of which increase with increasing concentration of surfactant. Several authors have taken point A to represent the maximum amount of solubilization and the maximum temperature of solubilization [197–200].

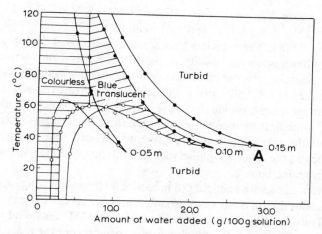

Figure 5.29 The effect of temperature and surfactant concentration on the solubility behaviour of water by di(2-ethylhexyl) sodium sulphosuccinate in cyclohexane solutions. From Kon-no and Kitahari [197] with permission.

The solubility behaviour of water in non-aqueous solutions of ionic surfactants has been widely studied. Kon-no and co-workers reported an increase in the solubilizing capacity for water by salts of the anionic dialkyl sulphosuccinates in various hydrocarbon solvents with increase in the alkyl chain length [202], bulkiness of the hydrocarbon group [197], and decrease in the valency of the counterion [197, 201]. Frank and Zografi [196] also noted a greater solubilization by the bulky di(2-ethylhexyl) sodium sulphosuccinate (Aerosol OT) compared with di(n-octyl) and di(n-hexyl) sodium sulphosuccinate in octane. The effect on solubilizing behaviour of addition of electrolyte to Aerosol OT in various solvents has been studied [202, 203]. Kon-no and Kitahara [203] showed that all electrolytes decreased the solubilization, the order of effectiveness of the cations in decreasing solubilization being $\frac{1}{2}Ca^{2+} \simeq \frac{1}{2}Mg^{2+} > Cs^+ \simeq K^+ > NH_4^+ \simeq Na^+ > Li^+$. The effects due to different anions were virtually indistinguishable. Solubilization was decreased by the addition of a strong base but increased by addition of strong acids. The lowering of solubilization capacity was attributed to a decrease of the repulsive charge between the ionic head groups and the squeezing out of the solubilized water.

Palit and Venkateswarlu investigated the solubilization of water by a series of organic acid salts of n-dodecylamine in xylene and reported maximum solubilization with butyrate salts [204]. A similar trend was observed with salts of n-octadecylamine in xylene, maximum solubilization being observed with the propionate salt. The amount of water solubilized by n-dodecylamine and n-octadecylamine salts of several carboxylic acids in benzene or cyclohexane was greatest for the shortest chain carboxylate [205]. The effect of electrolyte in decreasing the minimum temperature of solubilization and increasing the maximum amount of water solubilized in dodecylammonium carboxylates in hydrocarbon solvents was for cations $Cs^+ > K^+ \simeq Na^+ > Li^+ \simeq NH_4^+$ $> \frac{1}{2}Ca^{2+} \simeq \frac{1}{2}Mg^{2+} > \frac{1}{3}Al^{3+}$, and for anions $SCN^- > I^- > NO_3^- > Br^-$ $> Cl^- > F^-$ [198]. These orders of influence correspond with the order of weakness of Lewis acidity or basicity, respectively, i.e. they form a lyotropic series.

The solubilization of water in mixtures of ionic surfactants was studied by Palit and co-workers [206, 207] using dodecylamine derivatives and lauryl and hexadecylammonium bromide derivatives dissolved in various organic solvents. In general, solubilization is enhanced over the value for a single surfactant when mixtures of the two surfactants are present, provided that one surfactant is hydrophilic and the other lipophilic, such as a mixture of dodecylamine chloride and dodecylamine laurate.

The solvent can have a pronounced influence on the solubilizing capacity of an ionic surfactant due, in many cases, to differences in aggregate size in the various solvents. Thus the solubilization capacity of Aerosol OT was found to decrease with increase of chain length of hydrocarbon solvents over the range heptane to octadecane [196, 208].

Several authors have shown a general increase in the capacity of polyoxyethylated non-ionic surfactants to solubilize water with increase in the length of the oxyethylene chain [209, 210]. Fig. 5.30 shows the vapour pressure of water in 0.5% solutions of a series of dodecyl polyoxyethylene glycols in tetrachlorethylene plotted as a function of the water content of the system. The decrease in

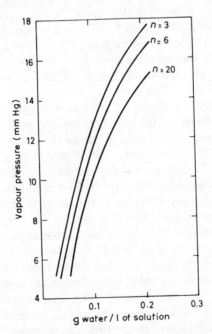

Figure 5.30 Water vapour pressure over 0.5% solutions of $C_{12}H_{25}O(CH_2CH_2O)_nH$ in tetrachloroethylene as a function of water content. From Wedell [209] with permission.

vapour pressure with increase in ethylene oxide chain length is indicative of an increased capacity to solubilize water. Decrease of the hydrocarbon chain length at a constant polyoxyethylene chain length has been reported by Wedell [209] to enhance the solubilizing capacity; other authors have, however, noted no significant effect [210–212].

Decrease of solubilizing power of non-ionics by the addition of electrolyte was far less pronounced than with anionic surfactants. The order of decrease of solubilization by a series of polyoxyethylene nonylphenol ethers in cyclohexane and benzene was: salts $\simeq$ base > acids [202]. The order of effectiveness of salts corresponds to that of increasing lyotropic numbers for anions and cations and this suggests that the effects of salts on solubilization by non-ionic surfactants may involve a salting-out action for hydrogen bonding between the ether oxygens and the solubilized water molecules. Shinoda and Ogawa [213] have demonstrated an increase in the optimum temperature for the solubilization of water (point A of Fig. 5.29) with increase in polyoxyethylene chain length. The optimum temperature for solubilization also depended on the solvent. Aromatic hydrocarbons such as toluene and *m*-xylene show about 40° C lower optimum temperature compared with saturated hydrocarbons such as methylcyclohexane and cyclohexane [213].

Solubilization studies on small organic molecules, such as methanol, acetic acid, and propylamine, by non-aqueous solutions of sodium and potassium dinonylnaphthalene sulphonates, and magnesium phenyl stearates have been reported [194, 214]. Bascom and Singleterry [213] indicated that in the initial stages of the uptake the solubilizate was co-ordinated by the cation. For methanol, which was solubilized to a larger extent than the other compounds at high concentrations, association with the hydrocarbon chains of the surfactant occurred. For this solubilizate there was some evidence of a decrease in micelle size on solubilization. A similar picture of a two-stage process in solubilization was apparent for acetic acid.

The naturally occurring phosphatide, lecithin, forms micelles in non-polar solvents like benzene, which have been suggested to be suitable models of the cell membrane. A potassium dye salt was solubilized to a greater extent than a sodium one by the large micelles present in this system [216]. The uptake of water has also been measured [217, 218], 0.33 g water per g lecithin being present in the lecithin–benzene system at the MAC; it was suggested that this water was associated with the polar head groups of the phosphatide molecules. The water did not significantly affect the number of monomers in the micelle, but did cause some changes in their shape.

The solubilization of dimethyl sulphoxide, imidazole, methanol, pyrazole, 2-pyridone and tetrabutylammonium perchlorate by dodecylammonium propionate in benzene, deuterochloroform and dichloromethane has been investigated using proton n.m.r. techniques [219]. The results suggest that hydrogen bonding is of primary importance for strong binding in these systems but weaker forces such as dipole–dipole interactions may also be of importance.

A comparison of the solubilization of water–formamide and water–*N*-

methlformamide mixtures by Aerosol OT in heptane [220] with the solubilization of pure water revealed a large increase in the total amount of these solubilizates incorporated into the micelles compared with water alone when the mole fraction of the amide in the mixture was low. Thus the amount of water–N-methylformamide mixture containing 0.02 mole fraction of the amide solubilized by a 5% w/v Aerosol OT solution was approximately 4.5 times that of pure water. The effect was attributed to a pronounced interaction between the amides, which have a very high permittivity, and the Na^+ counterions of the surfactant.

References

1. M. E. L. MCBAIN and E. HUTCHINSON (1955) *Solubilization and Related Phenomena,* Academic Press, New York.
2. H. B. KLEVENS (1950) *Chem. Rev.* **47**, 1.
3. C. ENGLER and E. DIEKHOFF (1892) *Arch. Pharm.* **230**, 561.
4. E. LESTER-SMITH (1932) *J. Phys. Chem.* **36**, 1401.
5. E. LESTER-SMITH (1932) *J. Phys. Chem.* **36**, 1672.
6. E. LESTER-SMITH (1932) *J. Phys. Chem.* **36**, 2455.
7. H. BRITZINGER and H. G. BEIER (1933) *Kolloid-Z.* **64**, 160.
8. J. W. MCBAIN and M. E. L. MCBAIN (1936) *J. Amer. Chem. Soc.* **58**, 2610.
9. F. VERZAR (1933) *Nutrit. Abs. Rev.* **2**, 441.
10. A. S. C. LAWRENCE (1937) *Trans. Faraday Soc.* **33**, 325.
11. A. S. C. LAWRENCE (1937) *Trans. Faraday Soc.* **33**, 815.
12. A. S. C. LAWRENCE (1940) *Ann. Reports* **37**, 107.
13. G. S. HARTLEY(1938) *J. Chem. Soc.* 1968.
14. J. W. MCBAIN and J. J. O'CONNOR (1940) *J. Amer. Chem. Soc.* **62**, 2855.
15. L. SJÖBLÖM (1956) *Acta Acad. Aboensis, Math. Phys.* **XX**, 14.
16. H. B. KLEVENS (1950) *J. Phys. Colloid Chem.* **54**, 283.
17. P. EKWALL, L. SJÖBLÖM and J. OLSEN (1953) *Acta Chem. Scand.* **7**, 347.
18. D. E. GUTTMAN, W. E. HAMLIN, J. W. SHELL and J. G. WAGNER (1961) *J. Pharm. Sci.* **50**, 305.
19. F. W. GOODHART and A. N. MARTIN (1962) *J. Pharm. Sci.* **51**, 50.
20. D. G. DERVICHIAN (1957) *Proceedings 2nd International Congress on Surface Activity,* Butterworths, London, Vol. 1, 327.
21. P. EKWALL, I. DANIELSSON and L. MANDELL (1960) *Kolloid-Z.* **169**, 113.
22. T. R. BATES, M. GIBALDI and J. L. KANIG (1966) *J. Pharm. Sci.* **55**, 191.
23. J. ROGERS (1966) M.Sc. Thesis, Toronto.
24. J. E. CARLESS and J. K. NIXON (1960) *J. Pharm. Pharmac.* **12**, 348.
25. W. HELLER and H. B. KLEVENS (1946) *J. Chem. Phys.* **14**, 567.
26. P. A. WINSOR (1950) *Trans. Faraday Soc.* **46**, 762.
27. W. J. O'MALLEY, L. PENNATI and A. N. MARTIN (1958) *J. Amer. Pharm. Ass. Sci. Edn.* **47**, 334.
28. N. A. HALL (1963) *J. Pharm. Sci.* **52**, 189.
29. D. GUERRITORE, L. BELLELLI and M. L. BONACCI (1964) *Ital. J. Biochem.* **13**, 222.
30. W. D. HARKINS, R. W. MATTOON and M. L. CORRIN (1946) *J. Colloid Sci.* **1**, 105.
31. W. D. HARKINS, R. W. MATTOON and M. L. CORRIN (1946) *J. Amer. Chem. Soc.* **68**, 220.
32. N. K. PATEL and H. D. KOSTENBAUDER (1958) *J. Amer. Pharm. Ass. Sci. Edn.* **47**, 289.
33. C. K. BAHL and H. D. KOSTENBAUDER (1964) *J. Pharm. Sci.* **53**, 1027.
34. N. K. PATEL and N. E. FOSS (1964) *J. Pharm. Sci.* **53**, 94.
35. R. A. ANDERSON and A. H. SLADE (1966) *J. Pharm. Pharmac.* **18**, 640.
36. A. G. MITCHELL and K. F. BROWN (1966) *J. Pharm. Pharmac.* **18**, 115.
37. H. MATSUMOTO, H. MATSUMURA and S. IGUCHI (1966) *Chem. Pharm. Bull.* **14**, 385.

38. S. J. A. KAZMI and A. G. MITCHELL (1973) *J. Pharm. Sci.* **62**, 1299.
39. S. J. A. KAZMI and A. G. MITCHELL (1967) *Canad. J. Pharm. Sci.* **11**, 10.
40. S. J. A. KAZMI and A. G. MITCHELL (1971) *J. Pharm. Pharmacol.* **23**, 482.
41. R. WITHINGTON and J. H. COLLETT (1973) *J. Pharm. Pharmacol.* **25**, 273.
42. M. DONBROW, E. AZAZ and R. HAMBURGER (1970) *J. Pharm. Sci.* **59**, 1427.
43. M. DONBROW and C. T. RHODES (1963) *J. Pharm. Pharmacol.* **15**, 233.
44. M. DONBROW and C. T. RHODES (1963) *J. Pharm. Pharmacol.* **17**, 258.
45. M. DONBROW and C. T. RHODES (1964) *J. Chem. Soc.* 6166.
46. M. DONBROW and J. JACOBS (1966) *J. Pharm. Pharmacol.* **18**, 92S.
47. M. DONBROW, P. MOLYNEUX and C. T. RHODES (1967) *J. Chem. Soc. A.* 561.
48. E. AZAZ and M. DONBROW (1976) *J. Colloid Interfac. Sci.* **57**, 11.
49. W. P. EVANS (1964) *J. Pharm. Pharmac.* **16**, 323.
50. E. AZAZ and M. DONBROW (1977) *J. Phys. Chem.* **81**, 1636.
51. R. W. ASHWORTH and D. D. HEARD (1966) *J. Pharm. Pharmac.* **18**, 98S.
52. E. HUTCHINSON and P. M. SCHAFFER (1962) *Z. Phys. Chem. N.E.* **31**, 397.
53. S. J. DOUGHERTY and J. C. BERG (1974) *J. Colloid Interfac. Sci.* **48**, 110.
54. T. SHIMAMOTO and Y. OGAWA (1975) *Chem. Pharm. Bull.* **23**, 3088.
55. T. SHIMAMOTO, H. MIMA and M. NAKAGAKI (1979) *Chem. Pharm. Bull.* **27**, 1995.
56. T. SHIMAMOTO and H. MIMA (1979) *Chem. Pharm. Bull.* **27**, 2602.
57. D. G. HERRIES, W. BISHOP and F. M. RICHARDS (1964) *J. Phys. Chem.* **68**, 1842.
58. B. BORGSTRÖM (1965) *Biochim. Biophys. Acta.* **106**, 171.
59. H. SUZUKI and T. SASAKI (1971) *Bull. Chem. Soc. Japan* **44**, 2630.
60. H. SUZUKI (1976) *Bull. Chem. Soc. Japan* **49**, 375.
61. A. GOTO, F. ENDO and K. ITO (1977) *Chem. Pharm. Bull.* **25**, 1165.
62. A. GOTO and F. ENDO (1978) *J. Colloid Interface Sci.* **66**, 26.
63. A. GOTO and F. ENDO (1979) *J. Colloid Interface Sci.* **68**, 163.
64. A. GOTO, M. NIHEI and F. ENDO (1980) *J. Phys. Chem.* **84**, 2268.
65. F. H. QUINA and V. G. TOSCANO (1977) *J. Phys. Chem.* **81**, 1750.
66. H. J. POWNALL and L. C. SMITH (1974) *Biochemistry* **13**, 2594.
67. R. R. HAUTALA, N. E. SCHORE and N. J. TURRO (1973) *J. Amer. Chem. Soc.* **95**, 5508.
68. E. LISSI, E. ABUIN and A. M. ROCHA (1980) *J. Phys. Chem.* **84**, 2406.
69. M. ALMGREN, F. GRIESER and J. K. THOMAS (1979) *J. Phys. Chem.* **101**, 279.
70. J. Y. PARK and E. G. RIPPIE (1977) *J. Pharm. Sci.* **66**, 858.
71. A. G. MITCHELL and J. F. BROADHEAD (1967) *J. Pharm. Sci.* **56**, 1261.
72. N. K. PATEL and N. E. FOSS (1965) *J. Pharm. Sci.* **54**, 1495.
73. J. H. COLLETT and L. KOO (1975) *J. Pharm. Sci.* **64**, 1253.
74. E. R. GARRETT (1966) *J. Pharm. Pharmacol.* **18**, 589.
75. G. SCATCHARD (1949) *Ann. N.Y. Acad. Sci.* **51**, 660.
76. P. MUKERJEE (1971) *J. Pharm. Sci.* **60**, 1531.
77. P. MUKERJEE (1979) in *Solution Chemistry of Surfactants* (ed. K. L. Mittal) Plenum, New York, Vol. 1, p. 153.
78. P. H. ELWORTHY and F. J. LIPSCOMB (1968) *J. Pharm. Pharmac.* **20**, 817.
79. B. A. MULLEY and A. D. METCALF (1956) *J. Pharm. Pharmac.* **8**, 774.
80. A. S. WAGGONER, O. H. GRIFFITH and C. R. CHRISTENSEN (1967) *Proc. Nat. Acad. Sci. USA.* **57**, 1198.
81. P. MUKERJEE (1971) *J. Phys. Chem.* **60**, 1528.
82. J. H. COLLETT, R. WITHINGTON and L. KOO (1975) *J. Pharm. Pharmac.* **27**, 46.
83. J. H. COLLETT and E. A. TOBIN (1979) *J. Pharm. Pharmacol.* **31**, 174.
84. W. PHILIPPOFF (1950) *J. Colloid Sci.* **5**, 169.
85. G. FOURNET (1951) *Disc. Faraday Soc.* **11**, 121.
86. G. S. HARTLEY (1949) *Nature* **163**, 767.
87. W. D. HARKINS and R. MITTLEMANN (1949) *J. Colloid Sci.* **4**, 367.
88. J. D. BERNAL (1946) *Trans. Faraday Soc.* **42B**, 197.
89. J. H. SCHULMAN and D. P. RILEY (1948) *J. Colloid Sci.* **3**, 383.

90. F. REISS-HUSSON and V. LUZZATI (1964) *J. Phys. Chem.* **68**, 3504.
91. F. REISS-HUSSON and V. LUZZATI (1966) *J. Colloid Interfac. Sci.* **21**, 534.
92. B. SVENS and B. ROSENHOLM (1973) *J. Colloid Interfac. Sci.* **44**, 495.
93. S. RIEGELMAN, N. A. ALLAWALA, M. K. HRENOFF and L. A. STRAIT (1958) *J. Colloid Sci.* **13**, 208.
94. M. DONBROW and C. T. RHODES (1966) *J. Pharm. Pharmac.* **18**, 424.
95. S. J. REHFELD (1970) *J. Phys. Chem.* **74**, 117.
96. S. J. REHFELD (1971) *J. Phys. Chem.* **75**, 3905.
97. J. H. FENDLER, E. J. FENDLER, G. A. INFANTE, P. S. SHIH and L. K. PATTERSON (1975) *J. Amer. Chem. Soc.* **97**, 89.
98. T. CORBY and P. H. ELWORTHY (1971) *J. Pharm. Pharmac.* **23**, 49S.
99. A. L. THAKKAR and P. B. KUEHN (1969) *J. Pharm. Sci.* **58**, 850.
100. J. R. CARDINAL and P. MUKERJEE (1978) *J. Phys. Chem.* **82**, 1614.
101. P. MUKERJEE and J. R. CARDINAL (1978) *J. Phys. Chem.* **82**, 1620.
102. S. G. BJAASTAD and N. A. HALL (1967) *J. Pharm. Sci.* **56**, 504.
103. E. M. KOSOWER (1958) *J. Amer. Chem. Soc.* **80**, 3253.
104. E. M. KOSOWER (1961) *J. Amer. Chem. Soc.* **83**, 3142.
105. A. L. THAKKAR and N. A. HALL (1967) *J. Pharm. Sci.* **56**, 1121.
106. J. C. ERIKSSON and G. GILLBERG (1966) *Acta. Chem. Scand.* **20**, 2019.
107. J. E. GORDON, J. C. ROBERTSON and R. L. THORNE (1970) *J. Phys. Chem.* **74**, 957.
108. J. C. ERIKSSON (1963) *Acta. Chem. Scand.* **17**, 1478.
109. J. J. JACOBS, R. A. ANDERSON and T. R. WATSON (1971) *J. Pharm. Pharmac.* **23**, 148.
110. E. J. FENDLER, C. L. DAY and J. H. FENDLER (1972) *J. Phys. Chem.* **76**, 1460.
111. J. H. FENDLER and E. J. FENDLER (1975) in *Catalysis in micellar and macromolecular system*, Academic Press, London, Ch. 2.
112. M. GRÄTZEL, K. KALYANSUNDARAM and J. K. THOMAS (1974) *J. Amer. Chem. Soc.* **96**, 7869.
113. C. A. BUNTON, M. J. MINCH, J. HIDALGO and L. SEPULVEDA (1973) *J. Amer. Chem. Soc.* **95**, 3262.
114. C. A. BUNTON and M. J. MINCH (1974) *J. Phys. Chem.* **78**, 1490.
115. J. ULMIUS, B. LINDMAN, G. LINDBLOM and T. DRAKENBERG (1978) *J. Colloid Interfac. Sci.* **65**, 88.
116. R. C. DORRANCE and T. F. HUNTER (1977) *J. Chem. Soc. Farad I.* **73**, 1891.
117. S. MIYAGISHI and M. NISHIDA (1980) *J. Colloid Interfac. Sci.* **73**, 270.
118. K. K. FOX, I. D. ROBB and R. SMITH (1972) *J. Chem. Soc. Farad. I* **68**, 445.
119. R. C. DORRANCE and T. F. HUNTER (1972) *J. Chem. Soc. Farad. I* **68**, 1312.
120. R. C. DORRANCE and T. F. HUNTER (1974) *J. Chem. Soc. Farad. I* **70**, 1572.
121. R. C. DORRANCE, T. F. HUNTER and J. PHILIP (1977) *J. Chem. Soc. Farad. I* **73**, 89.
122. A. NAKAJIMA (1971) *Bull. Chem. Soc. Japan* **44**, 3272.
123. A. S. WAGGONER, A. D. KEITH and O. H. GRIFFITH (1968) *J. Phys. Chem.* **72**, 4129.
124. J. OAKES (1971) *Nature* **231**, 38.
125. J. OAKES (1973) *J. Chem. Soc., Farad. II* **69**, 1464.
126. H. YOSHIOKA (1979) *J. Am. Chem. Soc.* **101**, 28.
127. T. NAKAGAWA and K. TORI (1964) *Kolloid-Z. u. Z.-Polymere* **194**, 143.
128. T. NAKAGAWA and H. JIZOMOTO (1972) *Kolloid-Z. u. Z. Polymere* **250**, 594.
129. T. NAKAGAWA and H. JIZOMOTO (1974) *Colloid Polymer Sci.* **252**, 482.
130. D. HALL, P. L. JOBLING, E. WYN-JONES and J. E. RASSING (1977) *J. Chem. Soc., Farad. II* **73**, 1582.
131. J. GETTINS, D. HALL, P. L. JOBLING, J. E. RASSING and E. WYN-JONES (1979) *J. Chem. Soc., Farad. II* **75**, 1957.
132. B. H. ROBINSON, N. C. WHITE and C. MATEO (1975) *Adv. Mol. Relaxation Processes* **7**, 321.
133. A. A. ISMAIL, M. W. GOUDA and M. M. MOTAWI (1970) *J. Pharm. Sci.* **59**, 220.
134. N. N. SALIB, A. A. ISMAIL and A. S. GENEIDI (1974) *Pharm. Ind.* **36**, 108.

Solubilization · 291

135. T. ARNARSON and P. H. ELWORTHY (1980) *J. Pharm. Pharmac.* **32**, 381.
136. T. ARNARSON and P. H. ELWORTHY (1981) *J. Pharm. Pharmac.* **33**, 141.
137. I. M. KOLTHOFF and W. STRICKS (1948) *J. Phys. Colloid Chem.* **52**, 195.
138. J. W. MCBAIN and A. A. GREEN (1946) *J. Amer. Chem. Soc.* **68**, 1731.
139. M. W. RIGG and F. W. LIU (1953) *J. Amer. Oil. Chemists' Soc.* **30**, 14.
140. P. T. JACOBS and E. W. ANACKER (1973) *J. Colloid Interfac. Sci.* **43**, 105. ibid., 1976, **56**, 255.
141. P. T. JACOBS and E. W. ANACKER (1976) *J. Colloid Interfac. Sci.* **56**, 255.
142. F. TOKIWA (1968) *J. Phys. Chem.* **72**, 1214.
143. H. SCHOTT (1967) *J. Phys. Chem.* **71**, 3611.
144. F. W. GOODHART and A. N. MARTIN (1962) *J. Pharm. Sci.* **51**, 50.
145. B. W. BARRY and D. I. D. EL EINI (1976) *J. Pharm. Pharmacol.* **28**, 210.
146. M. W. GOUDA, A. A. ISMAIL and M. M. MOTAWI (1970) *J. Pharm. Sci.* **59**, 1402.
147. G. S. HARTLEY (1955) in *Progress in the Chemistry of Fats and the other Lipids* Vol. III (ed. Holman, Lundberg and Malkin) Pergamon Press.
148. R. S. STEARNS, H. OPPENHEIMER, E. SIMON and W. D. HARKINS (1947) *J. Chem. Phys.* **15**, 496.
149. M. J. SCHWUGER (1972) *Kolloid-Z. u. z. Polymere* **250**, 703.
150. A. J. HYDE and D. J. M. ROBB (1964) *Proceedings 4th International Congress on Surface Active Substances*, Brussels, Gordon and Breach, New York.
151. M. SMITH and A. E. ALEXANDER (1957) *Proceedings 2nd International Congress on Surface Activity*, Butterworths, London, Vol. 1, 349.
152. T. NAKAGAWA and K. TORI (1960), *Kolloid Z. Z. Polymere*, **168**, 132.
153. J. W. MCBAIN and P. H. RICHARDS (1946) *Ind. Eng. Chem.* **38**, 642.
154. P. EKWALL, T. LUNDSTEN and L. SJÖBLOM (1951) *Acta. Chem. Scand.* **5**, 1383.
155. S. D. HOYES and L. SAUNDERS (1966) *Biochim. Biophys. Acta.* **116**, 184.
156. I. W. KELLAWAY and L. SAUNDERS (1967) *Biochim. Biophys. Acta.* **144**, 145.
157. M. M. GALE and L. SAUNDERS (1971) *Biochim. Biophys. Acta.* **248**, 466.
158. H. TOMIDA, T. YOTSUYANAGI and K. IKEDA (1978) *Chem. Pharm. Bull.* **26**, 2824.
159. H. TOMIDA, T. YOTSUYANAGI and K. IKEDA (1978) *Chem. Pharm. Bull.* **26**, 2832.
160. A. CAMMARATA, J. H. COLLETT and E. TOBIN (1980) in *Physical Chemical Properties of Drugs* (eds S. H. Yalkowsky, A. A. Sinkula and S. C. Valvani) Marcel Dekker, New York, ch. 8.
161. A. KAMINSKI and J. W. MCBAIN (1949) *Proc. Roy. Soc.* **A198**, 447.
162. K. KURIYAMA (1962) *Kolloid-Z* **180**, 55.
163. K. J. HUMPHREYS and C. T. RHODES (1968) *J. Pharm. Sci.* **57**, 79.
164. I. M. KOLTHOFF and W. F. GRAYDON (1951) *J. Phys. Colloid Chem.* **55**, 708.
165. J. W. MCBAIN, R. C. MERRILL and J. R. VINOGRAD (1941) *J. Amer. Chem. Soc.* **63**, 670.
166. P. EKWALL, L. MANDELL and K. FONTELL (1977) *J. Colloid Interfac. Sci.* **61**, 519.
167. A. GOTO, R. SAKURA and F. ENDO (1980) *Chem. Pharm. Bull.* **28**, 14.
168. H. SAITO and K. SHINODA (1967) *J. Colloid Interfac. Sci.* **24**, 10.
169. A. M. MANKOWICH (1955) *Ind. Eng. Chem.* **47**, 2175.
170. R. C. PINK (1939) *J. Chem. Soc.* 53.
171. J. WEICHERZ (1929) *Kolloid-Z.* **47**, 133.
172. J. WEICHERZ (1929) *Kolloid-Z.* **49**, 133.
173. P. A. WINSOR (1956) *Manuf. Chemist.* 89.
174. P. A. WINSOR (1956) *Manuf. Chemist.* 130.
175. J. W. MCBAIN and A. A. GREEN (1947) *J. Phys. Colloid Chem.* **51**, 286.
176. A. W. RALSTON and D. N. EGGENBERGER (1949) *J. Phys. Colloid Chem.* **52**, 1494.
177. A. W. RALSTON and C. W. HOERR (1946) *J. Amer. Chem. Soc.* **68**, 2460.
178. P. F. H. WARD (1940) *Proc. Roy. Soc.* **A176**, 412.
179. H. B. KLEVENS (1949) *J. Chem. Phys.* **17**, 1004.
180. H. B. KLEVENS (1950) *J. Amer. Chem. Soc.* **72**, 3780.
181. M. J. CROOKS and K. F. BROWN (1973) *J. Pharm. Pharmacol.* **25**, 281.

182. T. LÖVGREN, B. HEIKIUS, B. LUNDBERG and L. SJÖBLOM (1978) *J. Pharm. Sci.* **67**, 1419.
183. B. LUNDBERG, T. LÖVGREN and B. HEIKIUS (1979) *J. Pharm. Sci.* **68**, 542.
184. W. N. MACLAY (1956) *J. Colloid Sci.* **11**, 272.
185. M. H. J. WEIDEN and L. B. NORTON (1953) *J. Colloid Sci.* **8**, 606.
186. H. K. LIVINGSTON (1954) *J. Colloid Sci.* **9**, 365.
187. J. W. HADGRAFT (1954) *J. Pharm. Pharmacol.* **6**, 816.
188. M. DONBROW and E. AZAZ (1976) *J. Colloid Interface Sci.* **57**, 20.
189. J. NAKAGAWA, K. KURIYAMA and H. INOUE (1959) *Symposium on Colloid Chem.* *(Chem. Soc. Japan)*, 12th Symposium, p. 32.
190. T. NAKAGAWA, K. KURIYAMA and H. INOUE (1960) *J. Colloid Sci.* **15**, 268.
191. D. ATTWOOD, P. H. ELWORTHY and S. B. KAYNE (1971) *J. Pharm. Pharmac.* **23**, 775.
192. M. B. MATHEWS and E. J. HIRSCHHORN (1952) *J. Colloid Sci.* **8**, 86.
193. M. ZULAUF and H-F. EICKE (1979) *J. Phys. Chem.* **83**, 480.
194. S. KAUFMAN (1964) *J. Phys. Chem.* **68**, 2814.
195. K. KON-NO and A. KITAHARA (1971) *J. Colloid Interfac. Sci.* **35**, 409.
196. S. G. FRANK and G. ZOGRAFI (1969) *J. Colloid Interfac. Sci.*, **29**, 27.
197. K. KON-NO and A. KITAHARA (1971) *J. Colloid Interfac. Sci.* **37**, 469.
198. K. KON-NO and A. KITAHARA (1970) *J. Colloid Interfac. Sci.* **33**, 124.
199. K. KON-NO and A. KITAHARA (1970) *J. Colloid Interfac. Sci.* **34**, 221.
200. K. SHINODA and T. OGAWA (1967) *J. Colloid Interfac. Sci.* **24**, 56.
201. K. KON-NO, Y. VENO, Y. ISHII and A. KITAHARA (1971) *Nippon Kagaku Zasshi*, **92**, 381 [*CA* (1971) **75**, 80877c].
202. A. KITAHARA and K. KON-NO (1966) *J. Phys. Chem.* **70**, 3394.
203. K. KON-NO and A. KITAHARA (1972) *J. Colloid Interfac. Sci.* **41**, 47.
204. S. R. PALIT and V. VENKATESWARLU (1951) *Proc. Roy. Soc.* **A208**, 542.
205. A. KITAHARA (1956) *Bull. Chem. Soc. Japan* **29**, 15.
206. S. R. PALIT, V. A. MOGHE and B. BISWAS (1959) *Trans. Farad. Soc.* **55**, 463.
207. S. R. PALIT and V. VENKATESWARLU (1954) *J. Chem. Soc.* 2129.
208. W. I. HIGUCHI and J. MISRA (1962) *J. Pharm. Sci.* **51**, 455.
209. H. WEDELL (1960) *Int. Congr. Surfac. Activ.* 3rd, Vol. IV, p. 220.
210. M. NAKAGAKI and S. SONE (1964) *Yakugaku Zasshi* **84**, 151 [*CA* (1964) **61**, 5911A].
211. H. SAITO (1972) *Nippon Kagaku Kaishi* 491 [*CA* (1972) **77**, 77053S].
212. M. NAKAGAKI and S. SONE (1965) *Yakugaku Zasshi* **85**, 125 [*CA* (1965) **62**, 13392].
213. K. SHINODA and T. OGAWA (1967) *J. Colloid Interfac. Sci.* **24**, 56.
214. J. KAUFMAN (1962) *J. Colloid Sci.* **17**, 231.
215. W. D. BASCOM and C. R. SINGLETERRY (1958) *J. Colloid Sci.* **13**, 569.
216. I. BLEI and R.E. LEE (1963) *J. Phys. Chem.* **67**, 2085.
217. P. A. DEMCHENKO (1960) *Kolloid-Z* **22**, 297.
218. P. EKWALL, I. DANIELSSON and L. MANDEL (1960) *Proceedings 3rd International Congress on Surface Activity* Universitätsdruckerei Mainz, Vol. 1, 89.
219. O. A. EL-SEOUND, E. J. FENDLER and J. H. FENDLER (1974) *J. Chem. Soc. Farad. I* **70**, 450, 459.
220. D. ATTWOOD, C. MCDONALD and S. C. PERRY (1975) *J. Pharm. Pharmac.* **27**, 694.

6 *Pharmaceutical aspects of solubilization*

6.1 Introduction

Solubilization in surfactant solutions above the critical micelle concentration offers one approach to the formulation of poorly soluble drugs in solution form [1].

The objective of this chapter and of Chapter 7 is to review the state of the art of solubilization in surfactant systems with emphasis on the consequences of a surfactant presence in pharmaceutical formulations. In particular, emphasis will be placed on the effect of surfactants on bioavailability and the toxicity of formulations for neglect of these topics will, on the one hand, prevent the realization of the potential of surfactant systems and, on the other, might lead to the unwise use of surfactants in formulations. Some attempt will be made to place the topic in perspective and to answer the question as to the real value of surfactant solubilization in pharmaceutical formulation.

It is perhaps true that micellar solubilization has not made much impact on drug formulation. There are relatively few marketed products which could be considered to be isotropic solutions of drug and surfactant in either the UK or the USA, although surfactants are present in many formulations as minor adjuvants and to that extent their presence and influence is perhaps hidden.

The limiting factors in the use of solubilizers as effective formulation aids are (i) the finite capacity of the micelles for the drug, (ii) the possible short- or long-term adverse effects of the surfactant on the body, and (iii) the concomitant solubilization of other ingredients such as preservatives, flavouring and colouring matter in the formulation with consequent alterations in stability and effectiveness. Nonetheless, there is scope for development simply because there is a need for agents to increase the solubility of poorly soluble drugs even if only at the stage of pharmacological evaluation where, indeed, surfactants are used often without due regard to the implications. The use of co-solvents and surfactants to solve problems of low solubility has the advantage that the drug entity can be used without chemical modification and hence toxicological data do not have to be repeated as would be the case when alternative approaches are used to produce more soluble compounds. Some caution has, however, to be adopted in the interpretation of animal pharmacology and toxicology on formulations which differ from the final marketed product, especially if the final preparation contains surfactant but early test formulations do not or *vice versa*. Surfactants, as we will see in Chapter 7, are not inert substances some having distinctive pharmacological actions. There is a demonstrable need for the development of less toxic surfactants; the polyoxyethylene–polyoxypropylene block co-polymers which will be discussed later seem to have fewer side effects than conventional surfactants and seem to be worthy of further investigation.

With the development of new dosage form technology in which control of drug release is achieved, it is conceivable that micellar systems will find some place because of the ability of the micellar phase to alter the transport properties of solubilized drug molecules. One can envisage the deliberate addition of surfactants to drug reservoirs to control the exit rate of drugs from polymeric devices. This will be explained in Chapter 7.

This chapter is restricted mainly to aqueous systems and the solubilization of water-insoluble and poorly soluble drug entities and pharmaceutical additives, and, because of the lesser toxicity of non-ionic surfactants, it will concentrate on non-ionic surfactant systems. Wherever possible cited work refers to systems which have potential utility in pharmacy as there is a danger that all our knowledge is gained on model systems (of toxic ionic surfactants which are used because they are available in a pure state, benzene or similar well-defined solutes, and other unacceptable additives such as propanol) while we remain blissfully unaware of how to solve the real problems that arise [1].

As we have seen in Chapter 1, the range of available surfactants is wide, and so, too, are the mechanisms of solubilization and the effects the surfactants have on the solubilized material. Examples are known of enhanced drug activity and of inactivation, of increased stability, and instability; the interactions of the surfactants with components of the body must also be considered. In the case of insoluble drugs, the presence of micelles may enhance their activity through solubilization and transport to the site of action, a process which otherwise might have been a slow one. This has, of course, dire consequences in the case of carcinogens: normally insoluble carcinogenic substances which may be ingested may become very active in combination with surfactants, and, as the latter are taken in increasing amounts in food (non-ionics in bread is one example), this is a problem which warrants further study. Drugs which are meant to act on the intestinal mucosa, such as sulphaguanidine, might be inadvertently solubilized. There is the problem, especially with non-ionic surfactants, of interactions with preservatives in pharmaceuticals and consequent loss of biological action.

Some drugs themselves are surface-active and form micelles. While surface activity may not, in all cases, be the cause of their biological activity, it must in some way influence it and modify their interaction with the components of dosage forms or the components of the body. Surface-active drugs and surfactant molecules will interact to form mixed micelles at sufficiently high concentrations, a phenomenon which has implications for the thermodynamic activity and possibly the biological activity of the drug molecule.

Since 1964 there have been several comprehensive reviews of solubilization in surfactant systems, notably those by Swarbrick [2], Mulley [3], Sjöblom [4], Droseler and Voight [5], Elworthy *et al.* [6], and Florence [1]. These reviews together cite over a thousand sources primarily concerned with pharmaceutical applications. Other major publications which deal with micellar systems implicating solubilized species include Cordes [7], Fendler and Fendler [8] and the collections of papers edited by Mittal contain several contributions on the topic [9].

In this chapter the solubilization of a number of classes of drugs and pharmaceutical products will be dealt with; in some cases the division into sections has had to be somewhat arbitrary, but, as far as possible, compounds with similar structures, such as the steroids, have been dealt with as a group.

6.2 Solubilization of drugs

6.2.1 Antibacterial compounds

(A) PHENOLIC COMPOUNDS

Solutions of cresol with soap were early pharmaceutical examples of solubilized systems. Phenol itself is soluble in water to the extent of 7.7% (w/v), but it has disadvantages; the alternatives, cresol, chlorocresol, chloroxylenol, and thymol, are much less soluble in water, and their use as disinfectants has led to the need for formulation in surfactant solutions.

Solution of cresol with soap (lysol) is a saponaceous solution containing 50% v/v cresol. Its monograph specifies no particular soap, although activity of the preparation depends to a large extent on the type of soap employed. Although still used, the absence of strict standards for lysol, the widely varying phenol fractions used in its preparation, and the varying properties of the soaps make it an unsatisfactory solution. The high toxicity of phenol and the cresols has mitigated against their more widespread use. Emphasis is now being placed on their chlorinated derivatives, chloroxylenol and chlorocresol. Chloroxylenol is a potent, non-irritant bactericide of low toxicity. It has, however, a low solubility in water, 0.031 g ml^{-1} at $20°C$ [10]; the official preparation, Solution of Chloroxylenol B.P., contains 5% v/v chloroxylenol with terpineol in an alcoholic soap solution. A modification of this, claimed to be less alkaline, has been described by Lloyd and Clegg [11]. There are numerous commercial formulations with a wide spread of Rideal–Walker coefficients.

Mulley and Metcalf [12] have carried out detailed investigations of the phase behaviour of non-ionic detergent systems containing chloroxylenol (4-chloro-3,5-xylenol). Two of their phase diagrams are reproduced in Fig. 6.1. The surfactant $C_6H_{13}(OCH_2CH_2)_6OH$ is an efficient solubilizer above its CMC, which is approximately 3% w/w, but high concentrations are required to form isotropic liquids containing reasonable quantities of chloroxylenol. $C_6H_{13}(OCH_2CH_2)_2OH$ requires concentrations above 50% w/w to achieve an isotropic solution, and this compound probably acts more as a hydrotrope than as a micellar solubilizer in this concentration region. The former detergent forms only small aggregates of about 13 monomers in aqueous solution [13]. To obtain systems suitable for use it is essential to increase the alkyl chain length; $C_{10}H_{21}(OCH_2CH_2)_6OH$ has a low CMC and its micelles are reasonably large, containing 73 monomers at 25°C [14]. Isotropic micellar systems are formed at lower concentrations of detergent than for the shorter alkyl-chain homologues. However, there is a concomitant increase in the complexity of the phase diagram with the formation of liquid crystalline phases (Fig. 6.1).

There is apparently no evidence from these phase diagrams for the existence of simple phenol–glycol chain complexes: the liquid which separates at a solubility limit is a solution of variable composition. Ultraviolet spectroscopy shows that a hydrogen-bonded complex between the phenolic hydroxyl group and the other oxygens of cetomacrogol 1000 ($C_{16}H_{33}(OCH_2CH_2)_{23-24}OH$) is formed when chloroxylenol is solubilized by this commercial non-ionic surfactant. The

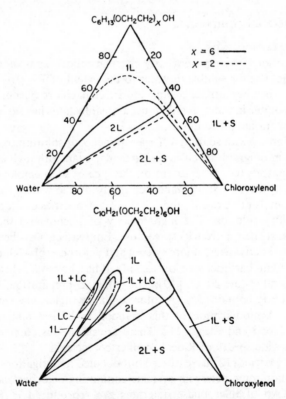

Figure 6.1 The upper phase diagram (after Mulley and Metcalf [12]) illustrates the phases existent in the system $C_6H_{13}(OCH_2CH_2)_xOH$: 4-chloro-3,5-xylenol: water at 20° C. The dotted line represents the behaviour when $x = 2$ and the solid lines where $x = 6$. The lower diagram shows the much more complex behaviour in the system: $C_{10}H_{21}(OCH_2CH_2)_6OH$: chloroxylenol: water. IL = isotropic liquid; LC = liquid crystalline; S = solid 2L = two liquids (immiscible.)

solubility of the chloroxylenol is directly proportional to the surfactant concentration, above the CMC [15]. However, rough determinations of the solubilities of resorcinol and phenol in cetomacrogol solutions, varying in concentration from 1 to 20%, showed these compounds to be less soluble than in water, although their solubility was proportional to detergent concentration [16].

Solutions of phenols in ionic systems exhibit similar behaviour. An initial fall in the solubility of 2-hydroxyphenol and 4-benzylphenol in potassium laurate solutions was noted below the CMC of the soap [17]. Few workers have commented on this 'insolubilization'; compounds with very low water solubility possibly do not show this property. That it is not restricted to phenols is shown by the results of Heller and Klevens [18] for ethyl benzene in potassium laurate. Ethyl benzene has a solubility in water similar to that of 4-benzylphenol.

The binding of series of phenols, cresols and xylenols to the non-ionic surfactant cetomacrogol 1000 can be described by a Langmuir adsorption isotherm [19]

$$x = K_1 K_2 c / (1 + K_1 c)$$

where x is the solute bound (mmol g^{-1} micelle), c is the concentration of free unionized solute (mmol), K_1 is the binding constant (1 mmol^{-1}) and K_2 the solute bound at hypothetical saturation (mmol g^{-1} micelle). The combined parameter $K_1 K_2$ is specific for each system and may be defined as the distribution coefficient of the solubilizate at infinitely dilute solubilizate concentration (P_0), Azaz and Donbrow [19] assert. Its value characterizes ideal behaviour both in the aqueous and micellar phases hence strictly would be subject to activity corrections. Binding capacity is inversely related to the water solubility of the phenol, cresol and xylenol, as can be seen in Table 6.1.

Values of P_0 in 0.1 M NaCl are also shown for a few compounds in this Table. A log–log plot of binding capacity and aqueous solubility yields a straight line. Azaz

Table 6.1 Aqueous solubility and distribution coefficient at infinite dilution (P_0) between cetomacrogol and water of phenols at 25° C

Compound	Solubility in water (mol l^{-1} × 10^3)	P_0 in water*	Solubility in 0.1 M NaCl (mol l^{-1} × 10^3)	P_0 in 0.1 M NaCl*
Phenol	1000	42.0	233	117
o-Cresol	240	79.5	188	80.6
p-Cresol	199	76.4	133	
m-Cresol	142	85.1		
2,4-Xylenol	51.0	125		
2,6-Xylenol	49.5	114		
3,5-Xylenol	40.0	132		190
3,4-Xylenol	39.0	151		
2,3-Xylenol	37.4	169		
2,5-Xylenol	29.0	197		

* Units: (1/g) × 10^3 = 1/1000 g or dimensionless units assuming density of cetomacrogol is unity at 25° C. Measured in 2% cetomacrogol. From Azaz and Donbrow [19].

and Donbrow's work has supported earlier work [20–23] which demonstrated that in unsaturated systems the binding 'constants' of solubilizates to surfactants are concentration-dependent and not, in fact, constant as some authors have assumed (e.g. [24–26]). This variation may have important practical implications in formulation.

A wider range of 34 benzoic acid derivatives has been studied in detail by Tomida et al. [27]. Using a solubility method these workers obtained saturation solubilities of the benzoic acid derivatives in Brij 35 (a polyoxyethylene lauryl ether) over a range of concentrations. Solubility ratios, calculated as the solubility in the surfactant solution/solubility in HCl, were a linear function of surfactant concentration allowing the calculation of a partition coefficient P_m which can be defined as

$$P_m = \frac{C_m}{C_a}, \tag{6.1}$$

where C_m and C_a are the concentrations in the micellar and aqueous phases, respectively. P_m is obtained from the solubility data as

$$\frac{S_t}{S_a} = (P_m - 1)\bar{v}C_s + 1 \tag{6.2}$$

where S_t is the total solubility of solubilizate in the presence of surfactant at concentration C_s, S_a is the solubility in the absence of surfactant and $\bar{v}$ is the partial molar volume of the surfactant. Some of the extensive data is reproduced in Table 6.2 for the ortho, para and meta substituents. The data are consistent with the findings of Azaz and Donbrow: the order of aqueous solubilities is always ortho > meta > para and the order of P_m is the opposite except for the hydroxybenzoic acids for which the ortho compound was solubilized most, followed by para and meta compounds. Patel and Foss [21] obtained the

Table 6.2 Aqueous solubilities, S_a, and partition coefficients of benzoic acids, P_m, between aqueous and micellar phases obtained from solubility method*

Substituent	ortho		meta		para	
	$S_a(\text{mol l}^{-1})$	P_m	$S_a(\text{mol l}^{-1})$	P_m	$S_a(\text{mol l}^{-1})$	P_m
H			2.61×10^{-2}	57.4		
F	4.05×10^{-2}	43.1	1.65×10^{-3}	91.2	4.98×10^{-3}	94.6
Cl	8.66×10^{-3}	99.8	1.92×10^{-3}	346	3.48×10^{-4}	446
Br	5.29×10^{-3}	150	1.36×10^{-3}	505	1.42×10^{-4}	634
I	1.75×10^{-3}	271	2.74×10^{-4}	1150	9.16×10^{-6}	908
CH$_3$	6.55×10^{-3}	120	6.13×10^{-3}	166	2.23×10^{-3}	163
OCH$_3$	2.50×10^{-2}	32.0	1.18×10^{-2}	72.8	1.30×10^{-3}	109
OH	1.08×10^{-2}	116	5.71×10^{-2}	38.3	4.17×10^{-2}	42.2
NO$_2$	2.53×10^{-2}	47.5	1.57×10^{-2}	96.8	1.01×10^{-3}	117
CN			2.35×10^{-3}	50.1	5.60×10^{-3}	57.0
COOH	2.57×10^{-2}	22.4	4.42×10^{-4}	155	6.50×10^{-5}	69.8

* From [27].

magnitude of interaction of hydroxy-, chloro- and aminobenzoic acids in polysorbate 80 and cetomacrogol 1000; hydroxy and amino derivatives showed the order of interaction to be ortho > para > meta. Substitution of a hydroxy group in the ortho position results in more affinity for any surfactant than para and meta substituents. This can be explained by the fact that the intramolecular hydrogen bonding increases the proton-donating nature of the carboxylic group. The greater the dissociation constant of the acid group the greater the hydrogen bonding to the oxyethylene groups in the micelle.

Plots of $\log P_m$ versus $\log P_{octanol}$ for the 34 compounds studied produced three groupings of results [27] to which the following equations applied.

$$\begin{array}{cccc} & n & r & s \end{array}$$

$$\log P_m = 0.921 \log P_{octanol} + 0.118 \quad 21 \quad 0.986 \quad 0.080 \tag{6.3}$$

$$\log P_m = 0.881 \log P_{octanol} + 0.392 \quad 5 \quad 0.999 \quad 0.014 \tag{6.4}$$

$$\log P_m = 0.968 \log P_{octanol} + 0.600 \quad 3 \quad 0.999 \quad 0.036 \tag{6.5}$$

where n is the number of points used in the regression, r is the correlation coefficient, and s is the standard deviation. Equation 6.3 applies to the majority of the compounds studied; Equation 6.4 to the nitro and cyano derivatives and the last equation to the compounds with dicarboxylic groups. The intercept values of the three groups are quite different; it is believed that the magnitude of the intercept is a reflection of the site of solubilization in the micelle. The closer the environment is to the nature of octanol used in the partitioning studies to obtain P_{oct}, the closer the intercept should be to zero. A negative intercept (for salicylic acid derivatives) has been identified with solubilization in the hydrocarbon core.

The site of solubilization while of little practical importance in the design of pharmaceutical formulations is of more than academic interest as the position of the solubilizate in the micelle may determine its stability and reactivity towards attacking species in the continuous phase (see Chapter 11).

To obviate the problem of the different affinities of ionized and unionized species for micelle, Tomida *et al.* [27] carried out their investigations at a pH such that ionization was suppressed. pH is rarely as low as that in this work and its influence on solubilization must be considered. Although the hydrogen ion concentration can influence the solution properties of non-ionic surfactants [28], the principal influence on uptake is exercised through the effect of pH on the equilibrium between ionized and unionized drug or solute species. This effect has been studied in most detail by Collett and Koo [29]. Increasing pH leads to a decrease in the micellar uptake of organic acids because of increasing solute solubility in the aqueous phase through increased ionization. This effect is clearly seen in Fig. 6.2 when the results of uptake of 4-chlorobenzoic acid between pH 3 and 4.40 are plotted as a ratio of its solubility in water of the appropriate pH, i.e. as the solubility ratio *R*. Considering the micellar species to form a phase or pseudophase allows a simple quantitative measure of the interaction between solubilizate and micelle. The concentration of solubilizate in the micelle is related to its concentration in the aqueous phase by a partition coefficient as defined in Equation 6.1.

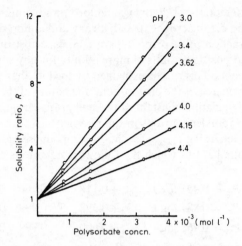

Figure 6.2 The influence of polysorbate 20 concentration and pH on the solubility ratio, R, of 4-chlorobenzoic acid. From Collett and Koo [29].

Curved Scatchard plots for the interaction of propyl *p*-hydroxybenzoate (propyl paraben) with four polyoxyethylene dodecyl ethers are shown in Fig. 6.3 [31]. The primary class of binding sites exhibited a high affinity and a low capacity for the preservative while the secondary sites had a low affinity and large binding capacity. Thus

$$r = \frac{n_1 K_1 D_f}{1 + K_1 D_f} + n_2 K_2 [D_f] \tag{6.6}$$

Analysis of these has allowed n_1, n_2, K_1 and K_2 to be estimated and related to the nature of the binding process in the micelles, especially in respect of the interaction with the hydrophilic polyoxyethylene layer [32].

Of great practical importance are the effects of additives on the binding of preservative molecules to surfactant micelles. Blanchard *et al.*, [33] have confirmed the negligible effect of sorbitol on the interaction of phenolic preservatives with polysorbate 80 using a Scatchard approach. The sorbitol is probably too hydrophilic to interact with the micelle and thus does not compete for binding sites. Similar conclusions were reached by Shimamoto and Mima [34] studying the effects of glycerol, propylene glycol and 1,3-butylene glycol on paraben–non-ionic surfactant interactions. These polymers had little effect on the binding of preservatives to the primary binding sites located at the core/PEG boundary of the micelle but they were thought to decrease binding at the secondary sites, 1,3-butylene glycol being most effective in displacing the preservatives. These secondary sites are reckoned to be non-specific and located in the PEG layer. As materials such as 1,3-butylene glycol may penetrate the PEG region they would probably displace solubilizate molecules. It would seem that displacement from the primary site would require much greater structural specificity (see discussion on interaction of preservative mixtures with micelles,

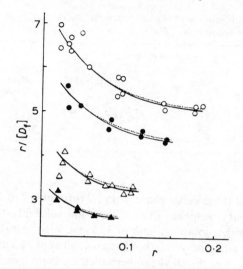

Figure 6.3 Scatchard plots for the interaction of propyl paraben with polyoxyethylene dodecyl ethers, n being the number of oxyethylene units.
○ $n = 15$; ● $n = 20$; △ $n = 30$; ▲ $n = 50$
From [31] with permission.

below). Any displacement of preservative from the micelle is likely to increase the preservative activity of the formulation.

The effects of added electrolytes on solubilized systems are discussed in Chapter 5. In Table 6.1 it can be seen that the addition of sodium chloride to a non-ionic system increases the P_m of the solubilizate. An electrolyte can have a dual effect, first on the properties of the surfactant and secondly on the solubilizate. If the electrolyte salts out the solubilizate P_m will increase, an effect observed with non-ionic surfactant systems whose micelles would be increased in size by such electrolytes. In ionic surfactant systems the effect can be more complex. The addition of electrolyte to an ionic surfactant results in a decrease in CMC, increase in micellar size and a decrease in effective charge per monomer, probably leading to a greater concentration of head groups and a more rigid micellar interior [35] which might result in decreased uptake of solubilizate into the micellar core. Uptake of methyl and ethyl paraben is increased by the addition of 10 mM NaCl to sodium lauryl sulphate [36] (see Table 6.3). As both electrolyte and the presence of paraben lowers the surfactant CMC, analysis of the results produced the unexpected conclusion that for all three compounds the partition coefficient to the micellar phase is reduced on addition of electrolyte. This is a problem which occurs and recurs in detailed studies of mechanisms of solubilization, being clearest when pH effects are studied. Generally the formulator is interested in total solubility which includes solubility in the aqueous and micellar phases. While the partitioning of a species to the micellar phase might be reduced, its increased solubility in the aqueous phase may compensate for this. In spite of the partition coefficient of the methyl, ethyl and butyl paraben

Table 6.3 Solubilization of alkylparabens in water and in 40 mM sodium lauryl sulphate solution at 27° C*.

Alkylparaben	Solubility (mmol l⁻¹) in		
	Water	40 mM NaLS	40 mM NaLS and 50 mM NaCl
Methylparaben	14.5	33.9	31.6
Ethylparaben	5.4	22.7	21.9
Butylparaben	1.1	24.3	26.7

* From [36, 38].

increasing towards the micellar phase from 1.2 through 3.2 to 21, respectively, in 40 mM sodium lauryl sulphate (NaLS) the total solubility is still highest for methyl paraben with a solubility limit of 33.9 mM. Ethyl paraben has the lowest solubility (22.7 mM) and butyl paraben has a solubility of 24.3 mM in 40 mM NaLS [38]. In a series such as the alkyl parabens their different locations in the micelle may be another factor complicating a ready understanding of the observation; the effect of the paraben on CMC which follows the order butyl > ethyl > methylparaben is of little importance when the total surfactant concentration is 40 mM as in the investigations in question but would obviously be important at surfactant concentrations close to the concentration (Fig. 6.4 shows this effect).

Uptake of solubilizate into surfactant micelles changes the physical state of the micelle (see Section 5.5). Sometimes the change in shape may result in drastic changes in the physical properties of the system as a whole – this may influence its use. The effect of additive on the cloud point of non-ionic surfactants

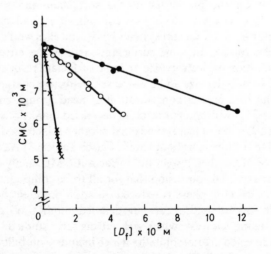

Figure 6.4 The CMCs of sodium lauryl sulphate solutions in the presence of alkyl parabens. ●, methylparaben; ○, ethylparaben; ×, butylparaben. From Goto and Endo [37] with permission.

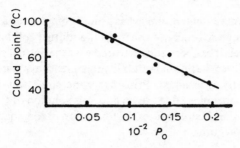

Figure 6.5 Relation between the cloud point and P_0 values for phenol and its homologues at concentrations of $0.05\,\mathrm{mol\,l^{-1}}$ in 2% cetomacrogol 1000. From Donbrow and Azaz [43]. Values of P_0 from Table 6.1.

[39–42] is of some practical importance as the cloud point may be lowered below room temperature. Fig. 6.5 shows the effect of a range of phenols on the cloud point of cetomacrogol solutions (cetomacrogol 1000 B.P. is $C_{16}H_{33}(OCH_2CH_2)_{22-24}OH$) where the relation between cloud point and the distribution coefficient of phenols, cresols and xylenols between micelles and water is demonstrated. The effect of a phenol on the cloud point is inversely related to its hydrophilicity. As the cloud point is thought to be due to the growth of the non-ionic micelles with increasing temperature, the binding of solute to the micellar structure could explain the lowering of the cloud point if the surfactant monomer–solute complex is more hydrophobic than the surfactant monomer itself.

Phenol–water systems display critical solution temperatures (CST). Addition of fatty acid soaps generally causes a lowering of CST. 3% sodium oleate lowers the CST of the phenol–water mixture from ~ 65 to 0° C [45]. 1% lowers it to 43° C and 1% sodium stearate lowers it to 49.1° C [46]. Prins [44] in investigations on cetyltrimethylammonium bromide (CTAB)–phenol–water mixtures, found striking effects caused by the detergent on the CST of the phenol–water system. Table 6.4 gives the concentrations of CTAB and phenol which have in admixture with water a CST of 20° C. The figures were obtained from a study of the phase diagram of the system.

Table 6.4 Concentrations of CTAB and phenol having a critical solution temperature of 20° C*

CTAB (% w/w)	Phenol (% w/w)
8.0	11.0
17.5	16.8
28.0	25.2
34.0	31.7
40.5	35.6
41.5	41.0
40.0	45.1

* From Prins [44].

Cetyltrimethylammonium bromide has a more complicated action than the fatty acid soaps. Although it generally enhances the mutual solubility of phenol and water, at some levels it has been shown to decrease the mutual solubility. Addition of more CTAB causes the mutual solubility to increase again until at 48 % CTAB, complete miscibility is attained. Prins has gone some way to explaining this behaviour. When phenol is added to an aqueous solution of CTAB consisting of spherical micelles the phenol induces the formation of rod-shaped aggregates (see simplified phase diagrams in Fig. 6.6).

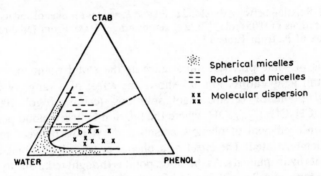

Figure 6.6 A diagrammatic representation of the cetyltrimethylammonium bromide–phenol–water system. (After Prins [44]). Horizontal arrow shows increasing phenol concentrations and passage from solutions containing spherical micelles, through solutions containing asymmetric micelles to a molecular dispersion on breakdown of the micelles.

The rod-shaped micelles can solubilize large quantities of phenol, but they reach a point (b) where they disintegrate, forming a molecular dispersion which results in a loss of mutual solubility. All systems containing more than 48 % CTAB are completely miscible. Examination of the interfacial tension curve of the co-existent phases in the water–phenol–CTAB system (phenol:water 40:60 by weight) shows that the curve mimics the solubility behaviour. A minimum at about 4 % is followed by an increase in interfacial tension, which falls again after 15 % concentration, falling to $0\ \mathrm{mN\,m^{-1}}$ at about 40 %.

(B) INTERACTION OF PRESERVATIVE MIXTURES WITH SURFACTANTS.
In many formulations more than one solute will be a potential solubilizate whether or not this is desired. As discussed in Chapter 5, the effect, if any, of one solute on the solubilization of another will depend on the mechanisms of solubilization. If solubilization of one solute occurs at specific 'sites' within the micelles then molecules with similar binding affinities might compete for the available sites leading to a decreased solubilization of each. In some cases one solute might induce a reorganization of the micelle structure and allow increased uptake; both mechanisms might operate such that maxima and minima are seen in the plots of solubility versus the concentration of second solubilizate [47] (see Fig. 5.23). Benzoic acid, for example, increases the solubility of methyl paraben in

cetomacrogol solutions, but dichlorophenol decreases its solubilization [47]. Chloroxylenol reduces the solubility of methylparaben and methylparaben reduces the solubility of chloroxylenol in cetomacrogol, there being no effect on mutual solubilities in the absence of surfactant.

The distribution of a solubilizate between micelles and the aqueous phase does not obey necessarily a simple partition law when a second solubilizate is present [48]. A non-linear increase in solubilization with increasing surfactant concentration has been found with a second solubilizate present. Fig. 6.7 shows the change in micellar partition coefficient of *o*-hydroxybenzoic acid in the presence of increasing levels of benzoic acid when polysorbate 80 is the solubilizer [48]. At 1 % surfactant there is a marked decrease in partition coefficient, but at 3 % there is little change. Nalidixic acid does not alter the micellar distribution coefficient of *o*-hydroxybenzoic acid but chloramphenicol reduces the distribution. Nalidixic acid has no detectable effects on the cloud point of the polysorbate solutions, suggesting that it does not alter micellar structure. Alhaique *et al.* [48] conclude that if the added compound does not induce significant changes in micellar structure it will not alter the distribution into the micelle of another solubilizate; this problem requires further and more detailed examination, primarily because of its importance in pharmaceutical systems and because of the potential importance of the phenomenon in altering drug bioavailability from micellar systems. Preliminary work on permeation through polymer membranes [48] has shown that the reduction in permeation caused by solubilization of a solute can

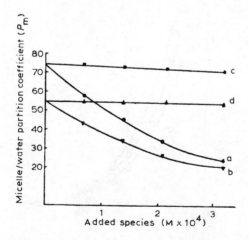

Figure 6.7 Changes in micelle/water apparent partition coefficient (P_m) of a solubilizate after progressive addition of a second species. Polysorbate 80 concentrations range from 1 to 3 % w/v. Each plot refers to a constant concentration of the surfactant. *o*-Hydroxybenzoic acid partition coefficients on addition of benzoic acid ((a) 1 % and (c) 3 % w/v polysorbate 80). Benzoic acid partition coefficients on addition of *o*-hydroxybenzoic acid ((b) 1 % and (d) 3 % w/v polysorbate 80). Temperature, 25° ± 0.1°, pH = 2.0, and the initial solubilizate concentration, 3.2×10^{-4} M. Redrawn from Alhaique *et al.* [48].

be minimized to some extent by the addition of a second solute which would decrease the value of P_m; thus while nalidixic acid had no effect on the permeation rate of o-hydroxybenzoic acid, benzoic acid increases the permeation of this compound through polydimethylsiloxane from 5.61×10^{-10} mol cm^{-2} s^{-1} to 6.5×10^{-10} mol cm^{-2} s^{-1}.

An attempt to characterize the interaction of mixtures of preservative molecules with non-ionic surfactants using the theory of competitive binding was unsuccessful [49]. In the presence of a second preservative (C) which may act as a competitor, the binding equation may be rewritten

$$r = \frac{n_1 K_1 [D_f]}{1 + K_1 [D_f] + K_{c_1}[C_f]} + \frac{n_2 K_2 [D_f]}{1 + K_2 [D_f] + K_{c_2}[C_f]} \tag{6.7}$$

where $[C_f]$ is the concentration of free competitor and K_{c_1} and K_{c_2} are the intrinsic association constants of the competitors for binding sites of class 1 and class 2, respectively. Scatchard plots for the interaction of chlorocresol and cetomacrogol in the absence and presence of a constant concentration of methyl paraben are shown in Fig. 6.8. On increasing the concentration of methyl paraben there is a downward displacement of the curve suggesting competition between the solubilizates for the same binding sites. Theoretical lines obtained using Equation 6.7 are shown; reasonable agreement is shown, but this does not apply to chlorocresol–cetomacrogol systems in the presence of propyl paraben or

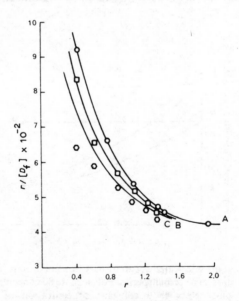

Figure 6.8 Scatchard plot for the interaction of chlorocresol with cetomacrogol in absence and presence of methyl paraben. Cetomacrogol concentration = 7.69×10^{-3} mol l^{-1}. Initial total methyl paraben concentration: O, 0.0; □, 8.54×10^{-2}; O, 13.14×10^{-2} mol l^{-1}. Points experimental, curves B and C calculated using Equation 6.7. From Kazmi and Mitchell [49].

perhaps surprisingly when the interaction of methyl paraben is measured in the presence of chlorocresol.

It is most likely that none of the preservative combinations used had exactly the same locus or are solubilized by the same mechanism, so that simple competition between the solubilizates in the micelle is unlikely; alternatively the interaction of some of the preservatives with the micelle (or monomers) leads to perturbations of micelle size and shape such that binding sites are altered in their capacity to accept solubilizate molecules.

For a mixture of two surfactants (S_I and S_{II}), Equation 6.11 becomes

$$r = 1 + \frac{n_1 K_1 [M_1]}{1 + K_1 [D_f]} + \frac{n_2 K_2 [M_1]}{1 + K_2 [D_f]} + \frac{n_1' K_1' [M_{11}]}{1 + K_1' [D_f]} + \frac{n_2' K_2' [M_{11}]}{1 + K_2' [D_f]}, \quad (6.8)$$

where M_1 and M_{11} are the molar concentrations of the two surfactants; n_1, n_2, K_1 and K_2 are constants for S_I and n_1', n_2', K_1' and K_2' are the corresponding constants for S_{II} as defined above. Kazmi and Mitchell [49] found that the extent of the interaction of a single preservative with a mixture of two surfactants can be predicted from a knowledge of the binding constants which characterize the interaction of the preservatives with the individual surfactants in the mixture. Variation of the free preservative concentration with total preservative concentration in mixtures of Texofor A16 and Texofor A60 are shown in Fig. 6.9a. The theoretical lines were calculated by substituting experimentally determined values of n_1, n_2 and K for each surfactant.

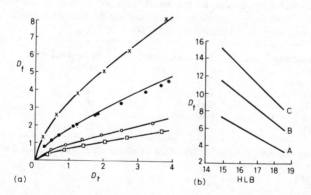

(a)

(b)

Figure 6.9(a) Variation of free preservative concentration $[D_f]$ with total preservative concentration $[D_t]$ for the interaction of chlorocresol with mixtures of Texofor A16 and Texofor A60. Concentration of Texofor A16 and Texofor A60 in a mixture (mol l^{-1}): X, 50.89×10^{-3} A16 + 2.75×10^{-3} A60; ●, 10.57×10^{-3} A16 + 2.75×10^{-3} A60; ○, 10.57×10^{-3} A16 + 13.26×10^{-3} A60; □, 2.11×10^{-3} A16 + 5.51×10^{-3} A60. Points experimental, curves calculated.

(b) $[D_t]$ versus HLB at constant $[D_f]$ for the interaction of chlorocresol with Texofor A16, Texofor A60 and mixtures of Texofor A16 and Texofor A60. Total concentration of surfactant or surfactant mixtures = 1 % (w/v). Concentration of free chlorocresol $[D_f]$: A, 1.0×10^{-3}; B, 2.0×10^{-3}; C, 3.0×10^{-3} mol l^{-1}. HLB Texofor A16 = 14.88; HLB Texofor A60 = 18.67. Curves calculated from Kazmi and Mitchell [49].

When, for a given value of D_f, D_t is plotted as a function of the HLB of the surfactant mixtures, it is clear that in the Texofor mixtures a smaller total preservative concentration is required to maintain a given free concentration of chlorocresol as the HLB is raised (Fig. 6.9b). With cetomacrogol–polysorbate 80 mixtures there is a slight increase in the overall concentration of chlorocresol required, but these two surfactants have almost identical binding characteristics towards chlorocresol which is thought to reside in the ethylene oxide layer which is of about equal size in the two surfactants in question. It seems, therefore, that direct experimental verification of antibacterial and antifungal activities are still required in mixtures of surfactants and preservatives, although by judicious choice of preservative and surfactant once the characteristics of individual systems are known, reasonable calculations can be made of possible changes in activity.

Mitchell [50] has shown that the activity of chloroxylenol in water and in solutions of cetomacrogol 1000 is related to the degree of saturation of the system. A saturated solution of chloroxylenol in water was found to have the same bactericidal activity as saturated surfactant solutions containing up to 100 times as much chloroxylenol. It is thus apparent that the activity depends on the amount of the bactericide free in the aqueous phase; the compound has apparently no action inside the micelles. Table 6.5 shows some of these results and should emphasize the importance of these factors in formulation.

Table 6.5 Dependence of the death time of *E. coli* in chloroxylenol–cetomacrogol solutions on cetomacrogol concentration at 20° C at a constant chloroxylenol concentration of 1.5%

Cetomacrogol conc. (mol l^{-1})	Saturation ratio*	Mean death time (min)
0.049	1.00	49
0.051	0.95	88
0.054	0.90	104
0.057	0.85	75–79 (h)

* The saturation ratio is the ratio of the amount of chloroxylenol present to its solubility in the solution.
From [50].

Of interest in this context are the results of Good and Milloy [51] (see Fig. 6.10). The partial pressure of phenol above CTAB–phenol–water solutions was determined as a function of phenol concentration by analysis of the gas phase above the solutions at 25° C. As P/P_0 is proportional to the activity of the phenol in the water, this, in accord with Ferguson's principle, should broadly determine its bactericidal activity. Studies of this nature on solutions with varying concentrations of surfactant would be valuable in interpreting the microbiological behaviour of such systems.

It has been suggested [52] that solubilization provides a means not only of

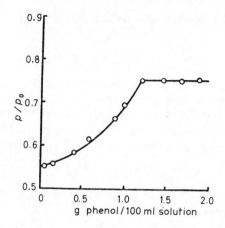

Figure 6.10 The partial pressure of phenol above solutions of phenol in 1 % cetyltrimethylammonium bromide at 25°C; p = partial pressure of phenol in the surfactant solution; p_0 = the partial pressure of phenol in water. After Good and Milloy [51].

modifying the biological activity of the phenols but also of providing a reservoir of materials in the micelles which would prolong the duration of their action. The apparent divergence of opinion between Mitchell [50] and Bean and Berry [53, 54] probably results from the properties of the solubilizers used. If Mitchell's findings are of universal application, then the only point in having solubilized preparations of phenols is to provide a suitable form in which they may be distributed and diluted for use.

Note should be taken of the fact that Mitchell's results were obtained with a non-ionic detergent of the polyoxyethylene ether class in which tendency to complex with phenolic hydroxyls has been noted before: all evidence of synergism between surfactants and phenols have been with ionic soaps or with non-ionics of other classes. We may cite briefly some examples. Soaps of coconut oil, castor oil, or linseed oil increased the germicidal activity of various phenols in the absence of organic matter [55]; sodium riconoleate, sodium linoleate, resinate, and oleate enhanced the activity of chlorothymol, chlorocarvacol, thymol, chlorophenyl phenol, resorcinol, *n*-hexylresorcinol [56]; and sodium dodecyl sulphate has a similar effect on 2,4,6-trichlorophenol [57]. Shafiroff [58] quotes the synergistic effect of sucrose mono-laurate on *p-m*-chlorocresol against *Staph. aureus*. Sodium dodecyl sulphate, it should be remembered, has cytostatic and sometimes lytic activity by itself.

Calculations of free preservative concentrations in surfactant systems have been attempted by several workers. Evans and Dunbar [59] attempted with a simple approach to quantify the effects by calculating the free concentrations of antibacterial. Their derivation reproduced below leads to the relationship between $[D_a]$ the concentration of antimicrobial in the water phase, C_s the total concentration of solubilizer, $[D_m]$ the concentration of antimicrobial in the micelle, R the ratio of antimicrobial to solubilizer, and P_m the partition coefficient of the antimicrobial substance towards the micelle phase

$$[D_a] = \frac{R[C_s]}{1 + P_m[C_s - \text{CMC}]} \qquad (6.9)$$

where the critical micelle concentration of the surfactant is denoted by CMC. The assumption made in the derivation is that the concentration of monomers is constant above the CMC and that the surfactant has no effect on the intrinsic biological activity of the antimicrobial, which as we will see later is not always so.

Equation 6.9 is obtained as follows. $R = [D_t]/[C_s]$ and $P_m = [D_m]/[D_a]$

$$[D_t] = [D_m][C_m] + [D_a], \qquad (6.10)$$

where $[D_t]$ is the concentration of the antimicrobial in the system in g g^{-1} and $[D_m]$ is the concentration in the micelle (g g^{-1} micelle). Using the product $[D_m][C]$ we convert this to g g^{-1} *system.*

$$\therefore \quad [D_a] = [D_t] - [D_m][C_m] \qquad (6.11)$$

or $$[D_a] = R.[C_s] - P_m.[D_a][C_m]. \qquad (6.12)$$

As $[C_m] \approx (C_s - \text{CMC})$ we obtain, on rearrangement, Equation 6.9.

Evans and Dunbar's equation predicts a maximum biological activity at the CMC when R is constant at high P_m. Using this model and different values of P_m,

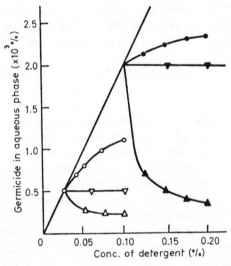

Figure 6.11 The effect of critical micelle concentration and solubilization on the concentration of germicide in aqueous phase, at a germicide: detergent ratio of 0.02. P_m is the distribution coefficient of germicide between aqueous and micellar phases

	P_m	CMC (%)		P_m	CMC (%)
○	1×10^3	0.025	●	7×10^2	0.10
▽	4×10^3	0.025	▼	1×10^3	0.10
△	1×10^4	0.025	▲	1×10^4	0.10

From Evans and Dunbar [59].

the concentration of active ingredient in the aqueous phase can remain the same, decrease, or increase slowly above the CMC (Fig. 6.11). The values of P_m used to construct Fig. 6.11 are reasonable. Methyl paraben has a $P_m = 3.24 \times 10^3$ and butyl paraben 7.29×10^4 in sodium lauryl sulphate solution [35]. Evans [60, 62] used a titration method to analyse preservative–non-ionic surfactant interactions, similar to that developed independently by Donbrow and Rhodes [61]. The method depends on the pH changes which occur when an acidic material is solubilized (see Chapter 5).

Knowing the total concentration of a substance, e.g. *p*-hydroxybenzoic acid, its dissociation constant, the concentration of detergent and P_m, it is a simple procedure to calculate the amount of acid dissolved in the aqueous and micellar phases at different pH values. The results of such a calculation assuming a total concentration of 0.1 % w/v *p*-hydroxybenzoic acid and 5.8 % w/v detergent are given in Table 6.6. If 0.1 % w/v is the optimum concentration of acid for a required preservative effect in water at pH 4, the concentration of the unionized acid – the active species – is $5.59 \times 10^{-3} \, \text{mol l}^{-1}$. Addition of 5.8 % of non-ionic detergent reduces this concentration to $1.14 \times 10^{-3} \, \text{mol l}^{-1}$ as is seen in column 3 of Table 6.6.

The practical importance of these surfactant-preservatives is that much greater quantities of preservative must be added to a formulation containing a non-ionic to have an equivalent action to a specified amount in an aqueous solution. This is

Table 6.6 Effect of pH on solubilization of *p*-hydroxybenzoic acid in 5.8 % octylphenyl E 8.5, and percentage acid required to be equivalent to 0.1 % w/v acid

pH	[HA]* water	[HA]* detergent solution	% w/v required
3.5	0.00662	0.00118	0.56
4.0	0.00559	0.00114	0.49
4.5	0.00376	0.00104	0.36
5.0	0.00220	0.00081	0.27

* Molar concentration of unionized *p*-hydroxybenzoic acid in the aqueous phase.
From [60].

Table 6.7 Inhibitory concentrations of methylparaben in presence of polysorbate 80, observed for one month

Organism	Detergent conc. (%)	Inhibitory conc. (%)
A. aerogenes	0	0.075–0.08
	2	0.18–0.20
	4	0.28–0.30
	6	0.40–0.42
Asp. niger	0	0.045–0.05
	7	0.32–0.34

From [63].

illustrated in Table 6.7 after Pisano and Kostenbauder [63] which shows the inhibitory concentration of methyl paraben in the presence of polysorbate 80 against two organisms increases with increasing detergent concentration. Solubility data of methyl *p*-hydroxybenzoate in polysorbate at 27° C show that in a 5.8% solution it is five times more soluble than in water (1 mole benzoate for each 4 glycol units of detergent monomer [64]), and this agrees with the results of Table 6.7.

(c) CHLORHEXIDINE

$$Cl\!\!-\!\!\langle\rangle\!\cdot NH\cdot \underset{\underset{NH}{\|}}{C}\cdot NH\cdot \underset{\overset{NH}{\|}}{C}\cdot NH\cdot (CH_2)_6\cdot NH\cdot \underset{\overset{NH}{\|}}{C}\cdot NH\cdot \underset{\underset{NH}{\|}}{C}\cdot NH\cdot \langle\rangle\!\!-\!\!Cl\cdot$$

Chemical structure of chlorhexidine available commercially as gluconate solution, acetate, and hydrochloride

(I)

Chlorhexidine possesses marked bactericidal action against a wide range of micro-organisms. The base has a low aqueous solubility (0.008% w/v); a wide range of salts have been prepared and their solubilities measured (Table 6.8). The dihydrochloride has a solubility of 0.06%, the diacetate 1.8% and as the gluconate has a solubility > 70% there would appear to be little need for the preparation of solubilized formulations. However, surfactants may be present in chlorhexidine formulations; because of the low solubility of chlorhexidine sulphate and related salts with inorganic ions present in water, extemporaneously prepared solutions diluted from concentrates may precipitate. Non-ionic and

Table 6.8 Chlorhexidine salts – water solubilities at 20° C

Salt	% w/v	Salt	% w/v	Salt	% w/v
(Base)	0.008	Diformate	1.0	Dilactate	1.0
Dihydriodide	0.1	Diacetate	1.8	Di-α-hydroxy*iso*butyrate	1.3
Dihydrochloride	0.06	Dipropionate	0.4	Digluconate	>70
Dihydrofluoride	0.5	Di-*iso*butyrate	1.3	Diglucoheptonate	>70
Diperchlorate	0.1	Di-*n*-valerate	0.7	Dimethanesulphonate	1.2
Dinitrate	0.03	Dicaproate	0.09	Di-isothionate	>50
Dinitrite	0.08	Malonate	0.02	Dibenzoate	0.03
Sulphate	0.01	Succinate	0.02	Dicinnamate	0.02*
Sulphite	0.02	Malate	0.04	Dimandelate	0.06
Thiosulphate	0.01	Tartrate	0.1	Di-isophthalate	0.008*
Di-acid phosphate	0.03	Dimonoglycolate	0.08	Di-2-hydroxynaphthoate	0.014*
Difluorophosphate	0.04*	Monodiglycolate	2.5	Embonate	0.0009*

* These are approximate values.
From [65].

quaternary ammonium surfactants serve to prevent this precipitation [65].

Surfactants may also be required as wetting agents and detergents and as emulsifiers in creams. Some non-ionic surfactant–chlorhexidine interactions are described by Senior [65]. 1 % and 3.3 % of polysorbate 80 reduces the activity of 0.1 % chlorhexidine acetate solution to 39 % and 14 %, respectively; corresponding figures for Lubrol W, a surfactant related to cetomacrogol, were 9 % and 5 %. Fig. 6.12 shows how the addition of ethanol to the formulation can reduce the interaction between the bactericide and the surfactant. The difference in surfactant uptake of two salts of chlorhexidine has been demonstrated by Wesoluch *et al.* [66] who conclude that the ion pair is solubilized into the micellar interior. The solubility of the diacetate is about six times that of the dihydro-chloride in both Brij 96 and Tween 80. One might have expected the least soluble salt to have been solubilized to a greater extent but Fig. 6.13 demonstrates that this is not so and implies that the salt rather than the chlorhexidine ion is solubilized, a suggestion supported by the fact that the solubility of both salts increases in solvents of decreasing polarity. The surfactant properties of chlorhexidine diacetate [67] may induce the formation of mixed micelles in which the diacetate molecule retains its counter ions by orientating radially in the micelle with the surfactant monomers.

Differences in micellar uptake would affect the bactericidal effect of the chlorhexidine formulation; the choice of salt and surfactant must, therefore, involve a careful analysis of intrinsic solubilities and activities of the salts and their percentage solubilization in surfactant micelles.

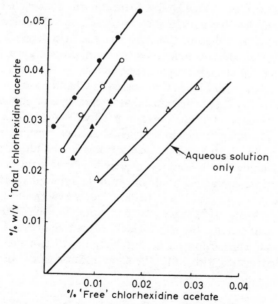

Figure 6.12 Effect of ethanol (v/v) on chlorhexidine acetate solubilization in 1 % w/v aqueous polysorbate 80. ●, 0 % ethanol; ○, 10 % ethanol; ▲, 20 % ethanol; △, 50 % ethanol. From Senior [65] with permission.

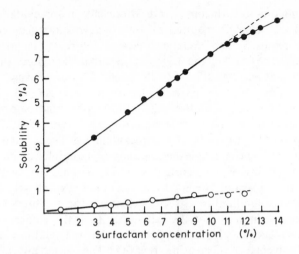

Figure 6.13 Solubility of chlorhexidine diacetate (●) and dihydrochloride (○) as a function of concentration of the decaoxyethylene oleic ether, Brij 96. Each point is the average of two or three independent determinations. From Wesoluch *et al.* [66] with permission.

(D) HEXACHLOROPHENE

The bactericidal properties of hexachlorophene in surfactant solutions have been studied by a number of workers [68–70]. Hexachlorophene is used in soaps for pre-operational scrubbing – the United States Pharmacopoeia has a hexachlorophene liquid soap which is a 0.225–0.26 % w/v solution of hexachlorophene in a 10 % potassium soap solution. Concern over the percutaneous absorption of hexachlorophene and its subsequent toxicity in infants has given fresh relevance to investigations of hexachlorophene–surfactant interactions. The effect of surfactants on skin permeability is discussed in Chapter 7.

Russell and Hoch [71] have claimed that the presence of non-ionic detergents in a number of shampoo formulations has no effect on the antibacterial action of bacteriostats (including hexachlorophene), but their results are difficult to interpret because of the presence in each formulation of additional surfactants. The biological action was, however, considered to be as great as or greater than that of preparations containing triethanolamine lauryl sulphate as the solubilizer and soap. The two detergents have, perhaps, some synergistic effects.

Anderson and Morgan [72] have related the solubilization of hexachlorophene with its biological activity, the bactericidal action being related to the concentration of unbound hexachlorophene, but the results of agar-plate diffusion tests could not be correlated with either the concentration of unbound agent or its total concentration.

(E) IODINE SYSTEMS

The term iodophor (phoros: bearer, carrier) is used to describe preparations of iodine in surfactant solutions. While all types of surfactant can be used to solubilize iodine, non-ionic polyoxyethylene derivatives have been found most suitable, as the iodophor can be formulated without instability in acid conditions in which antibacterial activity is enhanced. Iodine may be solubilized to the extent of 30 % by weight, of which three-quarters is released as available iodine when the iodophor is diluted. A cationic iodophor has, however, been used as an irrigant for the conjunctival sac and lachrymal passages.

Gershenfeld and Witlin [73] found that 1.49 % iodine was soluble in a 1:1 mixture of propylene glycol and water. In most cases the bactericidal efficiency of iodine–iodide solutions prepared in aqueous propylene glycol was identical with that of Iodine USP XIII, and a satisfactory non-irritant formulation was given containing 2 % iodine and 2.4 % sodium iodide in distilled water containing 25–50 % propylene glycol. Osol and Pines [74] investigated the solubility of iodine in aqueous ethylene, diethylene, triethylene, and propylene glycols and in glycerin, and cite evidence to support the presence of Lewis acid–base-type interactions in the solubilization process. To increase the solubility of the iodine to any great extent, large quantities of the glycols are required; this is a disadvantage.

Values for the solubility of iodine in water at 20° C vary between 0.335 and 0.285 g 1^{-1} [75]. The problem which occurs when such small quantities of iodine are dissolved results from the depletion of the solution through interaction of the iodine with the bacterial proteins. The use of non-ionic surfactants to produce systems with a high proportion of iodine was first described by Terry and Shelanski [76]. Unlike iodine–iodide systems, iodophors can be diluted without causing the precipitation of the iodine. Among other advantages claimed for iodine–surfactant systems are increased stability and decreased corrosion of metals, for example in instrument sterilization. Iodine is lost less readily from iodine–cetomacrogol solutions than from iodine solution (NF), as shown in Fig. 6.14 from Hugo and Newton [77]. It is an important consideration that the major proportion of iodine applied to a surface, for example, in the form of Strong Iodine Solution USP, is lost through sublimation [78]. Allawala and Riegelman [79] give evidence of the penetration of an iodophor solution (iodine–polyoxyethylene glycol nonylphenol) into the hair follicles of the skin, whereas iodine–iodide showed no such ability. This combination of less rapid sublimation and superior penetration results in enhanced activity. The product 'Wescodyne' (7.75 % polyoxyethylene polypropoxyethanol–iodine complex; 7.75 % nonylphenyl polyoxyethylene glycol ether–iodine complex and 0.1 % HCl) is highly fungicidal and lethal to tubercle bacilli, the bactericidal action of the iodine being enhanced in the iodophors [80].

Iodophors are used in the dairy industry for sterilizing equipment and for application to cows' udders. In addition to the advantages already listed, the iodophor prevents the accumulation of milkstone by solubilization of the salts which are associated with the formation of these deposits [81].

The mechanism of solubilization of iodine by non-ionic surfactants has been discussed by a number of workers [82, 83], who have concluded in favour of the formation of a complex rather than true micellar solubilization. Polyoxyethylene

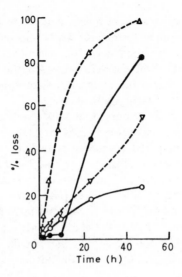

Figure 6.14 Changes in the weight and iodine content of iodine preparations stored in open beakers at room temperature. △ % weight of iodine solution; ▽ % weight of iodine–cetomacrogol complex; ● % weight of iodine lost from iodine solution; ○ % weight of iodine lost from iodine–cetomacrogol complex. From Hugo and Newton [77] with permission.

glycols increase the solubility of iodine in water, suggesting some complexation with the ether oxygens. Henderson and Newton [84] suggest that 1:1 charge transfer complexes are formed, characterized by a negative standard enthalpy change (ΔH°). Negative ΔH° values can also be observed for iodine–potassium iodide systems. If the reason for the increase in solubility is the formation of a complex with the ether oxygens one would expect little effect from the presence of micelles in solution. A comparison of Fig. 6.15a and b will show the great difference between the amount of iodine solubilized in cetomacrogol solutions (micellar) and in polyoxyethylene glycol solution (PEG 1540, 35 units) in the same concentration region. The molecular ratios of iodine to ether found for a series of monocetyl ethers and monolauryl ethers show that neither an ethylene oxide unit nor a molecule of ether associates with one molecule of iodine. The association must be more complex than the simple acid–base-type postulated for glycol–iodine interactions.

It should be remembered that evidence of complexation does not preclude the possibility of normal micellar solubilization.

Fig. 6.15 shows that the ratio of available iodine to total iodine in the cetomacrogol iodophor is approximately 0.86, a value which agrees well with those of Brost and Krupin [85] (0.775 and 0.80 for two non-ionic iodophors prepared with a nonyl phenylether).

The formation of solid polyoxyethylene glycol–iodine complexes is the cause of the incompatibility of potassium iodine–iodine solutions and certain glycol ointment bases [86]. A polyoxyethylene glycol 4000–iodine precipitate can be

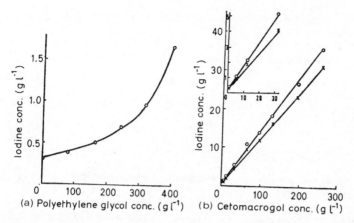

Figure 6.15(a) The solubility of iodine in aqueous solutions of polyoxyethylene glycol 1540 at 20° C. (b) The solubility of iodine in cetomacrogol solutions at 20° C is denoted by the open circles. The available iodine is shown by the crosses; the inset shows the effect at low cetomacrogol concentrations. The much higher solubility of iodine in the detergent solutions is evident. From Hugo and Newton [77] with permission.

assumed to have the iodine randomly distributed along its ether oxygens, not every oxygen being co-ordinated, according to Hiskey and Cantwell [87]. Their results are explicable in terms of a competing equilibrium for the iodine by the ether oxygens and iodide present in solution.

A complex between iodine and the micelles of non-ionic association colloids has been discovered in both aqueous and non-aqueous media [88]. Because of the similarity of the absorption spectra of the complex in aqueous and non-aqueous media – the ligand is shown to be the tri-iodide ion in both cases – Ross and Baldwin suggest that the site of the interaction between the tri-iodide species and the micelle is at the boundary between the hydrophobic and hydrophilic regions of the micelle. In aqueous and non-aqueous systems these regions are simply 'reversed', leaving the ions in identical environments.

(F) GLUTARALDEHYDE

Glutaraldehyde is an effective sporicide and chemisterilant. In alkaline solution it is effective but unstable; in acid conditions it is stable but weakly active. Cationic surfactants were suggested as stabilizing agents and the advantage of the addition of non-ionic surfactants has been demonstrated [89, 90]. As activity of glutaraldehyde has been enhanced by addition of divalent cations, various surfactant–cation combinations have been examined as possible potentiators and stabilizers of glutaraldehyde biocidal activity [91]. The magnesium salt of sulphated lauryl alcohol (Empicol ML 26A) was found to be effective in maintaining stability over 12 months and as a synergistic agent. Some of the effect may be on solution pH as Table 6.9 indicates. Table 6.9 compares bactericidal and fungicidal activity of two surfactant (Empicol) formulations with the activity of the simple solution and of an alkaline solution.

Table 6.9(a) Bactericidal activity of glutaraldehyde formations (0.01 % w/v) at 180° C (initial viable count: 1×10^8 ml^{-1})

Additive	Concentration (%)	Formulation pH	Time (min) for 99.9 % kill of:		
			E. coli	*Staph. aureus*	*Ps. aeruginosa*
Empicol	2.5	7.08	22	9	25
Empicol	10.0	7.60	15	7	20
None	—	4.6	120	100	65
NaHCO₃	0.3	7.9	20	12	35

(b) Fungicidal activity of glutaraldehyde formulations (0.5 % w/v) at 18° C (initial spore count: 1×10^6 ml^{-1})

Additive	Concentration (%)	Formulation pH	Time (min) for 99.9 % kill of:	
			A. niger	*T. mentagrophytes*
Empicol	2.5	4.8	110	75
Empicol	10.0	4.9	105	65
None	—	4.3	>180	>180
NaHCO₃	0.3	7.9	80	45

From [91].

6.2.2 Antibiotics and sulphonamides

The antibiotics and sulphonamides have been formulated in micellar solutions in the same way as other poorly soluble medicaments, but the range of chemical structures which exist would make it difficult to predict without experiment the solubilities of such drugs in surfactant solutions.

(A) CHLORAMPHENICOL

Chloramphenicol, soluble 1 in 400 of water at 20° C and 1 in 7 of propylene glycol, has been solubilized in Tween solutions [92, 93]. In spite of its superior solubility in propylene glycol, this compound cannot be used as a solvent for chloramphenicol in eye-drops or nasal preparations since it causes a marked burning sensation. Simple aqueous solutions of chloramphenicol lose about half their antibiotic activity by hydrolysis on storage for 290 days at 20 to 22° C [94].

Chloramphenicol 1 %, polysorbate 80 6 %, in water for injection has been suggested as an ophthalmic solution [95]. Other formulae have been given and some of these are collected in Table 6.10. A solution of the antibiotic has been prepared in 50 % *N,N*-dimethylacetamide as an intravenous injection. *N,N*-dimethylacetamide is a hydrotropic substance, a group of compounds whose actions are discussed in Section 6.7. Different crystal forms of chloramphenicol palmitate are soluble to differing extents in solutions of polysorbate 60. A detailed study of the solubilization of chloramphenicol in cetomacrogol solutions has been reported by Rogers [98]; Regdon-Kiss and Kedvessy [99] have studied the

Table 6.10 Solubilized chloramphenicol preparations

Chloramphenicol (%)	Solubilizer (%)	Reference
1	6% polysorbate 80	[95]
1.22	10% polysorbate 80	[96]
1.56	10% Brij 35	[96]
5.00	50% polysorbate 20	[97]
25.00	50% N,N-dimethylacetamide	
25.00	40% N-methyl-2-pyrrolidone	[98]

surface tension of polysorbate 20 solutions containing solubilized chloramphenicol.

(B) TYROTHRICIN AND RELATED SUBSTANCES

A mixture of gramicidin and tyrocidin, tyrothricin is stable in aqueous solutions of cationic and non-ionic surface-active agents. A 0.025% solution of the drug in 0.05% aqueous cetyltrimethylammonium bromide is stable for at least 6 months and the solution has a somewhat greater bactericidal action than solutions of either component alone. Levin [100] describes tyrothricin solutions containing 0.02% w/v of the antibiotic and employing 0.05% polyoxyethylene sorbitan monolaurate or 0.02% cetylpyridinium chloride as solubilizer. The latter solution is unstable in the presence of high concentrations of electrolytes, and a non-ionic should be used wherever there is the possibility of electrolyte contact. An isotonic solution for topical application (Soluthricin[R]) which contains 0.05% tyrothricin and 0.05% cetylethyldimethylammonium bromide is stable for at least 1 year at room temperature [101]. A concentrate containing 2.5% antibiotic and 2.5% surfactant which can be diluted for normal use has been marketed [102–104]. 4.98 g tyrothricin can be solubilized per gram 20% polysorbate 80 solution [105]. Such concentrates can be incorporated into jellies, emulsions, or ointments. In some manufacturing procedures it is convenient to evaporate off the aqueous phase, leaving a dried residue of surfactant and drug usable in tablet or ointment formulations; the requisite amount of solubilizer is then available when the dosage form is dissolved in the gastro-intestinal tract or in the body cavity and will promote the dissolution of the antibiotic.

Gillissen [104] finds that the antibacterial action of tyrothricin is influenced by the presence of solubilizers. Cationic surfactants have a synergistic effect (as mentioned above) on its activity against Gram-positive and Gram-negative bacteria, whereas polysorbate 80 inhibits its activity. It is thus important that these effects are borne in mind, and a compromise must be found between the stability and incompatibility characteristics of the solution and the activity of the product. The complexity of the situation is revealed by the fact that the activity of bacitracin is enhanced by the presence of cationic and non-ionic surfactants [106] but decreased by anionic agents. The effect of non-ionic surfactants on the bactericidal activity of tyrothricin has been measured [107]. Some results are shown in Table 6.11.

Table 6.11 Relation between HLB of polyoxyethylene glycol stearates and polyoxyethylene glycol sorbitan fatty acid esters and the minimum bactericidal concentration of tyrothricin. Surfactant concentration $2 \times 10^{-3}\,\mathrm{mol\,l^{-1}}$

Surfactant		HLB	Min. bactericidal concentration $(10^4\,\mathrm{mol\,l^{-1}})$
PEG-900-stearate		15.0	4.4
PEG-1800-stearate		16.9	6.8
PEG-4700-stearate		18.8	9.3
PEG-900-Sorbitan	laurate	16.7	4.3
	palmitate	15.6	5.6
	stearate	14.9	6.7

From [107].

Tyrothricin is surface active (see Fig. 6.16). Its size would seem to preclude significant interaction with micelles but the data in Table 6.11 show clearly that interactions occur, possibly by formation of mixed micelles.

Substances such as gramicidin J_1 and chloramphenicol, although they are only slightly soluble in water, have a high antibacterial activity. This presents little problem in *in vitro* antibacterial testing, but where study of a series of antibiotic derivatives is being made, low activity combined with low solubility can cause obvious difficulties. Such a problem was encountered during an investigation into the activity of a series of acyl derivatives of gramicidin J_1. Solubilization of the antibiotics in detergent solutions was possible – gramicidin J_1, aureothricin, and trichomycin were solubilized in a variety of non-ionic and cationic surfactants – but activity was affected in a variety of ways. Chloramphenicol,

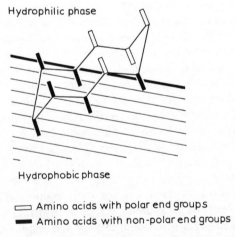

Hydrophilic phase

Hydrophobic phase

▭ Amino acids with polar end groups
▬ Amino acids with non-polar end groups

Figure 6.16 Schematic representation of the possible orientation of a tyrothricin molecule at the micellar interface between the hydrophobic core and the hydrophilic envelope of a non-ionic detergent. After Ullmann *et al.* [107].

dihydrostreptomycin, and colistin represent antibiotics suffering no, partial, and strong influence of surfactant, respectively. However, the activity of the antibiotics changed linearly with concentration and although polysorbate 80 and polyoxyethylene (15) octylphenol affect the colistin in opposite ways, extrapolation of results to zero surfactant concentration gives a value which agrees well with that determined in simple aqueous solution. This serves as a basis for the antibacterial test for poorly soluble substances: activities are determined at a number of surfactant concentrations and the minimum inhibitory concentration versus concentration (surfactant) plot is extrapolated to zero surfactant concentration [108].

Observations on the activity of antibiotic–surfactant combinations are not easy to collate. For example, although polysorbate 60 and Myri 52 (Polyoxyl 40 Stearate USP) do not impair the activity of bacitracin, oxytetracycline hydrochloride, polymixin B sulphate, or neomycin sulphate [109], Bliss and Warth [93] have concluded that polysorbate 80 potentiates the action of polymixins B and D and circulin. Brown and Winsley [110, 111] also reported that polymixin B and polysorbate 80 act synergistically against *Pseudomonas aeruginosa* on viability, cellular leakage and lysis, although the surfactant alone possessed little intrinsic activity. In attempts to elucidate this effect, spheroplasts of *Ps. aeruginosa* have been used as the test organisms for polysorbate 80-polymixin combinations [112]. The conclusion was that the synergism of action was probably due to the penetration of polysorbate 80 into the cytoplasmic membranes, facilitated by polymixin induced damage to the outer membrane and secondly to their combined action on the outer membrane structure and function. Concentrations of polysorbate 80 up to 10 % w/v are required to reduce the growth rate of *Ps. aeruginosa* cultures [113] while less than 0.01 % will lyse the corresponding spheroplast.

(C) GRISEOFULVIN

Orally administered griseofulvin is poorly and irregularly absorbed in rats and humans. Work has been directed towards increasing its absorption. Micronization of the drug has resulted in increased activity gram per gram [114, 115] but the effect of surface-active agents is not so obvious. Kraml *et al.* [116] believe that the addition of surfactants to either aqueous or corn-oil suspensions does not alter the levels of griseofulvin in the serum, yet Duncan *et al.* [117] observed that addition of butylated sodium naphthalene sulphonate (one of the compounds used by Kraml) to suspensions gave rise to higher serum levels of griseofulvin.

Griseofulvin has a low aqueous solubility – of the order of 1 mg per 100 ml – and solubilization as a means of improving its activity has been investigated [118]. In 2 % NaLS at 30° C, the solubility of griseofulvin reaches 171 mg per 100 ml, too low to be of practical use as the oral dose is of the order of 125 mg. A comparison of griseofulvin plasma levels following oral administration of a solution (0.5 % in polyoxyethylene glycol) and a suspension showed that higher levels were obtained with the solution. Sodium lauryl sulphate improved the blood levels of griseofulvin with a specific surface area 0.41 m² g⁻¹, but reduced the levels obtained with samples of specific surface area of 1 m² g⁻¹. It did not enhance the activity when given in multiple doses. Apparently, multiple dosing improved the efficiency of absorption as much as the surfactant could.

Bates *et al.* [119] have studied the solubilization of griseofulvin by bile-salt solutions in order to gain insight into the possibility that insoluble drugs may be absorbed by a mechanism involving preliminary solubilization of the drug by the bile salts which are normally present in the intestine. Hexoestrol and glutethimide were also studied; solubilization was found to increase in the order griseofulvin < glutethimide < hexoestrol.

Lysolecithin is also capable of solubilizing griseofulvin [120]. Solubilization in a range of non-ionic surfactants [121, 122] failed to achieve realistic levels of griseofulvin in an isotropic solution. Uptake increased in a homologous series with increasing oxyethylene chain length and in individual surfactants with increasing temperature but neither effect was dramatic [121]. Recent results [122] with surfactants based on erucyl and behexyl (C_{22}) alcohols (ErE_{24} and BE_{21}) indicated uptake of griseofulvin to the extent of 0.83×10^{-2} g g⁻¹ and 0.62×10^{-2} g g⁻¹ surfactant, respectively, compared with 0.95×10^{-2} g g⁻¹ of $C_{16}E_{20}$. Synthetic non-ionic surfactants with long-chain hydrocarbons (C_{32-35}) or long polyoxyethylene chains have been found by Arnarson and Elworthy [123] to solubilize less efficiently than $C_{16}E_{20}$. This perhaps suggests that manipulation of surfactant structure is not going to lead to systems with visibly increased solubilizing capacity. It may be that the subsidiary effects of surfactants such as their influence on dissolution rate may, for some drugs at any rate, be the sole advantage of inclusion in a formulation. The dissolution rate of griseofulvin is increased by a wide range of surfactants [119, 120, 124]. 1% of non-ionic surfactant can increase the rate of solution of griseofulvin by 2.5 to 3 times; higher surfactant concentrations increase this to up to 8 times. Further increase in

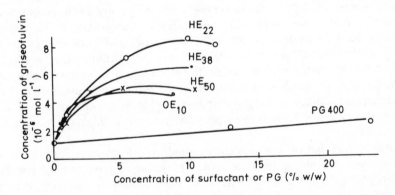

Figure 6.17 The amount of griseofulvin dissolved after 100 min in contact with various surfactant and polyoxyethylene glycol solutions, showing the decrease in rates of dissolution at higher surfactant concentrations. H = $C_{16}H_{33}$, from Elworthy and Lipscomb [124] with permission.

surfactant concentration is frequently not beneficial as the increased viscosity of the solution media will reduce solution rate giving the solution rate profile as shown in Fig. 6.17 an effect also noted in other work [125].

(D) PENICILLIN

Most of the penicillins are available as soluble salts and the need for solubilization is not great. Some of the less soluble derivatives, e.g. the *N*-benzylphenylethylamine salt of benzyl-penicillin, are used purposely, and this is the basis of their prolonged activity.

In aqueous solutions macrogols inactivate penicillin [126], as do many nonionic and ionic surfactants [127]. However, aqueous solutions of benzalkonium chloride and dioctylsulphosuccinate have been used as solvents for penicillin for topical instillation therapy of the sinus tract [128].

(E) STREPTOMYCIN

The solubility of dihydrostreptomycin sulphate, which is water soluble, is decreased in concentrated solutions of sorbitan monolaurate (E = 12) [129]. Streptomycin sulphate is also very soluble in water and is incompatible with sodium lauryl sulphate. Combination of streptomycin with polysorbate 20 produces a strong bacteriostatic effect against antibiotic resistant bacteria. Intrapleural injection of 0.5 g dihydrostreptomycin with 'one drop' of polysorbate 20 in 4 ml has been reported to result in a sterile pleural sample 1 week after injection [130].

(F) AMPHOTERICIN B

The polyene antibiotic amphotericin B is poorly soluble in water at neutral pH. Suspensions of the antibiotic when injected by subcutaneous or intramuscular routes cause pain and are poorly adsorbed. Solubilized preparations of amphotericin B, which are more active than the crystalline form are available [Fungizone Intravenous (Squibb), Amphotericin B for Injection USP; Amphotericin Injection BNF] employing sodium deoxycholate as solubilizer. However, intravenous injection of the colloidal preparation is likely to cause a greater incidence of nephrotoxicity and nausea, probably because of the ability of the solubilized antibiotic to persist in the circulation; this is offset by its greater activity and ease of handling [131]. The solubilized systems may be diluted with Dextrose Injection but precipitation has been reported within 6 hours of its addition. Procaine hydrochloride, lignocaine hydrochloride and chlorpromazine hydrochloride cause the precipitation of amphotericin [132, 133] possibly by complexing with the solubilizing agent. The solubilized preparation is also precipitated by addition to sodium chloride injection [134].

An intravenous solution of another antifungal agent, miconazole, is available as a 200 mg solution in 10% Cremophor EL (Daktarin, Janssen) some of whose properties as an intravenous solubilizer are discussed in the next chapter.

(G) SULPHONAMIDES

The less-soluble sulphonamides are liable to be deposited in the renal tubules or ureters after oral administration. The consequent renal damage may be prevented by maintaining an alkaline urine and a high liquid intake. The use of surfactants to prepare solubilized preparations or prevent the precipitation of excess compound in the tubules is a possibility that does not seem to have been investigated. Other potential sulphonamide–surfactant interactions possibly need investigation as phthalylsulphathiazole is almost insoluble in water and is used to treat infections of the intestine, from which it is only sparingly absorbed. The presence of surfactants, whether natural or ingested, could interfere with the absorption of this sulphonamide. Similarly, since sulphonamides when used on wounds penetrate the skin with difficulty, Hadgraft [135] states that the use of aqueous vehicles containing alkylbenzene sulphonates promotes the absorption of sulphonamides through the hair follicles. This could lead to toxic systemic effects. Absorption of insoluble drugs designed to act in the intestine has been suggested before, in the case of clioquinol. The occurrence of neurotoxic reactions following oral administration of halogenated hydroxyquinolines has been reported. Whether the toxic symptoms were due to genetic factors, duration of administration or formulation effects remain to be established. Clioquinol and di-iodohydroxyquinoline tablets may contain dispersing agents to aid wetting of the hydrophobic drugs. A brand of clioquinol tablets (Entero-Vioform tablets) contains a synthetic surfactant (sapamine) as a wetting agent. The systemic absorption in man of clioquinol, administered as a powder with 7% sapamine, has been confirmed [136]; Khalil and El-Gholmy [137] have shown the effectiveness of sapamine in increasing the dissolution rate of both clioquinol and di-iodohydroxyquinoline *in vitro*; a 0.2% NaLS solution causes an 18-fold increase in rate of clioquinol solution. While this does not necessarily imply a biological effect, the possible implications are clear.

Khawam *et al.* [138, 139] studied the solubilization of sulphanilamide in polysorbate 20, 60, and 80 solutions. The effect is not striking: a 4% solution of polysorbate 20 increases the solubility of the drug at 24° C from 7.17 g l^{-1} to only 9.81 g l^{-1}, and there are no great differences between the three detergents. It is doubtful if this is a micellar effect, as the solubility is also increased by 10% aqueous solutions of PEG 400, 4000, and 6000.

The behaviour of sulphisoxazole in surfactant and glycol solutions has been studied in a series of papers [140, 141]. In order to clarify the mechanism of the reduction in rectal absorption of this sulphonamide in the presence of PEG 4000 the effect of this compound on its physicochemical properties was examined. There is a linear relationship between the solubilities of sulphathiazole, sulphapyridine, and sulphisoxazole and the concentration of PEG 4000. The drugs apparently do not form complexes with the glycols; it is thought that the reduction in activity is due to a depression of the concentration of the drug in rectal lipid. The effect of non-ionic surfactants is to reduce the absorption of the sulphonamides through solubilization in micelles [141]. Fig. 6.18 indicates the extent of solubilization in polysorbate 80 solutions. Values of apparent distri-

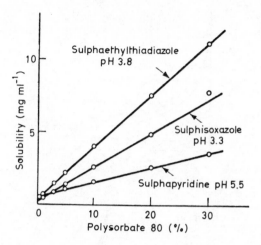

Figure 6.18 The solubility of sulphaethylthiadiazole, sulphisoxazole, and sulphapyridine in buffered solutions of polysorbate 80. From Kakemi *et al.* [141] with permission.

bution coefficients were obtained (K_m is the apparent distribution coefficient of the unionized drug between micelle and aqueous solution – the higher its value, the greater is the possibility of finding the drug in the micellar 'phase') in solutions of polyoxyethylene surfactants. Table 6.12 shows that the more hydrophobic the detergent, the greater the tendency of the sulphonamide to partition in favour of the micelle. Ionized sulphonamides are poorly solubilized in the micelles, but it is the unionized form which is biologically active. A correlation between K_m and the reduction in rectal absorption can be seen by comparing the values in Table 6.12 with the effect on absorption shown in Fig. 6.19a and b. The higher the K_m value, the greater is the reduction in absorption.

The diffusion of sulphanilamide from ointments has been increased by addition of either Tweens or Spans [142]. This, then, is another mechanism whereby surfactants can influence the effectiveness of formulation. For example, Sulphanilamide (5 %) in petrolatum does not inhibit sensitive *Staph. aureus*, yet in

Table 6.12 Values of the apparent distribution constant of unionized sulphisoxazole in presence of non-ionic surfactant

Non-ionic surfactant	K_m
Polyoxyethylene sorbitan monolaurate	0.90
Polyoxyethylene sorbitan monopalmitate	0.95
Polyoxyethylene sorbitan monostearate	1.20
Polyoxyethylene (30) – stearate	1.00
Polyoxyethylene (45) – stearate	0.70
Polyoxyethylene (10) – lauryl ether	1.50
Polyoxyethylene (15) – lauryl ether	1.00

From [141].

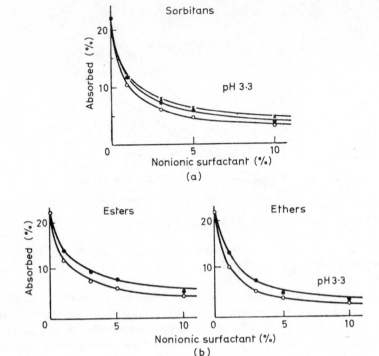

Figure 6.19(a) The effect of polyoxyethylene sorbitan alkyl esters on the rectal absorption of non-ionized sulphisoxazole.

    ×—× polyoxyethylene sorbitan monolaurate
    ●—● polyoxyethylene sorbitan monopalmitate
    ○—○ polyoxyethylene sorbitan monostearate

  (b) The effect of polyoxyethylene alkyl esters and ethers on the rectal absorption of non-ionized sulphisoxazole at pH 3.3.

               LH diagram
    ○—○ polyoxyethylene stearate-45
    ●—● polyoxyethylene stearate-30
               RH diagram
    ○—○ polyoxyethylene lauryl ether-10
    ●—● polyoxyethylene lauryl ether-15

From Kakemi *et al.* [141].

the presence of polysorbate 20 (15 %) some inhibition is observed; in the presence of the same concentration of PEG 400 large inhibition zones have been noted [143]. These results should be considered in the light of the previous remarks and the finding that at 1 % concentration levels polysorbates 20, 40, 60, and 80 markedly reduce the activity of sulphanilamide and it is likely that the surfactant increases release rate but decreases the activity of the drug.

    Span 60 (sorbitan monostearate) and Atlas G-2164 (Polyoxyethylene propylene glycol monostearate) increase the absorption of sulphathiazole from a

Table 6.13 Solubility of sulphacetamide sodium and sulphathiazole sodium in liquid petrolatum and cottonseed oil bases

Base	Surfactant	Sulphacetamide sodium solubility (mg %)	Sulphathiazole sodium solubility (mg %)
Liquid petrolatum	None	Negligible	Negligible
	1% Arlacel 83	2.18	1.83
	5% Arlacel 83	9.27	0.84
	10% Arlacel 83	17.14	1.52
Cottonseed oil	None	0.86	1.48
	1% Arlacel 83	0.79	0.92
	5% Arlacel 83	8.09	2.13
	10% Arlacel 83	19.95	2.04

From [145].

lanolin–petrolatum base [144]. It is not possible to decide whether this is due to a solubilization effect or a simple miscibility effect. However, it is known that surfactants increase the solubility of soluble sulphonamides in ointment bases. Whitworth and Becker's results [145] are shown in Table 6.13. Arlacel 83 increased the diffusion of both drugs from the cottonseed oil; the highest concentration of surfactant decreased the diffusion process from the petrolatum base. It is evident that the solubility of a drug in the vehicle is an important factor in the process. Solubilization will increase the saturation levels of the drug and will tend to promote its diffusion from the vehicle.

An ultracentrifugal study of polysorbate–drug interaction [146] has given values of apparent micellar partition coefficients for sulphapyridine and sulphisoxazole quite different from those quoted in Table 6.12 for sulphisoxazole, although no reference is made to this. A P_m of 79 ± 2 is quoted for sulphisoxazole in 1 to 4% polysorbate 80 at 0.001 and 0.01% solute levels.

(H) TETRACYCLINES

Naggar *et al.* [147] investigated the solubilization of the zwitterionic antibiotics tetracycline and oxytetracycline by polysorbate 20 and 80 at pH 5 and assumed that the interactions were due to some form of 'complexation' which seems unlikely. A wider range of tetracyclines and their interactions with a non-ionic, anionic and cationic surfactant were studied by Ikeda *et al.* [148] over a pH range of 2.1 to 5.6. It is unlikely that surfactant solutions of tetracycline are required, but the results are relevant in discussing tetracycline–surfactant interactions that could influence the activity of the antibiotic. Apparent partition coefficients of four tetracyclines in a polyoxyethylene (Brij 35) are shown in Table 6.14 obtained from dynamic dialysis measurements; the interactions with ionic surfactants (Table 6.15) appear more complex.

The tetracyclines have three macroscopic dissociation constants and thus their ionic behaviour is a complex function of pH. In the pH range of the solubilization

Table 6.14 Apparent partition coefficients of tetracyclines in Brij 35 solution at various pHs (25° C)

Substance	pH 2.1	3.0	3.9	5.6
Tetracycline	8.05	8.64	6.31	5.80
Oxytetracycline	8.01	7.61	6.54	5.68
Chlortetracycline	19.0	17.9	13.3	10.0
Minocycline	2.1	4.1	3.8	17.0

From [148].

Table 6.15 Apparent partition coefficients of tetracycline in sodium lauryl sulphate and dodecyltrimethylammonium chloride solutions at various pHs (25° C)

Solution	pH 2.1	3.0	3.9	5.4
Sodium lauryl sulphate	2860	2690	1130	390
Dodecyltrimethylammonium chloride	0	13	15	18

From [148].

study (2.1 to 5.6) the tetracycline molecules convert from cationic species to zwitterionic species. The zwitterionic form partitions most into a lypophilic phase and is the most active form biologically. The micellar partitioning results indicate that the cationic form is preferred for interaction with the micelle, perhaps because some of the tetracyclines are surface active, there having been a report that oxytetracycline aggregates in solution [149]. Minocycline shows the opposite trend to that displayed by tetracycline, oxytetracycline and chlortetracycline in Table 6.14. Minocycline has two $-N(CH_3)_2$ groups at position R_1 (IV) and III. At pH 2 the molecule will thus have little surface activity and would have difficulty orienting itself in a surfactant micelle. As the pH is increased, the protonation of the dimethylamino groups decreases and the molecule perhaps regains its amphipathic nature so allowing increased interaction with the surfactant. These results emphasize that many drugs do not obey the simple rules of micellar partitioning discovered with smaller and perhaps simpler molecules, and, indeed, that molecules within a given series do not always behave in the same manner.

Ikeda *et al.* [148] have analysed their results to obtain the partition coefficients for the cationic (P_c) and zwitterionic (P_z) tetracycline species. Fig. 6.20 should be consulted for the structures. The cationic form may be represented thus ($I°$, $II°$, III^+) and the zwitterionic (I^-, $II°$, III^+) or more simply $(0\ 0\ +)$ and $(-\ 0\ +)$, respectively.

If the apparent partition coefficient, P_m, is defined as

$$P_m = \frac{[D_m]/\phi}{[D_w]/(1-\phi)} = \frac{[D_m](1-\phi)}{[D_w]\phi} \tag{6.13}$$

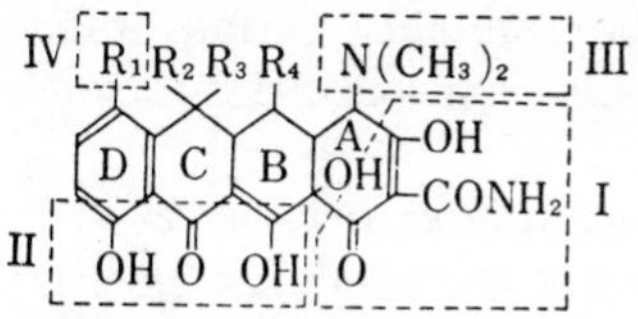

Figure 6.20 Three or four functional groups associated with macroscopic dissociation constants of tetracycline derivatives
tetracycline:
  $R_1 = H$, $R_2 = CH_3$, $R_3 = OH$, $R_4 = H$
oxytetracycline:
  $R_1 = H$, $R_2 = CH_3$, $R_3 = OH$, $R_4 = OH$
chlortetracycline:
  $R_1 = Cl$, $R_2 = CH_3$, $R_3 = OH$, $R_4 = H$
minocycline:
  $R_1 = N(CH_3)_2$, $R_2 = H$, $R_3 = H$, $R_4 = H$

where ϕ is the volume fraction of the micellar phase,

$$P_c = \frac{[(0,0,+)_m]}{[(0,0,+)_w]} \tag{6.14}$$

$$P_z = \frac{[(-0+)_m]}{[(-0+)_w]}. \tag{6.15}$$

If K_1 is the dissociation constant of the tetracycline, i.e.

$$K_1 = \frac{[(-0+)_w][H^+]}{[(00+)_w]} \tag{6.16}$$

$$P_m = \frac{[(0,0,+)_m] + [(-0+)_m](1-\phi)}{[(00+)_w] + [(-0+)_w]\phi}. \tag{6.17}$$

Substituting Equations 6.14, 6.15 and 6.16 into Equation 6.17 yields

$$P_m\{[H^+] + K_1\} = P_c[H^+] + K_1 P_z. \tag{6.18}$$

P_c and P_z can be estimated from a plot of $P_m\{[H^+] + K_1\}$ versus $[H^+]$.

The interaction between the tetracyclines and the ionic surfactants (Table 6.15) is of a different nature: a relatively small P_m being observed for tetracycline in DTAC and a large, presumably electrostatic interaction with NaLS at pH 2.1. The anionic–cation interaction would sufficiently alter transport properties so that dialysis rates would be altered; it is perhaps wrong to ascribe the notation K_m to the values obtained.

6.2.3 Steroids

The steroids have wide pharmacological applications, and there is a need for solutions of these compounds for topical and parenteral uses. Many of the steroid hormones are of low aqueous solubility. Various authors have reported the use of surface-active agents, proteins, and bile acids to solubilize these hormones.

(A) EFFECT OF STEROID STRUCTURE ON SOLUBILIZATION

The effect of steroid structure on solubilization has been briefly discussed in Section 5.3.2. Sjöblöm [150] has compared the maximum solubilizing powers of association colloids for a wide range of steroids (Table 6.16).

Table 6.16 Maximum solubilization of steroids in association colloid solutions

Steroid	Mol steroid/mol micelle		
	NaDS (40°C)	C_{14}TAB (20°C)	Polysorbate 20 (20°C)
Oestrone	0.014		0.0068
Oestradiol-17β	0.025		0.013
Oestradiol-17α	0.029		0.017
Oestriol	0.031		0.024
17α-ethynyloestradiol-17β	0.13		0.18
Oestrone-3-acetate	0.15		0.046
Oestradiol-3-benzoate	0.018		0.010
Oestradiol-3,17-dipropionate	0.051		0.013
Testosterone	0.18		0.027
17α-methyl testosterone	0.24	0.17	0.046
17α-ethynyl testosterone	0.0074	0.0044	0.0007
19-nortestosterone	0.27		0.13
Testosterone acetate	0.24		0.03
Testosterone propionate	0.22	0.094	0.044
Progesterone	0.24		0.037
11-hydroxy progesterone	0.30		0.026
17-hydroxy progesterone	0.090		0.0064
21-hydroxy progesterone (= desoxycorticosterone)	0.38		0.10
11,21-dihydroxy progesterone (= corticosterone)	0.42		0.14
17,21-dihydroxy progesterone	0.15		0.022
11,17,21-trihydroxy progesterone (= hydrocortisone)	0.30		0.057
11-desoxycorticosterone	0.38		0.11
11-desoxycorticosterone-21-acetate	0.16	0.070	0.013
Cortisone	0.20	0.14	0.023
Cortisone-21-acetate	0.071	0.050	0.009
Prednisone acetate	0.23	0.27	0.036
	0.087		0.012
Hydrocortisone	0.30	0.32	0.057
Hydrocortisone-21-acetate	0.026	0.025	0.0043
Prednisolone	0.22	0.21	0.047
Dexamethasone	0.16	0.27	0.041

From [150]. See also [151, 152–157].
NaDS = sodium dodecyl sulphate, C_{14}TAB = tetradecyltrimethylammonium bromide.

(II)

Uptake ranges from 7×10^{-4} mol steroid/mol polysorbate 20, for example, to $0.18 \, \text{mol} \, \text{mol}^{-1}$. If we consider two closely related structures, oestrone and oestradiol-17β, differing only in position 17; the marked difference this structural change induces is seen in Table 6.17. The oestradiol-17β is more hydrophilic with a hydroxyl group replacing the keto group of the oestrone. The results of Ekwall *et al.* [151] that in aqueous solutions of sodium lauryl sulphate the order of increasing solubilization follows the trend testosterone < progesterone < desoxycorticosterone confirms the findings that the substituent in position 17 determines the degree of solubilization. Progesterone has, in position 17, a $-\text{CO.CH}_3$ group, testosterone an $-\text{OH}$ group, and desoxycorticosterone a $-\text{CO.CH}_2\text{OH}$ group. A fair degree of correlation of solubilization parameters and the partition coefficients of a range of 19 steroids is displayed in Table 6.18 taken from the work of Tomida *et al.* [158].

When the data are plotted, two almost parallel lines are obtained which can be expressed by the following equations derived by least-squares:

$$\log P_m = 0.494 \log P_{\text{octanol}} + 1.24 \, (n = 12, \, r = 0.986, \, s = 0.066) \quad (6.19)$$

$$\log P_m = 0.523 \log P_{\text{octanol}} + 1.46 \, (n = 7, \, r = 0.995, \, s = 0.044). \quad (6.20)$$

The main selective structural feature is whether or not the steroid possesses a fluorine atom at carbon 9.

Those with fluorine are solubilized to a greater extent than would be predicted, this also being the conclusion of Barry and El Eini [160] who found that

Table 6.17 Maximum solubilizing power of surfactants for oestrone and oestradiol-17β

Surfactant	Conc. range (mol l^{-1})	Temp. (°C)	Mol micellar substance mol hormone^{-1}	
			Oestrone	Oestradiol-17β
Sodium caprate	0.1–0.5	20	202	99
Sodium lauryl sulphate	0.01–0.15	40	72.5	58.1
Tetradecyltrimethyl-ammonium bromide	0.005–0.08	20	44.6	13.3
			g mol^{-1}	
Polysorbate 20	1–20%	20	179×10^3	95.5×10^3

From [150].

Table 6.18 Solubilization parameters for the steroids in polyoxyethylene (23) dodecyl ether ($C_{12}E_{23}$) and partition coefficients between water and n-octanol, $P_{octanol}$, and water and ether, P_{ether} at 25 $\pm$ 1°C (from [158])

No.	Compound	Aqueous solubility (M)	R*	P^m_{app}	$P_{octanol}$	P_{ether}†
1	Hydrocortisone (XVIII)	1.08×10^{-3}	0.115	97.9	35.7	1.60
2	Corticosterone	5.79×10^{-4}	0.131	187	86.5	4.52
3	Deoxycorticosterone	3.55×10^{-4}	0.157	399	798	52.0
4	Cortisone	5.32×10^{-4}	0.0388	66.7	26.2	1.40
5	Hydrocortisone acetate (VII)	4.58×10^{-5}	0.0116	229	154	26.0
6	Cortisone acetate	6.18×10^{-5}	0.0140	205	126	25.1
7	Deoxycorticosterone acetate	2.35×10^{-5}	0.0187	718	1190	95.5‡
8	11-Hydroxy progesterone	1.53×10^{-4}	0.0459	271	227	35.5‡
9	Progesterone (XIX)	3.79×10^{-5}	0.0559	1330	7410	604
10	Testosterone (VI)	8.26×10^{-5}	0.0579	633	1960	87.3
11	Prednisolone	6.54×10^{-4}	0.0794	110	41.4	1.13
12	Prednisolone acetate	4.22×10^{-5}	0.0131	281	250	21.1
13	Triamcinolone	2.07×10^{-4}	0.0219	96.3	10.8	0.757
14	Triamcinolone acetonide (VIII)	4.95×10^{-5}	0.0312	569	205	14.6
15	Triamcinolone diacetate (XVII)	7.41×10^{-5}	0.0245	299	83.7	—
16	Dexamethasone	2.58×10^{-4}	0.0721	253	67.8	3.87
17	Betamethasone	1.71×10^{-4}	0.0535	283	87.7	4.76
18	Dexamethasone acetate	1.25×10^{-5}	0.0134	967	806	70.8†
19	Betamethasone 17-valerate	1.95×10^{-5}	0.0387	1790	3070	509

* The slope of the solubility versus surfactant concentration (M) plot. † Data taken from [159]. The experiment was carried out at 23 $\pm$ 1°C. ‡ Value estimated from the data in [159].

dexamethasone was solubilized to a greater degree than would be expected from the R_m value obtained by chromatography. This is probably due to the different sites of solubilization for the two groups of steroids. The R_m value, defined as

$$R_m = \log\left(\frac{1}{R_f} - 1\right)$$

(6.21)

is a useful quantitative expression of a molecule's polarity.

While general trends, e.g. between hydrophilicity or solubility in water and uptake into micelles of a given surfactant, have been demonstrated the other factors which specifically influence packing into a structured micelle are more difficult to quantitate. Perusal of the results in Tables 6.16 and 6.18 indicates some of the structural features of the steroids which increase or decrease solubilization. Introduction of an ethynyl group at C_{17} as in ethynyloestradiol enhances the solubility in ionic and non-ionic micelles. However, introduction of a 17-ethynyl group into the testosterone molecule results in decreased solubility. Steroids of the testosterone group are generally solubilized in much greater amounts than the oestrogens in ionic micellar solutions. According to Sjöblöm, this indicates that the solubilization of steroids is not uniformly influenced by a certain substituent, but that the whole of the steroid molecule determines its micellar solubility.

Sjöblöm concludes that: (i) hydrophilic substituents do not unconditionally increase the micellar solubility of the steroids, possibly because of the orientation of the steroids in the micelle, which might depend on the balance of hydrophilicity between rings AB and CD; (ii) the position of the hydrophilic substituents is of great importance; and (iii) suitable hydrocarbon substituents increase the solubility in the micelle [152].

A short side chain at C_{17} enhances the solubilization of steroids, especially when it contains a free hydroxyl group. A greater number of non-ionic molecules are required to solubilize one steroid molecule than is required by ionic micelles. In the oestrogens this difference is small, but in most of the other cases the difference is nearly 10-fold (see Table 6.16). This would indicate a different mechanism of solubilization; the shift of the absorption maximum and the depression of the molar extinction for all the steroids except the oestrogens are much more pronounced in Tween solutions than in ionic colloid solutions. This would suggest a unique mode of solubilization for the oestrogens, most of which are indeed poorly solubilized, only the 21-acetoxy steroid occupying an intermediate position.

Compare oestrone and testosterone in the diagram below:

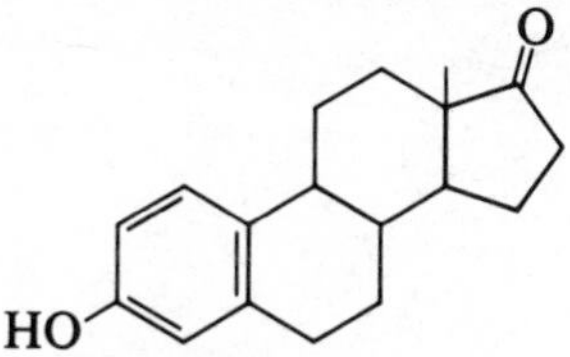

(III) Oestrone 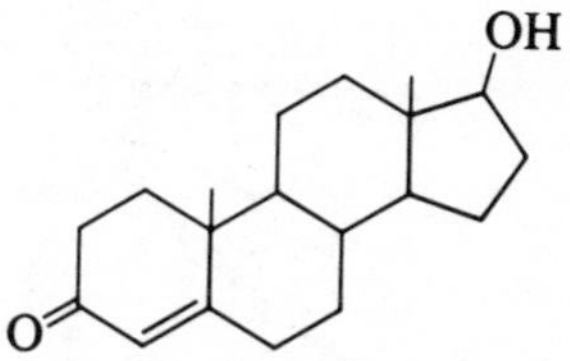   (IV) Testosterone

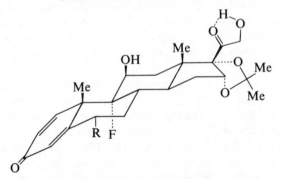

(V) Oestrone   (VI) Testosterone

(VII) Hydrocortisone acetate (R = H)
Fludrocortisone acetate (R = F)

(VIII) Triamcinolone acetonide (R = H)
Fluocinolone acetonide (R = F)

(IX) Betamethasone valerate

In both compounds rings B, C, and D have the same rigid configuration. In oestrone (V) ring A is rigid and planar, although in testosterone (VI) ring A is flexible. In the latter case this might have some influence on the packing of the testosterone molecule into the micelle. Ring A in oestrone has an ionizable group, while the keto group in testosterone on ring A is non-ionizable. The differences between these two rings may influence to some extent the orientation of the molecule in the micelle. The differences between the molecules and their derivatives would require a more critical examination before any conclusive results could be obtained.

Triton WR 1339 and Tween 80 have both been used in formulations of adrenal cortical hormone preparations for the eye [161, 162]. It was found that the solubility of prednisolone, methylprednisolone, and fluoromethalone in aqueous solutions of Triton WR 1339 was linearly dependent on the detergent concentration. The structures of the compounds are given below their solubility behaviour in Table 6.19. From this it can be seen that there are striking differences in the amounts solubilized, and it is obvious that the fluoro-derivative is more difficult to solubilize than the others mentioned.

Table 6.19 Solubilizing power of Triton WR 1339 for steroids

Steroid	Solubility in water (mg ml^{-1})	Steroid mg ml^{-1}/%w/w Triton WR 1339	Mol steroid mol^{-1} Triton WR 1339	Mol Triton mol^{-1} steroid
Prednisolone (X)	0.223	0.249	0.0486	20.6
Methylprednisolone (XI)	0.095	0.114	0.0214	46.7
Fluorometholone (XII)	0.003	0.00927	0.00173	578.0

From [161, 162].

(X) Prednisolone   (XI) Methylprednisolone   (XII) Fluorometholone

The solubility of non-steroidal oestrogens in ionic and non-ionic surfactant solutions has been investigated by Nakagawa [163, 164]. The compounds included dienoestrol, hexoestrol, diethylstilboestrol, and chlorotrianisene: the structures of which may be compared below. In both polysorbate 20 and 80 the order of increasing solubility (gram for gram) is dienoestrol, hexoestrol, and diethylstilboestrol. Chlorotrianisene, a bulkier molecule, is much less soluble, being approximately 1/20 as soluble in polysorbate 80 at 30° C than the others in this series and less soluble than many of the steroidal hormones. It has been

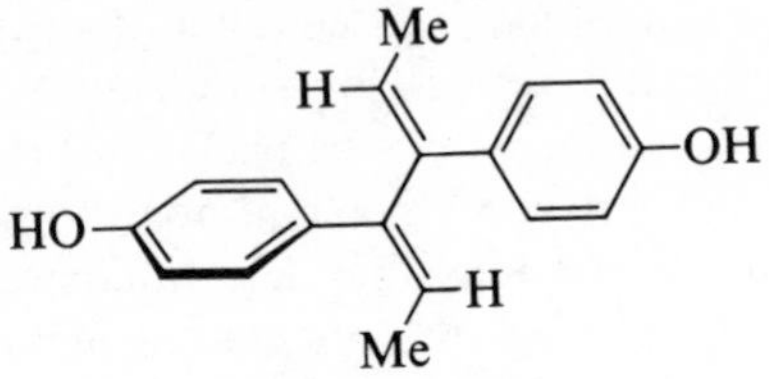

Me
H
OH
HO
H
Me
(XIII) Dienoestrol

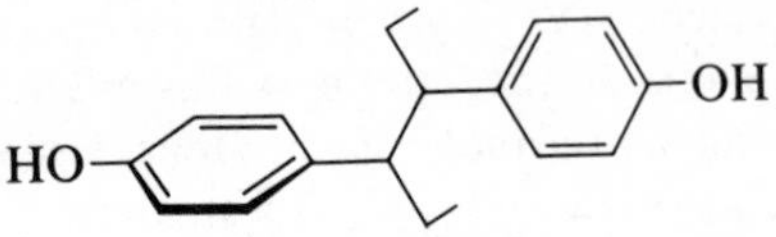

OH
HO
(XIV) *meso*-Hexoestrol

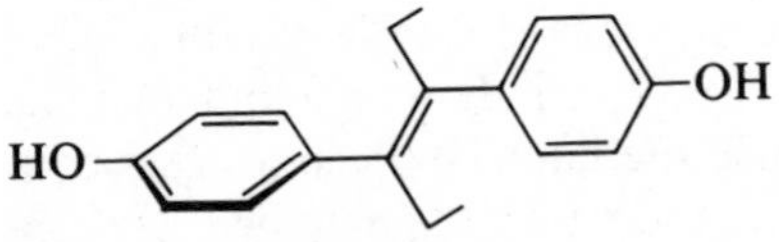

OH
HO
(XV) Stilboestrol

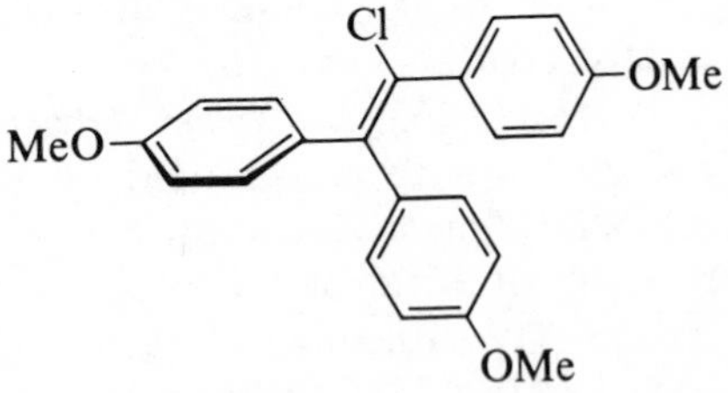

Cl
OMe
MeO
OMe
(XVI) Chlortrianisene

shown that the steroid hormones have 100 to 500 times the solubility of methylcholanthrene, a polycyclic aromatic hydrocarbon which resembles the steroids in structure. It is possible that in the solubilization process there is some interaction between the hydrophilic groups of these hormones and some portion of the surfactant.

(B) EFFECT OF SURFACTANT STRUCTURE ON UPTAKE OF STEROIDS

The solubilizing efficiency of a series of cetyl polyoxyethylene esters decreased as the polyoxyethylene chain length was increased [160] when surfactants were compared on a weight basis. Partition coefficients obtained by dialysis and solubility methods are shown in Table 6.20. P decreases with increasing polyoxyethylene chain length. As discussed in Section 5.3.1, although the number of steroid molecules per micelle is smaller for more hydrophilic surfactants, the total amount of steroid per mole of surfactant is greater, hence the observed increase in solubilizing efficiency with increased hydrophilic chain length when molar concentrations are compared. In practical terms comparison on a weight basis is more realistic and the results are clearly shown in Fig. 6.21.

(C) EFFECT OF TEMPERATURE

Increasing temperature decreases the solubilization capacity of non-ionics even though the micelles grow in size [160], but if the molar ratio of steroid to surfactant (and not micelle) is calculated this value increases for steroids in polysorbate 40 and tetradecylammonium bromide (TDABr) [165] (see Table 6.21). This topic is further discussed in Section 5.3.3.

(D) EFFECT OF ELECTROLYTE ADDITION

Very little work has been published on the effect of additives on steroid solubilization. Lundberg *et al.* [166] have, however, measured the uptake of three

Table 6.20 Partition coefficients of steroids between water and ether, P, and aqueous and micellar phases from solubility, P_s, and dialysis, P_d, at 25° C

Steroid	Surfactant									
	$C_{16}E_{17}$ $n = 99$		$C_{16}E_{32}$ $n = 56$		$C_{16}E_{44}$ $n = 39$		$C_{16}E_{63}$ $n = 25$			
	R_s	R_d	R_s	R_d	R_s	R_d	R_s	R_d	R	R_m
Hydrocortisone (XVIII)	110	110	101	103	86	87	68	66	1.63	0.27
Dexamethasone (XVII)	314	295	273	269	244	240	199	208	3.89	0.48
Testosterone (VI)	786	807	661	654	570	588	452	442	56.9	1.04
Progesterone (XIX)	2160	2230	1790	1730	1550	1400	1250	1000	613	1.46

From [160]. n = aggregation number of the surfactant and R_m is calculated from $\log\left(\dfrac{1}{R_f} - 1\right) = R_m$.

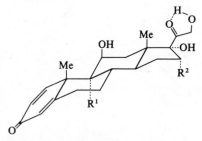

(XVII) Dexamethasone (R^1 = F; R^2 = Me)
    Triamcinolone (R^1 = F; R^2 = OH)
    Prednisolone (R^1 = R^2 = H)

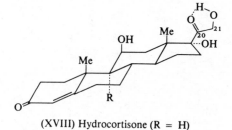

(XVIII) Hydrocortisone (R = H)

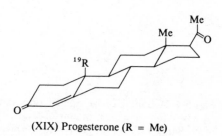

(XIX) Progesterone (R = Me)

steroids in TDABr. 0.2 M NaCl decreased the solubilization of testosterone and progesterone and increased that of oestrone confirming the notion that the first two steroids are solubilized in the polyoxyethylene layer of the micelle and that oestrone is solubilized in the hydrocarbon core. 0.1 M NaCl decreased the solubilizing capacity of sodium glycocholate for testosterone [167] by about 10%.

(E) SOLUBILIZATION OF STEROID MIXTURES

Of the two widely known solubilized preparations of intravenous anaesthetics on the market, one, Althesin (Glaxo, UK) contains a mixture of steroids. The more active anaesthetic is alphaxolone (9 mg ml^{-1}) (XX) and a less active alphadalone acetate (3 mg ml^{-1}) (XXI) has been added to improve the solubility of the alphaxolone in the Cremophor EL vehicle (20%).

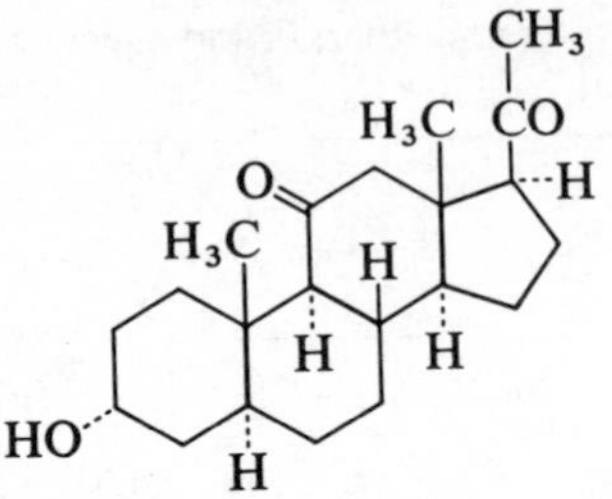

CH$_3$

H$_3$C CO

O

H$_3$C

H

H H

HO

H

(XX) Alphaxolone

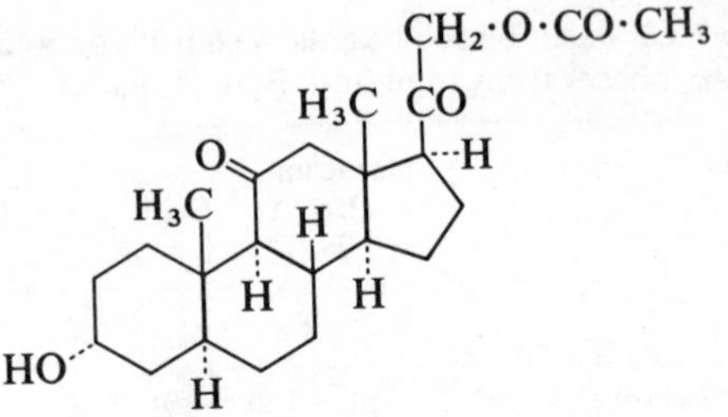

CH$_2$·O·CO·CH$_3$

H$_3$C CO

O

H$_3$C

H

H H

HO

H

(XXI) Alphadalone acetate

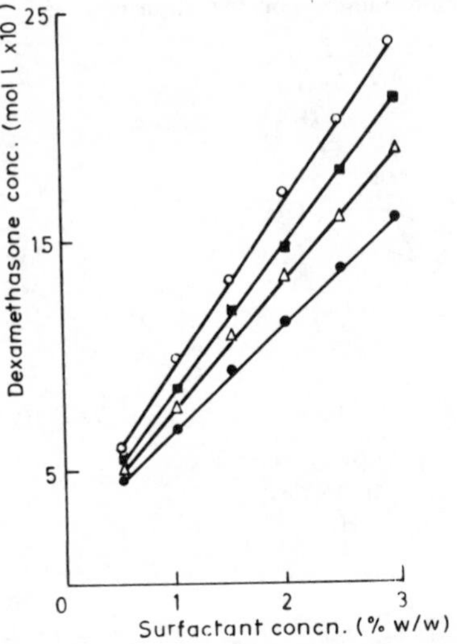

Figure 6.21 Solubility of dexamethasone in water (mol l^{-1} × 10^4), ▲, and as a function of % w/w aqueous concentrations of C$_{16}$E$_{17}$, ○, C$_{16}$E$_{32}$, ■, C$_{16}$E$_{44}$, △, and C$_{16}$E$_{63}$, ●. From Barry and El Eini [160] with permission.

The effect of alphadalone acetate on the solubility of alphaxolone is a phenomenon that remains to be explained. Simultaneous solubilization of steroid hormones has only been studied by Lövgren and co-workers [168, 169]. Oestradiol was solubilized independently of the C$_{21}$ steroids and testosterone studied, i.e. the capacity for oestradiol was unaffected by the solubilization of the latter. However, the solubilization of ethinyl oestradiol with progesterone and with testosterone was dependent on the presence of the other. The solubility of 11 α-hydroxyprogesterone was enhanced by ethinyl oestradiol – an effect akin to that of alphaxolone and alphadalone acetate. In several other pairs of steroids solubilization was reduced. When a progesterone-saturated solution of poly-sorbate 40 was equilibrated with an excess of ethinyl oestradiol, 96% of the solubilized progesterone precipitated while the oestrogen component was

Table 6.21 Solubilization capacities of surfactants for hormonal steroids a. _ function of temperature (K)

Surfactant	Steroid	Mol steroid/mol surfactant				
		293	300.5	308	315.5	323
Polysorbate 40	Oestradiol	0.013	0.016	.0.019	0.022	0.026
	Ethinyl oestradiol	0.18	0.23	0.27	0.32	0.37
	Testosterone	0.027	0.039	0.052	0.065	0.076
	Ethisterone	0.0007	0.0009	0.0012	0.0016	0.0018
	Progesterone	0.037	0.049	0.063	0.073	0.084
	17α-Hydroxyprogesterone	0.0072	0.0079	0.0085	0.0091	0.0091
Tetradecyltrimethyl-	Oestradiol	0.068	0.080	0.092	0.105	0.118
ammonium bromide	Ethinyl oestradiol	0.27	0.34	0.43	0.51	0.57
	Testosterone	0.13	0.19	0.25	0.29	0.35
	Ethisterone	0.0046	0.0055	0.0066	0.0074	0.0083
	Progesterone	0.16	0.15	0.16	0.16	0.16
	17α-Hydroxyprogesterone	0.043	0.060	0.082	0.098	0.114

From [165].

solubilized maximally. When the saturation was carried out in the opposite way, 81% of the ethinyl oestradiol precipitated and progesterone was solubilized maximally. If an excess of both steroids was added at the same time, progesterone was solubilized to its maximum extent, while the solubility of the ethinyl oestradiol dropped to 19% of its maximal value in agreement with the result

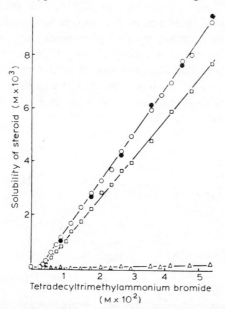

Figure 6.22 Solubility of progesterone in aqueous solutions of tetradecyltrimethylammonium bromide. ○, progesterone only; △, progesterone first and ethinyl oestradiol second; □, ethinyl oestradiol first and progesterone second; and ●, progesterone and ethinyl oestradiol at the same time. From Lövgren *et al.* [168].

obtained when it was added as the first component [168]. When solubilization occurs in different sites, if micellar structure is unaltered, solubilization of solute pairs should be independent and unaffected by mode of addition. No relationship between thermodynamic parameters and simultaneous solubilization behaviour has been found [165] and one is left with only the notion that there is independent solubilization if the sites of solubilization are different and if the same or adjacent, the solutes interfere with the others' accommodation in the micelle.

Some results of the solubility of progesterone in aqueous TDTMABr are shown in Fig. 6.22 in which differences in the order of addition of ethinyl oestradiol and progesterone are reflected in large differences in the solubility of one of the species.

Probably more work has been carried out on steroid solubilization than on most other classes of drug. In spite of the mass of data the behaviour of some steroid–surfactant systems, especially those containing two steroids, is by no means understood. The use of oil–water partition coefficients allows us to predict with a reasonable degree of precision the rank order solubility of a given steroid of a series in a surfactant, but not yet to relate surfactant properties to micellar capacity.

6.2.4 Fat-soluble vitamins

Of the six vitamins regarded as essential accessory food factors (vitamins A, B_1, B_2, nicotinamide, C and D), only vitamins A and D are insoluble in water. Presentation of fish-liver oils, rich in these two vitamins, as emulsions enhances the absorption of the vitamins, but such preparations are not always palatable. However, half a century ago Lester-Smith [170–172] observed the solubilization of vitamins A and D in soap solutions formed by the saponification of vitamin-containing oils. Vitamins E and K are also insoluble in water; vitamin E is used in the treatment of habitual abortion, and vitamin K is employed to combat hypoprothrombinaemia.

(XXII) Vitamin A

(XXIII) Vitamin D_2

Using sucrose mono-esters of fatty acids prepared according to Osipow et al. [173], Mima [174] solubilized vitamins A, D_2, and vitamin E acetate. The sucrose esters were employed in an attempt to overcome the problems encountered when polyoxyethylene glycol ethers are used, such as the sensitivity

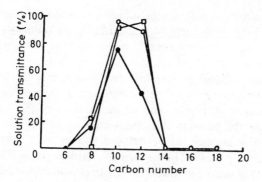

Figure 6.23 The solubility of vitamins in solutions of sucrose esters of varying alkyl chain length as shown by transmittance data, illustrating maximum solubilization at C_{10} and C_{12}. O—O vitamin A alcohol; ●—● vitamin D_2; □—□ vitamin A acetate. Ratio of vitamin : ester : water : 1 : 6 : 200. Drawn from the data of Mima [174].

of such systems to clouding (especially in the presence of non-polar materials) and their potential toxicity. Aqueous solutions of the vitamins prepared with the sucrose esters do not cloud and are very stable. Of sucrose mono-caprylate, caprate, laurate, myristate, palmitate, and linolenate, the caprate (C_{10}) ester was found to be most efficient for solubilizing vitamin D_2 and vitamin A alcohol [175] (Fig. 6.23). The haemolytic activity of some of these esters is unfortunately higher than for the established non-ionic detergents, and they must, therefore, be used with caution in injections. A Japanese patent [176] describes the use of 6-L-ascorbyl caprylate to solubilize the fat-soluble vitamins.

Aqueous injections of vitamins A, D, E, and K have been prepared in polysorbate 20, 40, 60, and 80 solutions [177]. Table 6.22 shows the solubility of these vitamins in 10 % polysorbate solutions, polysorbate 20 and 80 being the best two solubilizers.

It has been reported that the absorption of carotene (a precursor of vitamin A) is more rapid when presented solubilized in solutions of polysorbate 80 than when administered orally or intramuscularly in oil [178]. Sobel [179] has revealed improved absorption of vitamin A itself when in solubilized form; the transfer of the vitamin to the milk of nursing mothers is superior in such aqueous solutions [180].

Table 6.22 Solubility of fat-soluble vitamins in 10 % polysorbate solutions

Polysorbate	Vitamin D_2 (I.U. ml^{-1})	Vitamin E (mg ml^{-1})	Vitamin K_3 (mg ml^{-1})	Vitamin A alcohol (I.U. ml^{-1})
20	20 000	5.7	4.7	80 000
40	16 000	3.8	4.0	60 000
60	15 000	3.2	3.7	60 000
80	20 000	4.5	4.5	80 000

From [177].

The stability of vitamin A alcohol in neutral aqueous solution is enhanced by polysorbate 20, but not by polysorbates 40 and 60 [177] its stability in 20% polysorbate 20 is greater than its stability in cottonseed oil or in pure surfactant [181].

Many of these non-ionic substances are bitter and to minimize the amounts required, some solubilization studies have been undertaken in systems containing polygols. Coles and Thomas [182] observed that the addition of 30% glycerol makes it possible to halve the amount of surface-active agent required in the solubilization of vitamin A alcohol. More detailed studies have been made [183, 184] and the results are presented in the form of phase diagrams. The more comprehensive diagram (Fig. 6.24) illustrates the variety of phases possible in the system and the comparatively small region of isotropic liquid phase. The concentration of vitamin A palmitate is 6.6%. Increasing the concentration of glycerin reduces the amount of polysorbate required to form an isotropic phase.

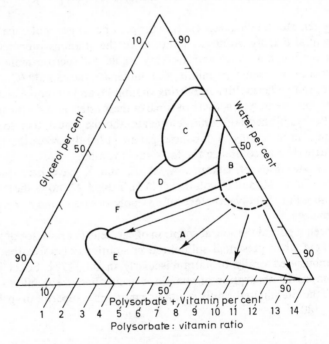

Figure 6.24 The vitamin A: polysorbate 80: glycerol: water system.

Zone	Description
A	Transparent, single phase
B	Semi-solid
C	Faintly opalescent
D	Markedly opalescent
E	Two transparent phases
F	Emulsions

From Boon *et al.* [183].

Hüttenrauch and Klotz [184] found the order of decreasing 'co-solubilization' saccharose > sorbitol > glycerin in the preparation of clear aqueous solutions of vitamin A with 15% polyoxyethylene sorbitan oleate. The polygol is thought to increase the proportion of hydrophile in the micelle, reducing the hydration of the oxyethylene chains, producing areas which would approximate to 100% glycol. If polygols increase the solubilization of vitamin A then it would be expected that the extent of solubilization would vary with the HLB of the surfactant. A relationship has been found between the HLB values of commercial polyoxyethylene glycol sorbitan mono-oleates and their solubilizing capacity for vitamin A palmitate [185, 186].

Table 6.23 shows that the solubility of vitamin A palmitate increases the larger the lyophobic chain and the smaller the polyoxyethylene radical, which is opposite

Table 6.23(a) Solubilizable amounts of vitamin A palmitate per mole of surface-active agent in 20% aqueous solution

Surfactant	E^*	Mean molecular weight	MAC (mol vitamin/mol surfactant)
Polysorbate 20	23.5	1,385	0.15
Polysorbate 40	20.1	1,286	0.54
Polysorbate 60	18.7	1,251	0.67
Polysorbate 80	19.1	1,270	0.68
PEG monolaurate	12.8	764	0.12
PEG monomyristate	13.4	851	0.31
PEG mono-oleate	13.3	867	0.67
PEG mono-oleate	8.7	586	0.16
PEG monolaurate	12.8	764	0.12
PEG monolaurate	23.2	1,223	0.08
PEG monolaurate	30.8	1,551	0.04

* E = mean number of ethylene oxide units per monomer (assay figure).

(b) Solubility of vitamin K_3 (2-methyl-1,4-naphthoquinone) in surfactant (20%) solutions

Surface-active agent	E^{1*}	MAC (mg ml^{-1})	MAC (mol/mol surfactant)
Polyoxyethylene sorbtan monolaurate	8.5	3.58	7.5×10^{-2}
	8.8	3.22	6.9×10^{-2}
	13.3	2.94	8.0×10^{-2}
	17.4	2.24	7.2×10^{-2}
Polyoxyethylene monolauryl ether	7.5	4.06	6.1×10^{-2}
	6.9	3.52	5.0×10^{-2}
	10.1	3.04	5.6×10^{-2}
	18.0	2.82	4.4×10^{-2}

* E^1: number of ethylene oxide units from assay.
From [186, 189].

to the expected findings derived from the co-solvent effects yet expected from predictions of micelle size. Some results showing similar effects are included in Table 6.23 for vitamin K_3. Indeed, Nakagawa [187] finds that the solubility of the fat-soluble vitamins in liquid paraffin is very much greater than in an 80% w/v PEG 300 solution, except for acetonaphthone, vitamin K_4, and vitamin K_3 (2-methyl-1,4-naphthoquinone). There are, however, optimum HLB values for the solubilization of vitamins A and D. Using transmittance data, as for the sucrose esters, Mima [188] determined these HLB values for vitamin A palmitate, vitamin A acetate, vitamin A alcohol, and vitamin D_2, these being, respectively 14.5 to 15.5, 15.8 to 16.2, and above 17.9 for the latter two. Vitamins obtained by purification (those with a greater number of international units of activity per gram) have a wider range of optimum HLBs, but the optimal HLB tends to be higher. Thus not only do we have to contend with variations in the properties of commercial surface-active agents but also with the degree of purity of the vitamin. Ito *et al.* [189] found that the range of HLB for solubilization of vitamin A palmitate was 15 to 17.

Considerable batch variation in the solubilizing properties of polysorbate 80, and a correlation between the assay for ethylene oxide content and solubilizing capacity has been found [183]. This emphasizes the need for careful analytical control of materials when experimental work is in progress. The areas of the phase diagram which gave clear solutions with all the surfactant samples represent about half the area shown in Fig. 6.24 obtained with one sample.

Formulae have been assembled for multi-vitamin syrup and multi-vitamin drops [190, 191] using polysorbate 80 to solubilize vitamin A palmitate and vitamin D with glycerin as co-solubilizer, enabling a high concentration of drug to be given per dose. The solubilizer serves the extra purpose of allowing the preparations to be diluted into water or milk for paediatric administration. The sorbitol, primarily added to enhance the taste of the preparation, but which increases the absorption of vitamin B_{12}, obviously also acts as a co-solubilizer [192]. A similar preparation is described by Whittet and Cummins [193].

Non-ionic surfactant solutions have, however, a tendency to cloud, and many substances lower the cloud point of the detergent solutions. The polarity of the solubilized substance affects the turbidity formation: vitamin A alcohol and vitamin D cause clouding, but vitamin A palmitate has practically no effect. With severe clouding, separation into two layers or precipitation may occur [190, 191], the preparation returning to the crystal state on cooling or on remixing the separated layers. The cloud point must be sufficiently high to prevent such separation through variation in storage temperatures, as such fluctuations must adversely affect the stability of the vitamins.

The solubility of vitamin A in surfactant solutions is utilized in an assay procedure for the estimation of naturally occurring vitamin A in chicken livers, synthetic vitamin A in powdered formulations for infants, and vitamin A in stabilized animal-feed supplements [192, 193]. The vitamin is solubilized by Triton X-100 and extracted with a mixed solvent.

6.2.5 Barbiturates

The search for suitable solubilizers is made necessary in the case of the barbiturates by the fact that the soluble sodium salts are unstable and less-soluble forms have therefore to be used. The increases in solubility brought about by 1 to 2% solutions of polysorbates 20–80 are not startling; more water-soluble derivatives show a smaller increment in solubility in surfactant solutions (see Table 5.7, Chapter 5).

The degradation of barbiturates in alkaline solution is well known. Stability may be increased by selection of a suitable co-solvent or solubilizing agent such as ethanol, propylene glycol or polysorbate 80 [194]. A polysorbate 80 solution was used to solubilize 0.4 g phenobarbitone but polysorbate 80 used alone in preparations of this type tends to impart an obnoxious taste to the product, and sweetening agents are required. The device of lowering pH to increase stability resulted in the precipitation of the free acid at pHs in the region of 7 to 8 and led to the investigation of methods to increase the solubility of the acidic form of the barbiturates [195, 196].

Uptake of phenobarbitone in a sodium paraffin sulphonate is not a linear function of surfactant concentration [197]. Above the critical micelle concentration there is an inflection, around 1%, which might result in problems in the dilution of the system.

6.2.6 Salicylates and related compounds

The insolubility of acetylsalicylic acid is a contributing factor to its irritant action on the gastric mucosa; it is hydrolysed in aqueous solution. It has, however, been incorporated in a suppository base with macrogols 1540 and 6000 which increase its solubility and appear to improve the absorption of the drug [198].

Representative classes of surfactants have been considered as solubilizers for aspirin [199]. Ranked in order of decreasing effectiveness were cetylpyridium chloride > polysorbate 20 > benzalkonium chloride > polysorbate 80 > di-octylsulphosuccinate, DOSS.

In a detailed study of the influence of non-ionic surfactant structures on the solubilization of salicylic acid it was found that as the alkyl chain length of polysorbate increases the molar ratio of solubilizate to surfactant increases [200]. As the oxyethylene chain length of a series of Myrj surfactants is increased this molar ratio also increases. A series of monohydric alcohols of decreasing dielectric constant, a group of polyhydroxy alcohols, PEG 400, and polysorbate 20 were investigated for their effect on the solubilizing power of the non-ionic surfactant polysorbate 80 [201]. Monohydroxy alcohols increased or decreased the solubilizing power of the detergent in order of their polarity while the polyglycol and the polyhydroxy alcohols had little effect. A surprising finding was that polysorbate 20 decreased the solubilizing power of polysorbate 80 for salicylic acid in a linear fashion; the results are given in Table 6.24. It might have been expected that the addition of another micelle-forming compound, especially

Table 6.24 Effect of polysorbate 20 on the solubilizing power of polysorbate 80 on salicylic acid

Polysorbate 80		Critical miscibility ratio: salicylic acid/polysorbate 80
No additive		0.150
Polysorbate 20	20%	0.145
	40%	0.140
	60%	0.135
	80%	0.130

From [201, 202].

one of a similar structure, would have increased the solubility of the salicylic acid in the solution but this is too simplistic a view.

Evidence of decreased solubility of both chlorhexidine diacetate and chlorhexidine dihydrochloride in Brij surfactant mixtures has been adduced [203] though this is only clear in the case of the dihydrochloride at concentrations higher than 5% when uptake is some 30% higher in Brij 96 than in Brij 92–96 mixtures if HLB = 11.

Nishikido has analysed uptake of a dye into mixed non-ionic surfactants [204] in terms of the solubilizing powers of the simple surfactants and the mixtures. Defining S_m^{ad} as the solubilizing power in the ideal state if each component forming mixed micelles contributes separately to the total solubilization, one can write

$$S_m^{ad} = \alpha_1 x_1^m + \alpha_2 x_2^m \tag{6.22}$$

where α_1 and α_2 are the solubilizing power of components 1 and 2 and x_1 and x_2 their mole fractions in the micelle. In an ideal system the ratio S_m/S_m^{ad}, where S_m is the actual solubilizing power of the system, would be equal to unity. In the alkyl polyether systems studied all expect the $C_{10}E_6$–$C_{12}E_6$ system, which is nearly ideal, showed negative deviation from ideality, thus solubilization is generally less than expected in mixed micelles because of the nature of the mixed micelle. A positive deviation in S_m/S_m^{ad} is expected in anionic–non-ionic polyether systems from which it is concluded that some interaction between polyoxyethylene chains and anionic surfactants contributes favourably to solubilization. Such beneficial effects have been measured in sodium lauryl ether sulphate–non-ionic systems in solubilizing perfume oils [205]. A range of glycols (hexylene, butylene, dipropylene and diethylene glycol) was measured: all had little solubilizing effect in the presence of sodium lauryl ether sulphate, except hexylene glycol which was as effective as polyoxyethylene (9) nonyl phenol, which was the most effective of the nonyl phenyl surfactants chosen as co-solubilizer.

6.2.7 Oils

For most pharmaceutical oils three to five parts of surfactant are sufficient to solubilize one part of oil in water; surfactants with an HLB in the region of 15 to

18 are ideal as solubilizers for this purpose, polysorbate 60 and polysorbate 80 being widely used in liquid oral preparations. The fish-liver oils, such as those of cod, halibut, and shark, can be made water miscible so that preparations can be diluted with flavoured vehicles for administration. Polysorbate 80, used to solubilize vitamins A and D in an aqueous vehicle of sorbitol and water [206], also acts as a carrier for flavouring oils.

Table 6.25 gives the amounts of cetomacrogol 1000 or polysorbate 20 required to solubilize 1% v/v of various oils used in flavouring. Polysorbate 20 has been used to prepare peppermint oil concentrates [208] and the phase diagram of a peppermint oil–water–polysorbate 20 system has been studied [209] (Fig. 6.25). A concentrate of 7.5% oil, 42.5% polysorbate 20, 50.0% water (represented by

Table 6.25 Amount of surfactants (cetomacrogol and polysorbate 20) required to solubilize 1% flavouring oils in water

Oil	Cetomacrogol, % w/v	Polysorbate 20, % v/v
Peppermint oil	4.5	5.0
Anise oil	7.0	9.0
Caraway oil	5.0	—
Dill oil	4.0	—
Cinnamon oil	7.0	12.0
Clove oil	—	6.0

From [207, 208].

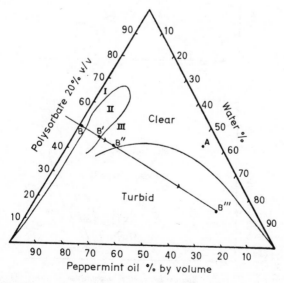

Figure 6.25 Partial phase diagram of polysorbate 20-peppermint oil–water system from O'Malley *et al.* [209]. Phases with compositions represented by the upper parts of the diagram are clear, those below, turbid. Point A represents 7.5% oil, 42.5% polysorbate 20, and 50.0% water.

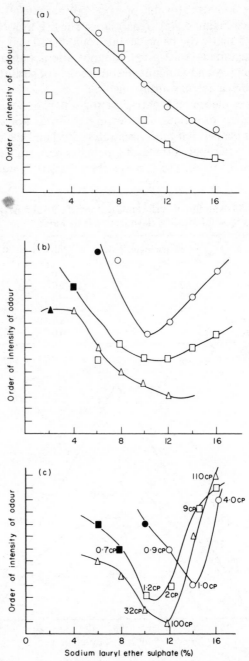

Figure 6.26 Comparison of the orders of intensity of odours of (a) 0.5 %, (b) 1 % and (c) 2 % of perfume W 100 in sodium lauryl ether sulphate solutions with 0 % (O), 2 % (□) and 4 % (△) added coconut diethanolamide (CDE). Solid points indicate opalescent solutions. From Blakeway *et al.* [205].

point A in Fig. 6.25) can be diluted ten times with water to give a satisfactory preparation of peppermint water, but care must be taken in choosing concentrates. Point B, for example, represents 49% oil, 50% polysorbate 20, and 1.0% water (clear solution). Dilution of 10 ml with 1.0 ml distilled water produces a cloudy solution. Addition of a further 1 ml causes the solution to become clear (B^{11}) and diluting the original solution three times with water produces a turbid mixture (B^{111}). Moderate changes in temperature (20, 30, 40° C) have little effect on the phase diagram.

Similar phase diagrams have been obtained for lavender, anise, clove and peppermint oils when the surfactant is a polyoxyethylene glyceryl fatty acid [210]; Ello [211] has solubilized a number of essential oils in polysorbate 20, 40, 60 and 80.

Many essential oils are subject to atmospheric oxidation. There is some evidence that solubilized benzaldehyde is more resistant to atmospheric oxidation than emulsified benzaldehyde [212, 213]. Suspensions of methyl linoleate in water in the absence of surface-active agents are oxidized at a very low rate; the presence of soap in all cases increases the rate of oxidation [212, 213], yet emulsions of the oil are oxidized more quickly than solubilized solutions. One may conclude from this evidence that if an insoluble oxidizable substance has to be formulated in an aqueous vehicle it is better solubilized rather than emulsified.

Solubilization of essential oils is unlikely to alter the perception of flavour; however, the intensity of odour perceived when perfumery oils are solubilized in aqueous surfactant systems can be a complex function of the solubilizer system [205]. The intensity of odour is proportional to the concentration of the perfume in the aqueous phase rather than to its concentration in the micelles. Increasing

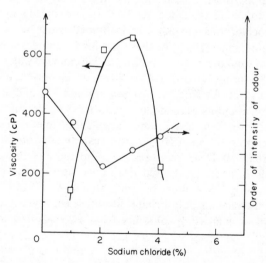

Figure 6.27 Effect of increasing NaCl content on viscosity (□) and rank order of odour intensity (○) of a solution of 2% perfume W 100 in 12% sodium lauryl ether sulphate and 2% coconut diethanolamide. From Blakeway *et al.* [205].

the concentration of surfactant decreases the odour intensity. This is seen clearly in Fig. 6.26 at low (0.5 %) concentrations of perfume W100 in sodium lauryl ether sulphate solution. With higher concentrations of perfume up to 2 %, the intensity profile takes on a biphasic aspect. One aspect of the system which influences the perfume intensity is its viscosity which when altered by the addition of NaCl causes the changes seen in Fig. 6.27. Non-polar oils such as liquid paraffin (mineral oil) have been used as model substances in investigation of solubilization [214–216]. The influence of the nature (e.g. polarity) of the oil on its solubilization has been studied by Lo *et al.* [216]; this work is discussed in more detail in Chapter 2. The kinetics of solubilization of non-polar oils by non-ionic surfactants has been the subject of a recent study [216a].

6.2.8 Miscellaneous drugs

It has been impossible to deal with all reports of solubilization of drug substances. Those chosen have tended to illustrate certain trends, such as the influence of drug structure or surfactant HLB on drug solubility. Some systems not considered above, perhaps because they deal with only one drug substance and one surfactant, will be referred to in Chapter 7 which considers the influence of surfactants on the bioavailability of solubilized drugs and other consequences of the addition of surfactants to pharmaceutical products. We consider below reports on the solubilization of a miscellaneous selection of drugs.

Some interest has been shown in solubilization of diuretics cyclopenthiazide, hydrochlorothiazide, hydroflumethazide and bendrofluazide [217] and frusemide [218,219]. The solubilities of the thiazide diuretics in water were not quoted. Micellar partition coefficients and the slopes of solubility–surfactant concentration plots were tabulated. At 35° C in polysorbate 80 the order of P_m values was hydrochlorothiazide (50) < hydroflumethazide (106) < bendrofluazide (186) < cyclopenthiazide (195). For these and for frusemide, polysorbate 80 was the most efficient solubilizer. The solubility of frusemide in water is 65 μg ml^{-1} at $35 \pm 0.5°$ C [218]. Its normal dose is 10 to 40 mg which must be accommodated in a liquid dose of 5 ml. As 20 % w/v of polysorbate 80 solubilizes only 7.2 mg frusemide per ml, attempts were made to reduce the surfactant concentration by using co-solvents propylene glycol, ethyl alcohol and dimethylacetamide (DMA), but DMA at a concentration of 50 % in water can dissolve only 8.3 mg frusemide per ml. Polysorbate 80–DMA mixtures are compared with the co-solvent–water mixtures in Fig. 6.28. If the desired drug concentration is 10 mg ml^{-1}, 10 % polysorbate 80, 40 % DMA must be used. Shihab *et al.* [219] manipulate the solubility of frusemide in polysorbate 80 with electrolytes, addition of 0.2 M potassium sulphate increasing solubility from 1.26 mg ml^{-1} to 1.61 mg ml^{-1}. NaCl, KCl, MgCl$_2$ and Na$_2$SO$_4$ were also used (Table 6.26).

The practicality of both approaches is open to question in view of the potential toxicity of the co-solvents and electrolytes. The order of interaction of prostaglandins with the non-ionic surfactant $C_{12}E_{23}$ was PGE$_1$ > PGE$_2$ > PGF$_{2\alpha}$ [220] which corresponds to the order of their cyclohexane/water partition coefficients.

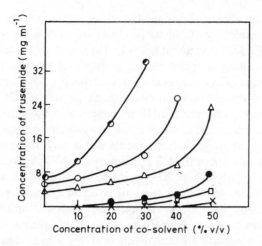

Figure 6.28 Effect of co-solvents and surfactants on the solubilization of frusemide at 35 ± 0.5° C.
●—●: polysorbate 80 (20% w/v) + DMA, ○——○: polysorbate 80 (15% w/v) + DMA
△——△: polysorbate 80 (10% w/v) + DMA, ●——●: DMA, □——□: ethyl alcohol
×——×: propylene glycol.
From Sivakumar and Mithal [218] with permission.

Table 6.26 Effect of electrolytes on the solubility of frusemide in 5% w/v polysorbate-80

Molar conc. of electrolyte	Solubility of frusemide (mg/100 ml)				
	NaCl	KCl	$MgCl_2$	Na_2SO_4	K_2SO_4
0.00	125.6	125.6	125.6	125.6	125.6
0.01	130.2	127.0	128.2	129.9	131.0
0.02	131.9	129.8	131.4	133.7	134.4
0.05	132.3	131.9	135.5	139.6	144.6
0.10	133.2	132.5	136.4	149.9	147.5
0.20	134.1	134.1	141.4	157.8	160.9

From [219].

Stable aqueous solutions of narcotic phenoxyacetamides for intravenous administration have been prepared with non-ionic solubilizing agents. The solutions are clear, can be sterilized, and show 'venous compatibility' [221]. Sucrose laurate is among the surfactants described in a similar patent for solubilizing narcotic amides [222]. Various formulations of tetrahydrocannabinol for intravenous administration have been suggested [223, 224].

Studies on the effect of solubilizer concentration on the biological activity of propanidid [225] and tetrahydrocannabinol [226] have been published and are discussed at length in Chapter 7.

Many active ingredients of ointments and lotions are not readily dispersible because of their insolubility. Coal-tar products are an example which have been successfully blended into ointment bases by the use of surfactants [227]. A 1 % crude coal-tar ointment in which the tar is dispersed by the addition of 0.5 % polysorbate 20 prior to its incorporation in the base, produces fewer adverse skin reactions than the normal preparations without surfactant [228]. Such preparations are also more readily removed from the skin with water. It has been stated, however, that incorporation of coal-tar into hydrophilic ointment bases allows the penetration of carcinogenic components which may be present in the tar. A clear transparent solution of the US Formulary Coal-Tar Solution can be made, provided that 10 % polysorbate 20 remains in the final dilution [229].

Spans and Tweens have been used to overcome similar problems in the formulation of medicines for internal use. The solubilization of resinous components of tinctures such as benzoin and myrrh in aqueous vehicles and the incorporation of water-soluble ingredients into oily vehicles has been discussed by Stoklosa and Ohmart [230]. Gerding and Sperandio [229] give examples of mixtures of tinctures and fluid extracts which, on addition of polysorbate 20, will not precipitate on dilution. Cetomacrogol 1000 added in small amounts to opiate linctus of squill, syrup of ginger, compound mixtures of camphor, and of lobelia and stramonium has a similar clearing action [207].

Cetomacrogol also prevents the precipitation of chlorophyll in mixtures containing tinctures of solanaceous drugs. Addition of non-ionics to preparations containing balsamic compounds, such as Gee's Linctus, renders the preparation clear.

6.3 Pharmaceutical aspects of solubilization in non-aqueous systems

Comparatively little material on solubilization of drugs in non-aqueous systems is available, yet the amount of water or aqueous solution which can be incorporated in organic solvents containing surfactants which form inverse micelles can be considerable. One of the limitations has, of course, been the availability of surfactants sufficiently soluble in non-aqueous solvents to reach a critical micelle concentration. Frank and Zografi [231] observed that 20 mol H_2O was solubilized by Aerosol OT (di(2-ethylhexyl) sodium sulphosuccinate) in octane. Most work has been carried out on systems such as these which are totally unsuitable for most pharmaceutical purposes. Some measurements on more "acceptable" non-aqueous solvents have been made [232] using a range of vegetable oils such as almond oil and olive oil as the non-aqueous phases. In this work the L_2 (isotropic non-aqueous phase) was identified along with the regions for L_1, M_1, G and related mesophases. The solubility of the non-ionic surfactants (Brij 92 and 96) in many of these oils is low thus limiting the formation of inverse micelles. Some preliminary work was carried out using Span 80 and Tween 80 mixtures and almond oil. Limited areas of L_2 phase formation occur (Fig. 6.29). Results [216] suggest that the nature of the surfactant aggregates is of

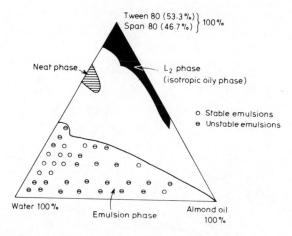

Figure 6.29 Partial phase diagram for a water–almond oil–surfactant system showing the absence of significant L_1 phase and the limited range of the L_2, neat and emulsion phases. From Kabbani *et al.* [232].

considerable importance in determining the uptake of water. While one might expect Brij 96 with the largest hydrophilic group to form the largest aggregates in a non-aqueous solvent, it might not sterically be suited to forming large aggregates except in the presence of the smaller Brij 92 molecules. An attempt to depict this is shown in Fig. 6.30. Palit *et al.* [233, 234] have found that, in general, solubilization in non-aqueous solvents is enhanced when mixtures of the two surfactants are present provided that one surfactant is hydrophobic and the other hydrophilic. The two surfactants in this study satisfy this requirement.

The considerable influence of the oil phase on L_2 phase formation has been noted.

Considerable data have been gathered on solubilization of water in non-aqueous liquids by Lin *et al.* in the course of work on emulsification [236] thus elucidating some of the factors influencing solubilization: optional ethylene oxide chain length, ratios of surfactants, the nature of the oil phase. Fig. 6.31 shows that

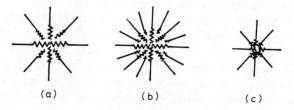

(a)   (b)   (c)

Figure 6.30 Diagrammatic representation of maximal micelle size and water uptake in non-aqueous solvents when the Brij 96 and Brij 92 are mixed to provide micelles as in (b). (a) Pure Brij 96, (c) pure Brij 92 micelle. In (b) better packing and exclusion of the polyoxyethylene core from non-aqueous solvent is made possible by alternating short and long polyoxyethylene chain components. From Lo *et al.* [235] with permission.

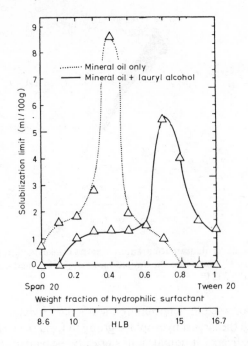

Figure 6.31 Shifting of optimum water solubilization by addition of lauryl alcohol. (Emulsions contain 30 % oil phase, 65 % deionized water, and 5 % surfactant mixtures. Surfactant mixtures consist of hydrophilic Tween 20 and lipophilic Span 20 at ratios and corresponding HLB values indicated by abscissa. Dotted lines represent data for pure mineral oil systems. Solid lines represent data for oil mixture consisting of 8 parts mineral oil and 2 parts lauryl alcohol). From Lin *et al.* [236].

in mixtures of Span 20 and Tween 20 in mineral oil, the optimal ratio for water uptake is shifted to a higher HLB on addition of lauryl alcohol to the oil phase, and the uptake of water considerably reduced. The addition of the polar oil has marked effect on the capacity of the system. Very little work has been published on the effects of drugs added to the solubilizate phase on the properties of the system although it is likely to be considerable [237].

Water-in-oil solubilized adjuvant formulations of vaccines containing *Clostridium welchii* type D toxoid as antigen were prepared first in 1968 and tested in laboratory animals by Coles *et al.* [238]. The adjuvant action of oil-in-water emulsions, multiple emulsions and water in gelled oil emulsions is well known but these varied systems have the disadvantages of high viscosity which makes injection physically difficult. Lin [236] quotes an HLB of 9.7 as the optimum value for water solubilization in mineral oil. Coles *et al.* [238] found a value of 10. While the addition of a small quantity of the lipophilic surfactant Arlacel 80 (sorbitan mono-oleate) to a system of Tween 81 (polyoxyethylene (5)-sorbiton mono-oleate) allowed increasing amounts of water to be solubilized, when toxoid solution was substituted for water the Arlacel decreased the amount which could

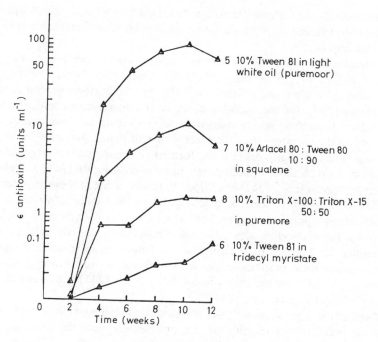

Figure 6.32 ε-Antitoxin titres in guinea-pig serum ($n = 6$) after 1 ml subcutaneous doses of vaccines 5–7 and a 0.2 ml dose of vaccine 8. From Coles *et al.* [238].

be solubilized. They also found that water was solubilized in paraffin oil and pure hydrocarbons, straight or branched, at lower concentrations in fatty alcohols and fatty acid esters and at extremely low concentrations in vegetable oils, pure triglycerides and fatty alcohol esters. This then limits non-aqueous solubilization for medicinal products. Vaccines in tridecyl myristate and squalene as well as mineral oil were examined and in one system (8) a Triton X-100/Triton X-15 mixture was used (unsuccessfully) as the solubilizer. ε-Antitoxin titres produced in rabbit serum on administration of four of these vaccine formulations are shown in Fig. 6.32. The tridecyl myristate system was unstable at 37° C with Arlacel and Tween mixtures but the solubilized systems are generally more stable than their emulsified counterparts, although not of course immune to destabilization in a biological environment. They are now more readily prepared than emulsions and have a lower viscosity.

6.4 Solubilization with block co-polymeric surfactants

So far in this chapter we have attempted to survey solubilization of pharmaceutical products by drug class. Here we diverge to discuss solubilization by a class of surfactant. For reasons of toxicity many ionic surfactants are excluded from serious contention as solubilizing agents for use in medicines. Not all non-ionic surfactants are without blemish in this regard, as we will see in Chapter 9, and there must still be scope for the investigation of new surfactants which can be used with impunity.

An interesting class of non-ionic surface-active agents are polyoxy-ethylene–polyoxypropylene–polyoxyethylene block co-polymeric surfactants, sold under the trade name Pluronic and also known by their generic name as poloxamers [239]. Of the available block co-polymeric surfactants, the polo-xamers have been most widely studied to date, yet there has been considerable confusion in the literature over the exact nature of their colloidal behaviour, in particular whether or not micelles are formed [240]. Recently, surface-tension measurements on a series of poloxamers in aqueous solution [241] and photon correlation spectroscopy [242] has helped to resolve some of these problems but as befits their structure their behaviour patterns tend to be complex. At low concentrations, approximating to those at which more conventional non-ionic detergents form micelles, the poloxamer monomers are thought to form monomolecular micelles by a change in configuration in solution. At higher concentrations these monomolecular micelles associate to form aggregates of varying size which have the ability to solubilize drugs [243] and to increase the stability of solubilized agents [244].

Table 6.27 lists approximate values of molecular weight and ethylene oxide and propylene oxide chain lengths for the poloxamers, and the designation of poloxamers and the commercial Pluronic surfactants.

Table 6.27 Approximate values of n, m and M for various polyoxy-ethylene–polyoxypropylene glycols (Pluronic or poloxamers)

Poloxamer designation	Pluronic* designation	Molecular weight of C_3H_6O-portion	m[†]	'Percent' C_2H_4O	Molecular weight of C_2H_4O-portion	n[†]	Total molecular weight, M
181	L61	1 750	23	10	194	4	1 944
182	L62	1 750	23	20	438	10	2 188
183	L63	1 750	23	30	750	17	2 500
184	L64	1 750	23	40	1 167	27	2 917
185	P65	1 750	23	50	1 750	40	3 500
188	F68	1 750	23	80	7 000	159	8 750
231	L81	2 250	30	10	250	6	2 500
234	P84	2 250	30	40	1 500	34	3 750
235	P85	2 250	30	50	2 250	51	4 500
237	F87	2 250	30	70	5 250	119	7 500
238	F88	2 250	30	80	9 000	205	11 250
331	L101	3 250	43	10	361	8	3 611
333	P103	3 250	43	30	1 393	32	4 643
335	P105	3 250	43	50	3 250	74	6 500
338	F108	3 250	43	80	13 000	296	16 250
101	L31	950	13	10	106	2	1 056
401	L121	4 000	53	10	444	10	4 444

* F denotes 'solid', P denotes 'pasty' and L denotes 'liquid' consistencies at 25° C.
[†] Molecular weight of C_3H_6O- is 76 and of C_2H_4O- is 44.

Some relationships between poloxamer structure and the solubilization of para-substituted acetanilides have been defined by Collett and Tobin [243]. The solubilities of the substituted acetanilides such as 4-hydroxyacetanilide, in aqueous poloxamer solutions increase with increasing oxyethylene content of the polymer although the more hydrophobic members of the series do not show this trend [243]. The results as expressed in Table 6.27 show that, for example, 4-nitroacetanilide is less soluble in the more hydrophilic poloxamers, and this is the general trend shown by the halogenated derivatives. These are apparently contradictory results. Some attempt was made to relate solubilization of the series to the π values of their functional groups. Thus in Table 6.28 we see solubilization expressed as the slope of the plot of mol drug solubilized mol^{-1} poloxamer against percentage ethylene oxide in the surfactant. Slope of the hydrophilic derivatives are thus positive and those of the more hydrophobic compounds, negative. A linear relationship is obtained for the solubilization of a hydrophobic acetanilide, 4-fluoroacetanilide and the propylene oxide–polyethylene oxide ratio of the solubilizer (Fig. 6.33a) but when the amount of drug solubilized by the hydrophobe is calculated it decreases as the hydrophobicity of the solubilizate increases, which is contrary to expectation (Fig. 6.33b). Collett and Tobin suggest some hydrophobic barrier in the micelle which seems unlikely, but there is no doubt that the micellar properties are not as predicted [241]. Apparent critical micelle concentrations determined from surface tension measurements decrease with increasing HLB. The fact that this is contrary to expectation might lead one to suspect that these are not true CMCs but are the consequence of interaction between the solubilizate and polymer. Methyl, ethyl, and propyl parahydroxy-benzoate, for example, interact with poloxamer co-polymers to no greater extent than they do with polyoxyethylene glycol 6000 which does not micellize; butyl parahydroxybenzoate, on the other hand, is solubilized to a greater extent in this Pluronic than by polysorbate 80. The flexibility of the chains at the air–water

Table 6.28 The slopes for plots of mol p-substituted acetanilide solubilized mol^{-1} poloxamer (pH 1.0, 37° C) against percentage oxyethylene in the poloxamer molecule and the π value of the substituent (from [247])

Substituent	Slope, $K \times 10^2$	π*
H	6.30	0
4-OH	15.0	−0.36
4-OMe	2.74	−0.133
4-OEt	0.31	0.367[†]
4-CHO	5.20	0.091
4-NO$_2$	−0.32	0.499
4-F	−1.30	0.309
4-Cl	−0.78	0.714
4-Br	−1.03	1.130
4-I	−0.83	1.303

* From [245]
† From [246].

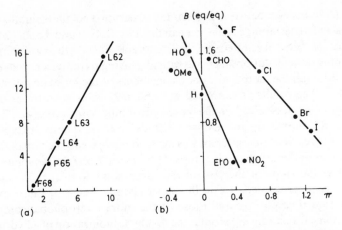

Figure 6.33(a) Solubilization of 4-fluoroacetanilide in aqueous solutions of poloxamers L62, L63, L64, P65 and F68 expressed as equivalents of drug per equivalent of ethylene oxide against the poloxamer mole ratio. Ordinate: $S/C_{EO} \times 10^2$ (equivalents of drug solubilized per equivalent of ethylene oxide). Abscissa: $C_R/C_{EO} \times 10^2$ (propylene oxide–ethylene oxide mol ratio). (b) The amount of p-substituted acetanilide solubilized (B) by the hydrophobe of poloxamer molecules as a function of the π value of the substituent group on the acetanilide molecule. From Collett and Tobin [247].

interface [241] suggests that the folding of the longer hydrophobic chains in bulk solution effectively decreases the exposed hydrophobic surface and this reduces the tendency to form polymolecular aggregates even though the monomer is calculated to be more hydrophobic through its HLB number. Another explanation of the trends may be that when the polyoxyethylene chains are short the molecules do not display sufficient amphipathy. Amphipathic properties increase with increase in the size of the hydrophile. Some evidence for this is that the addition of sodium chloride to a solution of poloxamer L64 causes a reduction in the measured mean radius of the aggregates in solution, suggesting that salting out of the hydrophile at both ends of the molecule converts it into a non-aggregating species, by making it more closely resemble a hydrocarbon chain [241].

Nuclear magnetic resonance has been used [248] to study the interaction of poloxamer F68 and phenol. Starting with low phenol concentrations, up to 2%, in a 10% aqueous poloxamer F68 solution, it was reported that the phenol was associated mainly with the polyoxypropylene chain. However, as the ratio of phenol to poloxamer increased, it appeared that the polyoxypropylene chain became saturated with phenol and relatively more phenol entered the polyoxyethylene chain.

A chlorhexidine gluconate–poloxamer 187 solution has been developed as an antiseptic skin cleansing formulation [249]. This contains 25% poloxamer 187, chosen to produce the greatest foaming capacity and also because the poloxamers as a class interfere with the activity of the chlorhexidine less than other non-ionic surfactants tested. An alcohol-based mouthwash has also been described. Choice of poloxamer rested on lack of noxious taste (cf. some other non-ionics) and its ability to solubilize aromatic flavours [250].

Marked increases in the dissolution rate of digitoxin and digoxin has been achieved by dispersing the drugs in solid poloxamer 188 (Pluronic F68) as a carrier [251] (see Fig. 6.34). Poloxamer 188, in concentrations equivalent to that in the digoxin co-precipitates studied, increased the solubility of the digoxin as shown in Table 6.29 in which results are compared with the effects of deoxycholic

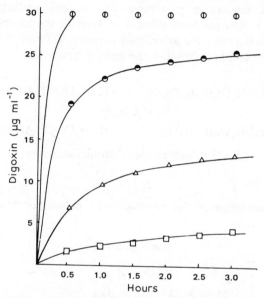

Figure 6.34 Dissolution of digoxin from poloxamer 188 test preparations. □, Untreated drug; △, 10 and 1 % physical mixtures; ●, 10 % co-precipitate; and ①, 1 % co-precipitate. From Neddy *et al.* [251] with permission.

Table 6.29 Effect of poloxamer 188 and deoxycholic acid on the solubility of digoxin in water at 37° C

Test system	Solubility mg/100 ml
Water	3.47
Poloxamer 188 in concentration equivalent to 10% co-precipitate	4.77
Poloxamer 188 in concentration equivalent to 1% co-precipitate	5.38
Deoxycholic acid in concentration equivalent to 10% co-precipitate	4.62
Deoxycholic acid in concentration equivalent to 1% co-precipitate	4.25

From [251].

acid. Enhanced dissolution could be due to the presence of the drug in an amorphous state in the co-precipitate, to surface-tension lowering and to increase in the bulk solubility of the dry substance (see Chapter 7).

Poloxamers have also been incorporated into white petrolatum USP ointment bases in the presence of dimethylsulphoxide to modify the absorption of drugs presented in the base [252]. Percutaneous absorption of salicylic acid was increased significantly by poloxamers 231 and 182 and absorption of sodium salicylate by poloxamer 182.

Sheth and Parrott [244], in their study on the hydrolysis of esters, measured the solubility of benzocaine in a range of non-ionic surfactants including poloxamer 188. It was the least efficient, a Tetronic co-polymeric surfactant (Tetronic 908) having twice the solubilizing capacity. Tetronic is the proprietary name for the poloxamine series with the general structure,

$$H(CH_2CH_2O)_a(C_3H_6O)_b \qquad (C_3H_6O)_b(CH_2CH_2O)_aH$$
$$NCH_2CH_2N$$
$$H(CH_2CH_2O)_a(C_3H_6O)_b \qquad (C_3H_6O)_b(CH_2CH_2O)_aH$$

(XXIV) Poloxamine (Tetronic) structure

Table 6.30 Nomenclature of the meroxapol and poloxamine block co-polymeric surfactants

Hydrophobe molecular weight	Meroxapol series							
3100	31R1	31R2	—	31R4	—	—	—	—
2500	25R1	25R2	—	25R4	25R5	—	—	25R8
1700	17R1	17R2	—	17R4	—	—	—	17R8
1000	—	—	—	—	10R5	—	—	10R8
% Ethylene oxide	10	20	30	40	50	60	70	80

Hydrophobe molecular weight	Poloxamine series							
6750	1501	1502	—	1504	—	—	—	1508
5750	1301	1302	—	1304	—	—	1307	—
4750	1101	1102	—	1104	—	—	1107	—
3750	901	—	—	904	—	—	—	908
2750	701	702	—	704	—	—	707	—
1750	—	—	—	504	—	—	—	—
750	—	—	—	304	—	—	—	—
% Ethylene oxide	10	20	30	40	50	60	70	80

From Schmolka [239].

The nomenclature of the poloxamers and the meroxapols (polyoxypropylene–polyoxyethylene–polyoxypropylene block co-polymers) 'reversed' poloxamers is explained in Table 6.30. Another class of block co-polymers which has no generic name has the name Pluradot (Wyandotte). These have three block co-polymer chains with the general formula,

$$R[O(C_3H_6O/CH_2CH_2O)_n - (CH_2CH_2O/C_3H_6O)_m H]_3$$

$$\left(\frac{C_3H_6O}{C_2H_4O}\right) > 1 \qquad \left(\frac{C_2H_4O}{C_3H_6O}\right) > 1$$

Pluradot structure

XXV

The solubilizing ability of these complex polymers has not been reported.

6.5 Polymer–surfactant interactions

Pharmaceutical formulations are rarely simple solutions. The increasing likelihood of the presence of polymers in formulations should alert us to the possibility of surfactant–polymer interactions which can influence the capacity of the surfactants to perform their function of increasing the solubility of drug substance. Polymer–surfactant interactions are of some interest in view of the use of polymers as viscosity modifiers and suspension stabilizers [253]. Interactions between surfactants and non-ionic polymers such as polyethylene oxides [254], polypropylene oxides [255], polyvinylpyrrolidone [256, 257] and polyvinylalcohol [260] have been studied [259]. An interesting property of some of these polymer–surfactant complexes, e.g. polyvinylpyrrolidone–NaLS, is the synergistic effect of the polymer on the capacity of the surfactant to solubilize oil-soluble dye [256, 257]. An instance of such synergism occurring in hydrocarbon media has also been reported [260]. Interactions between polymer and a given surfactant increase with the increasing hydrophobicity of the macromolecule; indeed it has proved possible to solubilize poorly soluble hydrophobic polymers by the addition of surfactant [261, 262]. Polyelectrolytes form precipitation complexes with oppositely charged surfactants which can in many cases be completely re-solubilized by the addition of excess surfactant [259]. Maximum precipitation has been found to occur when a single layer of adsorbed surfactant formed on the polymer chains; the resolubilized form appearing when a double layer of surfactant was achieved. Goddard and Hannan's detailed study [259] has revealed that optimal interactions between polymer and surfactant occurred when the surfactant had a long, straight hydrocarbon chain with the polar group terminal to the alkyl chain. Departure from this structural constraint reduces the extent of the interaction and also renders the resolubilization difficult, the latter being difficult to achieve if the charge density on the polymer is also high [259]. As might be expected, the complex formed between some surfactants and polymers has a solubilization capacity which is different from that of the surfactant alone. Fig. 6.35 shows the effect of PVP on the solubilization of Yellow

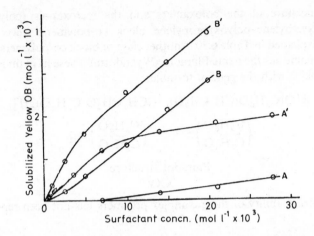

Figure 6.35 Effect of PVP addition on the solubilization of an oil-soluble dye, Yellow OB, in surfactant solutions at 30° C: A, NaDS; B, dodecyl-(oxyethylene)-ether. The primes refer to the addition of 0.1 % PVP to the corresponding surfactant solutions. From Saito [261].

OB in solutions of NaDS, a non-ionic detergent and sulphated non-ionics [263]. The latter seem to form complexes only when the number of oxyethylene groups is small; on the other hand, $C_{12}E_4$ with no sulphate groups shows no sign of complexation with PVP (Fig. 6.35 DD'). It is evident that the sulphate group contributes to the binding giving rise to enhanced solubilization. The mechanism of interaction of ionic surfactants and polyglycol ethers has been discussed by Schwuger [255]. In water only weak hydrophobic bonding between cationic surfactants and polyoxyethylene glycol ethers is evident. Complexing of anionic surfactants is very marked, the interactions being the result of both hydrophobic interactions and, more importantly, the lone pair on the ether oxygen confers a slight positive charge on the ether linkage which favours interaction with sulphate ions. With increasing pH the positive charge of the ether oxygens is reduced and the tendency to complex formation also is reduced.

Studies on the interaction between surfactants and styrene–ethylene oxide block co-polymers, however, indicate that the polymers exhibit, in the presence of surfactant, typical polyelectrolyte character. This, it has been suggested [264], is due to interaction repulsions between like charges of the NaDS ions adsorbed onto the polyoxyethylene blocks. Investigating the interaction of the same detergent with methylcellulose and poly(vinyl alcohol), Lewis and Robinson [265] also observed the polyelectrolyte character of the polymer–surfactant complexes. A complex between non-ionic surfactants and a polycarboxylic acid in water can solubilize oil-soluble dyes below the surfactant CMC [268]. The complex containing the solubilizate can be precipitated; the solubilizate remains in the precipitated complex and is leached out only slowly on placing the precipitate in fresh solvent. This has potential pharmaceutical implications. Halothane uptake by coacervate systems of gelatin–benzalkonium [269] has been considered to indicate the formation of a highly structured phase with non-polar domains similar to surfactant micelles in a polar medium. Consideration of these complex surfactant systems is essential if progress is to be made in developing solubilizers for pharmaceutical use. Solubilizers are rarely if ever used alone in a formulation, and, as indicated above, the 'capacity' of simple micellar systems is only rarely sufficient for practical formulations except for a handful of drugs of high potency.

In most of the work on solubilization of drugs and other solutes by single surfactant species a linear relationship between solubility and surfactant concentration has been noted. Chlorhexidine uptake into non-ionic micelles was one example where the solubility plot deviated from linearity because of an assumed change in micellar structure (see p. 313). As ionic surfactant is added to a macromolecule the conformational change that will be observed will almost certainly lead to deviations from linearity. Fig. 6.36 shows one example [266]. Solubilization of Orange OT occurs below the CMC of NaDS in the presence of additive. Two transition points are seen, the transition which occurs at the highest NaDS concentration designated as the 'second transition', should correspond to the concentration at which all the adsorption sites along the polymer backbone become saturated with surfactant molecules, and it was shown that a reasonable estimate of the NaDS concentration at the second transition could be obtained from the contour length of the polymer if it was assumed that the DS$^-$ ions were fully extended and bound linearly [266]. Interaction of NaDS with a cationic substituted polymer leads to maximum precipitation at ratios of polymer to surfactant at which the charge of the polymer is balanced by that of the surfactant. As the surfactant level is increased above this ratio the precipitated polymer is solubilized [267].

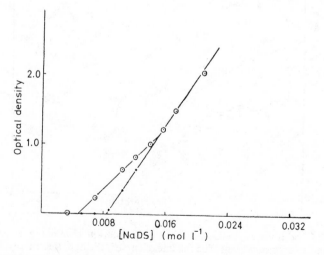

Figure 6.36 Optical density versus [NaDS] in the presence of Orange OT: ●, in the absence of polymer, ⊙, PEO concentration 0.071 %. From Jones, [266].

Significantly increased surface activity in some cationic polymer–anionic surfactant systems is observed in the low surfactant concentration range even when the added polymer is only weakly surfactive. It has been suggested [267] that this effect arises from the adsorption of surfactant or ions onto each cationic site rendering the polymer more surface active. Progressive addition of surfactant leads to further adsorption until the polymer now acts as an anionic polymer. The changes are represented diagrammatically in Fig. 6.37. A non-ionic surfactant (Tergitol 15–5.9) and a C_{14} betaine had no effect on the same cationic polymer.

Both ionic and non-ionic surfactants influence the rheological behaviour of gum arabic solutions [253]. Brij 96 increases the relative viscosity of gum arabic up to a 5% surfactant concentration at concentrations of gum up to 10%. NaLS also increases the viscosity of the gum but beyond 1% NaLS the viscosity is reduced (Fig. 6.38).

As we have seen the rheological properties of polymer–ionic surfactant systems suggest that the individual polymer chains adsorb a large number of the detergent anions so that they behave as polyanions with chains highly expanded because of the mutual electrostatic repulsion of the adsorbed species. Although the gum arabic molecules are polyanions, sodium lauryl sulphate may adsorb on the hydrophobic residues of the side chain and, in spite of the relatively high concentration of gegenions, the increased charge arising through adsorption would maintain a more expanded molecule than in the presence of an equivalent amount of NaCl. The effectiveness of the cetyltrimethylammonium bromide in reducing the intrinsic viscosity at low concentrations undoubtedly arises from the

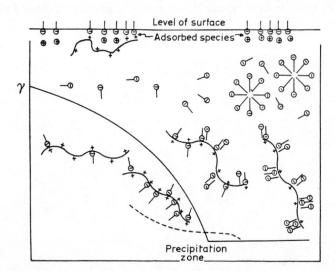

Figure 6.37 Conditions in bulk and surface of solution containing a polycationic electrolyte and anionic surfactant. The full line of curve is the hypothetical surface tension concentration plot of the surfactant alone; dashed line is that of mixture with polycation. Simple gegen-cations are depicted only in the surface zone. From Goddard *et al.* [267].

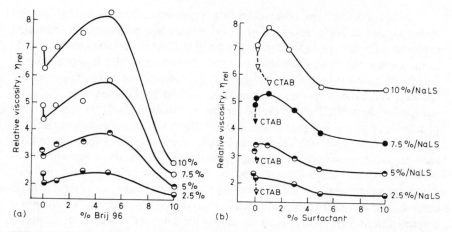

Figure 6.38 Relative viscosity of gum arabic solutions (a) in the presence of increasing concentrations of Brij 96, concentrations of gum arabic as shown on plots, and (b) in the presence of sodium lauryl sulphate (NaLS) and cetyltrimethylammonium bromide (CTAB) at the concentrations of gum shown, from 2.5% to 10%. From [253].

cumulative effect of the counterions and adsorption of the positive species, possibly directly with the anionic groups. At higher concentrations hydrophobic adsorption of further cetyltrimethylammonium bromide molecules could increase the positive charge on the macromolecule such that expansion occurs again. However, in this system, physical changes (e.g. coacervation) occur such that measurements cannot be made in the intermediate concentration range. If adsorption of NaLS occurs, undoubtedly the non-ionic polyoxyethylene-oleyl ether Brij 96 is also adsorbed. The increase it induces in the relative viscosity of the system may thus be due to the increased hydration of the surfactant–gum complex, water being trapped in the adsorbed polyoxyethylene layers.

6.6 Surfactant interactions with oppositely charged species

The possible interactions that may occur between a surfactant ion and an oppositely charged organic ion are outlined in Fig. 6.39.

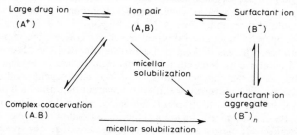

Figure 6.39 Equilibria possible in mixtures of oppositely charged organic ions when one is a surface-active agent (B^-) and the other a solute or drug molecule (A). From Tomlinson [270] with permission.

At low concentrations complexation occurs between the ions and usually turbidity occurs as a result, leading to phase separation of the so-called coacervate. On increasing the concentration of the surfactant ion above the CMC in the system, the coacervates may be solubilized resulting in a loss of turbidity. The interaction between the di-anionic drug disodium cromoglycate (cromolyn sodium) and cationic surfactants has been studied by Tomlinson *et al.* [271]; equilibrium between the ions can be represented quantitatively by the solubility product K_s and the ion-pair association constant K_{ip}. In aqueous solution, the value of K_{ip} increases with an increase in the carbon number of the ions forming the ion pair. A composite phase diagram showing the primary phase boundaries between sodium cromoglycate and a homologous series of alkylbenzyl-dimethylammonium chlorides is shown in Fig. 6.40. As is seen, an increase in alkyl chain length causes a shift in the phase boundaries to lower anion and cation concentrations, that is there is an increasing tendency for coacervation to occur. On the other hand, the CMC of the surfactant ion decreases with increasing alkyl

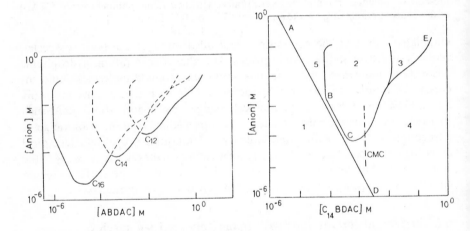

Figure 6.40 Composite diagram showing primary phase boundaries between sodium cromoglycate and an homologous series of alkylbenzyldimethylammonium chlorides at 25°C. The diagram for the C_{14} compound is shown with details of the phases observed. Region 1 represents that area in which complexation does not occur, and has as its boundary the solubility product line A-D. In Regions 2 and 3 visual evidence of complexation is apparent. In Region 2 this is observed as a grey/white complex, somewhat milky in appearance, whereas in Region 3 a brown viscous oil is observed. Although Regions 4 and 5 are above the theoretical solubility product line, no evidence of turbidity can be seen. In area 4 the concentrations of surface-active agent used are above its measured critical micelle concentration, and it is apparent that up to a limiting amount, (represented by line C-E), formed complex is solubilized within the surfactant micelles and also that above the critical micelle region insufficient surfactant monomer is available to ensure complexation. It needs to be appreciated here that the presence of free cromoglycate ion will tend to depress the measured surfactant critical micelle concentration. At high sodium cromoglycate concentrations and low surfactant concentrations (Region 5), no turbidity can be observed. From Tomlinson *et al.* [271] with permission.

chain length and the point of solubilization thus is reduced. Solubilization of the coacervate is achieved as these are hydrophobic species and can partition to the micellar phase. The solubilization of dye–detergent salts in excess detergent was first observed by Mukerjee and Mysels [272]. There are thus regions where anion and cation form an isotropic solution in spite of being incompatible at lower concentrations.

6.6.1 Solubility product

Tomlinson *et al.* [271] gave the following derivation of K_s for an interaction between an anion A^{x-} and a cation B^{m+} thus

$$A^{x-} + B^{m+} = A_m B_x. \tag{6.23}$$

The equilibrium constant is given by

$$K = \frac{a^m [A^{x-}] a^x [B^{m+}]}{a [A_m B_x]} \tag{6.24}$$

where a terms are the activities. As the coacervates form a separate phase $a = 1$ so that the solubility product, K_s is given by

$$K_s = a^m [A^{x-}] \, a^x \, [B^{m+}]. \tag{6.25}$$

As $a = c\gamma$ we can write

$$K_s = [A^{x-}]^m [B^{m+}]^x (\gamma_{A^{x-}})^m (\gamma_{B^{m+}})^x. \tag{6.26}$$

At low concentrations $\gamma_i = 1$ so that

$$K_s = [A^{x-}]^m [B^{m+}]^x. \tag{6.27}$$

Complexation has frequently been correlated with the hydrophobic character of one (or both) of the interacting ions [273–279]. Details of the interaction between a series of dyes and alkyltrimethylammonium bromides have been published [277]. The structure of the dyes used, tartrazine (XXVI) amaranth (XXVII) carmoisine (XXVIII) and erythrosine (XXIX) are shown below. These are all important colours used in the food, drug and cosmetic industries. Phase separation diagrams were constructed to indicate the relationship between surfactant concentration and the anisotropic solution–coacervate boundary. Differences between the interactions of a hydrophilic dye, tartrazine and amaranth, carmoisine and erythrosine which have both hydrophobic and hydrophilic moieties were exhibited. Tartrazine appears to behave like a simple electrolyte interacting simply with the charged groups at the micellar surfaces while the other dyes complexed and were solubilized as a complex in addition to interacting with the micelle surface [277]. These dyes also induced the formation

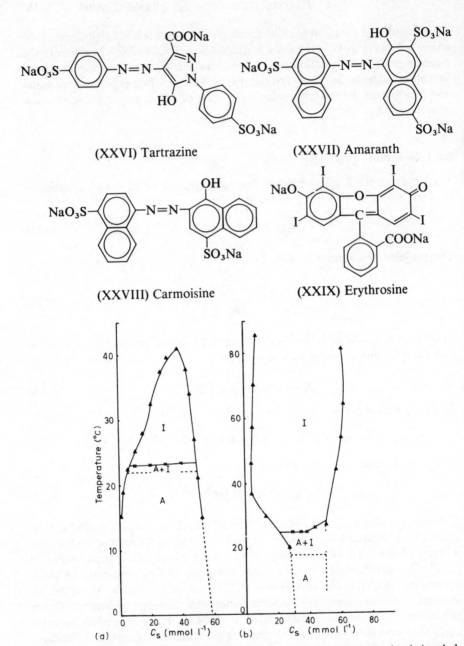

Figure 6.41 Diagrams showing temperatures at which dye–tetradecyltrimethyl-ammonium bromide interaction products changed phase, as determined by microscopy: (a), 15 mmol l^{-1} tartrazine–tetradecyltrimethylammonium bromide; (b) 15 mmol l^{-1} amaranth–tetradecyltrimethylammonium bromide; in each case the surfactant concentration is C_s, mmol l^{-1}. I, separated phase is isotropic; A, separated phase is anisotropic; and A + I, separated phase is mixed anisotropic and isotropic. From Barry and Gray [277] with permission.

of surfactant micelles at concentrations well below the curve of the surfactants in water. The quite different phase diagrams for tartrazine–tetradecyltrimethyl-ammonium bromide and amaranth–tetradecyltrimethylammonium bromide systems are shown in Fig. 6.41. Warming suppressed coacervation in the tartrazine–detergent system and a maximum temperature for coacervation occurred in each system while heating increased the area of coacervation in the other systems. Surfactant chain length effects are shown in Fig. 6.42.

The viscosity of these mixed systems can vary considerably due to the complex interactions and the formation of colloidal particles. The effect of surfactant concentration and alkyl chain length on the specific viscosity of amaranth solutions is illustrated in Fig. 6.43. Viscosity increases sharply as the ratio of dye to surfactant increases up to the point where the system coacervates. Ratios of surfactant to dye at the maximum agree with the ratios for compatability in these systems. Dye solutions which contain short-chain homologues have a smaller viscosity maximum than those which contain long-chain homologues.

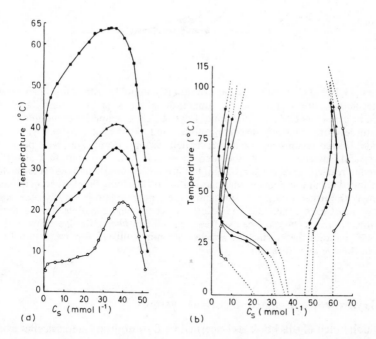

Figure 6.42 Diagrams showing temperatures and surfactant concentration (C_s) at which (a) tartrazine and (b) amaranth–alkyltrimethylammonium bromide solutions separate into two or more phases. Dye concentration 15 mmol l^{-1}. ■; hexadecyltrimethylammonium bromide system ▲ tetradecyltrimethylammonium bromide; ●: cetrimide; ○: dodecyltri-methylammonium bromide. From Barry and Gray [277] with permission.

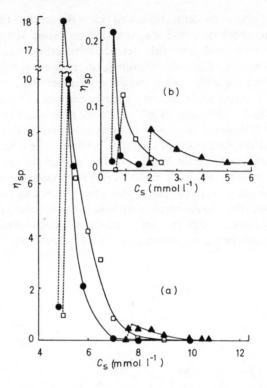

Figure 6.43 Effect of surfactant concentration (C_s, mmol 1^{-1}) and alkyl chain length on the specific viscosity (η_{sp}) of aqueous amaranth solutions. (a) surfactant above and (b) surfactant below the CMC and 1.5 mmol/l dye, ▲ dodecyltrimethylammonium bromide; □, tetradecyltrimethylammonium bromide; and ●, hexadecyltrimethylammonium bromide. Dotted regions represent coacervated systems. From [266] with permission.

Barry and Russell explain 'At the viscosity maximum the concentration of surfactant required to reach reversal of charge point has just been exceeded and the excess surfactant acts as added salt and suppresses the coacervation by screening the charges on the reacting species. With further addition of surfactant, the ζ-potential progressively rises, colloidal particles repel each other and the solutions become more mobile. In the limit, the viscosities approach those of the dye-free surfactant solutions. Results above and below the CMC indicate that surfactant micelles are not necessary for interaction to occur between the dye and surfactant.'

6.7 Hydrotropy in pharmaceutical systems

Although much of this book is concerned with solubilization in micellar systems, there is a need to discuss the phenomenon of hydrotropy, as there is now a considerable body of literature on the pharmaceutical aspects of the subject. As has been discussed, hydrotropy is the term reserved for the action of increasing the solubility of a solute by a third substance which is not highly surface active – at least one which does not form micelles at low concentrations. The mechanism of action of hydrotropes is varied, and we shall deal first with hydrotropes which exert their action through complexation. Caffeine is such a compound.

The presence of hydrotropes in drug formulations may be expected to influence the activity of the drug *in vivo*, although little work seems to have been done on this point. This may be a fruitful topic of research, as in these systems the lack of surface activity makes experimentation and interpretation of results less difficult. Higuchi and Drubulis [278] suggest that data on the complexing of drugs is important because the action of drugs is often the result of complex formation in an aqueous environment, and because the thermodynamic activities of the drugs in that environment may be modified by this phenomenon.

Because of the limited solubility of the xanthine derivatives in water, their solubilization by hydrotropes has been the subject of much interest. Such diverse compounds as piperazine [279], sodium salicylate [280], adenosine [281], and diethanolamine [282] have been used to solubilize theophylline. The action of each is presumably due to complex formation. Neish [283] found that caffeine, a xanthine derivative, increased the solubility of a number of aromatic amines, including sulphapyridine, sulphathiazole, and certain dyes.

It is perhaps relevant to note here that in clinical use sulphonamide mixtures are preferred to single drugs, to minimize the formation of crystal deposition in the kidney. The use of mixtures of related sulphonamides results in enhanced mutual solubility [284]. This is perhaps a case of hydrotropy. Lehr [285] has determined the solubility of a number of sulphonamide combinations, and Frisk *et al.* [286] have reported on the solubility of the combination sulphadiazine, sulphamerazine, and sulphathiazole. The influence of sulphanilamide on the solubility of sulphathiazole indicated that a 1 : 1 molecular complex was formed [287].

Apart from the possible prevention of unwanted physiological effects, hydrotropes can have a direct action on efficacy. Theobromine is soluble in water to the extent of 1 in 2000; an equimolar mixture with sodium acetate has a solubility of 1 in 1.5, and a mixture with sodium salicylate, 1 in 1. Clinical evaluation of various theophylline and theobromine preparations in the treatment of angina pectoris [288] showed a theobromine–sodium acetate mixture to be the most effective. Ergotamine levels have been shown to be enhanced when the drug is administered in combination with caffeine, which increases the solubility and rate of solution of the ergotamine [289].

Description of some of the hydrotropes used in the solubilization of riboflavin will illustrate the diversity of compounds employed. Nicotinamide increases the solubility of riboflavin in polar liquids [290], and it has been observed [291] that ascorbic acid also has a hydrotropic effect on riboflavin. A 20 % aqueous ascorbic acid solution increases the aqueous solubility of riboflavin at room temperature by 4.5 times. This is important in the formulation of multi-vitamin preparations. Concentrated solutions of (−) −, or (+) − tryptophan dissolve riboflavin to the extent of 4 mg ml^{-1} at pH 6.8, and the addition of nicotinamide markedly increases the solubility, giving stable solutions, suitable for injection [292]. *N*-(2-hydroxyethyl) gentisamide may also be employed as solubilizer [293], as can water-soluble salts of benzoic acid and its amino- or hydroxyl-substituted derivatives [294]. Riboflavin in concentrations of 1, 2, and 120 mg ml^{-1} requires 15, 25, and 625 mg, respectively, of sodium salicylate per ml to effect its solution.

A comprehensive study of the solubility of riboflavin in a range of hydrotropes [295] revealed some interesting differences due to structural changes in the hydrotrope (Table 6.31). The effect of a range of nicotinic acid and *isonicotinic* acid derivatives was not affected by the structure of the side chain to any great extent, but alterations to the hydrophobic radical proved more startling: 3-hydroxy-2-naphthoic acid and its mono-ethanolamide has approximately ten times the effect of salicylic acid and its mono-ethanolamide.

Table 6.31 Solubility of riboflavin in solutions of hydrotropes (solubility in mg ml^{-1})

Hydrotropic compound	Conc. of hydrotrope (%)				
	1	2	3	5	10
⬡—COONa	0.38	0.48	0.64	1.10	3.10
HO, HO—⬡—COONa, HO	0.67	0.92	1.46	2.85	7.34
OH ⬡—COONa	0.53	0.96	1.30	2.90	7.70
OH ⬡⬡—COONa	5.60	12.50	19.60	34.30	97.80
N⬡—COONa	0.37	0.50	0.66	1.10	2.40

From [298].

In an investigation of the solubility of theophylline, hydrocortisone, prednisolone, and phenacetin in a range of hydroxybenzoic acids and their sodium salts and several hydroxynaphthoates, it was concluded [278] that the major factor in the interactions were the donor–acceptor interactions; hydrophobic bonding and hydrogen bonding were thought to play a less-important role. Table 6.32 presents some of these results; the figures represent first-order constants, i.e. apparent stability constants. Complex formation between riboflavin and caffeine has been demonstrated and the solubilizing properties of caffeine and theophylline and dimethyluracil have been studied [296]. The marked difference in the solubilizing powers of the caffeine and dimethyluracil

Table 6.32 Solubility of hydrocortisone, prednisolone, and phenacetin in solutions of naphthoates and hydroxy naphthoates: apparent stability constants K ($l\,mol^{-1}$)

	Hydrocortisone	Prednisolone	Phenacetin
1-Naphthoate	8.2	8.0	2.5
2-Naphthoate	19.0	20	5.8
2-Hydroxynaphthoate	21	23	6.6
1-Hydroxy-2-naphthoate	35	35	7.0
2-Hydroxy-2-naphthoate	32	39	6.3

$$A + B \rightleftharpoons AB$$
$$\updownarrow$$
$$[B]\ \text{solid}$$

$$K = \frac{[AB]}{[A][B]}$$

The results are listed as first-order interaction constants, since the amount of solubilizate increased linearly with concentration of aromatic hydrotrope. From [278].

suggests that the imidazole ring of the xanthine nucleus is involved in the interaction which, it is thought, leads to the formation of 1:1 complexes. The molecular interaction of caffeine with benzoic acid, methoxybenzoic acid, and nitrobenzoic acid [297] has been investigated in detail. Fig. 6.44 shows the typical picture with an initial increase in the solubility of the acid with increasing caffeine concentration followed by a plateau region which results because of the limited solubility of the caffeine. Similar behaviour is exhibited by theophylline–acetylsalicylic acid mixtures [298]. Of the three xanthines studied, caffeine has the greatest solubilizing power towards acetylsalicylic acid. While 1:1 complexes are formed between caffeine and sulphathiazole (1 % w/v caffeine increases the solubility of sulphathiazole by 50 %), sulphadiazine, benzocaine, and *p*-aminobenzoic acid, and the barbiturates are capable of forming 2:1 complexes [298].

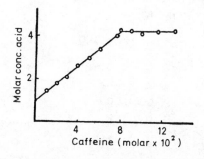

Figure 6.44 Solubility of *p*-methoxybenzoic acid in aqueous solutions of caffeine showing the plateau region caused by precipitation of the caffeine. From Donbrow and Jan, [297].

Some rationalization of drug complexation is possible by the use of Hückel frontier molecular orbitals [300]. These confirm π-donor-π-acceptor mechanisms for the interactions with niacinamide of the compounds shown below in aqueous solution. Phase-solubility diagrams all showed positive deviations from linearity, behaviour attributed to first- and second-order interactions between the solute (S) and the hydrotrope (H):

$$S + H \rightleftharpoons SH \tag{i}$$

$$K_{1:1} = \frac{[SH]}{[S][H]}$$

$$SH + H \rightleftharpoons SH_2 \tag{ii}$$

$$K_{1:2} = \frac{[SH_2]}{[SH][H]}.$$

Table 6.33 gives values for these two equilibrium constants [300].

Table 6.33 Equilibrium constants and calculated stabilization energies for niacinamide* complexes

Compound	$K_{1:1}$	$K_{1:2}$	ΔE	$\log K_{1:1}$	Calculated $\log K_{1:1}$
XXX	12.31	14.76	0.598	1.090	1.022
XXXI	7.36	0.33	0.561	0.861	0.876
XXXII	18.45	0.51	0.660	1.266	1.267
XXXIII	9.88	2.26	0.607	0.995	1.058

* Niacinamide concentration was 0.0–2.0 M for II and 0.0–0.2 M otherwise [300].
The ΔE value is a partial measure of stabilization.

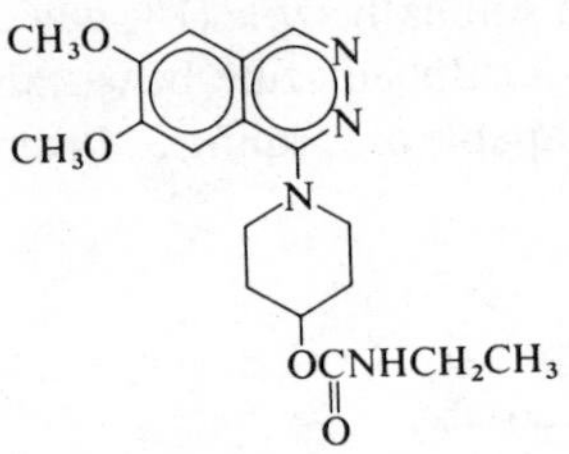

(XXX)

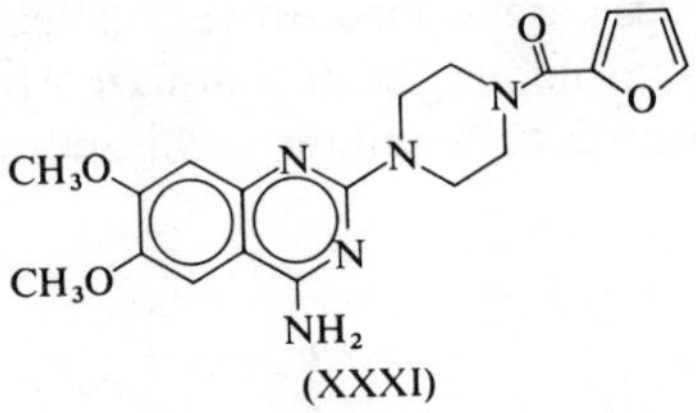

(XXXI)

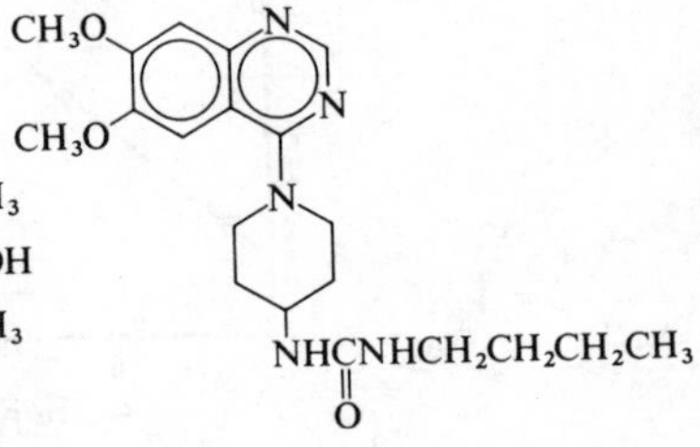

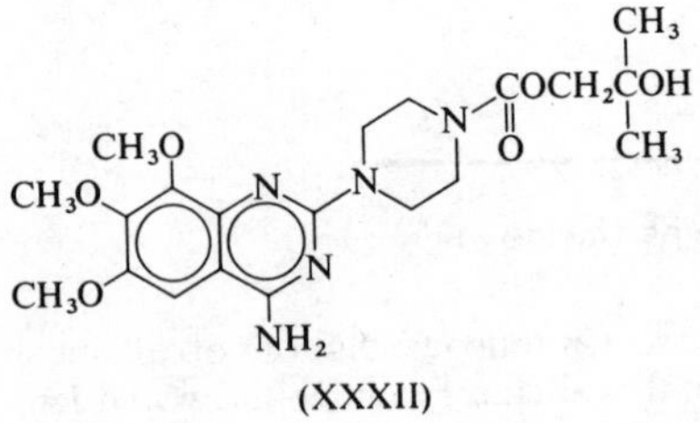

(XXXII)

(XXXIII)

The complexation of (XXXI) is shown in the diagram below and may be compared with the complex formed by compound (XXXII). The small change in structure gives rise to a change in the topology of the complex to give maximum degree of interaction

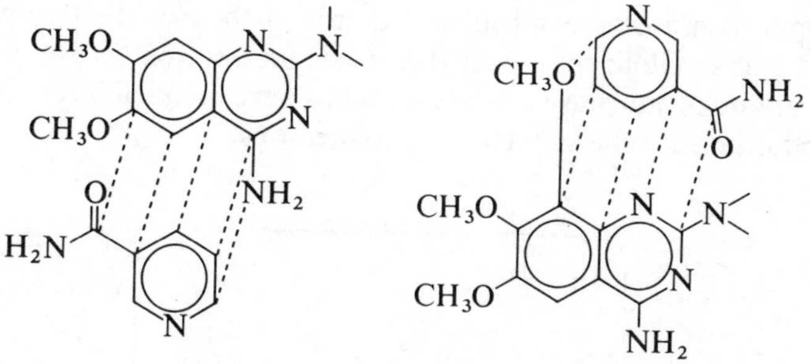

Schematic representation of interaction between niacinamide and (XXXI) and (XXXII) from Fawzi *et al.* [300] with permission.

The difficulty in using the xanthines in pharmacy is that they have a pharmacological action of their own and limited solubility. However, a number of medicaments contain hydrotropes for their therapeutic effect and not their solubilizing action, such as in aspirin, phenacetin, and caffeine mixtures; it is unlikely that the caffeine in this preparation is inactive.

Gentisic acid (XXXIV) and its alkyl derivatives have been reported to form complexes with several solutes including acronine (XXXV) [299].

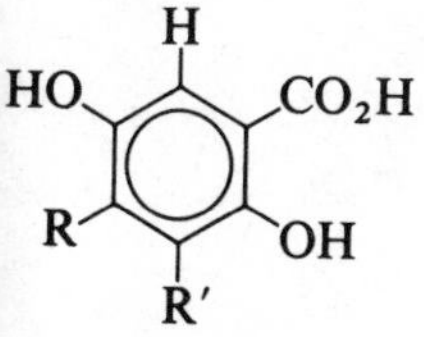

(XXXIV) (a) R = R' = H; gentisic acid
(b) R = H, R' = CH₃; 3-methyl gentisic acid
(c) R = H, R' = CH₂CH₃; 3-ethyl gentisic acid
(d) R = CH₃, R' = H; 4-methyl gentisic acid
(e) R = CH₂CH₃, R' = H; 4-ethyl gentisic acid

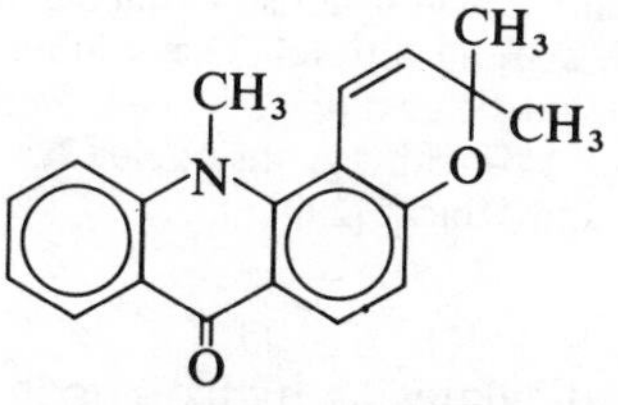

(XXXV) Acronine

The complexation of acronine with gentisate, and 3-methyl, 4-methyl, 3-ethyl, and 4-ethyl-substituted gentisates was studied by solubility techniques in aqueous solutions at 25° C. In all cases both 1:1 and 1:2 (acronine:gentisate) complexes were found and apparent increases in the solubility of acronine were observed with 3-methyl gentisate bringing about the largest increase. Both the nature of the substituent and its position were important in the complexation and the effects appear to be due to opposing steric and hydrophobic contributions [299]. In none of the systems was there evidence of precipitate formation. While a methyl substituent at the 3 or 4 position increases the solubility, a further increase in the chain length of the substituent decreases the extent of complexation. (Fig. 6.45). The suitability of 3-methyl gentisate for intravenous use has not been reported, although aqueous gentisate solutions have been studied (Cradock, see [299]) and deemed to be suitable for parenteral use.

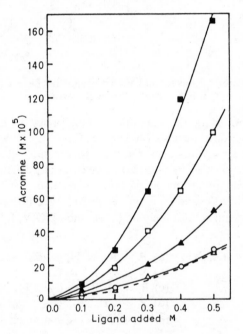

Figure 6.45 Plot of the apparent solubility of acronine at $25°$ C as a function of the concentration of various gentisic acid ligands in aqueous buffer (0.1 M succinate, pH 5.5, ionic strength ~ 0.6 M with sodium nitrate). The solubility of acronine in the absence of ligands is 8.5×10^{-6} M. Gentisic acid (○, ———); 3-ethyl gentisic acid (□, ———); 4-ethyl gentisic acid (△, – – –); 3-methyl gentisic acid (■, ———); 4-methyl gentisic acid (▲, ———). From Repta and Hincal [299].

6.7.1 Benzoates and salicylates as hydrotropes

Solutions of sodium *p*-toluene sulphonate enhance the solubility of phenolic compounds in general; however, of sodium benzoate, sodium *p*-toluene sulphonate, and sodium salicylate the last compound is the best hydrotrope [301]. Sodium benzoate is used to solubilize chlorocresol in solution of sodium benzoate and chlorocresol and to increase the solubility of the haemostatic adrenochrome monosemicarbazide [302]. Injection of caffeine and sodium benzoate is the parenteral form in which caffeine is usually administered. A 25:1 salicylic acid–adrenochrome complex dissolves in water up to 25 mg mol⁻¹ [303]. Iwao [304] found that a 25:1 mixture can be diluted with water to any degree without precipitation.

The solubility of the antitubercular drug, pyrazinamide, is directly proportional to the concentration of sodium *p*-aminosalicylate (sodium PAS) or sodium hydroxybenzoate in aqueous solution [305]. Thermal analysis has confirmed the complex formation between the drug and sodium PAS, but the authors [305] list the alternatives of complexation and normal increase in solubility in the presence of additive as the causes of solubilization. As the antituberculars are always used clinically in combination – sodium PAS is also an effective drug – this study may have some bearing on the efficacy of the combinations. It might be of interest for an investigation to be carried out on the solution properties of isoniazid, streptomycin, and sodium PAS mixtures, especially as streptomycin is thought to have some colloidal electrolyte properties of its own.

Saleh *et al.* [306] have studied the solubility of diazepam in sodium salicylate solution as a potential parenteral formulation. At present diazepam, which is practically insoluble in water, is formulated with propylene glycol as the main solvent component but undesirable clinical effects following intramuscular and intravenous injection have been attributed to the propylene glycol. The solubility of diazepam increases significantly at concentrations of salicylate greater than 15 to 20%. At 30%, the solubility has increased to over 16 mg ml⁻¹ (Fig. 6.46). The

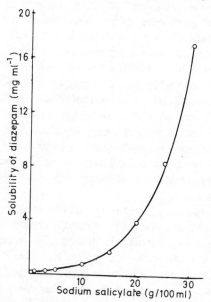

Figure 6.46 Effect of sodium salicylate on the solubility of diazepam in water at 37° C. From Saleh *et al.* [306].

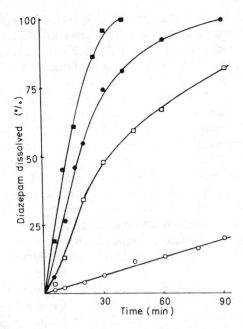

Figure 6.47 Effect of various concentrations of sodium salicylate on the dissolution rate of diazepam in water at 37° C. Water, ○; and 4%, □, 8%, ●; and 12%, ■; sodium salicylate. From El-Khordagui *et al.* [307].

concomitant increase in rates of solution can be seen in Fig. 6.47. As sodium salicylate decreases the surface tension of water the effect on dissolution may be partly due to this. There is, however, some evidence that at concentrations greater than 20% w/v, sodium salicylate associates in solution. The compatability and stability of diazepam in 30% sodium salicylate solutions following dilution with 5% dextrose and normal saline has been studied by El-Khordagui *et al.* [309] who found that 1:1 and 1:100 dilutions remained clear for up to 3 h, although microcrystal formation was noted at longer times. The diazepam–sodium salicylate combination also induced higher degrees of haemolysis *in vitro* than a commercial diazepam injection containing Cremophor EL. The interpretation of the haemolytic activity may be complicated by the fact that Cremophor EL has previously been implicated in the inhibition of haemolysis at concentrations of 0.8 to 4 mg ml^{-1} [308].

Sodium salicylate has been found to enhance rectal absorption of drugs [309] but contrary to the action of some surfactants the absorption promotion was not found to be the result of a permanent change in the rectal mucosa.

The effect of the caffeine–sodium salicylate molar ratio in the distribution of caffeine to chloroform from an aqueous phase has been studied by Blake and Harris [310] and is reproduced in Fig. 6.48. This effect might in itself cause a change in the physiological action of the caffeine. Sodium salicylate does not affect the biological availability of riboflavin in solutions containing 25% sodium

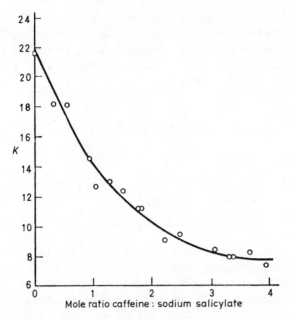

Figure 6.48 The effect of sodium salicylate on the distribution coefficient (K) of caffeine between chloroform and water. From Blake and Harris [310].

salicylate and 1.2% vitamin, the riboflavin activity being measured by two biological procedures. No difference was detected in response between riboflavin powder and solubilized preparations [311].

It is possible that the use of sodium salicylate in doses of 4 to 9 g daily in chronic gout and its ability to lower serum uric acid levels is due to a hydrotropic action on the acid, preventing its reabsorption by the kidney tubules. According to Lieber [312] the concept of hydrotropy allows some understanding of the etiology of metabolic diseases, such as atherosclerosis, diabetes, lithiasis, and gout. Lieber suggests that the efficacy of the oral hypoglycaemic agents used in the treatment of diabetes is due to their hydrotropic properties, by which they liberate insulin from its protein complex and thereby activate it. Tolbutamide is a derivative of *p*-toluene sulphonic acid (see below), and other agents used orally in diabetes are similar.

$$H_3C-\!\!\left\langle\bigcirc\right\rangle\!\!-SO_2NH\cdot CONH\cdot C_4H_9 \qquad H_2N-\!\!\left\langle\bigcirc\right\rangle\!\!-SO_2NH\cdot CONH\cdot C_4H_9$$

       (XXXVI) Tolbutamide   **(XXXVII) Carbutamide**

$$Cl-\!\!\left\langle\bigcirc\right\rangle\!\!-SO_2NH\cdot CONH\cdot C_3H_7$$

                (XXXVIII) Chlorpropamide

The tetracyclines are a group of drugs which form soluble complexes with typical hydrotropes. Oxytetracycline dihydrate and tetracycline dihydrate complex with sodium salicylate, sodium saccharin, sodium *p*-aminobenzoate, and *N*-methylpyrrolidone [313] which is hardly surprising considering the polar, multifunctional character of the drugs:

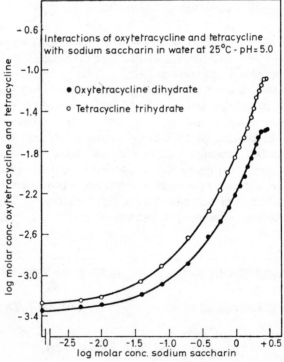

(XXXIX) Oxytetracycline

Since many hydrotropic agents possess strong negative groups, it seemed logical to ascribe the solubilization to a displacement reaction whereby the acidic hydrogens of the solute co-ordinate with the negative centres of the additive, replacing water molecules. This interaction may be rendered more favourable by

Interactions of oxytetracycline and tetracycline with sodium saccharin in water at 25°C - pH = 5.0

● Oxytetracycline dihydrate

○ Tetracycline trihydrate

Figure 6.49 The effect of sodium saccharin on the solubility of tetracycline and oxytetracycline in water at 25°C. From Gans and Higuchi [313].

the formation of hydrophobic bonds between the hydrocarbon parts of the interacting molecules. The large aromatic nucleus of the tetracycline antibiotics makes this a possibility. Fig. 6.49 shows the effect of sodium saccharin on the solubility of the two tetracyclines. The typical hydrotropic solubility curve is not, apparently, a linear function of concentration of the hydrotrope, but shows considerable sigmoidal character [314]. Relatively weak complexing tendencies exist between saccharin and various substances in aqueous solution [315], a 1:1 complex being formed with theophylline but there being no interaction with *N*-methyl pyrrolidone or γ-butyrolactone.

References

1. A. T. FLORENCE (1982) in *Techniques of Solubilization of Drugs* (ed. S. Yalkowsky) Marcel Dekker, New York, ch 2.
2. J. SWARBRICK (1965) *J. Pharm. Sci.* **54**, 1229.
3. B. A. MULLEY (1964) in *Advances in Pharmaceutical Sciences*, (eds H. S. Bean, A. H. Beckett, and J. E. Carless) Academic Press, London.
4. L. SJÖBLOM (1967) in *Solvent Properties of Surfactant Solutions*, (ed. K. Shinoda), Marcel Dekker, New York.
5. N. DROSELER and R. VOIGHT (1967) *Die Pharmazie.* **22**, 699.
6. P. H. ELWORTHY, A. T. FLORENCE and C. B. MACFARLANE (1968) *Solubilization by Surface-Active Agents* Chapman and Hall, London.
7. E. H. CORDES (ed.) (1973) *Reaction Kinetics in Micelles* Plenum Press, New York.
8. J. H. FENDLER and E. J. FENDLER (1975) *Catalysis in Micellar and Macromolecular Systems* Academic Press, New York.
9. K. L. MITTAL (1977) *Micellisation, Solubilization and Microemulsions* Vols 1 and 2, Plenum Press, New York.
10. A. G. MITCHELL (1964) *J. Pharm. Pharmacol.* **16**, 533.
11. J. B. LLOYD and B. W. CLEGG (1954) *J. Pharm. Pharmacol.* **6**, 797.
12. B. A. MULLEY and A. D. METCALF (1964) *J. Colloid Sci.* **19**, 501.
13. P. H. ELWORTHY and A. T. FLORENCE (1965) *Kolloid-Z.* **204**, 105.
14. R. R. BALMBRA, J. S. CLUNIE, J. M. CORKILL and J. F. GOODMAN (1962) *Trans. Faraday Soc.* **58**, 1661.
15. B. A. MULLEY and A. D. METCALF (1956) *J. Pharm. Pharmacol.* **8**, 774.
16. J. W. HADGRAFT (1954) *J. Pharm. Pharmacol.* **6**, 816.
17. H. BERRY, A. M. COOK and B. A. WILLS (1956) *J. Pharm. Pharmacol.* **8**, 425.
18. W. HELLER and H. B. KLEVENS (1946) *J. Chem. Phys.* **14**, 567.
19. E. AZAZ and M. DONBROW (1976) *J. Colloid Interface Sci.* **57**, 11.
20. M. DONBROW, E. AZAZ and R. HAMBURGER (1970) *J. Pharm. Sci.* **59**, 1427.
21. N. K. PATEL and N. E. FOSS (1965) *J. Pharm. Sci.* **54**, 1495.
22. C. T. RHODES and M. DONBROW (1965) *J. Pharm. Sci.* **54**, 1130.
23. M. J. CROOKS and K. F. BROWN (1974) *J. Pharm. Pharmacol.* **26**, 235.
24. B. A. MULLEY and A. J. WINFIELD (1970) *J. Chem. Soc. A*, 1459.
25. A. G. MITCHELL and K. F. BROWN (1966) *J. Pharm. Pharmacol.* **18**, 115.
26. K. J. HUMPHREYS and C. T. RHODES (1968) *J. Pharm. Sci.* **57**, 79.
27. H. TOMIDA, T. YOTSUYANAGI and K. IKEDA (1978) *Chem. Pharm. Bull.* **26**, 2824.
28. H. SCHOTT and S. K. HAN (1975) *J. Pharm. Sci.* **64**, 658.
29. J. H. COLLETT and L. KOO (1975) *J. Pharm. Sci.* **64**, 1253.
30. E. R. GARRETT (1966) *J. Pharm. Pharmacol.* **18**, 589.
31. T. SHIMAMOTO, H. MIMA and M. NAKAGAKI (1979) *Chem. Pharm. Bull.* **27**, 1995.
32. T. SHIMAMOTO, H. MIMA and M. NAKAGAKI (1979) *Chem. Pharm. Bull.* **27**, 2557.

33. J. BLANCHARD, W. T. FINK and J. P. DUFFY (1977) J. Pharm. Sci. 66, 1470.
34. T. SHIMAMOTO and H. MIMA (1979) Chem. Pharm. Bull. 27, 2602.
35. M. GRATZEL and J. K. THOMAS (1973) J. Amer. Chem. Soc. 95, 6885.
36. A. GOTO, R. SAKURA and F. ENDO (1980) Chem. Pharm. Bull. 28, 14.
37. A. GOTO and F. ENDO (1978) J. Colloid Interface Sci. 66, 26.
38. A. GOTO, F. ENDO and K. HO (1977) Chem. Pharm. Bull. 25, 1165.
39. H. SCHOTT (1969) J. Pharm. Sci. 58, 1443.
40. A. T. FLORENCE, F. MADSEN and F. PUISIEUX (1975) J. Pharm. Pharmacol. 27, 385.
41. W. N. MACLAY (1956) J. Colloid Sci. 11, 272.
42. K. SHINODA (1967) J. Colloid Interface Sci. 24, 10.
43. M. DONBROW and E. AZAZ (1976) J. Colloid Interface Sci. 57, 20.
44. A. PRINS (1962) Doctoral Dissertation, Technische Hogeschool, Eindhoven.
45. C. R. BAILEY (1923) J. Chem. Soc. 2579.
46. J. H. PURNELL and S. T. BOWDEN (1954) J. Appl. Chem. 4, 648.
47. M. J. CROOKS and K. F. BROWN (1973) J. Pharm. Pharmacol. 25, 281.
48. F. ALHAIQUE, D. GIACCHETTI, M. MARCHETTI and F. M. RICCIERI (1977) J. Pharm. Pharmacol. 29, 401.
49. S. J. A. KAZMI and A. G. MITCHELL (1976) Canad. J. Pharm. Sci. 11, 10.
50. A. G. MITCHELL (1964) J. Pharm. Pharmacol. 16, 533.
51. W. GOOD and M. H. MILLOY (1956) Chem. Ind. 872.
52. A. E. ALEXANDER (1949) Surface Chemistry, Butterworths, London, p. 299.
53. H. S. BEAN and H. BERRY (1951) J. Pharm. Pharmacol. 3, 639.
54. H. S. BEAN and H. BERRY (1953) J. Pharm. Pharmacol. 5, 632.
55. J. M. SCHAFFER and F. W. TILLEY (1930) F. Agric. 41, 137.
56. A. R. CADE (1935) Soap 11 (9), 27.
57. E. J. ORDAL and F. DEROMEDI (1943) J. Bact. 45, 293.
58. I. SHAFIROFF (1961) Proc. Chem. Spec. Mfrs. Assoc., 47th Meeting, p. 142.
59. W. P. EVANS and S. F. DUNBAR (1965) Surface Activity and the Microbial Cell Society for Chemical Industry, London, 1965.
60. W. P. EVANS (1964) J. Pharm. Pharmacol. 16, 323.
61. M. DONBROW and C. T. RHODES (1965) J. Pharm. Pharmacol. 17, 258.
62. W. P. EVANS (1965) J. Pharm. Pharmacol. 17, 462.
63. F. D. PISANO and H. B. KOSTENBAUDER (1959) J. Amer. Pharm. Assoc. 48, 310.
64. N. K. PATEL and H. B. KOSTENBAUDER (1958) J. Amer. Pharm. Assoc. 47, 289.
65. N. SENIOR (1973) J. Soc. Cosmetic Chem. 24, 259.
66. F. WESOLUCH, A. T. FLORENCE, F. PUISIEUX and J. T. CARSTENSEN (1979) Int. J. Pharmaceutics 2, 343.
67. D. D. HEARD and R. W. ASHWORTH (1968) J. Pharm. Pharmacol. 20, 505.
68. M. FROBISHER (1927) J. Bact. 13, 163.
69. A. L. ERLANDSON and C. A. LAWRENCE (1953) Science 118, 274.
70. R. BERTHET (1947) Schweiz, Apoth. Ztg. 85, 833.
71. K. L. RUSSELL and S. G. HOCH (1965) J. Soc. Cosmetic Chem. 16, 169.
72. R. A. ANDERSON and K. J. MORGAN (1966) J. Pharm. Pharmacol. 18, 449.
73. L. GERSHENFELD and B. WITLIN (1950) J. Amer. Pharm. Assoc. 39, 489.
74. A. OSOL and C. C. PINES (1952) J. Amer. Pharm. Assoc. 41, 634.
75. A. SEIDELL (1965) Solubility of Inorganic Compounds 4th edn, Van Nostrand, New York.
76. D. H. TERRY and N. SHELANSKI (1952) Modern Sanitation 4 (1), 61.
77. W. B. HUGO and J. M. NEWTON (1964) J. Pharm. Pharmacol. 16, 273.
78. W. NYIRI and M. JANNITTI (1932) J. Pharmacol. Exptl. Therap. 45, 85.
79. N. A. ALLAWALA and S. RIEGELMAN (1953) J. Amer. Pharm. Assoc. 42, 396.
80. C. A. LAWRENCE, C. M. CARPENTER and A. W. C. NAYLOR-FOOTE (1957) J. Amer. Pharm. Assoc. 46, 500.
81. N. E. LAZARUS (1954) J. Milk Tech. 17, 144.

82. V. ROSSETTI (1959) *Ann. Chim. Appl.* **49**, 923.
83. W. B. HUGO and J. M. NEWTON (1963) *J. Pharm. Pharmacol.* **15**, 731.
84. G. HENDERSON and J. M. NEWTON (1966) *Pharm. Acta Helv.* **41**, 228.
85. G. A. BROST and F. KRUPIN (1957) *Soap Chem. Spec.* **33** (8), 93.
86. T. KORIYA, A. D. MARCUS and B. E. BENTON (1953) *J. Amer. Pharm. Assoc. Pract. Ed.* **14**, 297.
87. C. F. HISKEY and F. F. CANTWELL (1966) *J. Pharm. Sci.* **55**, 166.
88. S. ROSS and V. H. BALDWIN (1966) *J. Colloid Sci.* **21**, 284.
89. R. W. SIDWELL, L. WESTBROOK, G. J. DIXON and W. F. HAPPCH (1970) *Appl. Microbiol.* **19**, 53.
90. L. J. WILKOFF, G. J. DIXON, L. WESTBROOK and W. F. HAPPCH (1971) *Appl. Microbiol.* **21**, 647.
91. S. P. GORMAN and E. M. SCOTT (1979) *Int. J. Pharmaceutics* **4**, 57.
92. H. OLDBERG (1958) *Arzneimittel-Forsch.* **8**, 143.
93. E. A. BLISS and P. T. WARTH (1950) *Ann. N.Y. Acad. Sci.* **53**, 38.
94. A. BRUNZELL (1957) *Svensk Tidskr.* **6**, 129.
95. ANON (1952) *Bull. Amer. Soc. Hosp. Pharm.* **9**, 56.
96. H. MATSUMURA et al. (1958) *Yakuzaigaku* **18**, 124–6 [*CA* (1959) **53**, 6535].
97. H. LEHMANN and J. CROT (1957) *Schweiz. Apoth. Ztg.* **95**, 367.
97a Dutch Patent 89,788 (1958).
98. J. A. ROGERS (1966) MSc Thesis, University of Toronto.
99. E. REGDON-KISS and G. KEDVESSY (1963) *Pharmazie* **18**, 131.
100. R. LEVIN (1952) *Pharm. J.* **168**, 56.
101. *Extra Pharmacopoeia* Pharmaceutical Press, London, 24 edn, Vol. 1.
102. British Patent 633,175 (1949).
103. US Patent 2,472,640 (1949).
104. G. GILLISSEN (1955) *Arzneimittel-Forsch.* **5**, 460.
105. T. NAKAGAWA (1956) *J. Pharm. Soc. Japan* **76**, 1113.
106. C. B. BRUCE and L. MITCHELL (1952) *J. Amer. Pharm. Assoc.* **41**, 654.
107. E. ULLMANN, K. THOMA and L. PATT (1978) *Tenside* **15**, 9.
108. H. WATANABE, S. OTANI, T. UEHARA and R. UEHARA (1961) *J. Antibiotics (Tokyo) Ser. A.* **14**, 264.
109. M. BARR and L. F. TRICE (1955) *Amer. J. Pharm.* **127**, 260.
110. M. R. W. BROWN and B. E. WINSLEY (1969) *J. Gen. Microbiol.* **56**, 99.
111. M. R. W. BROWN and B. E. WINSLEY (1971) *J. Gen. Microbiol.* **68**, 367.
112. M. R. W. BROWN, E. M. GEATON and P. GILBERT (1979) *J. Pharm. Pharmacol.* **31**, 168.
113. M. R. W. BROWN and R. M. E. RICHARDS (1964) *J. Pharm. Pharmacol.* **16**, 51.
114. R. M. ATKINSON, C. BEDFORD, K. J. CHILD and E. G. TOMICH (1962) *Nature* **193**, 588.
115. R. M. ATKINSON, C. BEDFORD, K. J. CHILD and E. G. TOMICH (1962) *Antibiot. Chemother.* **12**, 232.
116. M. KRAML, J. DUBUC and D. BEALL (1962) *Canad. J. Biochem.* **40**, 1449.
117. W. A. M. DUNCAN, G. MACDONALD and M. J. THORNTON (1962) *J. Pharm. Pharmacol.* **14**, 217.
118. J. R. MARVEL, D. A. SCHLICHTING, C. DENTON, E. J. LEVY and M. M. CAHN (1964) *J. Invest. Dermatol.* **42**, 197.
119. T. R. BATES, M. GIBALDI and J. L. KANIG (1966) *J. Pharm. Sci.* **55**, 191.
120. T. R. BATES, S. L. LIN and M. GIBALDI (1967) *J. Pharm. Sci.* **56**, 1492.
121. P. H. ELWORTHY and F. J. LIPSCOMB (1968) *J. Pharm. Pharmacol.* **20**, 817.
122. T. ARNARSON and P. H. ELWORTHY (1980) *J. Pharm. Pharmacol.* **32**, 381.
123. T. ARNARSON and P. H. ELWORTHY (1981) *J. Pharm. Pharmacol.* **33**, 141.
124. P. H. ELWORTHY and F. J. LIPSCOMB (1968) *J. Pharm. Pharmacol.* **20**, 923.
125. E. L. PARROTT and U. K. SHARMA (1967) *J. Pharm. Sci.* **56**, 1341.
126. R. R. SHERWOOD and A. M. MATTOCKS (1951) *J. Amer. Pharm. Assoc.* **40**, 90.
127. G. WOODARD (1952) *J. Pharm. Pharmacol.* **4**, 1009.

128. R. N. MITRA and E. J. GRACE (1956) Antibiotic Ann. 6, 455.
129. T. NAGAKAWA and R. MUNEYUKI (1953) J. Pharm. Soc. Japan 73, 1106.
130. U. COCCHI (1955) Schwr. Med. Wochr. 86, 916.
131. J. E. BENNETT (1964) Ann. Internal Med. 61, 335.
132. D. A. WHITING (1967) Br. J. Den. 79, 345.
133. B. B. RILEY (1970) J. Hosp. Pharm. 28, 228.
134. V. T. ANDRIDE and H. M. KRAVETZ (1962) J. A. M. A. 80, 269.
135. J. W. HADGRAFT Extra Pharmacopoeia, Pharmaceutical Press, London, 24 edn, p. 249.
136. D. B. JACK and W. RIESS (1973) J. Pharm. Sci. 62, 1929.
137. S. A. H. KHALIL and Z. A. EL-GHOLMY (1977) J. Pharm. Pharmacol. 29, Suppl. 21P.
138. M. N. KHAWAM, R. TAWASHI and H. V. CZETSCH-LINDENWALD (1964) Sci. Pharm. 32, 271.
139. M. N. KHAWAM, R. TAWASHI and H. V. CZETSCH-LINDENWALD (1965) Sci. Pharm. 33, 153.
140. K. KAKEMI, T. ARITA and S. MURANISHI (1965) Chem. Pharm. Bull. 13, 965.
141. K. KAKEMI, T. ARITA and S. MURANISHI (1965) Chem. Pharm. Bull. 13, 976.
142. R. T. YOUSEF and M. N. KHAWAM (1966) Arch. Mikrobiol. 53, 159.
143. R. T. YOUSEF, M. N. KHAWAM, R. TAWASHI and H. V. CZETSCH-LINDENWALD (1966) Arzniemittel-Forsch 16, 515.
144. T. SHIRAHIGE (1953) Folia Pharmacol. Japan 49, 282.
145. C. W. WHITWORTH and C. H. BECKER (1966) Amer. J. Hosp. Pharm. 23, 574.
146. J. Y. PARK and E. G. RIPPIE (1977) J. Pharm. Sci. 66, 858.
147. V. NAGGAR, N. A. DAABIS and M. M. MOTAWI (1974) Pharmazie 29, 122.
148. K. IKEDA, H. TOMIDA and T. YOTSUYANAGI (1977) Chem. Pharm. Bull. 25, 1067.
149. T. I. RAZHANSKAYA, L. V. DMITREKO, G. B. SELEKHOVA and G. V. SAMSONOV (1974) Kolloid-Z. 36, 58.
150. L. SJÖBLOM (1958) Acta Acad. Aboensis. Math. Phys. 21 (7).
151. P. EKWALL, L. SJÖBLOM and L. OLSEN (1953) Acta Chem. Scand. 7, 347.
152. L. SJÖBLOM (1965) in Surface Chemistry (eds. P. Ekwall et al) Munksgaard, Copenhagen.
153. L. SJÖBLOM and N. SUNDBLOM (1964) Acta Chem. Scand. 18, 1996.
154. C. BLOMQUIST and L. SJÖBLOM (1964) Acta Chem. Scand. 18, 2404.
155. P. EKWALL, T. LUNDSTEN and L. SJÖBLOM (1951) Acta Chem. Scand. 5, 383.
156. A. NYLANDER (1953) Farm Notis blad. 62, 183.
157. A. THAKKAR and N. HALL (1967) J. Pharm. Sci. 56, 1121.
158. H. TOMIDA, T. YOTSUYANAGI and K. IKEDA (1978) Chem. Pharm. Bull. 26, 2832.
159. G. L. FLYNN (1971) J. Pharm. Sci. 60, 345.
160. B. W. BARRY and D. I. D. EL EINI (1976) J. Pharm. Pharmacol. 28, 210.
161. US Patent 2,880,130 (1959).
162. US Patent 2,880,138 (1959).
163. T. NAKAGAWA (1954) J. Pharm. Soc. Japan 74, 1116.
164. T. NAKAGAWA (1956) J. Pharm. Soc. Japan 76, 1113.
165. B. LUNDBERG (1980) J. Pharm. Sci. 69, 20.
166. B. LUNDBERG, T. LÖVGREN and C. BLONQUIST (1979) Acta Pharm. Suec. 16, 144.
167. L. MARTIS, N. A. HALL and A. L. THAKKAR (1972) J. Pharm. Sci. 61, 1757.
168. T. LÖVGREN, B. HEIKIUS, B. LUNDBERG and L. SJÖBLOM (1978) J. Pharm. Sci. 67, 1419.
169. B. LUNDBERG, T. LÖVGREN and B. HEIKIUS (1979) J. Pharm. Sci. 68, 542.
170. E. LESTER-SMITH (1928) Analyst 53, 632.
171. E. LESTER-SMITH (1930) Biochem. J. 24, 1942.
172. E. LESTER-SMITH (1932) J. Phys. Chem. 36, 1401.
173. L. OSIPOW, F. D. SNELL, W. C. YORK and A. FINCHLER (1956) Ind. Eng. Chem. 48, 1459.
174. H. MIMA (1957) Pharm. Bull. Tokyo 5, 496.

175. H. MIMA (1958) *J. Pharm. Soc. Japan* **78**, 988.
176. Japanese Patent 12,287 (1962).
177. F. GSTIRNER and PH. S. TATA (1958) *Mitt der Deutsch Pharm. Gesell* **28**, 191.
178. R. M. TOMARELLI, J. CHARNEY and F. W. BERNHART (1946) *Proc. Soc. Exptl. Biol. Med.* **63**, 108.
179. A. E. SOBEL (1956) *Arch. Dermatol.* **73**, 388.
180. A. E. SOBEL (1949) *Fed. Proc.* **8**, 253.
181. C. J. KERN and T. ANTOSHKIW (1950) *Ind. Eng. Chem.* **42**, 709.
182. C. L. J. COLES and D. F. W. THOMAS (1952) *J. Pharm. Pharmac.* **4**, 898.
183. P. F. G. BOON, C. L. J. COLES and M. TAIT (1961) *J. Pharm. Pharmac.* **13**, 200 T.
184. R. HÜTTENRAUCH and L. KLOTZ (1963) *Arch. Pharm.* **296**, 145.
185. A. WATANABE, T. KANAZAWA, H. MIMA, N. YAMAMOTO and T. SHIMA (1955) *J. Pharm. Soc. Japan* **75**, 1093.
186. T. NAKAGAWA and C. R. MUNEVUKI (1954) *J. Pharm. Soc. Japan* **74**, 856.
187. T. NAKAGAWA (1956) *J. Pharm. Soc. Japan* **76**, 1118.
188. H. MIMA (1958) *J. Pharm. Soc. Japan* **78**, 983.
189. A. ITO, K. INAMI and A. OHARA (1956) *Ann. Rept. Takamine Lab.* **6**, 41.
190. H. MIMA (1958) *J. Pharm. Soc. Japan* **78**, 381.
191. *Formulary of Liquid Oral Products*, Atlas Chemical Industries, (1962) 29–30.
192. E. T. GADE and J. D. KADLEC (1956) *J. Agric. Food Chem.* **4**, 426.
193. T. D. WHITTET and M. CUMMINS (1955) *Pharm. J.* **174**, 271.
194. R. W. APPLEWHITE, A. P. BUCKLEY and W. L. NOBLES (1954) *J. Amer. Pharm. Assoc. Pract. Ed.* **15**, 1641.
195. T. HIGUCHI and R. KURAMOTO (1954) *J. Amer. Pharm. Assoc.* **43**, 398.
196. W. J. TILLMAN and R. KURAMOTO (1957) *J. Amer. Pharm. Assoc.* **46**, 211.
197. C. VAUTION, J. PARIS, F. PUISIEUX and J. T. CARSTENSEN (1978) *Int. J. Pharmaceutics* **1**, 349.
198. A. F. CACCHILLO and W. H. HASSLER (1954) *J. Amer. Pharm. Assoc.* **43**, 683.
199. J. K. LIM and C. C. CHEN (1974) *J. Pharm. Sci.* **63**, 559.
200. J. H. COLLETT, R. WITHINGTON and L. KOO (1975) *J. Pharm. Pharmacol.* **27**, 46.
201. N. A. HALL (1963) *J. Pharm. Sci.* **52**, 189.
202. N. A. HALL and R. A. SOUDAH (1966) *Amer. J. Pharm.* **138**, 245.
203. F. WESOLUCH (1978) DEPS Thesis, Université de Paris.
204. N. NISHIKIDO (1977) *J. Colloid Interface Sci.* **10**, 242.
205. J. M. BLAKEWAY, P. BOWDEN and M. SEN (1979) *Int. J. Cosmetic Sci.* **1**, 1.
206. US Patent, 2,417 229 (1947).
207. Pharmaceutical Society Report (1956) *Pharm. J.* **i**, 383.
208. A. J. MONTE-BOVI (1950) *J. Amer. Pharm. Assoc. Pract. Ed.* **11**, 107.
209. W. O'MALLEY, L. PENNATI and A. N. MARTIN (1958) *J. Amer. Pharm. Assoc.* **47**, 334.
210. K. THOMA and G. PFAFF (1976) *J. Soc. Cosmetic Chem.* **27**, 221.
211. I. ELLO (1964) *Mitt. der Deutsch Pharm. Gesell.* **34**, 193.
212. J. E. CARLESS and J. R. NIXON (1957) *J. Pharm. Pharmacol.* **9**, 963.
213. J. E. CARLESS and J. R. NIXON (1960) *J. Pharm. Pharmacol.* **12**, 348.
214. L. S. C. WAN and P. F. S. LEE (1975) *Canad. J. Pharm. Sci.* **10**, 69.
215. J. P. TREGUIER, I. LO, M. SEILLER and F. PUISIEUX (1975) *Pharm. Acta Helv.* **50**, 421.
216. I. LO, A. T. FLORENCE, J. P. TREGUIER, M. SEILLER and F. PUISIEUX (1977) *J. Colloid Interface Sci.* **59**, 319.
216a B. J. CARROLL, B. G. C. O'ROUKE and A. J. I. WARD (1982) *J. Pharm. Pharmacol.* **34**, 287.
217. A. E. ABOUTALEB, A. A. ALI and R. B. SALAMA (1977) *Indian J. Pharm. Sci.* **39**, 145.
218. K. SIVAKUMAR and B. M. MITHAL (1978) *India J. Pharm. Sci.* **40**, 157.
219. F. A. SHIHAB, A. R. EBIAN and R. M. MUSTAFA (1979) *Int. J. Pharmaceutics* **4**, 13.
220. S. OGURI, T. YOTSUYANAGI and K. IKEDA (1980) *Chem. Pharm. Bull.* **28**, 1768.
221. British Patent, 941,694 (1967).

222. Belgium Patent, 624,258 (1963).
223. J. C. CRADOCK, J. P. DAVIGNON, C. L. LITTERST and A. M. GUARINO (1973) *J. Pharm. Pharmacol.* **25**, 345.
224. R. D. SOFIA, R. K. KUBANA and H. BARRY (1974) *J. Pharm. Sci.* **63**, 939.
225. H. F. ZIPF and L. KUHLMANN (1967) *Arzneimittel Forsch* **17**, 1021.
226. S. H. ROTH and P. J. WILLIAMS (1979) *J. Pharm. Pharmacol.* **31**, 224.
227. S. C. PFLAG and L. C. ZOPF (1951) *US Armed Forces Med. J.* **2** (8), 1177.
228. O. CARNEY and L. C. ZOPF (1955) *A.M.A. Arch. Dermatol* **72**, 266.
229. P. W. GERDING and G. J. SPERANDIO (1952) *Amer. Profess. Pharmacist* **18**, 888.
230. M. J. STOKLOSA and L. M. OHMART (1951) *J. Amer. Pharm. Assoc. Pract. Ed.* **12**, 23.
231. S. G. FRANK and G. ZOGRAFI (1969) *J. Colloid Interface Sci.* **29**, 27.
232. B. KABBANI, F. PUISIEUX, J. P. TREGUIER, M. SEILLER and A. T. FLORENCE (1977) *Proc. 1st. Int. Congr. Pharmaceutical Technology* **1**, 53.
233. S. R. PALIT, V. A. MOGHE and B. BISWAS (1959) *Trans. Faraday Soc.* **55**, 467.
234. S. R. PALIT and V. VENKATESWARLU (1954) *J. Chem. Soc.* 2129.
235. I. LO, F. MADSEN, A. T. FLORENCE, J. P. TREGUIER, M. SEILLER and P. PUISIEUX (1977) in *Micellization, Solubilization and Microemulsions* (ed. K. Mittal) Vol. 1, Plenum, New York, p. 455.
236. T. J. LIN, H. KURIHARA and H. OHTA (1977) *J. Soc. Cosmet. Chem.* **28**, 457.
237. D. ATTWOOD, C. MCDONALD and S. C. PERRY (1975) *J. Pharm. Pharmacol.* **27**, 692.
238. C. L. J. COLES, J. R. HEPPLE, M. L. HILTON and C. A. WALTON (1968) *J. Pharm. Pharmacol.* **20** Suppl., 26S.
239. I. R. SCHMOLKA (1977) *J. Amer. Oil Chemists Soc.* **54**, 110.
240. A. AL-SADEN, A. T. FLORENCE, T. L. WHATELEY, F. PUISIEUX and C. VAUTION (1982) *J. Colloid Interface Sci.* **86**, 51.
241. K. N. PRASAD, T. T. LUONG, A. T. FLORENCE, J. PARIS, C. VAUTION, M. SEILLER and F. PUISIEUX (1979) *J. Colloid Interface Sci.* **69**, 225.
242. A. AL-SADEN, A. T. FLORENCE and T. L. WHATELEY (1979) *J. Pharm. Pharmacol.* **31**.
243. J. H. COLLETT and E. A. TOBIN (1979) *J. Pharm. Pharmacol.* **31**, 174.
244. P. B. SHETH and E. L. PARROTT (1967) *J. Pharm. Sci.* **56**, 983.
245. J. C. DEARDEN and E. TOMLINSON (1971) *J. Pharm. Pharmacol.* **23**, 735.
246. T. FUJITA, J. IWASA and C. HANSCH (1964) *J. Amer. Chem. Soc.* **86**, 5175.
247. J. H. COLLETT and E. A. TOBIN (1977) *J. Pharm. Pharmacol.* **29**, 19P.
248. J. JACOBS, R. A. ANDERSON and T. R. WATSON (1972) *J. Pharm. Pharmacol.* **24**, 586.
249. M. BARNES, M. R. DILLANY and A. J. SANDOE (1973) *Mfg. Chem.* **44**(10), 29.
250. US Patent 3,639,563 (1972).
251. R. K. NEDDY, S. A. KHALIL and M. W. GONDA (1976) *J. Pharm. Sci.* **65**, 1753.
252. W. W. SHAW, A. G. DANTI and F. N. BRUSCATE (1976) *J. Pharm. Sci.* **65**, 1780.
253. L. BELLOUL, M. SEILLER, A. T. FLORENCE and F. PUISIEUX (1979) *Acta Pharm. Tech.* **25**, 133.
254. M. N. JONES (1967) *J. Colloid Interface Sci.* **23**, 36.
255. M. J. SCHWUGER (1973) *J. Colloid Interface Sci.* **43**, 491.
256. S. SAITO (1957) *Kolloid-Z* **154**, 19.
257. S. SAITO (1958) *Kolloid-Z* **158**, 120.
258. M. N. BREUER and I. D. ROBB (1972) *Chem. Ind.* 530.
259. E. D. GODDARD and R. B. HANNAN (1977) *J. Amer. Oil Chem. Soc.* **54**, 561.
260. B. J. FONTANA (1968) *Macromolecules* **1**, 139.
261. H. ARAI and S. HOVIN (1969) *J. Colloid Interface Sci.* **30**, 373.
262. T. ISEMURA and A. IMANISHI (1958) *J. Polymer Sci.* **33**, 337.
263. S. SAITO (1960) *J. Colloid Sci.* **15**, 283.
264. K. NAKEMURA, R. ENDO and M. TAKEDA (1977) *J. Polymer Sci. Polymer Physics Ed.* **15**, 2087.
265. K. E. LEWIS and C. P. ROBINSON (1970) *J. Colloid Interface Sci.* **32**, 539.

266. M. N. JONES (1968) *J. Colloid Interface Sci.* **26**, 532.
267. E. D. GODDARD, T. S. PHILLIPS and R. B. HANNAN (1975) *J. Soc. Cosmetic Chem.* **26**, 461.
268. S. SAITO and Y. MATSUI (1978) *J. Colloid Interface Sci.* **67**, 483.
269. W. F. STANASZEK, R. S. LEVINSON and B. ECANOW (1974) *J. Pharm. Sci.* **63**, 1941.
270. E. TOMLINSON (1980) *Pharmacy International* **1**, 156.
271. E. TOMLINSON, S. S. DAVIS AND G. I. MUKHAYER (1979) in *Solution Chemistry of Surfactants* Vol. 1 (ed K. L. Mittal) Plenum, New York.
272. P. MUKERJEE and K. J. MYSELS (1955) *J. Amer. Chem. Soc.* **77**, 2937.
273. B. W. BARRY and G. F. J. RUSSELL (1972) *J. Pharm. Sci.* **61**, 502.
274. G. ZOGRAFI, P. R. PATEL and N. D. WEINER (1964) *J. Pharm. Sci.* **53**, 544.
275. E. TOMLINSON and S. S. DAVIS (1978) *J. Colloid Interface Sci.* **66**, 335.
276. B. W. BARRY and G. M. T. GRAY (1975) *J. Colloid Interface Sci.* **52**, 327.
277. B. W. BARRY and G. M. T. GRAY (1974) *J. Pharm. Sci.* **63**, 548.
278. T. HIGUCHI and A. DRUBULIS (1961) *J. Pharm. Sci.* **50**, 905.
279. German Patent, 224,981 (1908).
280. German Patent, 340,744 (1922).
281. Belgium Patent, 447,975 (1943).
282. German Patent, 583,054 (1934).
283. W. J. P. NEISH (1948) *Rec. Trav. Chim.* **67**, 361.
284. A. R. BIAMONTE and G. H. SCHNELLER (1952) *J. Amer. Pharm. Assoc.* **41**, 341.
285. D. LEHR (1945) *Proc. Soc. Expt. Biol. Med.* **58**, 11.
286. A. R. FRISK *et al.* (1947) *Brit. Med. J.* **i**, 7.
287. K. ITO and K. SEKIGUCHI (1966) *Chem. Pharm. Bull.* **14**, 255.
288. M. G. BROWN and J. E. F. RISEMAN (1937) *J. Amer. Med. Assoc.* **109**, 256.
289. M. A. ZOGLIO (1969) *J. Pharm. Sci.* **58**, 222.
290. D. V. FROST (1947) *J. Amer. Chem. Soc.* **69**, 1064.
291. R. HÜTTENRAUCH (1965) *Pharmazie* **20**, 243.
292. R. A. HARTE and J. L. CHEN (1949) *J. Amer. Pharm. Assoc.* **38**, 568.
293. P. E. BRUMFIELD and H. M. GROSS (1955) *Drug. Cosmet. Ind.* **77**, 46.
294. US Patent 2,395,378 (1946).
295. YAMAMOTO, FUJISAWA and TANAKA (1955) *Ann. Repts. Shionogi Labs* no. 5, 95.
296. D. E. GUTTMAN and M. Y. ATHALYE (1960) *J. Amer. Pharm. Assoc.* **49**, 687.
297. M. DONBROW and Z. A. JAN (1965) *J. Pharm. Pharmac.* **17**, 1295.
298. T. HIGUCHI and J. L. LACH (1954) *J. Amer. Pharm. Assoc.* **43**, 527, 349.
299. A. J. REPTA and A. A. HINCAL (1980) *Int. J. Pharmaceutics* **5**, 149.
300. M. B. FAWZI, E. DAVISON and M. S. TUTE (1980) *J. Pharm. Sci.* **69**, 104.
301. L. KNAZKO (1966) *Farm. Obzor* **35**, 298.
302. Belgian Patent 525,542 (1954).
303. US Patent 2,581,850 (1952).
304. J. IWAO (1956) *Chem. Pharm. Bull.* **4**, 247.
305. H. NEGORO, T. MIKI and S. UEDA (1959) *Chem. Pharm. Bull.* **7**, 91.
306. A. M. SALEH, S. A. KHALIL and L. K. EL-KHORDAGUI (1980) *Int. J. Pharmaceutics* **5**, 161.
307. L. K. EL-KHORDAGUI, A. M. SALEH and S. A. KHALIL (1980) *Int. J. Pharmaceutics* **7**, 111.
308. K. REBER (1965) *Nature* **208**, 195.
309. T. NISHIHARA, J. H. RYTTING and T. HIGUCHI (1980) *J. Pharm. Sci.* **697**, 44.
310. M. BLAKE and L. E. HARRIS (1952) *J. Amer. Pharm. Assoc.* **41**, 521.
311. W. C. GEWANT and H. K. LANE (1965) *Proc. Penn. Acad. Sci.* **38**, 111.
312. I. I. LIEBER (1963) *Semana Medica* **123**, 1810.
313. E. H. GANS and T. HIGUCHI (1957) *J. Amer. Pharm. Assoc.* **46**, 458.
314. H. S. BOOTH, H. E. EVERSON (1950) *Ind. Eng. Chem.* **42**, 1536.
315. J. R. MARVEL and A. P. LEMBERGER (1960) *J. Amer. Pharm. Assoc.* **49**, 417.

7 Biological implications of surfactant presence in formulations

7.1 Introduction

The use of surfactants as emulsifying agents, solubilizers, suspension stabilizers and as wetting agents in formulations intended for administration to human subjects or to animals can lead to significant changes in the biological activity of the active agent in the formulation. A drug is seldom administered as such but as a complex formulation. Surfactant molecules incorporated into the formulation can exert their multifarious effects in several ways, e.g. by influencing the deaggregation and dissolution of solid dose forms (Fig. 7.1), by controlling the rate of precipitation of drugs administered in solution form, by increasing membrane permeability and affecting membrane integrity. Complex interactions occur between surfactants and proteins and thus there is the possibility of a surfactant-induced alteration of drug metabolizing enzyme activity. There has also been the suggestion that surfactants may influence the binding of the drug to the receptor site. The determinants of the effectiveness of surfactants on drug absorption are several. Drugs in which dissolution and not membrane transport is the rate-limiting step in absorption and drugs in which the latter is the rate-determining step may be affected differently. Water-soluble drugs will not and water-insoluble drugs will interact with surfactant micelles thus high concentrations of surfactants are likely to affect lipophilic and hydrophobic drugs to differing degrees. Some surfactants have direct physiological activity of their own and in the intact animal can thus affect the physiological environment, e.g. by altering gastric residence time such that without physico-chemical intervention, a surfactant-effect will be seen. It is only possible to isolate some of these effects and to examine the effect of surfactants in each. Studies in whole animals have sometimes given what appear to be contradictory results.

Numerous studies on the influence of surfactants on drug absorption have shown them to be capable of increasing, decreasing, or exerting no effect on the transfer of drugs across biological membranes [1]. Perhaps the earliest report of the effect of soap on drug activity is that of Billard and Dieulafe [2], who noted that the toxic effect of curare injected intraperitoneally into guinea-pigs could be increased by the addition of low concentrations of soap and decreased by high concentrations. This biphasic action of surfactants has been noted several times

388

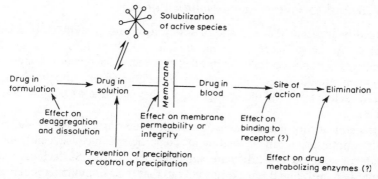

Figure 7.1 Possible sites of surfactant influence on drug absorption and activity. Utilization of a drug involves its release from the formulation, its solution in the body fluids, and its passage through barrier membranes into the systemic blood stream before transport into tissues and eventual arrival at the target organ. Release of poorly soluble drugs from tablets and capsules for oral use may be increased by the presence of surfactants, which may decrease the aggregation of the drug particles and therefore increase the area of particle available for dissolution. The lowering of surface tension may also be a factor in aiding the penetration of water into the drug mass; this wetting effect is operative at low concentrations. Above the critical micelle concentration (CMC) the increase in the saturation solubility of the drug substance by solubilization in the surfactant micelles can result in more rapid rates of drug solution. Where dissolution is the rate-limiting step in the absorption process, as it is with many poorly soluble drugs, an increase in rate of solution will increase the rate of drug entry into the blood and may affect peak blood levels. Very high concentrations of surfactant can decrease drug absorption by decreasing the chemical potential of the drug. This results when surfactant is present in excess of that required to solubilize the drug.

since, but nonetheless the literature tends to be confused. The observed influences of surfactants depend on the concentration of the agent used (which is difficult to assess when the formulation has been administered to man or intact animal) and even in model systems this leads to complications in elucidating effects especially when the surface-active agent exerts several actions simultaneously. Much of the confusion in the literature on this subject arises from discussion of the influence of different concentrations of surfactant, and from attempts to generalize on the action of varied surfactants on many different types of biological membrane. As with the physical effects noted above, distinct changes in the activity of the surfactant can frequently be observed on increase of surfactant concentration. This can be demonstrated by experiments in model systems, for example, in goldfish immersed in solutions of drug and surfactant [3–5]. Low concentrations of polysorbate 80 increase the absorption of secobarbitone; concentrations above the CMC decrease absorption. Similarly, the influence of surfactant structure and properties on drug absorption can also be demonstrated with the goldfish; some of these experiments will be discussed later in this chapter.

7.2 Effect of surfactants on dissolution of drugs

It is readily apparent that the rate of solution of poorly soluble drugs can be increased by the presence of surfactants in the dissolution medium. Most experiments have been carried out *in vitro*; the effect *in vivo* is more complex with the concomitant dilution of the surfactant by a complex medium, the absorption of the surfactant itself and the adsorption of other substances onto the dissolving particles.

Surfactant adsorption on to hydrophobic drug particles below the critical micelle concentration can aid wetting of the particles and consequently increase the rate of solution of particulate agglomerates [6–10]. Surfactants may be incorporated into solid dosage forms [11] so that their solubilizing action comes into play as the disintegration process starts and water penetrates to form a concentrated surfactant through lowering of surface tension solution around the drug particles or granules. Both facilitation of wetting and solubility increase will aid dissolution of the drug. Finholt and Solvang's results [12] on the dissolution *in vitro* of phenacetin and phenobarbitone in the presence of polysorbate 80 show clearly the influence of surface tension (Fig. 7.2). The solubility of phenacetin is little affected by the concentrations of polysorbate 80 used and thus enhanced wetting is the primary cause of improved dissolution rates, a result in accord with the finding that sodium lauryl sulphate (NaLS) increased the rate of solution of salicylic acid from compressed tablets owing to better solvent penetration into the tablets and granules [13]. Finholt and Solvang [12] determined the pH and surface tension of gastric juice from 27 patients. Surface tension ranged between 35 and 50 mN m^{-1} and pH between 1 and 7.5, and was independent of secretion rate. Such are the complications of the *in vivo* environment and the problems of determining the effect of synthetic surfactants on dissolution rates *in vivo*; the rate of solution of a drug such as phenobarbitone is significantly higher in diluted gastric juice than in 0.1 N HCl because of the

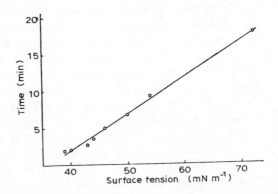

Figure 7.2 Relationship between the surface tension of the dissolution medium and the time necessary for dissolution of 100 mg phenacetin. Dissolution media: 0.1 N HCl containing different amounts of polysorbate 80. From Finholt and Solvang [12] with permission.

difference in surface tension. In addition, the amount of a soluble salt such as phenobarbitone sodium dissolved in diluted gastric juice at 1 h has been shown to be considerably increased, presumably because the precipitation of the free acid is reduced by components in gastric fluid. Nevertheless increased absorption of paracetamol has been observed *in vivo* [14]. Enhanced absorption of digoxin and digitoxin [15] and sulphadiazine and sulphisoxazole [16] have been ascribed to increased dissolution rates of these drugs brought about by the incorporation of surfactants into the formulation. The effect of poloxamer 188 and dioctyl sulphosuccinate (DOSS) on absorption of sulphisoxazole from rat intestinal loops is shown in Table 7.1, and the influence of these surfactants on dissolution rate shown in Fig. 7.3. Poloxamer 188 and DOSS are both used below their critical micelle concentrations, at concentrations likely to be found *in vivo* where they are used as faecal softeners in laxative products. In some systems negligible effects are noted below the surfactant CMC. Such is the case with hydrocortisone [17]; neither polysorbate 80 nor two Solulan surfactants (Solulan 25 and 16, American Cholesterol Products Inc., USA) increased the dissolution rate of this steroid until their respective CMCs were exceeded. However, the solubility of hydrocortisone was increased much less than the increased solution rate would imply suggesting that the solubility increase was not of major importance in this case. Short *et al.* [8] have also considered the effect of surfactant on hydrocortisone dissolution. An increased dissolution rate constant below the CMC of polysorbate 80 is observed, this decreasing just above the CMC; Short *et al.* suggest that this might be related to a surface tension effect, the maximum in dissolution rate constant coinciding with the surface tension minimum of the polysorbate. A minimum surface tension around the CMC value implies the presence of surface-active impurities [18] which may adsorb preferentially on the drug particles decreasing dissolution rate.

Concentrations of polysorbate 20 well in excess of the CMC have been used by Collett and Rees in their studies on salicylic acid dissolution [10, 19]. Dissolution rates were measured over a pH range from 1.0 to 4.0; the dissolution rate increases very slowly above 12% surfactant (Fig. 7.4) but there was no evidence of a decreased dissolution rate such as found by Parrott and Sharma [20], e.g. above

Table 7.1 Effect of poloxamer 188 and dioctyl sodium sulphosuccinate on the absorption of sulphisoxazole from rat intestinal loops*

Surfactant	Concentration, % w/v	Dose absorbed, % ± S.D.
Control	—	45.3 ± 6.5
poloxamer 188	0.01	56.1 ± 3.9
	0.10	57.3 ± 10.1
Dioctyl sodium	0.01	53.9 ± 9.4
sulphosuccinate	0.10	55.0 ± 8.4

* Values represent mean of 6 animals.
From [16].

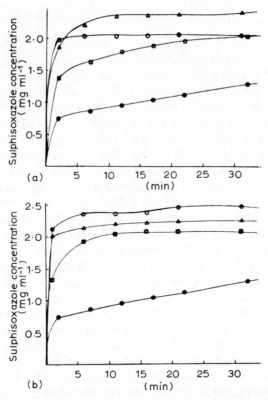

Figure 7.3(a) Effect of poloxamer 188 on sulphisoxazole dissolution ●, control; □, 0.001 %; △, 0.01 %; and ○, 0.1 %. (b) Effect of dioctyl sodium sulphosuccinate on sulphisoxazole dissolution. ●, control; □, 0.001 %; △, 0.01 %; and ○, 0.1 %. From Reddy *et al.* [16] with permission.

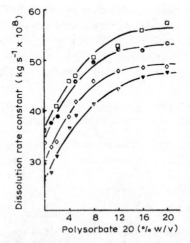

Figure 7.4 Plot of dissolution rate constants ($kg\,s^{-1} \times 10^8$) of salicylic acid against concentration of polysorbate 20 at several pH values ▼pH 10, ◇pH 2.0,● pH 3.0, □pH 4.0. From Rees and Collett [10] with permission.

12% polysorbate 80 with benzoic acid. Collett and Rees [19] suggest that the decreased dissolution rates are not a function of the viscosity of the dissolution medium but rather an artefact due to lack of pH control in the system, the decreased pH resulting from the dissolution of benzoic acid leading to decreased solubility and thus solution rate. However, such an explanation cannot be put forward to discuss the decreased rate of solution of griseofulvin [21] at high concentrations of non-ionic surfactant.

7.2.1 Theoretical approaches to dissolution rates in high concentrations of surfactant

Higuchi [22] has analysed the dissolution process in the presence of micellar solutions. His equations predict that the effect of surfactant on dissolution rate will be less than predicted by the Noyes–Whitney equation on the assumption of increased bulk solubility. The Noyes–Whitney relation in the form

$$\frac{dc}{dt} = kA(c_s - c) \tag{7.1}$$

shows the rate of change of concentration of solute, c, related to its surface area, A, and its saturation solubility, c_s. When $c_s \gg c$ there is a direct proportionality between the rate of solution, dc/dt and c_s. The studies discussed above have shown that this is frequently not observed, as clearly demonstrated in Fig. 7.5.

Higuchi [24] assumes that an equilibrium exists between the solute and the solution at the solid–liquid interface and that the rate of movement of solute into the bulk is governed by the diffusion of the free and solubilized solute across a stagnant diffusion layer. Drugs solubilized in micelles will have a lower diffusion coefficient than free drug so that the effect of additive on dissolution rate will be related to the dependence of dissolution rate on the diffusion coefficients of the diffusing species, and not to their solubilities, as suggested by simple interpretation of Equation 7.1. The effective diffusion coefficient (D_{eff}) is given by [24]:

$$D_{eff} = \frac{D_f c_f + D_m c_m}{c_s + c_m}, \tag{7.2}$$

subscripts f and m referring, respectively, to the free and micellar drug; c_m is thus the *increase* in solubility due to the micellar phase. This leads to the following equation for dissolution of a solid at constant area A and under sink conditions, i.e. $c_s \gg c$,

$$\frac{dc}{dt} = \left[\frac{D_f c_f}{h} + \frac{D_m c_m}{h}\right] \tag{7.3}$$

where h is the diffusion layer thickness. Substituting Equation 7.2 into Equation 7.3 gave, where c_t is the total solute concentration,

$$\frac{dc}{dt} = D_{eff} c_t / h. \tag{7.4}$$

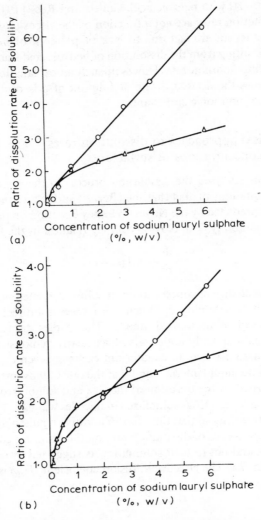

Figure 7.5(a) Ratio of dissolution rates and solubilities of sulphamethizole in surfactant solution to those in distilled water. (b) Ratio of dissolution rates and solubilities of sulphadiazine in surfactant solution to those in distilled water.

△: ratio of dissolution rate constant.

○: ratio of solubility.

From Watari and Kaneniwa [23] with permission.

However, both Collett and Rees [19] and Gibaldi *et al.* [25] find that dissolution rate is proportional to the effective diffusion coefficient raised to the power 0.5 to 1.0, thus placing in some doubt the diffusion coefficients of salicylic acid calculated assuming Equation 7.4 to hold [20]. The lack of agreement between the dissolution data and the predictions of Equation

7.4 leads to the conclusion that alternative models are required. A 'film-penetration' model incorporating the surface renewal concepts of Danckwerts [26] has been proposed [27]. In this, mass transfer from the surface is believed to occur by two simultaneous processes – one involving a stagnant film in which steady state molecular transfer occurs, and the other encompassing non-steady state mass transfer by eddy formation in the surface layer. The film-penetration model predicts a dependence of dissolution rate on diffusion coefficient with an exponent between 0.5 and 1.0 [25, 27].

Predictions of dissolution rate may be made using diffusion coefficients of the solutes in their solubilized state by applying the Stokes–Einstein equation.

$$D = \frac{RT}{6\pi\eta N_A} \sqrt[3]{\left(\frac{4\pi N_A}{3M\bar{v}}\right)}, \tag{7.5}$$

where D is diffusion coefficient, R is the molar gas constant, T is the absolute temperature, η is the viscosity of the solvent in poise, $\bar{v}$ is the partial specific volume of the micelles, M is the micellar molecular weight, and N is Avogadro's number. More direct measurements of D_m are now possible by photon correlation spectroscopy and this should lead to a better analysis of dissolution models for solubilizing systems.

Elworthy and Lipscomb [28] considered dissolution to consist of two processes occurring simultaneously:

(1) a zero order reaction for the transfer of griseofulvin molecules from the solid surface into the solution, with rate constant k_1;
(2) a first order reaction for the deposition of solute from solution to solid surface, with rate constant k_2.

The rate of increase of concentration in solution:

$$\frac{dc}{dt} = k_1 - k_2 c. \tag{7.6}$$

The solution to this equation with the condition that at $t = 0$, $c = 0$ is

$$c = \frac{k_1}{k_2}(1 - e^{-k_2 t}). \tag{7.7}$$

Expanding the exponential term and rearranging gives

$$\frac{c}{t} = k_1 - \frac{k_1 k_2 t}{2} + \frac{k_1 k_2^2 t^2}{6} - \frac{k_1 k_2^3 t^3}{24} + \ldots$$

At fairly early times in the dissolution process, terms in t^2 and t^3 etc. can be neglected giving:

$$\frac{c}{t} = k_1 - \frac{k_1 k_2 t}{2}. \tag{7.8}$$

A plot of c/t versus t will have an intercept k_1, and a slope $k_1 k_2/2$, enabling both constants to be evaluated. Trial calculations show that Equation 7.8 gives 1%

error in c compared to the exact Equation 7.7 provided that the $k_2 t$ term does not exceed 0.25.

Equation 7.8 reduces to the Noyes–Whitney equation. When equilibrium is reached, i.e. a steady state between dissolution and redeposition,

$$\frac{dc}{dt} = 0 = k_1 - k_2 c_s,$$

where c_s is the saturation solubility,

$$c_s = k_1/k_2, \tag{7.9}$$

and from Equation 7.7

$$c = c_s(1 - e^{-k_2 t})$$

or,

$$k_2 = \frac{1}{t} \ln\left(\frac{c_s}{c_s - c}\right), \tag{7.10}$$

which is the more usual form of the Noyes–Whitney equation. The rate constant of Equation 7.6 thus appears to be the first order constant arising in the consideration of the dissolution–redeposition process. Equation 7.8 is useful if the saturation solubility is not known; when it is, Equation 7.9 can be used to evaluate one constant when the other has been determined from Equation 7.8 or 7.10.

A result of this analysis is shown in Fig. 7.6 for the cetomacrogol–griseofulvin system [29]. The considerable effect of stirring rate on the dissolution rate of the powdered drug is seen, leading to the conclusion that it is necessary to choose

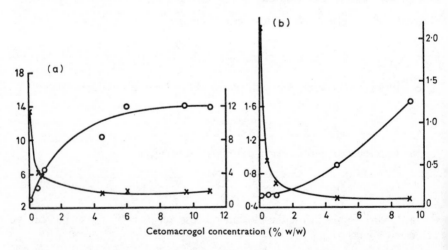

Figure 7.6 Effect of cetomacrogol concentration on k_1 (O) and k_2 ($\times$) at a stirring rate of (a) 200 rev min^{-1} (b) 60 rev min^{-1}. Left hand ordinates $10^7 k_1$. Right hand ordinates $10^3 k_2$. The solute griseofulvin, is in powdered form. From Elworthy and Lipscomb [29] with permission.

carefully the rate of stirring in attempts to obtain *in vitro–in vivo* correlations. It has been found [30] that *in vitro* rates of methyl prednisolone, for example, correlated with *in vivo* absorption rates only when the rate of stirring employed in the dissolution test was low.

It seems likely [28] that the presence of surfactants facilitates the transfer of drug molecules from the crystal surface into solution as the activation energy for this process was found to be lower in surfactant than in water. In the case of k_2, the activation energy increases in the surfactant solution which probably reflects the viscosity increase and also the possibility that a layer of adsorbed surfactant molecules interferes with the redeposition process.

Chan *et al.* [31] have presented a theory of solubilization kinetics and its relation to the flow of dissolution medium, based on an analysis of five steps depicted in Fig. 7.7. Surfactant molecules diffuse to the surface as micellar species (step 1). These molecules are adsorbed on the surface of the solid (step 2) and on the surface the surfactant and solubilizate form a mixed micelle (step 3). In step 4 the mixed micelle is dissolved and it diffuses away into the bulk solution in the last step (step 5). The solubilization rate is assumed to be controlled by steps 4 and 5 in Fig. 7.7. If these steps are rate controlling

$$\frac{d[M]}{dt} = k_i A[M_i] \tag{7.11}$$

where $[M]$ is the concentration of mixed micelles in the bulk solution, and $[M_i]$ is the concentration of micelles at the interface. A is the surface area per volume, k_i is the forward reaction rate constant for step i.

$$\frac{d[M_i]}{dt} = k_4[M_s] - k_{-4}[M_i][S] - k_5 A[M_i] = 0 \tag{7.12}$$

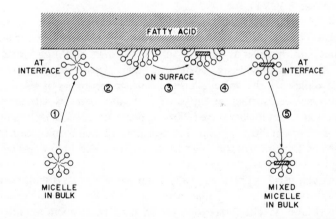

Figure 7.7 Schematic mechanism for initial solubilization. Mixed micelle desorption and diffusion (steps 4 to 5) are assumed to control stearic acid solubilization. From Chan *et al.* [31].

where $[M_s]$ is the concentration of mixed micelle on the surface and $[S]$ is the number of free sites for micelle adsorption

$$[M_s] = K_3[B_s] \tag{7.13}$$

$$[B_s] = K_2[B][S] \tag{7.14}$$

$$[S_0] = [S] + [B_s] + [M_s]. \tag{7.15}$$

$[B]$ is the concentration of surfactant micelles in bulk, $[B_s]$ at the surface and $[B_i]$ in the interface. $[S_0]$ are the total number of sites in the surfaces. K_i is the equilibrium rate constant for step i.

Combining Equations 7.11 to 7.15 we obtain

$$\frac{d[M]}{dt} = \frac{\{k_4 K_3[S_0]/(1+K_3)\}[B]}{\{k_{-4}[S_0] + k_5 A/k_5 AK_2(1+K_3)\} + [B]}. \tag{7.16}$$

$d[M]/dt$ is difficult to measure. It is assumed that the solubilizate concentration $[F]$ is proportional to $[M]$ and that $d[F]/dt \propto d[M]/dt$.

Obtaining $[F_{sat}]$ and $[B]$ by experiment, Equation 7.16 can be rewritten in the form,

$$\left(\frac{d[F]}{dt}\right)^{-1} = \left\{\frac{1+K_3}{nk_4 K_3[S_0]}\right\}$$

$$+ \left\{\left(\frac{[F_{sat}]}{nK_2 K_3[B]}\right)\left[\frac{1}{k_4[S_0]} + \frac{1}{k_5 AK_4}\right]\right\}\frac{1}{[F_{sat}]}. \tag{7.17}$$

This equation predicts that, providing steps 4 and 5 are rate controlling, a plot of $(d[F]/dt)^{-1}$ versus $[F_{sat}]^{-1}$ will be linear; the intercept of the plot is independent of k_5 and hence independent of flow; the slope of the plot is flow dependent, being dependent on k_s. In experimental studies of fatty acid dissolution into NaLS solutions the validity of the first two predictions was established (see Fig. 7.8).

The model on which the above derivations are based is by no means unequivocal. There is no proof that micelles diffuse to the surface and adsorb, or, indeed, that hemi-micelles as depicted in Fig. 7.7 form, although Somasundaran et al. [32] have previously postulated their existence. The transfer of solute molecules to the micelle at the surface probably involves complex interactions between surfactant, fatty acid and water perhaps with liquid crystal formation as an intermediate stage following penetration of surfactant molecules. As the earlier steps in the process are not rate limiting their formulation is perhaps less important. Diffusion of the solubilizate-laden micelle is a process which must occur.

Higuchi's analysis [24] predicts that substantial effects on dissolution rate will only be evident when the drug concentration in solution approaches or exceeds saturation solubility. The dissolution model used by Higuchi assumes that an equilibrium exists between the solid and the solution at the interface and that the rate is controlled by the diffusion of free and solubilized solute across the diffusion layer which has a thickness δ.

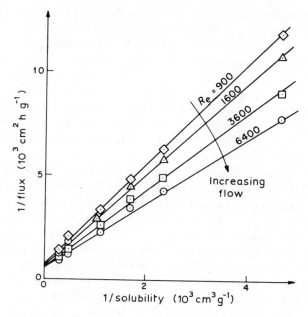

Figure 7.8 Solubilization kinetics of stearic acid. These data support the hypothesis that mixed micelle desorption and diffusion are rate controlling. From Chan *et al.* [31] with permission. R_e is the Reynolds number.

Provided sink conditions obtain (i.e. $c < 0.1 c_s$);

$$dc/dt = A[(Dc_s/\delta) + (D_m c_m/\delta)] \tag{7.18}$$

where c_m is the increase in solubility due to the surfactant and D_m is the diffusion coefficient of the drug in the micelle, it being assumed that δ is the same for both.

7.2.2 Dissolution from drug–surfactant mixtures

The work on dissolution rate, rather than solubility, tends to be of rather academic interest as a drug is rarely to be found dissolving into concentrated surfactant solutions. It is of more practical interest to consider dissolution from intimate mixtures of drugs and surfactants into water [34]. Application of the technique of formation of solid dispersions by fusing poorly soluble drugs with water-soluble carrier has been shown to increase the solution rate of drugs; carriers used include polyoxyethylene glycol, polyvinylpyrrolidone [34] but also surfactants [33, 35] in their solid or waxy state. The enhanced rate of dissolution of testosterone [35] from Myrj 51 (but also from polyoxyethylene glycol 1000 and PVP 11500 dispersions) was attributed to the small particle size of the drug in the solidified melt and to a lesser degree to the increased solubility in the carrier solution which formed. Ford and Rubinstein [33] made a more detailed study of a glutethimide–non-ionic surfactant system using Renex 650, a nonylphenyl-polyoxyethylene condensate. Phase diagrams showed the presence of a eutectic at 21% of the drug, 79% surfactant with a eutectic temperature of 35° C. Solid solutions of the drug in the surfactant and of Renex in the drug also existed. When placed in water, drug and carrier do not dissolve at rates directly proportional to their concentration in the dispersion and the dissolution rate of the drug is maximal when the drug concentration reaches about 25% in the disc (Fig. 7.9). Dissolution of digitoxin from co-precipitates of the drug with poloxamer 188 or deoxycholic acid has been shown to be enhanced over dissolution from physical mixtures and administration of the co-precipitates to mice significantly increased the oral toxicity [15] (Table 7.2).

Other techniques involving attempts to utilize the properties of surfactants have included crystallization of poorly soluble drugs such as sulphathiazole, prednisone and chloramphenicol in the presence of small amounts of surfactants [36]. Increases in the rate of solution were observed in each case when polysorbate was used as a 2.5% solution as the crystallization medium. While the result might be partly ascribed to adsorption of surfactant molecules on to the hydrophobic crystal surface, differential thermal analysis also suggests that some surfactant is incorporated into the crystal structure. Interference of a surfactant in the crystallization process could lead to defect formation. Model studies with

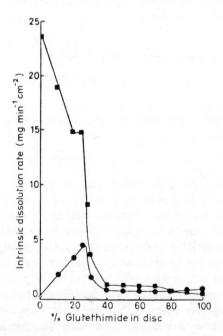

Figure 7.9 Dissolution rate–composition profile. Effect of glutethimide-Renex composition on the intrinsic dissolution rates of 1 h old resolidified melts into distilled water at 30°C. ■ Renex 650. ● Glutethimide. From Ford and Rubinstein [33].

Table 7.2 Oral toxicity of various digitoxin preparations in mice*

Test system	Number of animals dead[†]	Mortality (%)
Digitoxin	6	20
Digitoxin–poloxamer 188[‡] co-precipitate	29	97
Digitoxin–deoxycholic acid[‡] co-precipitate	30	100
Digitoxin–poloxamer 188[‡] physical mixture	11	37
Digitoxin–deoxycholic acid[‡] physical mixture	9	30
Poloxamer 188[§]	0	0
Deoxycholic acid[¶]	0	0

* A dose of 70 mg of digitoxin/kg was administered as a suspension in 0.5% methylcellulose. Thirty animals were used for each test system. † Animals were observed for 7 days post-administration. ‡ A 700 mg/kg dose was administered containing 10% (w/w) digitoxin. § A 2.7 g/kg dose was used. ¶ A 630 mg/kg dose was used.
From [15].

adipic acid have shown that surfactant adsorption on to growing crystal faces can change crystal habit [37, 38] (see Chapter 9).

7.3 Effect of surfactants on membrane permeability

Before we discuss some of the work which has been carried out on surfactant effects on drug absorption in whole animals, we review in this section some of the work which has been done using model systems. Foremost amongst these has been the goldfish *Carassius auratus*. In choosing this system Levy *et al.* [39] explain: 'Most of the studies of surfactant effects on drug absorption have been carried out on microbial systems. The results thus obtained may have limited applicability to multicellular organisms, since the latter are able to maintain homeostasis much more effectively. Moreover, the presence of enzymes and other vital cell constituents in the cell membrane makes unicellular organisms particularly sensitive to direct effects of surfactants.'

Use of small animals or humans presents great difficulties, not the least being the difficulty of maintaining a constant, known concentration of surface-active agent and drug. The major advantage of the fish system is that large quantities of test solution can be used, permitting the maintenance of constant concentration gradients across the membranes, which behave, as far as passive diffusion characteristics are concerned, in a similar way to human membranes. Fig. 7.10 shows the effect of polysorbate 80 on the time of death of goldfish immersed in sodium secobarbitone solution. The results show an enhancement of activity of the barbiturate at low concentrations and a decrease at higher concentrations, in common with other studies using alternative systems.

The end point in the experiment is the turnover time or death time of the fish.

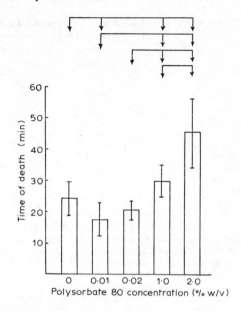

Figure 7.10 The effect of polysorbate 80(I) on the time of death of goldfish immersed in 0.02 % sodium secobarbitone solution at pH 5.9 and 20° C. Mean values of 10 fish are shown. Vertical bars indicate ± 1 standard deviation. Arrows connect values which differ significantly ($p < 0.05$) from one another. From Levy *et al.* [39].

The reciprocal death time (T^{-1}) is proportional to the rate of absorption of the drug, k_1

$$\frac{1}{T} = k_1 c_B/c_F - k_2/2, \qquad (7.19)$$

where c_B and c_F are the concentrations in the bathing solution and the threshold concentration in the fish, respectively, and k_2 is the rate of elimination of the drug.

A range of non-ionic surfactants has been studied for their effect on absorption of drugs in goldfish. Not all surfactants do increase absorption [40–42] some exhibiting only an inhibiting effect as seen in Fig. 7.11. Three main types of activity have been noted [43] when surfactant concentration is increased (Fig. 7.12), namely (a) the increase and decrease depicted in Fig. 7.12 when a drug is solubilized in the surfactant micelles (e.g. thioridazine–Renex 650 mixtures); (b) an overall decrease in activity when solubilization occurs, the surfactant having no influence on membrane permeability (e.g. thioridazine–Cremophor EL 120 (Fig. 7.12b), and (c) (Fig. 7.12c) an overall increase in activity when the surfactant increases the flux through the membrane and the drug is not associated with the micelles (e.g. paraquat–non-ionic surfactant systems) [44].

In some systems where the drug concerned interacts to a small degree with a surfactant which has a significant effect on permeability, only the increase in absorption is detectable. This is the case with thiopentone and a series of non-

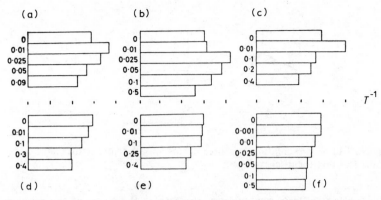

Figure 7.11 Absorption of thioridazine in goldfish in the presence of increasing concentrations of various non-ionic detergents, the rate of absorption being proportional to the reciprocal of the death time of the fish, reciprocal death time is plotted on the ordinate concentrations of surfactants (% w/v) are marked. Lack of enhancement of absorption by some surfactants is probably due to poor ability to penetrate lipid membranes because of shape factors. Decrease in absorption is due to non-ionic micelle formation. From Florence and Gillan [41] with permission. The surfactants are all Atlas products (Honeywill-Atlas, UK).
(a) Atlas G2162 (II); (b) Renex 650 (III); (c) Atlas G1790; (d) G1295 (IV); (e) G1300 (IV); (f) Cremophor EL.

Polysorbate 80

$$
\begin{array}{l}
\text{CH}_2\text{---} \\
\text{HCO (CH}_2\text{CH}_2\text{O)}_x\text{H} \\
\text{H(OCH}_2\text{CH}_2\text{)}_y \text{ OCH} \\
\text{HC---} \\
\text{HCO(CH}_2\text{CH}_2\text{O)}_z\text{H} \\
\text{H}_2\text{CO(CH}_2\text{CH}_2\text{O)}_w\text{OCR}
\end{array}
$$

$x + y + z + w = 20$

(I)

G2162

$$
\begin{array}{l}
\text{CH}_3 \\
\text{RCOOCH}_2\text{---CHO(CH}_2\text{CH}_2\text{O)}_{25}\text{H}
\end{array}
$$

(II)

Renex 650

R
⬡
O(CH$_2$CH$_2$O)$_{30}$H

(III)

G1295 and G1300

$$
\begin{array}{l}
\text{CH}_2 \text{ O(CH}_2\text{CH}_2\text{O)}_x\text{R} \\
\text{CH} \text{ O(CH}_2\text{CH}_2\text{O)}_y\text{R} \\
\text{CH}_2 \text{ O(CH}_2\text{CH}_2\text{O)}_z\text{R}
\end{array}
$$

(IV)

for G1295 $x + y + z = ca$ 150

for G1300 $x + y + z = ca$ 200

ionic surfactants studied in goldfish [45] using mean reciprocal overturn time as an index of the rate of absorption. Some results are shown in Table 7.3.

There are several competing mechanisms for surfactant-induced effects when solid oral dosage forms are administered. When solutions are administered,

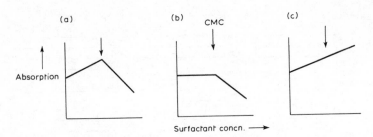

Figure 7.12 Representation of three forms of absorption–surfactant concentration profile (see text for discussion).

Table 7.3 Reciprocal turnover times (min^{-1}) ($\pm$ S.D.) of thiopentone in presence of surfactants

Surfactant	HLB	0.0005 %	0.1 %
None	—		0.09 $\pm$ 0.03
POE (4) lauryl ether	9.7	0.11 $\pm$ 0.02	—*
POE (10) lauryl ether	12.0	0.54 $\pm$ 0.02	—*
POE (23) lauryl ether	16.9	0.25 $\pm$ 0.09	0.18 $\pm$ 0.02
POE (2) stearyl ether	4.9	0.07 $\pm$ 0.01	(0.07 $\pm$ 0.03)†
POE (10) stearyl ether	12.4	0.15 $\pm$ 0.02	0.41 $\pm$ 0.07
POE (20) stearyl ether	15.3	0.21 $\pm$ 0.01	0.31 $\pm$ 0.02
POE (2) oleyl ether	4.9	0.11 $\pm$ 0.02	(0.11 $\pm$ 0.01)†
POE (10) oleyl ether	12.4	0.17 $\pm$ 0.03	0.34 $\pm$ 0.14
POE (20) oleyl ether	15.3	0.19 $\pm$ 0.04	0.29 $\pm$ 0.04

* At this concentration these surfactants were toxic. † Cloudy dispersion.
From [45].

provided the drug is maintained in solution in the gut and is not solubilized in the surfactant micelles, the only effect will be that of the surfactant on membrane permeability, if the surfactant does not itself alter the physiological status of the GI tract. When a drug is partly in solution, as in a suspension, the results are perhaps perplexing.

Fig. 7.13 illustrates the complexity of surfactant effects. It shows the influence of increasing surfactant concentration on drug absorption at several pH values. At pH 7.4 (where all the drug is in solution) increasing the concentration of polysorbate 80 from 0.01 to 0.1 % decreases the absorption, as the saturation of the system is reduced by incorporation of free drug in the micellar reservoirs. On increase of pH to 8.6 and 9.0 there is a decrease in absorption of drug as some of the drug is now in its insoluble form and has precipitated from solution. However, at these pH values increasing the surfactant concentration increases the rate of absorption, as solubilization increases the solubility of the drug and thereby increases its concentration gradient across the membrane [41].

There is no simple explanation of the absorption-promoting effect of the surfactant. Penetration of the surfactant into the liquid membrane seems to be

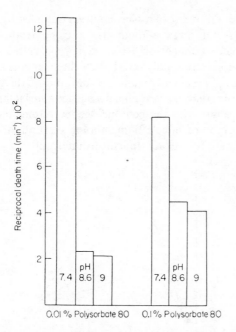

Figure 7.13 The influence of pH and polysorbate 80 concentrations on the absorption of solutions or suspensions containing 0.08 % thioridazine as shown by reciprocal death times of goldfish. From Florence and Gillan [41] with permission.

one step in the action as the interpolation of a foreign hydrocarbon chain certainly would result in an increase in the fluidity of the hydrocarbon interior of the membrane. This should lead to decreased resistance to passage of solutes through the membrane. In some experiments it appears that there is a decrease in permeability at higher surfactant concentrations which is not fully explained by the interaction of the permeant with the surfactant micelles. This suggests that a physical blocking mechanism is operating, perhaps in the manner suggested by Smith *et al.* [46] to explain the decreased penetration of pesticides into plants. Kameda *et al.* [47] have also noted inhibition of the absorption of species which did not interact with the micellar phase of polysorbate 80.

7.3.1 Influence of surfactant structure on membrane permeability

Until more is known of the molecular interactions of surfactants with membrane components and the factors controlling such interactions it will remain virtually impossible to predict which surfactants will be capable of enhancing permeability of membranes without causing damage. Careful choice of solubilizer both of appropriate structure and optimum concentration is obviously paramount. In experiments on goldfish the effects of polyoxyethylene non-ionic surfactants on the absorption of various barbiturates is dependent on the surfactant hydrophobic and hydrophilic chain lengths and possibly also on the size of the surfactant molecule. The range of molecular areas obtained from surface-tension measurements was not large enough for a categorical statement on the importance of surfactant dimension. However, it appears that surfactants having C_{12}–C_{16} hydrocarbon chains, polyoxyethylene chain lengths between 10 and 20, and molecular areas of between 1.00 and 1.60 nm^2 induce the greatest increase in absorption. Fig. 7.14 reveals that the effectiveness of a surfactant depends also on the solute, as the results with thiopentone, secobarbitone and phenobarbitone show a different order of effect. If membrane disruption occurs as well as increased fluidity these results are difficult to interpret.

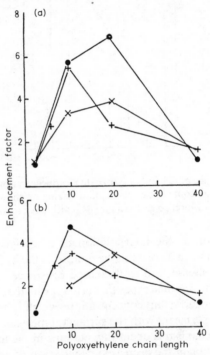

Figure 7.14 The mean effect of two groups of nonionic surfactants: (a) of the Brij 50 series and (b) of the Brij 70 series compared in goldfish ($n = 6$) on the absorption of ● thiopental, + secobarbital, and × phenobarbital. All surfactants at 0.1 % level, from [53].

Experiments in whole animals are even more difficult to decipher. The relative effects of different surfactants on intestinal absorption have been studied recently by Whitmore *et al.* [48] using the everted sac preparation from rat small intestine. A range of anionic, cationic and non-ionic surfactants were used and a relationship was found between the absorption of salicylates and L-valine and the release of protein and phospholipid from the preparation as a result of membrane disruption. Sodium lauryl sulphate increased the rates of uptake of salicylate and L-valine but cetyltrimethylammonium bromide had no effect on L-valine absorption and in fact decreased salicylate absorption. The non-ionic surfactant on the other hand, increased valine absorption but had no effect on salicylate transfer. CTAB has been shown before to decrease absorption of glucose, methionine and acetylsalicylic acid [49, 50]. Whitmore's data showed that CTAB appeared to prevent the loss of a protein of molecular weight of about 39 000 which is perhaps crucial in determining permeability. But it is almost impossible to obtain a consistent picture of action from different publications; one reason is that concentrations of surfactants studied are so different and the complications of surfactant interactions with drugs and with tissue components can obscure mechanisms of action. This is especially true when intact animals are used, although these are the most crucial test of solubilizer effects in relation to clinical bioavailability and to the use of solubilizers in pharmaceutical systems. One type of surfactant effect on drug absorption that can occur which would only be detected *in vivo* is the indirect effect such as that reported recently [51]. An increase in the absorption of tripalmitate by a detergent was attributed to the increase in gastro-intestinal motility induced by the surfactant.

Isolated tissue work should not suffer from such complications. In measuring the effect of a range of alkyl polyoxyethylene ethers on paraquat transfer across rat stomach epithelium, Walters *et al.* have found [44] no simple correlation between surfactant structure and transport. However, with a group of surfactants such as these with a given polyoxyethylene chain length a certain dependency on alkyl chain length can be seen (Fig. 7.15).

Permeability changes have been observed in reconstituted cell 'membranes' following treatment with surface-active agents [52] producing a selective permeability for cations. Addition of surfactant lowered the initially high resistance by several decades.

Investigation of the interaction of polyoxyethylene alkyl ethers with cholesterol monolayers [53] interestingly reveals a biphasic effect when surface pressure

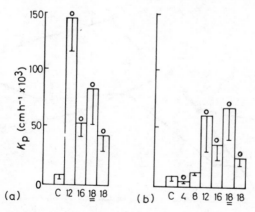

Figure 7.15 Values of K_p for paraquat obtained at 1.0 % surfactant levels with isolated rabbit gastric mucosa as a function of alkyl chain: C_{12}, C_{16}, C_{18} and oleyl, marked $\underline{18}$ on the abscissa. (a) compounds with 10 ethylene oxide units and (b) compounds with 20 ethylene oxide units. Results which are statistically significantly different ($P < 0.05$) from the control values without surfactant are marked (O). From [44] with permission.

is measured as a function of surfactant concentration (Fig. 7.16). Increasing penetration of the surfactant into the cholesterol monolayer causes increased surface pressure; at higher concentrations solubilization of the cholesterol molecules results in a decrease in surface pressure. Seeman [54, 55] has clearly demonstrated – with surface-active drug molecules – stabilization of erythrocyte membranes at low drug concentrations and labilization of the membrane at higher concentrations. We did not find [53] that decreased surface pressures of cholesterol monolayers coincided with the surfactant CMC, suggesting perhaps that micelle formation was occurring in the interfacial region and solubilization taking place in the concentrated interfacial surfactant layer (Fig. 7.16). While penetration and labilization of the membrane are undoubtedly factors in enhanced permeability, some workers have implicated solubilization of membrane components [48]. The release of protein and phospholipid from rat jejunal tissue has been related to the absorption of salicylate and L-valine (Figs 10.5(a) and (b)). Fig. 7.17 shows the extent of membrane protein extraction from rat gastric mucosa incubated with three surfactants. Of these, only Brij 76 and 78 had any significant effect on permeability. It is perhaps relevant here to discuss work which was carried out in a biochemical context aimed at selective solubilization of membrane components for further study. Since, as we have just discussed, simple penetration is unlikely to be involved in permeability enhancement, these results may illuminate the problem.

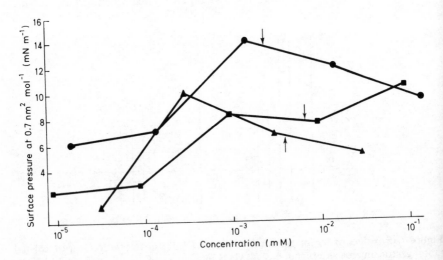

Figure 7.16 Plot of surface pressure of cholesterol monolayers at constant molecular area against surfactant concentration. ▲ Brij 72; ● Brij 76; ■ Brij 78. Arrows denote the CMC for each surfactant. From K. A. Walters *et al.* [53].

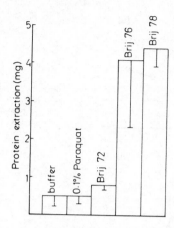

Figure 7.17 Protein extraction (mg) in bovine serum albumin equivalents occurring in 3 h following incubation in the medium as marked. Surfactant concentration 0·1 %.

The solubilization of mitochondrial components by non-ionic Tritons was found to depend on the ethylene oxide chain length; longer-chain detergents are less effective on a molar basis and volume basis in their clearing action on mitochondrial suspensions. Swanson *et al.* [56] have made a detailed study of the use of solubilizers in the extraction of the constituents of cerebral microsomes. The effect of polyoxyethylene chain length is shown in Fig. 7.18.

Certain features of the solubilization of microsomes by detergents suggest that the mechanism involved may be similar to that postulated as occurring in the lysis of erythrocytes by detergents. In this mechanism haemolysis was pictured as resulting from breakdown of a lipoprotein–detergent complex formed by penetration of the detergent into the erythrocyte membrane. Analogous penetration of microsomal membranes by detergents is suggested by the marked effects on the activity of the Na^+ ion-stimulated adenosine triphosphatase even at concentrations of these agents too low to bring about solubilization [57]. For maximal removal of protein, cholesterol and phospholipid, an ethylene oxide chain length of 10 to 13 units is required when the alkyl chain is a C_{16} hydrocarbon. These results should be compared with those in Fig. 7.17 above.

Some non-ionic surfactants have been found [58] to inhibit transmucosal absorption of water from the gut containing hypotonic solutions. As water flow affects drug movement, this is an additional factor, as yet not widely studied, which might have a bearing on the interpretation of surfactant effects on absorption.

Further aspects of surfactant and membrane interactions will be discussed in Chapter 10.

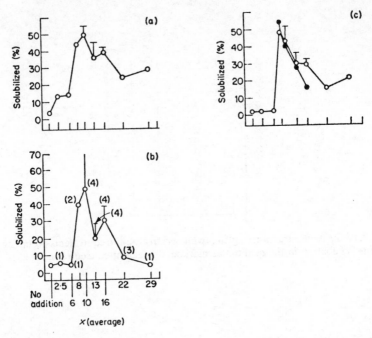

Figure 7.18 Extraction of microsomes with non-ionics of general formula $R(OCH_2CH_2)_xOH$ where R = cetyl for x = 16 to 29 and a mixture of oleyl and cetyl for x average 2.5 to 13. The detergent concentration was 1.3 mM in each case. (*a*) Percentage solubilization of protein; (*b*) protein solubilization in presence of 100 mM NaCl; (*c*) percentage solubilization of cholesterol (●) and phospholipid (○). From Swanson *et al.* [56].

7.3.2 Effect of surfactants on transfer of solutes across membranes and interfaces *in vitro*

It is pertinent to discuss the effect that surfactants have on permeation of solutes across artificial membranes as the solute–micelle interaction which is reflected in reduced transport rates, will be obtained generally without the complication of alteration to the permeability of the membrane.

The presence of polysorbates 20 and 80 decreases the transfer rate constant of salicylic acid across cellophane membranes at low pH [59]. Ionized salicylic acid does not partition into the micelles and thus at pH values above 5 polysorbate has little effect on permeation as cellophane membranes with small pore size are regarded to be impermeable to surfactant micelles [60]. The effect of solubilization is to reduce the concentration gradient of the solute across the membrane. Taking into account distribution of drug between the aqueous and micellar phases, Juni *et al.* [61] have derived an equation to describe the permeation profiles of drugs from systems containing micelles. A simplified model is shown in Fig. 7.19.

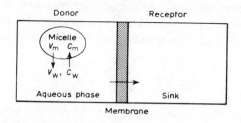

Figure 7.19 Schematic representation of sustained release of drug through membrane from a system containing micelle. From Juni *et al.* [61].

The distribution coefficient, P_m, of the drug between these two phases is given by

$$P_m = \frac{C_m}{C_w} = \frac{M_m/V_m}{M_w/V_w},$$ (7.20)

where C, M, and V denote the concentration and the amount of the drug in each phase, and the volume of each phase, respectively, and the subscripts m and w indicate a micellar phase and an aqueous phase, respectively. Under these conditions, if sink conditions are maintained in the receptor side, the permeation rate of drug is given by Fick's law:

$$\frac{dM_r}{dt} = \frac{APC_w}{l},$$ (7.21)

where M_r is the amount of drug in the receptor solution at time t, A, the area of the membrane available for permeation, l, the membrane thickness, and P, the permeability. The total amount of drug in the donor solution, M_t is given by:

$$M_t = M_m + M_w = M_t^\circ - M_r,$$ (7.22)

where M_t° is the total amount of drug initially introduced into the system. Rearrangement of Equations 7.20 to 7.22 leads to:

$$M_r = M_t^\circ \left[1 - \exp \left\{ -\frac{APt}{l(K_p V_m + V_w)} \right\} \right].$$ (7.23)

By definition:

$$M_t^\circ = V_m C_m^\circ + V_w C_w^\circ$$ (7.24)

$$C_m^\circ = P_m C_w^\circ,$$ (7.25)

where C_m° and C_w° are the initial concentrations of drug in the micellar phase and in the aqueous phase, respectively. When the volumes of the donor and receptor compartments are equal and represented by V, rearrangement of Equations 7.23 to 7.25 leads to Equation 7.26

$$-\ln\left(1 - \frac{C_r}{C_t^\circ}\right) = \frac{APC_w^\circ}{l V C_t^\circ} t$$ (7.26)

where C_t° denotes the total initial concentration of drug in the donor solution. The ratio C_w°/C_t° is equal to C_s°/C_s at equilibrium, where C_s° and C_s are the solubility of drug in water and that in a surfactant solution, respectively. Then Equation 7.26 becomes:

$$-\ln\left(1-\frac{C_r}{C_t^\circ}\right) = \frac{APC_s^\circ}{lVC_s}t. \tag{7.27}$$

Release profiles of the drug butamben (*n*-butyl *p*-amino benzoate) shows how release can be controlled by the presence of surfactant (Fig. 7.20). Slight increases in permeation of this solute from suspensions were caused by 0.5 NaLS and 0.5 % dodecyltrimethylammonium chloride because of melting effects and promotion of dissolution; such effects require the solutions to be saturated.

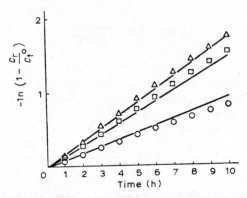

Figure 7.20 Release profiles of butamben through the silicone membrane from its initially saturated solutions in ○, 0.5 % sodium lauryl sulphate; □, 0.5 % polysorbate 80 and △, 0.4 % dodecyltrimethylammonium chloride solution at 30° C.
—, theoretical profile.
From Juni *et al.* [61] with permission.

The approach described above simplifies the *in vitro* situation and it has been found that the assumption of drug absorption from the aqueous phase only, underestimates the extent of absorption occurring *in vivo* [62] presumably because the enhanced membrane permeability is only detected in biological membranes.

The concentration of salicylic acid in the aqueous phase of a 1.0 % w/v polysorbate 20 solution at pH 1.0 can be calculated if the total salicylic acid concentration is known [62] using the equation,

$$C_a = C_t\left[\frac{V_t}{(P^\circ V_m/V_a) + P^- V_m/[(1/f_i)-1]V_a + 1 + 1/[(1/f_i)-1]V_a}\right] \times$$

$$\left[1 + \frac{1}{(1/f_i)-1}\right] \tag{7.28}$$

where C represents concentration, V volume and P partition coefficient; subscripts a, m denote aqueous and micellar phases and t is the sum of aqueous and micellar phases. Superscripts $^\circ$ and $^-$ denote unionized and ionized salicylic acid, respectively. f_i is the fraction of salicylic acid ionized at any pH. If only aqueous salicylic acid was available for absorption and provided that the surfactant did not influence absorption by mechanisms other than solubilization, an average of 18.4 % and 26.6 % of a 0.011 mg ml^{-1} salicylic acid solution should be absorbed from polysorbate 20 solutions after 15 and 30 min, respectively. These figures are equivalent to 10.1 % and 14.6 % of the total salicylic acid concentration and are lower than the corresponding experimentally determined values (*in vivo*) of 14.6 % and 19.1 %, respectively [63].

Salicylic acid–polysorbate 80 mixtures have been used in several other investigations including that of Hikal *et al.* [64]. In this work the apparent partition coefficient of the salicylic acid between chloroform and phosphate buffer (pH 6.5) was measured with polysorbate 80 in the aqueous phase at and above its CMC. Their results show that the surfactant increased the partitioning of the drug into the non-aqueous phase, perhaps indicating inverse micelle formation in the chloroform layer or, as the authors suggest 'complexation' between surfactant and salicylate forming a more lipid soluble species. At pH 6.5 little salicylate would be present in the micellar phase; these data require further investigation. More extensive investigations of mass transfer between aqueous and non-aqueous phase in the presence of surfactants have been carried out by Brodin [65]. Most workers have shown that surfactants reduce the rate of mass transfer either through their influence in reducing the circulation of bulk phases (especially in small droplets) or by the barrier of surfactant molecules aligned at the interface. Very low concentrations of surfactant inhibit circulation by the formation of a monolayer at the droplet–water interface. Thus Garner and Hale have observed [66] the rate of extraction of diethylamine from toluene droplets by water was reduced to 45 % of its normal value by the addition of 150 ppm of Teepol. Kitler and Lamy [67], however, have reported that lecithin can increase the transport of some phenothiazine into cyclohexane. Lecithin forms micelles in organic solvents [68, 69] an event which may explain these results. Brodin found that cetyltrimethyl ammonium bromide, sodium lauryl sulphate and the non-ionic Pluronic F68 caused decreases in the rate constants for transfer of a range of drugs, the maximum decrease in the pH-dependent values being about 10 fold. In this work, the partition coefficients of the solutes were not affected by the low concentrations of surfactants used; Brodin concludes that the decreases are the result of changes in the effective area available for transport when the surfactant molecules block the interface. Owing to the special nature of the biological membrane–water interface it is unlikely that this effect is in operation *in vivo*; at the higher concentrations used pharmaceutically the effect on partition coefficient is much more likely to be paramount. Brodin demonstrated the effect of concentration of CTAB on the partitioning of phenylbutazone between cyclohexane and water (Fig. 7.21). The rate of transfer $(\mathrm{d}m/\mathrm{d}t)$ of a solute across a

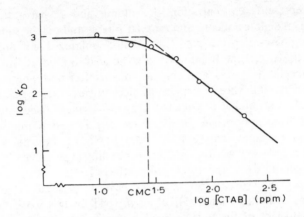

Figure 7.21 The effect of cetyltrimethylammonium bromide on the partitioning of phenylbutazone between cyclohexane and water, plotted as the logarithm of K_d the partition coefficient of the uncharged solute species. From Brodin [65].

membrane can be estimated from

$$\frac{dm}{dt} = K(\Delta c), \tag{7.29}$$

where Δc is the difference in concentration across the membrane and K is the dialysis rate constant. The integrated form of this equation is

$$\log(\Delta c) = -Kt + \text{constant}. \tag{7.30}$$

Experimental rate constants K_{app} can be obtained from plots of $\log(\Delta c)$ versus time. From Equations 7.30 and 7.28, Collett and Koo [70] obtained Equation 7.31 to calculate the theoretical dialysis rate constant from information on $K°$ and $P°$ the dialysis rate constant and micellar partition coefficient of unionized solute molecules, respectively.

$$K_s = [K°(1-f_i)]\left[\frac{V_t}{[(P°V_m/V_a)+1+((1/f_i)-1)^{-1}]V_a}\right]\left[1+\frac{1}{(1/f_i)-1}\right]. \tag{7.31}$$

According to Equation 7.31 the dialysis rate constant should increase with decreasing $P°$. As $\log P°$ is linearly dependent on π, the hydrophilic–lipophilic constant of the corresponding substituent in a homologous series, this equation allows one to predict the dialysis rate constants of a series of compounds. The equation should also predict the effect of surfactant concentration through its effect on $P°$. The values of K_{app} and K_s for a series of para-substituted benzoic acids in 1% polysorbate 20 at pH 1 are shown in Fig. 7.22 as a function of π.

The effect of surfactants on diffusion of substances through gels has biopharmaceutical overtones; one paper [71] has suggested that at or about the surfactant CMC the diffusion of malachite green through gelatin gels is increased.

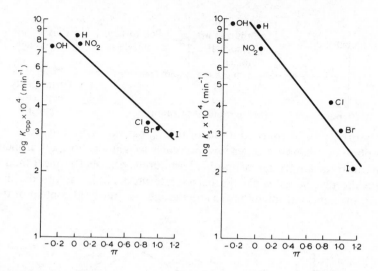

Figure 7.22 Relationships between π and experimental (left hand diagram) and theoretical (right hand diagram) dialysis rate constants, K_{app} and K_s respectively of para-substituted benzoic acids from 1 % polysorbate 20 solutions at pH 1.2. From Collett and Koo [70].

Post-CMC, the diffusion coefficients fall, as anticipated, although the maximum in diffusion coefficient does not coincide with known values of CMC. It might be that some ionic interactions either between dye and surfactant or surfactant and gel are complicating the interpretation. *In vitro* experiments using artificial membranes can define the physicochemical interactions between drug and micelle. So far we have seen the effects of the nature of the solute exerted through P_m, the pH of the solution and the capacity of the micelle for the drug, and their influence on dialysis rate. The reduction in free drug and the consequent reduction in transport rate can be quantified; it is not possible to quantify the increase in permeability caused by the surfactant monomers at low concentrations, or the increased permeability which can arise through solubilization of membrane components. It is unlikely that the membrane–surfactant–water interface bears much relationship to an oil–water interface and one can anticipate that we have much to learn about the nature of the interactions that occur.

It has been shown [72] that some biological membranes have a dissociating effect on certain types of complexes. Since the absorption-retarding effect of polysorbates 80 on secobarbitone was evident during rapid stirring of the solution and in the quiescent state, Levy *et al.* [73] concluded that the fish membrane does not have a dissociating effect on secobarbitone–non-ionic micelle complexes. 1:1 complexes formed between drugs and hydrotropes are probably broken because of the greater contact between drug and membrane. The following scheme was put forward by these workers to describe the effect of non-ionics on secobarbitone absorption:

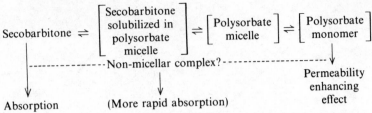

Polysorbate 80 in concentrations of 0.01% has no significant effect on the absorption of ethanol or other low-molecular-weight alcohols, but it increases significantly the absorption of another barbiturate, pentobarbitone. Ethanol can diffuse through pores, while the barbiturate must diffuse across the lipoidal barrier; the non-ionic might have a specific effect on the lipid content of the cell

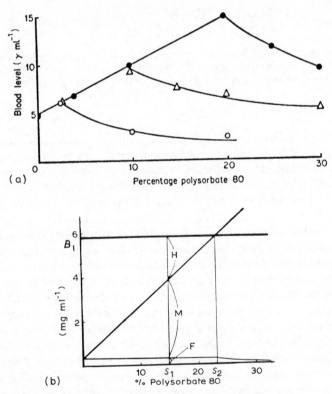

Figure 7.23 (a) The effect of polysorbate 80 on blood levels of sulphisoxazole. Concentration of sulphonamide solutions and suspensions administered: ●—● $5 \, mg \, ml^{-1}$; △—△ $2.5 \, mg \, ml^{-1}$; –O—O–: $1 \, mg \, ml^{-1}$. Redrawn from Kakemi *et al.* [75]. (b) Diagram representing sulphisoxazole in solution of polysorbate 80
H : solid sulphisoxazole
M : sulphisoxazole in micelles
F : free sulphisoxazole in solution.

membrane, Kay [74] having found evidence for this in studies on the effects of polysorbate 80 on the *in vitro* metabolism of the Ehrlich–Lettre Ascites carcinoma.

The effect of solubilization on absorption, so far evidenced to reduce absorption of the drug, can be beneficial if the system is saturated. Solubilization, while reducing the amount of drug absorbed when it is present in solution, allows larger concentrations of drug to be administered. When suspensions consisting of free sulphisoxazole, solubilized sulphisoxazole, and solid drug were placed in the rectal sac it was found that the blood levels of the sulphonamide increased with increasing surfactant concentration as shown in Fig. 7.23. Fig. 7.24 which shows the effect on blood levels of increasing the drug concentration in the administered solution, should also be consulted. In Fig. 7.23 it is evident that when 2.5 mg drug is presented per ml solution its activity increases with increasing concentration of polysorbate 80 until at 10% polysorbate the blood level falls. At 10% polysorbate 80 the drug is completely dissolved and the normal reduction in activity due to solubilization takes place.

An equation relating the total absorption rate (A_T) to the observed absorption rate of free drug (A_f), in which S is the concentration of surfactant (g/100 ml), A_m the absorption rate of micellar drug, and P_m the distribution constant, was

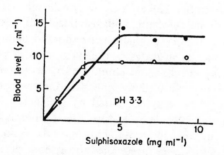

Figure 7.24 The effect of the amount of sulphisoxazole administered at two different polysorbate 80 levels (○—○ 10% polysorbate 80; ●—● 20% polysorbate 80) on blood levels achieved. From Kakemi *et al.* [75].

According to Kakemi *et al.* [75] the sulphisoxazole–polysorbate system can be explained by Fig. 7.23(b). B_1 is a concentration of sulphisoxazole, just solubilized in water at the concentration S_2 of polysorbate 80. When the surface-active agent is incorporated at the concentration S_1, a suspension is considered as the three-phase system consisting of the free sulphisoxazole in the solution (F), the drug entrapped in micelles (M), and the solid form of the drug dispersed in the solution (H). M increases with increasing polysorbate 80 up to S_2, and F is constant. Above S_2, the drug is completely solubilized and therefore M/F increases with increasing polysorbate 80; the free drug concentration decreases. Among three components, the free sulphisoxazole is readily absorbed, and the solid form of sulphisoxazole is not absorbed. If the drug in micelles is absorbed a little, it would be expected that the absorption rate increases as the concentration of surface-active agent below S_2, and decreases above S_2. This is demonstrated in Fig. 7.23a.

derived by Kakemi and co-workers and described their experimental results closely:

$$A_T = \frac{A_f}{1 + P_m S} + \frac{A_m P_m S}{1 + P_m S}. \tag{7.32}$$

This equation has a term for the absorption of the drug enclosed within micelles. It is unlikely that this is significant, a standpoint suggested by the experimental work of Kakemi *et al.* [75] and recently supported by Mysels' analysis [76] of the mechanism of transport of Orange OT through a membrane in the presence of sodium dodecyl sulphate. The rate of dialysis of solubilized dye was estimated at 3.6×10^{-5} h^{-1}, which is negligible compared to the 1.0 h^{-1} of the free drug. This is probably the general case, although electron micrographs of the intestinal microvilli seem to suggest that micellar particles can penetrate far into this specialized membrane during fat absorption.

7.4 Effect of surfactants on drug absorption

7.4.1 Effect of surfactants on intestinal absorption

In this section is reviewed a selection of the available evidence on this topic. Typical of the confusion that still exists is the conclusion arrived at in one paper [77] in which tetracycline, sulphanilamide, isoniazid and salicylic acid were used as test drugs and sodium lauryl sulphate, benzethonium chloride, polysorbate 80 and sucrose mono- and di-stearates as the surfactants. A perfusion technique involving the rat small intestine was employed. It was found that: (i) the ionic nature of the surfactants substantially influenced the absorption; (ii) the rate of absorption of tetracycline was accelerated by the presence of sodium lauryl sulphate, benzethonium chloride or sucrose esters; (iii) polysorbate 80 caused a marked reduction in the absorption of salicylic acid and tetracycline; (iv) benzethonium chloride reduced the absorption of salicylic acid; and (v) sucrose esters within the concentrations tested did not decrease the absorption of all the drugs tested [77]. Following oral administration of some of the solutions to adult human subjects the urinary excretion results supported the view that sucrose esters greatly enhanced, while polysorbate 80 markedly reduced the absorption of tetracycline and that the absorption of sulphanilamide was not affected significantly by the presence of sucrose esters. Here polysorbate 80 is having a negligible or detrimental effect on absorption. Similarly no significant change in the absorption of salicylic acid from the *in situ* rat intestine has been effected by polysorbate 80 at concentrations of 0.001 %, 0.01 % and 2 % [64]. When polysorbate 80 is used as the vehicle for dicoumarol, griseofulvin and sulphisoxazole acetyl, there are significant increases in absorption in the rat [77]. It is important that formulation approaches, other than addition of surfactants, are considered in contemplation of the viability and wisdom of a surfactant-containing formulation. The results shown in Fig. 7.25 are of some significance in this regard. In these experiments the drugs in polysorbate are in solution form;

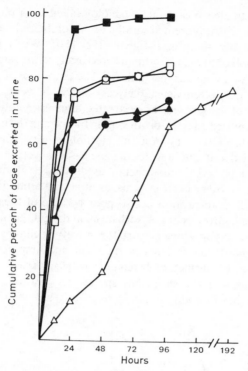

Figure 7.25 Cumulative urinary excretion of free sulphisoxazole (expressed as a percent of the administered dose on a molar basis) following oral administration of a 100-mg/kg dose sulphisoxazole acetyl in lipid vehicles and water. Each point represents the average of six animals. ●, hexadecane; △, oleyl alcohol; ■, polysorbate 80 (solution); □, trioctanoin; ○, triolein; and ▲, water (with 0.5% methylcellulose). From Bloedow and Hayton [78].

Percent drug dissolved in the suspension dosage forms

	% dissolved*		
Vehicle	Sulphisoxazole acetyl	Dicoumarol	Griseofulvin
Hexadecane	<0.01	1.4	<0.01
Oleyl alcohol	1.7	12	3.6
Polysorbate 80	100†	100†	94
Trioctanoin	6.8	19	13
Triolein	3.1	13	3.4

* Calculated from solubilities. † Solutions.

the others in suspension in varying degrees (see legend of Fig. 7.25). Although polysorbate 80 increases the bioavailability of dicoumarol when in solution and in suspension, Bloedow and Hayton's results on griseofulvin suspensions indicate that the release rate of drug from suspension is the primary factor in enhancing absorption (Fig. 7.26).

In warfarin-pretreated Wistar rats the biological effect of phytomenadione is greatly increased when presented orally as a solubilized aqueous solution compared to an oily solution (Miglyol 812) [80]. When phytomenadione (30 mg kg^{-1}) dissolved in oil, was administered orally to the rats, the effect of the drug on prothrombin time was insignificant. The same dose of phytomenadione solubilized with polyoxyethylene(20)glyceryloleate, however, completely abolished the effect of warfarin in the pretreated animals: 3 hours after administration of the solubilized vitamin, prothrombin time was in the normal range. The effect of the surfactants is not on the clotting process according to other experiments carried out but due to the increased absorption of the phytomenadione. It is unlikely that the surfactant used would be absorbed significantly to exert its effect on other body systems, as after oral administration the ester bond is cleaved [81]; absorption is also poor because of their high molecular weight. Following hydrolysis of polysorbate 80 in the gut the oleic acid moiety is absorbed and the polyoxyethylene sorbiton moiety is eliminated in the faeces [82]. Thus the model experiments *in vitro* can only impinge slightly on the complex influence of surfactant behaviour *in vivo*. Apart from the largely unknown behaviour of the surfactant species, its location, absorption* and degradation and thus its ability to retain an effectiveness over a given period of

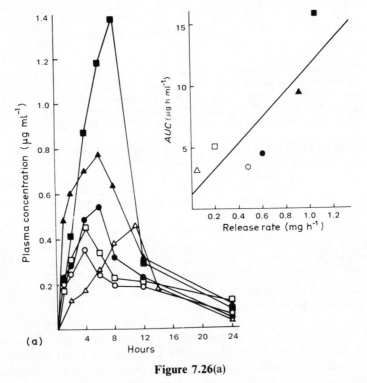

Figure 7.26(a)

---

* Recently the absorption of iodine-labelled polysorbate 80 from the rat gut has been studied [82a].

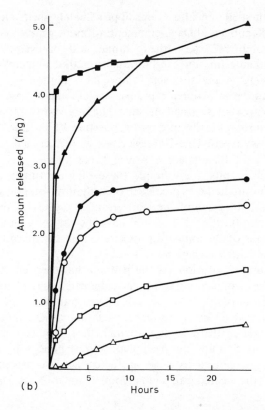

(b)

Figure 7.26(a) Representative plasma concentrations of griseofulvin following oral administration of 50 mg/kg of griseofulvin suspended in lipid vehicles and water. Each curve, representing data from one animal, has a peak plasma concentration and t_{max} closest to the mean values for each group. For key, see Fig. 7.25. Inset: correlation of the area under the plasma concentration–time curve (AUC) with the average 0 to 4 h rate of release of griseofulvin *in vitro*. (b) Release of griseofulvin into water from suspensions containing 5 mg drug in lipid and water. Each point represents the mean of three experiments. Key as above. From Bloedow and Hayton [79] with permission.

time, there are the normal problems of defining the bioavailability of a drug and the influence of formulation. Reddy *et al.* [15] show that poloxamer 188 and sodium sulphosuccinate increase the absorption of sulphadiazine from rat intestinal loops, there is no significant effect on bioavailability when the drug is administered with these surfactants to rats, if the total urinary excretion of the sulphonamide is measured over 24 h. Absorption rate might well have been affected but 24 h bioavailability was not.

Absorption of normally non-absorbed or poorly absorbed water soluble drugs from a Thomas gastric fundic pouch of the dog is greatly increased by certain surfactants [83]. Vitamin B12 absorption from both stomach and intact gastrointestinal tract of the rat is similarly enhanced [84]. As might be anticipated, while blood levels of cephaloridine are elevated several fold when surfactant is added to the ligated stomach, their influence in the intact GI tract is diminished and confined to the first 30 min, after which approximately normal levels of drug are observed [85]. Kreutler and Davis commenting on their results conclude that the absorption promoters exert their rapid and transient effect in the duodenum–small intestine and exert little effect in the stomach: 'This may be due to rapid emptying of the stomach followed by dilution of the dose in the duodenum, or to the subsequent rapid passage of a liquid dose out of the more absorptive upper part of the small intestine. The comparatively poor results in the intact animals also raise the interesting question of possible specific incompatibilities of polyoxyethylene-20-oleyl ether with intestinal secretions in the intact GI tract' [85]. Results are shown in Table 7.4.

From previous results, for example those of griseofulvin and dicoumarol suspensions, one might have predicted that addition of polysorbate 80 to a suspension of a poorly soluble anticonvulsant, a piperazine derivative with a solubility less than 0.1 mg ml^{-1}, would have increased its bioavailability. But, responses in terms of the animal's protection from convulsant challenges was decreased by 0.8 % polysorbate 80 (Fig. 7.27).

Recently, insulin absorption via the jejunum has been effected by administration of insulin–cetomacrogol solutions to diabetic rats [87]. Results presented in Table 7.5 are most likely to be due to a membrane effect rather than a surfactant-protecting effect on insulin degradation, as insulin administered $\frac{1}{2}$ h after cetomacrogol elicited a hypoglycaemic effect (see Section 7.4.2 below on rectal absorption). Sodium lauryl sulphate (0.75 %) and sodium taurocholate (3.2 %) have been reported to cause an increase in the percentage of insulin absorbed from the ligated rat jejunal loop from 0.4 % to 3.2 % and 3.4 %,

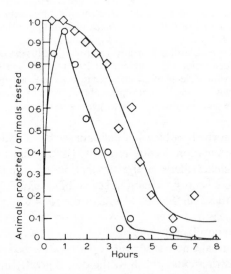

Figure 7.27 Time course of pharmacological activity of 1-diphenyl-4-[6-methyl-2-pyridyl methyleneamino] piperazine [I] after oral administration of a suspension of I without (◇) and with (○) 0.8 % polysorbate 80 solution in saline. From Sanvordeker and Bloss [86].

Table 7.4 Normal and promoted absorption from doubly ligated duodenum–small intestine

Drug combination	Number of animals	Minimum detectable levels ($\mu g\,ml^{-1}$)	Whole blood levels ($\mu g\,ml^{-1}$) and standard errors (time post-dosing, min)					
			20	40	60	80	120	180
Cephaloridine only	5	0.25	0.41 ±0.13	0.95 ±0.19	1.90 ±0.35	2.03 ±0.37	1.66 ±0.46	1.54 ±0.38
Cephaloridine with polyoxyethylene-20-oleyl ether	5	0.25	3.44 ±0.72	9.36 ±1.14	10.52 ±0.62	8.36 ±1.71	3.58 ±0.64	0.95 ±0.18
Cephalothin only	4	0.10	0.17 ±0.11	0.34 ±0.04	0.40 ±0.05	0.35 ±0.04	0.18 ±0.01	—
Cephalothin with polyoxyethylene-20-oleyl ether	5	0.10	1.39 ±0.10	5.90 ±0.69	5.96 ±0.97	2.25 ±0.37	0.76 ±0.20	—

Normal and promoted absorption from intact rat gastro-intestinal tract

Drug combination	Number of animals	Minimum detectable levels ($\mu g\,ml^{-1}$)	Whole blood levels ($\mu g\,ml^{-1}$) and standard errors (time post-dosing, min)					
			20	40	60	80	120	180
Cephaloridine only	15	0.25	0.33 ±0.09	0.82 ±0.23	1.12 ±0.28	1.35 ±0.30	0.79 ±0.17	0.45 ±0.12
Cephaloridine with polyoxyethylene-20-oleyl ether	11	0.25	1.16 ±0.37	1.10 ±0.28	1.04 ±0.21	0.93 ±0.13	0.79 ±0.17	0.57 ±0.12
Cephalothin only	4	0.10	0.21 ±0.02	0.33 ±0.06	0.33 ±0.05	0.34 ±0.10	—	—
Cephalothin with polyoxyethylene-20-oleyl ether	5	0.10	0.50 ±0.06	0.30 ±0.04	0.25 ±0.06	0.17 ±0.05	—	—

From [85].

Table 7.5 Effect of cetomacrogol on intrajejunal absorption of insulin in diabetic rats.

Sample administered*	Initial blood glucose conc. mg % (mean ± S.E.M.)	Blood glucose at times after insulin administration[†] (% of initial content) (mean ± S.E.M.)		
		1 h	2 h	4 h
Saline (n = 4)	326 ± 17.0	97 ± 2.9	96 ± 5.0	90 ± 4.3
Insulin (n = 5)	296 ± 18.8	104 ± 3.0 $P\ddagger < 0.2$	100 ± 3.5 $P\ddagger < 0.6$	95 ± 2.0 $P\ddagger < 0.4$
Insulin-cetomacrogol (n = 9)	309 ± 14.3	56.5 ± 2.8 $P\S < 0.001$	21.0 ± 2.0 $P\S < 0.001$	37.7 ± 4.1 $P\S < 0.001$

* Each run was carried out on a different animal; n = no. of rats. For sample composition see text.
† Blood glucose content (% of initial) after interperitoneal injection of 4 i.u. of insulin was 36.1 ± 2.2, 28.7 ± 1.0, 39.4 ± 3.6 at 1, 2 and 4 h respectively (4 rats). Initial blood glucose concentration in mg %: 315 ± 32 (mean ± S.E.M.).
‡ Insulin versus saline.
§ Insulin-surfactant versus saline or insulin (same P values).
From [87].

respectively, while a W/O/W emulsion system increased absorption to 30.6% [88].

Polysorbate 20 has been found to enhance the gastro-intestinal absorption of iron-59, but the mode of action was not clear [89]. The absorption of barium chloride ingested by cats was promoted by both polysorbate 20 and sodium lauryl sulphate at low concentrations and inhibited at high concentrations [90]. Non-toxic doses of sodium lauryl sulphate greatly increased the rate of glucose absorption in rabbits [91]. It has been claimed that sodium lauryl sulphate inhibits gastric motility in certain doses and Nissim concludes that large doses of ionic surfactants lead to structural damage, while small doses reduce the functional efficiency of mucosal cells [92]. Both sodium lauryl sulphate and dioctyl sodium sulphosuccinate, but not pluronic F68, increase the absorption of phenol red from the colon [93], as shown in Table 7.6. This table also shows the effect of administered drugs on the absorption of the dye. The fact that pharmacologically active agents can markedly affect the absorption rate suggests that the effect of the detergent may not be wholly physical. No data were quoted for the effect which the two drugs, atropine and chlorisondamine, had on absorption in the absence of surfactant. These results are of importance in pharmacy, as it is evident that drugs taken concomitantly with the solubilized preparation can seriously affect the theoretical performance of the formulation.

Preliminary data obtained by Lish and Weikel [93] indicate that dioctyl sodium sulphosuccinate increases the absorption of sulphathalidine. None of the surfactants studied influenced the absorption of the cationic dye, methyl violet; presumably there was interaction with the ionic detergents. The fact that the non-ionic surfactant has no effect with either phenol red or methyl violet may be the

Table 7.6 Effect of drugs on the enhancement of absorption of phenol red by surfactants (1 % in normal saline)

Solvent	Drug treatment	% Dye absorbed
Normal saline	—	6
Dioctylsulphosuccinate	—	58
	Chlorisondamine* (0.8 mg kg^{-1})	13
	Atropine sulphate (2.0 mg kg^{-1})	36
Sodium lauryl sulphate	—	77
	Chlorisondamine	45
	Atropine sulphate	65
Pluronic F68	—	1
	Chlorisondamine	5

* Ethylene-1-(4,5,6,7-tetrachloro-2-methyl*iso*indolinium)-2-trimethylammonium dichloride, a ganglion blocking agent used in severe hypertension. From [93].

result of mixed micelle formation. Mixed micelle formation and the phenomenon of therapeutic interference [94] will be discussed later.

(A) EFFECT OF BILE SALT

The bile salts have been studied for their effect on drug absorption. They are obvious objects of interest in view of their presence in the intestine and their involvement in fat absorption.

The bile salt concentration is 100 to 300 mM in human bile, and about one-tenth of this in human intestinal content and as the CMCs are in the region of 2 to 3 mM both bile and intestinal fluids contain bile salt micelles [95]. The ability of bile salt solutions to solubilize insoluble drugs such as griseofulvin and hexoestrol [96] suggests that the bile salts may be involved in the solubilization of drugs prior to absorption. Bates *et al.* [97] have found that physiological concentrations (0.04M) of sodium cholate and sodium deoxycholate enhance the rate of solution of hexoestrol and griseofulvin over their rate of solution in water, as shown in Table 7.7. This effect strengthens the view of their supposed action. It is interesting to note that the oxidized bile salt dehydrocholate does not micellize and does not enhance fat absorption [95].

Table 7.7 Relative dissolution rates for griseofulvin and hexoestrol at 37° C [97] in sodium deoxycholate and sodium cholate solutions

Dissolution medium	Griseofulvin time (min)			Hexoestrol time (min)		
	2	5	10	2	5	10
Water	1.0	1.0	1.0	1.0	1.0	1.0
Sodium deoxycholate	7.5	6.6	6.0	14.6	18.6	24.0
Sodium cholate	7.2	6.1	5.5	20.3	30.4	36.0

The enhanced absorption of medicinals on administration with deoxycholic acid may be due to reduction in interfacial tension or micelle formation. The inefficient absorption of reserpine promoted an investigation into its absorption in combination with deoxycholic acid [98]. Deoxycholic acid was found to increase the rapidity of absorption of reserpine and to increase its potency. The solubility of reserpine is increased in hydro-alcoholic deoxycholic acid solutions [99], it being suggested that both micellar solubilization and inclusion formation is responsible. A combination of these effects may facilitate the absorption of the reserpine.

The administration of quinine and other Cinchona alkaloids in combination with bile acids has been claimed to enhance their parasiticidal action [100, 101]. Quinine, taken orally, is considered to be absorbed mainly from the intestine. A considerable quantity of bile salts is required to maintain a colloidal solution of quinine; one might argue that an efficient supply of bile salts was therefore a prerequisite of quinine absorption [102]. It is interesting to note in this context that recurrent attacks of malaria are sometimes found to be accompanied by hepatic disturbances which may prejudice the normal flow of bile [103].

A mechanism has been suggested to explain the enhancement of drug absorption following a meal of high fat content. Triglycerols and similar materials increase the flow of bile into the small intestine, which results in the increased solubility of any drugs present. The effect of lipid additives on the solubilization of glutethimide, hexoestrol, and griseofulvin in a simulated bile salt mixture is small; the lipids would simply increase the concentration of bile, and although they themselves will be solubilized, would not preclude the solubilization of other drugs [104].

A comparative study of the effect of sodium cholate, sodium deoxycholate, sodium chenodeoxycholate and taurodeoxycholate on the absorption of quinalbarbitone sodium by goldfish [105] has shown no correlation between their effectiveness as absorption promoters and their relative hydrophobicity or ability to lower interfacial tension. In general terms their ability to increase absorption is predictable because of their surface activity and their ability to abstract lipid from erythrocyte ghosts indicating their freedom to interact with biological membranes [106, 107].

20 mM sodium taurocholate increases the absorption of procaineamide from rat small intestine [108]; an effect abolished by the addition of 18 mM oleic acid. Feldman and Gibaldi [109] reported that the addition of lecithin and fat digestion products to solutions containing sodium taurodeoxycholate produced a pronounced decrease in the permeability of the everted rat intestine to salicylate. The oleic acid perhaps decreases the ability of the bile salt to solubilize membrane components; more likely the oleic acid forms mixed micelles with the bile salt and thus solubilizes more procaineamide. Sodium taurocholate has no effect on the absorption of 2-allyloxy-4 chloro-N-(2-diethyl aminoethyl) benzamide (ACDB). Mixtures of this bile salt with lauric acid, palmitic acid or oleic acid reduced the intestinal absorption of ACDB [108].

In contrast to the findings on procaineamide absorption in the presence of mixed micelles, the absorption of amino glycosides has been found to be little affected by bile salt alone yet significantly increased in the presence of mono-olein-*N* oleic acid–bile salt mixed micelles [110]. Pretreatment of the gut with mixed micellar solution, 1 h prior to administration of the drug has no effect on absorption; there was no evidence of membrane damage. The mixture was much more effective in the large intestine rather than in the small intestine. To obtain an effect, concentrations of 40 mM of the mixed micelle are required hence as dilution rapidly occurs following oral administration in the intact animal, unspectacular results may be obtained. For this reason rectal absorption of the aminoglycoside–mixed micellar system was suggested [110] and shown to improve the absorption of both gentamicin and streptomycin.

Bile salts have achieved improved absorption of urogastrone, a glycoprotein with gastric antisecretory activity. Urogastrone alone or bile salts alone failed to inhibit H^+ secretion when administered intrajejunally in the rat, but presented together, a strong inhibitory response of gastric acid secretion was observed [111]. Sodium taurocholate was the least effective of the three bile salts studied and urogastrone administered in 0.2% polysorbate 80 was ineffective. Since EDTA facilitates intestinal absorption of heparin, presumably by chelation of membrane calcium and magnesium, and as bile salts possess an EDTA-like effect on the intestinal membrane [112, 113], it was suggested that they increase permeability to urogastrone by increasing the permeability of the absorptive membrane.

The importance of bile salt concentration on transport rates is clearly shown in Fig. 7.28 from the work of Feldman and Gibaldi [114] which suggests that two mechanisms are operating, one below and one above the CMC, leading to the

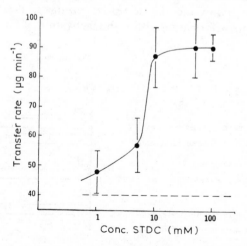

Figure 7.28 Effect of mucosal concentrations of sodium taurodeoxycholate STDC (log scale) on mean steady-state transfer rates of salicylate across the everted intestine of the rat. Bars denote ± 1 s.d. Dashed line indicates mean control value. From Feldman and Gibaldi [114] with permission.

question as to whether monomeric and micellar bile salt species are absorbed in a different manner. Absorption of the bile salts has been studied [115] and it appears from this work that micellar taurocholate moves across intestinal membranes twice as fast as monomeric bile salt. This does not necessarily mean that micelles are involved in the transport process but as Feldman and Gibaldi rightly observe, it is likely that the higher concentrations of bile salt alteration of membrane permeability will enhance the possibility of transport of the bile salt across the membrane. Some researchers have found that the effect of bile salts on the permeability of the intestinal barrier is not readily reversible [116]. Direct comparison of sodium taurocholate, polysorbate 80 and an alkyl ether non-ionic surfactant [117] indicated that the presence of 10 mM oleic acid in 0.2% surfactant solution was essential for significant absorption of heparin (see Fig. 7.29). Thus it appears that the bile salts are not unique in their action. The role of the oleic acid or mono-oleic in these mixed micellar systems has yet to be elucidated.

A logical extension of this type of study, because of the natural presence of bile salts in the intestine and the presence of synthetic surfactants in formulations, is the consideration of bile salt–surfactant mixtures. One such study [118] has considered the effect of sodium glycocholate and its mixtures with NaLS and polysorbate 80 on the absorption and metabolism of a thiamine disulphide derivative, in rats (see Scheme 7.1). It had previously been shown that surfactants altered the reaction rates of the thiol-disulphide exchange reaction that these compounds undergo [119] and that *o*-benzoyl thiamine disulphide interacts with the lauryl sulphate anion to form a 1:2 complex; this complex is broken up by sodium glycocholate to form new mixed micelles of thiamine derivatives and the surfactants. NaLS decreases k_A and k_D promoting the conversion of V to VII. The reduction in absorption and the decreased enzymatic deacylation are both explained by complex formation, although inactivation of intestinal esterase by

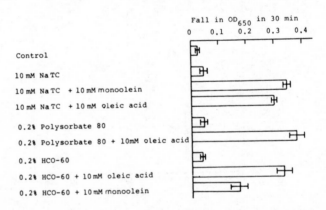

Figure 7.29 Plasma clearing factor activity after the administration into the large intestine. Several types of mixed micelles containing oleic acid were tested. Each value is the mean ± SEM of 4 to 5 animals. NaTC = sodium taurocholate; HCO-60 = hydrogenated castor oil based non-ionic surfactant with 60 ethylene oxide units. From Taniguchi *et al.* [117] with permission.

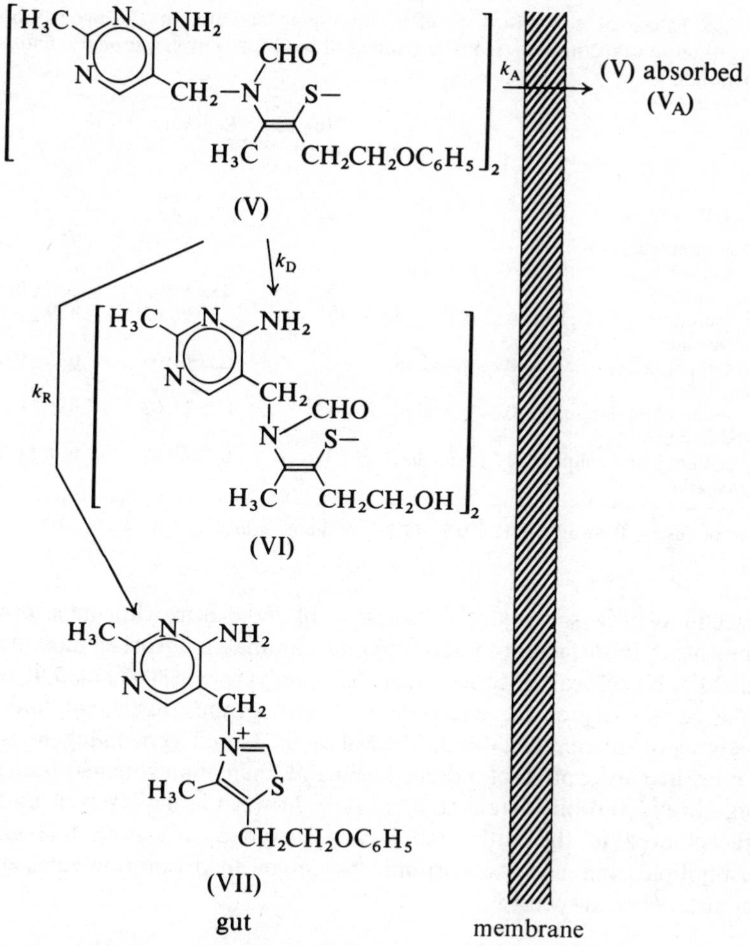

Scheme 7.1

(V) *o*-benzoyl thiamine disulphide
(VI) thiamine disulphide
(VII) *o*-benzoyl thiamine

Kinetic model for absorption and metabolism. In the intestinal tract, (V) is absorbed in an intact form but is partially metabolized by two processes (Scheme 7.1). One process is a reduction, being considered nonenzymatic, to form *o*-benzoylthiamine (VII), the other is an enzymatic hydrolysis of ester linkage (deacylation) to form the less absorbable (VI). From [118].

the surfactant is possible. The increase in k_R has also been observed *in vitro*. Sodium glycocholate at concentrations above 0.1 % increased k_A and at 0.015 % and above decreased k_D. Polysorbate 80 (0.5 %) reduced absorption, the enzymatic deacylation and the reduction. The effect of concentration of sodium glycocholate on k_A and k_R in the presence of 0.1 % sodium lauryl sulphate is shown in Table 7.8. The bile salt is seen to cancel out the effect of the anionic

Table 7.8 Effect of concentration of sodium glycocholate on the absorption of o-benzoylthiamine disulphide from mixtures of sodium lauryl sulphate and sodium glycocholate*

System	Number of rats	Rate constants (h^{-1})	
		k_A	k_R
Control[†]	3	0.40 ± 0.03	0.37 ± 0.05
0.1 % sodium lauryl sulphate	3	0.18 ± 0.03	0.76 ± 0.14
0.1 % sodium lauryl sulphate–0.1 % sodium glycocholate	3	0.22 ± 0.03	0.50 ± 0.07
0.1 % sodium lauryl sulphate–0.17 % sodium glycocholate	3	0.23 ± 0.01	0.39 ± 0.03
0.1 % sodium lauryl sulphate–0.25 % sodium glycocholate	3	0.37 ± 0.05	0.36 ± 0.06
0.1 % sodium lauryl sulphate–0.34 % sodium glycocholate	3	0.43 ± 0.07	0.39 ± 0.03
0.1 % sodium lauryl sulphate–0.5 % sodium glycocholate	3	0.38 ± 0.07	0.38 ± 0.07

* V in perfusate: $10 \mu g \, ml^{-1}$, pH 6.4, 37° C.   † Without surfactant.
From [118].

surfactant (which is to reduce absorption of V). o-Benzoylthiamine disulphide administered in surfactant solutions to intact animals is absorbed most efficiently from 0.1 % NaLS, least effectively from 0.5 % polysorbate 80 (Table 7.9), pointing to the complexity of the effects *in vivo*; endogenous surfactant interactions sometimes of an unpredictable nature will occur. The effect of endogenous bile on the intestinal absorption of indomethacin and phenylbutazone has been studied using normal and bile fistulated rats [120]. Lower plasma levels of both drugs were achieved in the latter and it was surmised that both bile salts and phospholipids influenced absorption via enhanced dissolution rates of drugs administered in suspension.

Table 7.9 Urinary excretion of thiamine after administration of o-benzoylthiamine disulphide micellar solutions*

Micellar solution	Urinary excretion of thiamine (%)
Control[†]	18.1 ± 1.8
0.1 % sodium lauryl sulphate	30.4 ± 7.7
0.5 % polysorbate 80	14.4 ± 3.5
0.5 % sodium glycocholate	24.7 ± 4.2

* Dose: $200 \, \mu g/2 \, ml$ (pH 6.4).   † Without surfactant.
From [118].

7.4.2 Surfactants and rectal absorption

The rectal route offers an alternative to the oral route of administration and offers some advantages in experimental work. Drug absorption from the rectum (of the rat) has been said to be more consistent with the pH-partition hypothesis than is absorption from the small intestine [121]. Results shown in Fig. 7.23 earlier in this chapter were obtained by administration of solutions and suspensions of sulphisoxazole into the sac of the rectum of rats. This is probably one of the most significant experiments on surfactant effects, possibly because of the relatively stable conditions in the rectum compared with the rest of the gastro-intestinal tract. Polysorbate 80 produced no histological damage to the rectal mucosa [75].

Bioavailability of gentamicin after rectal administration of the drug as a 20 mM mixed bile salt–mono-oleic micellar solution is approximately 45% whereas without surfactant absorption it is negligible [110]; bioavailability is further improved by installation of a freeze-dried gentamicin–mixed surfactant powder presumably because of the high concentration of surfactant that is achieved (see Table 7.10). The effect of surfactant in the rectum can be compared with that in the duodenum. The influence of a surfactant ion on absorption may not be directly related to surface activity or solubilization but, when solute ions of opposite charge are involved, to lipophilic ion pair formation. Such is the conclusion [122] drawn from observations of the effect of sodium lauryl sulphate on the rectal absorption of a variety of amines (Table 7.11), although as the anion is not transported in equal amounts this cannot be the whole explanation of increased absorption. It has thus been postulated that the binding of the species to the mucosal tissue perhaps offers a better explanation.

Saccharin sodium produces qualitatively similar results yet is not known to be a surfactant ion [123]. At pH 7.4 NaLS has been found to increase the binding of ephedrine and quinine to rectal mucosal preparations by an unknown mechanism. It is presumed that Kakemi *et al.* [122] propose that the evidence of increased affinity for mucosal tissue signifies an increased concentration gradient of the drug at the interface but this is a concept that requires much more data and experimental evidence before being acceptable. The increased surface activities of the ion-pair may be a factor in increasing the interfacial concentration of the drug

Table 7.10 Bioavailability of gentamicin in various preparations and routes of administration

Preparation	Bioavailability (% ± S.E.M.) from AUC
Intravenous injection	100
Duodenal instillation	
None	4.1 ± 1.6
40 mM mixed micellar solution	6.8 ± 0.5
Rectal instillation	
None	0.1 ± 0.1
10 mM mixed micellar solution	18.4 ± 3.4
20 mM mixed micellar solution	44.6 ± 4.9
Rectal insertion	
Powdered mixed micelles	58.1 ± 8.2

From [110].

Table 7.11 Effect of NaLS on the rectal absorption of various amines at pH 7.4

Drug	pK_a	% absorbed in 1 h		Apparent partition coefficient			
				Chloroform		Benzene	
		Alone	With NaLS	Alone	With NaLS (0.4 mM)	Alone	With NaLS (0.4 mM)
Aminopyrine	5.0	21.2 (2)	26.5 (2)	—	—	6.77	7.00
ACDB*	7.9	22.8 (2)	27.8 (2)	∞	∞	37.10	102.18
Quinine	8.4	14.9 (3)	30.7 (3)	∞	∞	1.45	12.60
Procaine	9.0	0.8 (3)	8.9 (3)	11.17	44.64	—	—
Ephedrine	9.6	2.9 (5)	7.1 (5)	0.07	0.22	—	—
Fuchsin 'Basic'	—	7.0 (4)	13.4 (4)	0.77	31.49	—	—
Homatropine	10.4	7.2 (4)	24.5 (4)	0.72	3.50	0.01	0.11

Numbers in parentheses represent number of experiments.
* 2-Allyloxy-4-chloro-N-(2-diethylaminoethyl)benzamide hydrochloride
Apparent partition coefficient is given by the following equation.

$$\text{apparent partition coefficient} = \frac{\left(\begin{array}{c}\text{drug concentration in water phase}\\\text{before the distribution is carried out}\end{array}\right) - \left(\begin{array}{c}\text{equilibrium concentration}\\\text{in water phase}\end{array}\right)}{(\text{equilibrium concentration in water phase})}.$$

Adapted from [122].

species in the interfacial region as suggested by Fiese and Perrin [124] and Patel and Zografi [125].

Compartmental kinetic analyses by the method of Doluisio et al. [126] of data for the in situ rectal absorption of quinine revealed that the rate constant of absorption from the gut lumen to the absorptive membrane was increased by four- to five-fold in the presence of the anions investigated. 'This phenomenon', state Suzuki et al. [123], 'can be interpreted as an increase of the binding tendency of the drug to the absorptive membrane. Binding or 'accessibility' seems to be favoured by the ion-pair formation'. They continue, 'This view was further substantiated by the results of the experiments in which the amount of intravenously administered drug being taken up by perfusion of the gut was measured. The apparent rate that the drug entered into the intestinal lumen was hardly affected by the intestinal perfusion of NaLS solution, thus ruling out the possibility of a general increase in permeability caused by NaLS, an anionic component of the ion-pair'.

These explanations are difficult to reconcile with the known effects of NaLS on membrane permeability.

Following on reports (referred to above) of macromolecule absorption facilitated by surfactants, there have been successful attempts to achieve insulin absorption per rectum [127–129, 126]. Non-ionic ethers, anionic, cationic and amphoteric surfactants, as well as bile acids, increased absorption. The optimal effect has been obtained with 1% polyoxyethylene (9) lauryl ether [128], the effect of both polyoxyethylene chain length and alkyl chain having been

Table 7.12 Effects of polyoxyethylene (POE)(n) fatty alcohol ethers in insulin suppositories on blood glucose level in rabbits. Insulin suppositories contained 0.5% polyoxyethylene (n) fatty alcohol ethers and 1 U kg^{-1} insulin in corn oil. The initial blood glucose concentration was 118.3 ± 6.2 mg/100 ml. Each value represents the blood glucose concentration at 30, 60, 90 and 120 min after rectal administration of insulin suppositories and mean of three rabbits $\pm$ s.e.m.

Surfactants POE (n)—alcohol ethers	Decrease in blood glucose %			
	30 (min)	60 (min)	90 (min)	120 (min)
(3) lauryl	-8.8 ± 7.7	-3.0 ± 3.5	-11.2 ± 2.3	-9.8 ± 4.6
(6) lauryl	-12.3 ± 0.6	-23.8 ± 6.9	-20.6 ± 6.5	-11.2 ± 6.1
(9) lauryl	-12.7 ± 8.5	-47.9 ± 5.6	-47.1 ± 7.4	-32.6 ± 11.2
(25) lauryl	$+0.6 \pm 1.2$	-4.2 ± 2.6	-4.0 ± 2.8	-0.9 ± 1.0
(40) lauryl	$+17.3 \pm 2.9$	$+18.5 \pm 2.5$	$+14.9 \pm 3.1$	$+13.5 \pm 8.2$
(9) octyl	$+3.9 \pm 5.5$	$+12.8 \pm 8.0$	$+13.6 \pm 9.2$	$+13.0 \pm 6.2$
(9) decyl	-21.6 ± 4.8	-36.2 ± 3.7	-16.2 ± 5.1	-12.6 ± 6.8
(9) cetyl	-28.4 ± 3.6	-43.1 ± 2.6	-35.9 ± 5.7	-16.8 ± 5.4
(9) stearyl	-22.0 ± 6.2	-22.2 ± 3.2	-19.8 ± 4.8	-26.2 ± 7.9

From [128].

determined (Table 7.12). The dose of insulin in suppositories requires to be two to three times the intravenous dose to produce the same order of hypoglycaemia.

7.4.3 Surfactants and intramuscular injections

Drugs administered intramuscularly are absorbed after diffusion of the soluble molecular species across capillary walls. Molecular size and charge and protein binding of the drug are influences, and as diffusion of a soluble species is involved, as well as membrane transport, it is not surprising that surfactants can affect absorption of drugs from muscle. Both promotion and reduction in absorption have been detected.

The basic polypeptide antibiotic, enduracidin, isolated from *Steptomyces fungicidus* has a molecular weight of about 2500 and is poorly absorbed following i.m. administration [130]. Non-ionic surfactants promote absorption; polysorbate 80 and a series of surfactants based on hydrogenated castor oil (HCO) with $n = 30, 50$, and 120 used at 5% levels enhanced absorption, an optimal effect being obtained with the surfactant HCO-50 (HLB = 13.4). The surfactant might both increase capillary permeability to the drug or prevent precipitation of the antibiotic in the tissues (it precipitates on the addition of sodium chloride in the absence of surfactant). Some results are shown in Fig. 7.30. An opposite effect on the absorption of some water-soluble drugs has been observed [131, 132].

Low concentrations of the polysorbate surfactant series reduced absorption of isonicotinamide, insulin, procaineamide and sulphanilamide [131]. Several mechanisms for the reduction in absorption and plasma levels of these drugs were considered. Micellar interaction was ruled out as a negligible effect in these

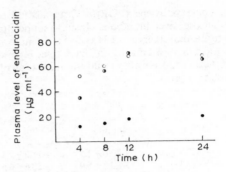

Figure 7.30 Effect of HCO-50 on blood levels of enduracidin following intramuscular injections to rats. Concentration of Enduracidin; 2.5 % dose; 6.25 mg/rat (0.25 ml injected)
●: without HCO-50
O: with 2.5% HCO-50
◑: with 5% HCO-50
Each value is an average of four rats. From Matsuzawa *et al.* [130] with permission.

systems. Histological investigation showed no relationship between the inhibition of absorption and the inflammation caused by polysorbate 80 at the site of injection [132]. Neither was the effect of the surfactant exerted on the capillary walls. In the process of studying this problem the disappearance of polysorbate 80 from rat thigh muscle was measured (Fig. 7.31). Obviously for the surfactant to exert some physical effect it must be at the site of injection; isonicotinamide escapes faster than the surfactant administered as a 5.0% solution. After 6 h 10.51 ± 0.79% of surfactant remained, while after 24 h less than 5% of the surfactant remained in the thigh muscle. Kobayashi and his colleagues have concluded that the polysorbates exert their action by reducing the rate of transport of the drugs

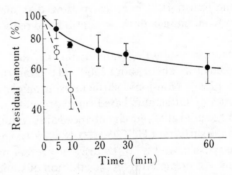

Figure 7.31 Semilogarithmic plots of the disappearance of polysorbate 80 and isonicotinamide from the rat thigh muscle.
A 50 μl of 5.0% polysorbate 80 containing 50 mM isonicotinamide was injected intramuscularly. Vertical bars indicate standard deviation.
—●—: polysorbate 80
—O—: isonicotinamide
From Kobayashi *et al.* [132] with permission.

through the extracellular space and connective tissue. 5.0% polysorbate reduces transport through the extracellular space by about 17%, 10 min after administration using insulin as a marker. Intradermal injection of polysorbate reduces the spread of a dye administered at the same time [133]. While this might be due to solubilization of the dye it could also be ascribed to a reduction in dermal tissue permeability.

7.4.4 Surfactants and percutaneous absorption

Skin permeability is increased by contact with a variety of substances, soap and detergents being deemed to be among the most damaging of all substances routinely applied to the skin [134]. The increased permeability of the human epidermis can be measured in the presence of very low concentration of anionic and cationic surfactants [135], although non-ionic surfactants are less damaging [136]. Detailed investigations of the interaction of surfactants with skin have been undertaken [137] showing that typical cationic and non-ionic surfactants are weak penetrants of skin unlike sodium lauryl sulphate which readily penetrates and destroys the integrity of the stratum corneum in hours. The addition of polyoxyethylene glycols or polyoxyethylated non-ionic surfactants to NaLS solutions reduces the rate of permeation of the ionic surfactant (see Fig. 7.32) possibly by complexing or forming mixed micelles with NaLS (see Chapter 6). Such effects are termed 'anti-irritation' phenomena and may generally be the result of three separate mechanisms [138]:

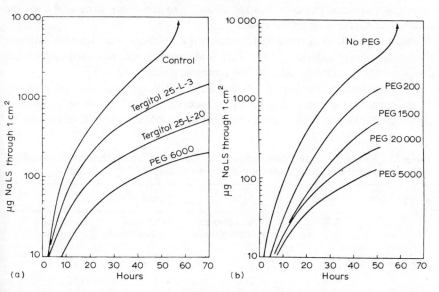

Figure 7.32 Permeation curves for 10% sodium lauryl sulphate (control) containing (a) 5% various ethoxylated compounds, PEG6000, Tergitol 25-L-3 ($C_{11-15}E_3$) and Tergitol 25-L-20 ($C_{11}E_{20}$) and (b) 5% of a series of PEG homologues. The membranes used were neonatal rat stratum corneum membranes. From Faucher *et al.* [137].

(1) prevention of intimate physical contact of irritant and skin;
(2) complexation; and
(3) blocking of otherwise reactive sites on the skin.

The first would occur by occlusion caused by film formers. The last can only be speculated upon but it is possible that competition for sites does occur.

Provided that the surfactants used in topical formulations are not toxic a more pressing biopharmaceutical concern is the effect of surfactants on the absorption of active ingredients. The nature of topical cream and ointment formulations is such that the unravelling of surfactant influences is not easy. Surfactants will affect the stability of an emulsified vehicle, the solubility of drugs in the vehicle and the spreadability of the formulation and thus directly influence drug release. This can be demonstrated *in vitro*; for example non-ionic surfactants increase the diffusion of sulphanilamide from oily bases and solubilization of the drug in the vehicle retards its release compared to release from systems in which the drug is suspended [139]. Such effects are to be anticipated from the prediction of Poulsen and colleagues [140] of the factors affecting release of drugs from topical formulations. The problem is similar to that encountered in Fig. 7.23 with sulphonamide suspensions and solutions, but less well defined. Inhibition zones on agar plates obtained with different concentrations of solubilized and suspended sulphanilamide in ointments containing 15 % surfactant are shown in Table 7.13. Increasing the sulphanilamide concentration increases the amount solubilized in both polysorbate 85 and 20; in the latter at 3 % sulphonamide, inhibition zones are maximal for reasons that are not clear. Obviously in the suspension systems the base has the capacity to solubilize drug and thus the continuous phase is never saturated.

The incorporation of emulsifying agents into ointments was shown to improve the release of sulphadiazine [141]. Such findings were also obtained using other drugs such as hexetidine, and yellow mercuric oxide [142, 143].

An *in vivo* method has been developed for monitoring the effect of polysorbate 85 on epidermal permeability [144] by measurement of moisture loss over a period of several days. After treatment for this period of time with surfactant the

Table 7.13 Inhibition zones (mm) obtained with different concentrations of solubilized and incorporated sulphanilamide (in ointments containing 15 % surfactant)

Sulphanilamide (%)	Tween 85		Tween 20	
	Solubilized sulphanilamide	Incorporated sulphanilamide	Solubilized sulphanilamide	Incorporated sulphanilamide
0.5	2.06	3.75	3.25	5.50
1	3.50	4.88	4.50	4.75
2	6.62	7.18	4.75	4.33
3	7.56	7.87	5.16	5.63
4	—	8.31	4.00	5.50
5	—	8.54	4.31	5.50

From [139].

skin can become irregular due to sloughing of the epidermis [145] when the surface is occluded with the vehicle.

Only sodium lauryl sulphate and sodium laurate increased the permeation of naproxen from aqueous gels through excised human abdominal skin, hexadecyl pyridinium chloride, polysorbate 60 and polyoxyethylene (23) lauryl ether decreasing permeation or having little effect [146]. Methyl decylsulphoxide, a surfactant derivative of dimethylsulphoxide (DMSO) has at 1% levels a considerable effect on flux, increasing it in excised human skin by ten times when naproxen was presented as an O/W cream formulation. Its mode of action is not known. These cationic surfactants are thought to bind to α-protein causing a reversible denaturation and uncoiling of the filaments. Membrane expansion, 'hole' formation and loss of water binding capacity are said to be consistent with the reversible $\alpha \rightleftharpoons \beta$ conversion of keratin [147] induced by surfactant binding. A more extensive range of non-ionic surfactants was incorporated into white petrolatum USP ointment base containing 10% salicylic acid or sodium salicylate (11.5% w/v) with dimethyl sulphoxide [148] (Fig. 7.33). These formulations were applied to rabbits and percutaneous absorption found to increase significantly in the presence of several of the non-ionic surfactants, even when DMSO was present. The nature of the effects is not yet clear.

Salicylic acid is absorbed faster from two ointment bases containing surfactants when applied to oral mucous membranes than from bases containing none but the complexity of the formulations used prevents detailed analysis. Some results are shown in Fig. 7.33c.

Using a more restricted range of components, an attempt has been made to optimize a steroid formulation containing propylene glycol or polyoxypropylene (15)-stearyl ether [150]. The partition coefficient between skin and vehicle, P_s, the solubility of the drug diflorasone diacetate in the vehicle and the percutaneous absorption were measured. Data for P_s and solubility are presented in Fig. 7.34a. The solubility increases with increasing surfactant concentration, a break occurring at about 0.2 weight fraction. As the solubility in the vehicle increases, P_s naturally falls. The steady state flux of ^3H diflorasone diacetate from various formulations (Fig. 7.34b) decreases with increasing surfactant beyond 0.2 weight fraction, the results agreeing moderately with those predicted from the following analysis [150].

The data obtained for the *in vitro* percutaneous penetration kinetics of diflorasone diacetate in vehicles consisting of propylene glycol–water and polyoxypropylene (15)-stearyl ether–mineral oil suggest that the skin is the rate-determining barrier for this compound. In this case, the appropriate relationship is represented by:

$$-\frac{dC_F}{dt} = \frac{P_s C_F D_s}{V_F h_s},$$ (7.33)

where the equation refers to unit area (1 cm^2).

C_F = concentration of dissolved diflorasone diacetate in the vehicle ($\mu g\,cm^{-3}$)
D_s = diffusion coefficient of diflorasone diacetate through the skin (cm^2 s^{-1})

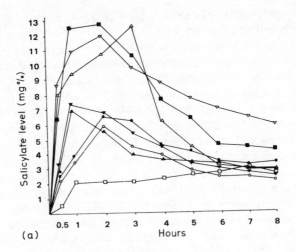

(a)

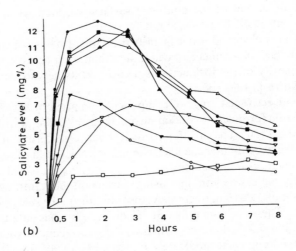

(b)

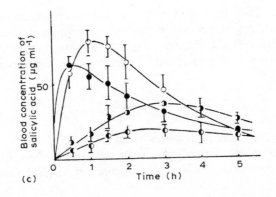

(c)

h_s = thickness of the skin barrier (cm)
P_s = diflorasone diacetate skin-vehicle partition coefficient
V_F = volume of formulation applied (cm^3).

The thickness of the skin barrier and the diffusion coefficient are combined and defined as a resistance, $R_s = h_s/D_s$. The resistance has units of time per length. Equation 7.33 can be simplified to:

$$-\frac{dC_F}{dt} = \frac{P_s C_F}{V_F R_s}.$$ (7.34)

Figure 7.33(a) Effect of sorbitan and polysorbate surfactants on percutaneous absorption of salicylic acid in the presence of dimethyl sulphoxide. △, 10% sorbitan monolaurate plus 10% dimethyl sulphoxide plus 10% salicylic acid; ▽, 10% sorbitan monopalmitate plus 10% dimethyl sulphoxide plus 10% salicylic acid; ■, 10% sorbitan trioleate plus 10% dimethyl sulphoxide plus 10% salicylic acid; ●, 10% polysorbate 20 plus 10% dimethyl sulphoxide plus 10% salicylic acid; ▲, 10% polysorbate 40 plus 10% dimethyl sulphoxide plus 10% salicylic acid; ▼, 10% polysorbate 60 plus 10% dimethyl sulphoxide plus 10% salicylic acid; ○, 10% dimethyl sulphoxide plus 10% salicylic acid; and □, 10% salicylic acid.
(b) Effect of poloxamer and polyoxyethylene surfactants on percutaneous absorption of salicylic acid in the presence of dimethyl sulphoxide. △, 10% poloxamer 182 plus 10% dimethyl sulphoxide plus 10% salicylic acid; ▽, 10% poloxamer 184 plus 10% dimethyl sulphoxide plus 10% salicylic acid; ■, 10% poloxamer 231 plus 10% dimethyl sulphoxide plus 10% salicylic acid; ●, 10% polyoxyethylene (2) oleyl ether plus 10% dimethyl sulphoxide plus 10% salicylic acid; ▲, 10% polyoxyethylene (4) lauryl ether plus 10% dimethyl sulphoxide plus 10% salicylic acid; ▼, 10% polyoxyethylene (20) oleyl ether plus 10% dimethyl sulphoxide plus 10% salicylic acid; ○, 10% dimethyl sulphoxide plus 10% salicylic acid; and □, 10% salicylic acid. From Shen *et al.* [148] with permission.
(c) Blood concentrations of salicylic acid following application of four different ointments to the cheek pouch of the hamster. ○ absorption ointment, ● hydrophilic ointment, ◑ macrogol ointment, ◐ white petrolatum. Each symbol represents the mean of five determinations with different animals. Bars indicate the standard error. The formulation of the ointments is given below

	Absorption ointment	Hydrophilic ointment	Macrogol ointment	White petrolatum
White petrolatum	40.0	25.0	—	98.0
Cetyl alcohol	18.0	—	—	—
Stearyl alcohol	—	22.0	—	—
Hexadecyl alcohol	—	—	—	—
Oleyl alcohol	—	—	—	—
Lanolin	—	—	—	—
Beeswax	—	—	—	—
Sorbitan monooleate	5.0	—	—	—
Sorbitan monostearate	—	—	—	—
Propylene glycol	—	12.0	—	—
Sodium lauryl sulphate	—	1.5	—	—
Macrogol 400	—	—	49.0	—
Macrogol 4000	—	—	49.0	—
Salicylic acid	2.0	2.0	2.0	2.0

From Tanaka *et al.* [149].

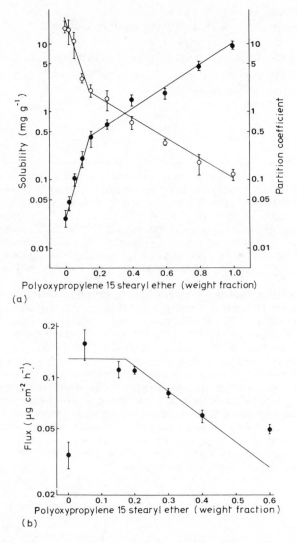

Figure 7.34 (a) Solubility and partition coefficients of diflorasone diacetate as a function of the weight fraction of polyoxypropylene (15) stearyl ether in mineral oil; average ($\pm$ s.D.) of four determinations. ●, solubility, and ○, partition coefficient. (b) Steady-state flux of 0.05% ³H-diflorasone diacetate formulations containing various weight fractions of polyoxypropylene (15) stearyl ether in mineral oil. The solid line was generated using Equation 7.33. The points are experimental values obtained from penetration studies. From Turi *et al.* [150] with permission.

The concentration of dissolved diflorasone diacetate, C_F, the partition coefficient, P_s and possibly the resistance, R_s, are influenced by the quantity of solvent or surfactant in a given vehicle. Under certain conditions, the solubility of a drug in a co-solvent system can be represented by the following expression:

$$C_F = C_0 \, e^{\alpha(f_s)} \tag{7.35}$$

where C_0 is the solubility of the drug in the formulation when the weight fraction of the solvent is zero, α is a constant, and f_s is the weight fraction of the solvent.

In a similar manner, the partition coefficient of a drug between the skin and the vehicle can be expressed as:

$$P_s = P_0 \, e^{-\beta(f_s)}, \tag{7.36}$$

where P_0 is the partition coefficient of the drug between the skin and the vehicle when the weight fraction of the solvent is zero and β is a constant. Inserting Equations 7.35 and 7.36 into Equation 7.34 leads to

$$-\frac{dC_F}{dt} = \frac{\left[P_0 \, e^{-\beta(f_s)}\right]\left[C_0 \, e^{-\alpha(f_s)}\right]}{V_F \, R_s}. \tag{7.37}$$

During the steady state period of penetration, the following relationship is valid:

$$V_R \frac{dC_R}{dt} = -\frac{V_F \, dC_F}{dt}, \tag{7.38}$$

where C_R is the concentration of diflorasone diacetate in the receptor compartment of the diffusion apparatus and V_R is the volume of the receptor compartment.

Equation 7.38 states that the amount of diflorasone diacetate leaving the vehicle per unit time is equal to the amount entering the receptor solution of the diffusion apparatus. With this relationship, Equation 7.37 can be written as:

$$V_R \frac{dC_R}{dt} = \frac{\left[P_0 \, e^{-\beta(f_s)}\right]\left[C_0 \, e^{\alpha(f_s)}\right]}{R_s}. \tag{7.39}$$

Integration of Equation 7.39 gives

$$Q_R = \frac{1}{R_s}\left[P_0 \, e^{-\beta(f_s)}\right]\left[C_0 \, e^{\alpha(f_s)}\right] t \tag{7.40}$$

where Q_R is the amount of diflorasone diacetate in the receptor compartment at time t. Equation 7.40 predicts that the addition of a solvent to a formulation could increase, decrease, or have no effect on the amount of drug diffusing through the skin. The result depends on the magnitudes of α and β and whether or not the drug solution is saturated or unsaturated.

Consideration of Equation 7.33 shows immediately that the vehicle has an influence on the absorption of the drug; if the vehicle is changed so that the drug becomes less soluble in it, P increases so that permeability increases. The vehicle is more dominant in topical therapy than in most routes of administration because the vehicle remains at the site, although not always in an unchanged form. Evaporation of water from the base would leave drug molecules immersed in the oily phase. Oil-in-water emulsion systems may invert to water-in-oil systems, such that the drug would have to diffuse through an oily layer to reach the skin. Non-volatile components of the formulation increase in concentration as the volatile components are driven off; this may alter the state of saturation of the drug and hence its activity. Drug may precipitate due to lack of *remaining solvent*. These changes mean that theoretical approaches very much represent the ideal case.

The thermodynamic activity of the drug is obviously the determinant of biological activity. If the solubility of the drug in the base is increased by addition of propylene glycol then its partition coefficient towards the skin is reduced. On the other hand, the increasing amount which can be incorporated in the base is an advantage and the concentration gradient is increased. It is apparent that there is an optimum amount of solubilizer. The optimum occurs at the level of additive which just solubilizes the medicament. Addition of excess results in desaturation of the system, and therefore a decrease in thermodynamic activity.

Other aspects of formulation such as the nature of the binary or ternary vehicle (oil–surfactant, water–surfactant or oil–water–surfactant, respectively) have been considered recently [151]. Addition of polysorbate 80 to the aqueous phase has no significant effect on the epidermal transport of ethanol, but a significant reduction in the transport of the less soluble octanol results, in line with the arguments presented above; in isopropyl myristate, octanol transport is not affected by the solubilizer while that of ethanol is decreased. In the ternary systems identified in Fig. 7.35, the results in Table 7.14 were obtained indicating a general decrease in permeability constants for ethanol, butanol and octanol. The viscosity of the vehicles was not a factor although this varied from 1 to 39×10^3 cP. In the ternary systems a surfactant will distribute itself between the aqueous and non-aqueous phase; quantitative prediction of permeation is made difficult even with data on the transport properties of the permeants in the individual phase. The results indicate that the percutaneous absorption of the

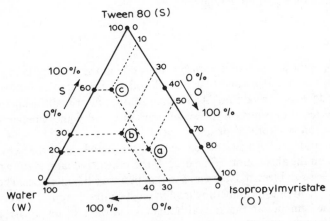

Figure 7.35 Phase diagram of the isopropylmyristate, polysorbate 80, water system. From Garcia *et al.* [151].

Table 7.14 Epidermal permeability constants for three alkanols in ternary vehicles

Alcohol	K_p (cm h^{-1} × 10^3)		
	a	b	c
Ethanol	1.8 ± 0.36	1.4 ± 0.13	1 ± 0.3
Butanol	3.6 ± 0.21	3.5 ± 1.2	1.1 ± 0.12
Octanol	0.1 ± 0.02	0.1 ± 0.02	0.01 ± 0.001

a, b, c as in Fig. 7.35.
From [151].

alcohols from these two-phase mixtures is a function of their affinity for the aqueous phase in contact with the stratum corneum.

Non-ionic surfactants affect the local anaesthetic intensity and duration of tetracaine when present at 5% levels in a vehicle containing propylene glycol (10%) [152]; corroborative results on bupivacaine have been published [153] indicative of increased penetration of the anaesthetic (Table 7.15). However, an increased toxic response was observed due both to increased drug levels in the tissues and to the vasodilating effect of the surfactant (a C_{12} polyoxyethylene ether). The surfactant increased the toxicity of bupivacaine when administered together into the trachea or the bladder of rabbits. The somewhat complicated relationship between solubility and toxicity in the bladder is demonstrated by the finding that racemic bupivacaine with a higher solubility than the D (+) and L (−) isomers does not precipitate in the bladder and is consequently absorbed to a greater extent.

Table 7.15 Topical anaesthesia in man by application of test solutions on the medial part of the upper lip. The maximal pain thresholds are expressed in per cent of the normal pain thresholds (NaCl) of the individual volunteers ($n = 6$.)

Compound	Concentration (%)	Pain threshold (%)	Duration of anaesthesia (min) ($\bar{x} \pm$ S.E.M.)
Surfactant (SA)	1.0	100 ± 12	< 5
Bupivacaine	1.0	130 ± 9	8 ± 3.1
Bupivacaine	2.0	175 ± 10	18 ± 4.4
Bupivacaine + SA	2.0 + 0.25	190 ± 8	22 ± 3.2
Bupivacaine + SA	2.0 + 0.50	190 ± 14	28 ± 3.6
Bupivacaine + SA	2.0 + 1.0	270 ± 19	41 ± 6.4
Tetracaine	2.0	250 ± 16	58 ± 7.4
NaCl	0.9	100	—

From [153].

7.4.5 Surfactants and corneal permeability

Marsh and Maurice [154], in their paper on the influence of non-ionic detergents on human corneal permeability, list previous attempts at increasing corneal penetration by application of surfactants, dating back to 1942. Ionic surfactants increase drug penetration in both man and animals [155–160] and in studies on non-ionic detergents [158–162], polysorbate appeared to be the most effective agent. As with many of the studies we have discussed, the range of studies of the surfactants was too wide to allow a better understanding of the processes involved. Marsh and Maurice [153], however, have attempted to relate the HLB of the surfactant to the corneal permeation of fluorescein in human subjects. No clear relationship with HLB was adduced although surfactants with HLB values in the range 16 to 17 including polysorbate 20 and Brij 35 caused the greatest increase in permeability, Myrj 52 had little effect in spite of an HLB value in this range, Brij 58 was also effective but caused 'alarming epithelial changes' [154]. The concentration dependency of the effect of polysorbate 20 and Brij 35 are shown in Fig. 7.36. Both substances are non-irritating to the rabbit eye [163]. Maximum comfortable levels are shown in Fig. 7.36.

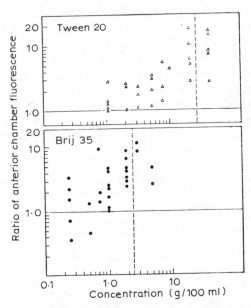

Figure 7.36 Effect of concentration of surfactants, HLB 16-17, on penetration of fluorescein from 1 drop into the anterior chamber of the eye. Ordinate: Log ratio of fluorescein in experimental aqueous humour to that in control. Vertical line: maximum comfortable concentration. From Marsh and Maurice [154] with permission.

7.5 Miscellaneous formulations and the influence of surfactants

Evidence for increased transport of ^{14}C-labelled nitrogen mustard N-oxide through rat ascites hepatoma cell membranes by polysorbate 80 [164] and the recent observation that this surfactant can increase methotrexate uptake into the brain [82a] gives rise to hope that surfactants can play a useful role in modification of drug action in specialized forms of treatment. This is a largely unexplored field. This finding of enhanced entry of anticancer drugs into tumours is a clear indication of the influence of solubilizer but the solubilizer may have other effects which might preclude their use in medicines. At least one should be aware of the potential. Especially is this true in biochemical and pharmacological experiments when disregard of the surfactant's potential for altering pharmacological response is scientifically dangerous. As an example one can cite several papers which have discussed the difficult problem of formulating $^9\Delta$-tetrahydrocannabinol (THC), a very lipid-soluble molecule with a high octanol/water partition coefficient [165]. A mixed solvent system of ethanol, polyoxyethylated non-ionic surfactant, and physiological saline (5:5:90) previously used for two antineoplastic nitrosoureas [166] was evaluated as a solvent for THC. The utility of the solvent depends on its pharmacological effects in the test system and the results seemed equivocal. The more detailed study of Roth and Williams [165], in particular in relation to the interaction of THC with specific receptor sites or membrane components and the effect of solubilizers on this is valuable. The membrane/solvent partition coefficient of the THC was reduced to almost zero at levels of cremophor EL of 0.4 mg ml^{-1}, the effects of polysorbate 80 being qualitatively similar. Ethanol also decreased the partition coefficient but at 5% v/v the reduction was not as significant. Membrane concentrations of the THC are estimated to be considerably reduced by the presence of solubilizers although this effect can be compensated to some extent by the increased concentrations that can be applied. Nevertheless the conclusion reached by Roth and Williams [165] was that the use of solubilizers to increase the water solubility of THC did not increase the membrane concentration to levels in excess of those which would occur in the absence of solubilizer; the only apparent advantage of adding solubilizer was 'to decrease the loss by adsorption of THC on to glassware and other apparatus!' At least one recent paper [167] has acknowledged the potential biological problems with solubilizing agents and reports ethanol as a substitute in an examination of THC on contractions of the isolated rat vas deferens.

The manner in which the formulation of THC affects the biological performance has been examined by two groups [168, 169]. Polysorbate 65–sorbitan monolaurate mixtures appeared to confer a longer duration of action by the intraperitoneal or subcutaneous route when compared with a PVP suspension and polysorbate 80 dispersion. Contrary to these findings, Sofia *et al.* [169] find 1% polysorbate 80 to be a poor vehicle for oral, s.c. i.p. or i.v. administration producing inadequate results when compared to a PVP dispersion or a dispersion in propylene glycol. The most suitable vehicle did, however, contain 1% polysorbate. The divergent opinion no doubt arises from the fact that the preparations are dispersions or emulsions and their mode of manufacture differs from laboratory to laboratory. Particle size and stability will thus vary.

Propylene glycol and 20% Cremophor EL have been compared as vehicles for diazepam [170] in view of the number of reports of thrombophlebitis associated with intravenous diazepam. The Cremophor vehicle caused significantly less post-injection thrombophlebitis possibly because it prevents the precipitation of the drug substance at the site of injection by its solubilizing effect. Ease of injection was also improved owing to the lower viscosity of the aqueous Cremophor (Table 7.16). These are relevant factors in the choice of a formulation and can override advantages which might be gained in drug absorption.

It is often assumed that drug solutions must be the most bioavailable form of the drug, but solutions of drugs poorly soluble at tissue pH will precipitate at the site of injection and subsequently release from the site might be slow. Precipitation may also occur if the drug is solubilized in a mixed solvent. In the presence of surfactant, as solubility is in most cases a linear function of surfactant

Table 7.16(a) Frequency (%) of thrombophlebitis after i.v. diazepam dissolved in propylene glycol or Cremophor EL

Complication	Propylene glycol (right hand)	Cremophor (right hand)	Other anaesthetic agents (left hand)
Swelling			
None	80.0	93.3	95.2
Moderate	20.0 { 8.9	6.7 { 6.7	4.8 { 3.8
Marked	{ 11.1	{ 0	{ 1.0
Erythema			
None	86.7	98.3	96.1
Moderate	13.3 { 8.9	1.7 { 1.7	3.9 { 2.9
Marked	{ 4.4	{ 0	{ 1.0
Phlebitis			
None	37.8	96.6	91.3
Moderate	62.2 { 17.8	3.4 { 1.7	8.7 { 6.7
Marked	{ 44.4	{ 1.7	{ 2.0

(b) Symptoms after i.v. injection of diazepam dissolved in propylene glycol or Cremophor (%)

Pain or functional disturbance	Propylene glycol (right hand)	Cremophor (right hand)	Other anaesthetic agents (left hand)
None	34.1	91.5	91.4
Moderate	65.9 { 34.1	8.5 { 6.8	8.6 { 4.8
Severe	{ 31.8	{ 1.7	{ 3.8

(c) Ease of injection of diazepam dissolved in propylene glycol or Cremophor (%)

Solvent	*n*	Good	Fair	Poor
Propylene glycol	82	35.4	37.8	26.8
Cremophor	97	62.9	32.0	5.1

From [170].

concentration, precipitation should not occur, although it will occur if the surfactant is diluted to below its CMC. A comparison of plasma levels achieved by two commercial formulations of diazepam (available in Finland) showed that diazepam solubilized in cremophor EL (Stesolid[R]) produced peak levels twice those produced by a formulation in propylene glycol (Diapam[R]) which is the same solvent as used in Valium[R] injection (Fig. 7.37) [171].

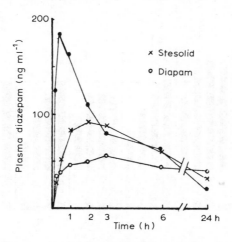

Figure 7.37 Plasma diazepam levels following intramuscular administration of diazepam; O in propylene glycol vehicle; × in a Cremophor EL vehicle compared with ● intravenous administration of the propylene glycol preparation. From Kanto [171] with permission.

7.6 Surfactants and antibacterial activity

As surfactants alter the permeability of mammalian cells it is not too surprising that, in spite of the differences between bacterial cell walls and mammalian cell membranes, some surfactants have the ability to increase the permeability of the bacterial cell wall or to act synergistically with antibacterial agents. There are several unique facets to discussion of this topic; some surfactants have antibacterial properties and some antibacterial agents have surface-active properties. In considering the subject one has to be aware, as before, of surfactant antibacterial interactions, the influence of surfactant on the performance of the dosage form or formulation, and surfactant–cell wall interactions.

Thus the antimicrobial effectiveness of the range of substances presented in Fig. 7.38a and b at a range of surfactant concentration will not be a simple function. The inactivation that occurs is frequently preceded at lower concentrations by an enhancement of activity. This is brought out well in Fig. 7.38a and b.

Since Dubos and Davis [174] first recommended the use of a polysorbate–albumin medium for cultivation of tubercle bacilli there have been a number of reports on the effect of these compounds on antibacterial activity. Reduction in activity has been pointed out by Forrest *et al.* [175], Youmans and

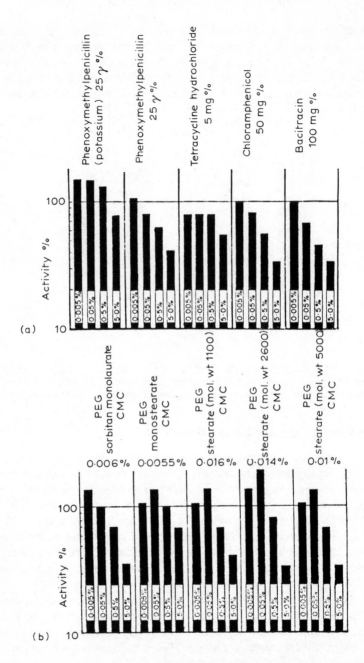

Figure 7.38 (a) and (b) The effect of non-ionic detergents on the activity of antibiotics as shown by the effect of: (a) polyoxyethylene lauryl ether (CMC 0.011 %) on penicillin, tetracycline, chloramphenicol, and bacitracin in concentrations ranging from 0.005 to 5.0 %, and (b) the effect of five different non-ionics on the activity of chloramphenicol (50 mg %). The critical micellar concentrations of the detergents are shown [172].

Youmans [176], and Fenner [177]. This inactivation is utilized in microbiological tests where sterility tests are being carried out in the presence of antibacterial agents [174].

Natori [178] found that the addition of 0.05 % polysorbate 80 decreased the activity of isoniazid, 4,4'-diaminodiphenylsulphone, oleic acid, and 3-aminoben-zofuran, but had no significant effect on streptomycin sulphate. The activity of isoniazid and 4,4'-diaminodiphenylsulphone was, however, reported to be increased by the addition of surfactants under different conditions [176].

The decrease of antibacterial activity caused by the addition of surfactants has been broadly related to solubilization of the antibacterial in the detergent micelles. The apparent increase in solubility in the presence of surfactants, does not exactly parallel the decrease in biological activity. Undoubtedly the effect of the surfactant on bacterial permeability and viability will be one factor causing deviation from strictly mathematical relationships. Correlation of mathemati-cally predicted preservative availability in solubilized and emulsified systems with the measured antimicrobial activity has been attempted by Kazmi and Mitchell [179]. Fig. 7.39 shows some of their results for the bacterial activity of chlorocresol against *E. coli* and theoretical estimates based on an equation relating free bactericide concentration and activity. The difference between the slopes of curves B and C in Fig. 7.39 according to Kazmi and Mitchell suggests that increasing the cetomacrogol concentration may decrease bactericidal activity. This decrease could be due to a stimulation of microbial growth or protection of the organism by the non-ionic surfactant. Although solubilized and emulsified dispersions with the same D_t are equitoxic, the present results indicate that they do not have the same activity as a solution of the preservative in water with the same D_f. The antimicrobial activity of chlorocresol in each surfactant solution was less than that of the solution in water. However, increasing the surfactant concentration in the range shown in Fig. 7.39 is not important provided that the concentration of 'unbound' or 'free' chlorocresol is the same.

Others [180] have asserted that theories equating antimicrobial activity to the concentration of non-micellar preservatives are inadequate. Comparison of systems containing benzoic acid with and without surfactant, for their antifungal activity versus *Schizosaccharoryces pombé* at equivalent values of free benzoic acid, demonstrated a significant increase in activity in the presence of surfactant (see Table 7.17). If no synergism between surfactant and antifungal agent occurs then systems containing the same free concentration should have identical activities regardless of total active agent present.

Figures illustrate the effect of non-ionic detergents on the activity of a range of antibiotics, showing the effect of concentration and of detergent structure on chloram-phenicol activity as shown by zone-inhibition assay. Whether this mode of assay bears any relationship to actual conditions *in vivo* is a matter for debate. See also Ullmann and Moser [173].

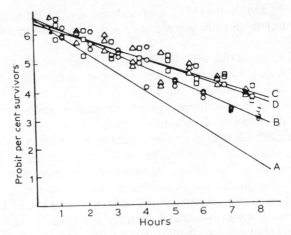

Figure 7.39 Probit % survivors as a function of time for the bactericidal activity of chlorocresol in aqueous cetomacrogol solutions against *E. coli*. [Cetomacrogol concentration (%)]: A, 0.0; B, ○, 1.0; C, □, 3.0; and D, △, 5.0 [Total preservative concentration, $[D_t]$ (%)]: A, 0.0350; B, 0.1743; C, 0.4528; and D, 0.7314. The initial free preservative concentration, $[D_f]$, = 0.035%. The points are experimental; the lines were fitted using a linear model. The equation

$$Y = \beta_0 + \beta_1 x_1 + \beta_2 x_2 + \beta_3 x_3 + \beta_4 t + \beta_5 x_1 t + \beta_6 x_2 t + \beta_7 x_3 t + \varepsilon$$

where β_0 is the intercept of curve A; β_1, β_2, and β_3 are the differences between the intercepts of curves B and A, C and A, and D and A, respectively; β_4 is the slope of curve A; β_5, β_6, and β_7 are the differences in slope between curves B and A, C and A, and D and A, respectively; x are the dummy variables where $x_1 = 1$ if curve B and otherwise is zero, $x_2 = 1$ if curve C and otherwise is zero, and $x_3 = 1$ if curve D and otherwise is zero; ε is a random variable; and t is time. From Kazmi and Mitchell [179] with permission.

Phenylethanol and polysorbate 80 when used in combination with benzalkonium chloride show enhanced activity against *Pseudomonas aeruginosa* [181–183]. Benzalkonium-sensitive cells grown in 0.5% polysorbate 80 appear to have normal cell walls [184]. Resistant cells grown in benzalkonium chloride solutions are also normal, but when grown in the presence of polysorbate 80 (0.02%) or benzalkonium chloride plus polysorbate 80 exhibits evidence of cytoplasmic damage. At high concentrations, polysorbate 80 totally inactivates the antipseudomonal activity of the benzalkonium chloride. At lower concentrations it undoubtedly increases the permeability properties of the cell and enables the benzalkonium to reach its site of action more efficiently. No evidence of synergism between hexadecyl pyridinium chloride or dodecyl pyridinium chloride and a series of non-ionic surfactants against *E. coli* above or below the surfactant CMC, was noted [185].

Polysorbate 80 was found to be more effective as an inactivating medium for hexachlorphene than serum albumin, but the Spans were found to be devoid of inactivating activity [186]. The importance of finding inactivators for hexachlorophene arises because its greatest area of usefulness has been in soaps

Table 7.17 Comparison of fungicidal activity of benzoic acid systems, with and without surfactant, at the same $[D_w]$*

Mixture	Exposure time (h)	% survival in replicate determinations				
		$1\frac{3}{4}$	$3\frac{1}{2}$	$3\frac{1}{2}$	$1\frac{1}{2}$	$1\frac{3}{4}$
	$[D_t]$	26.0 mM	26.0 mM	26.0 mM	29.0 mM	29.0 mM
	$[D_w]$	13.5 mM	13.5 mM	13.5 mM	15.0 mM	15.0 mM
B Benzoic acid + surfactant		12.8	0.8	0.1	0.04	0.09
		12.7	0.7	0.1	0.07	0.02
		13.1	0.8	0.2	0.00	0.1
		11.2	0.9	0.1	0.03	0.09
		15.3			0.00	0.07
	Mean	13.0	0.8	0.1	0.03	0.07
	$[D_t] = [D_w]$	13.5 mM	13.5 mM	13.5 mM	15.0 mM	15.0 mM
C Benzoic acid alone		24.9	5.3	0.2	0.6	2.8
		23.4	7.5	0.2	0.9	3.2
		27.2	6.3	0.2	0.4	3.0
		26.1	7.5		0.4	2.5
			3.8		0.5	2.8
	Mean	25.4	6.1	0.2	0.6	2.9

Saturation solubility, C_s (benzoic acid) in N/1000 HCl = 26.5 mM. C_s (benzoic acid) in 2% w/v surfactant = 51.0 mM.* $[D_w]$ is the concentration of benzoic acid in non-micellar phase. From [180].

and shampoos, and the consequent need to test their efficacy. Inactivation prevents false results caused by the retention of the bactericide on the bacterial cell.

An apparent potentiation of hexachlorophene by polysorbate 80 in high concentration in a hydrophilic ointment base, observed by a zone-inhibition method, agrees with the finding of Berthet [187] that the ionic Aerosol OT in a concentration of 1:2000 increased the phenol coefficient of hexachlorophene four-fold. However, the two concentration levels are different and the method of testing different. It is dangerous to rely on one method alone. The surface-active agents may, in some way, facilitate the diffusion of the antibacterial agent through the agar, yet in solution may solubilize it and reduce its activity. It is safe to assert that most drugs act in solution, so the agar plate method would give a misleading indication of surfactant effect. Certainly Anderson and Morgan [188] observed that their agar plate diffusion results for hexachlorophene–non-ionic surfactant systems bore no relationship to their solubilization or dialysis data. Attempts to determine minimum inhibitory concentrations of hexachlorophene in the presence of solubilizing agents have been prevented by the interaction of the surfactants with both components [188].

In actual use antibacterial agents will be lost from the system by interaction with bacteria, skin, foreign substances and the 'capacity' of the system to compensate for such losses has to be considered. In a solubilized system the 'capacity' would depend on the degree of saturation of the system or more precisely the change in total saturation of the system as a function of the saturation of the aqueous phase [188]. Anderson and Morgan have attempted to measure this by applying the results of dialysis experiments (Fig. 7.40). From this diagram it can be seen that if a saturated solution of hexachlorophene in 1% macrogol at pH 8 is used under conditions when the phenol is being lost, a fifth of the phenol can be removed and the residual activity is equivalent to that of an aqueous solution 85 to 90% saturated with respect to hexachlorophane; on the other hand, a similar loss of a fifth of the total phenol from a saturated solution of 1% Brij 35 reduces the activity to that of a 60% saturated aqueous solution. The uptake of antibacterial agents has been quantified. Hugo and Newton [189] when comparing the uptake of iodine by micro-organisms and serum from an iodide solution and an iodophor, found that there was a greater uptake of iodine from the former (see Fig. 7.41) which suggested that the iodine may 'be absorbed from the cetomacrogol system in the form of a complex, or that there is a greater affinity of the iodine for the cetomacrogol than for the ethanol–potassium iodide solution'. Interfacial tension will also play a part, as Freundlich considered that adsorption is greatest where the interfacial tension between solvent and substrate is high. The surfactant, of course, lowers the interfacial tension.

The uptake of hexylresorcinol by *E. coli* in the presence and absence of cetomacrogol exhibits the same trend as the uptake of iodine from aqueous solution and surfactant mixtures, with a marked reduction in the presence of detergent [190]; the rate of uptake is not affected. Beckett *et al.* [190] consider that the phenol–cetomacrogol complex probably prevents cell-wall penetration. The amount of hexylresorcinol bound per organism is less than the theoretical

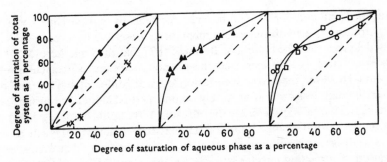

Figure 7.40 Distribution of hexachlorophene in solutions of various agents in aqueous 0.05M *tris* buffer, pH 8.0 at 25° C. ○, sucrose laurate 1%. □, polysorbate 20 1%. △, lauromacrogol 0.1%. ▲, lauromacrogol 1%. ×, macrogol 1%. ●, poloxamer 188 1%. Lauromacrogol = Brij 35; macrogol = PEG 4000; sucrose laurate = sucrose mono-laurate; poloxamer 188 = Pluronic F68.
From Anderson and Morgan [188] with permission.

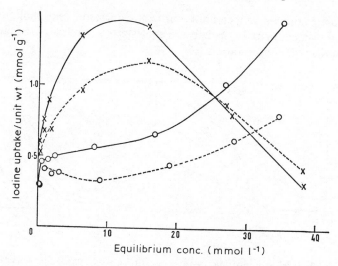

Figure 7.41 Adsorption isotherms for the uptake of iodine by *E. coli* (—) and *Staph. aureus* (- - -) from iodine formulations, after 5 h. *E. coli*, dry weight 2690 μg ml^{-1} for iodine solution and 3190 μg ml^{-1} for the iodine : cetomacrogol complex. *Staph. aureus*, dry weight 3430 μg ml^{-1} for iodine solution and 3060 μg ml^{-1} for the iodine : cetomacrogol complex. × Iodine solution. ○ Iodine : cetomacrogol complex (iodophor). From Hugo and Newton [189] with permission. Similar differences were obtained from the adsorption isotherms when the substrate was yeast or serum [189].

amount required to form a monomolecular layer around the organism in the absence of additive. It is possible [191] that hexylresorcinol becomes bound to the bacteria in the presence of excess cetomacrogol (i.e. in micellar solutions) in the form of a phenol–non-ionic complex. When cationic antibacterials or cationic surfactant antibacterials such as cetyl pyridinium chloride are involved, the anionic groups on the surface of most cells will be implicated in the antibacterial surfactant–cell interaction [192]. In yeast suspensions at pH values between 3.5 and 6.0 the cationic surfactants cetrimide and cetyl pyridinium chloride have strong cytolytic effects above certain critical concentrations. Sodium dodecyl sulphate, on the other hand, was only cytolytically active below pH 3.2 [193]. However, the strong binding of surfactant cations by the cell surface suggests that van der Waals' interactions were also operative between the surfactant alkyl chains and hydrophobic groups on the cell surface.

Riemersma [192] suggests that phosphate groups belonging to phosphoinositides, phosphatidic acid and other anionic lipids were involved in the ionic interaction of the surfactant head groups while the alkyl chain actually penetrated the membrane bilayer. At a certain concentration the membrane would form 'mixed micelles' with the surfactant cations leading to higher permeability and cytolysis. However, both anionic and cationic surfactants induce lysis and their mode of action cannot be identical. Bradford *et al.* [194] examining the solubilization of microsomal constituents observed that both CTAB and deoxycholate solubilized membrane protein, cholesterol phospholipid and an enzyme in a similar manner, protein and cholesterol solubilization occurring at a critical surfactant concentration (Fig. 7.42).

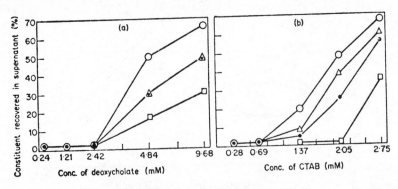

Figure 7.42 Solubilization of microsomal constituents by increasing concentrations of: (a) deoxycholate (b) cetyltrimethylammonium bromide; Protein (O), cholesterol (△), phospholipid phosphorus (●), and Na$^+$-ion-stimulated adenosine triphosphatase (□). Diagrams show the percentage of this constituent in the supernatant of the microsomal suspensions after treatment, as described by Bradford et al. [194].

In relation to surfactant influences on antibacterial activity some information is required on how surfactants will influence antibacterial binding to cell components. Some interactions between surfactants and body components including proteins are discussed in Chapter 10 but here we can consider some of the effects, discussed by Alhaique et al. [195], which might be fundamental to our understanding of this complex problem. They had found [196] that an allosteric transition could be easily effected by reaction of surfactant monomers with a protein in the presence of a ligand, chloramphenicol. The interaction of the anionic detergent, NaDS, with the protein resulted in an increase in the free antibiotic; in diffusion experiments this complex interaction was paralleled by an increase in the transfer rate of the antibiotic [197]. NaDS and CTAB both increase the amount of chloramphenicol bound to an albumin–lecithin complex when present in concentrations below their CMCs; polysorbate has little effect (see Fig. 7.43). This is most likely to be due to the surfactant causing the dissociation of the protein–phospholipid complex into surfactant–phospholipid and serum–albumin–surfactant complexes each capable of bonding the antibiotic to a greater extent than the original complex. As polysorbate 80 had negligible effects on ligand binding the dissociative process must be associated with adsorbed surfactant ions. Such interactions have been shown by Alhaique et al. [197] to alter the transport of chloramphenicol across a barrier prepared from aqueous dispersions of phospholipid or phospholipid–albumin complexes. In the latter case permeability coefficients are decreased and lag times increase on addition of low concentrations of NaDS (< 1 $\times 10^{-7}$ M) CTAB has the same effect and polysorbate 80, none.

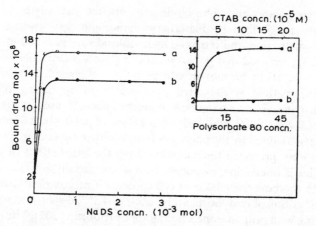

Figure 7.43 Association curves of chloramphenicol with the albumin–lecithin complex (3 % w/v) in phosphate buffer (pH = 6.8) at 25° C, in the presence of increasing amounts of sodium dodecyl sulphate (NaDS). Plots a (○) and b (●) refer to different initial concentrations of the antibiotic, i.e.—3.1 × 10⁻⁵ and 6.2 × 10⁻⁵ M, respectively. Inset. Association curves of chloramphenicol with the albumin–lecithin complex (3 % w/v) in phosphate buffer (pH 6.8) at 25° C, in the presence of increasing amounts of surfactants. Plots a′ and b′ refer to cetyltrimethylammonium bromide (●) (CTAB) and polysorbate 80 (○) (concn in mg ml⁻¹ × 10²), respectively. In all cases, the initial concentration of the antibiotic was 3.1 × 10⁻⁵ M. From Alhaique *et al.* [195] with permission.

While we can observe these and other effects in isolated systems they allow us, at this stage, simply to appreciate the variety of interactions that occur when foreign surfactant molecules insinuate themselves into membranes. We are probably not much further along the road to a complete understanding of the specific proteins or sites that are involved. Nor indeed has sufficient work been done to differentiate one surfactant's effects on a variety of membranes of known composition. The day of prediction is still a long way off.

7.6.1 Other observations on interactions of solubilizers and antibacterials

The formation of mixed micelles of quaternary ammonium compounds and non-ionic surfactants has been suggested as a possible mechanism for the association of antibacterials with polysorbate 20 [198], for such an interaction the degree of binding would be expected to increase with increasing length of hydrocarbon chain of the cation below its normal CMC. DeLuca and Kostenbauder [199] deduce that according to the treatment of the process by the law of mass-action, a maximum should occur in the concentration of monomeric long-chain ions as the total surfactant concentration is increased, and they were able to find experimentally a maximum in the adsorption isotherm for the interaction of cetyl pyridinium chloride and 0.2 % polysorbate 80 at 30° C. One could regard this process as a form of solubilization where this term is taken to mean 'interaction with micellar component'. The binding of organic electrolytes by non-ionic detergents is not limited to quaternary ammonium derivatives, as chlorpromazine, promethazine, and tetracine hydrochloride are also bound.

Ansel [200] reported that polyoxyethylene glycols prevent the haemolysis of rabbit erythrocytes by haemolytic concentrations of phenol, m-cresol, p-chlorophenol. The method is suggested as a means of appraising phenol–PEG interaction in preservative systems. A number of phenols cause leakage of the cell contents of E. coli. Judis [201] studying the effect of polysorbate 80 on the release of cell contents caused by the phenolics found further evidence of complexation, as the non-ionic protected the bacterium from the lethal effects of p-chloro-m-xylenol. This is interesting, as polysorbate is interfacially active and would be expected to promote the release of cell contents if no complex formation took place. Lytic effects of lysophosphatidyl choline (LPC) dispersions were reduced by saturating with progesterone, cholesterol or trioolein [202]. The lytic activity of a mixed LPC–phosphatidyl choline (PC) dispersion is completely abolished by incorporation of progesterone, suggesting that the co-operativity of the mixed micellization reduces the escaping tendency of the LPC reducing its ability to interact with the membrane. Progesterone itself is, in certain concentrations, haemolytic yet mixtures of this steroid with a haemolytic phospholipid can be devoid of such activity. Co-solubilization and mixed micelle formation can thus complicate an already complex picture. The inactivation of preservative esters in surfactant solutions in which other oil substances are solubilized has been investigated. The results are of special relevance in formulation studies. Propyl paraben was most subject to interference. For example, where its effective preservative level in the presence of surfactant is 0.162 %, this rises to 1.30 % when 2 % of isopropyl myristate is solubilized in the same surfactant solution [203].

The results of Matsumoto and Aoki [204] have been recalculated in terms of solubilities in 0.01 M solutions. This presents a picture exactly opposite to the one obtained on a percentage basis (see Fig. 7.44). This evidence should make the selection of a preservative for use in non-ionic systems less empirical. It is suggested that where no information is available for the interaction of the preservative with the detergent the least hydrophobic compound is used to minimize solubilization, or where a compound such as the butyl paraben must be employed a surfactant with a long hydrophilic group is chosen. Often a balance between the two will have to be made.

Chlorobutanol, benzyl alcohol, and phenylethyl alcohol–non-ionic systems have been studied at two temperatures [205], as was the binding of benzoic acid by polyoxyethylene stearates [206]. The latter confirmed that solubilization was greatest in the least hydrophilic surfactant (on percentage basis) and there was no evidence of interaction below the CMC.

The results of Anderson and Slade [207] suggest that hydrogen-ion concentration has little effect on the amount of benzoic acid solubilized. In spite of the fact that they estimate that only one in every 90 ether oxygens in non-micellar glycols are associated with benzoic acid in solution, the authors conclude that solubilization takes place in the PEG region of the micelles.

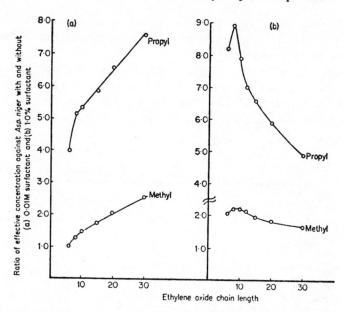

Figure 7.44 Effect of ethylene oxide chain length on the effective concentrations of the parabens versus *Asp. niger*, shown as a ratio of the concentration with and without surfactant. The results are shown: (a) on a molar basis, and (b) a percentage basis (0.01 M and 1.0% surfactant), respectively. Drawn from data calculated from Matsumoto and Aoki [204].

The increase of solubilization (mole per mole) with increasing glycol chain length may be explained by the fact that the ethylene oxide chains in the micelle form an environment exactly like a concentrated polyoxyethylene glycol solution, and hence the solubility of the solute increases. It remains true, however, that the ratio of solubilizate to ethylene oxide is 0.011 for PEG 3000, whereas it is 0.048 to 0.057 for non-ionics of polysorbate and Myrj and Brij type [207]. It is doubtful if there is a specific interaction. A large increase in the solubility of the paraben esters has been noted in solutions of carboxymethylhydroxyethyl cellulose and PEGs 200 and 400, and inactivation was noted in a bacteriological study. This finding is contrary to earlier investigations which suggested that there was no decrease in activity in the presence of glycols [208, 209]. Bolle and Mirimanoff [210] found that Crills, Spans and Tweens, but not Carbowax 1500 inhibited the activity of antiseptics against *Asp. niger*.

Antibacterial and antifungal agents have been used increasingly in shampoos and skin-cleansing agents. Russell and Hoch [211] have discussed the solubilization of typical materials (3,4,4′-trichlorocarbanilide and diaphene) in surfactant mixtures. The presence of lanolin – a common additive – did not interfere with the solubilization of the bacteriostats and in fact appeared to increase the amount solubilized. The presence of the non-ionics added as solubilizers did not appear to reduce the antibacterial activity of the mixture; indeed, in some cases there is a suspicion of enhanced activity, e.g. 3,4,4′-trichlorocarbanilide solubilized by

Igepal CO-630 in comparison with its activity in polysorbate 80 and Nimcolan 2. Banks and Huyck [212] indeed state that hexachlorophane must be solubilized to produce its maximal germicidal effect.

The problem of inactivation in emulsions and solubilized systems has received much attention; the observation that Millipore filters contain up to 3% of their dry weight of Triton X-100 [213], and the fact that these are used to filter solutions containing bacteriostats shows that the subject has many manifes-

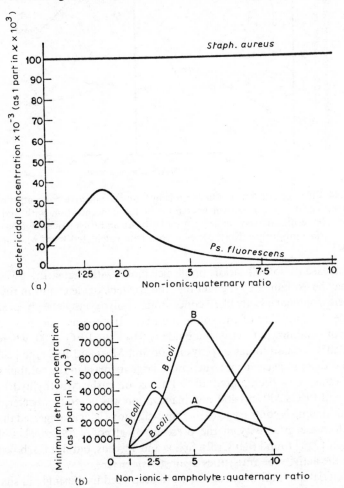

Figure 7.45(a) Interactions between detergents and quaternary ammonium compounds showing variation of bactericidal conentrations (1 part in $x \times 10^3$) with alteration in the ratio between non-ionic and quaternary showing initial increased activity, followed by a decrease in activity. *Staph. aureus* shows no variation, as it is so sensitive to the quaternary compound. From [220]. (b) Bactericidal activity in non-ionic:ampholyte:quaternary systems. (From [220]). A = 75% non-ionic, 25% sodium dodecyl amino propionate; B = 50% non-ionic, 50% ampholyte; C = 25% non-ionic, 75% ampholyte.

tations and is by no means exhausted. The danger of inadvertent inactivation must be avoided. A proprietary steroid cream diluted routinely to one-quarter of its strength with a cetomacrogol emulsifying wax resulted in a reduction of the chlorocresol content to 0.1 %, which was, in the presence of the detergent, insufficient to prevent the growth of *Ps. aeruginosa* [214]. Not all antimicrobials interact with non-ionics of course. Phenyl mercuric nitrate is not inactivated by 2 % polysorbate 80 [215]. Cases of increased activity are less apparent than cases of inactivation. Certain substances at appropriate concentrations should, in view of their membrane activity, be able to enhance activity. Polysorbate 80 has been shown to increase the action of polymixins B and D and of circulin [216]; the action of polymixin B sulphate, benzalkonium chloride, and chlorhexidine against *Ps. aeruginosa* is substantially increased in the presence of polysorbate 80. The inhibitory effect of the polymixin is enhanced at all concentrations of polysorbate from 0.004 to 0.5 %, the effect increasing with increasing concentration [217].

Synergism has also been noted between dodecyl hexahydroxyethylene glycol ester (Emulgen 106) and benzoquinone, the activity of which was increased 100 times against *M. pyogenes* var. *aureus* [218]. The bactericidal action of neomycin is claimed to be increased over forty times by cationic surfactants [219]. In admixture with small amounts of a non-ionic detergent the action of some quaternary compounds is increased, but with greater concentrations the activity shows a gradual decrease. It is obvious from Fig. 7.45a that the ratio of non-ionic to quaternary compound must not be greater than 4:1 if inactivation is to be avoided. The steep rise in the activity curve indicates that synergism is possible but close control of concentration is required. Ternary mixtures of Morpans (ampholytic surfactants) and non-ionics can also be used, the interactions becoming even more critically dependent on concentration, as is shown in Fig. 7.45b.

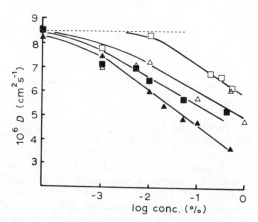

Figure 7.46 Diffusion coefficients of thioridazine (0.4 %) as a function of concentration of various non-ionic surfactants, measured by dialysis from Visking cellophane bags. ■ Cremophor EL; △ Atlas G2162; ▲ Renex 650 and □ Atlas G1295. From Florence [221].

7.7 Utilization of solubilization in drug delivery systems

Results of dialysis of solubilized systems using cellophane or polydimethyl-siloxane membranes indicate the possibility of using the solubilized state to control transport rates of drugs from reservoirs bounded by such inert membranes, permeable to drug and impermeable to micelles. The required degree of control over the free drug concentration in the reservoir is achieved by altering the surfactant concentration or the surfactant itself (Fig. 7.46). In Fig. 7.46 the diffusion coefficient of thioridazine has been reduced from 8.3×10^{-6} cm^2 s^{-1} aqueous solution to 4×10^{-6} cm^2 s^{-1} [221] in less than 1% of the non-ionic surfactant Renex 650. Micellar solutions, emulsions and co-solvent systems have been compared for their ability to control the release of butamben from silicone capsules [222]. Micellar solutions and emulsions provide reservoirs to maintain a more constant concentration of drug on the donor side of the membrane. With simple solutions the permeation rate falls as the solution concentration decreases. Emulsions and suspensions have problems of instability. In assessing a simple model for drug release the following equation was used to calculate the release profile

$$M(t) = M_\infty [1 - \exp(-APC_s^\circ t/lVC_s)], \qquad (7.41)$$

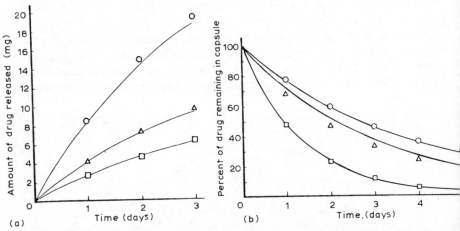

Figure 7.47(a) Release profiles of butamben from micellar systems in 10% sodium lauryl sulphate solution at three loading levels in a silicone capsule at 37° C. Drug loading levels were ○, 30 mg; △, 15 mg; and □, 10 mg. ——, theoretical profile predicted from Equation 7.7. (b) Effect of concentration of sodium lauryl sulphate on the release of butamben from micellar systems containing 15.5 mg butamben in silicone capsules at 37° C. Sodium lauryl sulphate concentrations were ○, 15%; △, 10%, and □, 5%. ——, theoretical profile predicted from Equation 7.41. In the theoretical calculations, the following parameters were used: $P = 1.24 \times 10^{-5}$ cm^2 s^{-1} (determined from a permeation study using the same silicone rubber membrane at 37° C), $A = 6.25$ cm^2, $l = 0.155$ cm, $C_s^\circ = 1.7$ mM, $C_s = 227$ mM (solubility of butamben in 10% sodium lauryl sulphate at 37° C). From Juni *et al.* [222] with permission.

where M_∞ and $M(t)$ are the amount of drug initially introduced and the cumulative amount of drug released at time t, respectively; A and l are the available area and thickness of the membrane, respectively; P is the permeability; V is the volume of intracapsular solution; and C_s^0 and C_s are the solubilities of drug in water and surfactant solution, respectively. P is dependent on the membrane used, and A, l and V depend on the size of the device. Control of release is achieved through C_s. The theoretical terms for release of half the drug in the silicone capsules in the system shown in Fig. 7.47b were 0.99, 2.14 and 2.74 days for the 5, 10 and 15% surfactant systems respectively [222]. The systems were tested after subcutaneous implantation in rabbits; close agreement between *in vitro* and *in vivo* release rates was found.

Mixed micelle formation and complex formation will achieve similar alterations in the transport properties of drug molecules. When the drug molecules themselves can aggregate, this will result in a reduction in permeability above their critical micelle concentration [221, 223].

7.7.1 Theoretical considerations of transport and release from solubilized systems

In order to evaluate the effect of micellar solubilization on the rate of transport of a solubilized drug, Matsumoto and co-workers compared experimental dialysis results with theoretical values based on an analysis of the system [224, 225] which considered the diffusion of the free drug. When the degree of interaction between drug and surfactant was low, theoretical predictions were accurate, but measured rates frequently exceeded the predicted rates of dialysis. Matsumoto ascribed the discrepancy to transport of some drug directly from the micelles following 'coalescence' with the membrane as the measured rates exceeded those predicted. A detailed analysis has been carried out by Goldberg and Higuchi [226, 227] of transport from a solubilized aqueous phase to an oil phase. They consider the possibility of transport of micellar-solubilized drug. In order for this to occur the micellar drug must diffuse 'to some point close to the oil, and then leave the micelle, unless the micelle itself enters the oil' [226]. The greater the solubilization the more important this process will be.

If the transport rate of the drug is diffusion controlled then the rate of uptake by the oil phase will be equal to the rate of transport through the aqueous phase. The steady-state rate of transport through the aqueous phase to the oil droplet, G, is given by

$$G = \left(AD_f \frac{dC_f}{dh} + AD_m \cdot \frac{dC_m}{dh} \right). \tag{7.42}$$

Subscripts f = free drug and m = drug in the micelle, as before. Other symbols have their usual meaning; h is the diffusion layer thickness. The oil is regarded as a 'perfect sink' and the following relation assumed

$$C_m = P_m \cdot C_f \cdot C_{SA} \tag{7.43}$$

where C_{SA} is the surfactant concentration. It was shown that for the planar case

$$\ln\left(\frac{\alpha}{\alpha - \beta C_0}\right) = \left(\frac{\beta A}{lv}\right)\left(\frac{D_f}{P_m C_{SA}} + D_m\right)t \qquad (7.44)$$

Where v is the volume of oil (ml ml^{-1}), C_0 is the concentration of drug in the oil, α is defined as

$$\alpha = \frac{C_t}{(1 + 1/P_m C_{SA})(1 - v)} \qquad (7.45)$$

and

$$\beta = \frac{v}{(1 + 1/P_m C_{SA})(1 - v)} + \left(\frac{1}{P_{app}} - \frac{1}{P_{O/W}}\right) \qquad (7.46)$$

where C_t is the total amount of drug present and P_{app} and $P_{O/W}$ being the apparent and true partition coefficients respectively. It can be shown that

$$\ln\left(\frac{\alpha}{\alpha - \beta C_0}\right) = \left(\frac{\beta A}{lv}\right)\left(\frac{D_f}{P_m C_{SA}} + D_m\right)t$$

C_0 = concentration of drug in oil.

The equivalent equation for the spherical case (diffusion to a sphere of oil) is

$$\ln\left(\frac{\alpha}{\alpha - \beta C_0}\right) = \frac{\beta A}{v}\left(\frac{D_f}{P_m C_{SA}} + D_m\right)t, \qquad (7.47)$$

the only difference being the disappearance of the diffusion layer thickness from the equation. The equation can be rewritten to give the concentration of drug in the oil as a function of time:

$$C_0 = \frac{\alpha}{\beta}\left\{1 - \exp\left[-\frac{\beta A}{v}\left(\frac{D_f}{P_m C_{SA}} + D_m\right)t\right]\right\}, \qquad (7.48)$$

in which every term can be independently determined so that predictions of transport can be made. The model becomes more complex with additional considerations of micellar charge and charged liquid interface. In this case the free drug and the drug in the micelle diffuse freely to some distance from the oil droplet where an electrical barrier then permits only free drug to diffuse. Interfacial barriers to transport other than electrical barriers may also exist [227], for example, as a result of adsorbed surfactant or polymer. An interfacial barrier constant, Γ, can be introduced into Equation 7.48, thus giving

$$C_0 = \frac{\alpha}{\beta}\left\{1 - \exp\left[-\frac{\beta \Gamma A}{v}\left(\frac{D_f}{P_m C_{SA}} + D_m\right)t\right]\right\}. \qquad (7.49)$$

As $\Gamma \to 1$ the model approaches the simple diffusion model. The smaller Γ the lower the rate of transport from the aqueous phase into the oil. The simple diffusion approach was deemed by Goldberg and Higuchi to be inadequate. Rate data for the indoxole–isopropyl myristate–polysorbate 80 system are shown in Fig. 7.48. The points are experimental, the solid lines based on Equation 7.49 when $\Gamma = 1.27 \times 10^{-4}$ for 2% polysorbate 80 and 1.85×10^{-4} for 1% poly-

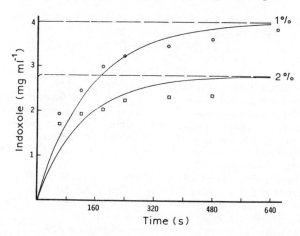

Figure 7.48 The appearance of indoxole in the oil phase as a function of time. Points, experimental data; dashed line, the theoretical rate of transport based on diffusion theory; solid line, the theoretical rate based on the interfacial barrier theory. Results for 1 % and 2 % polysorbate 80 are shown. The oil phase is isopropyl myristate. From Goldberg and Higuchi [227] with permission.

sorbate 80; the dashed lines are those calculated for single diffusion. Predicted rates are too fast, the magnitude of the barrier is sufficient to reduce rates of transport by up to several thousand times according to this analysis. Such observed discrepancies could not be attributed to an electrical barrier which reduced the calculated rates by at most a factor of ten for an oil droplet with a surface potential of 100 mV. Brodin's experimental results on the influence of surfactants on mass transfer between an aqueous phase and an oil phase (discussed earlier) confirm these impressions that interfacial barriers to transport exist. Whether these considerations apply at artificial membranes is another matter. Adsorption of surfactant onto membranes (especially multilayer adsorption) is likely to influence transport of drugs considerably. Adsorption of nonionic surfactants onto the intestinal membrane may be one of the factors contributing to the absorption inhibiting effect of high concentrations of surfactant [47] when there is little interaction between drug and micelles, a phenomenon postulated to explain reductions in transfer of substances into leaves [228] and goldfish [40].

References

1. O. BLANPIN (1958) *Prod. Pharm.* **13**, 425.
2. G. BILLARD and L. DIEULAFE (1904) *Compt. Rend. Soc. Biol.* **56**, 146.
3. G. LEVY and S. P. GUCINSKI (1964) *J. Pharmac. exp. Therap.* **146**, 80.
4. G. LEVY, K. E. MILLER and R. H. REUNING (1966) *J. Pharm. Sci.* **55**, 394.
5. G. LEVY and J. A. ANELLO (1968) *J. Pharm. Sci.* **57**, 101.
6. I. MORAN, J. GILLARD and M. ROLAND (1971) *J. Pharm. Belg.* **26**, 115.
7. S. SCHRANG and P. FINHOLT (1970) *Meddr. norsk farm Selsk.* **31**, 101.

464 · *Surfactant systems*

8. M. P. SHORT, P. SHARKEY and C. T. RHODES (1972) *J. Pharm. Sci.* **61**, 1732.
9. L. S. C. WAN (1972) *Canad. J. pharm. Sci.* **7**, 25.
10. J. A. REES and J. H. COLLETT (1974) *J. Pharm. Pharmacol.* **26**, 956.
11. E. CID and F. JAMINET (1971) *J. Pharm. Belg.* **26**, 369.
12. P. FINHOLT and S. SOLVANG (1968) *J. Pharm. Sci.* **57**, 1322.
13. G. LEVY and R. H. GUMTOW (1963) *J. Pharm. Sci.* **52**, 1139.
14. L. PRESCOTT, R. F. STEEL and W. R. FERRIER (1970) *Clin. Pharmacol. Therap.* **11**, 496.
15. R. K. REDDY, S. A. KHALIL and M. W. GOUDA (1976) *J. Pharm. Sci.* **65**, 1753.
16. R. K. REDDY, S. A. KHALIL and M. W. GOUDA (1976) *J. Pharm. Sci.* **65**, 115.
17. B. R. HAJRATWALA and H. TAYLOR (1976) *J. Pharm. Pharmacol.* **28**, 934.
18. K. J. MYSELS and A. T. FLORENCE (1970) in *Clean Surfaces* (ed. G. Goldfinger) Marcel Dekker, New York. pp. 227ff.
19. J. H. COLLETT and J. A. REES (1975) *J. Pharm. Pharmacol.* **27**, 647.
20. E. L. PARROTT and V. K. SHARMA (1967) *J. Pharm. Sci.* **56**, 1341.
21. P. H. ELWORTHY and F. J. LIPSCOMB (1968) *J. Pharm. Pharmacol.* **20**, 923.
22. W. I. HIGUCHI (1964) *J. Pharm. Sci.* **53**, 532.
23. N. WATARI and N. KANENIWA (1976) *Chem. Pharm. Bull.* **24**, 2577.
24. W. I. HIGUCHI (1967) *J. Pharm. Sci.* **56**, 315.
25. M. GIBALDI, S. FELDMAN and N. D. WEINER (1970) *Chem. Pharm. Bull.* **18**, 715.
26. P. V. DANCKWERTS (1951) *Industr. Eng. Chem.* **43**, 1460.
27. H. L. TOOR and J. M. MARCHELLO (1958) *A. I. Chem. E. J.* **4**, 97.
28. P. H. ELWORTHY and F. J. LIPSCOMB (1968) *J. Pharm. Pharmacol.* **20**, 923.
29. P. H. ELWORTHY and F. J. LIPSCOMB (1969) *J. Pharm. Pharmacol.* **21**, 273.
30. W. E. HAMLIN, E. NELSON, B. E. BALLARD and J. G. WAGNER (1962) *J. Pharm. Sci.* **51**, 432.
31. A. F. CHAN, D. F. EVANS and E. L. CUSSLER (1976) *A. I. Chem. E. J.* **22**, 1006.
32. P. SOMASUNDARAN, T. W. HEALY and D. W. FUERSTENAU (1966) *J. Colloid Interface Sci.* **22**, 599.
33. J. L. FORD and M. H. RUBINSTEIN (1978) *J. Pharm. Pharmacol.* **30**, 512.
34. A. HOELGAARD and N. MØLLER (1975) *Arch. Pharm., Chemi. Sci. Ed.* **3**, 65.
35. A. HOELGAARD and N. MØLLER (1975) *Arch. Pharm., Chemi. Sci. Ed.* **3**, 34.
36. W. L. CHIOU, S-J. CHEN and N. ATHANIKAR (1976) *J. Pharm. Sci.* **65**, 1702.
37. A. S. MICHAELS and A. R. COLVILLE (1960) *J. Phys. Chem.* **64**, 13.
38. A. S. MICHAELS, P. L. T. BRIAN and W. F. BECK (1967) *Proc. 4th Int. Conf. Surface Active Agents (1964)* **2**, 1053.
39. G. LEVY, K. E. MILLER and R. H. REUNING (1966) *J. Pharm. Sci.* **55**, 394.
40. A. T. FLORENCE and J. M. N. GILLAN (1975) *J. Pharm. Pharmacol.* **27**, 152.
41. A. T. FLORENCE and J. M. N. GILLAN (1975) *Pesticide Sci.* **6**, 429.
42. A. T. FLORENCE (1977) in *Micellization, Solubilization and Microemulsions* (ed. K. L. Mittal) Vol. 1, Plenum Press, New York, pp. 55–74.
43. A. T. FLORENCE (1981) in *Techniques of Solubilization of Drugs* (ed. S. Yalkowsky) Marcel Dekker, New York.
44. K. A. WALTERS, P. H. DUGARD and A. T. FLORENCE (1981) *J. Pharm. Pharmacol.* **33**, 207.
45. K. A. WALTERS, A. T. FLORENCE and P. H. DUGARD (1982) *Int. J. Pharmaceutics* **10**, 153.
46. L. W. SMITH, C. L. FOY and D. E. BAYER (1966) *Weed Research* **6**, 233.
47. A. KANEDA, K. NISHIMURA, S. MURANISHI and H. SEZAKI (1974) *Chem. Pharm. Bull.* **22**, 523.
48. D. A. WHITMORE, L. G. BROOKES and K. P. WHEELER (1979) *J. Pharm. Pharmacol.* **31**, 277.
49. J. A. NISSIM (1960) *Nature* **187**, 308.
50. E. CID (1971) *Pharm. Acta. Helv.* **46**, 377.
51. B. ISOMAA and G. SJOBLÖM (1975) *Fd. Cosmet. Toxicol.* **13**, 517.
52. W. D. SEUFERT (1965) *Nature* **207**, 174.

53. K. A. WALTERS, A. T. FLORENCE and P. H. DUGARD (1982) *J. Colloid Interface Sci.* in press.
54. P. SEEMAN (1966) *Biochem. Pharmacol.* **15**, 1737.
55. P. SEEMAN (1969) *Biochim. Biophys. Acta* **183**, 490.
56. P. D. SWANSON, H. F. BRADFORD and H. MCILWAIN (1964) *Biochem J.* **92**, 235.
57. H. F. BRADFORD, P. D. SWANSON and D. B. GAMMACK (1964) *Biochem J.* **92**, 247.
58. S. KITAZAWA, M. ISHIZU and K. KIMURA (1977) *Chem. Pharm. Bull.* **25**, 590.
59. R. WITHINGTON and J. H. COLLETT (1973) *J. Pharm. Pharmacol.* **25**, 273.
60. H. MATSUMOTO, H. MATSUMURA and S. IGUCHI (1966) *Chem. Pharm. Bull.* **14**, 385.
61. K. JUNI, T. TOMITSUKA, M. NAKANO and T. ARITA (1978) *Chem. Pharm. Bull.* **26**, 837.
62. J. H. COLLETT, R. WITHINGTON and B. COX (1974) *J. Pharm. Pharmacol.* **26**, 34.
63. J. H. COLLETT and R. WITHINGTON (1973) *J. Pharm. Pharmacol.* **25**, 723.
64. A. H. HIKAL, L. DYER and S. W. WONG (1976) *J. Pharm. Sci.* **65**, 621.
65. A. BRODIN (1975) *Acta. Pharm. Suec.* **12**, 41.
66. F. H. GARNER and A. R. HALE (1953) *Chem. Engn. Sci.* **2**, 157.
67. M. E. KITLER and P. LAMY (1971) *Pharm. Acta. Helv.* **46**, 483.
68. P. H. ELWORTHY (1959) *J. Chem. Soc.* 813 and 1951.
69. P. H. ELWORTHY and D. S. MCINTOSH (1964) *J. Phys. Chem.* **68**, 3448.
70. J. H. COLLETT and L. KOO (1975) *Acta. Pharm. Suec.* **12**, 81.
71. A. A. EL-SAYED and M. S. MOHAMMED (1976) *Bull. Fac Pharm., Cairo University* **15**, 217.
72. G. LEVY and T. MATSUZAWA (1965) *J. Pharm. Sci.* **54**, 1003.
73. G. LEVY, K. E. MILLER and R. H. REUNING (1966) *J. Pharm. Sci.* **55**, 394.
74. E. R. M. KAY (1965) *Cancer Res.* **25**, 764.
75. K. KAKEMI, T. ARITA and S. MURANISHI (1965) *Chem. Pharm. Bull.* **13**, 976.
76. K. J. MYSELS (1969) *Adv. Chem. Ser.* **86**, 24.
77. K. KAKEMI, T. ARITA, H. SEZAKI and I. SUGIMOTO (1964) *Yakugaku Zasshi* **84**, 1210.
78. D. C. BLOEDOW and W. L. HAYTON (1976) *J. Pharm. Sci.* **65**, 334.
79. D. C. BLOEDOW and W. L. HAYTON (1976) *J. Pharm. Sci.* **65**, 328.
80. K. THOMAS, G. PFAFF and K. QUIRING (1978) *J. Pharm. Pharmacol.* **30**, 270.
81. B. L. OSER and M. OSER (1957) *J. Nutr.* **61**, 149.
82. P. J. CULVER, C. S. WILCOX, C. M. JONES and R. S. ROSE (1951) *J. Pharmacol. Exp. Therap.* **103**, 377.
82a M. N. AZMIN, A. T. FLORENCE, J. F. B. STUART and T. L. WHATELEY (1982) *J. Pharm. Pharmacol.* **36**.
83. W. W. DAVIS, R. R. PFEIFFER and J. F. QUAY (1970) *J. Pharm. Sci.* **59**, 960.
84. W. W. DAVIS and C. J. KREUTLER (1971) *J. Pharm. Sci.* **60**, 1651.
85. C. J. KREUTLER and W. W. DAVIS (1971) *J. Pharm. Sci.* **60**, 1835.
86. D. P. SANVORDEKER and J. BLOSS (1977) *J. Pharm. Sci.* **66**, 82.
87. E. TOUITOU, M. DONBROW and A. RUBINSTEIN (1978) *J. Pharm. Pharmacol.* **32**, 108.
88. M. SCHICHIRI, R. KAWAMORI *et al.* (1978) *Acta Diabetologica Latina* **15**, 175.
89. R. W. WISSLER, W. F. BETHARD, P. BARKER and H. D. MORI (1954) *Proc. Soc. Exp. Biol. Med.* **86**, 170.
90. J. SOEJIMA (1955) *Wagasaki Igakki kassi* **30**, 219 [*CA* **49**, 10524C).
91. V. KOZLIK and B. MOSINGER (1956) *Die Pharmazie* **11**, 22.
92. J. NISSIM (1960) *Nature* **187**, 308.
93. P. M. LISH and J. H. WEIKEL (1959) *Toxicol Appl. Pharmacol.* **1**, 501.
94. A. ALBERT (1979) *Selective Toxicity* 6th edn, Chapman and Hall, London.
95. A. F. HOFMANN (1965) *Gastroenterology* **48**, 484.
96. T. R. BATES, M. GIBALDI and J. L. KANIG (1966) *J. Pharm. Sci.* **55**, 191.
97. T. R. BATES, M. GIBALDI and J. L. KANIG (1966) *Nature* **210**, 1331.
98. M. H. MALONE, H. I. HOCHMAN and K. A. NIEFORTH (1966) *J. Pharm. Sci.* **55**, 972.
99. J. L. LACH and W. A. PAULI (1966) *J. Pharm. Sci.* **55**, 32.
100. British Patent 378, 935 (1931).

101. Swiss Patents 126, 502; 130, 091–3 (1948).
102. U. P. BASU, S. MUKHERJEE and R. P. BANERJEE (1947) *J. Amer. Pharm. Assoc.* **36**, 266.
103. *J. Amer. Med. Assoc.* (1945) **128**, 495.
104. T. R. BATES, M. GIBALDI and J. L. KANIG (1966) *J. Pharm. Sci.* **55**, 901.
105. C. MARRIOTT and I. W. KELLAWAY (1976) *J. Pharm. Pharmacol.* **28**, 620.
106. M. GIBALDI and C. H. NIGHTINGALE (1968) *J. Pharm. Sci.* **57**, 1354.
107. C. H. NIGHTINGALE, R. J. WYNN and M. GIBALDI (1969) *J. Pharm. Sci.* **58**, 1005.
108. K. INUI, M. SHINTOMI, R. HORI and H. SEZAKI (1976) *Chem. Pharm. Bull.* **24**, 2504.
109. S. FELDMAN and M. GIBALDI (1969) *Proc. Soc. Exp. Biol. Med.* **132**, 1031.
110. S. MURANISHI, N. MURANISHI and H. SEZAKI (1979) *Int. J. Pharmaceutics* **2**, 101.
111. R. HORI, K. OKUMURA, K. INUI, N. NAKAMURA, A. MIYOSHI and T. SUYAMA (1977) *Chem. Pharm. Bull.* **25**, 1974.
112. T. KIMURA, K. INUI and H. SEZAKI (1971) *Yakuzaigaku* **31**, 167.
113. K. KAKEMI, H. SEZAKI, R. KONISHI, T. KIMURA and A. OKITA (1970) *Chem. Pharm. Bull.* **18**, 1034.
114. S. FELDMAN and M. GIBALDI (1969) *J. Pharm. Sci.* **58**, 425.
115. J. M. DIETSCHY (1968) *J. Lipid Res.* **9**, 297.
116. E. G. LOVERING and D. B. BLACK (1974) *J. Pharm. Sci.* **63**, 671.
117. K. TANIGUCHI, S. MURANISHI and H. SEZAKI (1980) *Int. J. Pharmaceutics* **4**, 219.
118. I. UTSUMI, K. KOHNO and Y. TAKEUCHI (1974) *J. Pharm. Sci.* **63**, 676.
119. I. UTSUMI, K. KOHNO and Y. TAKEUCHI (1973) *Chem. Pharm. Bull.* **21**, 1727.
120. S. MIYAZAKI, T. YAMAHIRA, H. INOUE and T. NADAI (1980) *Chem. Pharm. Bull.* **28**, 323.
121. K. KAKEMI, T. ARITA and S. MURANISHI (1965) *Chem. Pharm. Bull.* **13**, 861.
122. K. KAKEMI, H. SEZAKI, S. MURANISHI and Y. TSUJIMURA (1969) *Chem. Pharm. Bull.* **17**, 1650.
123. E. SUZUKI, M. TSUKIGI, S. MURANISHI, H. SEZAKI and K. KAKEMI (1972) *J. Pharm. Pharmacol.* **24**, 138.
124. G. FIESE and J. H. PERRIN (1969) *J. Pharm. Sci.* **58**, 599.
125. R. M. PATEL and G. ZOGRAFI (1966) *J. Pharm. Sci.*
126. J. T. DOLUISIO, W. G. CROUTHAMEL, G. H. TAN, J. V. SWINTOSKY and L. W. DITTERT (1970) *J. Pharm. Sci.* **59**, 72.
127. Y. YAMASAKI, M. SCHICHIRI, R. KAWAMORI, Y. SHIGETA and H. ABE (1978) *Diabetes* **27**, Suppl. 2, 514.
128. K. ICHIKAWA, I. OHATA, M. MITOMI, B. KAWAMURA, H. MAENO and H. KAWATA (1980) *J. Pharm. Pharmacol.* **32**, 314.
129. E. TOUITOU, M. DONBROW and E. AZAZ (1978) *J. Pharm. Pharmacol.* **30**, 662.
130. T. MATSUZAWA, H. FUJISAWA, K. AOKI and H. MIMA (1969) *Chem. Pharm. Bull.* **17**, 999.
131. H. KOBAYASHI, T. NISHIMURA, K. OKUMURA, S. MURANISHI and H. SEZAKI (1974) *J. Pharm. Sci.* **63**, 580.
132. H. KOBAYASHI, T. PENG, M. FUJIKAWA, S. MURANISHI and H. SEZAKI (1976) *Chem. Pharm. Bull.* **24**, 2383.
133. H. KOBAYASHI, T. PENG, R. KAWAMURA, S. MURANISHI and H. SEZAKI (1977) *Chem. Pharm. Bull.* **25**, 569.
134. R. SCHEUPLEIN (1978) in *The Physiology and Pathophysiology of the Skin*, (ed. A. Jarrett) Vol. 5, Academic Press, London.
135. P. H. DUGARD and R. J. SCHEUPLEIN (1973) *J. Inert. Derm.* **60**, 263.
136. R. J. SCHEUPLEIN and L. ROSS (1970) *J. Soc. Cosmetic Chem.* **21**, 853.
137. J. A. FAUCHER, E. D. GODDARD and R. D. KULKARNI (1979) *J. Amer. Oil. Chem. Soc.* **56**, 776.
138. R. L. GOLDEMBERG and L. SAFRIN (1977) *J. Soc. Cosmet. Chem.* **28**, 667.
139. R. T. YOUSEF and M. N. KHAWAM (1966) *Archiv. für Mikrobiologic* **53**, 159.
140. B. J. POULSEN (1973) in *Drug Design* (ed. A. J. Ariens) Vol. 4, Academic Press, New York, p. 149.
141. B. LEVY and C. L. HUYCK (1949) *J. Amer. Pharm. Assoc. Sci. Ed.* **38**, 611.

142. W. SASKI and S. G. SHAH (1965) *J. Pharm. Sci.* **54**, 277.
143. L. H. MACDONALD and R. E. HIMELICK (1948) *J. Amer. Pharm. Assoc., Sci. Ed.* **37**, 368.
144. K. J. RYAN and M. MEZEI (1975) *J. Pharm. Sci.* **64**, 671.
145. M. MEZEI, R. W. SAGER, W. D. STEWART and A. L. deRUYTER (1966) *J. Pharm. Sci.* **55**, 584.
146. Z. T. CHOWHAN and R. PRITCHARD (1978) *J. Pharm. Sci.* **67**, 1272.
147. R. J. SCHEUPLEIN and I. H. BLANK (1971) *Physiol. Rev.* **51**, 202.
148. W. W. SHEN, A. G. DANTI and F. N. BRUSCATO (1976) *J. Pharm. Sci.* **65**, 1780.
149. M. TANAKA, N. YANAGIBASHI, H. FUKUDA and T. NAGAI (1980) *Chem. Pharm. Bull.* **28**, 1056.
150. J. S. TURI, D. DANIELSON and J. W. WOLTERSORN (1979) *J. Pharm. Sci.* **68**, 275.
151. B. GARCIA, J. P. MARTY and J. WEPIERRE (1980) *Int. J. Pharmaceutics* **4**, 205.
152. E. CID (1968) *Il Farmaco. Ed. Rrat.* **23**, 474.
153. G. ÅBERG and G. ADLER (1976) *Arzneimittel Forsch.* **26**, 78.
154. R. J. MARSH and D. M. MAURICE (1971) *Exp. Eye. Res.* **11**, 43.
155. C. S. O'BRIEN and K. C. SWAN (1942) *AMA, Arch. Ophthalmol.* **27**, 253.
156. I. H. LEOPOLD and H. G. SCHEIE (1943) *AMA, Arch. Ophthalmol.* **29**, 811.
157. J. G. BELLOWS and M. GUTMANN (1943) *AMA, Arch. Ophthalmol.* **30**, 352.
158. M. GINSBURG and J. M. ROBSON (1945) *Brit. J. Ophthalmol.* **29**, 185.
159. M. GINSBURG and J. M. ROBSON (1949) *Brit. J. Ophthalmol.* **33**, 574.
160. L. VON SALLMANN and K. MEYER (1944) *AMA, Arch. Ophthalmol.* **31**, 1.
161. K. A. REISER (1952) *Klin. Monatsb. Augenheik* **121**, 257.
162. H. FUJINO (1953) *Acta Soc. Ophthalmol. Jap.* **57**, 1347.
163. L. W. HAGLETON (1952) *Proc. Sci. Sect. Toilet Goods Ass.* **17**, 5.
164. T. YAMADA, Y. IWANAWA and T. BABA (1963) *Gann* **54**, 171.
165. S. H. ROTH and P. J. WILLIAMS (1979) *J. Pharm. Pharmacol.* **31**, 224.
166. J. P. DAVIGNON, H. B. WOOD and J. C. CRADDOCK (1973) *Cancer Chemotherapy Reports.* **4**, 7.
167. M. NICOLOU, A. J. LAPA and J. R. VALLE (1978) *Arch. Int. Pharmacodyn.* **236**, 131.
168. L. A. BORGEN and W. M. DAVIS (1973) *J. Pharm. Sci.* **62**, 479.
169. R. D. SOFIA, R. K. KUBENA and H. BARRY (1974) *J. Pharm. Sci.* **63**, 939.
170. M. A. K. MATTILA, M. RUOPPI, M. KORHONEN, H. M. LARNI, L. VALTOMEN and H. HEIKKINEN (1979) *Br. J. Anaesth.* **51**, 891.
171. J. KANTO (1974) *Brit. J. Anaesthesia* **46**, 817.
172. E. ULLMANN (1961) *Proceedings of the XXI International Pharmaceutical Congress,* Pisa.
173. E. ULLMANN and B. MOSER (1962) *Arch. Pharm.* **295**, 136.
174. R. J. DUBOS and B. D. DAVIS (1946) *J. Expt. Med.* **83**, 409.
175. H. S. FORREST, P. D'ARCY HART and J. WALKER (1947) *Nature* **160**, 94.
176. A. YOUMANS and G. YOUMANS (1948) *J. Bact.* **56**, 245.
177. O. FENNER (1954) *Arzneimittel Forsch.* **4**, 368.
178. S. NATORI (1958) *Chem. Pharm. Bull.* **6**, 94.
179. S. J. A. KAZMI and A. G. MITCHELL (1978) *J. Pharm. Sci.* **67**, 1260.
180. K. J. HUMPHREYS, G. RICHARDSON and C. T. RHODES (1968) *J. Pharm. Pharmacol.* **20**, 4S.
181. R. M. E. RICHARDS and R. J. MCBRIDE (1972) *J. Pharm. Pharmacol.* **24**, 145.
182. M. R. W. BROWN and R. M. E. RICHARDS (1964) *J. Pharm. Pharmacol.* **16**, Suppl 5T.
183. R. M. E. RICHARDS (1971) *J. Pharm. Pharmacol.* **23** Suppl 136S.
184. R. M. E. RICHARDS and R. H. CAVILL (1976) *J. Pharm. Pharmacol.* **28**, 935.
185. K. THOMAS and K. WILL (1975) *Pharmazeutische Zeit* **120**, 1013.
186. C. A. LAWRENCE and A. L. ERLANDSON (1953) *J. Amer. Pharm. Assoc.* **42**, 352.
187. R. BERTHET (1947) *Schweiz Apoth, Ztq.* **85**, 833.
188. R. A. ANDERSON and K. J. MORGAN (1966) *J. Pharm. Pharmacol.* **18**, 449.
189. W. B. HUGO and J. M. NEWTON (1964) *J. Pharm. Pharmacol.* **16**, 49.

190. A. H. BECKETT, S. J. PATKI and A. E. ROBINSON (1959) *J. Pharm. Pharmacol.* 11, 367.
191. A. H. BECKETT, S. J. PATKI and A. E. ROBINSON (1959) *J. Pharm. Pharmacol.* 11, 421.
192. J. C. RIEMERSMA (1966) *J. Pharm. Pharmacol.* 18, 657.
193. J. C. RIEMERSMA (1966) *J. Pharm. Pharmacol.* 18, 602.
194. H. F. BRADFORD, P. D. SWANSON, D. B. GAMMACK (1964) *Biochem. J.* 92, 247.
195. F. ALHAIQUE, D. GIACCHETTI, M. MARCHETTI and F. M. RICCIERI (1975) *J. Pharm. Pharmacol.* 27, 811.
196. F. ALHAIQUE, M. MARCHETTI, F. M. RICCIERI and E. SANTUCCI (1975) *Experientia* 31, 215.
197. F. ALHAIQUE, M. MARCHETTI, F. M. RICCIERI and E. SANTUCCI (1972) *Il Farmaco, Ed. Sci.* 27, 145.
198. C. D. MOORE and R. B. HARDWICK (1956) *Mfg. Chem.* 27, 306.
199. P. DELUCA and H. B. KOSTENBAUDER (1960) *J. Amer. Pharm. Assoc.* 49, 430.
200. H. ANSEL (1965) *J. Pharm. Sci.* 54, 1159.
201. J. JUDIS (1962) *J. Pharm. Sci.* 51, 261.
202. I. W. KELLAWAY and L. SAUNDERS (1969) *J. Pharm. Pharmacol.* 21, Suppl, 189S.
203. M. MATSUMOTO and M. AOKI (1962) *Chem. Pharm. Bull.* 10, 260.
204. M. MATSUMOTO and M. AOKI (1962) *Chem. Pharm. Bull.* 10, 251.
205. C. K. BAHAL and H. B. KOSTENBAUDER (1964) *J. Pharm. Sci.* 53, 1027.
206. F. W. GOODHART and A. N. MARTIN (1962) *J. Pharm. Sci.* 51, 50.
207. R. ANDERSON and A. H. SLADE (1965) *Austral. J. Pharm.* 46, S53.
208. G. V. STORZ, H. G. DEKAY and G. S. BANKER (1965) *J. Pharm. Sci.* 54, 92.
209. M. G. DE NAVARRE (1957) *J. Soc. Cosmetic Chemists* 18, 371.
210. M. BOLLE and A. MIRIMANOFF (1950) *J. Pharm. Pharmacol.* 2, 685.
211. K. L. RUSSELL and S. G. HOCH (1965) *J. Soc. Cosmetic Chem.* 16, 169.
212. C. J. BANKS and C. L. HUYCK (1962) *Amer. J. Hosp. Pharm.* 19, 132.
213. R. D. CAHN (1967) *Science* 155, 195.
214. W. C. NOBLE and J. A. SAVIN (1966) *Lancet* i, 347.
215. M. S. PARKER, M. BARNES and T. J. BRADLEY (1966) *J. Pharm. Pharmacol.* 18, Suppl. 103S.
216. E. A. BLISS and P. T. WARTH (1950) *Ann. N.Y. Acad. Sci.* 53, 38.
217. R. M. E. RICHARDS and M. R. W. BROWN (1964) *J. Pharm. Pharmacol.* 16, 360.
218. M. AKAGI, K. HIROSE, Y. OSE and J. AMANO (1954) *Ann. Proc. Gifu. Coll. Pharm.* no. 4, 41.
219. US Patent, 3,069,320 (1958).
220. Glover (Chemicals) Ltd (1965) *Morpans.*
221. A. T. FLORENCE (1977) in *Micellization, Solubilization and Microemulsions* (ed. K. L. Mittal) Vol. 1, Plenum Press, New York, p. 55.
222. K. JUNI, K. NOMOTO, M. NAKANO and T. ARITA (1979) *J. Membrane Sci.* 5, 295.
223. D. ATTWOOD, A. T. FLORENCE and J. M. N. GILLAN (1974) *J. Pharm. Sci.* 63, 988.
224. H. MATSUMOTO, H. MATSUMURA and S. IGUCHI (1966) *Chem. Pharm. Bull.* 14, 398.
225. H. MATSUMOTO (1966) *Yakugaku Zasshi* 86, 590.
226. A. H. GOLDBERG, W. I. HIGUCHI, N. F. H. HO and G. ZOGRAFI (1967) *J. Pharm. Sci.* 56, 1432.
227. A. H. GOLDBERG and W. I. HIGUCHI (1969) *J. Pharm. Sci.* 58, 1341.
228. L. W. SMITH, C. L. FOY and D. E. BAYER (1966) *Weed Research* 6, 233.

8 Emulsions

8.1 Introduction

Emulsions, traditionally prepared with naturally occurring gums such as acacia and tragacanth have been used in pharmacy for centuries as means of administration of oils or vitamins. In the last decade, however, there has been renewed interest in the emulsion as a vehicle for delivering drugs to the body as it has been found to have several advantageous characteristics, frequently enhancing the bioavailability of the drug substance. Concentrated emulsions are used in topical therapy as semi-solid vehicles. In these systems the delicate interplay between the components of the emulsion, the preservatives and the drug molecules, make this an intriguing though theoretically almost intractable topic. The replacement of the natural gums with surfactants has led to the advantages of a more rigorous and fundamental approach to the formulation of different systems, but the presence of surfactants has introduced extra facets to complicate their behaviour. The amphipathic nature of the surfactant leads to a potential biological effect on membranes; the presence of surfactant micelles in the continuous phase leads to the potential solubilization of components such as preservatives, flavours and drug. We will deal with some of these problems in this chapter and with what is still the foremost problem with emulsions – their physical stability.

In the first part of the chapter oil-in-water (O/W) emulsions will be dealt with, followed by consideration of water-in-oil (W/O) emulsions and multiple emulsions either as water-in-oil-in-water (W/O/W) or O/W/O systems. We do not intend here to deal exhaustively with emulsion stabilization, as specialist texts have recently tackled this. Carroll [1] has reviewed stability and mechanisms of emulsion breakdown, as have Vincent and Davis [2]. The literature up to 1971 was considered by Florence and Rogers [3, 4] particularly in respect of non-ionic surfactant-stabilized emulsions, those that are most frequently encountered in pharmacy because of the generally lower toxicity of non-ionic surfactants. Becher's classic text [5] should still be consulted for a wealth of information on all aspects of emulsions. The aim in this chapter is to concentrate on the factors affecting the formulation and use of emulsions, especially the influence of additives, and the way the emulsion systems perform *in vitro* and *in vivo*.

8.2 Aspects of emulsion stability

Emulsions have been defined [5] as heterogeneous systems of one liquic dispersed in another in the form of droplets usually exceeding 0.1 μm in diameter The two liquids are immiscible, chemically unreactive, and form system: characterized by a minimal thermodynamic stability.

Unless the free energy of the oil–water interface is zero, an emulsion cannot b(a thermodynamically stable system, since reduction of the area of oil in contac with the water will always result from coalescence of the droplets. 'Stability' i: therefore a relative term, but the degree of stability can be assessed by observin; the rate of change of a parameter such as interfacial area or droplet diameter Unstabilized emulsions, referred to as oil hydrosols by King [6], coalesce rapidly while stabilized emulsions can retain a highly dispersed internal phase for month: or years. As the free energy of the interface is the driving force for coalescence emulsions can be stabilized by the inclusion of a surface-active substance in th(system which concentrates at the oil–water interface. The role played by th(surface-active material may be three-fold, depending on the chemical nature o: the adsorbing material and the adsorbent (Fig. 8.1). The adsorbed surfactan molecules:

(a) can decrease the free energy of the system. Although at one time this effec was considered to be largely responsible for the stability, emulsions havin; the same interfacial free energy may have widely differing stabilities However, statements can still be found ascribing stability loosely t(interfacial tension;
(b) can form a barrier delaying the coalescence of the globules. This barrie represents a combination of steric, viscous and elastic properties [6, 8–10 depending on the emulsifier;
(c) may affect the electrostatic charge of the dispersed particles.

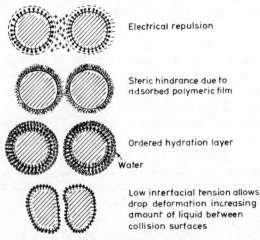

Electrical repulsion

Steric hindrance due to adsorbed polymeric film

Ordered hydration layer

Water

Low interfacial tension allows drop deformation increasing amount of liquid between collision surfaces

Figure 8.1 Schematic representation of mechanisms of emulsion stabilization due to th formation of interfacial films. Redrawn from Karel [7].

8.2.1 Instability

Emulsion instability is manifested in changes in the physical properties of the dispersion such as its droplet size distribution, its rheological properties or other parameters which are a consequence of the coalescence of globules or their flocculation, that is, of the alteration in the real or effective mean globule diameter, respectively. Flocculation, which is often the precursor of coalescence can affect the appearance of both liquid and solid emulsions. It accelerates the rate of creaming or settling which in itself is regarded as a form of instability.

Inversion of emulsion type is rare in practice although localized inversion may occur through interaction of the components of the systems with packaging materials. This type of instability is dealt with later in relation to phase volume and emulsifier type.

The problem is to prevent instability, not only to maintain the appearance of the emulsion, but so that the characteristics of the emulsion and of medicaments dissolved in the emulsion are as little changed on ageing as possible. As an example, ageing might alter the absorption of heparin from O/W emulsions where absorption of heparin appears to be directly related to the particle size and total surface area of the oil droplets [11]. Fat emulsions are used extensively in intravenous feeding [12] where it is vital that particles remain below 1 μm in diameter to avoid thrombophlebitis and other complications, but the state of the art is exemplified by the statement [13], that 'the emulsions must be stored in a refrigerator and no antibiotics, vitamins or potassium supplements added because they may break the emulsions'. Lynn [14] reports some experiments on the addition of disodium carbenicillin and sodium cloxacillin to intravenous lipid emulsions which verify this statement. The special case of intravenous emulsions is dealt with in Section 8.7.2.

8.2.2 HLB, PIT and emulsion stability

Before dealing with theoretical approaches to the stability of emulsion systems it is as well to consider the practical approaches to emulsion formulation, which, in spite of greatly increased understanding of the theoretical basis for stabilization and destabilization, remain the first avenue in development laboratories. We will consider the hydrophile–lipophile balance (HLB) system devised by Griffin [15] as a means of selecting the most effective non-ionic surfactant stabilizer for a given oil. Only now are we beginning to understand why it is that the HLB approach is a reasonable first approach to the choice of single or mixed emulsifiers, but the greater stabilizing power of mixed surfactants cannot yet be placed on a theoretical footing.

In this system one calculates the hydrophile–lipophile balance (HLB) of surfactants and matches the HLB of the surfactant mixture, in the case of O/W systems, to that of the oil being emulsified. The HLB number of a surfactant is calculated according to certain empirical formulae and for non-ionic surfactants the values range from 0 to 20 on an arbitrary scale (see Fig. 8.2). At the higher

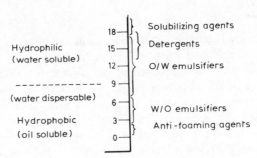

Figure 8.2 HLB scale and approximate range into which solubilizing agents, detergents, emulsifiers and antifoaming agents fall.

end of the scale the surfactants are hydrophilic and act as solubilizing agents, detergents and O/W emulsifiers. Oil-soluble surfactants with a low HLB act as W/O emulsifiers. In the stabilization of oil globules it is essential that there is a degree of hydrophilicity to confer an enthalpic stabilizing force and a degree of hydrophobicity to secure adsorption at the interface. The balance between the two will depend on the nature of the oil and the mixture of surfactants, hence the need to apply the HLB system. The HLB number of a non-ionic surfactant of the polyoxyethylene class is calculated from the equation

$$\text{HLB} = (\text{mol \% hydrophilic group})/5. \tag{8.1}$$

Polyoxyethylene glycols therefore have an HLB of 20. Other surfactant types can be treated as follows. The HLB of polyhydric alcohol fatty acid esters such as glycerol monostearate may be obtained from the equation

$$\text{HLB} = 20\left(1 - \frac{S}{A}\right), \tag{8.2}$$

where S = saponification number of the ester and A is the acid number of the fatty acid. The HLB of polysorbate 20 calculated using this formula is 16.7, S being 45.5 and A, 276. The polysorbates (Tweens) have HLB values in the range 9.6 to 16.7; the sorbitan esters (Spans) have HLBs in the lower range of 1.8 to 8.6.

For those materials for which it is not possible to obtain saponification numbers, e.g. beeswax and lanolin derivatives, HLB is calculated from:

$$\text{HLB} = (E + P)/5; \tag{8.3}$$

E is the percentage by weight of oxyethylene chains, P is the percentage by weight of polyhydric alcohol groups (glycerol or sorbitol) in the molecule.

The HLB system has been put on a more quantitative basis by Davies and Rideal [16] who calculated group contributions to the HLB number such that the HLB was calculable from

$$\text{HLB} = \sum (\text{hydrophilic group numbers}) - \sum (\text{lipophilic group numbers}) + 7 \tag{8.4}$$

Some Group numbers are given in Table 8.1.

Table 8.1 HLB group numbers for hydrophilic and lipophilic groups*

Hydrophilic groups	Group number	Lipophilic groups	Group number	Derived groups	Group number
$-SO_4-Na^+$	+38.7	$-CH-$	-0.475	$-(OCH_2CH_2)-$	+0.33
$-COO-K^+$	+21.1	$-CH_2-$	-0.475	$-(OCH_2CH_2CH_2)-$	-0.15
$-COO-Na^+$	+19.1	$-CH_3$	-0.475		
$-SO_3-Na^+$	+11.0	$=CH-$	-0.475		
N (tertiary amine)	+ 9.4	$-CF_2-$	-0.870		
Ester (sorbitan ring)	+ 6.8	$-CF_3$	-0.870		
Ester (free)	+ 2.4				
$-COOH$	+ 2.1				
$-OH$ (free)	+ 1.9				
$-O-$ (ether group)	+ 1.3				
$-OH$ (sorbitan ring)	+ 0.5				

* According to Davies and Rideal [16] and Lin [17]; Lin and Somasundaran [18]; Lin [19].

The appropriate choice of emulsifier or emulsifier mixture can be made by preparing a series of emulsions with a range of surfactants of varying HLB. It is assumed that the HLB of a mixture of two surfactants containing fraction f of A and $(1-f)$ of B is an algebraic mean of the two HLB numbers.

$$HLB_{mixture} = f\,HLB_A + (1-f)\,HLB_B. \tag{8.5}$$

For reasons not explained by the HLB system, mixtures of surfactants give more stable emulsions than single surfactants. In the experimental set up, creaming of the emulsion is observed and is taken as an index of stability. The system with the minimum creaming or separation of phases is deemed to have an optimal HLB. It is therefore possible to determine optimum HLB numbers required to produce stable emulsions of a variety of oils.

Each oil corresponds with an HLB number which will provide a stable O/W emulsion; for example, for liquid paraffin it is 10 to 12. A single surfactant or mixture of surfactants which provides this HLB number will stabilize the liquid paraffin dispersion. Estimations of stability in order to assess the 'required' or 'critical' HLB are carried out visually by observation of creaming of a series of emulsions prepared with a range of emulsifying agents. While this number has been invaluable for rapid choice of an emulsifier, its use still involves empirical standards. The HLB system neglects the dependence of stability on the concentration of surfactant as has been pointed out by Elworthy and Florence [20]. Riegelman and Pichon [21] have drawn attention to other drawbacks in its use; for example, while creaming is a criterion of instability in commercial formulations it is by no means the only one. It is imperative, they point out, to recognize that stability towards creaming is dependent on the rheological character of the emulsion far more than on the interfacial characteristics of the interfacial film. The influence of surfactants on the viscosity of the continuous phase is therefore of primary importance in this case. In spite of there being optimal HLB values for forming O/W emulsions, it is possible to formulate stable systems with mixtures of surfactants well below the optimum. This is because of the formation of a viscous network in the continuous phase. The viscosity of the medium surrounding the droplets prevents their collision and this overrides the influence of the interfacial layer and barrier forces due to the presence of the absorbed layer. Richards and Whittet [22] were able to obtain stable liquid paraffin-in-water emulsions with surfactant combinations having an HLB as low as 3.9. The stable emulsions were all thixotropic, indicating that the surfactants were contributing to the structural viscosity of the system and thereby contributing to stability by preventing creaming.

Hadgraft [23] also has obtained stable liquid paraffin emulsions with cetyl alcohol-cetyl polyoxyethylene ether combinations having HLB values as low as 1.9; this stability undoubtedly arises from the viscous nature of both interface and bulk phases. Rheological measurements on similar systems confirm this view [24].

The conclusion is that the HLB system will give a quick answer to a practical problem, but will offer little scope for basic improvements on the formulation. Vold [25] has written 'it is intriguing that HLB numbers of mixed non-ionics are additive according to the proportions of each present. One is left with the conviction that the HLB number has a rational interpretation and with a sense of frustration in not being able to show its origin conclusively'.

A linear relationship has been established between the heats of hydration of non-ionic surfactants and their calculated HLB numbers [26]. The dependence of the critical HLB on oil type suggests the importance of solubility and interfacial adsorption. The solubility and aggregation characteristics of surfactants have been shown to be related to solubility parameters (δ) of the surfactant (δ_s) and of the solvents in which they are dispersed [27, 28]. We define the solubility parameter as the square root of the ratio of the molar energy of vaporization to its molar volume. Little [29] has been able to obtain a correlation between δ_s and HLB when the solubility parameter was calculated with the aid of the Hildebrand rule relating molar energy of vaporization to boiling point. δ_s was computed from the contributions of the hydrocarbon (HC) and polar groups (P) according to

$$\delta_s = \phi_{HC}\delta_{HC} + \phi_P\delta_P \tag{8.6}$$

where ϕ is the appropriate volume fraction. Using solubility parameters for individual chemical groups (e.g. 9.6 for the polyethylene oxide group) the following relation held reasonably well:

$$HLB = \left\{\frac{\delta_s - 8.2}{\delta_s - 6.0}\right\} 54 \tag{8.7}$$

or

$$\delta_s = \left\{\frac{118.8}{54 - HLB}\right\} + 6.0. \tag{8.8}$$

It is because of the dependence of HLB on the solution properties of the surfactant molecules that HLB numbers cannot always be realistic measurements of actual HLB, which will depend on the nature of the solvent, the temperature and presence or absence of additives. The relationship between surfactant CMC and HLB has been discussed by Lin and Marszall [30]. CMC is, of course, dependent on the properties of the solvent and this explains the variable nature of the HLB for any given surfactant.

This has been pursued in work by Florence, Madsen and Puisieux [31] who investigated the influence of sodium chloride and sodium iodide on liquid paraffin-in-water emulsions stabilized by mixed non-ionic surfactants. The stability of the emulsions was assessed in several ways. A series of liquid paraffin emulsions was prepared with Brij 92/96 mixtures with a range of HLB values, with and without additives. The particle size of the resulting emulsions, their rheological properties, evidence of creaming, inversion and emulsion type were recorded. In this way the HLB value at which a change in property, e.g. emulsion type, occurred was noted and recorded as the 'critical' HLB (Fig. 8.3). Salting-out of the surfactant led to an increase in the critical HLB values and salting-in resulted in a decrease in the critical HLB. On addition of 15 % NaCl to the liquid paraffin-in-water emulsion the HLB for maximum stability rises from 7.8 to 9.5; the cloud point of the water-soluble surfactant decreases by 35° C in this concentration of sodium chloride. From Fig. 8.4 it is seen that a decrease of 35° C in the cloud point is equivalent to a decrease in critical HLB of 1.6, which is in agreement with the experimentally determined change.

The lesser effect of sodium iodide on the cloud point is reflected in the less pronounced effect on optimal or critical HLB value (Fig. 8.3).

The mechanism of action of the additives which alter the solution properties of the non-ionic surfactants is not yet clear. The measurable hydration of the

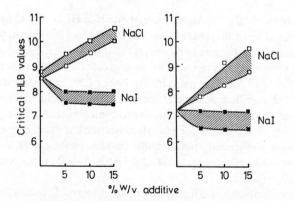

Figure 8.3 The change in critical HLB values as a function of added salt concentration, where the salt is either NaCl or NaI. Results were obtained from measurements of particle size, stability, viscosity and emulsion type as a function of HLB for liquid paraffin-in-water emulsions stabilized by Brij 92–Brij 96 mixtures. Data from different experiments showed different critical values hence on each diagram hatching represents the critical regions while data points actually recorded are shown. Results on the left-hand diagram show, respectively, particle size and stability data: those on the right-hand diagram show the HLB at transition from pseudoplastic to Newtonian flow properties and emulsion type (O/W → W/O) transitions. From Florence *et al.* [31] with permission.

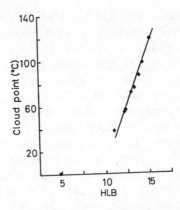

Figure 8.4 Variation of cloud point of surfactant solutions as a function of calculated HLB value. ◆ values for linear primary dodecanol-polyoxyethylene condensates from Schott [32]. ● Brij 96 and Brij 98, the latter point being obtained by extrapolation to zero salt concentration of cloud points in a series of NaCl solutions. From Florence *et al.* [31] with permission.

polyoxyethylene chains is not increased by the compounds which salt-in the surfactant molecules, as the viscosity results demonstrate. These substances must so alter the structure of water that the thermodynamic properties of the system are altered. The relation proposed by Racz and Orban [26] between the heat of hydration Q and HLB is,

$$HLB = 0.42Q + 7.5 \tag{8.9}$$

where Q has units of cal g^{-1}. An increase in intrinsic HLB caused by the addition of, say sodium iodide or thiocyanate, should result in an increase in Q in Equation 8.9. This expectation is borne out by the fact that the heat of dilution of non-ionic surfactant in sodium thiocyanate solution is greater than the heat of dilution in sodium chloride [33]. An increase in the heat of hydration caused by agents which salt-in the stabilizing chains of the surfactants should result in an increase in the enthalpic element of the steric barrier to coalescence. All the evidence points to the possibility of increasing the enthalpic barrier but experimental evidence is less easily gained. Preliminary results show that the phase-inversion temperature of xylene-in-water emulsions stabilized by Triton X-100 is increased by ethanol and propanol.

The apparent inability to effect larger or demonstrable increases in emulsion stability by the addition of substances which raise the cloud point of the stabilizing molecules may be due to several factors. Increase in aqueous solubility may decrease the concentration of emulsifier at the interface, although interfacial tension measurements do not indicate that this is so; as there is an optimal HLB for each oil to achieve stability, any change in HLB by addition of agents which salt-in or salt-out the stabilizing molecules will shift the stability index from the optimum position.

The determination of the factors influencing the critical HLB has been the preoccupation of Puisieux and his colleagues. In the course of new work they have found that the particle size of the oil phase is at a minimum at the critical HLB value, that the critical HLB is dependent on the concentration of surfactant present and have made a detailed study of the influence of the oil phase on the critical HLB value [34–37]. In Table 8.2 the critical HLB is compared with the dielectric constant of the oil phase and Fig. 8.5 shows a linear relationship between HLB and the logarithm of the dielectric constant in the case of the

Table 8.2 The dielectric constant and critical HLB for a number of oil phases stabilized by Brij 92–Brij 96 mixtures

Oil phase	Dielectric constant at 20° C	Critical HLB
Hexane	1.890	10.5–11
Heptane	1.924	10–10.5
Octane	1.948	9.5–10
Dodecane	2.014	9–9.5
Tetradecane	2.0368	8.5
Liquid paraffin	2.16	8.5
Perhydrosqualene		8
Cyclohexane	2.023	12
Decalin	2.26	10–11
Benzene	2.26	> 12
Toluene	2.28	> 12
Xylene	2.568	12
Ethyl oleate	3.17	10–12

From [37].

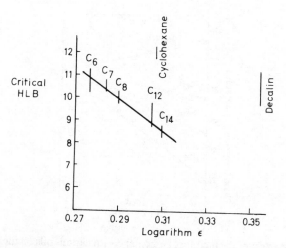

Figure 8.5 Critical HLB as a function of log dielectric constant (ε) of the C_6–C_{14} alkanes and cyclohexane and decalin. From Lo *et al.* [37] with permission.

straight chain alkanes. This finding has been confirmed by other work [38]. The reason for the lack of additivity may be the non-ideality of the oil mixtures. If one species is more surface active than the other in the mixture then there will be a surface excess concentration not reflecting the bulk concentration.

An investigation by the same group of the critical HLB values for mixtures of oils showed that the critical HLB values were not additive as they are for surfactants, as Table 8.3 demonstrates.

O/W emulsions stabilized with non-ionic surfactants tend to form W/O emulsions at elevated temperatures as the surfactant molecules dehydrate and become more lipophilic. The phase inversion temperature (PIT) can thus be ascertained by experiment. Arai and Shinoda [39] have found that the PIT of emulsions in which the oil phase consists of oil mixtures can be expressed as

$$PIT_{mixture} = PIT(A)_{\phi_A} + PIT(B)_{\phi_B} \qquad (8.10)$$

Table 8.3 Critical HLB values for stabilization of mixtures of oils using perhydrosqualene as reference

Mixture of oils	Critical HLB determined by experiment	Critical HLB calculated on assumption of additivity
Perhydrosqualene–xylene 80:10	8	8
Perhydrosqualene–xylene 75:25	8.5	10
Perhydrosqualene–xylene 50:50	9.5–12	11–16
Perhydrosqualene–heptane 50:50	9	10
Perhydrosqualene–heptane 75:25	8.5–9	10–12
Perhydrosqualene–heptane 25:75	9.5–10	10–10.6

From [37]. See Table 8.2 for critical HLB for individual oils.

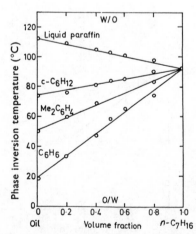

Figure 8.6 The effect of the mixture of *n*-heptane with various oils on the phase inversion temperatures of emulsions stabilized with 3 % w/w in water of polyoxyethylene (9.6) nonyl phenyl ether. From Arai and Shinoda [39] by permission of Academic Press.

where ϕ_A and ϕ_B are the volume fractions of oils A and B and PIT(A) and PIT(B) are the phase inversion temperature of emulsions of the oils alone. Fig. 8.6 shows the linear relationship between the PIT of mixtures and the volume fraction of heptane added to the oil phase. Shinoda [40] has suggested using the PIT values instead of HLB values in the selection of emulsifiers as the PIT (or 'HLB temperature') is 'an experimental value that takes into account the nature of the oils, the interaction between the surfactants and oils and the nature of the water'.

A correlation between PIT and calculated HLB value of a heptane–water emulsion stabilized by 3% surfactant mixtures is shown in Fig. 8.7.

Arai and Shinoda's conclusion that, as the PITs and the HLB values change linearly for a narrow range of temperature, the required HLB value of an oil mixture can be calculated from the volume or weight average of the respective values for the oils, is at variance with some of the findings discussed above. However, Arai and Shinoda did not determine experimental critical HLB values for their emulsion systems.

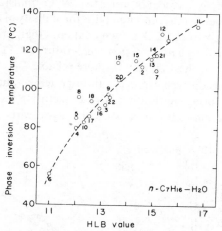

Figure 8.7 The correlation between the calculated HLB-value and the PITs of *n*-heptane–water emulsions stabilized with various surfactants (3 wt % for water). Weight ratios are used for mixtures.

(1) $R_9C_6H_4O(CH_2CH_2O)_{17.7}H$. (2) $R_9C_6H_4O \cdot (CH_2CH_2O)_{14.0}H$.
(3) $R_9C_6H_4O(CH_2CH_2O)_{9.6}H$. (4) $R_9C_6H_4O(CH_2CH_2O)_{7.4}H$.
(5) $R_{12}C_6H_4O \cdot (CH_2CH_2O)_{9.0}H$. (6) $R_9C_6H_4O(CH_2CH_2O)_{6.2}H$.
(7) $R_{12}O(CH_2CH_2O)_{13.0}H$. (8) $R_{12}O(CH_2CH_2O)_{6.5} \cdot H$. (9) Polyoxyethylene (13.9) styrenephenylether. (10) Polyoxyethylene (11.4) styrenephenylether. (11) Tween 20. (12) Tween 40. (13) Tween 60. (14) Tween 80.
(15) $R_9C_6H_4O(CH_2CH_2O)_{15.8}H/R_9C_6H_4O(CH_2CH_2O)_{7.4}H = 7/3$.
(16) $R_9C_6H_4O \cdot (CH_2CH_2O)_{9.6}H/R_9C_6H_4O(CH_2CH_2O)_{7.4}H = 7/3$.
(17) $R_9C_6H_4O(CH_2CH_2O)_{9.6}H/R_9C_6H_4O \cdot (CH_2CH_2O)_{7.4}H = 3/7$.
(18) $R_{12}O(CH_2CH_2O)_{6.5}H/R_9C_6H_4O(CH_2CH_2O)_{9.6}H = 1$.
(19) $R_{12}O \cdot (CH_2CH_2O)_{6.5}H/R_9C_6H_4O(CH_2CH_2O)_{15.8} = 1$.
(20) $R_{12}O(CH_2CH_2O)_{13.0}H/R_9C_6H_5O(CH_2CH_2O)_{7.4} = 1$.
(21) $R_{12}O(CH_2CH_2O)_{13.0}H/R_9C_6H_4O(CH_2 \cdot CH_2O)_{15.8}H = 1$.
(22) $R_9C_6H_5O(CH_2CH_2O)_{7.4}H/Tween\ 80 = 1$. From Arai and Shinoda [39] with permission.

(A) INVERSION OF EMULSIONS

In stable emulsion formulations, inversion can be induced by increasing the disperse phase volume, ϕ, to around 0.7. This is a phenomenon not accounted for in theoretical treatments. The exact value of ϕ at the inversion point depends on the surfactant present and its concentration. The viscosity of a series of chlorobenzene-in-water emulsions has been determined as a function of ceto-macrogol 1000 concentration and ϕ. Some results are shown in Fig. 8.8 which indicate that the higher the concentration of surfactant in the system the lower the phase volume at inversion. Becher [41] obtained this trend with low HLB emulsifiers (e.g. sorbitan mono-esters), but, in general, with polyoxyethylated compounds the inversion point increased with increasing concentration. In a recent paper, Shinoda and Saito [42] described emulsification by a phase-inversion method, this being the preparation of a stable, finely dispersed emulsion by rapid cooling of an emulsion at its phase inversion temperature. This differs from the normal emulsification by inversion which involves alteration of phase volume, for example, by addition of water to a W/O emulsion to form an O/W type. It was concluded that the optimum HLB for stability of an emulsion cannot be obtained accurately from HLB–stability data but that, as stability is sensitive to temperature near the phase-inversion temperature (PIT), the selection of an emulsifier according to the PIT may be more reliable.

Inversion of emulsion type can occur through temperature changes such as those encountered during sterilization or manufacturing procedures. The more soluble a non-ionic emulsifier in a particular hydrocarbon, the lower is the

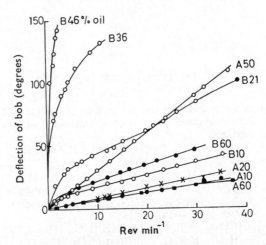

Figure 8.8 Viscosity results obtained with a Couette viscometer on a series of emulsions of chlorobenzene stabilized A with 5% cetomacrogol 1000 and B with 10% cetomacrogol 1000. Results are given in arbitrary units: deflection in degrees of inner bob versus Rev min^{-1} of outer container. Phase volumes for both series are appended to lines as percentage oil. The diagram shows the inversion of A and B at a phase volume about 0.60 (see low viscosity of this one). B inverts before A. From Florence and Rogers [4] with permission.

phase inversion temperature of the emulsion [12]. Hence, emulsifiers in systems which have to withstand elevated homogenization or sterilization temperatures must be more hydrophilic than those found satisfactory at normal temperatures. However, the phase-inversion temperature can be manipulated to a considerable extent by altering the composition of the oil phase. This is strikingly illustrated in Fig. 8.8.

It is laborious to determine the critical HLB experimentally and as can be seen from Figs. 8.3 and 8.5, it is often impossible to define a single value, as the stability against coalescence will not change rapidly with change in HLB. However, the change in stability of an emulsion is sensitive to temperature near the PIT and hence this phase inversion temperature tends to be more precise even if not less empirical. Stable emulsions will obviously be achieved if the emulsions are rapidly cooled when prepared by the PIT method and when the storage temperature is well below the PIT. Relatively stable systems are obtained when the PITs of the emulsions are about 20 to 65° C higher or 10 to 40° C lower than the storage temperature for O/W and W/O emulsions, respectively [42, 43].

8.2.3 Droplet size and emulsion stability

Alternative approaches to the selection of emulsifiers have been investigated. Prediction of optimum emulsifier mixtures has been made by way of solubilization measurements [44]. Lin *et al.* [44] found a correlation between the maximum amount of aqueous phase that could be solubilized in the oil phase containing the surfactant and the average droplet size of the emulsion subsequently formed (Fig. 8.9). The relationship held even when ionic–non-ionic mixtures of surfactants were used and thus displays an advantage over the PIT method as ionic surfactants do not produce PIT values. It is telling that the method works in the presence of additives such as lauryl alcohol. Fig. 8.9 shows the shift in optimum surfactant ratio when lauryl alcohol is added to the oil phase, in this case mineral oil. The addition of a polar oil to a non-polar oil will result in a predictable shift in required HLB. However, a shift of no more than 1 HLB unit would be expected from the linear additivity rule, while Fig. 8.9 shows a shift of some 2.4 HLB units. In some systems that Lin and his colleagues [44] investigated, the position of maximum solubilization did not coincide with the optimal O/W emulsion, an effect believed to be due to phase inversion at the point of maximum solubilization.

It is evident from several techniques, that the optimum emulsions form when the surfactant ratio is such that the globule size is at a minimum [45]. Sedimentation or creaming and attraction between the globules is minimal under these conditions. As ternary phase diagrams indicate (see Chapter 2) the phase behaviour of oil–non-ionic surfactant–water systems is complex, and highly dependent on the ratio of surfactants used [46]. A phase diagram of the water–dodecane system containing 20% non-ionic surfactant $C_{12}E_5$ is shown as a temperature–composition plot in Fig. 8.10. In the transition between O/W emulsion and W/O emulsion as temperature is raised, the system passes through

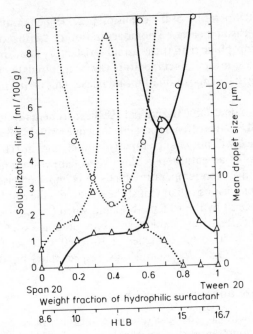

Figure 8.9 Shift of optimum emulsification peak by addition of lauryl alcohol (emulsions contain 30% oil phase, 65% deionized water, and 5% surfactant mixtures. Surfactant mixtures consist of hydrophilic Tween 20 and lipophilic Span 20 at ratios and corresponding HLB values indicated by abscissa). Dotted lines represent data for pure mineral oil systems. Solid lines represent data for oil mixture consisting of 8 parts mineral oil and 2 parts lauryl alcohol. ○ mean droplet size; △ solubilization limit. From Lin *et al.* [44] with permission.

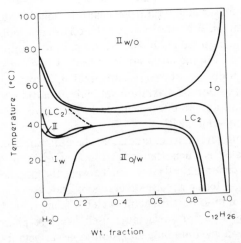

Figure 8.10 The phase diagram of the water–dodecane system containing 20 wt% pentaoxyethylene dodecyl ether, $C_{12}E_5$. From Harusawa and Mitsui [47] with permission.

the I_W region when dodecane is solubilized in aqueous micelles. LC is a liquid crystalline phase possibly with lamellar structure. I_o is a dodecane solution of water-swollen inverse micelles [47]. At the transition point, i.e. at the PIT there thus exists a third phase sometimes known as the surfactant phase. Van der Waals interactions in three-phase emulsions have been examined by Madaric and Friberg [48]. The third phase may be a liquid crystalline phase referred to above, whose presence causes a pronounced stabilization of emulsions when it exists in the vicinity of the emulsion droplets.

8.2.4 Interaction forces between dispersed particles and emulsion stability

The theory of stability of lyophobic colloids by Derjaguin and Landau (1941) and by Verwey and Overbeek (1948) (the 'DLVO theory') is still the most successful treatment of stability. The theory was developed to deal with the stability of inorganic sols, which possess intrinsic electrostatic charges, but over the years, the theory and modifications of it, have been applied to suspensions of clays and to emulsions. In general, the theory predicts the energy requirements of the system which will lead to stability against flocculation or coagulation of the particles – this step being the final one for solid dispersions in a liquid medium but the initial stage of instability in emulsions, as flocculation can occur without coalescence. However, as the average proximity of the particles will be governed by the forces discussed by the theory, if the barrier to flocculation is small, there will require to be a good resistance to coalescence if the emulsion is to remain stable.

Particles of all types possess a net electrostatic charge when suspended in a simple aqueous medium. The origin and magnitude of this charge is dependent on the nature of the medium, the nature of the particle surfaces and the presence or absence of other components in the system.

The DLVO theory of stability takes into account the interaction of two kinds of long-range forces which determine the closeness of contact of two particles approaching as a result of Brownian movement. The forces concerned are (i) the London–van der Waals' forces of attraction, and (ii) the electrostatic repulsion between electrical double layers.

Since the origin of one force is completely independent of the other, each force may be evaluated separately and the net result of their interaction obtained by summation. However, in addition to these forces of interaction, a free energy of interaction force arising from a third source, steric hindrance, involving the free energy of mixing solvated adsorbed layers, becomes of importance in systems containing non-ionic surfactants. This is particularly true for the polyoxyethylene type of non-ionic surfactants; to a lesser extent true for alkoxymethyl ethers of sucrose (I) in which the hydrophilic chain is neither long nor very flexible. The repulsion between two particles originating in this manner has been termed entropic repulsion because of the loss of configurational entropy of the adsorbed molecules on mixing. This entropy loss is manifested as a repulsive force. This

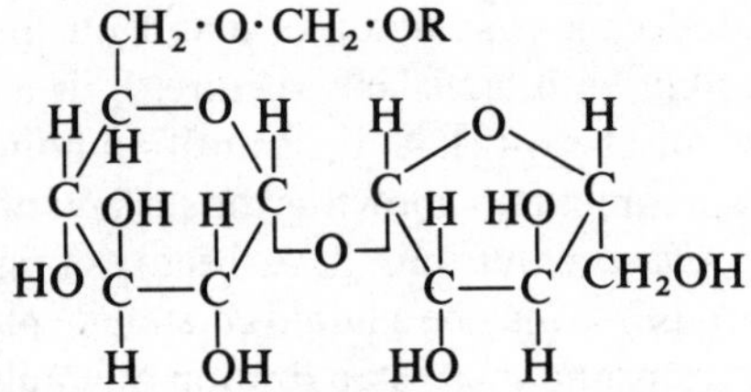

CH$_2$·O·CH$_2$·OR

(I)

force has far greater importance in emulsion systems than in conventional lyophobic colloids because of its short-range nature, which means it is usually operative in already flocculated or coagulated systems. However, conclusions about the stability of disperse systems may be reached only if the short- and the long-range forces affecting the interaction of the particles are assessed. The DLVO theory can be applied without difficulty only to monodisperse particles and therefore, its application to systems having wide particle-size distributions is limited. Hogg *et al.* [49] and Ho and Higuchi [50] assess the effect of heterodispersity on colloid stability.

(A) THE LONDON–VAN DER WAALS' FORCE OF ATTRACTION

The attractive forces which exist between like molecules in a vacuum have been quantitatively identified by London [51]. The relation for a pair of equal spheres composed of finite particles was derived by Hamaker [52]. For values of $H/a \ll 1$ the relation for two spherical particles of radius a can be reduced to

$$V_A = -\frac{Aa}{12H} \qquad (8.11)$$

here the distance of separation is H and A is the Hamaker constant.

When the particles 1 are not in a vacuum but are embedded in a medium of substance 2, the total interaction of two particles is dependent on the net interactions of all molecules. Hamaker deduced this interaction to be

$$A = A_{11} + A_{22} - 2A_{12} \qquad (8.12)$$

where $A_{22} = \pi^2 q_2 \beta_2$ for material 2 and $A_{12} = \pi^2 q_1 q_2 \beta_{12}$ (q is the number of molecules per cm^3 material and β is related to the square of the polarizability) for the corresponding interaction between material 1 and 2. If it is assumed that A_{12} can be taken as the geometric mean of A_{11} and A_{22}, then

$$A = (A_{11}^{\frac{1}{2}} - A_{22}^{\frac{1}{2}})^2. \qquad (8.13)$$

Thus, A is always positive. Consequently, emulsion droplets are always subject to an attraction, which has the same value for droplets in a particular emulsion whether the dispersion is O/W or W/O. Values of the Hamaker constant differ for different pairs of liquids, ranging from 1×10^{-20} J for paraffin-in-water to

2×10^{-19} J for carbon tetrachloride–water. The accuracy of estimating A is low, often by an order of magnitude [53].

In systems of dispersed particles containing adsorbed layers of surface-active substances, the effective Hamaker constant between the particle and the medium may be radically altered. If the Hamaker constant of the solvation sheath is close to that of the dispersion medium, the sheath simply acts as a mechanical barrier preventing the close approach of the dispersed particles in the range where attractive forces become strong. This layer would then contribute to stability.

Vold [54] has analysed the effect of adsorption on the attraction of spherical colloidal particles of radius a. In the case of a homogeneous adsorbed layer of thickness δ, having a Hamaker constant, A_{33}, the potential energy of attraction is given by

$$V_A = -\frac{1}{12}\left[(A_{22}^{1/2} - A_{33}^{1/2})^2 \left(\frac{a+\delta}{\Delta}\right) + (A_{33}^{1/2} - A_{11}^{1/2})^2 \left(\frac{a}{\Delta+2\delta}\right) \right.$$
$$\left. + 4a(A_{22}^{1/2} - A_{33}^{1/2})(A_{33}^{1/2} - A_{11}^{1/2})(a+\delta)(\Delta+\delta)(2a+\delta) \right] \quad (8.14)$$

where Δ is the distance between the surfaces of the adsorbed layers [55].

Using Vold's equation and estimating A_{33} from refractive index measurements, Elworthy and Florence [20] obtained a result which is surprising in the light of Ottewill's calculations: the adsorbed layer *increased* the attraction between chlorobenzene particles dispersed in water. Both theoretical and experimental derivation of A_{33} assumed that the surfactant layer is homogeneous. This is an over simplification.

Fig. 8.11 illustrates the effect of different thicknesses of adsorbed layers of a polyoxyethylene ether $C_{16}H_{33}(O.CH_2.CH_2.)_6OH$ on the calculated attraction

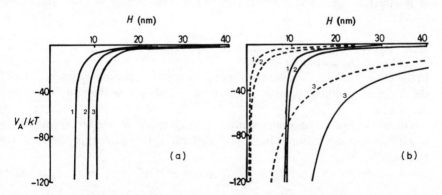

Figure 8.11(a) The influence of the thickness (δ) of an adsorbed layer of $C_{16}E_6$ and the attraction between two chlorobenzene droplets for a particle of size $a = 0.5 \times 10^{-4}$ cm. 1, $\delta = 2.5$ nm. 2, $\delta = 4.0$ nm. 3, $\delta = 5.0$ nm. (b) Influence of particle size a on the attraction between two chlorobenzene droplets: - - - uncoated; —— coated with a layer of $C_{16}E_6$ with $\delta = 4.0$ nm. 1, $a = 0.5 \times 10^{-4}$ cm. 2, $a = 1 \times 10^{-4}$ cm. 3, $a = 10 \times 10^{-4}$ cm. From Florence and Rogers [3] with permission.

energy between two chlorobenzene droplets 1 μm in diameter. As the thickness of the homogeneous layer of $C_{16}E_6$ increases, the attraction between the droplets increases. At a separation of 12.5 nm, the attraction energy is approximately doubled by increasing the layer thickness from 2.5 to 4.0 nm. A thickness of 5 nm results in a four-fold increase in the attractive forces. In this particular system, the Hamaker constant of the adsorbed layer was greater than that of the dispersed oil $(A_{33} = 6.7 \times 10^{-20}\,\text{J},\ A_{22} = 6.3 \times 10^{-20}\,\text{J},\ A_{11} = 3.78 \times 10^{-20}\,\text{J})$. Likewise, the attraction increases with increasing particle size for a given adsorbed layer thickness (Fig. 8.11) and a significant change is observed on the introduction of an adsorbed layer onto uncoated particles. It appears, therefore, that large particles will be more difficult to stabilize by adsorbed layers than small ones if only the effect on V_A is considered. In the development of the theory, Vold [54] explains that a reduction in the inter-particle attraction occurs when the density of the particle is greater than that of the medium, and the particle is coated with a weakly interacting adsorbed layer. Furthermore, it is pointed out that adsorbed layers on dispersed particles can never change the interparticle attraction into a repulsion. Derjaguin [56, 57] has doubted whether attenuation of attractive forces could contribute significantly to stability. The application of the Vold correction in this work has been criticized [58] as interparticle attraction was calculated keeping the centre-to-centre distance constant rather than the outer surface-to-surface distance constant [59].

At first sight it would appear that stability cannot be explained by the electrostatic repulsive forces which operate in many dispersions. Pure hydrocarbon droplets dispersed in water possess a net negative charge, but are unstable. The problems which arise in the theory of non-ionic dispersions are these [45, 60]. To what extent do electrostatic forces contribute to stability? How is any charge on the dispersed particles affected by the presence of non-ionic detergents or polymer molecules at the interface? What is the nature of other stabilizing forces?

(B) REPULSIVE FORCES ARISING FROM INTERACTION OF
 ELECTRICAL DOUBLE LAYERS

In the DLVO theory, the distribution of counterions in the immediate vicinity of the charged colloidal particle is assumed to obey the Poisson–Boltzmann distribution.

The work required to bring together identical spherical particles from infinity to a distance of separation H in a liquid medium is given in an approximate form by [61]:

$$V_R = \frac{\varepsilon a \dot{\psi}_0{}^2}{2} \ln(1 + e^{-\kappa H}). \tag{8.15}$$

This formula is valid only for low potentials ($\dot{\psi}_0 < 25$ mV), for spheres that are large in radius, a, compared to the thickness of the double layer, (i.e. $\kappa a \geqslant 1$), and for a large separation, H, between the spheres compared with the double layer thickness (i.e. $H > 1/\kappa$). However, it can be used as an approximation for most practical systems, particularly emulsions.

In Equation 8.15 the surface potential, $\dot{\psi}_0$, is usually equated to ζ, the zeta potential as determined from measurements of electrophoretic mobility. The two quantities are by no means always identical, but in view of the experimental difficulties of assessing $\dot{\psi}_0$, zeta potential may be used as a close approximation. ζ is the experimentally accessible potential difference between the bulk solution and the electrokinetic 'slipping plane' or 'plane of shear' which is situated in the diffuse layer close to the immobile Stern layer. (The meaning of these planes in molecular terms when a long chain hydrophilic non-ionic surfactant is adsorbed at the globule surface is not clear).

In emulsions, as opposed to solid dispersions, the potential energy barrier to contact of two globules can increase if distortion of the globules occurs as a result of their mutual repulsion. The significance of this effect is not estimable in unstabilized emulsions. When adsorbed layers of surface-active material are present, it is less likely that distortion occurs on collision. Minute amounts of surfactant retard circulation in droplets because adsorption increases the interface viscosity. Trace amounts of surfactant cause the motion of small bubbles and drops through a liquid to resemble that of rigid bodies [62].

Diffuse double layers extending into the disperse phase from the interface have been considered by Verwey [63] but they do not appear to influence the external electrical layers to any significant extent. If the disperse phase is polar (e.g. water), then the concentration of counterions in the disperse phase would serve to reduce the net electrokinetic potential attributed to the particle.

The degree of stability of many dispersions cannot be explained solely on the basis of V_a and V_R. Elworthy and Florence [56] have treated the stability of emulsions of chlorobenzene and anisole stabilized with a series of synthetic polyoxyethylene ethers in the light of colloid theory and have shown that electrical stabilization alone cannot explain the stability observed. The nature of this 'other force' which is invoked to explain discrepancies between theory and experiment is not fully worked out. Nevertheless, much interest has been shown in this alternative mechanism of stabilization, which for 'non-ionic' emulsions appears to play the major role [64]. Results have indicated that the thickness and degree of solvation of adsorbed layers is critical [65]. Thus, the particular conformation and length of the polyoxyethylene chains of non-ionic surfactants at interfaces is likely to be an important factor in the stabilization of emulsified droplets.

The overall free energy change when the adsorbed layers on identical spherical particles mix on contact is assumed to arise from the additive contributions of several energy changes. This overall free energy, being positive, results in repulsion between the particles because work is required to overcome these energy barriers.

If ΔG is the net free energy change, then

$$\Delta G = \Delta G_m + \Delta G_v + \Delta G_s + \Delta G_e, \tag{8.16}$$

where ΔG_m is the free energy change of mixing of the adsorbed layers which results in an increased concentration and density of chains in the overlap region. When coated particles are separated by a distance less than 2δ, restrictions are imposed on the volume occupied per chain, governed by the particular conformations of the chains. This excluded volume effect is manifested as a free energy change, ΔG_v. Considerations of interacting adsorbed layers on emulsion globules must include the tendency towards desorption of the molecules at the interface as the compression due to overlapping increases. This tendency will cause an increase in the local interfacial tension which constitutes an increase of the surface free energy, ΔG_s. Collisions between emulsion particles do not always result in the formation of an aggregate or in coalescence. In fact, the dispersed droplets have been observed to exhibit a certain degree of elasticity on collision. This phenomenon may be considered as an elastic energy of repulsion, ΔG_e.

(C) STERIC OR ENTROPIC REPULSION
The term 'entropic' has been applied somewhat indiscriminately to the forces of repulsion arising from interaction of the adsorbed surfactant layers. This term arises because the long chains of the surfactant being restricted on contact suffer a loss of entropy and contribute to the positive free-energy change, in addition to change in solute–solvent interactions in the overlap region.

Various theoretical approaches have been discussed in some detail by Lyklema [66]. For this reason they are not treated here. Instead we concentrate on the theory of Fischer [67] whose final equation contains parameters more amenable to substitution with experimental data than many others.

When adsorbed layers form on dispersed spherical particles, a region of high concentration of macromolecule extends outward from the particle surface into a region of lower concentration of macromolecule. The layer forms a gradient of polymer concentration which is a function of the length of the adsorbed chains, the degree of surface coverage and the conformation of the polymer chains in the environment. If two identical spherical particles (Fig. 8.12) having identical adsorbed layers of polymer chains collide in a common medium the overlap volume of their adsorbed layers is dV. The concentration of polymer chains is increased and one can express the excess chemical potential change of the chains in the overlap volume $(\Delta\mu_i)_E$ as the difference between the observed chemical potential change, $\Delta\mu$, and the change expected under ideal conditions $(\Delta\mu_i)_{ideal}$, i.e. with no solute–solvent interactions, thus

$$(\Delta\mu_i)_E = \Delta\mu - (\Delta\mu_i)_{ideal}. \tag{8.17}$$

The increase in chemical potential as a result of this process generates an excess osmotic pressure arising from solvent flow to the region of high concentration. This pressure is manifested as an energy of repulsion to mixing and its magnitude is determined by the partial molar volume of the solute, $\overline{V}_1$, the initial concentration of chains in the adsorbed layer, C, and the degree of polymer–solvent interaction, related to B, the second virial coefficient. Equation 8.17 may then be written following Ottewill's derivation [55, 67].

$$(\Delta\mu_i)_E = -RTB\overline{V}_1C^2 = -\pi_E\overline{V}_1, \tag{8.18}$$

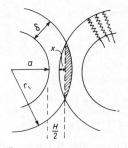

Figure 8.12 The model used in the derivation of Equations 8.25 and 8.30. Particles of radius a with absorbed layer of thickness δ approach to a distance H between the particle surface. $r_1 = (a + \delta)$. x is the distance between the particle surface and the line bisecting the volume of overlap.

Calculation

Volume of hatched section $= \dfrac{\pi x^2}{3}(3r_1 - x)$

i.e.
$$V_s = \frac{\pi}{3}\left(\delta - \frac{H}{2}\right)^2\left(3a + 3\delta - \delta + \frac{H}{2}\right)$$

$$= \frac{\pi}{3}\left(\delta - \frac{H}{2}\right)^2\left(3a + 2\delta + \frac{H}{2}\right)$$

$$V = \text{Total volume of overlap} = 2 \times V_s$$

whence
$$\pi_E = RTBC^2. \tag{8.19}$$

The free energy of mixing, dG_m, in the volume element dV is related to C^2 by

$$dG_m = \frac{2dV}{\overline{V}_1} \cdot RTB\overline{V}_1C^2 = 2d\overline{V}RTBC^2, \tag{8.20}$$

for the volumes of each adsorbed layer such that $dV_1 = dV_2 = dV$, where dV_1 and dV_2 are the volume elements in the adsorbed layers of the first and second particles, respectively. Therefore,

$$\Delta G_m = \int_0^{dV} 2d\,VRTBC^2 = 2\int_0^{dV} \pi_E dV = 2\pi_E dV \tag{8.21}$$

The total energy of repulsions over the whole overlap region between the spheres may be calculated by integrating the volume elements, dV.

The volume of a segment of a sphere is given by

$$V_s = \frac{\pi h^2}{3}3(r_1 - h). \tag{8.22}$$

In Fig. 8.12, $h = (\delta - H/2)$ and $r_1 = (a + \delta)$, hence

$$V_s = \frac{2\pi}{3}\left[\delta - \frac{H}{2}\right]^2\left[3a + 2\delta + \frac{H}{2}\right]. \tag{8.23}$$

If $V_s = V$, then from Equation 8.21,

$$\Delta G_m = \frac{4\pi . \pi_E}{3} \left[\delta - \frac{H}{2} \right]^2 \left[3a + 2\delta + \frac{H}{2} \right].$$ (8.24)

Combining Equations 8.24 and 8.19 gives

$$\frac{\Delta G_m}{kT} = \frac{BN_A C^2 4\pi}{3} \left[\delta - \frac{H}{2} \right]^2 \left[3a + 2\delta + \frac{H}{2} \right].$$ (8.25)

where N_A is Avogadro's number. If the dispersion medium is water, the greater the hydrophilicity of the adsorbed layer, the larger is B, and the higher the free energy of mixing. ΔG_m may be readily estimated from Equation 8.25 providing a suitable value for B can be obtained. This parameter has been evaluated for solutions of non-ionic detergents [68] and glycols [69] but discrepancies exist, depending on the method of measurement, and data are only available for micellar solutions and the behaviour of detergents at oil–water interfaces may, or may not, be identical to that in the micelles. However, Napper [70] found that incipient flocculation of polymer latexes occurred in dispersion media which were θ solvents for the stabilizing molecules in free solution as predicted by theory, i.e.

$$B \propto \left(1 - \frac{\theta}{T} \right),$$ (8.26)

where θ = Theta temperature. Thus when $T = \theta$ deviations from ideality vanish. Using the critical flocculation temperature as a criterion of stability, Napper [70] observed that aqueous dispersions stabilized by two non-ionic polymers differing by a factor of 3 in molecular weight possessed similar stability, suggesting an insensitivity of the steric stabilization to molecular size. However, with amphipathic compounds such as the $C_{16}E_x$ series of non-ionic detergents, changes in molecular weight result in different adsorption properties and, hence, differing steric stabilizing powers [56, 71].

Fischer [67] has evaluated ΔG_m for molecular overlap when the stabilizer segment density in the overlap volume is constant. However, an equation of the form given by Meier [72] should be employed for polymeric stabilizers. Thus, for $L < H < 2L$ where L represents the contour length of the longest stabilizing moiety,

$$\Delta G_m = 2kT(\psi_1 - \chi_1) \frac{V^2}{\overline{V}_1} (\rho_j \rho_k)_H dV.$$ (8.27)

and for $H < L$

$$\Delta G_m = 2kT(\psi_1 - \chi_1) \frac{V^2}{\overline{V}_1} \left[\int (\rho_j)_H^2 dV - \int (\rho_k)_H^2 dV + \int (\rho_j \rho_k)_H dV \right].$$ (8.28)

The relation between B, the second virial coefficient, and χ_1, the solvent–polymer (surfactant) interaction parameter, is

$$B = RT \frac{(\psi_1 - \chi_1)}{\overline{V}_1 . \rho_2^2}$$ (8.29)

where ρ_2 is the density of the adsorbed layer. ψ_1 is an entropy parameter which is given the ideal value of 0.5. Equation 8.28 represents the total change in Gibbs' free energy on bringing the particles from infinite separation to a distance, H, apart, where ρ_j and ρ_k are the segment density contributions in the volume element dV of the adsorbed chains on each particle surface.

Ottewill and Walker [73] have used Fischer's Equation 8.25 but substituted for B giving

$$\frac{\Delta G_m}{kT} = \frac{4\pi N_A C^2}{3\overline{V}_1 \rho_2^2} [0.5 - \chi_1] \left[\delta - \frac{H}{2}\right]^2 \left[3a + 2\delta + \frac{H}{2}\right]. \tag{8.30}$$

The calculated stabilizing contribution of adsorbed layers is represented in Fig. 8.13.

The derivation of these equations and their use involves the assumption that the stabilizing layer remains intact during collision whereas desorption can occur.

The loss of entropy resulting from the interaction of the chains should cause some of the molecules to desorb and to gain entropy by returning to the bulk solution [74]. In these cases the free-energy change corresponding to this process will be largely determined by the free energy of desorption. Therefore, a surfactant should have a high free energy of adsorption if it is to give rise to high stability. This is a function of its solubility in the disperse and continuous phases. When the surfactant is soluble in the disperse phase (as, for example, $C_{16}E_3$ in chlorobenzene) it should be free to desorb at the point of collision and diffuse away into the globule. A more water-soluble detergent (e.g. $C_{16}E_{25}$) will be prevented from doing so and must remain in the interface [75] and hence should contribute more to the stability.

Because the entropic force, however it is formulated, is a short-range force, it will be more directly related to coalescence behaviour than V_A or V_R.

Although an estimate of the required interaction parameters can be made, e.g. from osmotic pressure or light scattering measurements [3], the significance of these bulk values in respect of an oriented adsorbed layer is somewhat doubtful, especially in emulsion systems when the hydrophobic portion of the surfactant (which contributes to the interaction parameter in bulk water) will be embedded in the oil phase. However, the value of these theoretical approaches is that they give guidelines for sensible estimates of the likely effects of changing individual parameters.

Since the calculated free energy of interaction is largely dependent on the value chosen for χ_1, then evidence of the effect of salts on χ_1 could lead to significant conclusions on the stability of emulsions in the presence of salts. Several reports of non-ionic surfactants and polyethylene glycols bear out the contention that electrolytes dehydrate the ethylene oxide chains and promote their 'salting out'. This is what the study of the effect of NaCl and NaI referred to earlier (see Fig. 8.3) aimed to display – that 'salting in' and 'salting out' would have an effect on stability as predicted by Equation 8.30.

Carroll [1] has criticised Elworthy and Florence's attempts to rationalize the stability of chlorobenzene- and anisole-in-water emulsions stabilized by a series

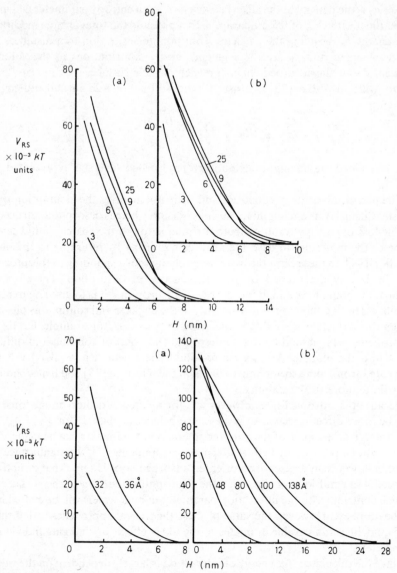

Figure 8.13 *Upper* The entropic stabilizing force, V_{RS}, as a function of distance of approach of the particles, H, (a) for anisole, (b) for chlorobenzene emulsions for four non-ionic detergents $C_{16}E_3$, $C_{16}E_6$, $C_{16}E_9$, $C_{16}E_{25}$. Ethylene oxide chain length marked. *Lower* The effect of V_{RS} versus H plots on altering for (a) $C_{16}E_6$ and (b) $C_{16}E_{25}$. δ as marked on diagram. From Elworthy and Florence [56] with permission.

of polyoxyethylated non-ionic surfactants based on a C_{16} alkyl chain, especially in relation to the use of zeta potential in place of surface potential in Equation 8.15. Indeed the use of this equation which presupposes constant charge is probably incorrect since the charge originates from the adsorption of extremely mobile hydroxyl ions. However, the trends noted are unlikely to be contradicted as adsorption of non-ionic surfactants lead to a decrease in the negative charge. It is unlikely that the mobile charges have much influence on stability anyway. The entropic stabilization calculated and drawn in Fig. 8.13 does display the correct trend of increasing stabilizing force as chain length is increased although the values of V_{RS} are insensible to changes in the concentration of bulk surfactant above the CMC.

Mobility of the adsorbed surfactant layer may well be a factor in the failure of established, though imperfect, theories to explain stability. Movement of the surfactant molecules from regions of contact of the droplets would be hindered if the surface viscosity was increased by the addition of a second component. In order to test the influence of surface change, a sulphated analogue (II) of cetomacrogol (III) was prepared and its stabilizing power compared with that of the non-ionic parent compound [76]. Fig. 8.14 shows the zeta potential of chlorobenzene globules as a function of concentration of (II) and (III). The rate of coalescence of chlorobenzene-in-water emulsions was $5.2 \times 10^{-7} \, s^{-1}$ when stabilized with cetomacrogol and $0.93 \times 10^{-7} \, s^{-1}$ when stabilized with the non-ionic surfactant, yet the interfacial tensions were similar (6.4 and $5 \, mN \, m^{-1}$). Addition of hexadecanol to chlorobenzene-in-water emulsions stabilized by $C_{16}E_6$ resulted in an increase in stability [77, 78], an observation in line with results of other workers [79, 80].

$$C_{16}H_{33}(OCH_2CH_2)_{21}OSO_3^- Na^+$$
(II)
(Cetomacrogol sulphate)

$$C_{16}H_{33}(OCH_2CH_2)_{21}OH$$
(III)
(Cetomacrogol 1000)

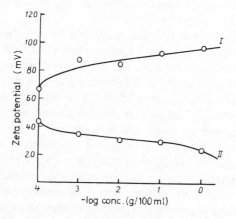

Figure 8.14 Change of zeta-potential with concentration of surfactant for chlorobenzene globules in aqueous solutions of (I) cetomacrogol sulphate and (II) cetomacrogol. From Attwood and Florence [76] with permission.

The most likely mechanism for the enhanced stability of xylene-in-water emulsions on the addition of cetyl alcohol to cetyltrimethylammonium bromide-stabilized systems is probably the formation of a coherent interfacial film which acts as a barrier preventing coalescence, possibly by virtue of its dilational viscoelasticity [81] which would have the effect of restricting the local dilation of the film as two drops approach each other. It has been suggested that when films such as the CTAB film carry an electrical charge the main importance of viscoelasticity is the maintenance of repulsion by preventing the lateral displacement of the adsorbed surfactant [82]. As Attwood and Florence [76] showed with cetomacrogol sulphate the presence of charged surfactant alone at the interface does not in itself lead to stability, nor indeed does the existence of a rigid interfacial film regardless of its hydrophilicity. The magnitude of the surface rheological parameters is not a measure of the overall stability of the droplets but only an indication that some stability will be present; the nature of the film is of more importance than its high viscosity. For example, rigid films of hexadecanol do not produce stable emulsions of chlorobenzene [79]. Sonntag *et al.* have confirmed this view in their statement [83] that 'in monomolecular adsorbed layers of surfactants, agreement between coalescence stability on the one hand and the mechanical properties, measured parallel with the interface, on the other, would occur only in a few selected instances'.

(D) STABILITY OF COMPLEX SYSTEMS

The addition of extra components such as long-chain alcohols or simply the use of mixed surfactants complicates an already complicated system. Emulsions are more complex than suspensions because of the ability of the surfactant to diffuse in and out of the disperse and continuous phases and because of the potential mobility of the surface layer itself. Migration of surfactant from one phase to another will take place in the initial moments of the life of the emulsion system after preparation. Lin has made extensive studies of the effect of surfactant location on emulsion properties [84, 85].

Probably more significant is the formation of structured phases in the continuous phase of the emulsion system. Such structures which may be in addition to any structured interfacial film, can lead to extremely stable systems. In the extreme case of the inability of the disperse phase droplets to move freely in the highly structured vehicle, forces of attraction will have no impact. Where a structured phase exists around the particle in an otherwise unchanged system, the forces of attraction and repulsion will be modified.

Friberg and his co-workers have drawn attention to the importance of mesomorphic phases to emulsion stability (e.g. [86, 87]). Fig. 8.15 shows a simplified phase diagram for an oil–water emulsifier in which the region of the mesomorphous phase (C) is shown. When the concentration of emulsifier is increased to the point where the composition of the system passes from the two-phase area A + B to the three-phase area A + B + C the stability of the emulsion increases suddenly and the viscosity rises. Both of these changes are due to mesophase C distributed in solution B which is the continuous phase of the

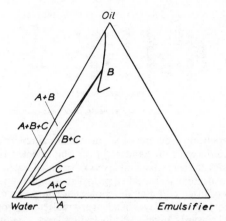

Figure 8.15 The phase diagram of a system water–oil–emulsifier. A, + water solution of emulsifier: B, + oil solution of emulsifier; C, + mesomorphous phase present in the system. From Friberg *et al.* [86] with permission.

emulsion. On further increase in surfactant concentration several alternatives are possible. If the composition corresponds to a point in B + C the viscosity will be very high; if it corresponds to a point in C the gel-like mesomorphic phase will assume the continuous phase [86]. The marked effect of change of the nature of the oil phase in such phase diagrams [46, 87] gives one clue as to the influence of the oil phase on emulsion stability, i.e. the oil phase not only affects surfactant adsorption [88, 89] but also the formation of mesomorphous phases which influence stability.

The most effective emulsion and foam stabilizers are aerosol systems containing fluorocarbon propellants as surfactants. These are believed to form an oriented polymolecular structure at the propellant–water interface; for optimum stability Sanders has found [90] that the surfactants must have a low solubility in both phases and have the ability to remain in the interfacial region. Hydrocarbon and fluorocarbon chains are not freely miscible and this perhaps explains the unusual behaviour of the surfactants in these systems. Addition of long-chain alcohols or acids enhance stability of the fluorocarbon emulsions and a hypothetical structure of the interfacial region has been proposed (Fig. 8.16). Davis *et al.* [91] have investigated the stability of fluorocarbon emulsions intended as artificial blood substitutes. Perfluorocarbon oils tended to produce unstable emulsions while oil phases such as perfluorotributylamine or per-fluorotetrahydrofuran formed more stable systems. These authors also refer to the possibility that as fluorocarbon–hydrocarbon mixtures have positive excess free energies, cohesive and adhesive forces between surfactant and oil phase will result.

Many pharmaceutical emulsions are semi-solid oil-in-water systems, prepared with mixed emulsifiers [92]. Often combinations of anionic, cationic or non-ionic surfactants with a fatty alcohol are used as the emulsifying system. Emulsifying

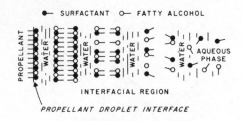

Figure 8.16 Hypothetical structure of a molecular complexed interfacial film at a propellant water interface. From Sanders [90]. 'The oriented liquid crystal nature of molecular complexes with their attendant layers of oriented water molecules suggests that the interfacial region around an emulsified propellant droplet can be viewed as consisting of alternating shells of oriented water and molecular complex molecules. The propellant interface would consist of a monolayer of adsorbed molecular complex molecules with the polar heads oriented towards an adjacent hydration layer. The hydration layer of water molecules in turn would be surrounded with a bimolecular shell of complex molecules with the polar heads on one side of the shell oriented towards the inner hydration layer and the polar heads on the other side oriented towards an outer hydration shell. This configuration of alternating layers of oriented water and bimolecular complex molecules would extend into the bulk phase with diminishing orientation until it disappeared.'

Wax BP, Cetomacrogol Emulsifying Wax BP and Cetrimide Emulsifying Wax (BP) are ready prepared mixtures for extemporaneous use. When these or similar mixtures are used the resultant emulsions are mobile at low emulsifier concentrations and semi-solid at moderate concentrations (about 10% or less of the total weight of the emulsion). The process whereby the surfactant mixture imparts the semi-solid characteristics over a period of time has been termed 'self-bodying' [92–94], the essential feature of which is the introduction of a significant elastic

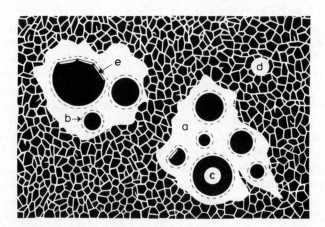

Figure 8.17 Theoretical diagram of structures which may occur in emulsions (viewed between crossed polars) (a) frozen liquid crystal; (b) L_2 phase; (c) oil globule; (d) gel network of frozen liquid crystal dispersed in L_1 phase; (e) crystal in globule (b and d not visible in light microscope). From Barry [93] with permission.

component into the rheology of the system. These systems are extremely complex structures. In liquid paraffin emulsions stabilized by cetrimide–cetostearyl alcohol mixtures the globules are aggregated, many are non-spherical and deformed by crystals of cetostearyl alcohol [94]. Not surprisingly they display complex flow properties in continuous shear experiments. Creep testing showed that when the emulsions contained less than 5% mixed emulsifier they decreased in consistency on storage while emulsions containing more than 5% emulsifier increased in consistency on ageing. Many of the systems could be considered to be essentially ternary cetrimide–cetostearyl alcohol–water systems with dispersed droplets of liquid paraffin. A diagrammatic representation of the system is given in Fig. 8.17. The rheological characteristics of these complex semi-solid emulsions have been comprehensively reviewed by Barry [95–99]. His group's extensive work in this area has allowed them to define the requirements for self-bodying action in the components of the emulsifier mixture, one of which will be lipophilic and the other hydrophilic [92]. These are as follows:

(i) Lipophilic component
1. This should be an amphiphilic compound which by itself promotes water-in-oil emulsions and is capable of complexing with the hydrophilic component at the oil–water interface.
2. Its concentration should at least be sufficient to form a close-packed mixed monolayer with the hydrophilic component. To promote semi-solid emulsions at room temperature it should be near or above its saturation concentration in the oil.
3. Excess material should diffuse readily from the warm oil phase into the warm aqueous micellar phase, and there be solubilized.
4. The melting point should be sufficiently high to precipitate solubilized material at moderate temperature.

(ii) Hydrophilic component
1. This should be a surface-active agent which by itself promotes oil-in-water emulsions and is capable of complexing with the lipophilic component at the oil–water interface.
2. Its concentration should at least be sufficient to form a close-packed mixed monolayer with the lipophilic component. To promote semi-solid emulsions it should be in excess of its critical micelle concentration in the aqueous phase.
3. It should be capable of solubilizing the lipophilic component when warm.

In the above requirements, the term solubilization implies incorporation of the lipophilic component into any type of micelle, whether spherical (L_1 and L_2 phase) or lamellar liquid crystalline phase.

In a paper on the self-bodying action of alkyltrimethylammonium bromide–cetostearyl alcohol mixed emulsifiers, Barry and Saunders [100] discuss the sequence of events in the preparation of liquid paraffin-in-water emulsions containing these components. The gel structure in the continuous phase is clearly responsible for the self-bodying. During mixing of a ternary system the quaternary ammonium surfactant penetrates the molten cetostearyl alcohol to form liquid crystals which as aqueous penetration proceeds, form smectic structures consisting of spherules, sheets and filaments. When the ternary system cools below the penetration temperature interaction is reduced and the system precipitates to form a three-dimensional viscoelastic gel network of frozen smectic phase. In the emulsion studies it was estimated that the mixed emulsifiers were present in concentrations in excess of those required to form condensed monomolecular films at the globule interface. In the preparation of emulsions, as the ingredients cool cetostearyl alcohol diffuses to the continuous phase and forms liquid crystals. A small amount of liquid paraffin is solubilized [100]. The kinetics of structure build-up in self-bodied emulsions have been studied [101]. Hexagonal oil droplets are detectable in emulsions in which the frozen smectic network is not extensive, i.e. does not prevent close packing of the particles due to flocculation in the secondary minimum.

Deformed particles would normally only be seen in highly concentrated emulsions such as those considered by Princen [102]. It is generally accepted that the volume fraction of the disperse phase in an emulsion can be increased up to a certain critical value before inversion takes place. If the disperse phase was monodisperse (and so far no emulsion has been prepared in a monodisperse form despite claims [103] and attempts to do so [104]) the maximum packing of spheres occurs at a phase volume, ϕ, of 0.7405. However, quite stable emulsions which have higher volume fractions have been prepared. Heterodisperse systems can readily exceed $\phi = 0.7405$ as particles which are small enough can infiltrate the spaces between the larger particles. In fact Lissant [105–107] has reported systems with $\phi = 0.99$, explained by the formation of polyhedral particles. Princen [102] has treated these systems from a thermodynamic viewpoint, considering the equilibrium thickness of the thin surfactant films separating the deformed droplets and the contact angle between solution and oil phase as factors affecting stability. The thermodynamic properties treated by Princen include the pressure exerted on the films, the 'osmotic pressure' of the emulsion, and the relative vapour pressure of the continuous phase; the resultant theory is capable of predicting the maximum attainable volume fraction as a function of drop size. Princen maintains that the presence of a finite contact angle leads to the spontaneous deformation of the drops and contraction of the emulsion to a volume fraction in excess of that corresponding to hexagonally close-packed cylinders [102].

Such factors as these discussed above, in self-bodied and in concentrated emulsions, obviously complicate the treatment of emulsion stability according to the concepts of classical DLVO colloid stability theory. There are other factors at work which work against a simple understanding of stability through extant theories. Prigorodov et al. [108] claim that in a manner analogous to liquid crystal formation in Friberg's system, 'supermolecular structures' form in the surface of droplets in concentrated emulsions stabilized by non-ionic surfactants. Although not well defined, this system is likely to be a microemulsion which forms spontaneously at the oil–water interface.

Deviations from theory result because of some special properties of emulsion systems. DLVO theory explains stability by describing forces preventing the close approach of the disperse particles, and hence where instability can arise from mechanisms other than coalescence and coagulation, one can expect deviation between experiment and theory. Such a mechanism of particle growth occurs in emulsions through diffusion of minute portions of oil through the continuous phase in micellar form. The more efficient the solubilizing capacity of the non-ionic stabilizer the more important will be diffusional growth of large particles in the emulsion at the expense of smaller particles. A theory pertaining to this mechanism of emulsion breakdown has been proposed by Higuchi and Misra [109]. It has been suggested [110] that oils which form less stable emulsions may be those which are more soluble in the aqueous phase. Ostwald ripening effects will occur in small particle colloidal systems such as emulsions or suspensions owing to operation of the Kelvin effect [111] the relationship between vapour pressure and radius of curvature may be written in the form

$$RT\ln\frac{p}{P_\infty} = \frac{\Delta V\gamma}{pd} \qquad (8.31)$$

when p is the vapour pressure of a small particle of diameter d, P_∞ is the vapour pressure of the same substance of infinite radius of curvature, V is the molar volume and σ the surface free energy of the material. For an oil droplet in water in which it has a low solubility we can write

$$RT\ln\frac{C_r}{C_\infty} = \frac{2\gamma V}{r} \qquad (8.32)$$

where C_r is the solubility of a droplet of radius r and C_∞ is the solubility of the bulk material, σ is its interfacial tension. Very small emulsion droplets will thus be more soluble than very large droplets and this results in the instability of small particles and their tendency to disperse, the larger droplets consequently growing.

For the case where there are n_A and n_B particles of radii a_A and a_B, respectively, the rate of change of the radius of B is given by:

$$\frac{da_B}{dt} = \frac{DC_\infty K}{\rho a_B^2}\left[\frac{n_A(a_B - a_A)}{n_A a_A + n_B a_B}\right] \qquad (8.33)$$

where $K = (2\gamma M/\rho RT)$, γ = interfacial tension, D = molecular diffusion coefficient, C_∞ = miscibility of an infinitely large drop, and ρ is the density of the disperse phase.

When the two initial particle radii are 0.5 and 1.0 μm, degradation by diffusional processes is as shown in Fig. 8.18. Decrease in the diffusion rate of the oil can be achieved by increasing the viscosity of the external phase, or by addition of a third component to the dispersed phase if the additive has a sufficiently low rate of diffusion.

Deformation of the globules on collision can feasibly result in the dissipation of some of the attractive energy but, if the DLVO theory reasonably predicts the

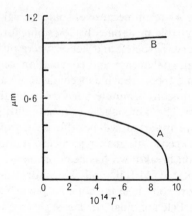

Figure 8.18 The degradation of an emulsion initially composed of a mixture of 0.5 μm and 1 μm radius droplets of equal number concentration, showing the droplet diameters versus a function (T^1) of time. $T^1 = \dfrac{DC_\infty Kt}{\rho}$. At $T^1 = 4 \times 10^{-14}$ there is about a 10% change in a_A, after $t = 3 \times 10^7$ s (1 year), using $D = 5 \times 10^{-6}$ cm^2 s^{-1}, $\rho = 1$, even though C_∞ is only 3×10^{-8} g ml^{-1}. A, mean size of smaller droplets; B, mean size of larger droplets. From Higuchi and Misra [109] with permission.

behaviour of suspensions of asymmetric particles, then it is unlikely that the unknown shape of the deformed particles in an emulsion will give rise to much discrepancy. More likely to be a source of deviation is the possible dissolution of the surfactant in the disperse phase when particles come together, with resultant depletion of the surface concentration of stabilizer, unless the diffusion of surfactant from the continuous phase is rapid enough to counteract this tendency. Presumably in systems with high bulk concentrations of emulsifier, replenishment of the surface is rapid; on the other hand, where the solubility of the surfactant in the oil phase is high and the total concentration is high, this inhibits diffusion of the surfactant into the oil. Therefore, in both cases, high concentrations of surfactant will increase stability. It has been shown that O/W emulsions made from or containing small quantities of long-chain alkanes are more stable than those made from short-chain alkanes. Two possible explanations of this behaviour have been suggested, one involving the concept that the long-chain component inhibits Ostwald ripening, the other that it inhibits coalescence. Buscall *et al.* [111] have shown that the former explanation is the more likely. In their work with sodium dodecyl sulphate-stabilized emulsions of hexane, addition of dodecane had little effect on the critical coalescence pressure determined by centrifugation (Table 8.4) but the significant effect of the alkanols was readily demonstrated due to interfacial complex film formation. As the Hamaker constants for the alkanes lie within a small range (3.4 to 4.4 $\times$ 10^{-21} J) and for most oils (3.4 to 5.6 $\times$ 10^{-21} J), it is unlikely that the alkanes influence coalescence by altering van der Waals' attraction.

Table 8.4 The effect of additives on the critical coalescence pressure (p_c) for hexane/sodium dodecyl sulphate emulsions

Oil phase	$p_c(\mathrm{N\,m^{-2}})$
Hexane	0.70×10^5
Hexane + 1 % dodecane	0.63×10^5
Hexane + 1 % hexadecane	0.83×10^5
Hexane + 1 % octanol	36.2×10^5
Hexane + $\frac{1}{2}$ % dodecanol	11.8×10^5
Hexane + 1 % dodecanol	47.0×10^5
Hexane + 1 % hexadecanol	16.5×10^5

From [111].

(E) STABILITY OF WATER-IN-OIL EMULSIONS

Water-in-oil emulsions form when the phase volume of oil reaches a critical value which is dependent on surfactant type and concentration and the particle-size distribution of the oil phase. Generally, water-in-oil emulsions will be formed when using surfactants of low HLB, the oil soluble surfactants. Water-in-oil creams are widely found in the cosmetic, food and pharmaceutical industries but have received less attention than O/W systems, although Albers and Overbeek over 20 years ago considered the correlation between electrokinetic potential and the stability of W/O emulsions [112], extending earlier work on the stability of suspensions in nonaqueous media [113]. Schulman and Cockbain [8] in 1940, noting the irregularity of water droplets in stabilized W/O emulsions, attributed the shape to the rigidity of the interfacial film and from many experiments concluded that W/O emulsions form only when the stabilizer molecules at the O/W interface are rigid as a result of complex formation. As oil is a non-ionizing medium, electrical forces of repulsion would, they argued, be unlikely. Molecules would be unionized and there would be no electrical diffuse double layer. In fact some slight degree of ionization can occur in a polar liquid. At the low ionic concentrations involved, the repulsion, V_R, is described in good approximation by [112]

$$V_R = \psi_0^2 \varepsilon a^2 / H \qquad (8.34)$$

where ψ_0 is the surface potential, ε the dielectric constant, a the radius of the drop, and H is the distance between the centres of the particles. Experiments conducted by Albers and Overbeek on W/O emulsion stabilized by oil-soluble ionizing stabilizers showed that no correlation existed between stability against flocculation and electrokinetic potential. All emulsions flocculated rapidly even in the presence of a surface potential considerably higher than 25 mV; this is explained as a consequence of the free mobility of the stabilizers in the surface. With non-ionic stabilizers there is little doubt that entropic stabilizing forces are responsible for stabilization. However, in these inverse systems, as in O/W systems, solubilized water and under some conditions, mesomorphic phases are present in the bulk phase.

Shinoda and Sagitani [114] have considered the question of emulsifier selection for W/O emulsions. W/O emulsions have been found to be most stable against coalescence when the storage temperatures were higher than the PIT values. As the PIT of suitable emulsifiers is lower than 0° C these values cannot be ascertained. Thus correlation between PIT, stability and temperature and HLB of emulsifiers have been studied so that some method of selection can be devised. Figs 8.19 and 8.20 show some of the phase diagrams obtained by Shinoda *et al.* In

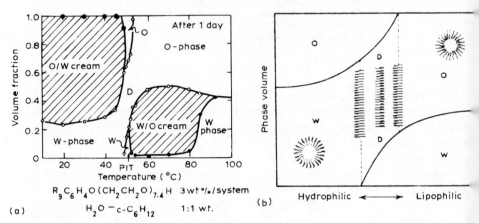

Figure 8.19(a) The effect of temperature on the types of emulsions, and the volume fractions of oil, cream, and water phases 1 day after agitation. (O) Drained phase-cream boundary; (●) coalesced phase-cream boundary. (b) Schematic diagram of water, surfactant, and oil phases after the complete phase separation of the system. Surfactant is lipophilic at higher temperature. From Shinoda and Sagitani [114] with permission.

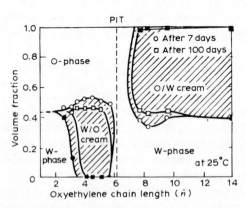

Figure 8.20 The effect of hydrophilic chain length of emulsifiers on the types and stability of: cyclohexane–water emulsions stabilized with 3% w/w system of $C_9H_{19}C_6H_4O(CH_2CH_2O)_nH$ at 25° C. (O, □) drained phase-cream boundary; (● ■) coalesced phase-cream boundary. From Shinoda and Sagitani [114] with permission.

Fig. 8.19 a non-ionic surfactant $(R_9C_6H_4O(CH_2CH_2O)_{7.4}H)$ is used in a cyclohexane–water system. The effect of temperature and volume fraction on the systems formed are shown. A diagrammatic representation of this system shows the transfer that occurs at the PIT. As no W/O emulsion exists in this system below the PIT (about 50° C) the particular surfactant is of no practical interest as a stabilizer of W/O emulsion. Fig. 8.20 explains in detail the influence of surfactant hydrophilic chain length on cyclohexane–water emulsion formation at 25° C. W/O emulsions form when the oxyethylene chain length of the nonyl-phenyl ethoxylate is in the approximate range 3.5 to 6. Shinoda and Sagitani conclude from their work that W/O emulsions are most stable when the oxyethylene chain length of the emulsifier was 0.7 to 0.88 that of the surfactant whose PIT is equal to the storage temperature of the emulsion, i.e. $R_9C_6H_4O(CH_2CH_2O)_{3.2-4.1}H$ for liquid paraffin-in-water and $R_9C_6H_4O(CH_2CH_2O)_{4.3-5.6}H$ for cyclohexane–water emulsions at 25° C.

The influence of the HLB of Span–Tween mixtures on the stability of W/O and O/W systems has been studied by Boyd *et al.* [115]; the rate of droplet coalescence was determined using a centrifugal photosedimentometer. Fig. 8.21 shows the sensitivity of coalescence rate to HLB near the critical HLB for inversion. The schematic representation (Fig. 8.22) of the polysorbate 40 and Span

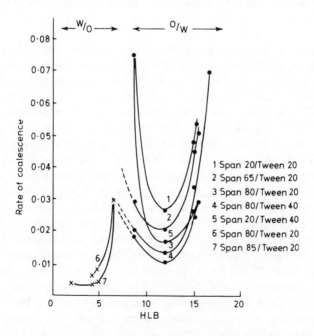

Figure 8.21 Influence of HLB on emulsion stability. Emulsions were prepared with liquid paraffin (Nujol) and contain 50% disperse phase. The total emulsifier concentration in each emulsion was 10^5 mol l^{-1}. The rate of coalescence is defined here as $d(\ln N_t)/dt$ where N_t is the number of globules at time t. From Boyd *et al.* [115] with permission.

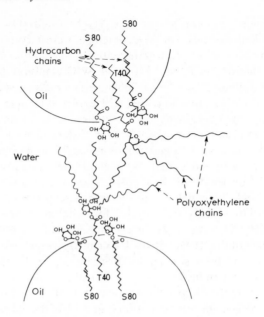

Figure 8.22 Schematic representation of orientation of Tween 40 and Span 80 molecules in mixed films adsorbed at the oil–water interface. From Boyd *et al.* [115].

80 molecules in mixed films adsorbed at the O/W interface is helpful in understanding stabilization of both O/W and W/O emulsions by these molecules and their analogues. In the O/W systems stabilization occurs through the interactions between the polyoxyethylene groups of the Tween molecules; the presence of Span molecules at the surface allows a more congenial geometric packing of the molecules and indeed allows more intimate contact and thus greater cohesion between the lyophobic portions of the Span and Tween molecules. In W/O systems the same holds, but stabilization is conferred by the entropic interaction between the hydrocarbon chains Blending Span 20 (sorbitan monolaurate, HLB 8.6) which can stabilize O/W or W/O emulsions with Span 80 (HLB 4.3, sorbitan mono-oleate) or Span 85 (HLB 1.8, sorbitan trioleate) does not improve the stability of W/O emulsions containing only Span 80 or Span 85 [115]. This is thought to be due to the fact that the sorbitan rings of these molecules are similar and the hydrocarbon chains of alternating Span 80 and Span 20 (or Span 85 and Span 20) molecules pack more closely together than when one species is present. Span 85–Span 20 blends produce more stable emulsions than Span 80–Span 20 blends because, according to Boyd *et al.* [115], the 'trioleate radicals develop a more complex structure in the oil phase'.

The O/W stabilizing ability of Tweens and Spans, widely used in the food and pharmaceutical industries, have been studied by a Japanese group [116, 117]. O/W emulsions prepared with Tween 20–Span 80 mixtures were unstable over the whole HLB range possible with these surfactants (4.3 to 16.9); the authors suggest that the great difference in HLB of the surfactants is the cause. The high water solubility of the one and the high oil solubility of the other may well result in too high concentrations of surfactant in the bulk phase. Friberg and Wilton [118] have postulated that when the emulsifier can form micelles in both the aqueous phase and the oil phase the resultant emulsion will have poor stability characteristics. The reason for this is that when the disperse phase droplets approach each other the emulsifier molecules can desorb readily into the bulk oil or water phase and thus interfacial stability is lost.

Studies of W/O emulsions and thin aqueous surfactant films between oil phases using Span 80 and other low HLB surfactants by Sonntag and Klare [119] have drawn attention to the importance of the stability of the thin film between the water droplets when flocculation occurs. Addition of electrolyte causes dehydration of the surfactant molecules and promotes de-emulsification. As with O/W emulsions, addition of electrolyte causes a shift in the HLB of the surfactant molecules. Increase in temperature results in an increased rate of flocculation because of surfactant desorption and an increase in the rate of coalescence. Using water droplets in octane the temperature for coalescence increased with Span 80 concentration from $42°$ C at $0.1 \, g \, l^{-1}$ to $68°$ C at $1 \, g \, l^{-1}$ and $> 75°$ C at $5 \, g \, l^{-1}$. The equilibrium non-aqueous black films which form between the water droplets are unaffected by temperature but their thickness is controlled by the nature of the oil phase. Span 80 in octane forms a film 3.9 nm thick, while in xylene the black film has a thickness of 28 nm [119]. Sonntag and Netzel have carried out similar measurements on this film relevant to O/W stability [120]. Because of the complexities of the real emulsion system such studies, discussed in some more detail in [4], have many uses in allowing the experimenter to isolate some of the variables in the system. The reader is referred to the literature on these films (see [121]) for further details and to Sonntag's paper in particular for information concerning non-ionic surfactant behaviour in aqueous and non-aqueous films (see [122–124]).

Structured W/O systems similar to those obtained in O/W systems have been reported in which the oil phase consists of liquid paraffin and a microcrystalline wax, the emulsifier being a polyoxyethylene alkyl ether [125] alerting us to the potential complexities of inverse emulsions.

(F) THE EFFECT OF ADDITIVES ON THE STABILITY OF NON-IONIC EMULSIONS

Non-ionic surfactants are said to have several advantages over ionic surfactants as emulsifiers. In general they are less toxic and less sensitive to electrolytes and pH. Emulsions stabilized with Triton X-400 were stable over the pH range 2.5 to 11 and were compatible with a wide range of substances used in dermatology [126].

Although non-ionic surfactants are often chosen for emulsified systems because of their relatively low toxicity, several problems in formulation have arisen with their use. Most emulsions of edible fats and oils must be preserved against microbial attack. The most effective preservatives have often been the *p*-hydroxybenzoic acids. Unfortunately, these acids are inactivated by interaction with polyoxyethylene compounds in the system, and, in addition, changes in emulsion stability have occurred. The formation of complexes between phenolic compounds and ethylene oxide condensates has been frequently demonstrated (see Chapter 6). Complex formation reduces the solubility of the polyoxyethylene non-ionic surfactants and renders them less effective as emulsifiers. Nevertheless, an example has been quoted [126] where the increase in viscosity caused by the addition of a phenol (IV) to emulsions stabilized by a polyoxyethylene non-ionic increased the resistance of the emulsion to separation by centrifugation.

$$HO-\bigcirc-CH_2O-\bigcirc$$

(IV)

It would be wrong to assume that non-ionic stabilized emulsions are immune to the effect of added electrolytes. The addition of electrolytes to non-ionic stabilized emulsions can cause pronounced effects on stability. In solutions of non-ionic surfactants, the addition of electrolytes generally causes a dehydration of the ethylene oxide chains by disruption of hydrogen bonds. Selected salts have been shown to exhibit interaction with polyethylene oxide ethers, reducing their solvation and producing more compact molecular conformations [127, 128].

The importance of the hydrated layer in maintaining stability is indicated by Levi and Smirnov [129] who showed that monoglycerides became efficient emulsifiers only when they possessed sufficient hydroxyl groups. These authors also noted that the introduction of ionizable carboxylic acid groups into the non-ionic stabilizer molecule containing many hydroxyl groups greatly increased its emulsifying properties. Stability of soybean oil-in-water emulsions stabilized by mono- and diglycerides has recently been studied [130]. A critical molar ratio of fatty acid groups in the monoglyceride compound to the fatty acid groups in the diglyceride was found when the stability was maximal; the mixed films show some viscoelasticity. The effect of pH change was investigated, electrophoretic mobility increasing as the pH value increased with the addition of 0.01 M NaOH (Fig. 8.23). Electrophoretic mobility has also been measured as a function of pH in liquid paraffin emulsions stabilized by Brij 92 and Brij 96 mixtures [45] (Fig. 8.24). A similar increase in zeta potential was noted with increasing pH at all ratios of the two non-ionic surfactants, the zeta potential reaching a maximum at a pH of just less than 10. There was no evidence from these results that the zeta potential reached a maximum value at the critical HLB as Becher and colleagues have suggested [131, 132]. However, maximal stability in the systems studied by Depraétere *et al.* [45] coincides with minimal particle size which in these systems falls below 0.1 μm (Fig. 8.25). There is a noticeable shift in critical HLB with increasing surfactant concentration which is thought to reflect changes in the interfacial layer composition as the bulk concentrations of the two surfactants increase. The zeta potential of the emulsion droplets decreases as the hydrophilicity of the surfactant mixtures increases, following the trends observed with

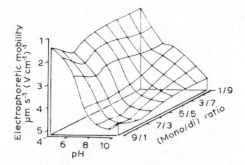

Figure 8.23 Influence of the ratio of stearyl mono- to stearyl di-glyceride and of pH on the electrophoretic mobility of soybean oil droplets. From Kako and Kondo [130] with permission.

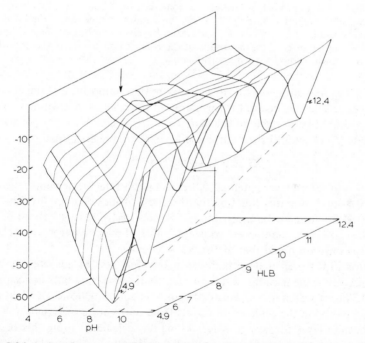

Figure 8.24 A plot of zeta potential of emulsions (ordinate, mV) as a function of pH of the aqueous phase and surfactant HLB. The arrows mark the critical HLB of the system in the absence of added electrolyte. Surfactant, Brij 92 and Brij 96. From Depraétere *et al.* [45] with permission.

single surfactants [56]. This confirms the expectation that the more hydrophilic surfactants with longer polyoxyethylene groups move the plane of shear further out into the bulk phase. The effect of increasing surfactant concentration is to decrease electrophoretic mobilities; as this occurs with single surfactant species

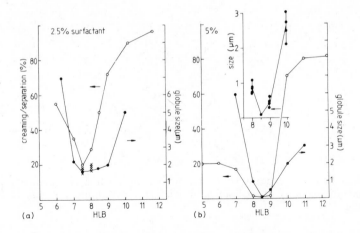

Figure 8.25 Plots of percentage creaming or coalescence ($\bigcirc$) and mean globule size ($\bullet$) in O/W emulsions stabilized by Brij 92–Brij 96 mixtures as a function of the HLB of the emulsifier mix at (a) 2.5 % and (b) 5 % surfactant. Results from the Coulter Nanosizer are shown ($\times$). In the inset in (b) the details of the particle size change in the critical HLB region are shown. From [45] with permission.

also [56, 64] this cannot be due to a change in the composition of the interfacial film. More likely it is due to the formation of multilayers. According to Tadros [133] it should be possible to calculate from the decrease in zeta potential the thickness (Δ) of an adsorbed non-ionic layer using the equation,

$$\tanh e\zeta/kT = \tanh e\psi/4kT \exp\left[-\kappa(\Delta - \delta)\right] \tag{8.35}$$

where ψ is the Stern layer potential, δ the thickness of the Stern plane, taken to be about 0.4 nm. Using this equation the thickness of the adsorbed layer has been calculated and found to be approximately 50 nm at HLB 12 (pH 5.5) and 20 nm at HLB 4.9 at 5 % surfactant levels, while the length of the polyoxyethylene chain of Brij 92 is only 1 nm and that of Brij 96, 4.5 nm.

Tadros [133] predicted an increase in the thickness of adsorbed polyvinyl alcohol layer at the paraffin–water interface with increasing bulk concentration from 3.5 nm at 1 ppm to 67.7 nm at 20 ppm. It is doubtful, however, whether it is possible solely on the evidence of results from application of Equation 8.35, to confirm multilayer formation. Kayes [134] has calculated using this equation, adsorbed layer thicknesses on solid particles for alkyl polyoxyethylene mono-ethers of 3.2 and 9.2 nm for derivatives with 30 and 60 ethylene oxide residues, respectively, which approximates to monolayer coverage.

Photomicrography was used to follow gross changes in the liquid paraffin emulsions on change of pH over the range 4 to 9.5 pH units [45]. In emulsions with surfactant HLB from 8 to 12.4 there was no significant change in appearance on changing the pH. In systems of lower HLB (4.9, 6 and 7) the globules aggregate at pH 4 and are separate at pH 9.5, reflecting the influence of surface charge. When the pH of the latter emulsion is decreased from 9.5 to 4 aggregation of the globules is seen to occur. Viscosity measurements on undiluted emulsions in the range HLB 7 to 12.4 at pH 4 and pH 9.5 indicate that at HLBs greater than, or equal to, 8 the viscosity of the emulsions is higher at higher pH values, although the situation is dramatically reversed at HLB 7. The very high viscosity of the emulsions at pH 4 and at lower pH values is attributable to the formation of floccules. The maximum viscosity in non-flocculated systems occurs at HLB 8.5 where the particles are smallest. It is erroneous, therefore, to suggest that pH has no effect on non-ionic stabilized emulsions, as has been claimed [135].

8.3 Multiple emulsions

Multiple emulsions are emulsion systems in which the disperse phase contains dispersed droplets of the external phase. Water-in-oil-in-water (W/O/W) type multiple emulsions are oil-in-water emulsions in which the dispersed oil drops themselves contain smaller dispersed aqueous droplets. These systems were observed as long ago as 1890 [136] but there has been increased interest recently in the use of W/O/W emulsions in such diverse fields as the separation of hydrocarbons [137], treatment of waste water [138], immobilization of enzymes [139], prolongation of drug release [140], and the treatment of drug overdosage [141]. O/W/O emulsions can also be formed. Two recent reviews have dealt with the topic of multiple emulsions, their preparation, pharmaceutical uses and stabilization [142, 143].

Sherman [144] reports that multiphase globules of increasing number, size and complexity appear on increasing the concentration of sorbitan monolaurate and oil in liquid paraffin-in-oil emulsions. At $\phi = 0.73$ and 6% surfactant concentration, inversion takes place to a W/O emulsion, the relative viscosity falling from 78.6 to 1.46. The resultant emulsion contains many multiple phase globules. Mulley and Marland [145] discussed conditions in non-ionic stabilized emulsions leading to multiple drop formation. It is to be expected that as an emulsion inverts the systems formed near the inversion boundary will have some dual characteristics. With the correct choice of emulsifying agents multiple emulsions of reasonable stability can be formed, although the systems have an inherent instability due to their very nature.

8.3.1 Preparation and stability of W/O/W emulsions

A stable W/O emulsion (the primary emulsion) is emulsified in water using surfactants appropriate to the stabilization of an oil-in-water emulsion. The emulsification of the primary emulsion is a critical stage as excess turbulence will cause the coalescence of the multiple droplets. Ultrasound and high shear mixers cannot, therefore, be used at this stage. A typical size distribution of internal (aqueous) and multiple (oil) drops is shown in Fig. 8.26. One of the difficulties in recording the size data of these systems is that oil droplets may be empty or contain one, two or more droplets. Davis *et al.* [146] used a Coulter counter to size the external oil droplets in multiple systems and the same group [147] used a

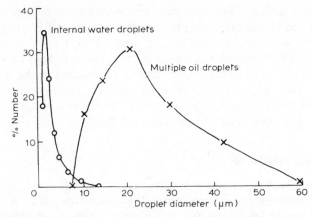

Figure 8.26 Droplet size distributions of the internal aqueous and multiple (filled oil) drops of a W/O/W emulsion. From Whitehill [142].

freeze-etching electron microscope technique to obtain both external and internal droplet size distribution.

The 'yield' of multiple droplets in the system is of great importance in assessing techniques of preparation and suitability of surfactants. It has been suggested [148] that if a yield of more than 90% of filled droplets is required the lipophilic surfactant used to prepare the primary emulsion must be present in amounts ten times greater than the hydrophilic surfactant. If migration of surfactant occurs in simple emulsions after preparation, the scope for migration of surfactants from their original sites in multiple emulsions is higher. If a mixture of surfactants is used to prepare the primary emulsion then it seems likely that the hydrophilic surfactant will diffuse to the aqueous continuous phase on re-emulsification, altering the ratio at the internal droplet–oil interface. This may well be one cause of instability. Fig. 8.27 shows one of the effects on yield as Span 80 and hydrophilic surfactant concentrations are adjusted. In Fig. 8.27a the Span 80: Tween 20 ratio is altered; in Fig. 8.27b the results obtained using four other surfactants in place of the Tween 20 are shown. No explanation of these effects has been proposed, but as there has been a suggestion that liquid crystalline phases are implicated in the greater stability of some multiple emulsions [149] then it is perhaps naive to consider looking for an explanation of stability on the basis of interfacial molecular structure.

Florence and Whitehill [150, 151] have described three types of multiple W/O/W emulsions, categorized according to the predominance of the multiple droplet type (Fig. 8.28). Using isopropyl myristate as the oil phase, 5% Span 80 to prepare the primary W/O emulsion and various surfactants to prepare the secondary emulsion, three main emulsion types were observed.

(a) Type 'A' droplets contained one large internal droplet, of the type described by Matsumoto [148] Brij 30 (polyoxyethylene (4) lauryl ether) (2%) was used as secondary emulsifier.

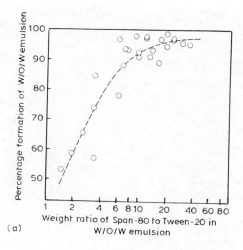

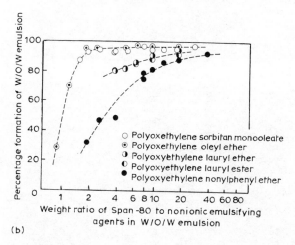

Figure 8.27(a) Plot of W/O/W emulsion formation against the weight ratio of Span-80 to Tween-20 in W/O/W systems, calculated from the constitution of various W/O/W emulsion samples. (b) Plot of W/O/W emulsion formation against the weight ratio of Span-80 to the nonionic emulsifying agents. From [158] with permission.

(b) Type B droplets contained several small internal droplets. These were prepared with 2% Triton X-165 (a polyoxyethylene (16.5) nonylphenyl ether).

(c) Type C droplets entrapped large numbers of small internal droplets and were prepared using a 3:1 Span 80:Tween 80 mixture.

Although all multiple emulsions contain a range of droplet types, the predominance of one type led the authors [150, 151] to categorize the emulsions as types A, B or C. The characteristics of the inclusion in the dispersed phase of emulsions has also been considered by Kessler and York [152] who have found that the

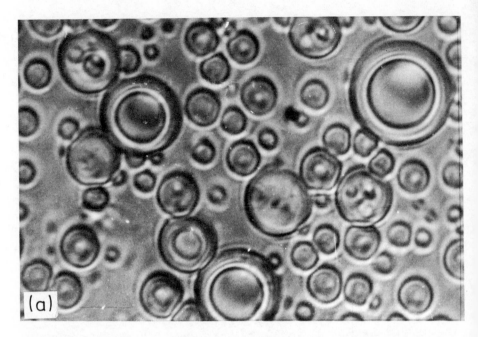

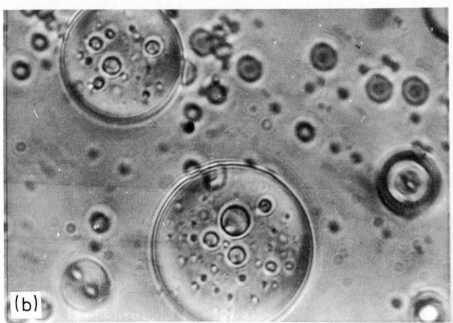

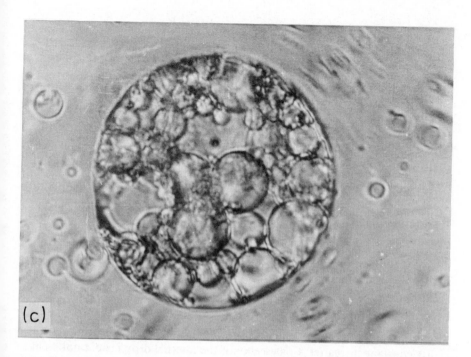

(c)

Figure 8.28 Photomicrographs of three types of W/O/W multiple-emulsion droplet, identified by Florence and Whitehill [150]. (a) Type A; (b) type B; and (c) type C systems obtained using isopropyl myristate as the oil phase with different surfactant mixtures.

distribution of large external droplets and that of the included drops is represented satisfactorily by a log-normal probability function. Large dispersed phase drops contained proportionally many more included droplets than did smaller drop sizes in systems which contained no stabilizer. Thus the efficiency of droplet entrapment in the presence of a secondary emulsifier would seem to suggest that those surfactants that are least effective in lowering interfacial tension will produce more droplets of type C (which are probably the most useful systems pharmaceutically). Yet if instability results from use of such a stabilizer the internal droplet system distribution might soon alter through coalescence.

Pathways of breakdown of multiple emulsions are shown in Fig. 8.29. Fig. 8.29 is oversimplified; in practice the number of possible combinations is larger. Consideration of such pathways, however, should make it possible to predict which basic mechanisms are responsible for instability. A number of factors will determine the breakdown mechanism in a particular system, but one of the main driving forces behind each step will be the reduction in the free energy of the system brought about by the reduction in the interfacial area.

Coalescence of the oil drops would result in a large change in the free energy of the system and this appears to occur frequently in the first few weeks of the lifetime of the emulsions. The mean diameter of the multiple drops of isopropyl

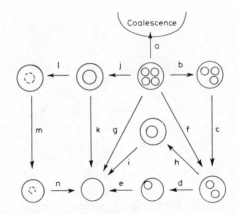

Figure 8.29 Some breakdown pathways that occur in W/O/W multiple drops. The external oil drop may coalesce with other oil drops, (which may or may not contain internal aqueous droplets) as in (a), the internal aqueous droplets may be expelled individually (b, c, d, e) or more than one may be expelled (f), or they may be less frequently expelled in one step (g); the internal droplets may coalesce before being expelled (h, i), (j, k); or water may pass by diffusion through the oil phase gradually resulting in shrinkage of the internal droplets (l, m, n). From Florence and Whitehill [150, 151].

myristate emulsions appeared to increase over the first three weeks but remained fairly constant thereafter. Coalescence of the internal droplets did not seem to occur to any extent. This would not be expected to be a major route of breakdown since coalescence of these droplets would not result in large decreases in free energy. On the other hand, coalescence of these small droplets would not result in a substantial change in the mean droplet diameter and thus changes may be masked by experimental error.

Another main breakdown mechanism is expulsion of the internal aqueous droplets through rupture of the oil layer. A calculation of the reduction in free energy, ΔG, as a consequence of the loss of the internal aqueous droplet from a type A multiple drop based on an internal droplet radius (r_1) of $4 \mu m$ and a multiple drop radius (r_2) of $4.25 \mu m$ is as follows: the total interface area ($4\pi[r_1^2 + r_2^2]$) is $4.28 \times 10^{-10} m^2$. The volume of the multiple oil drop is $3.215 \times 10^{-16} m^3$ and the volume of the internal aqueous droplet, $2.681 \times 10^{-16} m^3$. Thus calculation of the volume of the simple oil drop formed ($5.347 \times 10^{-17} m^3$) gives its radius ($2.337 \times 10^{-6} m$) from which one obtains the total interfacial area of the simple oil drop which is $6.863 \times 10^{-11} m^2$. The calculated change in interfacial area, ΔA, is thus $-3.594 \times 10^{-10} m^2$. For simplicity one assumes that the O/W interfacial tensions are equal at the different interfaces. An interfacial tension has to be assumed. Using the equation $\Delta G = \gamma \Delta A$ and assuming a constant, low interfacial tension, γ, of $5 \times 10^{-3} N m^{-1}$, the free energy change ΔG is $-1.797 \times 10^{-12} J$. Assuming a figure of $2.5 \times 10^{-19} m^2$ for the area per molecule of oil, $\Delta G = -632 J mol^{-1}$.

The loss of a smaller internal droplet would result in a smaller change in the free energy and so would be less likely to occur. Similar calculations can be made for the total loss of internal drops from type B and type C drops. From this, emulsion B may be expected to be more stable than emulsions A and C. The multiple drops of emulsion A certainly appear to be less stable, losing their internal droplets more readily than emulsion B.

Time-lapse studies [150, 151] showed that under normal conditions, in which distilled water without additives formed both internal and external aqueous phases, internal aqueous droplets were seen to collide frequently without coalescence. When electrolyte was present in the internal aqueous phase, emulsions A and B appeared to be stable on dilution with water, although changes may have occurred before filming commenced. Emulsion type C on dilution exhibited marked internal droplet coalescence with release of the internal phase, this activity ceasing after a few minutes, presumably after equilibrium had been reached.

8.3.2 Estimation of stability

Quantitative estimates of stability in multiple emulsions are complicated by the fact that the systems are not to be considered simply as water-in-oil emulsions dispersed as an oil-in-water emulsion. This approach is doomed to failure: the presence of water droplets in an oil drop suspended in water leads to the oil layer acting as a semi-permeable membrane. Osmotic factors thus lead to swelling or contraction of the inner water droplets by passage of water molecules across the oil layer [153].

(A) ANALYSIS OF OSMOTIC FLOW
Shrinkage of the internal droplets indicates osmotic flow of water. This analysis assumes that true osmotic flow occurs in $W/O/W$ systems, i.e. that the oil is impermeable to solute but permeable to solvent molecules. The volume flow of water, J_w, may be equated to the change in droplet volume with time (dv/dt) and we can write

$$J_w = \frac{dv}{dt} = -L_p \cdot A R T (g_2 c_2 - g_1 c_1) \tag{8.36}$$

where L_p is the hydrodynamic coefficient of the oil 'membrane', A is its cross-sectional area, R is the gas constant, T absolute temperature and g, the osmotic coefficient of electrolyte solutions of concentration c.

The flux of water ϕ_w is

$$\phi_w = J_w / \overline{V} \tag{8.37}$$

where $\overline{V}$ is the partial molar volume of water. If we define an osmotic permeability coefficient [21], P_o, as $L_p R T / \overline{V}$ we can obtain from Equations 8.36 and 8.37 the relation

$$\phi_w = -L_p A R T (g_2 c_2 - g_1 c_1) / \overline{V}, \tag{8.38}$$

and thus write for ϕ_w,

$$\phi_w = -P_o A (g_2 c_2 - g_1 c_1). \tag{8.39}$$

(B) CALCULATION OF AN APPROXIMATE DIFFUSION COEFFICIENT FOR WATER IN THE OIL

The diffusion coefficient of water, D_w, may be obtained from the value of P_o as

$$-P_o = \frac{D_w}{\Delta x},\qquad(8.40)$$

where Δx is the diffusion thickness, 8.2×10^{-6} m. D_w was found to be 5.15×10^{-8} cm^2 s^{-1} in isopropyl myristate emulsions [150, 151]. This value is extremely low in comparison with literature values. Schatzberg [154] for example, quotes values of 1.73×10^{-5} cm^2 s^{-1} and 4.16×10^{-5} cm^2 s^{-1} for the diffusion coefficients of water in hexamethyl tetracosane and m-hexane respectively at $25°$ C. The diffusion coefficient of inverse micelles of mixed non-ionic surfactants in dodecane ($\eta = 1.35$ cP at $20°$ C) has recently been measured by photon correlation spectroscopy to be 2.7×10^{-7} cm^2 s^{-1} [155]. The viscosity of isopropyl myristate is 4.15 times that of dodecane thus the same micellar unit might be expected to have a diffusion coefficient of approximately 6.5×10^{-8} cm^2 s^{-1} close to the value we have obtained for the diffusion coefficient of water in isopropyl myristate. This strongly suggests that transport of the water through the oil layer is by way of inverse micellar species.

(C) FORCES OF ATTRACTION IN W/O/W SYSTEMS

Some progress towards an understanding of these systems is possible by considering the influence of the dispersal of water in the oil droplets on the interactions between the 'multiple' drops and by consideration of the influence of the size of the water droplets on their internal stability and on the possibility of coalescence with the external phase. It is premature to consider all this in detail as the application of colloid stability theory to simpler emulsions has not been particularly successful. For type A systems the approach of Vold [156] may perhaps be used if the oil layer is thought of as the homogeneous 'adsorbed' layer. Alternatively, the effect of the internal phase on the size of the oil droplet is perhaps worth considering. The approach of Vold [156] has been adopted by Florence and Whitehill [150, 151] and, in the first instance, interactions between identical droplets were considered. In the simplest case two type A droplets may be considered.

For two spherical particles of radius $r_1 < r_2$ of composition P in a medium m with a separation Δ between the surfaces,

$$-12 V = (A_p^{\frac{1}{2}} - A_m^{\frac{1}{2}})^2 H\left(\frac{\Delta}{2r_1}, \frac{r_2}{r_1}\right)\qquad(8.41)$$

where the quantities A_p and A_m are constants depending on the composition of particle and medium, and the function $H(x, y)$ is positive, decreasing from its limiting value for $x \ll 1$ to 0 as y gets very large

$$\lim_{x \to 0} (x, y) = \frac{y}{(1 + y)}.\qquad(8.42)$$

The net interaction for two particles covered with an adsorbed layer is given by

$$V = V_f - 2V_D \tag{8.43}$$

where V_f is the sum of the interaction of two solvated particles in contact and the interaction energy of two imaginary particles with the composition of the medium, and V_D is the interaction of an imaginary particle and a real solvated particle in contact.

Modifying Vold's approach to type A multiple drops, the total interaction energy controlling flocculation is given by the equation

$$-12V = (A_w^{\frac{1}{2}} - A_o^{\frac{1}{2}})^2 H_o + (A_o^{\frac{1}{2}} - A_{wi}^{\frac{1}{2}})^2 H_w$$
$$+ 2(A_w^{\frac{1}{2}} - A_o^{\frac{1}{2}})(A_o^{\frac{1}{2}} - A_{wi}^{\frac{1}{2}}) H_{wo} \tag{8.44}$$

where the subscripts wi, o and w refer to the internal aqueous phase, oil phase and external aqueous phase, respectively. Since the internal phase and external phases are both water this simplifies to

$$V = -\frac{1}{12}[(A_w^{\frac{1}{2}} - A_o^{\frac{1}{2}})^2 H_o + (A_o^{\frac{1}{2}} - A_w^{\frac{1}{2}})^2 H_w$$
$$+ 2(A_w^{\frac{1}{2}} - A_o^{\frac{1}{2}})(A_o^{\frac{1}{2}} - A_w^{\frac{1}{2}}) H_{wo}]. \tag{8.45}$$

The symbol H_o represents two drops of radius $(r + \delta)$ and separation Δ and H_{wo} represents the H function for a sphere of radius r and one of radius $(r + \delta)$ separated by a distance $(\delta + \Delta)$. H_w represents the function for two spheres of radius r separated by a distance $(\Delta + 2\delta)$. δ is the thickness of the oil layer. Since the Hamaker constant for the oil phase, isopropyl myristate $(4.52 \times 10^{-20} \text{ J})$, is close to that of water $(3.66 \times 10^{-20} \text{ J})$ the effect of replacing part of the oil droplet with water is not great; in fact in a typical type A droplet the effect of a large internal aqueous droplet on energies of attraction is found to be insignificant. Only when the internal droplet grows to almost fill the whole diameter of the multiple droplet is the influence of the internal phase noticeable. A more significant influence on van der Waals' forces of attraction between multiple droplets appears to reside in the reduction in size which follows from expulsion of the internal droplets. The resultant reduction in diameter leads to a reduction in the force of attraction as V_A is related to globule radius, r, by

$$V_A = \frac{-(A_o^{\frac{1}{2}} - A_w^{\frac{1}{2}})^2 r}{\Delta}. \tag{8.46}$$

A full analysis of interactions in multiple emulsions would obviously have to take account of forces of repulsion. The systems are too complex to allow any reasonable estimate of repulsive forces at this stage, although simplified models are being developed to allow an approach along this route.

8.3.3 Alternatives to multiple emulsions

Perhaps because of the stability problems with conventional multiple emulsion systems, some effort has gone into devising alternative systems which will have the same potential advantage of isolating the internal phase from an external phase by way of an oil layer which acts as a transport barrier. When a microphase-in-oil dispersion is emulsified the resultant system is a microphase-in-oil-in-water dispersion. Such systems have been described [157] and the release of 5-fluorouracil from the system compared with release from the internal phase of W/O/W emulsion *in vitro* (Fig. 8.30).

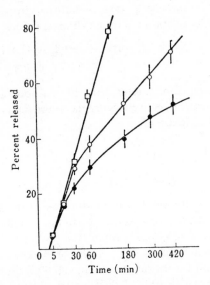

Figure 8.30 *In vitro* release of 5-FU (5 fluorouracil) from a gelatin microsphere in oil-in-water emulsion (S/O/W) and a W/O/W emulsion.
● S/O/W emulsion; ○ W/O/W emulsion;
□ Aqueous solution.
Results are expressed as the means ±S.D. of four experiments. From Hashida *et al.* [157] with permission.

W/O/W emulsions stabilized with soy lecithin–Span 80 mixtures have been used as the basis for the preparation of phospholipid vesicles [158]. A water-in-*n*-hexane emulsion was first prepared and the bulk of the hexane removed, the concentrate being dispersed in aqueous solution using a low concentration of hydrophilic surfactant which itself could then be removed leaving the phospholipid vesicles.

In an attempt to improve stability and to further retard the release of drugs two different methods have been used to prepare W/O/W emulsions both involving the formation a polymeric gel in the internal or external (continuous) aqueous phases [159]. The first of these methods was based on the production in the external aqueous phase of a cross-linked, hydrophilic gel by γ-irradiation of the emulsifier, a polyoxyethylene–polyoxypropylene ABA block co-polymer (poloxamer). These compounds are surface-active (promoting O/W emulsification) and are capable of being cross-linked [160]. O/W emulsions may be prepared which contain the poloxamer in the continuous aqueous phase. After emulsification, the surfactant molecules can be cross-linked at the O/W interface and in the continuous phase by γ-irradiation, forming a network of surfactant molecules which link the dispersed oil globules. Multiple emulsions prepared by re-emulsification of primary W/O emulsion in water saturated with nitrous oxide and containing poloxamer (types F68, F87, F88 or a combination of these) are exposed to ^{60}Co γ-irradiation; the viscosity of the emulsions increased gradually up to the gel point which is a function of the type and concentration of the poloxamer. These systems which are W/O/gel systems do not cream as ungelled multiple emulsions do. On mixing with water the gel swells and W/O droplets are released from the gel matrix. As the hydrophilic nature of the poloxamer compounds prevents their use as stabilizers of the primary W/O emulsion and as the more lipophilic members of the series which might be used degrade on irradiation [160,161], an alternative approach, a modified emulsion polymerization method, has also been used to gel the internal phase [159]. A W/O emulsion, formed with acrylamide (6% w/v) and *N*, *N*-methylene*bis*acrylamide (2% w/v) in water previously flushed with nitrogen as the aqueous phase was subjected to γ-irradiation at doses of up to 0.25 Mrad. The gelatinous polyacrylamide-in-oil dispersion produced was then redispersed in hydrophilic surfactant solutions (e.g. 2% w/v Triton X165 or Span 80:Tween 80 3:1) to produce a W/O/W emulsion containing a cross-linked polyacrylamide gel in the internal aqueous phase. The resultant system is an acrylamide gel/O/W emulsion with similarities to the gelatin microsphere/O/W system described above.

One drawback of these approaches is the possible adverse effects of γ-irradiation on any entrapped drugs [see, for example, 162], particularly with the first method which requires relatively high doses of radiation. It is possible, however, that the oil may exert a protective effect on labile drugs and this has still to be investigated in detail. Other block copolymeric surfactants have also been gelled by irradiation, increasing the scope for the use of these surfactant systems [163]. Three-ply walled microcapsules formed from a multiple emulsion have recently been described [164].

8.4 Microemulsions

Microemulsions or swollen micellar systems represent an intermediate state between micellar solutions and true emulsions as shown in the hypothetical phase diagram in Fig. 8.31. They are readily distinguished from emulsions by their transparency and more fundamentally by the fact that they represent single thermodynamically stable solution phases. The term microemulsion was introduced to describe systems first identified by Hoar and Schulman [166]. Interest in these fluid translucent isotropic dispersions of oil or water has grown rapidly; a book devoted to the theory and practice of microemulsions was published in 1977 [167].

Although the dividing line between a micelle swollen with solubilizate molecules and microemulsions is a narrow one, Prince [167] has identified

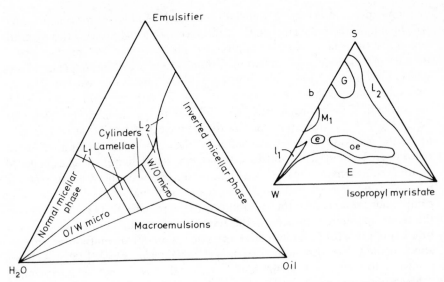

Figure 8.31 Hypothetical phase equilibria diagram showing regions of microemulsions and micellar solutions, after Prince [167]. This simplified phase diagram has been criticised by Rance and Friberg [165] but diagrams with the construction shown above have been obtained by Lo *et al.* [146] as shown in the insert to the figure, which illustrates the Brij 96–isopropyl myristate–water system at room temperature (oe represents a viscous phase of surfactant containing dispersed oil).

microemulsions as having droplet diameters in the range of 250 nm down to 10 nm and micellar solution as having diameters less than 10 nm. Like emulsions, microemulsions may be of the W/O or O/W type and the system can invert from one type to the other by addition of one phase or by altering the surfactant type. Inversion of a water-in-oil microemulsion takes place on addition of water via a viscoelastic gel stage which is composed of a hexagonal array of water cylinders at lower water ratios and a lamellar array of swollen bimolecular leaflets close to the O/W microemulsion boundary (Fig. 8.32) in the transitional stage the systems are turbid and birefringent.

Formation of microemulsions is encouraged by the addition of a co-solubilizer such as a long-chain alcohol, possibly because of the geometric requirements for the appropriate curvature in the interfacial region. The stability and structure of microemulsion particles is still a subject of debate [167, 171].

Schulman emphasized that micellar emulsions are systems in true equilibrium, it being proposed that the components of the surface films in these systems produce a negative interfacial tension at the hydrocarbon–water interface [172]. On mixing, a spontaneous interfacial area increase occurs until zero interfacial tension is attained. In Adamson's [173] model for micellar W/O emulsions, stability is accounted for by a balance of the Laplace pressure ΔP, related to the micellar radius r and interfacial tension γ by

$$\Delta P = 2\gamma/r \tag{8.47}$$

Water sphere in oil

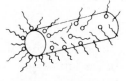

Water cylinder in oil

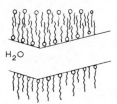

H_2O

Oil solubilized in surfactant bilayer

H_2O

Oil droplet in water

Surfactant   Oil

Figure 8.32 Mechanism of phase inversion of microemulsions. Adapted from Shah *et al.* [168] per Zajic and Panchal. [169]. The lamellar phase ('middle phase' microemulsions) co-exist with bulk aqueous and non-aqueous phases. These systems and their solubilizing ability has been treated theoretically by Huh [170].

and the osmotic pressure difference $\Delta\pi_{os}$ between the inner and outer regions of the micelle which arises from the difference in ionic concentration. The osmotic pressure difference is positive, hence in the presence of water an indefinite swelling of the micellar units is produced until equilibrium is reached when

$$\Delta\pi_{os} = 2\gamma/r. \tag{8.48}$$

Prince [167] considered the monolayer at the microemulsion droplet interface to have a real thickness thereby allowing the assignment of two interfacial

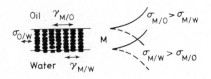

Figure 8.33 The direction of the curvature depends on the relative magnitudes of $\gamma_{M/W}$ and $\gamma_{M/O}$. After Prince [173].

tensions $\gamma_{M/O}$ and $\gamma_{W/M}$ representing the tensions at the two monolayer surfaces against oil and water (Fig. 8.33). The formation of curved surfaces is caused by different interfacial tensions $\gamma_{M/O}$ and $\gamma_{W/M}$ at a *flat* interface; the resultant geometry of the droplet is crucial. If the interfacial layer is 2.5 nm thick, Table 8.5 gives the ratio of the volume of the interface to the volume of the droplet core V_r/V_c. There is an abrupt change in V_r/V_c when the droplet size reaches about 10 nm in diameter, emphasizing the importance of the interfacial interactions in microemulsions. Robbins' theory for the phase behaviour of microemulsions [174] is consistent with the concept that interactions in the mixed film are responsible for the direction and extent of curvature and thus in the type and size of the droplets of the resulting microemulsion. The kind and degree of curvature is imposed by the different tendency of water to swell the hydrophilic head group and oil molecules to swell the hydrophobic regions of the amphiphiles. Prince [167] describes the situation in this way for a W/O system:

'When $V_r/V_c > 7$, the behaviour of the system depends on the molecular interactions at both sides of a well-defined 2.5 nm thick interphase. As the ratio of the number of tails to oil molecules increases, the orientation of the oil molecules becomes less random. Since the tails are anchored in the water, orientation is induced among the oil molecules, resulting in a liquid condensed film. This is particularly so if there is good association among the oil molecules and tails. It

Table 8.5 The geometry of droplets having a 2.5 nm thick interphase

Diameter (nm)		V_r/V_c^*
Core	Droplet	
1	6	215
1	6.5	80
3	8	18
5	10	7
6	11	5
8	13	3.3
10	15	2.375
100	1.05	0.158
1000	10.05	0.015

* Volume interphase/volume core.
From [167].

is reasonable to assume that such an interphase becomes a duplex film, possessing different tensions at each of its sides as the core diameter of the aggregate approaches 5 nm. It is this that differentiates the microemulsion from the micelle;' and for O/W microemulsions:

'When the amount of oil in the system is very small, the interaction of the surfactant heads with the water phase, aggregates the surfactant into normal micelles. The oil is now intercalated among closely packed tails and the volume solubilized in this manner will depend on association with the tails. For a normal micelle to become an O/W microemulsion by increasing in size, there must be strong interactions at both sides of a well-defined interphase. In this case, the interactions are much more specific. Because of space limitations in the centre of the droplet, sharp wedge formation must be achieved by long alcohol tails squeezing oil molecules away from the core and towards the head side of the interphase. Expansion of the water side of the interphase by increasing the size and number of polar groups associated with the surfactants abets this wedge formation.'

The hydrodynamic interface of the droplets has been shown to contain a significant amount of the continuous phase and a quantity of co-surfactant which increases as the amount of solubilizate increases [176]. Small-angle neutron scattering indicates that there is a portion of the interfacial film, 0.9 nm thick, which is not penetrated by the continuous phase [176] in water-in-cyclohexane microemulsions stabilized by sodium dodecyl sulphate and 1-pentanol.

While Schulman and Prince have both inferred that negative or zero interfacial tensions are required for the spontaneous formation of microemulsions, Ruckenstein has assumed a more realistic positive but small interfacial tension [177, 178] and the free energy of formation was found to have a negative minimum value at a certain radius representing the size of stable globules. In a note on the thermodynamic stability of microemulsions Ruckenstein has pointed out that as the surfactant and co-surfactant accumulate in the interface – which is extensive – their bulk concentration decreases and thus their chemical potentials decrease and the free energy of the system is decreased by this so-called dilution effect [179]. Stable microemulsions thus form not because of negative interfacial tension but because the negative free energy change due to dilution of the surfactant in the bulk overcomes the energy due to the small positive interfacial tension.

The preparation and physical properties of oil/water microemulsions containing liquid paraffin, glycerol, water and blends of Tween 60 and Span 80 have been examined [180]. The decrease in micellar size as the surfactant/alcohol ratio was increased is similar to the situation observed with solubilized micellar solutions formed by non-ionic surfactants. Turbidity spectra methods of particle sizing have shown that an increase of temperature of preparation over the range 25 to 80° C led to a gradual decrease in the modal diameter and the half-width of the size distribution curve. Phase diagram studies on micellar solutions prepared at 70° C have indicated a pronounced dependence of the area of existence of microemulsions on the ratio of Tween to Span in the system and on the oil content. Light-scattering investigations on these systems have indicated micelle diameters of between 13 and 45 nm. In general, micellar size was increased by increasing the oil content and by decreasing the surfactant/glycerol ratio. No significant effect on micellar diameter was detected as the Tween/Span molar ratio was varied between 1.0 and 1.6 [180]. As the amount of surfactant in the system is increased, a greater interfacial area is possible and the oil is thus distributed among a greater number of micelles, which are consequently smaller in size. This reasoning is no longer applicable to high surfactant/alcohol ratios, and clear micellar solutions fail to form presumably because there is an insufficient quantity of alcohol in the system to enable an interfacial film of the required composition to be formed. The same authors [181] in a related study used ionic surfactants. Light-scattering measurements have also been carried out on hexadecane–water–alcohol–non-ionic surfactant microemulsions, where the non-ionic surfactants studied have been Brij 96 and Tween 60 [182], the results suggesting that the fraction of alcohol co-surfactant in the interphase region decreases with increasing oil content, which is in some conflict with the work of Attwood *et al.* [181].

One could ask the relevance of the study of such fine points. They can be justified because these systems should possess a high volume fraction of disperse phase and have a very high interfacial area (approximately 10^9 cm^2 l^{-1}) and are useful *inter alia* in the study of chemical reactions and interactions at O/W boundaries. It is thus essential that the composition of the interfacial region is known and its behaviour understood. Attempts have been made to study by electron microscopy the structure of microemulsions using a carbon replica technique [183]. In most cases evidence of spherical structures were seen and in samples of high surfactant and co-surfactant concentrations, lamellar structures could also be detected.

8.5 Viscosity and rheological characteristics of emulsions

The main factors which affect the viscosity of emulsions are listed in Table 8.6. The properties of the disperse phase, the continuous phase and the emulsifying agent or agents all influence the emulsion viscosity. Each factor does not act independently and the interpretation of emulsion viscosity data is complicated by this fact and the fact that particles can deform under shear depending on the nature of the interfacial film. As we have also discussed, emulsions are complex systems, often highly structured, and at phase boundaries or on the point of inversion are very sensitive to small perturbations in the system. We will deal here first with mobile emulsions and then consider briefly the semi-solid state.

8.5.1 Mobile emulsions

Of practical interest is how the viscosity of an emulsion changes with concentration of disperse phase. If the particles in a suspension are rigid (i.e. non-deformable) and in a dilute state ($\phi = 0.05$) Einstein's equation gives the

Table 8.6 Factors which influence emulsion viscosity

1. *Internal phase*
   (a) Volume concentration (ϕ); inter-particle interference; flocculation; aggregation.
   (b) Viscosity (η_i).
   (c) Particle size, and size distribution; technique used to prepare the emulsion; interfacial tension; particle deformation.
   (d) Chemical constitution.

2. *Continuous phase*
   (a) Viscosity (η_0).
   (b) Chemical constitution, and polarity; effect on the potential energy of interaction between particles.

3. *Emulsifying agent*
   (a) Chemical constitution, and concentration.
   (b) Solubility in continuous and internal phases; pH of liquid phases.
   (c) Physical properties of film around the particles; thickness of film; particle deformation; fluid circulation within the particles; influence on the attraction forces between particles.
   (d) Electroviscous effect; electrolyte concentration in aqueous continuous media.

4. *Additional stabilizing agents*
   Pigments, hydrocolloids, hydrous oxides, etc.

From P. Sherman [186]

relationship between the relative viscosity of the system η_{rel} and ϕ,

$$\frac{\eta}{\eta_0} = \eta_{rel} = 1 + 2.5\phi, \tag{8.49}$$

where η_0 is the viscosity of the continuous phase. Modifications of this equation account for the deformability of the globules and the internal viscosity of the droplet, η_i. For example [184]:

$$\frac{\eta}{\eta_0} = \eta_{rel} = 1 + 2.5\left(\frac{\eta_i + 2/5\eta_0}{\eta_i + \eta_0}\right)\phi \tag{8.50}$$

This equation reduces to Equation 8.49 when $\eta_i = \eta_0$. Equation 8.50 was modified to take account of more concentrated systems and the following equation [185] has been found to agree with experimental data up to $\phi = 0.4$.

$$\ln \eta_{rel} = 2.5\left(\frac{\eta_i + 2/5\eta_0}{\eta_i + \eta_0}\right)(\phi + \phi^{5/3} + \phi^{11/3}). \tag{8.51}$$

Attempts have also been made to take into consideration the influence of the emulsifier layer around the globules. All equations show that η increases with increase of disperse phase volume fraction, ϕ. When ϕ exceeds 0.4 to 0.5 the emulsions tend to become pseudo-plastic and the viscosity increases significantly for small changes in ϕ. Like all equations which deal with emulsions higher orders of ϕ appear in the equation when $\phi > 0.05$. A more general equation for concentrated systems takes the form [186, 187]:

$$\eta = \eta_0(1 + a\phi + b\phi^2 + c\phi^3 + \ldots), \tag{8.52}$$

when a, b and c are constants, in general a having the value 2.5. When $\phi > 0.05$, particles interact with each other during flow and thus the nature of the globular interface and forces of attraction and repulsion are reflected in the viscosity. The particle size of globules is obviously a factor and because of the difficulty in obtaining monodisperse emulsions, interpretation of viscosity data is hazardous. Decreasing particle size increases the viscosity of the emulsion. Fig. 8.34 shows this effect for W/O emulsions stabilized by sorbitan sesquioleate. All had a narrow distribution of particle size. Emulsions with a broad particle size distribution have a lower viscosity than emulsions which are more monodisperse.

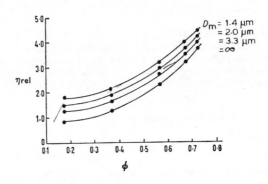

Figure 8.34 The relative viscosity of W/O emulsions stabilized by sorbitan sesquioleate as a function of D_m, the mean particle diameter, and ϕ. From Sherman [186] with permission.

Flocculation or aggregation of particles in these emulsions will give rise to significant changes in the rheological properties of the dispersion, but the effect decreases at higher rate of shear in non-Newtonian systems. At sufficiently high shear rates a limiting viscosity, η_∞, is measured when all the aggregates are disrupted and the limiting viscosity is thus due solely to hydrodynamic interaction. Flocculated particles trap quantities of continuous phase which results in the high viscosity of the flocculated emulsion at rest. An equation has been devised for deflocculated systems in which the increase, h, in the effective volume of the globules due to the hydration of the emulsifier is considered,

$$\frac{\eta_\infty}{\eta_0} = \left[\frac{1}{1 - 3\sqrt{(h\phi)}} \right]. \tag{8.53}$$

Although large variations in the values of h have been reported, many emulsions exhibit a value of 1.3 for h [184]. When the continuous phase is immobilized, and H' is a measure of this volume, the free volume of the continuous phase is thus $1 - H'\phi$ and several equations have taken the form

$$\eta_{rel} - 1 = \left[\frac{a\phi}{1 - H'\phi} \right] \tag{8.54}$$

to account for this.

Because of the influence of particle size and floccule structure, the viscosity of emulsions will change on ageing. The rheological changes in W/O emulsions, which as we have seen, are prone to flocculation, have been studied in depth by Sherman [187, 188] in which estimates of H have been made. The change in viscosity over 600 h of water-in-Nujol emulsions stabilized by 1.5% sorbitan mono-oleate is shown in Fig. 8.35.

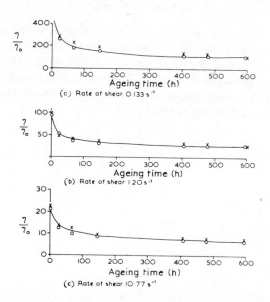

Figure 8.35 Comparison of experimental and calculated changes in η/η_0 during ageing of water-in-oil (Nujol) emulsions. O–O Experimental data; × – × theoretical data. Stabilizer 1.5% sorbitan mono-oleate. From Sherman [187] with permission.

8.5.2 Viscosity of multiple emulsions

A viscometric method for estimating the stability of W/O/W multiple emulsions has been described [189]. The dispersed phase consists of the inner aqueous phase (ϕ_{wi}) and the oil phase (ϕ_o). Using Mooney's equation in the form for Newtonian flow, where λ is a crowding factor,

$$\ln \eta_{rel} = \left[\frac{a\phi}{(1 - \lambda\phi)} \right], \tag{8.55}$$

($\phi_{wi} + \phi_o$) is substituted for ϕ giving

$$\ln \eta_{rel} = \frac{a(\phi_w + \phi_o)}{\{1 - \lambda(\phi_{wi} + \phi_o)\}}. \tag{8.56}$$

Equation 8.55 can be rewritten to give

$$\frac{\phi_{wi} + \phi_o}{\log \eta_{rel}} = \frac{2.303}{a} - \frac{2.303\lambda}{a}\{\phi_{wi} + \phi_o\}. \tag{8.57}$$

The inter-relationship between the volume fraction of the inner phase and the relative viscosity of the W/O/W emulsion can be represented by Equation 8.58 noting that ϕ_o will remain constant after the loss of the internal water phase while ϕ_{wi} should decrease with the increasing incidence of breakdown

$$\phi_{wi} = \frac{a[\log \eta_{rel}\{(2.303/a) - (2.303\lambda/a)\phi_o\} - \phi_o]}{[a + 2.303 \log \eta_{rel}]}. \tag{8.58}$$

Coalescence of the external oil phase is not accounted for in Equation 8.57 nor is the heterogeneity of particle size distribution taken into account. Measured values of a range from 2.28 to 3.21 and of λ from 0.46 to 1.86 [189]. However, using Equation 8.58 the data in Figure 8.36 were collected showing the magnitude of the putative fall in internal aqueous droplet volume. The short-term stability of these systems is amply demonstrated.

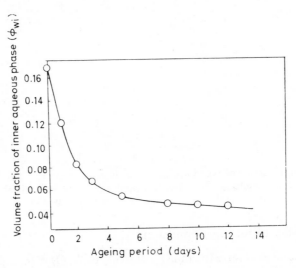

Figure 8.36 Decrease of the volume fraction of inner aqueous phase ϕ_{wi} as a function of the ageing period of W/O/W emulsions (where $\bigcirc$ = liquid paraffin) calculated from Equation 8.58. Hydrophobic surfactant Span 80; hydrophilic surfactant Tween 20. From Kita *et al.* [189] with permission.

8.5.3 Viscosity of microemulsions

The interpretation of the viscosity of macroemulsions is complicated; that of microemulsions is more problematical not because of uncertainties about the role of the interfacial region but because of the difficulty in obtaining measurements which are meaningful. Dilution of microemulsions to obtain data as a function of ϕ has to be carried out with due regard to the consequent alterations in concentration of all species present. Dvolaitzky *et al.* [175] have adopted a trial-and-error approach to dilution of their microemulsions, selecting the correct dilution medium as that which produced a linear relationship between η_{rel} and W, the concentration of water in the system. For their cyclo-hexane–pentanol–water (soap) system, they explain the possible non-linearity in η_{rel} versus W as follows. The initial W/O microemulsion has a high volume ratio (r_h, hydrocarbon radius r_w, radius of water droplet) and a water/soap ratio of about 16. When diluted by a phase without water a portion of the water from the

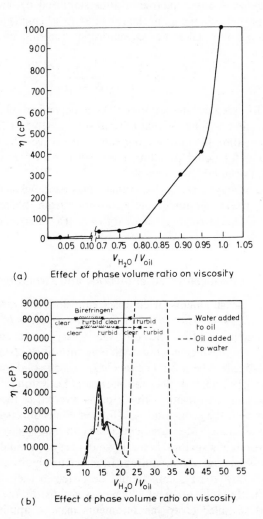

(a)   Effect of phase volume ratio on viscosity

(b)   Effect of phase volume ratio on viscosity

Figure 8.37(a) Changes in the viscosity of microemulsions due to the formation of cylindrical and lamellar structures upon increasing the water/hexadecane ratio above 0.8. (b) Changes in the viscosity of microemulsions and liquid crystals (cylindrical and lamellar structures) upon increasing the water/hexadecane ratio. Solid line represents the viscosity changes when water was added to hexadecane containing potassium oleate and hexanol; broken line represents the changes when hexadecane containing potassium oleate and hexanol was added to water. From Falco *et al.* [192] with permission.

droplets is solubilized leading to a lower water/soap ratio and to an increase in the volume ratio r_h/r_w. This leads to an increase in η because of the intake of continuous phase in the interfacial region due to the increased radius of curvature. This tendency has a limit and on further dilution of the system, the relative viscosity decreases, leading to the parabolic $\eta_{rel}-W$ relationship. Matsumoto and Sherman [190] have obtained an equation which describes the behaviour of benzene/water microemulsions stabilized by Tween 20/Span 20 mixtures with particle diameters in the range 54 to 125 nm. The volume (ϕ_s) of benzene solubilized in micelles of excess emulsifier is considered and the equation takes the form,

$$\eta_{rel} = \exp\left[\frac{a(\phi - \phi_s)}{1 - k(\phi - \phi_s)}\right], \tag{8.59}$$

where k is a hydrodynamic interaction coefficient dependent on the mean particle size, a, having a value well below that of rigid spheres (1.95 to 2.08). Others have also used this equation [191] obtaining values of between 2.5 and 3.8 and of k from 0.40 to 0.95 for liquid paraffin–glycerol–water microemulsions prepared with blends of Tween 60 and Span 80.

Microemulsions may behave as Newtonian dispersions when their particles are spherical. When the system undergoes a transition from spheres to cylinders or lamellae (Fig. 8.32) the viscosity changes abruptly at a water/oil phase ratio of 0.8 (Fig. 8.37) [192].

8.5.4 Rheology of concentrated emulsions and structured systems

Simple viscosity measurements of complex emulsions are rarely possible and almost meaningless, yet a knowledge of the flow properties of these systems is important in quality control and use. Only very dilute emulsions exhibit Newtonian flow; those which contain high concentrations of surfactant and co-surfactant may appear to be solid. The rheology of semi-solid emulsions has been reviewed by Barry [92, 93, 193, 194] and Barry and Eccleston [99] and the reader who wishes specialist treatment of this topic is directed to these sources.

The problem with the semi-solid, self-bodied emulsion systems has been to measure their consistency. Measurements from continuous shear measurements, as they are the result of structural breakdown in the systems under study, have to be treated with some degree of caution. In order to obtain a true measure of consistency or 'body' the systems should be tested in their native state, this requiring a method of measurement which does not disrupt the structures in the emulsion. Thus so-called creep measurements may be applied in which the emulsions are subject to only relatively minor deformities. In creep a shear is quickly imposed on the sample and maintained at a constant level; the time-dependent strain or compliance response to this steady stress provides the creep curve. A recovery curve is obtained on removal of the stress, a typical diagram showing the profile for creep and recovery is given in Fig. 8.38.

The instantaneous elastic deformation AB is associated with the uncoupled

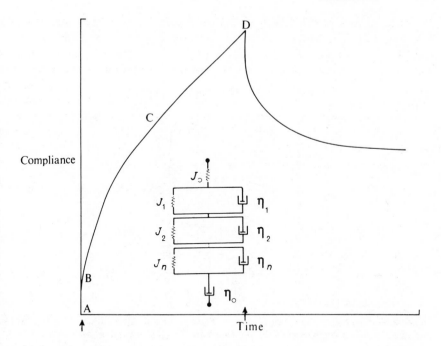

Figure 8.38 A typical creep and recovery curve for a visco-elastic material and an insert of the mechanical model used to describe the creep results.

Hookean spring. BC is the viscoelastic region represented by the Voigt units together with the Newtonian dashpot. Region CD is that of Newtonian flow and is associated with the residual dashpot, as the Voigt units are now fully extended and no longer affect flow. Regions AB and BC are recovered totally and in part, respectively, Region CD is not recovered. In molecular terms, the Hookean spring represents bonds stretching elastically and simulates the elasticity of the gel networks, and the residual dashpot represents viscous deformation of a solid dispersion in a liquid medium. The Voigt units represent that part of the structure in which secondary bonds break and reform. As all bonds do not do this at the same rate, a spectrum of retardation times exists [99]. $J(t)$ is the total creep compliance at any time t where J is the ratio of shear strain to shear stress. The creep compliance for each system is defined as

$$J(t) = J_o + \sum J_n (1 - e^{-t/\tau_n}) + \frac{t}{\eta_o} \qquad (8.60)$$

where J_o is the residual or instantaneous shear compliance, η_o is the residual shear viscosity, J_n is the shear compliance of the elastic part of the mechanical model used to simulate viscoelastic behaviour, $\tau_n (= J_n \eta_n)$ being the retardation time of this unit, due to the random breaking and reformation of bonds in the system randomly. Analysis of the creep–recovery curve gives J_o and η_o.

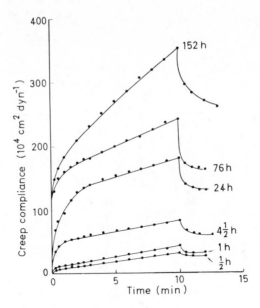

Figure 8.39 Creep compliance of 50% w/w W/O emulsions after different ageing times at a shear stress of 4.9 dyn cm^{-2}. From Sherman [188] with permission.

Creep compliance–time studies on water-in-liquid paraffin emulsions containing 30%, 50% and 65% w/w disperse phase stabilized by 1.5% w/w sorbitan mono-oleate show a marked change in behaviour during the first hours of ageing (Fig. 8.39), changes believed to be associated with the rapid coagulation of globules with diameters not exceeding 0.5 nm [188]. In the flocculated states coalescence leads, during ageing, to a more open disperse phase structure. The elastic moduli of the W/O emulsions presumably arise from the interlinking of the floccules in the system and thus will be sensitive to changes in the degree of flocculation. In concentrated W/O systems the magnitude of the elastic moduli will depend on the number of globules in close contact and on the attractive forces between globules and their neighbours. Viscoelastic moduli should decrease as globule coalescence results in a decrease in the number of globules and in the point of contact in the floccules. Thus creep compliance measurements can detect changes in the native state of the system. These results on simple formulations show the time dependence of properties arising from flocculation and coalescence. More dilute systems have been investigated by Thurston and Davis [195].

In self-bodied emulsions the structures take a finite time to reach a state of equilibrium and thus there is this additional ageing factor related to the continuous phase, creep compliances falling and viscosities rising [196]. An idealized diagram of the relationship between η and J and the mixed emulsifier concentration in oil-in-water emulsions is shown in Fig. 8.40.

To understand better the molecular interactions at work, the effect of the

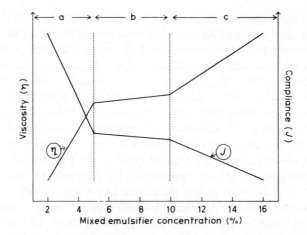

Figure 8.40 Idealized diagram showing the dependence of the viscosities (η) and compliances (J) on the mixed emulsifier concentration in oil-in-water emulsions. Region a – mobile emulsions, region b – semi-solid creams; region c – creams tend to be too stiff for ideal pharmaceutical use. From Barry [98] with permission.

nature of the mixed emulsifier system has to be studied. Barry and Eccleston have reviewed their work on this problem [99].

The effect of concentration of alkyltrimethylammonium bromides of C_{12} to C_{18} on η_0 and J_0 are shown in Fig. 8.41 for liquid paraffin-in-water emulsions

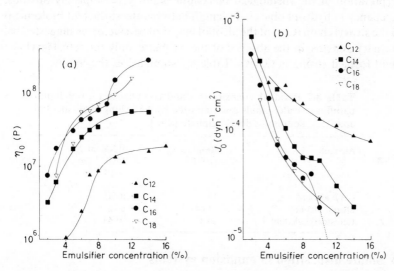

Figure 8.41 Variation of (a) residual viscosity, η_0 and (b) residual compliance, J_0 with approximate mixed emulsifier concentraton for liquid paraffin-in-water emulsions stabilized by cetostearyl alcohol and the alkyltrimethylammonium bromides dodecyl–C_{12}, tetradecyl–C_{14}, hexadecyl–C_{16} and octadecyl–C_{18}. From Barry and Eccleston [99] with permission.

stabilized by cetostearyl alcohol and the cationic surfactants. That it was the gel formation in the aqueous phase that was primarily responsible for the rheological behaviour was indicated by comparison of the results with measurements on the ternary systems (cetyl alcohol–surfactant–water). C_{12} networks in both ternary and emulsion systems are less rigid than networks formed from higher homologues; the C_{14} and C_{16} networks were extensive while the C_{18} networks, although rigid, were more diffuse than expected from the trends observed [99]. Barry and Eccleston attribute the results to the difference in chain lengths of the C_{12} surfactant and cetostearyl alcohol and to steric hindrance by the cationic head group which prevents the formation of a strong molecular complex. In higher homologues the greater opportunity for hydrophobic bond formation counteracts the steric hindrance of the cation. The comparatively low rigidity of the C_{18} network was ascribed to the high Krafft point of the C_{18} surfactant which was probably close enough to the storage temperatures to hasten network crystallization. If such interactions are important then it is obvious that one should be able to predict, for example, the influence of changing the chain length of the alcohol in a mixed stabilizing system. Increasing the chain length of the alcohol in a NaDS stabilized liquid paraffin-in-water emulsion showed compliances in the order stearyl alcohol (C_{18}) $\gg$ cetostearyl alcohol (50 to 70% stearyl alcohol; 20 to 35% cetyl alcohol) > cetyl alcohol (C_{16}), indicating that the network formed with the stearyl alcohol is much weaker. The molecular configuration of the cetostearyl alcohol–water binary phase is different from that of stearyl alcohol or cetyl alcohol/water phase: it thus seems that the molecular interpretation of the rheological behaviour is not yet completely obvious; the discrepancy in hydrophobic chain length between the surfactant molecule (C_{12}) and the stearyl chain (C_{18}) of the alcohol may well be a factor, as suggested before. In ternary systems, in the absence of the oil phase, only the cetostearyl alcohol system formed strong networks. Table 8.7 summarizes the results.

Table 8.7 Residual viscosity (η_o) and compliance (J_o) of liquid paraffin-in-water emulsions stabilized by 4.5% long-chain alcohol and 0.5% sodium dodecyl sulphate [196].

Alcohol	Residual η_o (10^6 P)	Residual J_o (dyn^{-1} cm^2 $\times 10^{-4}$)
Cetyl alcohol	13.6	0.745
Stearyl alcohol	1.64	6.72
Cetostearyl alcohol	14.4	0.944

8.6 Solute disposition in emulsion systems

The use of emulsions in medicine demands a knowledge of the distribution of added agents to the formulation. Flavouring agents, preservatives, stabilizers and, of course, therapeutically active substances are added to emulsion formulations and are distributed at equilibrium between the continuous phase and the disperse phase. As the continuous phase usually contains a micellar 'phase' the analysis of the distribution of solutes in emulsions is frequently complicated. A number of equations have appeared in the literature (for example [197, 198]) which attempt to quantify the distribution of organic solutes between the oil, aqueous and micellar phase of emulsified systems. These are of immediate importance in the rational formulation of systems which will maintain adequate concentrations of fungicidal and bactericidal agents in the aqueous phase, and in the design of systems which will maintain an optimal rate of release of active ingredient.

8.6.1 Preservative distribution

Many preservatives are weak acids which ionize in the aqueous phase,

$$HA \rightleftharpoons H^+ + A^- \tag{8.61}$$

so that the dissociation constant, K_a, is defined by

$$K_a = \frac{[H^+][A^-]}{[HA]}. \tag{8.62}$$

The following treatment has been given by Shimamoto and Mima [199]. If the volume of the aqueous phase is V_w the amount of ionized compound in the aqueous phase W_{A^-} is given by

$$W_{A^-} = [A^-]V_w = \frac{K_a[HA]V_w}{[H^+]}. \tag{8.63}$$

In a simple two-phase dispersion of oil and water the undissociated molecules partition such that the partition coefficient, P, is defined as

$$P = \frac{[HA_0]}{[HA]}, \tag{8.64}$$

where $[HA]$ is the concentration of undissociated molecules in the oil phase and $[HA]$ is that in the aqueous phase. If the volume ratio of oil to water is Φ, the volume of the oil phase is given by (ΦV_w). Hence the amount of preservative in the oil W_{HA_0} is

$$W_{HA_0} = [HA_0]\Phi V_w = P[HA]\Phi V_w. \tag{8.65}$$

Binding of the preservative with non-ionic surfactant micelles has been represented [199] by an equation of the form:

$$\frac{[D_b]}{[S]} = \frac{n_1 K_1[D_f]}{1 + K_1[D_f]} + n_2 K_2[D_f] \tag{8.66}$$

where $[D_f]$ is the concentration of free preservative in the aqueous phase and $[D_b]$ is the concentration of bound preservative, $[S]$ being the concentration of surfactant. The concentrations of $[D]$ refer to the total volume of solution. n_1, n_2 and K_1, K_2 are binding constants. The volume of the micellar phase is not accurately known but it can be considered proportional to surfactant concentration since CMCs of non-ionic surfactants are low. As the density of many non-ionic surfactants is approximately 1 g ml^{-1}, the volume of the micellar phase is hence considered to be approximated by the weight of surfactant, thus

$$[S] = \frac{W_s}{V_w + V_m} = \frac{W_s}{V_w + W_s}. \tag{8.67}$$

On the basis of the same approximation, W_{D_b} is

$$W_{D_b} = [D_b](V_w + W_s) = \left\{ \frac{n_1 K_1 [D_f]}{1 + K_1 [D_f]} + n_2 K_2 [D_f] \right\} \left\{ [S](V_w + W_s) \right\} \tag{8.68}$$

If $b = \{ (V_w + W_s)/V_w \}$, Equation 8.68 can be rearranged to give

$$W_{D_b} = b \cdot V_w [D_f][S] \left\{ \frac{n_1 K_1}{1 + K_1 [D_f]} + n_2 K_2 \right\}. \tag{8.69}$$

The total preservative concentration in the system is what must be known, or more importantly its relationship to the amount of preservative in free solution. The total amount of preservative in an emulsified system, W_{D_t} is given by

$$W_{D_t} = W_{D_f} + W_{A^-} + W_{HA_o} + W_{D_b}, \tag{8.70}$$

where W_{D_f} is the amount of free, undissociated preservative in the aqueous phase.

$$V_t = \Phi V_w + V_w + V_M \approx \Phi V_w + V_w + W_S = (\Phi + b)V_w. \tag{8.71}$$

As $[HA] = [D_f]$, and the amount of free undissociated preservative in the aqueous phase is $[D_f] \cdot V_w$, the total preservative concentration based on the total volume of the emulsion, $[D_t]$ is

$$[D_t] = \frac{W_{D_t}}{V_t} = \frac{[D_f]}{(\Phi + b)} \left\{ 1 + \frac{K_a}{[H^+]} + \Phi P + b[S] \left(\frac{n_1 K_1}{1 + K_1 [D_f]} + n_2 K_2 \right) \right\}. \tag{8.72}$$

When the required concentration of free undissociated preservative in the aqueous phase is known from the biological activity concentration profile, the total preservative concentration in the emulsified system can be obtained using Equation 8.72. Table 8.8 shows calculated and observed values of $[D_f]$ for methyl p-hydroxy benzoate in an emulsion of 30 g light mineral oil (light liquid paraffin) in water ($\simeq 100$ ml containing four non-ionic surfactants) with a total weight of 5.6 g.

Konning [200] has presented an equation to relate the quantity of preservative added to an emulsion to its free concentration in water which has the form,

$$[D_f] = \frac{W_{D_t}(\Phi + 1)}{R(\rho \Phi + 1)} \tag{8.73}$$

Table 8.8 Comparison of calculated with observed free methyl *p*-hydroxy benzoate concentrations in the aqueous phase of the emulsion

% w/w of methyl *p*-hydroxy benzoate in emulsion $[D_t]$	% w/w of free methyl *p*-hydroxy benzoate in aqueous phase $[D_f]$	
	Observed*	Calculated
0.0049	0.0010	0.0010
0.0195	0.0043	0.0040
0.0488	0.0116	0.0108
0.0971	0.0242	0.0234
0.193	0.0566	0.0522
0.292	0.0958	0.0846
0.484	0.196	0.150

* Shimamoto *et al.* [205].

where R is the preservative–surfactant interaction ratio, or the total/free preservative ratio, and if the surfactant concentration is $[S]$ we can write

$$\frac{W_{D_t}}{[D_f]} = R = 1 + k[S]. \tag{8.74}$$

Equations 8.73 and 8.74 have been criticized by Mitchell and Kazmi [201]. Equation 8.74 is a special case of the relation found by Mitchell and co-workers [202, 203], i.e.

$$\frac{[D_b]}{[S]} = \frac{n_1 K_1 [D_f]}{1 + K_1 [D_f]} \tag{8.75}$$

which is a form of Equation 8.66 neglecting class II binding sites. n_1 is the maximum number of independent binding sites of class I in the surfactant molecule, K_1 being the preservative surfactant association constant. The conditions under which Equation 8.74 is valid are unlikely as $R = 1 + k[S]$ when $k = n_1 K_1$, and this is only true as $[D_f] \rightarrow 0$. Kazmi and Mitchell [203, 204] have concluded that the equation which satisfactorily characterizes the interaction of several preservatives with non-ionic surfactants is

$$\frac{[D_b]}{[S]} = \left\{\frac{n_1 K_1 [D_f]}{1 + K_1 [D_f]}\right\} + \left\{\frac{n_2 K_2 [D_f]}{1 + K_2 [D_f]}\right\}, \tag{8.76}$$

where K_2 is the intrinsic binding constant for preservative to class II site. If $K_2[D_f]$ is small, Equation 8.76 reduces to Equation 8.66; Mitchell and Kazmi's final equation for emulsion systems takes the form

$$W_{D_t} = \frac{[D_f]}{\Phi + 1}\left\{1 + \frac{n_1 K_1 [S]}{1 + K_1 [D_f]} + \frac{n_2 K_2 [S]}{1 + K_2 [D_f]} + P\Phi\right\}, \tag{8.77}$$

which is similar in form but not identical to Equation 8.72. Where the emulsion is complex, determination of the parameters in Equation 8.72 or 8.77 may be tedious and direct experimental measurement of the distributions may be an essential step. Konning [200] has put forward the view that the selection of a suitable concentration for a preservative in the aqueous phase of an emulsion is not to be based on the minimum inhibitory concentration of the compound in simple aqueous solution. Kazmi and Mitchell have agreed with this view [201] and emphasize the fact that many factors neglected in the strictly mathematical approach would contribute to the success or failure of the preservative, decomposition, or interaction of the preservative with some additive, adaptation of micro-organisms might all occur. Surfactants might well act synergistically with the preservative although it is more likely that at the high concentrations found in emulsions, activity will be diminished, but the magnitude of the decrease may well be exaggerated by the mathematical models [204].

The above treatments apply equally to other organic solutes, the concentration of which in the continuous phase would determine release or activity [205]. Flavour release from O/W emulsions in the mouth has been described by a model which assumes that flavour molecules are transferred from oil to water when the equilibrium of the emulsion is disturbed by dilution with saliva, and that only the aqueous concentration stimulates perception [206, 207]. The release equations predict high release of flavour when the value of P for the flavouring agent is high, when dilution is high, and/or when the initial value of ϕ is high.

A dilution factor, f_e, is defined as

$$f_e = \frac{\phi_i}{\phi} = \frac{[F_i]}{[F]} \tag{8.78}$$

where ϕ_i is the initial and ϕ the final volume fraction of oil phase when the emulsion is diluted with water, and F_i and F refer to the corresponding concentrations of flavour in the emulsion. The initial aqueous concentration F_{wi} is reduced by the aqueous dilution factor, f_w, defined as

$$f_w = \frac{V_w}{V_{wi}}, \tag{8.79}$$

where V_w is the final volume of aqueous phase. Now,

$$[F_{wd}] = \frac{[F_{wi}]}{f_w} \tag{8.80}$$

where $[F_{w_d}]$ is the concentration of flavour in the aqueous phase immediately after dilution. As a result of dilution the compound is transferred from oil to water until a new equilibrium is stabilized to give $[F_{we}]$.

Release equations were arrived at as follows [206]. By mass balance, and assuming P to be independent of concentration the total amount of compound in the emulsion, $V_t[F_t]$, is given by

$$V_t[F_t] = V_o[F_o] + V_w[F_w] \tag{8.81}$$

and noting that $\phi = V_o/V_{wd}$ one obtains from Equations 8.79 and 8.81

$$[F_{we}] = \frac{[F_i]}{\phi_i(P-1)+f_e} \qquad (8.82)$$

and from Equations 8.78, 8.79, and 8.81:

$$[F_{wd}] = \frac{[F_{wi}](1-\phi_i)}{f_e - \phi_i}. \qquad (8.83)$$

McNulty and Karel define a dimensionless parameter $[F_{we}]/[F_{wd}]$ which is always greater than 1.0 in their model and is a measure of the potential extent of release, defined as the potential extent of increase in aqueous concentration. The release potential is determined by the difference between $[F_{wd}]$ and $[F_{we}]$. Fig. 8.42 shows the influence of partition coefficient P and ϕ_i on the aqueous concentration of flavour immediately after emulsion dilution $[F_{wd}]$ and at equilibrium $[F_{we}]$ when $f_e = 2$, and $[F_{wi}] = 100\,\text{ppm}$. Uptake curves $[F_{we}]/[F_{wd}] < 1.0$ are obtained for the model when an emulsion initially containing no flavour is diluted with an aqueous phase containing an initial

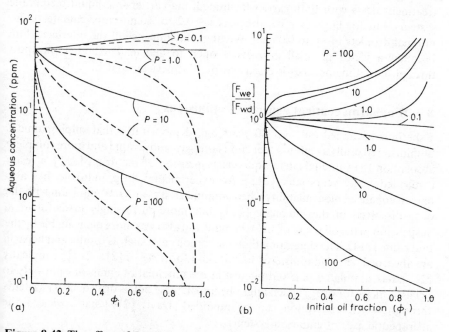

Figure 8.42 The effect of P and ϕ_i on (a) the aqueous concentration immediately after emulsion dilution ($-----$) and at equilibrium (———) following 1:1 dilution with water ($f_e = 2$) and (b) the results plotted as $[F_{we}]/[F_{wd}]$ which when > 1.0 represent flavour release and when < 1.0 represent uptake of flavour from the aqueous phase. From McNulty and Karel [206]. (b) shows that as ϕ_i increases the potential extent of flavour release and uptake increases at any value of P, and that as P increases, the potential extent of flavour release and uptake increases at any value of ϕ_i.

concentration of flavour. These equations might have relevance in the consideration of dilution of pharmaceutical ointments and creams and the consequent effect on drug release and preservative action.

8.6.2 Mass transfer in multiple emulsion droplets

Osmotic flow was previously discussed (see Section 8.3.2a)

A model which describes mass transfer in the liquid surfactant membranes which exist in multiple emulsion droplets has been proposed by Kopp *et al.* [207]. The multiple emulsion droplet is considered as a sphere in which there is embedded a great number of immobile reaction sites (internal dispersed droplets). When liquid membranes are used to detoxify systems, transfer of active agents occurs from the experimental phase to the internal droplets. To avoid early saturation of the inner phase the diffusing species is reacted with a substance in the receiving phase to form a product incapable of diffusing back through the membrane. Li [209] gives the example of phenol diffusing across a hydrocarbon membrane into sodium hydroxide solution; the sodium phenolate is insoluble in the liquid membrane. Specific 'carrier' molecules may also be used. Kopp's treatment deals with both carrier-facilitated and carrier-free liquid membrane systems, but for large external droplets (~ 0.2 to 2.0 nm approximately) and internal droplets of up to 0.05 nm. Whether the influence of the interfacial film becomes a factor in small droplets is open to debate. Facilitated transport through liquid membranes is discussed by Matulevicius and Li [210].

8.7 Biopharmaceutical aspects of emulsions

Experiments carried out over 40 years ago demonstrated that sulphonamides administered orally as oil-in-water emulsions gave more rapid and more complete absorption than in equivalent aqueous suspension of the drugs [211], a result confirmed many years later [212]. An experimental drug, indoxole, has also shown enhanced bioavailability when administered as an O/W emulsion with the drug dissolved in the oil phase [213] but more surprisingly griseofulvin in suspension in the oil phase of an O/W emulsion also was more bioavailable by the oral route [214]. The suggestion that macromolecules such as insulin and heparin are absorbed when administered in emulsion form *per os* [215–217] or topically [218], has stimulated renewed interest in the potential of emulsions not only to act as vehicles for drugs which might be impalatable but as drug delivery systems through which absorption can be modified. Davis [219] has reviewed the therapeutic uses of emulsion systems.

The modifying influence of the emulsion dosage form on bioavailability of a drug substance contained in the emulsion arises from several sources: the distribution of the drug between the oil phase and water phase will control availability to some extent; the oil phase might influence biological processes or the surfactant present as stabilizer might alter response through micellar entrapment of the drug, or through a direct effect on membrane permeability.

Kakemi and co-workers have studied the mechanisms of intestinal absorption of drugs from O/W emulsions extensively using *in vitro* and *in vivo* models. In their first paper [220] on the topic they use as their starting point the equation of Bean and Heman-Ackah [221] derived for a preservative–emulsion system in which the concentration of active agent in the aqueous phase was related to the overall concentration in the system

$$[D_w] = [D_t] \left\{ \frac{\Phi + 1}{P\Phi + 1} \right\}, \tag{8.84}$$

which is redolent of those equations discussed earlier, but considers only partition between the oil phase and the external water phase, neglecting micellar interactions. In view of the other complexities and from their finding for salicylamide that the same order of results was obtained in the presence and absence of emulsifier, Kakemi *et al.* proceeded with Equation 8.85. They noted too that micellation can occur in both aqueous and non-aqueous phase and the resulting mathematical relation which took all of these factors into account would be too unwieldy. Equation 8.84 can be written in terms of the amount of drug in the aqueous phase and in the emulsion:

$$W_w = W_t \frac{1}{P\Phi + 1}, \tag{8.85}$$

as $W_w = [D_w] V_w$ and $W_t = [D_t] [V_o + V_w]$ and $\Phi = V_o / V_w$.

The authors argue that not only the concentration of drug but also the absolute amount has to be taken into account in considering the absorption process. The absorption of acetanilide from isopropyl palmitate emulsions ($P = 1.05$) decreased with increase in Φ, $[D_w]/[D_t]$ is almost unity over the whole range of Φ while W_w/W_t decreased with increase of Φ. Comparison of percentage absorption from an aqueous solution and an emulsion using the ratio (A_e/A_s) indicated, as absorption falls with increasing Φ, that the amount of drug in the aqueous phase determined absorption and secondly that drugs were absorbed mainly via the aqueous phase and not directly from the oily phase. Fig. 8.43 shows salicylamide absorption from isopropyl palmitate emulsions ($P = 2.37$). As Equation 8.85 is an equilibrium equation and absorption is a dynamic process, absorption will be greater than predicted from the equation, because drug from the aqueous phase which is absorbed is replenished from the oil phase. Both salicylamide and acetanilide have values of $P > 1$. Investigation of some drugs with $P < 1$ showed an interesting phenomenon, namely an enhanced absorption for the emulsion. However, if the volume of the aqueous phase was maintained constant and equal to the comparison solution, increase in Φ gave rise to different absorption profiles, as shown in Fig. 8.44. These indicate that the absolute volume of the aqueous phase is one of the critical factors determining absorption of poorly oil-soluble drugs from O/W emulsions.

It would be wrong to consider the oil phase as an inert carrier or even simply as a reservoir. Recent work has shown that the oil phase exerts some biological

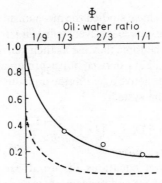

Figure 8.43 Salicylamide absorption from rat intestine from an isopropyl palmitate emulsion shown (————) as a ratio of percentage absorbed from the emulsion (A_e) and percentage absorbed from a solution (A_s); (— — — —) the ratio of W_w/W_t as a function of Φ. From Kakemi *et al.* [220] with permission.

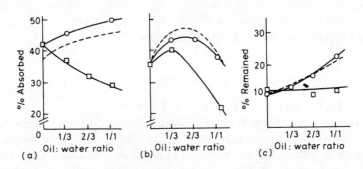

Figure 8.44 The effect of aqueous phase volume on drug absorption from emulsions when the drug has a partition coefficient of less than unity.
(a) Sulphapyridine absorption from isopropyl palmitate emulsion ($P = 0.089$).
(b) Sulphapyridine absorption from ethyl laurate emulsion ($p = 0.25$)
(c) Sulphanilamide absorption from isopropyl palmitate emulsion ($P = 0.034$)
-□- by instillation of constant aqueous phase volume.
-O- by instillation of constant overall emulsion volume.
---- by calculation using values of constant aqueous phase volume procedures using Nogami's [222] equation,

$$\text{Apparent absorption rate constant} = \left(\frac{1}{2.303} \cdot \frac{k}{V}\right),$$

where k is the intrinsic rate constant independent of V the volume of the perfusion fluid. The values calculated agree well with the observed values when the overall emulsion volume is kept constant supporting the view that the absolute volume of the aqueous phase is a critical factor. From Kakemi *et al.* [220].

effects of its own. The absorption of diazepam administered in medium chain triglyceride (MCT) was faster and significantly less variable than absorption of a suspension of drug, it being suggested that the diazepam was emptied from the stomach while retained in the lipid and thus its absorption was affected by the movement of the MCT in the gastro-intestinal tract and its rapid dispersion and metabolism [223]. MCT disappears almost completely from the small intestine (in a rat gastro-intestinal loop) in 30 min. Little metabolism of MCT occurs in the stomach although lingual lipase is said to exert a lipolytic activity towards long-chain triglycerides in the stomach [224]. The presence of emulsifiers may well alter the handing of MCT and other oils. Indeed it has been demonstrated [225] that the natural formation of monoglycerides by hydrolysis of MCT in the stomach will facilitate emulsification of the MCT.

Modification of gastric residence times in determining absorption from oils and emulsions has also been implicated by Bates and Sequeira [226], a view recently confirmed by Palin *et al.* [227].

Oral administration to rats of DDT in solution in three oils of different chemical composition was found to yield significantly different plasma-concentration time curves (see Table 8.9). Emulsification of the oils with 6 % v/v Tween 80 had different effects on both the rate and extent of absorption and was dependent upon the nature of the oil. The effect of each oil on the total gut transit time of a co-administered ^{99m}Tc-sulphur colloid was investigated. The time taken for 50 % of the marker to be excreted was determined from faecal recoveries and whole body gamma scintigraphy. The sulphur colloid was most rapidly cleared in the presence of liquid paraffin ($t_{50\%} = 9.8 \pm 3.6$ h). There was no significant difference in the total transit times in the presence of Miglyol 812 ($t_{50\%} = 15.5 \pm 2.0$ h) and arachis oil ($t_{50\%} = 14.1 \pm 1.1$ h). Therefore, the differences in DDT absorption may only be explained in part by the effect of oils on total gut transit time.

Other factors obtrude. Heman-Ackah and Konning [228] speculated that the activity of phenol in emulsions was higher than anticipated from calculation of its concentration in the aqueous phase because of its adsorption at the O/W interface at which micro-organisms were also adsorbed. Ogata *et al.* [228] have found evidence for the enhanced absorption of methyl orange (acting as a model drug) when adsorbed at the O/W interface in 'coexistence with surfactants'. As methyl orange does not partition to the oil phase it may be that the absorption of

Table 8.9 Pharmacokinetic data following oral administration of DDT (20 mg) to rats (mean ± s.d.) A = arachis oil, M = Miglyol 812 (fractionated coconut oil), P = liquid paraffin.

	AUC (μg ml^{-1} h)			T_{max} (h)			CP_{max} (μg ml^{-1})		
	A	M	P	A	M	P	A	M	P
Oil	118.3	57.9	23.9	7.0	6.0	4.0	11.1	4.7	1.7
(1 ml)	±6.6	±10.2	±4.2	±2.0	±0.0	±0.0	±1.9	±0.7	±0.1
Emulsion	105.8	54.5	54.3	4.0	5.5	5.0	9.2	3.6	4.3
(2 ml)	±12.5	±3.4	±8.2	±0.0	±1.0	±2.0	±2.1	±0.6	±0.4

AUC – area under plasma concentration–time curve 0 to 24 h, T_{max} – time to peak concentration, CP_{max} – plasma concentration at peak
From [227].

heparin and insulin from emulsions is enhanced also by adsorption. The oil volume fraction is important, as it was found that adsorption had an inhibiting effect on methyl orange (MO) absorption at phase ratios greater than 0.15. The results with MO led to the conclusion that the adsorbed state of the compound was an 'activated' one – perhaps analogous to the adsorption of poorly soluble drugs such as digoxin on high surface area materials which enhances their bioavailability. Some of the competing influences are depicted in Fig. 8.45 which attempts to show the effect of adsorption of oil droplets and micelles to the adsorbing surface, perhaps neglecting the influence of drug adsorbed on to the oil droplets themselves. This picture was devised to explain experimental results but one has to assume that in one case (b), the drug is absorbed from the micellar phase while in (a) it is not. The role of the micelle-absorption has not been clarified. It has been suggested [231] that oleic acid could be absorbed directly from emulsion droplets without passing through an intermediary phase in the rat jejunum. Absorption of oil red XO from methyl oleate following intramuscular administration was enhanced by the presence of 6 % Span 80, while that of Sudan black-B was hardly affected, yet it has been suggested that the effect of Span is one on the dispersion of methyl oleate itself. There is a significant effect of Span 80 on the disappearance of methyl oleate, following i.m. injection – the surfactant increasing the rate of disappearance [232] which could hardly explain the quite different effects of the surfactant on the absorption of two drugs. The absorption of these two oil-soluble drugs has also been studied from the rat small intestine when they were administered as O/W emulsions. The authors' conclusions [233] are that 'from emulsions using tributyrin as an oil phase, oil red XO was absorbed

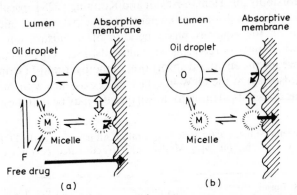

Figure 8.45 Schematic relation between absorption and hypothetical mode of drug distribution in O/W emulsion. O, M and F represent drug in oil droplets, in micelles, and in aqueous phase. In the case of (a), drug in the aqueous phase may be absorbed without interaction with micelles or oil droplets. On the other hand, in (b), as micelles are competing for binding sites with oil droplets on mucosal layer, one absorption rate of the drug will depend not only on the amount of drug in the micellar phase, but also on the degree of absorption of micelles on mucosal layer. Adsorption of micelles on mucosal layer is controlled by the oil concentration at constant concentration of surfactant. From Ogata *et al.* [230] with permission.

monoexponentially to the same extent as from polysorbate 80 micellar solution despite the fact that the dye is very lipophilic and is not considered to be localized in the aqueous phase. On the other hand, oil red XO was absorbed faster from emulsions using triolein as an oil phase, in the early stage and slower in the later stage than from tributyrin emulsions. These absorption characteristics of oil red XO were demonstrated to be connected with the absorption of the oil phase. Oil red XO does not seem to move into inner compartments with oil for it was not transported into intestinal lymph even from triolein emulsions', which added to the evidence above does not readily allow a composite theory to be applied to absorption from O/W emulsions in the presence of surfactants. Even *in vitro* experimentation is not at such a stage of refinement to allow us to progress too far with explanations of performance in the whole animal. For example, the inter-dependence of the parameters is much greater in emulsion systems and is well illustrated by the results obtained on the effect of surfactant concentration on

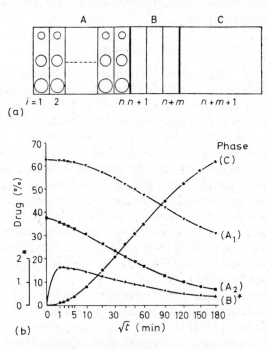

Figure 8.46(a) Simulation model of drug release from the sample (droplet matrix) in an *in vitro* cell with membrane filter separating A from C.
  A: sample phase (droplet matrix), n layers
  B: filter phase, m layers
  C: solvent phase
Sample phase is composed of the continuous phase (outer phase) and the droplets (inner phase) with a distribution of particle sizes.
(b) Model simulation: drug release from emulsion. (A_1) drug content in the droplets; (A_2) drug content in the continuous phase; (B) drug content in the filter phase; (C) drug content in the solvent phase. From Takehara and Koike [234].

release from an isopropyl myristate emulsion. The low release rate at the lowest surfactant concentration was ascribed to droplet coalescence because of the instability of the emulsion [234]. However, a theoretical approach to simulation of release from a polydisperse (i.e. real) emulsion has been made [234] in which polydispersity was treated by considering there to be groups of monosized particles; drug transport in the droplet was treated as radial diffusion. Theoretically, reasonable results were obtained from the models; drug release was faster for smaller droplets than from larger droplets (hence the experimental observation above on the unstable emulsion) and the concentration of drug in the droplets near the surface is lower than in the central part of the drop. The model is illustrated, along with some simulated release profiles in Fig. 8.46. Further work of this nature is essential if the importance of the resistance of the interfacial layer, drug adsorption on to the oil droplets, and micellar solubilization, are to be collated, although *in vitro* work and simulation cannot replace the vital physiological experiments required to elucidate the interactions between surfactant, oil and biomembranes.

8.7.1 Lymphatic transport of emulsions

Several studies have clearly shown that formulation techniques can increase the uptake of drugs into the lymphatic circulation, important in the treatment of cancers involving metastases via the lymph, and lymphomas. One study has shown enhancement of lymphatic transport of mitrocycin C to be accomplished by injection of O/W or W/O emulsions by the intraperitoneal and intramuscular route [235]. Bleomycin, a water soluble base, has been administered in several emulsion forms, including a 'drying' emulsion formulation which is a re-constitutable form designed with the aim of minimizing physical instabilities [236]. Following intramuscular injection and intraperitoneal administration, absorption into the lymph occurred in the order W/O emulsion > O/W emulsion > aqueous solution. Bleomycin being water soluble will be found mostly in the aqueous phase of O/W emulsions, yet is found to be bound to the extent of 40% to the oil droplets. This may be the reason for the enhanced absorption of O/W systems.

Lymphatic uptake of the lipid soluble dye Sudan blue and Vitamin A acetate from triolein-in-water emulsions requires the presence of both bile salts and phosphatidylcholine [237]. However, the overall contribution of lymphatic uptake for these two solutes is very small. In the absence of sodium taurocholate in rats with bile fistulae the administration of 20 ml polysorbate 80 did not stimulate lymphatic transport, although absorption from a 4% polysorbate 80 solution is equivalent to that for the triolein emulsions in intact amounts, but slightly more rapid.

8.7.2 Intravenous formulations

The insolubility of several drugs makes their formulation for intravenous use difficult. Alternatives to micellar solutions which as we have seen are not without

Table 8.10 Acute intravenous toxicity (LD 50) in mice of diazepam in different injection formulations and of some placebo formulations. Confidence limits at $P = 0.05$.

Preparation	LD 50 $mg\,kg^{-1}$	LD 5 $mg\,kg^{-1}$	LD 95 $mg\,kg^{-1}$	Injected amount at LD 50 $ml\,kg^{-1}$	LD 5 $ml\,kg^{-1}$
Apozepam*	52.2 (50.5–53.7) $n = 90$	45.0	60.5	10.4 (10.1–10.7) $n = 90$	9.0 (8.2–9.4) $n = 90$
Stesolid*	65.1 (61.0–69.0) $n = 90$	47.1	90.0	13.0 (12.2–13.8) $n = 90$	9.4 (7.0–10.6) $n = 90$
Valium*	83.4 (77.9–87.0) $n = 50$	70.4	98.7	16.7 (15.6–17.4) $n = 50$	14.1 (13.2–14.7) $n = 50$
Placebo Valium†	90.5 (87.0–94.1) $n = 69$	77.1	106.5	18.1 (17.4–18.8) $n = 69$	15.4 (14.8–16.0) $n = 69$
Diazepam emulsion	283.3 (263.5–312.6) $n = 60$	204.1	393.1	56.7 (52.7–62.5) $n = 60$	40.8 (29.3–45.8) $n = 60$
Placebo emulsion	—	—	—	—	> 64.0

From [240].
* This formulation contains a surfactant in micellar form.
† Vehicle without drug.

problems, are co-solvent systems or emulsions. Jeppsson has been foremost in investigating the intravenous use of emulsions using as his starting emulsion formulation, lipid emulsions such as soybean oil emulsions which have been used intravenously in parenteral nutrition. Barbituric acids [238], cyclandelate and nitroglycerine [239] have been incorporated into emulsions. Because of problems with diazepam injections which had a propylene glycol–ethanol–benzyl alcohol vehicle in relation to their toxicity (notably thrombophlebitis), alternative formulations have been sought. A soybean oil emulsion* of diazepam has been formulated and tested *in vivo*. It was shown to possess considerably lower toxicity than Valium injection or the equivalent solubilized product Stesolid (see Table 8.10). At the low dose level (2.5 mg kg⁻¹), Valium showed its peak activity at the first reading 4 min after injection, but the diazepam emulsion gave a maximum response after about 25 min. This may be due to a slower distribution to the brain because of a competition between the lipid particles and the CNS for the lipophilic drug.

More prolonged activity of lignocaine given as a fat emulsion compared with solution formulation has also been reported [241]. Obviously the fate of the lipid droplets in the body determines the pharmacokinetic · profile of the drug substance. This has been widely studied in view of its importance in parenteral

---

* The diazepam emulsion was of the following composition [240]:

Diazepam (WHO)	0.5 g
Soybean oil	15.0
Acetyl monoglycerides	5.0
Egg yolk phosphatides	1.2
Glycerol	2.5 ·
Distilled water ad	100.0 ml.

nutrition. Safe fat emulsions usually prepared from soybean oil or cottonseed oil for parenteral nutrition became available around 1960. There are two preparations containing fat as an emulsion mixed with amino acids and sorbitol or xylitol (see Table 8.11).

Electron microscope studies on Intralipid and chylomicrons showed that the size of the fat particles in Intralipid 10% (mean diameter 0.13 μm) was about the same as that of chylomicrons (the mean diameter varied between 0.096 and 0.21 μm [242, 243]). The particles in Intralipid 20% were somewhat larger (mean diameter 0.16 μm). These, and a number of other investigations, have shown many physical and biochemical similarities between chylomicrons and Intralipid. Groves and Yalabick [244] have found that the volume surface mean diameter of Intralipid globules was 0.46 μm.

Intralipid can be stored for at least 18 months at a temperature between 0 and

Table 8.11 Composition of fat emulsions and preparations

Composition of fat emulsions

	Intralipid[†] (g)	Lipiphysan[‡] (g)	Lipofundin S[§] (g)
Soybean	100 or 200		100 or 200
Cottonseed oil		150	
Egg yolk phospholipids	12		
Soybean lecithin	20		
Soybean phospholipids			7.5 or 15
Glycerol	25		
Sorbitol		50	
Xylitol			50
DL-tocopherol		0.5	
Distilled water to a volume (ml) of:	1000	1000	1000

[†] Vitrum, Stockholm, Sweden.
[‡] Egic, Loiret, France.
[§] Braun, Melsungen, Germany.

Fat–polyol-amino acid preparations

Constituents	Nutrifundin* (g)	Trivemil[†] (g)
Soybean oil fract.	38	38
Soybean lecithin		7
Soybean phospholipids fract.	3.8	
Sorbitol		100
Xylitol	100	
Amino acid mixture	60	60
Distilled water to a volume (ml) of:	1000	1000

* Braun, Melsungen, Germany.
[†] Egic, Loiret, France.
From [242].

4° C. There are no significant changes in any of the physical and chemical properties. Intralipid has also been stored for $2\frac{1}{2}$ years at $-4°$ C and at 20° C without any changes. Tests in both dogs and man did not show any decreased tolerance. The margin for the recommended storage conditions seems to be great [245]. Heparin and emulsions containing fat-soluble vitamins can be added to Intralipid. No other solutions are recommended for mixing with Intralipid. The effect of additives on the physical properties of Intralipid 10% have recently been studied [246] mainly by electrophoretic and viscometric techniques. Heparin had little effect on the zeta potential of the particles, but sodium and calcium chloride and calcium lactate had marked effects, reducing the zeta potential to zero at relatively low additive concentrations (Fig. 8.47).

The effect of glycerine solutions on the zeta potential of the droplets was dependent on the pH of the system, reducing the zeta potential at pH 3.4 for example, and increasing it at pH 7.0 due to the overall negative charge of glycerine at this point. As the pH of the emulsions decrease on ageing, presumably due to the formation of free fatty acid, the severity of the interaction or its consequences might thus be dependent on the age of the emulsion. Fig. 8.47a suggests that freshly prepared Intralipid might, however, be more sensitive to electrolyte than an emulsion nearing its expiry date. In spite of the significant effects on zeta potential there was no evidence of rapid flocculation of particles. It may be that the presence of cations increases the stabilizing properties of the phospholipid interfacial layer perhaps by altering the rigidity of the monolayer (or multilayer) at the interface.

Commercial lecithins which are used in the manufacture of fat emulsions and in emulsions used in foods and cosmetics vary in composition. The relationship between phospholipid composition and the phase behaviour of lecithin emulsion systems has been investigated [247]. Soybean oil emulsions were prepared with a series of commercial soybean lecithins (0.5 to 3% w/w) and emulsion stability and electrophoretic mobility and globule size determined. Addition of sodium stearate as a negatively charged component decreased globule size and increased stability. 1% sodium chloride was added to both phospholipid dispersions and to the emulsions. X-ray diffraction data on a phospholipid and a phospholipid–sodium stearate mixture show that sodium chloride compressed the lipid layer thickness or prevented the swelling of the layers by the soap. Calculation showed that the phospholipid forms 'at least one bilayer' (10 to 20 nm) at the oil–water interface; thus the direct influence of NaCl would be to compress this hydrated layer and reduce its efficiency as a stabilizing barrier.

Rydhag's results on mean globule diameter and electrokinetic mobility of soybean-in-water emulsions obtained with a series of commercial soybean lecithins listed in Table 8.12 are shown in Fig. 8.48. The anticipated increase in negative zeta potential is noted with increase in anionic lipid content; the approximate relationship between anionic lipid and mean globule diameter is not so predictable. The relatively large size of the emulsion globules ($\sim 5\,\mu m$) compared with i.v. fat emulsions should be noted.

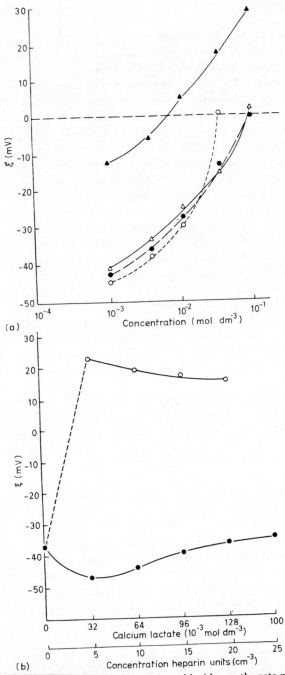

Figure 8.47(a) The effect of sodium and calcium chlorides on the zeta potential of various batches of Intralipid. ▲, Intralipid 10% (batch 196526), calcium chloride; △, Intralipid 10% (batch 191394), sodium chloride; ●, Intralipid 10% (batch 196526), sodium chloride; ○, Intralipid 20% (batch 298093), sodium chloride. (b) The effect of calcium lactate (○) and heparin (●) on the zeta potential of Intralipid 10% (batch 196526). From [246].

Table 8.12 The phospholipid composition of the commercially available soybean lecithins

Producer	Trade name	PC (mol %)	PE (mol %)	PI (mol %)	PS (mol %)	PA (mol %)	Lyso (mol %)	Anionic lipids (mol %)	P_{tot}* (wt %)
Ross & Rover Inc. N.Y., USA	YELKIN	41	34	19	—	6	—	25	49
Staley Manuf. Co, USA	STASOL 3318	34	10	29	—	12	15	41	65
Unimills, Hamburg, Germany	Bolec Z	33	29	24	—	14	—	38	51
Lucas Meyer, Hamburg, Germany	AZOL	45	19	11	25	—	—	36	41
Aarhus Oliefabr., Denmark	OKH	11	19	41	—	29	—	70	68
Unknown, Netherlands	SBP-BZ	33	32	21	—	14	—	35	78
Lucas Meyer, Hamburg, Germany	Epikuron-200, PC	100	—	—	—	—	—	—	98

* Total phospholipids in products calc. mol. wt = 750
PC = phosphatidyl choline, PE = phosphatidyl ethanolamine, PI = phosphatidyl inositol, PS = phosphatidyl serine, PA = phosphatidylic acid, Lyso = lysolecithin.
From [247].

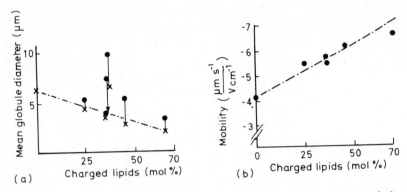

Figure 8.48(a) The mean globule diameter as a function of the amount of charged phospholipids in the emulsifiers in Table 8.12. × mean globule diameter, directly after the preparation. ● mean globule diameter after 5 days storage. (b) The electrophoretic mobility of the oil droplets as a function of the amount of negatively charged phospholipids in the emulsifiers shown in Table 8.12. From Rydhag [247].

8.7.3 Elimination of fat emulsions from the bloodstream

It is desirable that the fat particles in an artificial fat emulsion have the same biological properties as those of the natural chylomicrons in respect of their transport in the blood and distribution in the body. It is believed that the clearance of fat emulsions by the reticulo-endothelial system and their subsequent fate in the body is dependent on the surface characteristics of the droplets, in particular the nature of the emulsifying agent [248].

Scholler [249] reported that fat from some fat emulsions is accumulated in the reticulo-endothelial cells. This accumulation was found to depend on the composition of the emulsions and the size of the fat particles. The investigations were performed by electron microscope studies of liver biopsies from man and rats. After intravenous infusions of Lipofundin S the fat particles were taken up by the reticulo-endothelial cells by phagocytosis in both man and experimental animals. The reticulo-endothelial system was partly blocked, and antibody formation was significantly reduced. After infusion of Intralipid no accumulation of fat particles in the Kupffer's cells was observed.

No significant reduction in the formation of antibodies was found in guinea-pigs after Intralipid. Because of the impairment of the resistance of the body by the accumulations of particles in the reticulo-endothelial system, it has been stated that only those fat emulsions – such as Intralipid – which are not taken up by the reticulo-endothelial cells should be used clinically [249]. Hence the importance of the factors affecting uptake and clearance. Davis and Hansrani [250] have studied the phagocytosis of soybean oil emulsions prepared with emulsifiers which confer different surface characteristics onto the fat particles. They used model systems comparing the uptake of the fat particles by mouse peritoneal polymorphonuclear macrophages and by *Acanthamoeba castellanii*.

Addition of lysophosphatidylcholine (LPC) to an emulsion stabilized by a mixture of phosphatidylcholine and phosphatidyl ethanolamine (PC/PE) increased the charge on the droplets as well as the uptake by the phagocytosis systems. This uptake can be quantified by means of first order rate constant (Table 8.13). The results show clearly the importance of the nature of the surface layer. It is not clear whether the observed effects are due solely to surface charge, although there are correlations between uptake and microelectrophoretic mobility, or are a composite of surface layer properties and charge effects. However, it may be concluded that the manner in which emulsion droplets are handled by both phagocytic systems can be altered readily by small formulation changes.

Table 8.13 Phagocytosis of soybean oil emulsions (10% v/v)

Emulsifier (% w/v)	1st order rate constant (s^{-1})		Zeta Potential (mV)
	PMN	*Acanthamoeba*	
1.2% PC:PE (4:1)	1.56	0.14	−28.5
1.2% PC:PE (4:1) +0.1% LPC	7.80	0.33	−57.0
1% Pluronic F108	0.10	0.06	−12.2

From [250].

The Pluronic non-ionic block co-polymers (poloxamers) may be alternative emulsifiers to the phospholipids, although their nature can also alter the physical and biological behaviour of the fat emulsions. Fat emulsions stabilized by high molecular weight poloxamers were cleared more slowly by the blood than those prepared with low molecular weight polymers [251].

Jeppsson and Rossner [248] have also studied the influence of various emulsifying agents on the removal rate of soybean oil emulsions, containing egg yolk phosphatides, acetylated monoglycerides (Myvacets 9–40) and Pluronics (F68, F108). The addition of egg phosphatides to slowly eliminated emulsions promoted their elimination rates. Addition of Pluronic F68 did not significantly affect the elimination rate, whereas Pluronic F108 considerably impaired the removal of fat particles. It is possible that the surfactant prevents the lipid droplets from sticking to the blood vessel endothelium or interferes with lipoprotein–lipase activity which has been implicated in the removal of natural chylomicron triglycerides.

The influence of Myvacet on the course of particle elimination was not as dramatic as that of Pluronic F108, but a combination of Pluronic F108 and Myvacet produced emulsions with the lowest fractional removal rate, the results supporting the concept that the emulsifying agent is of major importance for the fractional removal rate of the emulsion from the blood stream. As with other investigations of emulsion systems, change of emulsifier leads to inevitable changes in particle size distribution and it is vital that this is taken into account in assessing the biological work, although the particle size distribution may well change *in vivo*, an added complication.

8.7.4 Biopharmaceutics of multiple emulsions and related systems

Herbert [252] made early use of multiple emulsions. Groups of mice were given ovalbumin in aqueous solution, W/O emulsion and W/O/W emulsion and their antibody response noted. It was found that mice treated with antigen in multiple emulsions exhibited a slightly better response than those treated with the same dose of antigen in the W/O emulsion and that this response was sustained. Herbert pointed out that W/O/W emulsions were easier to inject and formed diffuse depots after injection compared to W/O systems which, because of their generally high viscosity, were difficult to inject and formed discrete depots after injection, occasionally causing lesions.

Taylor et al. [253] extended the use of W/O/W vaccines to man and compared antibody levels and short-term reactions after administration of influenza vaccine in aqueous, W/O and W/O/W formulations. The W/O/W preparation was at least as effective as the W/O preparation. Short-term reactions to the multiple emulsion were more frequent than those associated with the aqueous and W/O preparations but these were found to be mild and the authors did not consider them to constitute a serious drawback to a more extensive use of the vaccine.

The potential of W/O/W systems for prolonged release of drugs rests in the presence of active ingredient in the disperse phase droplets. A water-soluble drug in an aqueous droplet thus has to diffuse out through the oil layer of the disperse droplet, although if the emulsion is not stable, drug will be released by the other mechanisms discussed earlier (Section 8.3). The presence of drugs and drug salts in the formulation may alter the stability of the emulsion and the appropriate surfactant concentrations and ratios have to be adjusted as not all of the primary emulsion is formed successfully into multiple droplets. In the second emulsification stage the efficiency of 'encapsulation' may vary. However, Collings [254] quotes examples of the possible entrapment of drugs including isoprenaline sulphate, sodium warfarin, chlorpromazine, methadone, benzylpenicillin and cephaloridine. Benoy et al. [255] incorporated various anti-cancer agents into the internal aqueous phase of W/O/W emulsions. It will be apparent that injection or ingestion of a W/O/W emulsion will rapidly result in the formation of a W/O emulsion in vivo; an O/W/O emulsion would be expected to retain its multiple phase characteristics. W/O/W and O/W/O emulsions as carriers for naltrexone and its hydrochloride have, in fact, been directly compared [256] although only in vitro.

In the case of W/O/W emulsions, a 70 % prolongation of release is obtained by adding sodium chloride or sorbitol to the internal aqueous phase. Compared to a solution of drug corresponding to the external aqueous phase of the system, no prolongation is seen in the multiple emulsion without any additives. Transfer of the drug through the external aqueous phase appears to be rate-limiting in a diffusion cell in which the formulation is separated from the sink compartment by a cellulose membrane.

In the release of naltrexone from O/W/O emulsions containing sodium chloride in the aqueous phase, a 70 % prolongation is also obtained compared to emulsions without sodium chloride. The effect decreased at higher concentrations of additive and is attributed to changes in the properties of the W/O interface [256].

Effective diffusion coefficients of the drugs in these emulsions and in their phases have been calculated from the equation

$$R = \frac{200A}{V} \sqrt{\frac{Dt}{\pi}} \tag{8.86}$$

where R is the percentage of drug released, A is the cross-sectional area of the small compartment in the diffusion cell, V is the volume of the formulation. Some results are presented in Table 8.14. The effective diffusion coefficient for the W/O emulsion is about 2.5 times greater than that for the O/W primary emulsion, but little different from that of the oil suspension. The O/W/O emulsion with the naltrexone dissolved in the internal oil phase has the same value of D as the corresponding primary O/W emulsion. From the results in Table 8.14, Brodin *et al.* [256] conclude that the release of naltrexone from multiple O/W/O systems is governed by transfer from the primary O/W emulsions and not by the external oil phase. It is apparent from these results that release is not significantly prolonged by any of these manipulations. Much will depend on the partition coefficient of the drug in question, but 70 % reductions in release rate are more easily obtained by other conventional pharmaceutical techniques. Additives can alter the observed values of D by changing the osmotic flow patterns, but again the scope is limited, the lowest value of D recorded by Brodin and his colleagues being approximately $0.4 \times 10^{-5} \, \text{cm}^2 \, \text{s}^{-1}$ for naltrexone in the presence of 2.5 % NaCl in the aqueous phase of a O/W/O emulsion. The diffusion coefficient of the drug in aqueous solution without surfactant is $1.62 \times 10^{-5} \, \text{cm}^2 \, \text{s}^{-1}$ and in the presence of 2 % of a propylene glycerol alginate it has been found to be $0.97 \times 10^{-5} \, \text{cm}^2 \, \text{s}^{-1}$. What then are the advantages of these systems? These probably lie not with prolongation of release but with protection of a sensitive component, e.g. insulin from gastric fluids, or with site directed therapy, e.g. preferential uptake of the system into the lymphatics.

Table 8.14 Effective diffusion coefficients of naltrexone in different systems. C_0 = initial concentration of drug.

No.	System	$D(10^5 \text{cm}^2 \text{s}^{-1})$	$C_0(10^2 \text{mol l}^{-1})$	Comments
1	oil suspension	1.3	3.6	no surfactant
2	oil suspension	1.8	1.8	5 % w/w Span 80
3	O/W emulsion	0.6	1.8	drug suspended in oil phase
4	W/O emulsion	1.5	7.3	4 % aq. phase, drug in oil phase
5	O/W/O emulsion	0.6	1.8	drug initially in internal oil phase, final emulsion formed by using the mixer
6	O/W/O emulsion	1.5	1.8	drug initially in both oil phases, final emulsion formed by using the mixer

From [256].

Any biologically active material in the innermost phase will be protected from the external environment. This opens up the possibility of delivering materials to the body which are normally inactivated when administered as solutions. The results of an interesting study by Shichiri *et al.* [257] indicate that multiple emulsions may be applied to the problem of the oral administration of insulin. W/O/W emulsions, *in vitro*, were found to be fairly resistant to the action of pepsin, chymotrypsin and trypsin, although they gradually lost their activity in the presence of pancreatic lipase. Definite responses were observed in about half of the animals tested following the administration of insulin in W/O/W emulsion. Apart from its role in the protection of the insulin from proteolytic destruction, the multiple emulsion appeared to facilitate the intestinal absorption of insulin (although the exact mechanism by which this occurred was not clear). The *in vivo* results were, however, rather variable (large variations in blood glucose and plasma insulin were found) and further work is required in this area. Rather different protective mechanisms appear to be operative in the effect of an emulsion formulation on the delivery of a bleomycin dextran complex to the lymphatics. Incorporation of the bleomycin dextran sulphate complex into W/O emulsions protected the complex from dissociation following intramuscular or intragastric administration; injection of a simple aqueous solution of the complex had not sufficient advantage over injection of the uncomplexed bleomycin [258]. Some results of lymph node uptake are shown in Fig. 8.49 showing the enhanced uptake of bleomycin as the complex administered in a W/O emulsion. One predicts a similar advantage for W/O/W emulsions but these were not studied. However, a comparison has been made by the same group, of water-in-oil and gelatin microsphere-in-oil emulsions [259], the latter showing a marked superiority over a W/O emulsion in its ability to transport iodohippuric acid to the regional lymph nodes after intragastric injection. It is difficult to understand why this should be so if the only difference between the W/O emulsion and the gelatin containing emulsion is that the aqueous phase has been replaced by microsphere but it is most likely due to the greater stability of the former. Both W/O and gelatin microsphere-in-oil emulsions, on injection to the body, will presumably disperse as multiple emulsions dependent on the concentration of surfactant present in the external phase.

Other evidence of the usefulness of multiple emulsions is present in the literature. Elson *et al.* [260] found the activity of methotrexate in leukaemic mice was enhanced when administered as a multiple emulsion.

Takahashi *et al.* [261] have used, with some success, intratumour injection of multiple emulsions of bleomycin and mitomycin in cases of inoperable carcinoma in human patients. Some of the impetus for the increased interest in multiple emulsions has come from the concept of liquid membranes for detoxification. Liquid membranes exist as part of large multiple oil droplets containing many internal dispersed droplets. The system was devised by Exxon Research and Engineering and they have coined the term liquid membrane capsule (LMC) (see Fig. 8.50). For use *in vivo* the LMCs must survive transit in the gut following oral administration. Asher *et al.* [262] have studied the mobility of LMCs through the

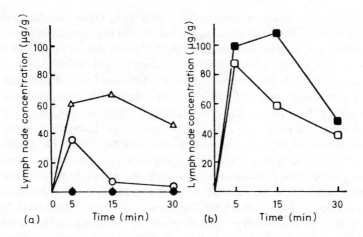

(a) Time (min)　　　(b) Time (min)

Figure 8.49 Lymph node concentration of bleomycin following intragastric injection of bleomycin or bleomycin–dextran sulphate in the form of (a) aqueous solution or (b) W/O emulsion.

(a) O: free bleomycin

　　△: bleomycin–dextran sulphate complex. (mol. wt d dextran sulphate, 5×10^5)

　　●: intravenous injection of aqueous solution of free bleomycin.

(b) □: free bleomycin

　　■: bleomycin–dextran sulphate complex

Injection volume was $50\,\mu l$ and concentration of injection vehicle was $30\,mg\,ml^{-1}$ for bleomycin and $100\,mg\,ml^{-1}$ for bleomycin–dextran sulphate complex. Results are expressed as the mean of two animal groups (10 rats). The water-in-oil emulsion comprised 77.8 % sesame oil, 1.5 % polyoxyethylated castor oil (E = 60) and 5.7 % sorbitan sesquioleate. From Muranishi *et al.* [258] with permission.

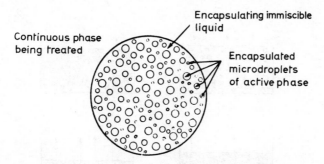

Figure 8.50 Liquid membrane capsule. The diameter of the capsule is 150 to 1000 μm. The diameter of encapsulated micro droplets is 1 to 5 μm. The encapsulated active phase can contain a catalyst, such as an enzyme; the LMC then function as 'reactors'. The reactants diffuse into the catalyst-active phase, and the products diffuse out. The encapsulated phase can also be a reagent. LMC encapsulating a reagent-active phase can be formulated to function as traps. Here the species to be removed diffuses from the solution being treated through encapsulating phase to the reagent-active phase. The reagent converts the material to a non-permeable species which cannot diffuse back through the encapsulating phase and becomes trapped [262].

upper gastro-intestinal tract into the colon of dogs by encapsulating meglumine diatrizoate in the LMC and following the progress of the capsules by fluoroscopy. The stability was studied by administering ten times the lethal dose of sodium cyanide in LMC to rats. LMCs were recovered unchanged in size; rats survived the lethal dose of NaCN. Intact non-coalesced LMC was found distributed in the stools. It seems surprising that LMCs can pass through the intestinal milieu with its phospholipids and bile salt and be unaffected. One clue might be in the use of isoparaffins as the oil phase, although this causes problems in relation to the transport of solutes such as urea through the liquid membrane. The consequence of this is that urease encapsulated in paraffin exhibits very low rates of hydrolysis. Replacing some of the paraffin with mono-olein increases the rate of hydrolysis as urea, the product of the reaction, can diffuse more readily away. Asher *et al.* [262] did not subject the mono-olein LMCs to the *in vivo* transit test, and it is doubtful if these would have survived as well. Triglyceride emulsification by amphipaths present in the intestinal lumen has been studied under conditions of low shear [263]. Bile salt alone can emulsify only small amounts of triolein but fatty acid salts and phosphatidyl choline alone produced emulsions of great stability. Mixtures of the three types of natural surfactant produced complex emulsification patterns. It is thus obvious that choice for the oil phase of LMCs is limited if the capsule is to be left untouched by interaction with natural surfactants. Emulsification of large LMCs would undoubtedly lead to loss of some percentage of the internal droplets and their contents.

8.7.5 Other factors affecting the use of emulsions

The optimal use of emulsion systems depends, as we have seen, on the maintenance of their stability in in-use conditions. One factor that has been

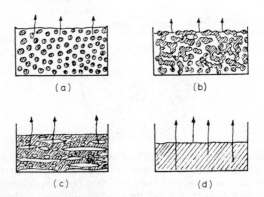

(a)　　　　(b)

(c)　　　　(d)

Figure 8.51 Possible stages of evaporation of thin emulsion layers (a) water evaporates freely from emulsion surface, (b) water evaporates from superficial pores while oil globules coalesce progressively, (c) water diffuses to the surface from droplets and/or lamellae beneath a superficial lipid layer, (d) evaporation of humectant- and/or surfactant-bound water. From Saettone *et al.* [264] with permission.

neglected in topical therapy is what changes occur in O/W emulsions spread onto the skin, for all theoretical approaches to topical therapy tend to assume that the system defined in the laboratory survives intact. Saettone and his colleagues [264] have recently investigated the changes in electrical impedance of thin layers of O/W emulsions during evaporation in controlled environments. Evaporation proceeds in distinct stages depicted in Fig. 8.51. Fig. 8.52 shows the influence of

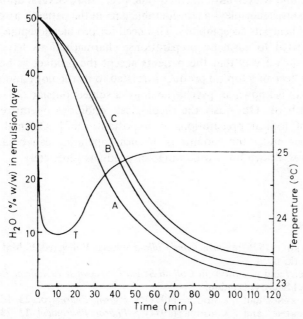

Figure 8.52 Water content versus time plots of the O/W emulsions under study (curves A, B and C) and temperature versus time plot of emulsion A (curve T). Temperature ordinate is on the right side. From Saettone *et al.* [264]. The emulsion composition is shown below.

Composition of the emulsion	A	B	C
Liquid paraffin, B.P.	15.10	14.60	14.60
Cetyl alcohol, B.P.	2.00	2.00	2.00
Isopropyl myristate	15.00	14.50	14.50
Glycerin monomyristate	2.00	2.00	2.00
Sorbitan monostearate	9.00	9.00	9.00
Polysorbate 60	6.50	6.50	6.50
Glycerin, B.P.	—	1.00	—
Sodium pyrrolidonecarboxylate¶	—	—	1.00
Methyl *p*-hydroxybenzoate	0.15	0.15	0.15
Propyl *p*-hydroxybenzoate *	0.25	0.25	0.25
Water**	50.00	50.00	50.00
Total	100.00	100.00	100.00

¶ Prepared in situ by neutralising an aqueous solution of the acid. ** Doubly distilled using an all-glass apparatus.

1 % glycerine (in emulsion B) 1 % sodium pyrrolidone carboxylate (in emulsion C) on water content as a function of time. Both additives are known humectants binding water and thus reducing the escaping tendency of the water molecules. Emulsion A differs from B and C only in the absence of these ingredients. This technique will prove useful in the study of the influence of other formulation factors on stability in thin films.

A not unrelated topic is that of the spreadability of cream and ointments to which Barry and Meyer have devoted time [265, 266]. Several authorities have noted that consistency plays a considerable part in the patient's assessment of a product and hence its acceptability. This need for product acceptability is by no means restricted to cosmetic preparations; pharmaceuticals have to be formulated in such a way that the patients accept the product as being of high quality. The 'feel' of a topical product is related to its rheology and this fact has spawned what is known as 'psychorheology' a subject outside the scope of this chapter and book. Obviously the rheological conditions operative during the spreading of topical preparations is important [267]. Optimal rheological characteristics under the conditions of shear operating can be assessed and 'master curves' drawn for a wide range of products [265, 266].

References

1. B. J. CARROLL (1976) in *Surface and Colloid Science*, Vol. 9, (ed. E. Matijevic), Wiley, New York, pp. 1–67.
2. B. VINCENT and S. S. DAVIS in *Colloid Science: Specialist Periodical Reports* Vol. 3, Chemical Society, London.
3. A. T. FLORENCE and J. A. ROGERS (1971) *J. Pharm. Pharmacol.* **23**, 153.
4. A. T. FLORENCE and J. A. ROGERS (1971) *J. Pharm. Pharmacol.* **23**, 233.
5. P. BECHER (1965) *Emulsions–Theory and Practice*, 2nd edn, Reinhold, New York.
6. A. KING (1940) *Trans. Faraday Soc.* **37**, 168.
7. M. KAREL (1975) in *Theory, Determination and Control of Physical Properties of Food Materials* (ed. C. Rha), Reidel, Boston, USA.
8. J. H. SCHULMAN and E. G. COCKBAIN (1940) *Trans. Faraday Soc.* **36**, 651; 661.
9. E. G. COCKBAIN and T. S. MCROBERTS (1953) *J. Colloid Sci.* **8**, 440.
10. E. G. COCKBAIN (1956) *J. Colloid Sci.* **11**, 575.
11. R. H. ENGEL and S. J. RIGGI (1969) *J. Pharm. Sci.* **58**, 3352.
12. R. P. GEYER (1960) *Physiol. Rev.* **40**, 150.
13. Todays Drugs (1970) *Br. Med. Journal* 352.
14. B. LYNN (1970) *J. Hosp. Pharm.* **28**, 71.
15. W. C. GRIFFIN (1949) *J. Soc. Cosmet. Chem.* **1**, 311.
16. J. T. DAVIES and E. K. RIDEAL (1963) *Interfacial Phenomena*, 2nd edn., Academic Press, London.
17. I. J. LIN (1971) *Trans AIME* **250**, 225.
18. I. J. LIN and P. SOMASUNDARAN (1971) *J. Colloid Interface Sci.* **37**, 731.
19. I. J. LIN (1972) *J. Phys. Chem.* **76**, 2019.
20. P. H. ELWORTHY and A. T. FLORENCE (1969) *J. Pharm. Pharmacol.* **21**, Suppl. 79S.
21. S. RIEGELMAN and G. PICHON (1962) *Am. Perfumer* **77**, 31.
22. R. M. E. RICHARDS and T. D. WHITTET (1955) *Pharm. J.* **175**, 141.
23. J. W. HADGRAFT (1954) *J. Pharm. Pharmacol.* **6**, 816.
24. F. A. J. TALMAN, P. F. DAVIES, E. M. ROWAN (1967) *J. Pharm. Pharmacol.* **19**, 417.

25. M. J. VOLD (1969) *J. Colloid Interface Sci.* **29**, 181.
26. I. RACZ and E. ORBAN (1965) *J. Colloid Sci.* **20**, 99.
27. R. C. LITTLE and C. R. SINGLETERRY (1964) *J. Phys. Chem.* **68**, 3457.
28. R. C. LITTLE (1975) *J. Colloid Interfac. Sci.* **51**, 200.
29. R. C. LITTLE (1978) *J. Colloid Interfac. Sci.* **65**, 587.
30. I. J. LIN and L. MARSZALL (1976) *J. Colloid Interfac. Sci.* **57**, 85.
31. A. T. FLORENCE, F. MADSEN and F. PUISIEUX (1975) *J. Pharm. Pharmacol.* **27**, 385.
32. H. SCHOTT (1969) *J. Pharm. Sci.* **58**, 1443.
33. A. DOREN and J. GOLDFARB (1970) *J. Colloid Interfac. Sci.* **32**, 67.
34. M. SEILLER, C. ARGUILLÈRE, P. DAVID, F. PUISIEUX and A. LE HIR (1968) *Ann. Pharm. Franc.* **26**, 557.
35. M. SEILLER, T. LEGRAS, F. PUISIEUX and A. LE HIR (1967) *Ann. Pharm. Franc.* **25**, 723.
36. M. SEILLER, T. LEGRAS and F. PUISIEUX (1970) *Ann. Pharm. Franc.* **28**, 425.
37. I. LO, T. LEGRAS, M. SEILLER, M. CHOIX and F. PUISIEUX (1972) *Ann. Pharm. Franc.* **30**, 211.
38. N. OHBA (1962) *Bull. Chem. Soc. Japan* **35**, 1021.
39. H. ARAI and K. SHINODA (1967) *J. Colloid Interfac. Sci.* **25**, 396.
40. K. SHINODA (1967) *J. Colloid Interfac. Sci.* **24**, 4.
41. P. BECHER (1958) *J. Soc. Cosmet. Chem.* **9**, 141.
42. K. SHINODA and H. SAITO (1969) *J. Colloid Interfac. Sci.* **30**, 258.
43. H. SAITO and K. SHINODA (1970) *J. Colloid Interface. Sci.* **32**, 647.
44. T. J. LIN, H. KURIHARA and H. OHTA (1977) *J. Soc. Cosmet. Chem.* **28**, 457.
45. P. DEPRAÉTERE, A. T. FLORENCE, F. PUISIEUX and M. SEILLER (1980) *Int. J. Pharmaceutics* **5**, 291.
46. I. LO, A. T. FLORENCE, J. P. TREGUIER, M. SEILLER and F. PUISIEUX (1977) *J. Colloid Interfac. Sci.* **59**, 319.
47. F. HARUSAWA and T. MITSUI (1975) *Progress in Organic Coatings* **3**, 177.
48. K. MADANI and S. FRIBERG (1978) *Prog. Colloid and Polymer Sci.* **65**, 164.
49. R. HOGG, T. W. HEALY and D. W. FUERSTENAU (1966) *Trans. Faraday Soc.* **62**, 1638.
50. N. F. H. HO and W. J. HIGUCHI (1968) *J. Pharm. Sci.* **57**, 436.
51. F. LONDON (1930) *Z. Physik* **63**, 245.
52. H. C. HAMAKER (1937) *Physica's Grav.* **4**, 1058.
53. J. I. GREGORY (1970) *Adv. Colloid Interfac. Sci.* **2**, 396.
54. M. J. VOLD (1961) *J. Colloid Sci.* **16**, 1.
55. R. H. OTTEWILL (1967) in *Non-Ionic Surfactants* (ed. M. J. Schick) Marcel Dekker, New York.
56. B. V. DERJAGUIN (1966) *Discuss. Faraday Soc.* **42**, 109.
57. B. V. DERJAGUIN (1966) *Discuss Faraday Soc.* **42**, 109; 317.
58. B. VINCENT (1973) in *Colloid Science: Specialist Periodical Reports* Vol. 1, The Chemical Society, London.
59. D. J. W. OSMOND, B. VINCENT and F. A. WAITE (1973) *J. Colloid Interfac. Sci.* **42**, 262.
60. P. BECHER, S. E. TRIFILETTI and Y. MACHIDA (1976) in *Theory and Practice of Emulsion Technology* (ed. A. L. Smith) Academic Press, London, p. 519.
61. B. V. DERJAGUIN (1939) *Trans. Faraday Soc.* **36**, 203.
62. D. M. NEWITT, R. K. DOMBROVSKI and F. KRELMAN (1954) *Trans. Inst. Chem. Eng.* **32**, 244.
63. E. J. W. VERWEY (1939) *Trans. Faraday Soc.* **36**, 192.
64. P. H. ELWORTHY and A. T. FLORENCE (1967) *J. Pharm. Pharmacol.* **19**, Suppl. 140S.
65. N. VAN DER WAARDEN (1950) *J. Colloid Sci.* **5**, 317.
66. J. LYKLEMA (1968) *Adv. Colloid Interfac. Sci.* **2**, 54.
67. E. W. FISCHER (1958) *Kolloidzeitschrift* **160**, 120.
68. P. H. ELWORTHY and C. MCDONALD (1964) *Kolloidzeitschrift* **195**, 16.
69. G. N. MALCOLM and J. S. ROWLINSON (1957) *Trans. Faraday Soc.* **53**, 921.
70. D. H. NAPPER (1968) *Trans. Faraday Soc.* **64**, 1701.
71. P. H. ELWORTHY and A. T. FLORENCE (1969) *J. Pharm. Pharmacol.* **21**, Suppl. 70S.

72. D. J. MEIER (1967) *J. Phys. Chem.* **71**, 1861.
73. R. H. OTTEWILL, T. WALKER (1968) *Kolloidzeitschrift* **227**, 108.
74. E. J. W. VERWEY (1966) *Trans. Faraday Soc.* **42**, 314.
75. F. MACRITCHIE (1967) *Nature*, **215**, 1159.
76. D. ATTWOOD and A. T. FLORENCE (1971) *Kolloidzeitschrift* **246**, 580.
77. P. H. ELWORTHY, A. T. FLORENCE and J. A. ROGERS (1971) *J. Colloid Interfac. Sci.* **35**, 23.
78. P. H. ELWORTHY, A. T. FLORENCE and J. A. ROGERS (1971) *J. Colloid Interfac. Sci.* **35**, 34.
79. S. W. SRIVASTAVA and D. A. HAYDON (1964) *Proc. IVth Int. Cong. Surface Activity*, Vol. 2, 1221.
80. B. BISWAS and D. A. HAYDON (1962) *Kolloidzeitschrift* **185**, 31.
81. TH. F. TADROS (1980) *Colloids and Surfaces* **1**, 3.
82. G. W. HALLWORTH and J. E. CARLESS (1970) in *Theory and Practice of Emulsion Technology* (ed. A. L. Smith) Academic Press, New York.
83. H. SONNTAG, J. NETZEL and B. UNTERBERGER (1970) *Spec. Discuss Faraday Soc.*
84. T. J. LIN, H. KURIHARA and H. OHTA (1975) *J. Soc. Cosmetic Chem.* **26**, 121.
85. T. J. LIN, H. KURIHARA and H. OHTA (1973) *J. Soc. Cosmetic Chem.* **24**, 797.
86. S. FRIBERG, L. MANDELL and M. LARSSON (1969) *J. Colloid Interfac. Sci.* **29**, 155.
87. S. FRIBERG and L. MANDELL (1970) *J. Pharm. Sci.* **59**, 1001.
88. B. R. VIJAYENDRAN and T. P. BURSH (1979) *J. Colloid Interfac. Sci.* **68**, 387.
89. P. BECHER (1963) *J. Colloid Sci.* **18**, 665.
90. P. A. SANDERS (1970) *J. Soc. Cosmet. Chem.* **21**, 377.
91. S. S. DAVIS, T. S. PUREWAL, R. BUSCALL, A. SMITH and K. CHOUDHURY (1976) *Colloid Interface Science*, Vol. 2, Academic Press, New York, p. 265.
92. B. W. BARRY (1971) *Rheol Acta* **10**, 96.
93. B. W. BARRY (1971) *Manufacturing Chemist*, April, 27.
94. B. W. BARRY and G. M. SAUNDERS (1970) *J. Colloid Interfac. Sci.* **34**, 300.
95. B. W. BARRY (1975) *Adv. Colloid Interfac. Sci.* **5**, 37.
96. B. W. BARRY (1974) *Adv. Pharm. Sci.* **4**, 1.
97. B. W. BARRY and B. WARBURTON (1968) *J. Soc. Cosmetic Chem.* **19**, 725.
98. B. W. BARRY (1970) *J. Texture Studies* **1**, 405.
99. B. W. BARRY and G. M. ECCLESTON (1973) *J. Texture Studies* **4**, 53.
100. B. W. BARRY and G. M. SAUNDERS (1971) *J. Colloid Interfac. Sci.* **35**, 689.
101. B. W. BARRY and G. M. SAUNDERS (1972) *J. Colloid Interfac. Sci.* **41**, 331.
102. H. M. PRINCEN (1979) *J. Colloid Interfac. Sci.* **21**, 55.
103. M. A. NAWAB and S. G. MASON (1958) *J. Colloid Sci.* **13**, 179.
104. J. A. ROGERS (1969) Ph.D. Thesis, University of Strathclyde.
105. K. J. LISSANT (1966) *J. Colloid Interfac. Sci.* **22**, 462.
106. K. J. LISSANT (1973) *J. Colloid Interfac. Sci.* **42**, 201.
107. K. J. LISSANT (1974) *J. Colloid Interfac. Sci.* **47**, 416.
108. V. N. PRIGORODOV, S. A. NIKITINA and A. D. TAUBMAN (1965) *Colloid J. USSR* **27**, 734.
109. W I. HIGUCHI and J. MISRA (1962) *J. Pharm. Sci.* **51**, 459.
110. S. S. DAVIS and A. SMITH (1976) in *Theory and Practice of Emulsion Technology* (ed. A. L. Smith), Academic Press, London p. 325.
111. R. BUSCALL, S. S. DAVIS and D. C. POTTS (1976) *Colloid and Polymer Sci.* **257**.
112. W. ALBERS and J. TH. G. OVERBEEK (1959) *J. Colloid Sci.* **14**, 501.
113. H. KOELMANS and J. TH. G. OVERBEEK (1954) *Discuss. Faraday Soc.* **18**, 52.
114. K. SHINODA and H. SAGITANI (1978) *J. Colloid Interfac. Sci.* **64**, 68.
115. J. BOYD, C. PARKINSON and P. SHERMAN (1972) *J. Colloid Interfac. Sci.* **41**, 359.
116. A. TAKAMURA, T. MINOWA, S. NORO and T. KUBO (1979) *Chem. Pharm. Bull.* **27**, 2921.
117. S. NORO, A. TAKAMURA and M. KOISHI (1979) *Chem. Pharm. Bull.* **27**, 309.
118. S. FRIBERG and I. WILTON (1970) *Am. Perfumer* **85** (12), 27.
119. H. SONNTAG and H. KLARE (1967) *Tenside* **4**, 104.

120. H. SONNTAG and J. NETZEL (1966) *Tenside* **3**, 296.
121. D. R. WOOD and K. A. BURRILL (1972) *J. Electroanalyt. chem.* **37**, 191.
122. H. SONNTAG, F. PÜSCHEL and BO. STROBEL (1967) *Tenside* **4**, 349.
123. H. SONNTAG, J. NETZEL and H. KLARE (1966) *Kolloid-Z.* **211**, 121.
124. H. SONNTAG (1968) *Tenside* **5**, 188.
125. H. KOMATSU, M. TAKAHASHI and S. FUKUSHIMA (1977) *Trans. Soc. Rheol.* **21**, 219.
126. S. CASADIO (1951) *Bull. Chim Farm* **90**, 277.
127. R. D. LUNDBERG, F. E. BAILEY and R. W. CALLARD (1966) *J. Polymer Sci.* **4**, 1563.
128. G. G. HAMMES and J. C. SWANN (1967) *Biochemistry* **6**, 1591.
129. S. M. LEVI and O. K. SMIRNOV (1959) *Colloid J. USSR* **21**, 315.
130. M. KAKO and S. KONDO (1979) *J. Colloid Interfac. Sci.* **69**, 163.
131. P. BECHER and S. TAHARA (1972) *Proc. IV th Int. Cong. Surface Activity* Zurich, Vol. 2(2), 519.
132. P. BECHER, S. E. TRIFILETTI and Y. MACHIDA (1976) in *Theory and Practice of Emulsion Technology* (ed. A. L. Smith), Academic Press, London, p. 235.
133. TH. F. TADROS (1976) in *Theory and Practice of Emulsion Technology* (ed. A. L. Smith), Academic Press, London.
134. J. B. KAYES (1976) *J. Colloid Interfac. Sci.* **56**, 426.
135. A. KAMEL, V. SABET, H. SADEK and S. N. SRIVASTAVA (1978) *Prog. Colloid and Polymer Sci.* **63**, 33.
136. W. SEIFRIZ (1925) *J. Phys. Chem.* **29**, 738.
137. N. N. LI and W. J. SOMERSET (1966) US Patent **3**, 410, 794.
138. N. N. LI and A. L. SHRIER (1972) in *Recent Developments in Separation Science*, CRC Press, Ohio, **1**, 163.
139. S. W. MAY and N. N. LI (1974) *Enzyme Eng.* **2**, 77.
140. A. F. BRODIN, D. R. KAVALIUNAS and S. G. FRANK (1978) *Acta Pharm Suec.* **15**, 1.
141. C. CHIANG, G. C. FULLER, J. W. FRANKENFELD and C. T. RHODES (1978) *J. Pharm. Sci.* **67**, 63.
142. D. WHITEHILL (1980) *Chem. Drug.* **213**, 135.
143. A. T. FLORENCE and D. WHITEHILL (1982) *Int. J. Pharmaceutics* **11**, 277.
144. P. SHERMAN (1963) *J. Phys. Chem.* **67**, 2531.
145. B. A. MULLEY and J. S. MARLAND (1970) *J. Pharm. Pharmacol.* **22**, 243.
146. S. S. DAVIS, T. S. PUREWAL and A. S. BURBAGE (1976) *J. Pharm. Pharmacol.* **28**, Suppl. 60P.
147. S. S. DAVIS and A. S. BURBAGE (1977) *J. Colloid Interfac. Sci.* **62**, 361.
148. S. MATSUMOTO, Y. KITA and D. YONEZAWA (1976) *J. Colloid Interfac. Sci.* **57**, 353.
149. D. R. KAVALIUNAS and S. G. FRANK (1978) *J. Colloid Interfac. Sci.* **66**, 586.
150. A. T. FLORENCE and D. WHITEHILL (1981) *J. Colloid Interfac. Sci.* **79**, 243.
151. D. WHITEHILL and A. T. FLORENCE (1979) *J. Pharm. Pharmacol.* **31**, Suppl. 3.
152. D. P. KESSLER and J. L. YORK (1970) *A.I Chem. E. J.* **16**, 369.
153. S. MATSUMOTO and M. KOHDA (1980) *J. Colloid Interfac. Sci.* **73**, 13.
154. P. SCHATZBERG (1965) *J. Polymer Sci.* **C10**, 87.
155. A. AL SADEN, A. T. FLORENCE and T. L. WHATLEY, unpublished results.
156. M. J. VOLD (1961) *J. Colloid Sci.* **16**, 1.
157. M. HASHIDA, T. YOSHIOKA, S. MURANISHI and H. SEZAKI (1980) *Chem. Pharm. Bull.* **28**, 1009.
158. S. MATSUMOTO, M. KOHDA and S. MURATA (1977) *J. Colloid Interfac. Sci.* **62**, 149.
159. A. T. FLORENCE and D. WHITEHILL (1980) *J. Pharm. Pharmacol.* **32**, Suppl. 64pp.
160. AL-SADEN, A. T. FLORENCE and T. L. WHATELEY (1980) *Int. J. Pharm.* **5**, 317.
161. A. A. AL-SADEN, A. T. FLORENCE and T. L. WHATELEY (1981) *Colloids and Surfaces* **2**, 49.
162. G. P. JACOBS (1980) *Int. J. Pharmaceutics* **4**, 299.
163. A. A. AL-SADEN, A. T. FLORENCE and T. L. WHATELEY (1980) *J. Pharm. Pharmacol.* **32**, Suppl. 5P.

164. N. J. MORRIS and B. WARBURTON (1982) *J. Pharm. Pharmacol.* **34**, 475.
165. D. R. RANCE and S. FRIBERG (1977) *J. Colloid Interface Sci.*, **60**, 207.
166. T. P. HOAR and J. H. SCHULMAN (1943) *Nature, Lond.* **152**, 102.
167. L. M. PRINCE (ed) (1977) *Microemulsions: Theory and Practice*, Academic Press, New York.
168. D. O. SHAH, A. TAMJEEDI, J. W. FALCO and R. D. WALKER (1972) *Amer. Inst. Chem. Eng. J.* **18**, 1116.
169. J. E. ZAJIC and C. J. PANCHAL (1976) *CRC Crit. Rev. Microbiology*, **39**.
170. C. HUH (1979) *J. Colloid Inter. Sci.* **71**, 408.
171. K. SHINODA and S. FRIBERG (1975) *Adv. Colloid Interfac. Sci.* **4**, 281.
172. J. H. SCHULMAN and J. D. MONTAGUE (1961) *Ann. NY Acad. Sci.* **92**, 366.
173. A. W. ADAMSON (1969) *J. Colloid Interface Sci.* **29**, 261.
174. L. M. PRINCE (1970) *J. Soc. Cosmetic Chem.* **21**, 192.
175. M. L. ROBBINS (1977) *Micellization, Solubilization and Microemulsions* (ed. K. L. Mittal) Vol. 2, Pharm. Press, New York, p. 755.
176. M. DVOLAITZKY, M. GUYOT, M. LAGUES, J. P. LE PESANT, R. OBER, C. SAVTERG and C. TAUPIN (1978) *J. Chem. Phys.* **69**, 3279.
177. E. RUCKENSTEIN and J. C. CHI (1975) *J. Chem. Soc. Faraday II* **71**, 1960.
178. E. RUCKENSTEIN (1977) *Micellization, Solubilization and Microemulsions* (ed. K. L. Mittal) Vol. 2, Pharm. Press, New York, p. 755.
179. E. RUCKENSTEIN (1978) *J. Colloid Interfac. Sci.* **66**, 369.
180. D. ATTWOOD, L. R. J. CURRIE, P. H. ELWORTHY (1974) *J. Colloid Interfac. Sci.* **46**, 249.
181. D. ATTWOOD, L. R. J. CURRIE and P. H. ELWORTHY (1974) *J. Colloid Interfac. Sci.* **46**, 255.
182. C. HERMANSKY and R. A. MACKAY (1980) *J. Colloid Interfac. Sci.* **73**, 324.
183. J. BIAIS, M. MERCIER, P. LALANNE, B. CLIN, A. M. BELLOCQ and B. LEMANCEAU (1977) *C. R. Acad. Sci. Paris*, C **285**, 213.
184. G. I. TAYLOR (1932) *Proc. Roy. Soc.* **A138**, 41.
185. A. LEVITON and A. LEIGHTON (1936) *J. Phys. Chem.* **40**, 71.
186. P. SHERMAN (1964) *J. Pharm. Pharmacol.* **16**, 1.
187. P. SHERMAN (1967) *J. Colloid Interfac. Sci.* **24**, 97.
188. P. SHERMAN (1967) *J. Colloid Interfac. Sci.* **24**, 107.
189. Y. KITA, S. MATSUMOTO and D. YONEZAWA (1977) *J. Colloid Interfac. Sci.* **62**, 87.
190. S. MATSUMOTO and P. SHERMAN (1969) *J. Colloid Interfac. Sci.* **30**, 525.
191. D. ATTWOOD, L. R. T. CURRIE and P. H. ELWORTHY (1974) *J. Colloid Interfac. Sci.* **46**, 261.
192. J. W. FALCO, R. D. WALKER and D. O. SHAH (1974) *A. I. Chem. E. J.* **20**, 510.
193. B. W. BARRY (1974) *Adv. Pharm. Sci.* **4**, 1.
194. B. W. BARRY (1975) *Adv. Colloid Interfac. Sci.* **5**, 37.
195. G. B. THURSTON and S. S. DAVIS (1979) *J. Colloid Interfac. Sci.* **69**, 199.
196. B. W. BARRY, G. M. SAUNDERS (1972) *J. Colloid Interfac. Sci.* **38**, 616.
197. B. W. BARRY (1970) *J. Colloid Interfac. Sci.* **32**, 551.
198. H. S. BEAN, G. KONNING and S. M. MALCOLM (1969) *J. Pharm. Pharmacol.* **21**, Suppl. 173S
199. T. SHIMAMOTO and H. MIMA (1979) *Chem. Pharm. Bull.* **27**, 2743.
200. G. KONNING (1974) *J. Pharm. Sci.* **9**, 103.
201. A. G. MITCHELL and S. J. A. KAZMI (1975) *Canad. J. Pharm. Sci.* **10**, 67.
202. A. G. MITCHELL and K. F. BROWN (1966) *J. Pharm. Pharmacol.* **18**, 115.
203. S. J. A. KAZMI and A. G. MITCHELL (1970) *J. Pharm. Pharmacol.* **23**, 482.
204. S. J. A. KAZMI and A. G. MITCHELL (1973) *J. Pharm. Sci.* **62**, 1299.
205. T. SHIMAMOTO, Y. OGAWA and N. OHKURA (1973) *Chem. Pharm. Bull.* **21**, 316.
206. P. B. MCNULTY and M. KAREL (1973) *J. Food. Technol.* **8**, 309.
207. P. B. MCNULTY and M. KAREL (1973) *J. Food Technol.* **8**, 319.

208. A. G. KOPP, R. J. MARR and F. E. MOSER (1978) *I. Chem. E. Symposium Series* **54**, 279.
209. N. N. LI (1978) *I. Chem. E. Symposium Series* **54**, 291.
210. E. S. MATULEVICIUS and N. N. LI (1975) *Separation and Purification Methods* **4**, 73.
211. W. H. FEINSTONE, R. WOLFF and R. O. WILLIAMS (1940) *J. Bact.* **39**, 47.
212. C. W. DAESCHNER, W. R. BELL, P. C. STIVRIUS, E. M. YOW and E. TOWNSEND (1957) *Med. Assoc. J. Dis. Child* **93**, 370.
213. J. G. WAGNER, E. S. GERARD and D. G. KAISER (1966) *Clin. Pharmac. Therap.* **7**, 610.
214. P. J. CARRIGAN and T. R. BATES (1973) *J. Pharm. Sci.* **62**, 1476.
215. R. H. ENGEL and M. J. FAHRENBACH (1968) *Proc. Soc. Exp. Biol. Med.* **129**, 772.
216. R. H. ENGEL and S. J. RIGGI (1969) *J. Pharm. Sci.* **56**, 1372.
217. R. H. ENGEL and S. J. RIGGI and M. J. FAHRENBACH (1968) *Nature* **219**, 856.
218. E. SCHRAVEN and D. TROTTNOW (1973) *Arzneim. Forsch* **23**, 274.
219. S. S. DAVIS (1976) *J. Chem. Pharm.* **1**, 11.
220. K. KAKEMI, H. SEZAKI, S. MURANISHI, H. OGATA and S. ISEMURA (1972) *Chem. Pharm. Bull.* **20**, 708.
221. H. S. BEAN and S. M. HERMAN-ACKAH (1964) *J. Pharm. Pharmacol.* **16**, Suppl. 58T.
222. H. NOGAMI, M. HANANO and H. YAMADA (1968) *Chem. Pharm. Bull.* **16**, 389.
223. Y. YAMAHIRA, T. NOGUCHI, T. NOGUCHI, H. TAKENAKA and T. MAEDA (1979) *Chem. Pharm. Bull.* **27**, 1190.
224. M. HAMOSH and R. O. SCOW (1973) *J. Clin. Invest.* **52**, 88.
225. Y. YAMAHIRA, T. NOGUCHI, T. NOGUCHI, H. TAKENAKA and T. MAEDA (1980) *Chem. Pharm. Bull.* **28**, 169.
226. T. R. BATES and J. A. SEQUEIRA (1975) *J. Pharm. Sci.* **64**, 791.
227. K. PALIN, S. S. DAVIS, A. J. PHILLIPS, D. WHATELAY and C. G. WILSON (1980) *J. Pharm. Pharmacol.* **32**, 611.
228. S. M. HEMAN-ACKAH and G. H. KONNING (1967) *J. Pharm. Pharmacol.* **19**, Suppl., 189S.
229. H. OGATA, K. KAKEMI, A. FURUYA, M. FUJII, S. MURANISHI and H. SEZAKI (1975) *Chem. Pharm. Bull.* **23**, 716.
230. H. OGATA, K. KAKEMI, S. MURANISHI and H. SEZAKI (1975) *Chem. Pharm. Bull.* **23**, 707.
231. J. D. HAMILTON (1971) *Biochim. Biophys. Acta* **239**, 1.
232. T. TANAKA, H. KOBAYASHI, K. OKUMURA, S. MURANISHI and H. SEZAKI (1974) *Chem. Pharm. Bull.* **22**, 1275.
233. T. NOGUCHI, C. TAKAHASHI, T. KIMURA, S. MURANISHI and H. SEZAKI (1975) *Chem. Pharm. Bull.* **23**, 775.
234. M. TAKEHARA and M. KOIKE (1977) *Yakugaku Zasshi* **97**, 780.
235. Y. NAKAMOTO, M. FUJIWARA, T. NAGUCHI, T. KIMURA, S. MURANISHI and H. SEZAKI (1975) *Chem. Pharm. Bull.* **23**, 2232.
236. Y. NAKAMOTO, M. HASHIDA, S. MURANISHI and H. SEZAKI (1975) *Chem. Pharm. Bull.* **23**, 3125.
237. T. NOGUCHI, Y. JINGUJI, T. KIMURA, S. MURANISHI and H. SEZAKI (1975) *Chem. Pharm. Bull.* **23**, 782.
238. R. JEPPSSON and S. LJUNGBERG (1972) *Acta Pharm. Suec.* **9**, 81; 199.
239. R. JEPPSSON and S. LJUNGBERG (1973) *Acta Pharm. Suec.* **10**, 129.
240. R. JEPPSSON and S. LJUNGBERG (1975) *Acta Pharm. Suec.* **36**, 312.
241. R. JEPPSSON (1975) *Acta Pharm. Toxic.* **36**, 299.
242. A. WRETLIND (1977) in *Current Concepts in Parenteral Nutrition* (eds J. M. Greep and P. B. Sveters) Nijhoff, The Hague.
243. G. I. SCHOEFL (1968) *Proc. Roy Soc. B* **169**, 147.
244. M. J. GROVES and H. S. YALABIK (1975) *Powder Technology* **12**, 233.
245. J. BOBERG and I. HAKANSSON (1964) *J. Pharm. Pharmacol.* **16**, 641.
246. C. R. T. KAWILARANG, K. GEORGHIOU and M. J. GROVES (1980) *J. Clin. Hosp. Pharm.* **5**, 151.
247. L. RYDHAG (1979) *Fette Seifen Anstrichmittel* **81**, 168.

248. R. JEPPSSON and S. ROSSNER (1975) *Act. Pharmacol. Toxicol.*, **37**, 134.
249. K. L. SCHOLLER (1968) *Z. prakt. Anästh Wiederbel* **3**, 193.
250. S. S. DAVIS and P. K. HANSRANI (1980) *J. Pharm. Pharmacol.* **32**, Suppl 61p.
251. R. P. GEYER (1974) *Bull Parenteral Drug. Assoc.* **28**, 88.
252. W. J. HERBERT (1965) *Lancet* **2**, 771.
253. P. TAYLOR, C. L. MILLER, T. M. POLLOCK, F. T. PERKINS and M. A. WESTWOOD (1969) *J. Hyg. Camb.* **67**, 485.
254. A. J. COLLINGS (1971) *Br. Patent* 1 235 667.
255. C. J. BENOY, L. A. ELSON and R. SCHNEIDER (1972) *Br. J. Pharmacol.* **45**, 135P.
256. A. F. BRODIN, D. R. KAVALIUNAS and S. G. FRANK (1978) *Acta Pharm. Suec.* **15**, 1.
257. M. SHICHIRI, Y. SHIMUZU, Y. YOSHIDA, R. KAWAMORI, M. FUKUCHI, Y. SHIZETA and H. ABE (1974) *Diabetologia* **10**, 317.
258. S. MURANISHI, Y. TAKAHASHI, M. HASHIDA and H. SEZAKI (1979) *J. Pharm. Dyn.* **2**, 383.
259. M. HASHIDA, Y. TAKAHASHI, S. MURANISHI and H. SEZAKI (1977) *J. Pharmac. Kinetics Biopharmac.* **5**, 241.
260. L. A. ELSON, B. C. V. MITCHLEY, A. J. COLLINGS and R. SCHNEIDER (1970) *Rev. Europe Etude Clin et Biol.* **15**, 87.
261. T. TAKAHASHI, S. VEDA, K. KONO and S. MAJIMA (1976) *Cancer* **38**, 1507.
262. W. J. ASHER, K. C. BOVÉE, J. W. FRANKENFIELD, R. W. HAMILTON, L. W. HENDERSON, P. G. HOLTZAPPLE and N. N. LI (1975) *Kidney International* **7**, 5409.
263. J. M. LINIHORST, S. B. CLARK and P. R. HOLT (1977) *J. Colloid Interfac. Sci.* **60**, 1.
264. M. F. SAETTONE, E. NANNIPIERI, L. CERRETTO, N. ESCHINI and I. V. CARELLI (1980) *Int. J. Cosmetic Sci.* **1**, 63.
265. B. W. BARRY and M. C. MEYER (1973) *J. Pharm. Sci.* **62**, 1349.
266. B. W. BARRY and M. C. MEYER (1975) *J. Texture Studies* **6**, 433.
267. B. W. BARRY and A. J. GRACE (1972) *J. Pharm. Sci.* **61**, 335.

9 *Surfactants in suspension systems*

9.1 Introduction

This short chapter deals primarily with the effect of surfactants on the properties of suspensions. Suspensions can be prepared without surfactants, there being an increasing use of polymers for stabilizing dispersions of solids in liquids. We will restrict discussion here to the effects of conventional surface-active agents but will deal also with the use of surface-active polymers such as the poloxamers, as these bridge the gap between two very large subjects. The emphasis of the chapter is towards pharmaceutical suspensions.

Pharmaceutical suspensions tend to be coarse rather then colloidal dispersions. The problems that arise when a solid drug is dispersed in a liquid include sedimentation and caking (leading to difficulty in resuspension), and flocculation and particle growth through dissolution and recrystallization. Adhesion of suspension particles to container walls has also been identified as a problem. The formulation of pharmaceutical suspensions requires that caking is minimized and this can be achieved by the production of flocculated systems. A flocculate or floc is a cluster of particles held together in a loose open structure. A suspension consisting of particles in this state is termed flocculated and we discuss states of flocculation and deflocculation. Unfortunately, flocculated systems clear rapidly and the preparation appears unsightly, so a partially deflocculated formulation is the ideal requirement. This viscosity of a suspension is obviously affected by flocculation.

9.2 Settling of suspended particles

In order to quantify the sedimentation of suspended particles, the ratio R of sedimentation layer volume (V_u) to total suspension volume (V_o)

$$R = \frac{V_u}{V_o} \approx \frac{h_\infty}{h_0} \tag{9.1}$$

may be used. A measure of sedimentation may also be obtained from the height of the sedimented layer (h_∞) in relation to the initial height of the suspension (h_0). In a completely deflocculated system the particles are not associated; particles settle under gravitational forces and the sediment layer builds up. The pressure on the individual particles can lead, in this layer, to close packing of the particles to such an extent that the secondary energy barriers are overcome and the particles become irreversibly bound together. In flocculated systems where the repulsive barriers have been reduced the particles settle as flocs and not as individual particles; the tendency is for the flocs to settle further; the supernatant clears but because of the random arrangement of the particles in the flocs the sediment is not compact. Caking, therefore, does not readily occur. In flocculated or concentrated suspensions, zone settling occurs.

9.3 Suspension stability

This is governed by the same forces as those described for other dispersion systems such as emulsions. There are differences, however, as coalescence obviously cannot occur in suspensions and the adsorption of polymers and surfactants also occurs in a different fashion. Flocculation, unlike coalescence, can be a reversible process and partial or controlled flocculation is attempted in formulation.

Caking of the suspension, which arises on close packing of the sedimented particles, cannot be eliminated by reduction of particle size or by increasing the viscosity of the continuous phase. Fine particles in a viscous medium settle more slowly than coarse particles but, after settling, they fit to a more closely packed sediment and frequently there is difficulty in redispersion. Particles in a close packed condition brought about by settling and by the pressure of particles above thus experience greater forces of attraction. Flocculating agents can prevent caking; deflocculating agents increase the tendency to cake. Use of flocculating and deflocculating agents is frequently monitored by measurement of the zeta potential of the particles in a suspension.

9.3.1 Zeta potential and its relationship to stability

Most suspension particles dispersed in water have a charge acquired by specific adsorption of ions or ionization of surface groups, if present. If the charge arises from ionization, the charge on the particle will depend on the pH of the environment. As with other colloidal particles, repulsive forces arise because of the interaction of the electrical double layers on adjacent particles. The magnitude of the charge can be determined by measurement of the electrophoretic mobility of the particles in an applied electrical field.

The velocity of migration of the particles under unit applied potential is μ_E. For non-conducting particles, the Henry equation is used to obtain the zeta potential, ζ, from μ_E. This equation can be written in the form,

$$\mu_E = \frac{\zeta \varepsilon}{4\pi\eta} f(\kappa a), \tag{9.2}$$

where $f(\kappa a)$ varies between 1, for small κa, and 1.5, for large κa. ε is the dielectric constant of the continuous phase and η is its viscosity. In systems with low values of κa the equation can be written in the form

$$\mu = \frac{\zeta \varepsilon}{4\pi\eta}. \tag{9.3}$$

After conversion to the appropriate units $\zeta = 14.08\,\mu$ (in millivolts). As we have seen, the zeta potential (ζ) is not the surface potential (ψ_0) discussed earlier but is related to it. Factors which alter ζ will affect ψ_0 and therefore ζ can be used as a reliable guide to the magnitude of electric repulsive forces between particles. Changes in ζ on the addition of flocculating agents, surfactants and other additives can then be used to predict the stability of the system. Martin [1] has described the changes in a barium subnitrate suspension system on addition of dibasic potassium phosphate as flocculating agent (Fig. 9.1). Bismuth subnitrate has a positive zeta potential; addition of phosphate reduces the charge and the zeta potential falls to a point where maximum flocculation is observed. In this zone there is no caking. Further addition of phosphate leads to a negative zeta potential and a propensity towards caking. Flocculation can therefore be controlled by the use of ionic species with a charge opposite to the charge of the particles dispersed in the medium.

The rapid clearance of the supernatant in a flocculated system is undesirable in a pharmaceutical suspension. The use of thickeners such as tragacanth, sodium carboxymethylcellulose or bentonite hinders the movement of the particles by

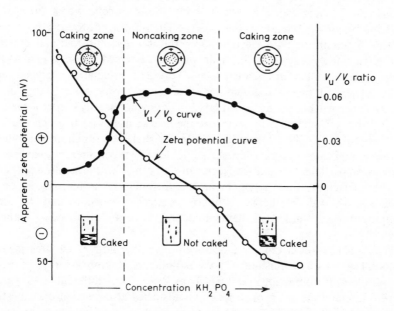

Figure 9.1 Diagram, from Martin [1], demonstrating the effect of the addition of the flocculating agent KH_2PO_4 on the flocculation of bismuth subnitrate, as measured by the sedimentation volume (V_u/V_o). The changes in zeta potential on addition of the flocculating agent are also shown as are the caking and noncaking zones.

production of a viscous medium, so that sedimentation is slowed down. The incompatibility of these anionic agents with cationic flocculating agents has to be considered. Martin has suggested a technique to overcome the problem – the conversion of the particle surfaces into positive surfaces so that they require anions and not cations to flocculate them. Negatively charged or neutral particles can be converted into positively charged particles by addition of a surface-active amine. Such a suspension can then be treated with phosphate ions to induce flocculation.

It is perhaps not surprising that with some complex systems the interpretation of behaviour differs and is open to debate. Consider the system shown in Fig. 9.2. One starts with a clumped suspension of sulphamerazine, a flocculated system which produces non-caking sediments. Addition of sodium dioctylsulpho-succinate confers a greater negative charge on the suspension particles and deflocculation results. Addition of aluminium chloride as a flocculating agent reduces the negative charge in a controlled way to produce the loose clusters illustrated in the diagram. These are the observable results of these procedures. The different interpretations (of Haines and Martin [2], on the one hand, and of Wilson and Ecanow [3], on the other) are diagrammatically realized. The difference lies in the manner in which the aluminium ions adsorb on to the sulphamerazine particles; in Haines' view they adsorb directly; in the other view they interact with the surfactant ions on the surface.

The attractive forces between suspension particles are considered to be exclusively London–van der Waals' interactions (except where interparticle bridging by long polymeric chains occurs). The repulsive forces, as discussed in Chapter 8, comprise both electrostatic repulsion and entropic and enthalpic forces. In aqueous systems the hydrophobic dispersed phase is coated with hydrophilic surfactant or polymer. As adsorption of surfactant or polymer (or, of course, both) at the solid–liquid interface alters the negative charge on the suspension particles, the adsorbed layer may not necessarily confer a repulsive effect. Ionic surfactants may neutralize the charge of the particles and result in their flocculation. The addition of electrolyte such as aluminium chloride can further complicate interpretation of results; electrolyte can alter the charge on the suspension particles by specific adsorption, and can affect the solution properties of the surfactants and polymers in the formulation. Some aspects of the application of DLVO theory to pharmaceutical suspensions and the use of computer programmes to calculate interaction curves are discussed by Schneider et al. [4].

There is not yet a complete and definitive thermodynamic treatment of colloidal systems which includes all the parameters of importance. In particular the problem of interactions between many particles of different sizes, particle solvent interactions in the presence of an adsorbed layer, etc. Several questions remain unresolved, e.g. the question of whether surface potentials or surface charges are modified as particles approach each other [5]. As the primary effect of surfactants on suspension stability is effected through adsorption and the modification of the properties of the interface, rather than on effects on bulk

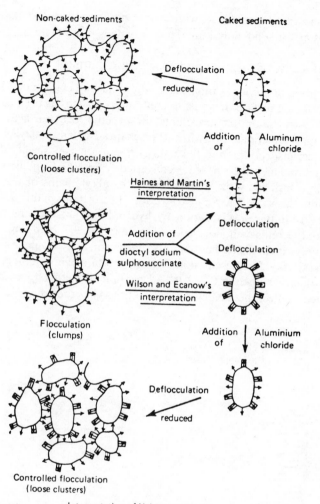

Non-caked sediments

Caked sediments

Controlled flocculation
(loose clusters)

Deflocculation
reduced

Addition
of

Aluminum
chloride

Haines and Martin's
interpretation

Addition of
dioctyl sodium
sulphosuccinate

Deflocculation

Deflocculation

Wilson and Ecanow's
interpretation

Flocculation
(clumps)

Addition
of

Aluminium
chloride

Deflocculation

reduced

Controlled flocculation
(loose clusters)

Interpretation of Haines and Martin

Bonds between particles in
flocculated sulphamerazine
suspension.

Aluminium adsorbed
on to sulphamerazine.

Controlled amount of alu
minium ions produces co
trolled flocculation.

Interpretation of Wilson and Ecanow

Dioctyl sodium sulpho-
succinate anions adsorbed
on to sulphamerazine.

Aluminium ions react with
dioctyl sodium sulphosuccinate anions.

Controlled amount of
aluminium ions produces
controlled flocculation.

Figure 9.2

viscosity (as can be the case with many emulsion systems) adsorption of surfactants at the solid–liquid interface is treated separately.

9.3.2 Surfactant adsorption and suspension stability

The adsorption of anionic, cationic and non-ionic surfactants on to hydrophilic and hydrophobic surfaces in aqueous continuous phases has been considered in Chapter 1. Adsorption of surfactants on to non-polar surfaces occurs via hydrophobic interactions, the hydrocarbon chain adsorbing and lying close to the solid surface; adsorption on to polar surfaces can occur by specific electrostatic interactions in which the surface is converted from a hydrophilic surface to a hydrophobic surface by the orientation of the alkyl chains of the surfactants outward into the water (Fig. 9.3). Adsorption of surfactants in this way frequently gives rise to multilayer adsorption by hydrophobic interactions between the primary and secondary monolayers, as shown in Fig. 9.3. Non-ionic surfactants based on polyoxyethylene ethers may also adsorb on to hydrophilic surfaces such as silica in this way. A representation of the orientation of non-ionic surfactants at a silica surface is shown in Fig. 9.3b. Adsorption isotherms for polar and non-polar systems reflect these different possibilities as has been discussed previously (section 1.4).

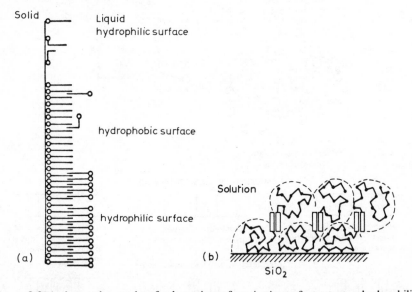

Figure 9.3(a) shows the mode of adsorption of an ionic surfactant at a hydrophilic surface, with its polar head groups close to the surface and non-polar chains pointing out into the aqueous phase. As one proceeds down the diagram, the bulk solution concentration of the surfactant is being increased until bilayers are formed by hydrophobic associations of the alkyl chains of the surfactant, rendering the surface hydrophilic. (b) shows the possible mode of association of non-ionic alkyl polyoxyethylene ethers at a hydrophilic silica surface at a concentration where bilayers are forming. From Rupprecht [6] with permission.

The influence of the change of surface charge on stability as the zeta potential goes through zero with increasing concentration of surfactant is shown in Fig. 9.4 along with the zeta potential data [7].

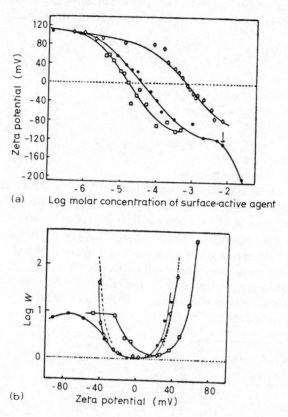

(a)

(b)

Figure 9.4(a) Zeta potential as a function of the logarithm of the concentration of anionic surfactants adsorbed on to polystyrene latex; □ sodium tetradecyl sulphate, ● NaDS, and ◇ sodium decyl sulphate. The CMC for NaDS in water is shown by the arrow. (b) Shows log W against zeta potential for these systems. The theoretical curve ---------- has been calculated from Equation (9.5 to 9.7) using the following parameters $a = 10$ nm, $\tau = 1.23$ and $A = 3 \times 10^{-19}$ J. From Ottewill and Watanabe [9] with permission.

The total energy V_t of interaction between the double layers on adjacent solid particles is the sum of the attractive (V_a) and repulsive (V_r) contributions. It is this potential V_t which retards the process of rapid coagulation by a factor W, the stability ratio of Fuchs [8], given by,

$$W = 2a \int_{2a}^{\infty} [\exp(V_t/kT)] \frac{dH}{H^2}, \tag{9.4}$$

where a is the particle radius, H is the distance between the particle centres, k is the

Boltzmann constant, and T is the absolute temperature. Notice that as W decreases, rapid coagulation occurs.

Following the treatment of Ottewill and Watanabe [9–11], the total energy V_t of interaction of double layers can be expressed in terms of the Stern potential ψ_δ to yield an expression for the stability ratio in terms of ψ_δ, namely,

$$\ln W = \frac{A}{24kT}\left[\frac{D^2\psi_\delta^2}{(1+2\kappa a)} - D\psi_\delta\right] - \tfrac{3}{4}\ln(D\psi_\delta)^2 + B, \qquad (9.5)$$

where

$$D^2 = \frac{12(2\kappa a + 1)\varepsilon a}{A} \qquad (9.6)$$

and

$$B = \ln 2\sqrt{\pi} - \frac{1}{2}\ln\frac{A}{(6kT)} + \frac{A}{24kT}, \qquad (9.7)$$

where A is the Hamaker constant.

Adsorption of ionic surfactants on to polar substrates will be affected by pH. The relationship between the amount of NaDS adsorbed on to colloidal alumina and the stability ratio, W, of alumina dispersions is shown in Fig. 9.5 at two different pH values, 6.9 and 7.2. The zero point of charge of alumina is pH 9.1 thus the higher surface charge at pH 6.9 ensures that hydrocarbon chain interactions occur at a lower surfactant concentration than at pH 7.2. As the pH is lowered further this effect is accentuated [12]. The effect of pH on adsorption of lauryl sulphate ions and tetradecylpyridinium ion is shown clearly in Rupprecht's [6] results using colloidal titanium dioxide as the adsorbate (Fig. 9.6).

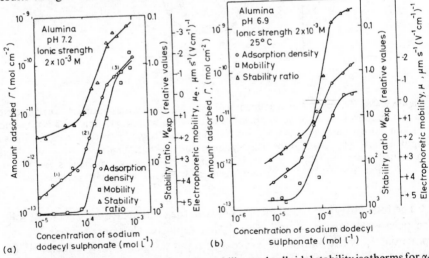

Figure 9.5(a) Adsorption, electrophoretic mobility, and colloidal stability isotherms for α-alumina at pH 7.2, 2×10^{-3} M ionic strength, and 25° C. as a function of the equilibrium concentration of sodium dodecyl sulphonate. (b) Adsorption, electrophoretic mobility, and colloidal stability isotherms for α-alumina at pH 6.9, 2×10^{-3} M ionic strength, and 25°C. as a function of the equilibrium concentration of sodium dodecyl sulphonate. From [12] with permission.

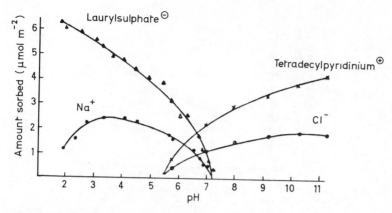

Figure 9.6 Adsorption of ionic surfactants on to colloidal TiO_2 as a function of the pH of the aqueous dispersion. From [6] with permission.

The nature of drug particle surfaces is rarely as clearly defined as those of alumina or titanium dioxide. The multiplicity of charged groups and the orientations of the molecules at the surface of disparately shaped particles leads to some difficulty in predicting the properties of drug suspensions.

With polymer dispersions a range of polarities can be achieved by suitable choice of monomer and the relation between surface polarity and surfactant adsorption can be studied [13]. Defining polarity, X_p, as γ^p/γ where γ^p is the polar contribution to polymer surface tension and γ is the surface tension of the polymer, Vijayendran [13] has found that the logarithm of the area per molecule of NaLS at the polymer surface increases with the increase in polarity of the polymer surface. Some of his data are shown in Fig. 9.7. These results account for, at least in part, the low aggregate stability of polar emulsions commonly encountered in practice, as the increased area per molecule on the more polar polymers indicates that there will be decreased stability against flocculation. In none of these systems were there ionizable groups at the surface; all are therefore hydrophobic systems where adsorption is likely to be monomolecular.

The degree of flocculation of sulphamerazine suspensions has been determined as a function of surfactant (sodium dodecyl polyoxyethylene sulphate) concentration and electrolyte concentration [14]. At low surfactant levels the system is flocculated, with sedimentation volume fractions of about 0.9; at higher levels the sedimentation volume falls to about 0.6. On the addition of 0.24 M NaCl a dramatic decrease in sedimentation volume is observed, indicating a dense sediment with cloudy supernatant. Further addition of NaCl results in re-establishment of a flocculated state perhaps by adsorption of ions on to the particles.

Specific interactions between surfactant ions and other additives in the system will sometimes significantly alter the behaviour of the particles. Starch and related materials even in concentrations as low as several parts per million can affect the adsorption of sodium oleate and other fatty acids on to minerals such as calcite

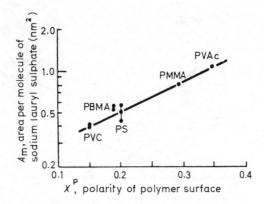

Figure 9.7 Plot of log A_m against calculated polymer polarity (X^P).
PVAc: poly(vinyl acetate)
PMMA: poly(methyl methacrylate)
PVC: poly(vinyl chloride)
PS: poly(styrene)
PBMA: poly(butyl methacrylate)
From [13] with permission.

[15]. Results obtained for the adsorption of oleate and starch on calcite as a function of the concentration of the other showed the existence of mutual enhancement of adsorption, possibly due to the formation of a helical starch–oleate clathrate with the oleate molecules held inside the helix [15]. The effect of the oleate would then be obscured even though its apparent adsorption was increased.

Interactions between an anionic surfactant used as a wetting agent for sulphamerazine and a cationic polymer have been the subject of investigations by Zatz *et al.* [16]. Surfactants have rarely been included in studies of flocculation by polymeric materials, yet they are usually present in formulations. Suspensions containing sufficient polysorbate 40 or dioctyl sodium sulphosuccinate to ensure complete wetting of the particles were deflocculated. In the presence of a commercial cationic flocculating polymer (Primafloc C-3, Rohm and Haas) flocculation was achieved over a limited range of polymer concentrations. The amount of polymer required to induce flocculation in systems containing the non-ionic surfactant was about 1 % of that needed in the suspensions with the anionic agent. Mixtures of the two surfactants at the 0.2 % level were used in different ratios, and a log–log plot of concentration of polymer at the point of maximum flocculation, C_p, versus the weight fraction of anionic surfactant was linear, demonstrating the feasibility of manipulating surfactant content to achieve the desired properties in the suspensions.

Cationic surfactants can reverse the charge on polystyrene latex particles which are initially negatively charged [17, 18]. One can expect the same principles to operate with cationic adsorbant molecules as with anionic agents. Some anomalous Langmuirian isotherms, however, have been reported for cetylpyridinium chloride and dodecyl-dimethyl-3,4-dichlorobenzylammonium chloride–colloidal silica systems [19]. Plots of amount adsorbed versus solution concentration reveal a levelling off of adsorption at the CMC of the surfactants. The adsorption of cetyltrimethylammonium bromide (CTAB) from mixed surfactant solutions (Fig. 9.8) on to kaolinite reveals a marked reduction in cation adsorption. Clay particles tend themselves to be more complex adsorbants because of the nature of their crystal structure (see Section 1.4.2). CTAB probably adsorbs by ion-exchange on to the negatively charged sites of the kaolinite crystal face with hydrophobic multilayer formation, as described before. However, at around 0.05 mol kg⁻¹ adsorbed CTAB, the discontinuity in the isotherm is probably due to further adsorption on the positive edges of the crystal. Competitive adsorption occurs in the presence of anionic surfactant, but the competition of non-ionic surfactants only becomes marked when the negative charge on the particle surfaces has been neutralized.

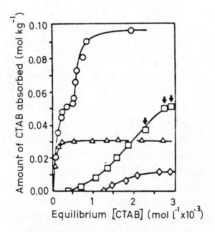

Figure 9.8 The adsorption of CTAB on to H-kaolinite at 298 K. ○ from water; □ from 0.01 mol dm⁻³ sodium dodecyl sulphate (NaDS); ◇ from 0.10 mol dm⁻³ NaDS; △ from 0.2 % w/v Triton X-100. Arrows indicate turbid mixtures. From [20] with permission.

(A) NON-IONIC SURFACTANTS

Flocculation and deflocculation of montmorillonite (Veegum) suspensions by a non-ionic surfactants has been studied by Ohno *et al.* [21]. Adsorption isotherms of the surfactant on to the clay showed points of inflexion due to bimolecular adsorption. Maximum flocculation occurred when the surface was covered by a layer of surfactant oriented with the alkyl chain pointing out into the aqueous phase. The deflocculation which occurs on further addition of surfactant is due to the hydrophilization of the surface by the second layer of surfactant molecules. The viscosity of the clay suspension reaches a maximum at the point of maximum monolayer coverage and then falls on further addition of surfactant. Fig. 9.9

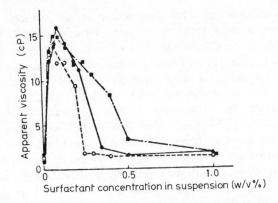

Figure 9.9 Apparent viscosities of 0.5 % veegum suspensions in the presence of non-ionic surfactants which are polyoxyethylated hydrogenated castor oil derivatives with the ethylene oxide chain lengths: ■ E = 30 ● E = 50 ○ E = 80 units. From Ohno *et al.* [21].

shows results for the apparent viscosity of 0.5 % Veegum suspensions in the presence of surfactants with increasing polyoxyethylene chain length. A number of other authors have dealt with the interaction of montmorillonites with non-ionic alkyl and alkylbenzene polyoxyethylene glycol ethers [22–24]. Depending on the surfactant concentration, monolayers or bilayers of molecules 0.38 to 0.43 nm and 0.84 to 0.85 nm thick were observed but analytically determined uptakes and X-ray data were not in agreement, an assumption having to be made that surfactant was adsorbed in the interstitial layers of the clay.

In the internal surfaces of the clay layers surfactant has to compete not only with solvent molecules (as it does on the externally exposed surfaces) but also with the attraction between the layers. The strong interactive forces between these surfaces will constrain the adsorption and orientation of the surfactant molecules [25]. A detailed study of the orientation of non-ionic surfactants on montmorillonite has been made by Platikanov *et al.* [25] in experiments where the interlayer cations were replaced by *n*-hexadecylammonium ions in order to decrease the electrostatic bonding energy between adjacent silicate layers.

Isotherms for the adsorption of non-ionic alkyl polyoxyethylene glycol ethers on polystyrene latex [26] or griseofulvin [27] show no evidence of bilayer formation, the surface reaching saturation around the region of the CMC of the surfactants. Adsorption of the surfactants onto griseofulvin decreases with increasing hydrophilic chain length and for a given surfactant, cetomacrogol, has been found to decrease with increasing temperature [27] (Fig. 9.10). The sensitivity of adsorption to temperature thus makes the interpretation of temperature-induced flocculation studies which assume a constant surface composition (see later) somewhat difficult. The opposite effect is noted for the adsorption of polyoxyethylene glycol 400. At first sight this is the expected result: that on increasing temperature the surfactant becomes more hydrophobic and more surface active. The results show that the area occupied by each molecule of

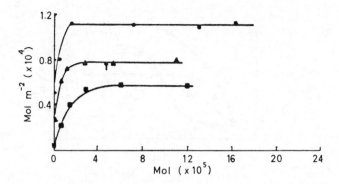

Figure 9.10 Effect of temperature on the adsorption of cetomacrogol on griseofulvin: ●, 288 K; ▲, 298 K; ■, 308 K. From Elworthy and Guthrie [27].

cetomacrogol and several other non-ionics increases as the temperature increases, in the case of the former compound from 1.48 nm² at 288 K to 3.43 nm² at 308 K. A non-ionic surfactant of a different class, a pentaerythritol mono-n-octyl ether, seems to form multilayers on griseofulvin. On polystyrene latex the molecules of $C_{12}E_6$ are vertically oriented with a saturation area per molecule of 0.4 nm² compared with a value of 0.6 nm² at the air–water interface. Sedimentation studies suggest an adsorbed layer thickness of about 5 nm ($\pm$ 1 nm) in agreement with the fully extended length of 4 nm. On graphon the area per molecule of this surfactant is somewhat greater, at 0.55 nm² [28].

Model systems such as polystyrene latex have found much favour among colloid scientists. Formulators have to deal with more complex materials with inhomogeneous surfaces. Kayes [29, 30] has extended the range of substances studied in detail using thiabendazole, nalidixic acid and betamethasone in addition to griseofulvin in an investigation of electrophoretic mobility, surfactant adsorption and suspension stability. Betamethasone shows little variation of zeta potential with pH as might be expected from its structure. As an example of the variation in charge with pH, nalidixic acid can be considered. At a pH of 3 the zeta potential is $+ 30$ m V; charge reversal occurs at pH 4.9 and at pH 7.0 has reached a 'steady state' value of -22 mV. Such changes would have some biological relevance as flocculation of suspensions *in vivo* can be a factor determining the bioavailability of suspended drugs. Above pH 8 the drug becomes quite soluble, a complication not encountered with latices. Fig. 9.11 shows the pH–zeta potential profiles obtained by Kayes.

Non-ionic surfactants were found to be more strongly adsorbed on to griseofulvin and thiabendazole than the other drugs studied. At the neutral pH studied all had a negative charge, griseofulvin, thiabendazole and betamethasone by adsorption of OH^- ions and nalidixic acid by ionization of the COO^- groups on the surface. There seems to be evidence that in the adsorption of cationic surfactant on to nalidixic acid particles the first stage is the coulombic attraction of the cationic heads to the ionic groups, this being followed by hydrophobic

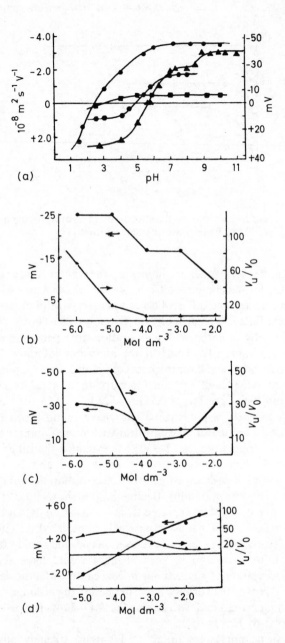

Figure 9.11(a) pH–mobility (10^{-8} m^2 s^{-1} V^{-1}) zeta potential (mV). ●—griseofulvin, ▲—thiabendazole, ●—nalidixic acid, ■—betamethasone. (b) Zeta potential (mV)—●, sedimentation volume (V_u/V_o %)—▲, thiabendazole—$\log_{10}$ concentration $C_{16}E_{30}$ (mol dm^{-3}). (c) Zeta potential (mV)—●, sedimentation volume (V_u/V_o %)—▲, griseofulvin—$\log_{10}$ concentration $C_{16}E_{30}$ (mol dm^{-3}). (d) Zeta potential (mV)—●, sedimentation volume (V_u/V_o %)—▲, thiabendazole—$\log_{10}$ concentration C_{12}TAB (mol dm^{-3}). From Kayes [29, 30] with permission.

adsorption resulting in reversal of charge, an effect which is moderated by the simultaneous adsorption of non-ionic surfactants. In the absence of non-ionic surfactant, the maximum zeta potentials found at 2.5×10^{-2} mol 1^{-1} $C_{12}TAB$ were griseofulvin $+55$ mV, betamethasone $+46$ mV, thiabendazole $+44$ mV, and nalidixic acid $+25$ mV. Plots of sedimentation volume for several of these systems do not at first sight accord with the idealized diagram in Fig. 9.11. Results for thiabendazole and griseofulvin are shown in Fig. 9.11b, c and d.

The effect of the non-ionic surfactant $C_{12}E_6$ on the stability of polystyrene latex as a function of electrolyte concentration and pH is shown in Fig. 9.12. At a constant surfactant concentration at pH 4.6 the value of $\log W_{exp}$ decreases gradually with increasing concentration of lanthanum nitrate until a concentration is reached where the curve becomes parallel to the concentration axis. This corresponds to the region of rapid flocculation. The portion below the critical concentration for rapid flocculation is the region of slow flocculation. The plots in Fig. 9.12 have been normalized to show more clearly the effect of increasing

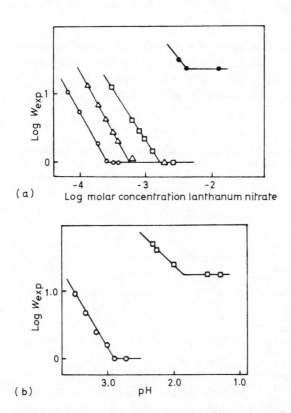

Figure 9.12(a) $\log W_{exp}$ against log molar concentration of lanthanum nitrate for latex A at pH 4.6. Temperature $20°$ C. $\bigcirc$, $C_{12}E_6$ absent; $\triangle$, 5×10^{-6} M $C_{12}E_6$; $\square$, 10^{-5} M $C_{12}E_6$; $\bullet$, 2×10^{-5} M $C_{12}E_6$ (b) $\log W_{exp}$ against pH for latex A at $20°$ C. $\bigcirc$, $C_{12}E_6$ absent; $\square$, 2×10^{-5} M $C_{12}E_6$. From Ottewill and Walker [26] with permission.

surfactant concentration. At the highest concentration studied the dispersion could not be flocculated, yet complete monolayer coverage had not been attained at that point. Lowering of pH did not induce flocculation even when the pH was decreased to 0.5 units. Table 9.1 lists the flocculation concentrations of the lattices at several $C_{12}E_6$ levels.

Table 9.1 Flocculation concentrations obtained with lanthanum nitrate at pH 4.6 (Latex A, number concentration $= 6.6 \times 10^{11}$ particles ml^{-1})

Total $C_{12}E_6$ concentr. (M)	Equilibrium $C_{12}E_6$ concentr. (M)	Flocculation concentr (M)
zero	zero	$2.8_3 \times 10^{-4}$
5.0×10^{-6}	2.0×10^{-6}	$6.9_0 \times 10^{-4}$
1.0×10^{-5}	3.0×10^{-6}	$1.7_4 \times 10^{-3}$
2.0×10^{-5}	7.0×10^{-6}	$4.2_0 \times 10^{-3}$
5.0×10^{-5}	2.8×10^{-5}	stable

From [26].

Adsorption of a series of non-ionic surfactants on to sulphathiazole and naphthalene has been compared [31]. The lower adsorption on to the naphthalene surface is reflected in the high areas/molecule given in Table 9.2. The packing of the surfactants on sulphathiazole is closer than at the air–water interface, perhaps suggesting that the values at the solid surface are apparent

Table 9.2 Areas per molecule at the solid–solution interface calculated from the saturation adsorption

Surfactant	Area per molecule, (nm^2) of solid:		CMC $\times 10^4$ (mol l^{-1})	Area per molecule at air–solution interface (nm^2)
	Sulphathiazole	Naphthalene		
Polyoxyethylated nonyl-phenol-5.8*	0.24	110	0.25	0.39
Polyoxyethylated nonyl-phenol-6.6	0.26	110	0.4	0.44
Polyoxyethylated nonyl-phenol-8.5	0.42	120	0.4	0.53
Polyoxyethylated nonyl-phenol-9.8	0.48	120	0.6	0.61
Polyoxyethylated nonyl-phenol-11.7	0.60	130	0.8	0.62
Polyoxyethylated octyl-phenol-7.7	0.29	180	2.7	0.59
Polyoxyethylated octyl-phenol-9.7	0.40	280	3.4	0.65

* Numbers indicate the average polyoxyethylene chain length.
From [31].

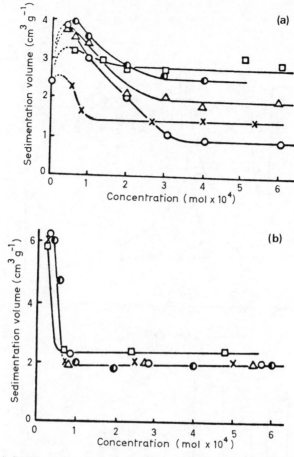

Figure 9.13(a) Sedimentation volume of sulphathiazole in the presence of poly-oxyethylated nonylphenols (Average polyoxyethylene chain length of polyoxyethylated nonylphenol): ○, 5.8; ×, 6.6; △, 8.5; ◐, 9.8; and □, 11.7. (b) Sedimentation volume of naphthalene in the presence of polyoxyethylated nonylphenols. (Average polyoxyethylene chain length of polyoxyethylated nonylphenol): ○, 5.8; ×, 6.6; △, 8.5; ◐, 9.8; and □, 11.7. From Otsuka *et al.* [31].

because of some multilayer formation. The sedimentation volumes of the sulphathiazole and naphthalene can be seen in Fig. 9.13.

The nonylphenyl ethoxylates have also been used by Kayes and Rawlins in their work with diloxanide furoate and polystyrene latex [32]. Adsorption isotherms show the well-established trend with a maximum adsorbed concentration of NPE_8 of around $2\,\mu mol\,g^{-1}$ and of NPE_{35} of about $0.35\,\mu mol\,g^{-1}$. Their results on the redispersibility of diloxanide suspensions, shown in Fig. 9.14, reflect therefore not the extent of adsorption of the surfactant molecules but their hydrophilic properties. The profiles can be compared with the corresponding sedimentation values for naphthalene and sulphathiazole above.

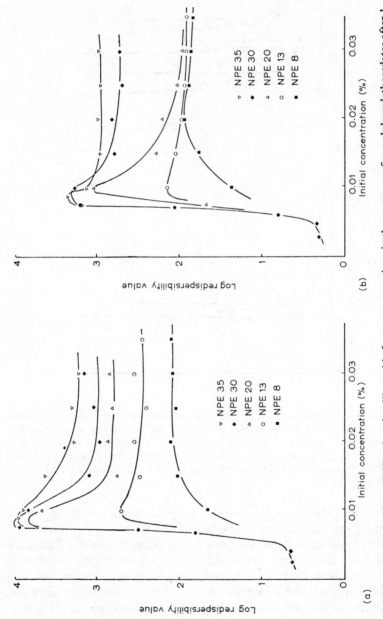

Figure 9.14(a) Redispersibility data for diloxanide furoate suspensions in the presence of nonylphenylethoxylates after 1 year's storage. (b) Redispersibility data for drug suspension in the presence of nonylphenylethoxylates after 3 days storage. From Rawlins [32].

Sulphamerazine suspensions are in a deflocculated state when prepared with polysorbate 40 [33] (2%). On standing, the sedimentation volume is around 0.15 and the system is impossible to redisperse after two weeks. Addition of high concentrations of propylene glycol (i.e. greater than 50%) flocculates the system and renders the suspension dispersible. Propylene glycol dehydrates the surfactant molecules (the cloud point is lowered) hence reducing its stabilizing properties.

9.3.3 Steric stabilization of suspensions

The term steric stabilization was used first by Heller and Pugh [34] in situations where uncharged particles were prevented from flocculating by adsorption of non-ionic polymeric molecules. We have discussed enthalpic and entropic components of steric stabilization in Chapter 8. The free energy of interaction of two spherical lyophobic particles of radius a coated with an adsorbed layer of thickness δ and surface layer concentration c, at a distance H_0 between the surfaces of the uncoated particle is given by

$$\Delta G_s = V = \frac{4kT\pi c^2}{3V_1\rho^2}(\psi - \chi)(\delta - H_0/2)^2 (3a + 2\delta + H_0/2), \qquad (9.8)$$

where V_1 is the molecular volume of the solvent molecules and δ is the density of the adsorbed film; χ characterizes the interaction of polymer and solvent and ψ is an entropy of mixing term. A positive value of ΔG_s is required for stability. Although some of the terms, as we have discussed before, are difficult to measure, the significance of δ and c in the equation is readily apparent. Addition of electrolyte and elevation of temperature can be assumed to affect both in certain ways, but as we have seen (e.g. with griseofulvin) one cannot always predict the effect of these changes in conditions. Ottewill and Walker [26] seemed to assume that addition of electrolyte did not affect adsorption of the non-ionic on to polystyrene. Tadros and Vincent [35] have shown clearly the effect of electrolyte on the adsorption of poly(ethylene oxide)–poly(propylene oxide) ABA block co-polymeric surfactants of the poloxamer series on to polystyrene. Electrolyte increases adsorption, and, contrary to the observations of Elworthy and Guthrie [27] with simpler non-ionics, increase in temperature also increases the adsorption of these co-polymers, although above 37° C there is no increase with further increase in temperature. Critical flocculation temperatures were measured as a function of KCl concentration. The lowest temperature is 75° C, much above the point at which adsorption has ceased to increase; flocculation is thus the result of the decreased solvency of the water for the polymer and the reduced stabilizing power of the adsorbed layer.

The systems flocculate below the θ-temperature of the free molecules (Fig. 9.15), the complex nature of the adsorption of the molecule at the surface may be one way to explain this behaviour. The molecules are highly folded at the air–water interface and also are likely to be so at a solid interface [36, 37]. An estimate of the adsorbed layer thickness of poloxamers and nonyl phenyl

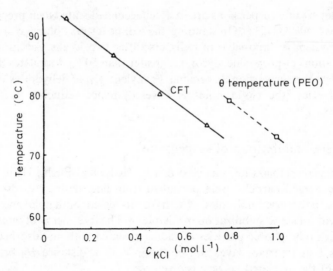

Figure 9.15 Critical flocculation temperature values and θ temperatures of polyethylene oxide (PEO) for polystyrene latex systems stabilized by adsorbed Pluronic in the presence of varying concentrations of KCl. Pluronic concentration, 200 ppm. From Tadros and Vincent [35].

Table 9.3 Adsorbed layer thickness of surfactants by microelectrophoretic measurement (a) on latex

Surfactant	Mobility $(10^{-8}\,\mathrm{m^2\,s^{-1}\,V^{-1}})$	Zeta potential (mV)	Absorbed layer thickness (nm)
Pluronic L62	-3.4 ± 0.10	-54.5 ± 2.0	0.9 ± 0.3 (1.8)*
Pluronic L64	-3.13 ± 0.14	-49.3 ± 1.8	1.7 ± 0.3 (2.6)
Pluronic F68	-1.55 ± 0.20	-22.8 ± 2.4	8.7 ± 1.0 (14.4)
Pluronic F38	-2.32 ± 0.13	-34.8 ± 1.8	4.8 ± 0.5 (7.7)
Pluronic F88	-1.18 ± 0.18	-17.3 ± 2.0	11.3 ± 1.0 (18.3)
Pluronic F108	-0.95 ± 0.20	-14.0 ± 3.0	13.5 ± 2.0 (26.1)
NPE 8	-3.50 ± 0.10	-56.8 ± 1.5	0.40 ± 0.35
NPE 13	-3.25 ± 0.05	-54.0 ± 1.5	0.95 ± 0.22
NPE 20	-3.05 ± 0.10	-48 ± 2.0	2.8 ± 0.4
NPE 30	-2.68 ± 0.23	-41.25 ± 3.8	3.28 ± 0.8
NPE 35	-2.55 ± 0.20	-39.5 ± 2.1	3.50 ± 0.4

* Extended length in parentheses [37].

(b) on diloxanide furoate B.P.

Surfactant	Mobility $(10^{-8}\,\mathrm{m^2\,s^{-1}\,V^{-1}})$	Zeta potential (mV)	Adsorbed layer thickness (nm)	Extended length (nm)
Pluronic L62	-2.60 ± 0.15	-33.4 ± 1.9	3.00 ± 0.52	1.74
Pluronic L64	-1.83 ± 0.18	-23.5 ± 2.3	6.30 ± 0.90	2.65
NPE 20	-1.5 ± 0.20	-19.27 ± 2.6	8.15 ± 1.25	5.4
NPE 30	-0.90 ± 0.20	-11.55 ± 2.6	13.15 ± 2.15	7.4

From Rawlins [32].

ethoxylates has been made (see Table 9.3). A problem in applying theories of steric stabilization is the question of whether or not desorption of the stabilizing chains occurs. A problem of great magnitude in the formulation of an adequate theory is that of the conformation of the molecules at the solid–liquid interface. In a critique of stearic stabilization theories, Osmond et al. [38] conclude that the only satisfactory approach is computer studies along the lines initiated by Clayfield and Lumb [39]. The explanation of incipient flocculation (phase separation type flocculation) in terms of Fischer's theory as applied by Ottewill and Walker [26] is, they suggest, fortuitous and they propose that a more realistic approach is to consider the theoretical basis of phase separation between adsorbed layers of the interacting particles. Clayfield and Lumb's approach was to consider that the repulsion necessary to prevent particle adhesion arises from the reduction in configurational entropy of the adsorbed polymer molecules when they are compressed on the close approach of neighbouring particles covered with polymer.

For the entropic repulsion calculations, the model system considered is that of a spherical particle, radius a, separated by a distance H from a plate, with both surfaces coated with adsorbed polymer chains of root-mean-square (rms) height l_r at a surface coverage θ. If pairs of chains of height l_1 on opposing surfaces just touch when those on the sphere lie on a circle which subtends, at the centre of the sphere, a semi-angle ϕ to the perpendicular to the plate, the entropic repulsion V_R for the system is given in kT units by Equation 9.9, where $W(l_1/l_r)$ is the proportion of the total number of configurations of a molecule for which the distance of the extremity of the chain from the adsorbing surface is less than l_1.

$$\frac{V_R}{kT} = 2\pi\theta \int_0^{\pi/2} \frac{(2a/l_r) + (H/2l_r)^2}{(1 + \cos\phi)^2} \times \sin\phi \cos\phi \left[-\ln W\left(\frac{l_1}{l_r}\right) \right] d\theta. \quad (9.9)$$

The following equation was proposed for the London–van der Waals' attraction energy V_A, in kT units, between the spherical particle and the plate, where A is the Hamaker constant:

$$V_A = -\frac{A}{6kT} \left[\frac{2a(H+a)}{H(H+2a)} - \ln\left(\frac{H+2a}{H}\right) \right]. \quad (9.10)$$

The theory was developed to deal with terminally adsorbed block co-polymer, and extended to deal with the more complex case of random co-polymer chains where the possibility exists of the adsorbed chain lying flat on the surface rather than retaining its original coiled conformation because of the placing of the polar adsorbing units along the chains [40]. Fig. 9.16 shows the interaction energy for three types of polymer molecule, an incompressible film, a random co-polymer and a 100-link terminally adsorbed chain.

Clayfield and Lumb [40] discuss the results as follows: 'The potential energy maximum is very high for the terminally adsorbed macromolecule and infinite for the other two, so that in all three cases irreversible adhesion should be prevented. The minima in the potential energy curves increase arithmetically in the order

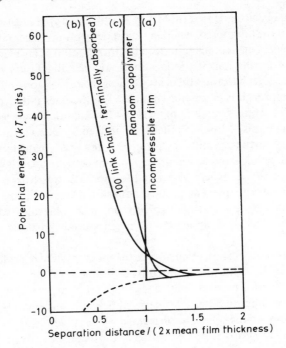

Figure 9.16 Interaction curves for (a) incompressible film, (b) film of terminally adsorbed chains, (c) film of random co-polymer chains, at equal root-mean-square film thicknesses, calculated from Equations 9.9 and 9.10 using $a/l_r = 5$, $\theta = 1.0$, $A/kT = 12.5$. From [40] with permission.

terminally adsorbed polymer, fluctuating random co-polymer, incompressible molecule, and thus the protection afforded against reversible adhesion decreases in the same order. Increase of height of each molecule, with the same attraction constant, increases the protection against adhesion by decreasing the potential energy minimum. An increase of height, however, will generally be achieved only with a corresponding increase in the area occupied by the macromolecule on the adsorbing surface and this will tend to reduce the repulsion.'

Adsorbed surfactant layers present a different problem for theoretical analysis, especially if there is multilayer adsorption. Configurational entropy would not seem to be high in monolayers of molecules such as $C_{12}E_6$ which are vertically adsorbed and close-packed at the interface. Further problems arise when considering additives in the bulk phase which are incompatible with the stabilizing molecules under some conditions. As far as we are aware no one has considered this problem with surfactants as stabilizers. However, the effect of free polymer on the stability of sterically stabilized dispersions in which a polymer is used as stabilizer has been considered [41]. When two dilute polymers in the same solvent are mixed, they are, as a rule incompatible and will exhibit phase separation. Addition of free polymer (type i) to a polymer (type ii)-stabilized dispersion is likely to lead to instability [42]. However, it has been found, for example, that dispersions of neutral polystyrene particles carrying terminally anchored polyoxyethylene chains (*MW*. 750 or 2000) flocculate on the addition of free, homo-polymer over a molar mass range of 200 to 300 000, i.e. the same polymer is leading to instability. The phenomenon has much in common with the phase separation that occurs on mixing two polymeric species. Three component phase diagrams can be constructed to show regions of instability. Fig. 9.17 is a phase diagram from Vincent *et al.* [41] for a polystyrene latex stabilized with grafted poly(ethylene oxide) chains of 750 molecular weight in the presence of poly(ethylene oxide) of two molecular weights 400 and 10 000. The continuous phase is water or 0.065 mol l^{-1} MgSO$_4$.

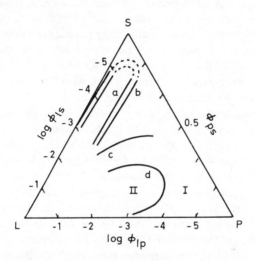

Figure 9.17 Experimental three-component phase diagram (25° C): latex (L) + polymer (P) + solvent (S). (a) PS-PEO 750 + PEO10 000 + water; (b) PS-PEO-750 + PEO 10 000 + 0.065 mol dm^{-3} MgSO$_4$; (c) PS-PEO-750 + PEO 1500 + water; (d) PS-PEO-750 + PEO 400 + water.
I = thermodynamically stable one-phase region.
II = floc phase plus disperse phase.
ϕ_{lp} = fraction latex/free polymer
ϕ_{ls} = fraction latex/solvent
ϕ_{ps} = volume fraction of free polymer.
From Vincent *et al.* [41] with permission.

A particle which is covered by anchored polymer chains can be regarded as a very large composite cross-linked polymer with a core replaced by the material of the particle; a dispersion of these particles may thus be expected to display some of the thermodynamic properties of the polymer [43].

Tadros [44] has considered the stability of aqueous dispersions of ethirimol in mixtures of poly(vinyl alchohol) and CTAB or NaDS.

9.3.4 Adhesion of suspension particles to containers and other substrates

Particles in a suspension not only have the opportunity to interact with each other but can interact with the walls of the container. The processes are not identical to those occurring in the bulk because adhesion often occurs during pouring and if during subsequent drainage particles begin to agglomerate and adhere to the container wall, a deposit will build up. Several types of wetting may be identified which might be the prelude to such adhesion. Where the suspension is in constant contact with the container (in the bulk), immersional wetting occurs, in which collisions occur between particle and walls in the continuous medium and particles may or may not adhere.

Above the liquid line spreading of the suspension during shaking or pouring may also lead to adhesion of the particles contained in the spreading liquid. Adhesional wetting occurs when a liquid drop remains suspended, like a drop of water on a clothes line. Obviously the surface tension of the suspension plays a part in the spreading and wetting processes (see Chapter 1). Adhesion increases with increase in suspension concentration, and with the number of contacts the suspension makes with the surfaces in question.

Additives, especially surfactants, will modify the adhesion of suspension particles. They will act in two ways: (i) by decreasing the surface tension; and (ii) by adsorption modifying the forces of interaction between particle and container. The example below refers to the addition of benzethonium chloride to chloramphenicol suspensions. Benzethonium chloride converts both the glass surface and the particles into positively charged entities. Adhesion in the presence of this cationic surfactant is concentration-dependent; the process is akin to flocculation. At low concentrations the surfactant adsorbs by its cationic head to the negative surface of the glass and to the suspension particle. The glass is thus made hydrophobic. At higher concentrations hydrophobic interactions occur between coated particle and surface. Further increase in concentration results in multilayer formation of surfactant, adsorbed on to the non-polar groups, rendering the surfaces hydrophilic. In this condition they repel, reducing adhesion.

This topic is dealt with in great detail by Uno and Tanaka [45–47]. Apart from the energies of repulsion and attraction due to adsorption of surfactants, adhesion may result from crystallization under the special conditions that obtain when particles are left behind after drainage of the bulk phase. Particles will be trapped by the surface tension of the liquid film (Fig. 9.18). As the particle dissolves there is supersaturation followed by deposition of the solid around the particle. According to Uno and Tanaka the solubility of the suspension particle increases at the particle–wall contact point due to the pressure exerted on the particle by the surface tension forces. Following deposition, evaporation of the continuous phase occurs as depicted in Fig. 9.18 [46]. Deposition due to cooling, to medium evaporation and to the adhesiveness of additives in the formulation can occur. The effect of polysorbate 80, benzethonium chloride, some anionic surfactants and poloxamers on the adhesion of chloramphenicol suspensions has

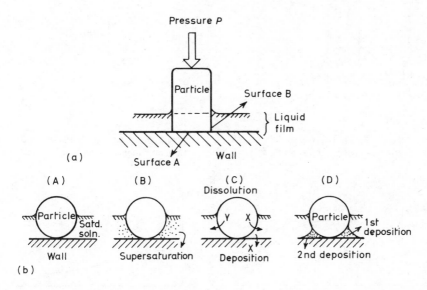

Figure 9.18(a) Solubility of solid increases under pressure which acts only upon solid and not on liquid. (b) representation of the processes involved in adhesion of particles. From Uno and Tanaka [46] with permission.

been reported [46]. Sodium lauryl sulphate and sodium cetyl sulphate decrease adhesion, but the other surfactants studied produce a maximum in the adsorption–surfactant concentration diagrams shown in Fig. 9.19.

High concentrations of non-ionic surfactant such as polysorbate 80 decrease adhesion to a low value. Cationic surfactants should be avoided as they either have little effect or enhance adhesion; anionics have to be avoided in pharmaceutical preparations for toxicological reasons.

In the process of contact with the wall of the container, spherical particles may become flattened at the point of contact, and this can have a considerable influence on the interactive forces [48]. Although most drug substances are hard and irregular, and hence unlikely to be deformed, this is not so with model systems such as lattices where this effect should be taken into account. Interaction forces between condensed bodies in contact have been discussed at length by van den Tempel [49].

The surfactant in suspension formulations might minimize adhesional deposition by solubilization. Precipitation of chloramphenicol palmitate from solutions containing polysorbate 80 has been studied by Moës [50]. Low concentrations of the surfactant give coarse particles and large compact aggregates which on a macroscale have low sedimentation volumes. Systems with low concentrations of polysorbate 80 have higher apparent viscosities because of the aggregation of the particles.

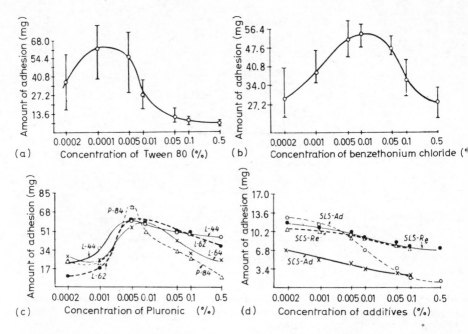

Figure 9.19 The relationship between (a) Tween 80 concentration, (b) benzethonium chloride concentration, (c) the concentration of several poloxamers as marked and adhesion from a 15% suspension of chloramphenicol in a container with 6.5 cm² glass wall, (d) the effect of anionic surface-active agents SLS (sodium lauryl sulphate) and SCS (sodium cetyl sulphate) on adhesion (Ad) and the amount of trapped and retained suspension (*Re*) × 0.25. Adhesion determined at 40° C after 48 h after a single wetting of the surface. From Uno and Tanaka [47] with permission.

9.3.5 Suspension stability in non-aqueous systems

In media of low dielectric constant, electrostatic stabilization is of little importance. Colloidal dispersions in non-aqueous media are thus more likely to be stabilized by steric barriers formed by adsorbed surfactants and polymers. Relatively little work has been done on the adsorption of surfactants on to solids from non-aqueous solvents, a limiting factor of course being the insolubility of many surfactants in solvents other than water. Non-ionic surfactants tend to be soluble in both aqueous and non-polar solvent systems. Rupprecht [6] has made a series of investigations of adsorption of non-ionic alkyl polyethers on to silica in various organic solvents. Fig. 9.20 shows some of the adsorption isotherms for nonylphenol $E_{8.5}$ from dichloromethane, *n*-butanol, *n*-propanol, ethanol, 1,4-dioxan and DMSO. As might be expected, adsorption is greatest from the dichloromethane and the effect of increasing polarity is clearly seen with the three alcohols.

Dioctylsulphosuccinate sodium is soluble in several organic solvents and some sorption isotherms have been obtained for its adsorption on to silica from carbon

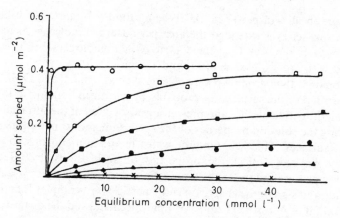

Figure 9.20 Sorption of NPE$_{8.5}$ on to hydrophilic silica from various organic dispersions
O——O Dichloromethane
□——□ *n*-Butanol
■——■ *n*-Propanol
●——● Ethanol
▼——▼ 1,4 Dioxan
×——× DMSO
From [6] with permission.

tetrachloride, chloroform and butanol (Fig. 9.21). The alkylpyridinium chlorides have also been studied by Rupprecht and some results for the tetradecyl derivative are shown in Fig. 9.21. Levels of adsorption in these solvents are less than from water, the area per molecule of the C$_4$ pyridinium

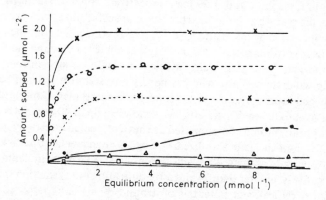

Figure 9.21 Sorption of ionic surfactants on to hydrophilic silica from organic solvents.
×——× CCl$_4$ ⎤
×- - - -× CHCl$_3$ ⎬ *n*-Dioctylsulphosuccinate
△——△ Butanol ⎦

O - - - - O CHCl$_3$ ⎤
●——● *i*-Propanol ⎬ Tetradecylpyridinium chloride
□——□ DMSO, Ethanol ⎦

homologue on silica dispersed in $CHCl_3$ being about 0.76 nm^2. As the alkyl chain length in this series is extended the area per adsorbed molecule increases to a maximum (1.10 nm^2) at C_{10}, the C_{12} to C_{18} homologues being identically adsorbed with areas/molecule of around 1.0 nm^2.

It is apparent that strongly polar solvents such as ethanol and DMSO reduce the adsorption of the surfactants, probably by hydrogen bonding to the surface groups on the colloidal silica and preventing adsorption of the surfactant, as well as affecting the solution properties of the surfactant and its natural tendency to seek out surfaces. Trace amounts of water in the non-aqueous solvents will obviously markedly affect the properties of suspensions, by affecting the adsorption process and the state of ionization of ionic surface-active agents. The magnitude and sign of the charge on rutile particles is a complex function of the concentration of Aerosol OT (dioctylsulphosuccinate) and the quantity of water in the system when the continuous phase is xylene. Rigorously dried systems exhibit a negative charge while those containing trace water are positive [51]. Investigation of the stabilization of rutile dispersions in xylene with Aerosol OT led to the conclusion that the DLVO theory provides a satisfactory explanation of behaviour. Romo [52] and Micale *et al.* [53] have also considered the effect of water on the stability of non-aqueous dispersions. Romo studied alumina and aluminium hydroxide suspensions in normal and iso-C_3, C_4 and C_5 alcohols. On addition of water the stability of the hydroxide suspensions increases whereas that of the alumina decreases. Progressive increases in the amount of water added to alumina results in a reversal of zeta potential. In anhydrous propanol the zeta potential is -26 mV, while with 40λ water/100 mg the potential changed to $+20$ mV. The values of zeta potential for aluminium hydroxide under the same conditions are $+25$ mV and $+29$ mV, respectively. Such interactions obviously complicate the picture as far as interpretation of surfactant effects are concerned. An early paper by Koelmans and Overbeek [54] concluded that the stability of suspensions in solvents of very low polarity was achieved by quite modest electric charges, and zeta potentials were sufficient to stabilize suspensions of coarse particles (greater than 1 μm) 'whereas hardly any stabilization can be expected from adsorbed layers of non-ionized long-chain molecules'. Among the surfactants used in stabilizing dispersions in xylene were Span 40 and Span 80. These, however, appeared to be contaminated as they conducted electricity on addition to xylene. Barium sulphate stabilized by Span 80 had a zeta potential of 60 mV, alumina stabilized by Span 40 a zeta potential of 33 mV. Sedimentation times were, respectively, greater than 24 and 5 h, confirming the prediction from DLVO theory that a critical zeta potential of around 30 mV was required to ensure stability. In view of later findings on the effect of water (especially in view of the positive charge on alumina) the results of Koelmans and Overbeek are undoubtedly complicated by traces of water and contaminants in the surfactants.

Koelmans and Overbeek's remarks about non-ionic stabilizers are probably explained by the choice of materials for their work. Useful light has been shed on the question of steric stabilization in non-aqueous systems by the work of Bagchi and Vold [55]. They studied dispersions of graphon in heptane using Triton X-35 or dodecylbenzene as stabilizer. Triton X-35 gave a stepped isotherm, while dodecylbenzene gave a Langmiurian isotherm resulting in decreasing rates of flocculation. The non-ionic surfactant Triton X-35 did not stabilize the dispersion, it being suggested [56] that even when a bilayer forms the thickness of the adsorbed layer is insufficient to stabilize the dispersion, only the short tertiary octyl group protrudes into the heptane. The second virial coefficient of the Triton in heptane is close to zero and as a result there would be no osmotic component in the repulsive force, although the evidence, from viscosity measurements, of a randomly oriented surfactant, is not conclusive [56]. The dodecylbenzene chains will afford a much greater monolayer thickness and stabilizing barrier if adsorbed perpendicularly to the surface. Two Tritons (Triton X-45 and X-102) have been used in work on the dispersibility of barium titanium oxide in water and methylisobutylketone [57]. These surfactants and Tween 20 decreased the settling rate in the non-aqueous solvent. Some of these results for specific sedimentation volume are shown in Table 9.4.

Table 9.4 Specific sediment volume of 0.68 μm BaTiO$_3$ with various surfactants

Surfactant	Specific sediment volume (cm^3 g^{-1})	
	Water	MIBK
None*	0.94	1.07
Triton X-45	0.97	0.83
Triton X-102	0.98	0.92
Tamol 731	0.47	1.52
Tamol 850	0.47	1.57
Darvan C	0.47	1.67
Tween 20	0.93	0.93
Oleic acid	1.33	0.57

* The dry specific sediment volume of the 0.68 μm sample was 0.77 cm^3 g^{-1}.

Triton X-45	Rohm and Haas	Alkylaryl polyether alcohol
Triton X-102	Rohm and Haas	Alkylaryl polyether alcohol
Tamol 731 (25% aqueous solution)	Rohm and Haas	Sodium salt of polymeric carboxylic acid
Tamol 850 (30% aqueous solution)	Rohm and Haas	Sodium salt of polymeric carboxylic acid
Darvan C	R. T. Vanderbilt Company	Condensation product of formaldehyde and naphthalene sulphonic acid
Tween 20	Atlas Chemical Industries	Polyoxyethylene sorbitan monolaurate
Oleic acid	Fischer Chemical Corp.	Certified reagent grade

From [57].

9.4 Effect of surfactants on adsorptive capacity of suspensions

Adsorbent materials are used medicinally to absorb toxins *in vivo*. The enterotoxin of *Pseudomonas aeruginosa* is implicated in the aetiology of diarrhoea; adsorbents such as kaolin are used to decrease the toxicity of such agents by removal of the toxin by adsorption. Armstrong and Clarke have investigated the adsorption sites on kaolin by measurement of electrophoretic mobility in the presence of surfactants and other materials as a function of pH [58]. In particular they studied the uptake of gentian violet on to koalin treated with anionic and cationic agents. Kaolin pretreated with cetrimide showed a marked decrease in adsorptive capacity. On the other hand, when pretreated with sodium hexadecyl sulphate the adsorptive capacity was increased (Fig. 9.22).

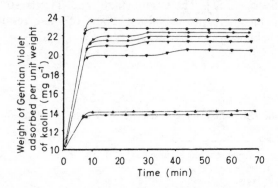

Figure 9.22 Variation with pH of the uptake of a 0.05 % (w/v) solution of Gentian Violet on natural kaolin and on kaolin treated with anionic and cationic materials. Kaolin, pH 7.4 ($\triangleright$) and 5.7 ($\blacktriangleright$); washed kaolin, pH 7.4 ($\triangledown$) and 5.4 ($\blacktriangledown$); kaolin plus sodium hexadecyl sulphate, pH 7.5 (O) and 5.6 ($\bullet$); and kaolin plus cetrimide, pH 7.6 ($\triangle$) and 5.4 ($\blacktriangle$). From Armstrong and Clarke [58] with permission.

At pH 7.4 sorbitan monostearate decreased the adsorptive capacity of 1 % kaolin suspensions or *P. aeruginosa* toxin by 70 %, sodium lauryl sulphate and cetrimide reducing absorption by 62 and 26 %, respectively [59]. Drugs such as lincomycin known to be strongly bound to kaolin reduced the absorption of the toxin by about 30 %. At pH values greater than 4.1 the toxin has a negative charge and will be adsorbed on to the edge of native kaolin particles. But at pH 7 to 8 the edges become positively charged and adsorption of toxin to the kaolin must be mediated through hydrophobic bonds. One might expect, therefore, that the adsorption of cationic molecules might increase adsorption of the toxin; the general effect of saturating a clay with organic cations is to render the clay surface hydrophobic. Kaolinite and montmorillonite saturated with alkyl-ammonium ions will separate from the aqueous phase because of this effect. However, the nature of the adsorption will change as the pH changes; if the negative sites disappear with increase in pH then there will be little opportunity for coulombic adsorption of cations. Adsorption studies might be complicated by changes in the degree of flocculation of the system which might change the available surface area of the dispersed clay.

9.5 Rheological characteristics of suspensions

Einstein considered a suspension of spherical particles which were far enough apart to be treated independently where ϕ is defined by

$$\phi = \frac{\text{volume occupied by the particles}}{\text{total volume of suspension}}.$$

The suspension could be assigned an effective viscosity, η_*, given by

$$\eta_* = \eta_0(1 + 2.5\phi) \tag{9.11}$$

where η_0 is the viscosity of the suspending fluid. As we have seen, the assumptions involved in the derivation of the Einstein equation do not hold for colloidal systems subject to Brownian forces, electrical interactions and van der Waals' forces.

A charged particle in suspension with its inner immobile Stern layer and outer diffuse Gouy (or Debye–Hückel) layer presents a different problem than a smooth and small non-polar sphere. Such particles when moving experience electroviscous effects which have two sources: (i) the resistance of the ion cloud to deformation; (ii) the repulsion between particles in close contact. When particles interact, for example, to form pairs in the system the new particle will have a different shape from the original and will have different flow properties. The coefficient 2.5 in Einstein's equation (9.11) applies only to spheres; asymmetric particles will produce coefficients greater than 2.5.

Other problems in deriving *a priori* equations result from the polydisperse nature of pharmaceutical suspensions. The particle size distribution will determine η. Experimentally a polydisperse suspension of spheres has a lower viscosity than a similar monodisperse suspension. It is obvious that a simple undifferentiated volume fraction ϕ cannot be expected to be of value in this situation.

Structure formation during flow is an additional complication. Structure breakdown occurs also and is evident particularly in clay suspensions, which are generally flocculated at rest. Under flow there is a loss of the continuous structure and the suspension exhibits thixotropy and a yield point. The viscosity decreases with increasing shear stress.

A very large number of equations, many of them empirical, have been proposed to relate the viscosity with the concentration of particles in suspensions. Fig. 9.23 shows the application of a typical empirical equation for suspensions on which measurements are made at sufficiently high shear rates that the system has become Newtonian. The limiting viscosity at these high shear rates reflects only hydrodynamic interactions. All other interactions, of course, lead to increased viscosity and it is never possible to discount multiple sources of interactions

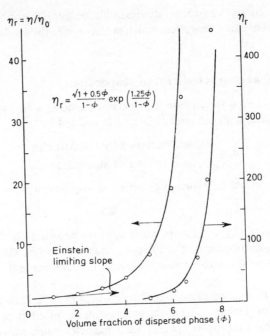

$$\eta_r = \eta/\eta_0$$

$$\eta_r = \frac{\sqrt{1 + 0.5\phi}}{1-\phi}\exp\left(\frac{1.25\phi}{1-\phi}\right)$$

Einstein
limiting slope

Volume fraction of dispersed phase (ϕ)

Figure 9.23 Relation between volume-fraction of dispersed phase and relative viscosity of well-stabilized dispersions. From van den Tempel [60].

especially in concentrated dispersions. The effect of non-hydrodynamic interactions outweighs nearly all other effects on the rheological properties of suspensions [60].

Attempts have thus been made, with partial success, to account for the effects of the formation of adsorbed layers of surfactants and stabilizers around the particles, hydration of particles and electrostatic interactions. Particle aggregation resulting from net attractive forces between the particles gives rise to the elastic and plastic flow properties observed in many systems. Electrostatic repulsive forces affect the viscosity of the particles but do not give rise to appreciable deviations from Newtonian behaviour [60]. High molecular weight surfactants adsorbed on to zinc oxide and other clays and pigments can alter the viscosity of concentrated (50%) dispersions dramatically, reducing apparent viscosities from values higher than 10 000 cP to about 10 cP [61]. The reduction in viscosity is closely related to the zeta potential of the particles, although quantitative relationships were not established. It seemed that an increase in zeta potential above a certain level was required to achieve a significant reduction in apparent viscosity. As the pH of the systems studied by Moriyama [61] were monitored along with the amount of additive adsorbed, the zeta potential and apparent viscosity, it is worthwhile reproducing some of these results, in spite of the lack of quantitation (see Fig. 9.24). For a comprehensive review of the role of colloidal forces in the rheology of suspensions readers should consult the article

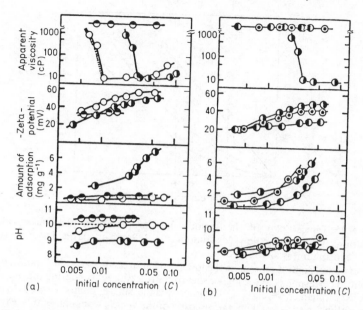

Figure 9.24(a) Apparent viscosity, zeta potential, amount of adsorption, and pH in zinc oxide–water (50/50) suspensions versus initial concentration, c, curves for Na salt of polyacrylic acid (PA) (○), Na salt of formalin condensate of β-naphthalene sulphonate acid (NSF) (◐), and sodium tripolyphosphate (◑). The initial concentrations of PA and NSF refer to the mole concentrations expressed/monomer unit of them. A dotted line refers to the apparent viscosity versus c curve for PA at a controlled pH of 10.1. C = initial mole concentration of surfactant and polyphosphate. (b) Apparent viscosity, amount of adsorption, and pH in zinc oxide–water (50/50) suspensions versus initial concentration, c, curves for Na salt of formalin condensate of alkyl (C_4) sulphonic acid (Al-NSF) (◑), sodium alkyl (C_4) naphthalene sulphonate (◐), and sodium dodecyl benzene sulphonate (⊙). The initial concentrations of Al-NSF refer to the mole concentrations expressed/monomer unit. C = initial mole concentration of surfactant. From [61] with permission.

by Russel [62] who, however, concludes 'no existing theories can cope quantitatively with the inevitable multiparticle hydrodynamic, Brownian and electrostatic or steric interactions in the disordered state normally encountered . . .'. With flocculated systems the position is worse as the interactions between flocs and the break-up of flocs is insufficiently understood. Tadros considers, in particular, the rheology of suspension concentrates [63]. The rheological characteristics of pharmaceutical suspensions are of some practical interest. Sedimentation of particles and their movement through the dispersion medium results in a low shear stress; shaking and pouring result in a high shear stress [58]. A pseudoplastic or plastic system is preferred. In such systems, falling particles exerting a low shear stress will experience a higher viscosity than the viscosity experienced under pouring conditions, as may be deduced from Fig. 9.25.

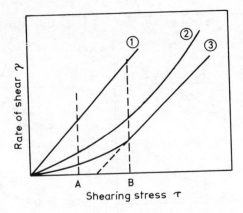

Figure 9.25 Typical plot of the various flow properties in pharmaceutical liquid formulations. 1, Newtonian flow; 2, pseudoplastic; 3, Bingham plastic. From Samyn [64]. Samyn's explanation is as follows:

'In this figure, point *A* represents the low shearing stress due to sedimentation (see text) and point *B* corresponds to the higher shearing stress caused by pouring the preparation. For the Newtonian curve, curve 1, the viscosity is the same at points *A* and *B*. In order to prevent settling in such a system, the viscosity of the vehicle would have to be excessively high and the product would be unpourable. For the pseudoplastic flow curve, curve 2, the viscosity at point *A* is greater than at point *B*. This means that at the low stresses induced by a falling particle, the effective viscosity of a pseudoplastic system is much greater than the effective viscosity under pouring conditions. From this viewpoint, it can readily be seen that a pseudoplastic vehicle is a better choice for a suspending agent than a Newtonian vehicle.

Consider now the case of the plastic suspending agent, curve 3. In this instance, at point *A* the shearing stress produced by a settling particle is less than the yield value of the suspension and no flow or sedimentation should occur. At point *B*, however, the yield value is exceeded and the preparation flows with a viscosity equal to the reciprocal of the slope at point *B*. A suspension with this flow curve should not settle, yet it should remain pourable. The essential feature of this hypothetical case is the position of the yield value of the suspension between the points *A* and *B*. If the yield value were higher than point *B*, the suspension would show poor pouring properties. If it were below *A*, there would be sedimentation. Further, since point *A* can be reduced to a lower shearing stress by employing small particles, minimizing density differences and having a viscous pseudoplastic media, it then becomes easier to obtain a yield value between *A* and *B*. In actual practice, then, the combined use of a pseudoplastic and a plastic suspending agent would appear to be an excellent media for suspending an insoluble material.'

For a Newtonian system the viscosity at levels A and B is the same (by definition). For the pseudoplastic system the viscosity at A is greater than that at B. In a plastic system with a yield value between A and B, no settling would occur. Samyn's own discussion of these cases is given in the legend to the figure.

Very few publications deal specifically with the effects of surfactants on the rheology of suspensions. Perhaps this is because it is not possible to isolate the effects of adsorption of surfactant *per se* and the flocculation–deflocculation behaviour that ensues. The influence of an organic salt, sodium benzoate and polysorbate 40 on 1 % dispersions of Veegum have been reported [64] as shown in Fig. 9.26. This type of action on clays is important in evaluating the possible use of the clay as a suspending agent. Sodium benzoate affects the hydration of the clay. Fully hydrated Veegum has a viscosity (at 30 rpm in a Brookfield viscometer) eight times that of an unhydrated specimen. 0.5 % sodium benzoate prevents hydration and alone results in a low viscosity preparation. In the presence of polysorbate 40 the effect is as seen in Fig. 9.26, the results showing the complex nature of the interactions. As the viscosity of the preparation is measured at a constant concentration of non-ionic surfactant the increased viscosity on increasing the sodium benzoate concentration cannot be due to solubilization and inactivation of this material.

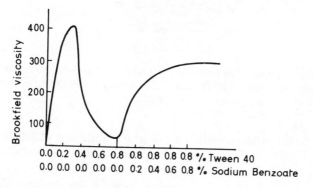

Figure 9.26 Influence of a salt (sodium benzoate) and a surfactant (Tween 40) on the viscosity of a clay. From Samyn [64].

9.6 Crystal changes in suspensions

Crystal growth in suspensions is not generally a serious problem, but it sometimes can occur with unstable polymorphic forms to the extent that a polymorphic transition can take place with time, with the possible attendant changes in the stability of the suspension due to particle size and shape changes and to an alteration in the hydrophobicity of the crystals themselves. Surfactants and other additives which adsorb on to solid surfaces can obviously affect this process in susceptible systems. Very low concentrations of surface-active materials can affect crystal growth. The effect of surfactants and the mechanism of their effect is best illustrated by the effect of anionic and cationic surfactants on adipic acid crystal form [65]. X-ray analysis showed that the linear 6-carbon dicarboxylic acid molecules were aligned end to end in parallel array in the crystal with their long axis parallel to the 0 1 0 faces so that the 0 0 1 face is made up entirely of —COOH groups while the 0 1 0 and 1 1 0 faces contain both COOH and hydrocarbon portions of the molecule (Fig. 9.27a).

The cationic surfactant, trimethyldodecylammonium chloride (TMDAC) was twice as effective in hindering the growth of the 0 0 1 face as that of the 1 1 0 and

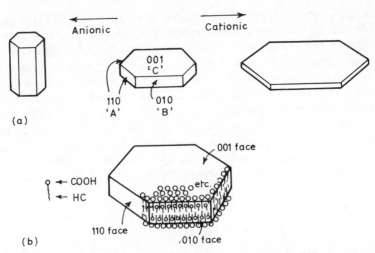

Figure 9.27(a) A representation of the effect of anionic and cationic surfactants on the habit of adipic acid crystals with (b) a diagrammatic representation of the arrangement of the adipic acid molecules at the crystal surface.

0 1 0 faces. In high concentrations it caused the formation of very thin plates or flakes. Levels of 50 ppm of the anionic surfactant sodium dodecyl benzene sulphonate (SDBS) were three times more effective in reducing growth rates of the 1 1 0 and 0 1 0 faces than of the 0 0 1 face. Higher levels of SDBS caused extreme habit modification, producing not hexagonal plates but long thin rods or needles.

The crystallographic faces whose growth rates were depressed most were those upon which surfactant adsorption was the greatest. Cationic additives adsorb on the face composed of carboxylic acid groups (0 0 1), anionic additives on the (1 1 0) and (2 0 0) faces which are hydrophobic. A coulombic interaction of the cationic head groups and the $-COO^-$ groups on the (0 0 1) faces has been suggested. The adsorption of the anionic surfactant, repelled from the anionic (0 0 1) faces, takes place amphipathically on the hydrophilic (1 1 0) faces and (1 0 0) faces (Fig. 9.27b).

Adipic acid has also been used as a model crystal by Fairbrother and Grant [66] in studies of the effect of alkanols and alkanoic acids on habit modification, during crystallization procedures. Crystallization of drugs in aqueous surfactant solutions can lead to enhanced dissolution rates of poorly soluble drugs such as prednisone, sulphathiazole and chloramphenicol [67, 68].

Carefully chosen surfactants can be used to inhibit crystal growth in solution, although there is little in the literature to substantiate this statement. Tadros [68] has compared the rate of growth of particles of the pesticide terbacil in the presence of two dispersing agents, Pluronic P75 and poly(vinyl alcohol). The block co-polymeric surfactant begins to solubilize the terbacil at concentrations above 10^{-2} %, whereas the PVA has little effect until it reaches 1 % levels. At these critical levels the growth rate obtained with the Pluronic was much higher than with the PVA at both polymer concentrations; Tadros attributes this to the solubilization of the terbacil by the former compound.

Crystal growth can be considered to be a reverse dissolution process. The diffusion theories of Noyes and Whitney and Nernst consider that matter is deposited continuously on a crystal face at a rate proportional to the difference of concentration between the surface and the bulk solution. So an equation for crystallization can be proposed in the form

$$\frac{dm}{dt} = A k_m (c_{ss} - c_s) \qquad (9.12)$$

where m is the mass of solid deposited in time t, A the surface area of the crystal, c_s the solute concentration at saturation, and c_{ss} the concentration at supersaturation. As $k_m = D/\delta$ where D is the diffusion coefficient of solute and δ is the diffusion layer-thickness, the degree of agitation of the system (which affects δ) influences crystal growth. As crystals generally dissolve faster than they grow, growth is not simply the reverse of dissolution. It has been suggested that there are two steps involved in growth; transport of the molecules to the surface and their arrangement in an ordered fashion in the lattice. Equation 9.12 may be written in a modified form

$$\frac{dm}{dt} = K_g A (c_{ss} - c_s)^n \qquad (9.13)$$

K_g being the overall crystal growth coefficient and n the 'order' of the crystal growth process. Solubilization, by increasing the saturation solubility of the solute, will increase crystal growth in a supersaturated medium such as a concentrated suspension.

Simonelli *et al.* found that crystal growth of sulphathiazole could be inhibited by poly(vinylpyrrolidone)[69]; Carless *et al.* [70] observed that cortisone alcohol inhibited crystal growth of cortisone acetate, suggesting that the alcohol is adsorbed on to particles of the stable form of the acetate, preventing further deposition of cortisone acetate on to the crystal nuclei. Growth when allowed to occur resulted in change of particle shape, the particles growing into long needles.

Crystal growth and dissolution are closely related. If the temperature does not remain constant on storage of suspensions continual dissolution and re-

Table 9.5 Increase of average particle size ($d = 45\,\mu m$) of salicylic acid within 10 days in non-aqueous media

Preparation	Increase (%)
Salicylic acid vaseline 5%	80
,,　　0.5% Span 20	111
,,　　3 %　,,	101
,,　　5 %　,,	107
,,　　0.5% Span 60	81
,,　　3 %　,,	96
,,　　5 %　,,	124

From [72].

deposition of molecules can occur; it is under these cyclical conditions that crystal growth occurs. The dissolution retarding effects of methylcellulose in prednisolone acetate systems has been ascribed tentatively to particle aggregation [71]. Hydroxypropylmethylcellulose appears to inhibit dissolution without affecting the aggregation of the particles in suspension. Changes in the particle size of suspended particles in ointment bases have long been recognized as a problem in

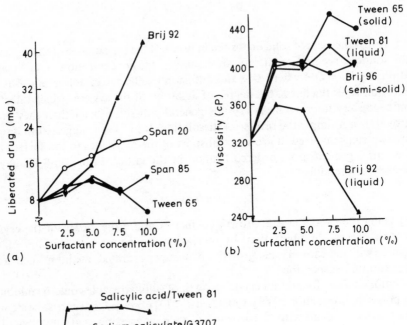

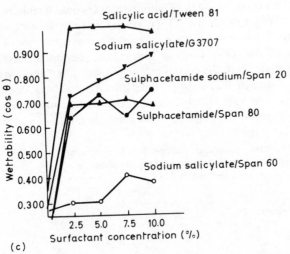

Figure 9.28(a) Effect of surfactant concentration on the liberation of sodium salicylate. (b) Quasi-viscosity of sodium salicylate ointments: dependence on surfactant concentration. (c) Wettability of preparations: dependence on surfactant concentration. From Voight [72].

pharmacy. Some attention has been given to this problem recently by Voigt [72]. Using salicylic acid dispersed in vaseline as an example, he showed the effect of addition of surfactants to the system. Surfactants such as the Spans which increase the solubility of the salicylic acid in the vehicle result in increased rates of particle growth, measured crudely by sizing the largest 20 particles in each preparation (Table 9.5).

In more extensive studies of changes in particle size distributions in these systems, the small particles were seen to disappear leaving crystals of up to

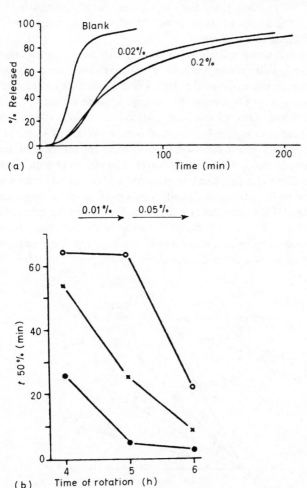

Figure 9.29(a) Influence of concentration of DOSS-Na (%) on the release of NaCl from dispersions in liquid paraffin after 4 h of rotation. Concentration of sodium chloride: about 5%; $n = 1$, blank = 0% DOSS-Na. (b) Influence of addition of water on the release with different concentrations of DOSS-Na. Concentration of sodium chloride: about 5%. ● = 0% DOSS-Na; × = 0.002% DOSS-Na; ○ = 0.2% DOSS-Na. From Crommelin and de Blaey [73].

200 μm in the system. No relation could be found between the HLB of the non-ionic surfactants used and crystal size changes, which must therefore be dependent not only on solubilization but on the viscosity of the vehicle which will also be a function of HLB. The apparent viscosity of some sodium salicylate ointments with added non-ionic surfactants as a function of surfactant concentration is shown in Fig. 9.28 along with the effect of surfactants on the wettability of some of these systems, two factors which might affect the release of active ingredient from topical preparations. In a computer analysis of many hundreds of experiments, Voigt finds that there is a pronounced correlation between the wettability of the ointment, as measured by the contact angle with water, and liberation of the active ingredient.

The release of a water-soluble agent (NaCl) from dispersions in liquid paraffin to an underlying water phase has been investigated by Crommelin and de Blaey [73]. Transport of the suspended particles to the interface by sedimentation was the rate-limiting step. Obviously the state of agglomeration of the system will affect the sedimentation. Low concentrations of di(2-ethylhexyl) sodium sulphosuccinate reduced agglomeration and trace amounts of water produced the opposite effect. Dissolution rates were measured in the presence of the surfactant and in the presence of surfactant and water. The effects of solubilization of the NaCl in inverse micelles, agglomeration of the sodium chloride by water and the deaggregation by the surfactant are all factors to be taken into account. Fig. 9.29 shows the effect of the surfactant on the release rates and the effect of addition of 0.01% and then 0.05% water on the outcome of the experiments.

A not dissimilar system is described by Gutteridge and Worthington [74] in the development of an oily antacid formulation containing calcium carbonate and

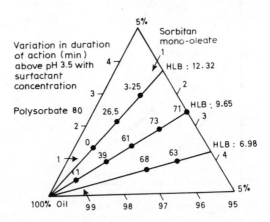

Figure 9.30 The *in vitro* test method adopted for this work is based on the original method of Fuchs [75]. The pH of a reaction mixture originally containing 100 ml of 0.05 M hydrochloric acid and the antacid under test, is continuously recorded while 0.1 M hydrochloric acid is continually added to the system at the rate of 2 ml min⁻¹. A constant level device maintains the total volume of the system at 100 ml. From Gutteridge and Worthington [74] with permission.

magnesium hydroxide. The oily suspension is placed in an aqueous medium for the testing of its neutralization capacity. It was found that the choice of surfactant in the oil had a considerable effect on the duration of action of the suspension. The effect of polysorbate 80 and sorbitan mono-oleate concentration are shown in Fig. 9.30. No explanation of these effects have been attempted, but the qualitative conclusions of the work were that the size and nature of the suspended particles and the viscosity of the suspension affected duration of activity.

9.7 Bacterial and other cell suspensions

Bacterial suspensions can be considered as colloidal systems and many aspects of their behaviour understood in terms of colloid stability theory. Factors such as adhesion of cells to surfaces, cell aggregation, spontaneous sorting out of mixed cell aggregates and other such phenomena depend to a large extent on interaction between the cells in an aqueous medium and on the nature of their surfaces. Although DLVO theory can be applied to gain an understanding of cell behaviour, the surfaces tend to be more labile than those encountered in drug suspensions or in model systems. The surface charge of some organism varies with the age of the cells; e.g. cells of *E. coli* have a lower negative charge when 'young' than at any other time in their growth cycle; the increase in the negative charge of *Strep. pyogenes* grown in liquid medium is due to accumulation of surface hyaluronic acid during the active growth phase [76]. The electrophoretic mobility of different cells, bacterial and mammalian, is given in Table 9.6 taken

Table 9.6 The electrophoretic mobility of different particles measured at pH 7

Species	Biological characters	$10^8 \times$ mobility/$m^2 s^{-1} V^{-1}$
Klebsiella aerogenes	fim +	−1.74
	fim −	−3.50
Staphylococcus aureus	methicillin-sensitive	−1.00
	methicillin-resistant	−1.48
	trained to methicillin	−1.50
Streptococcus pyogenes	Type 2G	−1.03
	Type 2M	−0.89
Human blood cells	Erythrocytes	−1.08
	Lymphocytes	−1.09
	Platelets	−0.85
Hamster kidney cells	Erythrocytes	−1.35
	Tissue cells	−0.65
	Tumour cells	−1.2
Erythrocytes	Chimpanzee	−1.18
	Chicken	−0.82
	Dog	−1.28
	Horse	−1.16
	Ox	−0.96
	Pig	−0.88
Chlorella cells		−1.70
Colloidal gold		−3.2
Oil droplets		−3.1

From [77].

from the review by James [77] of the molecular aspects of biological surfaces. All possess the same order of charge, although the nature of their surfaces differ considerably.

The contamination of pharmaceutical suspensions for oral use has recently caused concern. The behaviour of organisms in such suspensions is therefore of some interest before interpretation of experimental results can be undertaken. Under some conditions, bacteria may be strongly adsorbed and therefore more resistant to the effects of preservatives; in other cases, the bacteria may be free in suspension.

Two distinct phases of bacterial sorption on to glass have been observed [78]; the first reversible phase may be interpreted in terms of DLVO theory. Reversible sorption of a non-mobile strain (*Achromobacter*) decreased to zero as the electrolyte concentration of the media was increased, as would be expected. The second irreversible phase is probably the result of polymeric bridging between bacterial cell and the surface in contact with it. It is obviously not easy to apply colloid theory directly but the influence of factors such as ψ_0, pH and additives can be predicted and experimentally confirmed.

The flexibility of surfaces and the formation in the cells of pseudopodia with small terminal radii of curvature will obviously complicate the application of theory. In the agglutination of erythrocytes and the adsorption of erythrocytes to virus-infected cells, projections of small radius of curvature have been observed. Such highly curved regions could well account for local penetration of the energy barrier and strong adhesion at the primary minimum.

Sorption of microbial cells is selective and there is no obvious relation between gram-staining characteristics and attachment.

Addition of HCl and NaOH to drug suspensions alters the adhesion of *E. coli* to pyrophyllite and Kaolinite. The negative charge of the cell surfaces will decrease with decrease in pH; the isoelectric point of many bacteria lies between pH 2 and 3. At pH values lower than the isoelectric point the cell surface carries a positive charge. The flat surface of the clay also carries a negative charge, which also diminishes with decrease of pH; the positive charge localized on the edge of the clay platelet will be observed only in acidic solution.

In addition to the van der Waals' and electrical forces, steric forces resulting from protruding polysaccharides and protein affect interactions; specific interactions between charged groups on the cell surface and on the solid surface, hydrogen bonding or the formation of cellular bridges may all occur to complicate the picture. Pethica has summarized the possible forces acting between cells and surfaces:

(1) Chemical bonds between opposed surfaces;
(2) ion-pair formation;
(3) forces due to charge fluctuation;
(4) charge mosaics on surface of like or opposite over-all charge;
(5) electrostatic attraction between surfaces of opposite charge;
(6) electrostatic attraction between surfaces of like charge;

(7) van der Waals' forces;
(8) surface energy;
(9) charge repulsion;
(10) steric barriers.

Once attached to a surface, however, secretions from the cell often change and complicate the picture. Adhesion of organisms to teeth is but one instance where bacterial adsorption is a biologically important prelude to some biological problem, in this case caries. It is thus of interest to determine to what extent adhesion can be prevented, for example by the use of surface-active agents which modify the nature of both solid substrate and bacterial surface. Bacteria interact avidly with cationic surfactants, converting their negatively charged surfaces to

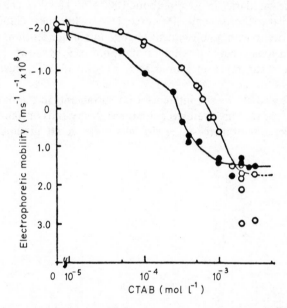

Figure 9.31 Electrophoretic mobility of *S. faecalis* as a function of cetyltrimethylammonium bromide concentration at two pH values and the constant ionic strength of 0.01 M. ○, pH 7.0; ●, pH 3.4. From Schott and Young [80]. Cetyltrimethylammonium bromide uptake by the cell wall resulted in reduced electrophoretic mobility as the negative carboxylate charges were neutralized by the positive charges of the surface-active cation. This is chemisorption via the ion-exchange reactions:

$$B\text{—}COO^-Na^+ + C_{16}H_{33}(CH_3)_3N^+Br^- \rightleftharpoons$$
$$B\text{—}COO^-{}^+N(CH_3)_3C_{16}H_{33} + Na^+Br^- \quad \text{(Reaction 1)}$$

at pH 7.0, and:

$$B\text{—}COOH + C_{16}H_{33}(CH_3)_3N^+Br^- \rightleftharpoons$$
$$B\text{—}COO^-{}^+N(CH_3)_3C_{16}H_{33} + H^+Br^- \quad \text{(Reaction 2)}$$

at pH 3.4, where B represents a portion of the cell wall.
Lysis of the cells at higher CTAB concentrations increased the positive zeta potential of the cell wall.

positively charged [80], the zeta potential being zero at about the CMC of the surfactants (see Fig. 9.31). CTAB completely inhibits the growth of *Streptococcus faecalis* at a concentration which produced only a 10 to 30 % fall in zeta potential, suggesting that the primary site of attack of this compound was not the cell wall [80].

Differences in the change of mobility of bacteria from widely different taxonomic groups in the presence of surfactants may be explained by variation in the nature of the surface chemical groups [81].

Blood is a non-Newtonian liquid showing a shear-dependent viscosity. At low rates of shear erythrocytes form cylindrical aggregates (rouleaux) which break up when the rate of shear is increased. Calculations show [82] that the shear rate (D) associated with blood flow in a large vessel such as the aorta is about $100\,s^{-1}$ but for flow in the capillaries it rises to about $1000\,s^{-1}$. The flow characteristics of blood are similar to those of emulsions except that, while shear deformation of oil globules can occur with a consequent change in surface tension, no change in membrane tensions occurs on cell deformation. Fig. 9.32 shows the viscosity of blood at low shear rates, measured in a Brookfield LVT micro cone-plate viscometer.

Figure 9.32 also shows the influence of a surfactant on the blood flow: low concentrations of sodium oleate decrease the viscosity and concentrations higher than 60 mg per 100 ml increase the viscosity. One would anticipate changes in

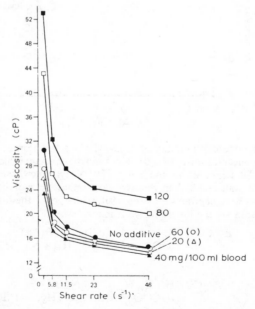

Figure 9.32 The influence of varying concentrations of sodium oleate on the viscosity of blood at different rates of shear. Each value represents an average of 8 subjects. From [84] with permission.

viscosity on addition of an anionic surfactant; interpretation is complicated by the fact that a reversible morphological effect takes place. Surfactants such as sodium oleate are also able to disaggregate erythrocytes, irrespective of rate of shear, and consequently would be expected to reduce the viscosity. At higher concentrations the increase in viscosity may be due to an electroviscous effect or the altered shape of the particle.

Phagocytosis has a cell-surface component which obviously may be affected by the presence of a surface-active agent. Van Oss records that if a detergent is added to polymorphonucleocytes (PMN) before incubation with bacteria, phagocytosis is enhanced due to the lowering of the contact angle of the PMNs without change in the bacterial surface. If the surfactant is added at the same time or immediately after the addition of bacteria the result is generally inhibition of phagocytosis. It is claimed that surfactants which induce the same contact angle in all cell types decrease the adhesion between bacteria and PMN and therefore make exocytosis 'almost as likely as phagocytosis' [83].

References

1. A. N. MARTIN (1961) *J. Pharm. Sci.* **50**, 513.
1a R. WOODFORD (1966) *Pharmacy Digest* **29**, 17.
2. B. A. HAINES and A. N. MARTIN (1961) *J. Pharm. Sci.* **50**, 228, 753, 756.
3. R. G. WILSON and B. ECANOW (1963) *J. Pharm. Sci.* **52**, 757.
4. W. SCHNEIDER, S. STAVCHANSKY and A. N. MARTIN (1978) *Amer. J. Pharm. Educ.* **42**, 280.
5. R. H. OTTEWILL (1976) *Prog. Colloid Polymer Sci.* **59**, 14.
6. H. RUPPRECHT (1978) *Prog. Colloid Polymer Sci.* **65**, 29.
7. P. SOMASUNDARAN, T. W. HEALY and D. W. FUERSTENAU (1966) *J. Colloid Interfac. Sci.*, **22**, 599.
8. N. FUCHS (1934) *Z. Phys.* **89**, 736.
9. R. H. OTTEWILL and A. WATANABE (1960) *Kolloid-Z.* **170**, 38, 132.
10. R. H. OTTEWILL and A. WATANABE (1960) *Kolloid-Z.* **171**, 33.
11. R. H. OTTEWILL and A. WATANABE (1960) *Kolloid-Z.* **173**, 7, 122.
12. P. SOMASUNDARAN and D. W. FUERSTENAU (1966) *J. Phys. Chem.* **70**, 90.
13. B. R. VIJAYENDRAN (1979) *J. Appl. Polymer Sci.* **23**, 733–742.
14. J. V. BONDI, R. L. SCHADARE, P. J. NIEBERGALL and E. T. SUGITA (1973) *J. Pharm. Sci.* **62**, 1731.
15. P. SOMASUNDARAN (1969) *J. Colloid Interfac. Sci.* **31**, 557.
16. J. L. ZATZ, L. SCHNITZER and P. SARPOTDAR (1979) *J. Pharm. Sci.* **68**, 1491.
17. P. CONNOR and R. H. OTTEWILL (1971) *Colloid Interfac Sci.* **37**, 642.
18. J. GREGORY (1977) *Effluent Water Treatment J.* **17**, 641 *et. seq.*
19. K. THOMA, E. ULLMAN and E. WOLFERSEDER (1966) *Archiv. der Pharmazie* **299**, 1020.
20. J. T. PEARSON and G. WADE (1977) *J. Pharm. Pharmacol.* **24**, Suppl. 132P.
21. Y. OHNO, Y. SUMIMOTO, M. O. SHIMA, M. YAMADA and T. SHIMAMOTO (1974) *Chem. Pharm. Bull.* **22**, 2788.
22. W. F. HOWER (1970) *Clays and Clay Minerals* **18**, 97.
23. H. SCHOTT (1964) *Kolloid-Z., Z. Polymere* **199**, 158.
24. H. SCHOTT (1968) *J. Colloid Interfac. Sci.* **26**, 133.
25. D. PLATIKANOV, A. WEISS and G. LAGALY (1977) *Colloid and Polymer Sci.* **255**, 907.
26. R. H. OTTEWILL and T. WALKER (1968) *Kolloid-Z, Z. Polymere* **227**, 108. For further analysis of the results in this paper see P. Bagchi (1975) in *Colloidal Dispersions and Micellar Behaviour* (ed. K. Mittal), ACS, Washington pp. 145 *et. seq.*

27. P. H. ELWORTHY and W. G. GUTHRIE (1970) *J. Pharm. Pharmacol.* **22**, Suppl. 1145.
28. J. M. CORKILL, J. F. GOODMAN and J. R. TATE (1966) *Trans. Farad. Soc.* **62**, 979.
29. J. B. KAYES (1977) *J. Pharm. Pharmacol.* **29**, 163.
30. J. B. KAYES (1977) *J. Pharm. Pharmacol.* **29**, 199.
31. A. OTSUKA, H. SUNADA and Y. YONEZAWA (1973) *J. Pharm. Sci.* **62**, 751.
32. D. A. RAWLINS (1979) PhD Thesis, University of Aston in Birmingham.
33. J. L. ZATZ (1979) *Int. J. Pharmaceutics* **4**, 83.
34. W. HELLER and T. L. PUGH (1954) *J. Chem. Phys.* **22**, 1778.
35. TH. F. TADROS and B. VINCENT (1980) *J. Phys. Chem.* **84**, 1575.
36. K. N. PRASAD, T. T. LUONG, A. T. FLORENCE, J. PARIO, C. VAUTION, M. SEILLER and F. PUISIEUX (1979) *J. Colloid Interfac. Sci.* **69**, 225.
37. J. B. KAYES and D. A. RAWLINS (1978) *J. Pharm. Pharmacol., Suppl.*, 75P.
38. D. W. J. OSMOND, B. VINCENT, F. A. WAITE (1975) *Colloid and Polymer Sci.* **253**, 676.
39. E. J. CLAYFIELD and E. C. LUMB (1966) *J. Colloid Sci.* **22**, 269; 285.
40. E. J. CLAYFIELD and E. C. LUMB (1968) *Macromolecules* **1**, 133.
41. B. VINCENT, P. F. LUCKHAM and F. A. WAITE (1980) *J. Colloid Interfac. Sci.* **73**, 508.
42. A. VRIJ (1976) *Pure Appl. Chem.* **48**, 471.
43. C. COWELL, F. K. R. LI-IN-ON and B. VINCENT (1978) *JCS Farad. I* **74**, 337.
44. TH. F. TADROS (1975) in *Colloidal Dispersions and Micellar Behaviour* (ed. K. L. Mittal), ACS Symp. Series 9 Washington, pp. 173, *et. seq.*
45. H. UNO and S. TANAKA (1970) *Kolloid-Z, Z. Polymere* **242**, 1186.
46. H. UNO and S. TANAKA (1972) *Kolloid-Z, Z. Polymere* **250**, 238.
47. H. UNO and S. TANAKA (1971) *Kolloid-Z, Z. Polymere* **245**, 519.
48. B. DAHNEKE (1972) *J. Colloid Interfac. Sci.* **40**, 1.
49. M. VAN DEN TEMPEL (1972) *Advan. Colloid Interfac. Sci.* **3**, 137.
50. A. MOËS (1970) *J. Pharm. Belg.* **25**, 409.
51. D. N. L. MCGOWN and G. D. PARFITT (1966) *Discuss. Faraday Soc.* **42**, 225.
52. L. A. ROMO (1966) *Discuss. Faraday Soc.* **42**, 225.
53. F. J. MICALE, Y. K. LUI and A. C. ZETTLEMOYER (1966) *Discuss. Faraday Soc.* **42**, 238.
54. H. KOELMANS and J. L. G. OVERBEEK (1954) *Discuss. Faraday Soc.* **18**, 52.
55. P. BAGCHI and R. D. VOLD (1970) *J. Colloid Interfac. Sci.* **33**, 405.
56. B. VINCENT (1973) in *Colloid Science* Vol. 1, Chemical Society, London 238.
57. J. A. BZDAWKA and D. T. HAWORTH (1980) *J. Disp. Sci. Tech.* **1**, 323.
58. N. A. ARMSTRONG and C. D. CLARKE (1976) *J. Pharm. Sci.* **65**, 373.
59. S. A. SAID, A. M. SHIBL and M. E. ABDULLAH (1980) *J. Pharm. Sci.* **69**, 1238.
60. M. VAN DEN TEMPEL (1971) in *Elasticity, Plasticity and the Structure of Matter*, 3rd edn (eds R. Houwink and H. K. de Decker), Cambridge University Press, Cambridge, pp. 123 *et. seq.*
61. N. MORIYAMA (1975) *J. Amer. Oil Chemists Soc.* **52**, 198.
62. W. B. RUSSEL (1980) *J. Rheology* **24**, 287.
63. TH. F. TADROS (1980) *Advan. Colloid Interfac. Sci.* **12**, 141.
64. J. C. SAMYN (1961) *J. Pharm. Sci.* **50**, 517.
65. A. S. MICHAELS and A. R. COLVILLE (1960) *J. Phys. Chem.* **64**, 13.
66. J. E. FAIRBROTHER and D. J. W. GRANT (1978) *J. Pharm. Pharmacol.* **30**, Suppl. 19P.
67. W. L. CHIOU, S. J. CHEN and N. ATHANIKAR (1976) *J. Pharm. Sci.* **65**, 1702.
68. TH. F. TADROS (1973) *Particle Growth in Suspensions* (ed. A. L. Smith) Academic Press, London p. 221 *et. seq.*
69. A. P. SIMONELLI, S. C. MEHTA and W. I. HIGUCHI (1970) *J. Pharm. Sci.* **59**, 633.
70. J. E. CARLESS, M. A. MOUSTAFA and H. D. C. RAPSON (1960) *J. Pharm. Pharmacol.* **20**, 639.
71. S. A. HOWARD, J. MANGER, J. W. HSIEH and K. AMIN (1979) *J. Pharm. Sci.* **68**, 1475.
72. U. VOIGT (1977) in *Formulation and Preparation of Dosage Forms* (ed. J. Polderman) Elsevier/North Holland, Amsterdam pp. 45 *et seq.*
73. D. J. A. CROMMELIN and C. J. DE BLAEY (1980) *Int. J. Pharmaceutics* **5**, 305.

74. M. C. GUTTERIDGE and H. E. C. WORTHINGTON (1980) *J. Pharm. Pharmacol.* **32** Suppl. 63P.
75. C. FUCHS (1949) *Drug and Cosmet. Ind.* **64**, 692.
76. M. J. HILL, A. M. JAMES and W. R. MAXTED (1967) *Biochim. Biophys. Acta.* **66**, 264.
77. A. M. JAMES (1979) *Chem. Rev.* **8**, 389.
78. A. S. G. CURTIS (1973) *Progr. Biophys. Mol. Biol.* **27**, 317.
79. B. A. DETHICA (1961) *Exp. Cell Res.* Suppl, **8**, 123.
80. H. SCHOTT and C. Y. YOUNG (1972) *J. Pharm. Sci.* **61**, 762.
81. A. M. JONES (1957) *Prog. Biophys. Biophys. Chem.* **8**, 98.
82. P. SHERMAN (1975) *Chemistry in Britain* **11**, 321.
83. C. J. VAN OSS (1978) *Ann. Rev. Microbiol.* **32**, 19.
84. A. M. EHRLY (1968) *Biorheology* **5**, 209.

10 *Aspects of surfactant toxicity*

10.1 Introduction

Toxicity is a manifestation of an unwanted biological activity. In this chapter aspects of surfactant-induced toxicity will be discussed. The nature of the subject demands that it is dealt with on a selective basis; no attempt will be made here to be comprehensive. Some effort will be made, however, to relate surfactant properties with their biological effects. These biological effects, which are occasionally put to therapeutic use, are here regarded as side effects or toxic effects when elicited on administration of a medicine containing surfactants. Biological responses leading to toxicity which are the result of surfactant-induced hyper-absorption of another active ingredient such as discussed in Chapter 7, will not be considered. The chapter is divided into several parts. In the first part interaction of surfactant with biological substances such as proteins, lipids, membranes and enzymes will be surveyed in the hope that this might aid our understanding of reported 'toxicities'. The second part of the chapter reviews reports on adverse reactions and unwanted effects derived from pharmaceutical products in animal and human subjects. The study of the absorption, distribution and metabolism of surfactant molecules is not a well developed art. We know too little of this topic, yet it is vital for our understanding of the interactions of amphipathic molecules with their biological environments. The impurity or heterogeneity of most commercial surface-active agents, however, does not assist us in gaining an understanding at the molecular level of toxic events which they might induce.

It is the ability of surfactants to adsorb at interfaces and bind through hydrophobic interactions to proteins and to solubilize components of membranes that implicates them in interactions with cell membranes and proteins – especially enzymes. Thus the physical act of interaction can lead to a disruption of normal cell function and to increased or decreased enzymic activity. Tissue damage is an overt manifestation of surfactant interactions with body components, as is haemolysis.

The toxicological properties of surfactants have been reviewed by Elworthy and Treon [1], in 1967, and by Gloxhuber [2] in 1974. Guidance on the suitability of surfactants for use in pharmaceuticals can be gained from lists of approved substances for use in foodstuffs (substances Generally Regarded As Safe, GRAS). Nine types of emulsifier are listed GRAS in the US Code of Federal Regulations,

Table 10.1 Alphabetical listing of FDA-approved emulsifiers*

FDA name (other names in parentheses)	CFR[†] section	HLB
Acetylated monoglycerides	121.1018	~ 3
Calcium stearoyl-2-lactylate	121.1047	2–3
Diacetyl tartaric acid esters	121.101	8–9
of mono- and diglycerides	(GRAS)	
Dioctyl sodium sulphosuccinate	121.1137	
Ethoxylated mono- and diglycerides	121.1221	13.1
Fatty acids	121.1070	1
Glyceryl-lacto esters of fatty acids	121.1004	7–12
Hydroxylated lecithin	121.1027	Med.
Lactylated fatty acid esters of glycerol and		
propylene glycol	121.1122	Med.
Lactylic esters of fatty acids	121.1048	8.8
Lecithin	121.101	Med.
	(GRAS)	
Mono- and diglycerides	121.101	~ 3
	(GRAS)	
Monosodium phosphate derivatives of		
mono- and diglycerides	121.101	Med.
	(GRAS)	
Polyglycerol esters of fatty acids	121.1120	Broad
Polyoxyethylene (20) sorbitan		
tristearate (polysorbate 65)	121.1008	10.5
Polysorbate 60 (polyoxyethylene (20) sorbitan		
monostearate)	121.1030	14.9
Polysorbate 80 (polyoxyethylene (20) sorbitan		
mono-oleate)	121.1009	15.9
Propylene glycol mono- and diesters of fats		
and fatty acids	121.1113	2–5
Sodium lauryl sulphate	121.1012	High
Sodium stearoyl-2-lactylate	121.1211	2–3
Sodium stearyl fumarate	121.1183	—
Sorbitan monostearate	121.1029	4.7
Stearyl monoglyceridyl citrate	121.1080	Low
Succinylated monoglycerides	121.1195	~ 3

* From Petrowski [4].
† Code of Federal Regulations, Food, Drug Administration, Dept of Health, Education and Welfare Title 21 Chapter 1 Sub Chapter B Food, Food Products Part 121 Food Additives Subpart B.

Sub part B; in addition to these another group of emulsifiers approved as food additives in sub part D [3] are declared safe when used in the quantities and for the applications mentioned in the lists. A listing of FDA-approved emulsifiers is given in Table 10.1. The safety of the mono- and diglycerides and polysorbates as food additives has been reviewed [5].

10.2 Metabolism of surfactants

10.2.1 Non-ionic surfactants

For a full understanding of the significance of *in vitro* studies on surfactant–membrane or surfactant–protein interactions, information on the metabolic fate of ingested surfactants is required. Studies on this topic are few. It has, however, been demonstrated that the ester link of polysorbate surfactants is split by intestinal lipase; the polyoxyethylene glycol moiety is not well absorbed and most is found thereafter in the faeces. Small amounts which are absorbed are excreted in the urine [6, 7]. Thus observed actions of polysorbates on cell membranes *in vitro* must be considered in the light of the subsequent breakdown of the molecule *in vivo*. Both the route of administration and the surfactant molecule influences the fate of the surfactant. Oral and parcutaneous absorption of a series of alkyl polyoxyethylene esters has been studied [18] in rats and humans. Rapid and extensive absorption occurred from the gastro-intestinal tract, more than 75 % of the dose being absorbed whereas intravenous doses were absorbed slowly, less than 50 % in 72 h. About 94 % of a 7 or 100 mg kg^{-1} dose of a polyoxypropylene–polyoxyethylene block co-polymer (poloxamer 108, Pluronic F38) administered i.v. to rats was excreted in 3 days. About 6 % of the polymer appeared in the faeces [9]. The rate of disappearance of the surfactant polymer from the plasma is relevant to the comprehension of the events following i.v. administration of a solubilized preparation; the relative rates of disappearance of drug and surfactant might determine the behaviour of the solubilizate species. 91 % of the administered dose of poloxamer 108 appears in the urine in the first 20 h, 22 % reaching the kidney in 3 min. (see Fig. 10.1). The distribution of the surfactant in the tissues 3 min and 20 h after administration is described in Table 10.2.

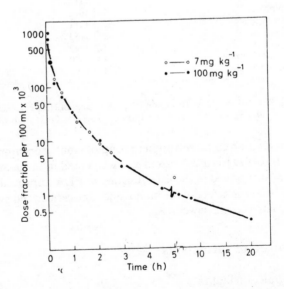

Figure 10.1 Time-course of ^{14}C-labelled F-38 (poloxamer 108) disappearance from rat plasma after intravenous administration of 7 or 100 mg kg^{-1}. From Wang and Stern [8] with permission.

Table 10.2 Distribution of ^{14}C-label in tissue and excreta after intravenous administration of $100 \, \text{mg kg}^{-1}$ of Poloxamer 108

Tissue	μg per g tissue		% administered dose	
	3 min	20 h	3 min	20 h
Carcass*	155	5	37.8 ± 2.6	1.1 ± 0.2
Kidney	10 660	177	22.5 ± 3.2	0.4 ± 0.1
Plasma	808	<1	$19.5 \pm 2.9^\dagger$	$0.0\S$
Small intestine	435	8	4.4 ± 0.7	0.1 ± 0.1
Liver	353	25	4.1 ± 0.5	0.4 ± 0.1
Lung	1 491	13	2.4 ± 0.4	0.0
Large intestine	203	11	1.6 ± 0.3	0.1
Pancreas	412	9	1.5 ± 0.2	0.0
Muscle	144	3	$1.2 \pm 0.3^\ddagger$	0.0
Salivary glands	606	9	0.9 ± 0.2	0.0
Stomach	246	3	0.8 ± 0.1	0.0
Heart	719	12	0.7 ± 0.1	0.0
Epididymal fat	97	2	0.4 ± 0.1	0.0
Gonads	137	5	0.4 ± 0.1	0.0
Bladder	853	12	0.3 ± 0.3	0.0
Spleen	346	9	0.3 ± 0.1	0.0
Thymus	320	6	0.2 ± 0.1	0.0
Diaphragm	322	6	0.2 ± 0.02	0.0
Brain	100	3	0.2 ± 0.04	0.0
Thyroid	610	7	0.1 ± 0.02	0.0
Urine		915		90.6 ± 4.7
Faeces		59		6.6 ± 4.0
Total recovery			99.5 ± 4.8	99.5 ± 8.8

* Defined as remainder of animal after bleeding and removing the whole organs or tissue samples described.
$\dagger$ Contained in approximately 5 ml plasma.
$\ddagger$ Contained in approximately 9 g wet muscle tissue sample.
$\S$ Less than 0.1 %.
From [9].

Table 10.3 Distribution of radioactivity in the gastro-intestinal tract of rats 8 h after oral intubation of $0.8 \, \text{mg} \, [^{14}\text{C}] \, \text{CTAB kg}^{-1}$

Region of gut	Radioactivity (% of administered dose*)
Stomach	$10.0 \pm 3.1^\dagger$
Small intestine, proximal half	$2.5 \pm 0.3^\dagger$
Small intestine, distal half	$13.5 \pm 1.5^\dagger$
Caecum and colon	$53.8 \pm 2.4^\dagger$
Complete gastro-intestinal tract: Total	$79.9 \pm 1.3^\dagger$
Contents	69.8 ± 2.4
Wall	10.1 ± 2.3

* Values represent means $\pm$ S.E.M. for five rats.
$\dagger$ Wall plus contents.
From [10].

10.2.2 Cationic surfactants

The absorption, distribution and excretion of a quaternary ammonium surfactant, cetyltrimethylammonium bromide (CTAB) has been observed in the rat [9] following oral administration. About 80% of the dose of ^{14}C-labelled CTAB was found in the gastro-intestinal tract 8 h after administration (Table 10.3). The low levels in serum and bile confirm the low level of intestinal absorption. After 3 days, 92% of the administered dose was excreted in the faeces and less than 1% in the urine. This poor absorption might be expected considering that monoquaternary ammonium salts with anticholinergic activity are absorbed to the extent of only 10 to 20% after 3 to 4 h in the intestine [11], and this probably explains why the quaternary ammonium surfactants are 10 to 100 times more toxic when administered intravenously [12]. While absorbed CTAB does not appear to concentrate in any particular organ, 0.8% of the administered radioactivity is discovered in the liver 8 h after administration. Being highly polar the surfactant is not readily metabolized but it is likely to influence enzyme activity in the liver. Cutler and Drobeck's review [13] on the toxicology of cationic surfactants should be consulted for more detailed information.

10.2.3 Anionic surfactants

The anionic alkyl sulphates appear to be rapidly absorbed from the gastro-intestinal tract. In experiments in which undecyl (^{35}S) sulphate was administered by intraperitoneal, oral or i.v. routes, most of the injected radioactivity was excreted in the urine of rats; a difference in the absorption of C_{10}(^{35}S) sulphate and C_{18}(^{35}S) sulphate has been demonstrated. Intraperitoneal injections of the former result in the recovery of 60% of the doses in the urine in 6 h while with the latter 5 to 26% is recovered in the same period. The C_{18} compound is excreted faster than the C_{10} surfactant when administered orally, this leading to the postulate of absorption of both monomers and micelles across the intestinal mucosa [14]. There appeared also to be a sex-linked difference in excretion patterns in rats, probably due to the greater capacity of the female rats to metabolize the sulphate ester. Both linear alkyl benzene sulphonates (LAS) and branched chain alkyl benzene sulphonates (ABS) are readily absorbed from the gastro-intestinal tract [15]. LAS and its metabolites (sulphophenylbutyric acid and sulphophenylvaleric acid) are excreted chiefly in the urine. ABS appear mostly in the faeces. Orally administered ^{35}S-sodium dodecyl sulphate is subsequently found in liver, kidney, blood, spleen, lung and brain [16]; 80.4% of 2-ethylhexyl sulphate appears in the urine along with its metabolite 2,3-dihydroxycaproic acid [17]. Attempts to assess surfactant effects in isolated systems should encompass, wherever possible, the effect of the principal metabolites.

Straight-chain anionic detergents are degraded in the rat by a ω, β oxidation, the metabolic products being carboxylic acid derivatives of short-chain alkyl sulphates. Anionic surfactants with even numbered chains are metabolized to butyric acid 4-sulphate and this is excreted in the urine [19–22]. The odd-chain length surfactant undecyl sulphate is eliminated in the urine as propionic acid 3-

sulphate with two other products [20]. Introduction of a terminal unsaturated linkage (as in 10-undecyl sulphate) or a terminal phenyl group (e.g. 10-phenyl decyl sulphate) does not prevent metabolism [21, 22] but the products of metabolism are not the same as that of the 'parent' compounds. The metabolism of the phenyl-substituted compound differs considerably from that of the decyl sulphate. ω, β oxidation is blocked. Biliary elimination of conjugated hydroxylated products occurs; so far this is the only alkyl sulphate ester which is known to be excreted to a major extent by this route.

Burke *et al.* [21, 22] conclude that 'the fact that both unsubstituted and substituted alkyl sulphate esters are concentrated and metabolized in the liver and that the route of elimination appears to be dependent on the nature of structural modifications, may have important implications in the design of commercial anionic detergents'.

10.3 Interactions of surfactants with membranes and membrane components

10.3.1 Membrane disruption by surfactants

The disruption of membrane integrity and function by surface-active compounds is at the centre of many of the observed biological effects of surfactants. Many surfactants have been studied to quantify their usefulness as solubilizing agents for membrane components for subsequent biochemical study. Non-ionic surfactants have been most widely used as 'chemically mild and efficient chaotropic agents' [23]. Steps in the solubilization of the components of biological membranes by a non-ionic detergent are shown in Fig. 10.2. The ratio of detergent to lipid is important in determining the exact nature of the interaction between amphipaths and membrane [24] and the nature of the biological membrane is also important. A study of the solubilization of mitochondrial inner membrane, microsomal and erythrocyte membrane components by Triton X-100, NaDS and sodium deoxycholate [26] has shown considerable differences between the surfactants in the percentage solubilization of protein and lipid phosphorus from the mitochondrial inner membrane. This is probably because of the high protein/lipid ratio in this membrane (see Table 10.4). Human erythrocyte ghosts contain a relatively high proportion of cholesterol. Loizaga *et al.* [26] find detergent/protein ratios more reliable than detergent/lipid ratios in assessing the extent of membrane solubilization, and these workers suggest that the systematic use of this parameter would make data from different laboratories more readily comparable. Chemical and ultrastructural studies suggest that the first step in the membrane solubilization process [25] involves solution of the protein constitutents associated with membrane structure, at surfactant concentrations up to 0.1 %. In the next stage lipids are solubilized after liberation from lipoprotein complexes (0.1 to 0.5 % surfactant), selective solubilization of components is possible with some surfactants and membranes. Other sequences have been suggested which probably are, in fact, the same as those depicted in Fig. 10.2. The

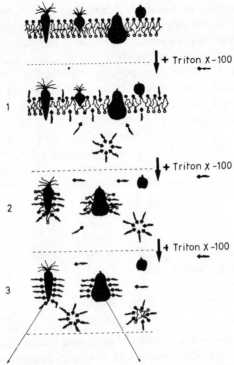

Figure 10.2 Solubilization of the biological membranes by non-ionic detergents. The membrane model is taken from [24]. Depending on the ratio of detergent–membrane lipids, different steps of solubilization may be obtained. In step 1, when a small amount of detergent is present, the molecules of detergent are incorporated into the membrane without breaking it. In step 2, the membrane is solubilized into micellar solution containing mixed protein–lipid–detergent micelles in equilibrium with detergent micelles and free detergent molecules. Finally (step 3), when enough detergent is added, pure protein–detergent micelles may be obtained in equilibrium with detergent–lipid and detergent micelles. From Gulik–Krzywicki [25] with permission.

Table 10.4 Detergent/protein (w/v) ratios for 90% maximal solubilization of membrane protein

Membrane	Detergent		
	Deoxycholate	Sodium dodecyl sulphate	Triton X-100
Mitochondrial inner	1.9	1.0	2.5
Human erythrocyte	5.0	0.9	4.7
Microsomal	2.9	2.2	3.3

From [26].

The membranes were obtained and purified by standard procedures. They were resuspended in 0.25 M-sucrose/10 mM-*Tris*/HCl buffer, pH 7.4, to a concentration of 1 mg protein. Detergents were then added to obtain final concentrations between 0.1 and 1.0% (w/v), and the suspensions were incubated at 20° C for 30 min. Each preparation was centrifuged for 1 h at 150 000 $g_{av.}$, at 4° C (Triton X-100 and deoxycholate treatments) or 18° C (sodium dodecyl sulphate). The supernatant was assumed to contain the solubilized fraction of the membranes.

results from Foster *et al.*'s [28] study of the action of non-ionic detergent on a kidney membrane fraction suggest that membrane disruption involves binding of detergent monomers to exposed polar segments of membrane protein followed by the formation of co-micelles of the detergent with segments of the membrane. Interactions below the CMC are evident and are shown to change only in magnitude at the CMC when data are analysed by Hill plots [28] as in Fig. 10.3a and b, where protein solubilization is recorded after treatment with a mixed detergent (LOC) and sodium dodecyl sulphate. Since both polar and hydrophobic forces are operating in membrane disruption by detergents, the former primarily in initial adsorption of ionic surfactants, the ionic strength of the system should affect the processes involved. An observed increase in (Na^+, K^+)-ATP-ase activity associated with increased ionic strength is in agreement with this proposal [28].

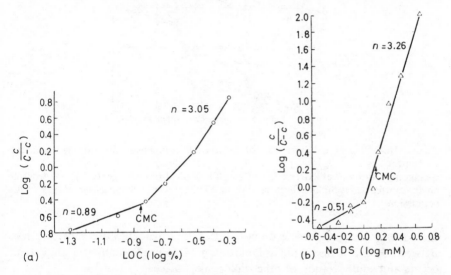

Figure 10.3(a) Hill plot of sup-20 protein with LOC treatment. *c*, protein concentration, $mg\,ml^{-1}$. *C*, maximal = 24%. The vertical arrow (↑) indicates the CMC. (b) Hill plot of erythrocyte membrane protein solubilized with NaDS treatment. *c*, protein concentration, $mg\,ml^{-1}$. *C*, maximal = 100%. The ratio of *n* for the part of the curve above the break to that below indicates an increased level of binding above the CMC, in (a) of 3.4 times, and in (b) of 7 times. From Foster *et al.* [28] with permission. LOC is composed of a diethanolamide/polyoxyethylene alcohol mixture.

Fitzpatrick *et al.* [29] have concluded that NaDS binds to all components of erythrocyte membranes uniformly; release of components approximately follows the order of their water solubility, lipid solubilization occurring around the CMC of NaDS. About all the lipid is dissolved at 10 mM. Between 2 and 10 mM NaDS (see Fig. 10.4) lipids are solubilized and form mixed micelles capable of solubilizing cholesterol. The exact sequence of events and nature of the co-micelles, etc., has still to be elucidated. The authors found the Triton X-100 behaviour difficult to interpret. Although its CMC is $1\,mmol\,l^{-1}$ it is possible

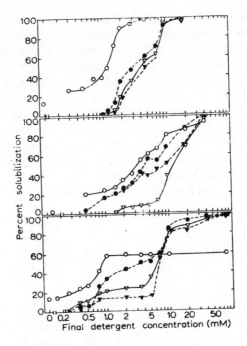

Figure 10.4 Percent solubilization of membrane components at various detergent concentrations. ○—○, total protein; ●---●, total lipid: △—△, cholesterol; ▲---▲, sphingomyelin. Point at extreme left of graph is control (no detergent). Top, dodecysulphate; centre, deoxycholate; bottom, Triton X-100. From Fitzpatrick *et al.* [29] with permission.

that, due to binding, micelles do not appear until five times this concentration. Between 5 and 10 mM rapid solubilization of sphingomyelin and cholesterol occurs and solubilization of other lipids takes place.

Such interactions are directly relevant to the effect of surfactants on drug absorption. Fig. 10.5a shows the relationship between absorption of salicylate or L-valine across rat jejunal tissue *in vivo* and release of protein and phospholipid by a series of surfactants. Fig. 10.5b shows the relative activity of a series of non-ionic surfactants of the Brij series and sodium taurodeoxycholic acid (NaTDC), sodium dodecyl sulphate (NaDS) and CTAB [30]. An association between protein release and increased absorption of solutes has also been reported by Feldman and Reinhard [31] and Walters *et al.* [32]. Anionic, non-ionic and cationic surfactants tended to accelerate the breakdown of the mucous layer covering the epithelium and at high concentrations were thought to interfere with the structure of the mucosal surface itself. Exposure of various anionic and non-ionic surfactants to rabbit intestinal mucosa causes epithelial desquamation and necrosis [33].

Perfusion of 1 % NaDS at pH 7.4 in phosphate buffer caused great histological damage to the rat intestinal mucosa: separation of the epithelium from the lamina

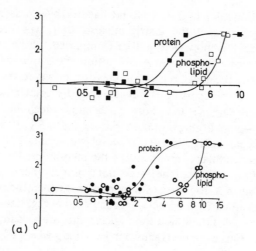

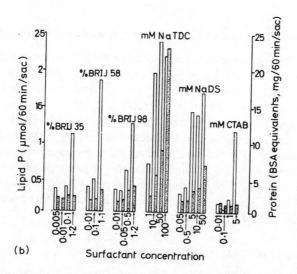

Figure 10.5(a) Relationship between absorption of salicylate (upper) or L-valine and the release of protein and phospholipid in the presence of the various surfactants. ●, ■, protein; ○, □, phospholipid. (Values pertaining to salicylate and the non-ionic surfactants have been omitted because the absorption rates were largely unaffected. Ordinates: Final serosal concentration of salicylate or L-valine (upper and lower figure respectively) relative to control value; upper jejunal values. Abscissa: Protein or phospholipid release relative to control value. (b) The release of protein and lipid phosphorus from the mucosal surface of rat everted jejunal sacs by incubation in the presence of various surfactants. Protein concentrations are shown by stippled lines and lipid phosphorus in outline. Release values by control preincubation solutions were 0.22 μmol lipid phosphorus per 60 min incubation per 3 cm sac and 1.6 mg protein per 60 min incubation per 3 cm sac. The surfactants gave negligible or zero levels of phosphorus by the assay method. From Whitmore *et al.* [30].

propria, destruction of blood vessels and haemorrhage was noted [34]. The histological appearance of the gastric mucosa after exposure to a range of surfactants for 0.25 h is shown in Fig. 10.6. Comparison of the treated specimens with the control shows in several cases disruption of the mucosal border. It is not possible to quantify the effect: it is our impression, however, that the surfactants which increased the transport of paraquat across the isolated gastric membrane caused membrane damage and those surfactants which had little effect on transport had a negligible effect. In the series shown in Fig. 10.6 sodium dodecyl sulphate (b) has a damaging effect at 0.1 % level, with loss not only of the mucosal border but deeper cellular components of the membrane. Polysorbate 80 (c) at 0.1 % levels results in the loss of the border but minimal other damage; at the same concentration Triton X-100 (d) causes more damage than the polysorbate. Two poloxamers (Pluronic P75 and F68) were compared in Fig. 10.6e and f. At higher concentrations (~ 1 %) these non-ionic surfactants are as damaging as sodium dodecyl sulphate. One example is shown in Fig. 10.6 g where 1 % Brij 35 has been applied to the mucosa.

10.3.2 Other effects of surfactants on the gastro-intestinal tract

Surfactants have other effects on gastro-intestinal function which might alter drug absorption: stimulation of mucus production has been suggested [33] and the influence of surfactant species on gastric emptying and intestinal transit is likely to affect drug absorption indirectly [35, 36]. CTAB in doses up to $8 \, mg \, kg^{-1}$ accelerates gastric emptying in the rat and hence increases the absorption rate of tripalmitate. At doses of $400 \, mg \, kg^{-1}$ the detergent retards gastric emptying [37]. Sodium deoxycholate decreases the rate of gastric emptying [38] as does dioctylsulphosuccinate [35] administered orally in a dose of 400 or $1600 \, mg \, kg^{-1}$ to rats. Inhibition of peristalsis of rat jejunum has been caused *in vitro* by nonionic surfactants of the Brij class, their potency increasing with increasing alkyl chain length [39].

10.3.3 Fusion of membranes

The stability and lack of coalescence of normal biological membranes is probably a consequence of the thermodynamic stability of the lipid bilayer structure. Lucy [40] suggested that membrane fusion, which can be observed on the addition of lysolecithin and other *fusogens* is the result of the transformation of parts of the bilayer structure into globular micellar form. Any substance which favours this transition from bilayer to micellar structure should facilitate membrane fusion. Lysolecithin, as a consequence of its surface activity and wedge shape, induces cell fusion, so too does retinol (vitamin A alcohol) which is also surface active. Triton X-100 increases the permeability of liposomes and planar lipid bilayers without disrupting their structures, and, not surprisingly, this non-ionic surfactant has been found to act as a fusogen [41] by, it is thought, the formation of inverted micelles with lipid molecules. Several fusogenic lipids behave as immunological

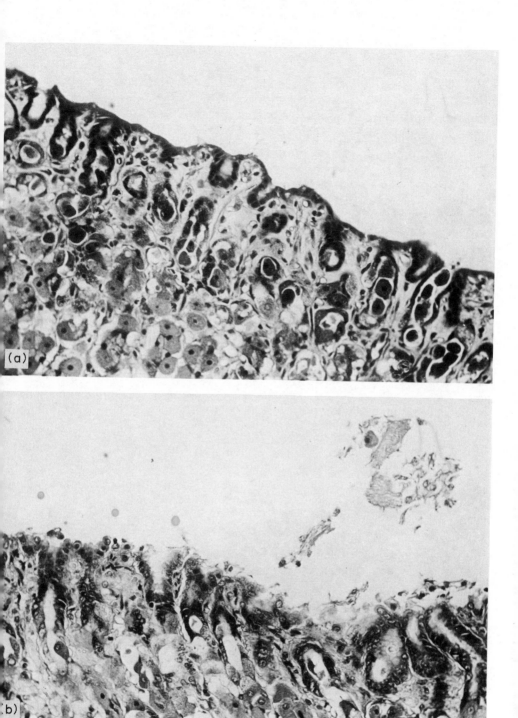

Figure 10.6 (a), (b) (see page 628 for caption)

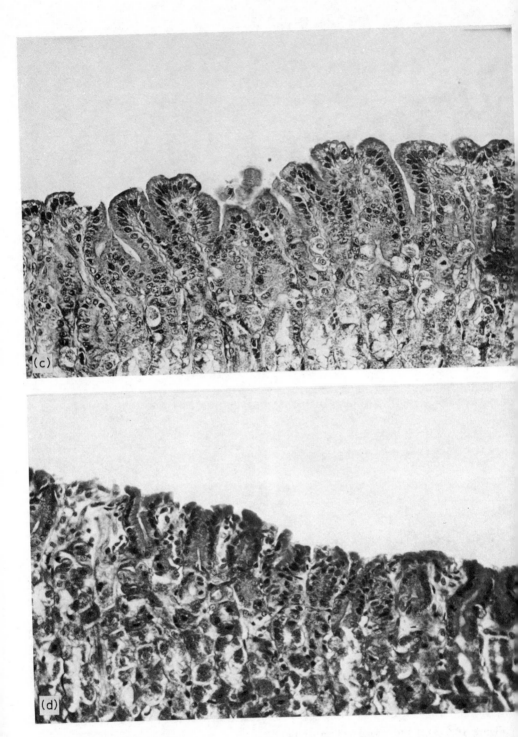

Figure 10.6 (c), (d)

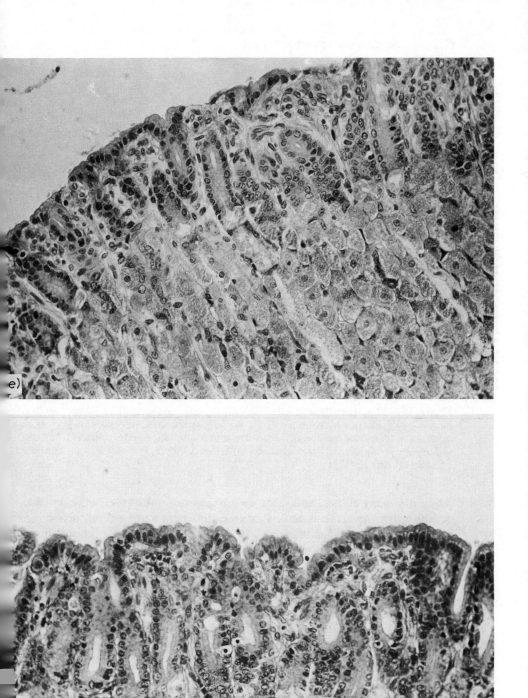

Figure 10.6 (e), (f)

Figure 10.6 Photomicrographs (× 312.5 in original photograph) of isolated gastric membrane of the rat, by courtesy of Dr K. A. Walters. (a) Control and (b)–(g) after exposure of the gastric mucosa to 0.1 % surfactant solutions for approximately 0.25 h; (b) sodium lauryl sulphate, (c) polysorbate 80, (d) Triton X-100, (e) Pluronic P75, (f) Pluronic F68 and (g) Brij 35 (1 %).

adjuvants; many of these adjuvants are surface-active substances which are able to labilize lysosomal membranes. It has therefore been suggested that actions on the cell or lysosomal membrane may be of some importance in adjuvant effects [42]. Components of Freund's adjuvants such as the emulsifier Arlacel A are active fusogens [42]. Span 80 and Span 20 have also been found to be able to fuse hen erythrocytes, but Span 40 and Span 60 were found to be inactive in this regard, consistent with previous findings that fusion occurs with esters of both medium-chain saturated fatty acids and with longer chain unsaturated fatty acids, but not in general with esters of palmitic and stearic acids.

10.3.4 Surfactant–protein interactions

Interactions between surfactants and proteins can lead to solubilization of insoluble membrane-bound proteins or to changes in the biological activity of enzyme systems. Obviously some understanding of such interactions may give some clues as to the possible interaction of surfactants in biological surroundings. Interactions between surfactants and water-soluble proteins also occur, many studies having been carried out using serum albumin.

It has been known for a long time that surfactants could precipitate, form complexes with or denature proteins at low concentrations. Early work with albumin suggested that the presence of a polar group in the surfactant molecule was a prerequisite to interaction with protein. The 'lack of interaction' of non-ionic surfactants and proteins, it was implied, had important consequences, since it divorced the non-ionics from biologically deleterious effects such as denaturation and inactivation of proteins. This view has since had to be modified as non-ionic surfactants are not the mild agents that they appeared to be. Hydrophobic interaction of the hydrocarbon chain of the non-ionic surfactant can result in a modification of the quaternary structure of proteins in solution. Possible modes of interaction of anionic surfactants are shown in Fig. 10.7.

Detergents produce conformational changes in proteins at low concentrations, the surfactants combining with native proteins in multiple equilibria (i.e. many mols per mol of protein) on unfolding, the protein will bind more detergents as more binding sites are exposed [45]. Dissociation constants and numbers of

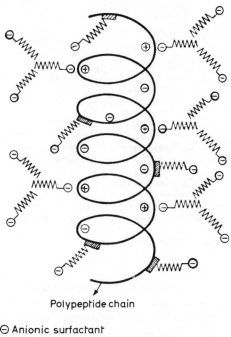

Polypeptide chain

MMMM⊖ Anionic surfactant

▨ Hydrophobic residue of the aminoacids

Figure 10.7 A diagrammatic representation of modes of binding of an anionic surfactant to a protein, after Dominguez [43]. Some modes of interaction, especially hydrophobic interactions with the amino acid hydrophobic residues would be equally appropriate for non-ionic and cationic surfactants. In addition, cationic surfactants could attach themselves electrostatically to the anionic sites. The hydrophobic interactions are supported by the results of Helenius and Simons [44] which demonstrated that lipophilic proteins bound up to about 70% of their weight of deoxycholate and Triton X-100 but that hydrophilic proteins bound little surfactant.

binding sites for bovine serum albumin (BSA), β-lactoglobulin and other proteins are shown in Table 10.5. The binding of detergent molecules to high-affinity sites does not result in denaturation [46] at low detergent concentration. Binding, a function of the free surfactant concentration in equilibrium with the protein, is influenced by pH, temperature, and ionic strength. The monomeric species attaches to the protein when both protein and ligand are together at low concentrations. Increasing surfactant concentration above the CMC thus results in little increase in binding. There is some evidence that proteins may adsorb on detergent micelles: pancreatic colipase, for example, is bound to bile salt micelles [56].

Table 10.5 High-affinity sites of proteins for detergents. K_d is the concentration required for 50% saturation of stray sites.

Protein	Detergent	Number of sites	K_d (mol l^{-1})	References
BSA*	Sodium dodecyl sulphate	8–10	9.1×10^{-7}	[47]
	Sodium tetradecyl sulphate	10–11	6.7×10^{-7}	[47]
	$C_{14}NMe_3Cl$	4	6.7×10^{-5}	[49]
	Triton X-100	4	5×10^{-5}	[50]
	DOC	4	1.5×10^{-5}	[50]
β-Lactoglobulin	Sodium dodecyl sulphate	2	—	[51–53]
Human Apo-A-I	Sodium dodecyl sulphate	4	5×10^{-5}	[54]
	$C_{14}NMe_3Cl$	4	5×10^{-5}	[55]
Human Apo-A-II	Sodium dodecyl sulphate	4	5×10^{-5}	[54]
	$C_{14}NMe_3Cl$	10	10^{-4}	[55]
Pyruvate oxidase†	Sodium dodecyl sulphate	2–4	$<5 \times 10^{-5}$	

* More references on serum albumin binding of detergents will be found in Steinhardt and Reynolds [48].
† Schrock and Gennis, unpublished work
DOC = deoxycholate
From Gennis and Jonas [46].

From Table 10.5 it is apparent that long-chain alkyl sulphates are bound to BSA in higher numbers and with higher affinity than cationic or non-ionic surfactants or bile salts. Helenius and Simons [57] believe this to be a natural consequence of the major biological role of serum albumin as a carrier of fatty acid anions in the circulation. The high-affinity sites seem to be hydrophobic in nature and exist as 'patches' or 'crevices' a feature not evident in other water-soluble polymers. However, it had been suggested [58] that the binding sites of carboxylates and sulphates or sulphonates are different.

Sodium dodecyl sulphate is bound co-operatively to most protein, the critical concentration for co-operative binding being about 25% of the CMC. Non-ionic surfactants and the bile salts do not usually induce co-operative binding and do not usually, therefore, denature proteins, although dissociation into inactive or less active subunits can occur. The cationic surfactants must reach a higher concentration before co-operativity is detected. The critical concentration for this is often close to the CMC, thus co-operative binding is limited. Helenius and Simons [57] speculate from this information that more proteins should be resistant to the denaturing effects of cationic detergents than to the effect of NaDS. Ribonuclease A, they point out, is denatured by the anionic surfactant but not by dodecyltrimethylammonium bromide [59]. The reversibility of the binding of cationic surfactants to myoglobin contrasts with the irreversible effects of NaDS on its conformation. NaDS also causes irreversible deactivation of trypsin [60]. Incubation of trypsin with benzyalkonium chloride (a mixture of C_8 to C_{18} alkyldimethylbenzylammonium chlorides) leads to inactivation below pH 3 but no effect above pH 5 [61]. This finding led to the suggestion that the protonation of a side chain carboxyl on the enzyme allowed the benzalkonium molecules to reach the hydrophobic bonding site; however, no such inactivation has been detected with cetyl dimethylbenzylammonium chloride below pH 3 [60] and the results have not been recorded. The effect of hydrocarbon chain length of the cation is shown in Table 10.6 and the influence of pH in Fig. 10.8. Studies on the inactivation of preparations of influenza and psittacosis viruses by surfactants have confirmed the importance of the ionic nature of the surfactant, the ratio of surfactant to protein and the pH of the solution [61].

As there is a considerable body of data on the binding of anions and cations to proteins [45, 46, 57] only non-ionic surfactant–protein interactions will be considered in detail here. Using pure alkyl glucosides rather than the hetero-geneous polyoxyethylene alkyl ethers, Wasylewski and Kozik [58] studied the binding of octyl and decyl glucosides to BSA and the influence of the octyl, decyl

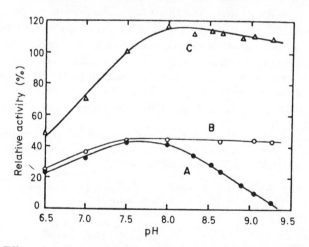

Figure 10.8 Effect of pH on the inhibition of tryptic activity by cetyl dibenzylammonium ions (curve A) or butylamine (curve B) and in the absence of the inhibitor (curve C) at 25 °C in 0.1 M phosphate buffer. The concentration of the inhibitors was 3.8 mM. The activity obtained in the absence of the inhibitor at pH 7.5 was taken as 100%. From Nakaya *et al.* [60] with permission.

Table 10.6 Inhibition of tryptic activity by various cationic detergents and amines which contain various lengths of hydrocarbon chains. The activity was measured in 3.8 mM of the reagents at 25° C and using BAPA* as substrate. (pH 7.5, 25° C)

Reagent		Relative activity (%)
	Control	100
(C₂)	Ethylamine	100
(C₄)	Butylamine	51
(C₆)	Hexylamine	77
(C₈)	Octylamine	99
(C₁₀)	Decylamine	—
(C₁₂)	Dodecylpyridinium chloride	100
(C₁₄)	Myristoylpyridinium chloride	95
(C₁₆)	Cetylpyridinium chloride	48
(C₁₆)	Cetyltrimethylammonium chloride	49
(C₁₆)	Cetyldimethylbenzylammonium chloride	49
(C₁₈)	Octadecylpyridinium chloride	60

From [60].
* BAPA = N^α benzoyl-L-arginine *p*-nitroanilide.

and dodecyl analogues on conformational changes in the albumin. The equation used by Reynolds *et al.* [47] was employed to interpret the isotherms. This takes the form:

$$\frac{1}{\bar{r}} = \frac{1}{nK} \cdot \frac{1}{C_d} + \frac{1}{n} \tag{10.1}$$

where $\bar{r}$ is the average number of detergent molecules bound per molecule of protein, K is the intrinsic association constant, n is the total number of sites with an association constant K and C_d is the free detergent concentration in equilibrium with the protein. The thermodynamic interaction parameters are shown in Table 10.7. The hydrophobic nature of the interaction is denoted by the value of $\Delta G^\ominus$ which is negative and increases with increasing hydrocarbon chain length. Competitive interaction with the sites occurs when sodium dodecyl sulphate and the non-ionic surfactants are present together. As NaDS is added, the number of bound decyl glucoside molecules decreases to zero when the molar ratio of NaDS/protein is 12 (Fig. 10.9). The higher affinity of the anionic

Table 10.7 Thermodynamic parameters for the interaction of alkyl glucosides with bovine serum albumin. Limits of error of n, K and $-\Delta G^\ominus$ were calculated from the equilibrium dialysis data by the least-squares method

Detergent	Temperature (°C)	Number of binding sites, n	Association constants, K (l mol⁻¹)	$-\Delta G^{\ominus *}$ (kJ mol⁻¹)
Octyl glucoside	7	12.0 ± 1.8	$(2.10 \pm 0.44) \times 10^2$	9.86 ± 0.52
	25	15.1 ± 1.9	$(3.05 \pm 0.54) \times 10^2$	13.54 ± 0.44
Decyl glucoside	7	10.2 ± 0.4	$(1.78 \pm 0.11) \times 10^3$	19.02 ± 0.15
	25	12.8 ± 1.6	$(1.88 \pm 0.35) \times 10^3$	19.46 ± 0.46

* $\Delta G^\ominus = RT \ln K$.
From [58].

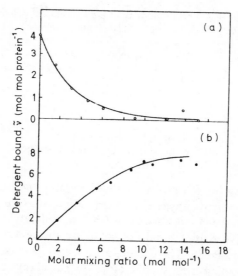

Figure 10.9 Competition of decyl sulphate for decyl glucoside binding sites of bovine serum albumin. (a) Solid curve for decyl glucoside binding as a function of molar mixing ratio of sodium decyl sulphate to bovine serum albumin. (b) Binding of sodium decyl sulphate to bovine serum albumin. From Wasylewski and Kozik [58] with permission.

surfactant might be due to specific interactions with the ε-amino groups of the lysine residues of albumin; the glucose residue in the non-ionic perhaps introduces a steric barrier. On the basis of infra-red and esr measurements it was concluded that in contrast to the anionic agents, the alkyl glucosides induced no detectable change in protein structure.

Colipase, a small protein which acts as a co-factor of pancreatic lipase, has been isolated in increased yields in the presence of Triton X-100. Addition of the surfactant not only results in increased extraction of the enzyme from pancreatic tissue but it results in the protection of the protein against proteolytic cleavage in particular at the Arg_5–Gly_6 bond [62]. This protection was thought to be due to interaction of colipase and detergent molecules. This study and an earlier one by Sari *et al.* [63] lend support to the conclusion that non-ionic and ionic detergents bind to a hydrophobic site on the colipase molecule which includes the three tyrosine residues of the protein. Such information is of value in interpreting data on the activity of enzymes in the presence of detergents. Triton X-100 and Brij 35 are strong inhibitors of the hydrolysis of water insoluble triglycerides by pancreatic lipase: complete inhibition occurs below the CMCs. The inhibition is reversed by addition of colipase in only a narrow range of Triton X-100 concentrations.

Five non-ionic detergents out of 18 studied have been found to increase the activity of L-glutamic acid dehydrogenase towards α-ketoglutaric acid; conformational changes induced by the detergent could lead to an increase in the binding of the substrates required for the reduction of α-ketoglutaric acid [52]. Several other surfactants in which the enzyme was soluble destroyed all activity

Table 10.8 Enzyme activity of L-glutamic acid dehydrogenase in the presence of surfactants (from [64])

Surfactant		Activity 100%*	Activity 100%[†]
Polysorbate	20	120	120
	40	118	97
	60	10	—
	80	—	—

* Measured as α-ketoglutaric acid reduction.
† Measured as α-monocarboxylic acid dehydrogenase.

perhaps by denaturing the protein or by disrupting the protein into inactive subunits. While polyoxyethylene sorbitan monopalmitate and monolaurate enhanced activity, the monostearate abolished it, and polyoxyethylene sorbitan mono-oleate (polysorbate 80) did not solubilize the enzyme and thus had no measured effect on activity. These trends are not easy to account for (Table 10.8).

Results forthcoming from the literature do not always assist in building a picture of events. The observation [65] that 2- and 4-hydroxylation of biphenyl was competitively inhibited by polysorbate 80 in the hamster contrasts with the finding that in the rat, polysorbate 80 (2 mM) had no effect on the linear microsomal demethylation of aminopyrine (also a 'Type I' substrate) [66]. The question arises whether surfactants such as polysorbate 80 produce their effects by interaction as an alternative substrate or by perturbation of a membrane bound enzyme system. Commercial samples of polysorbate 80, Brij 35 and Triton X-100 all enhance the activity of a sarcosine dehydrogenase isolated from a strain of *Pseudomonas*, due to the presence of free oleic acid in these non-ionic surfactants. Deoxycholate and a sarcosine surfactant (*N*-dodecanoyl *N*-methyl glycine) inhibit activity [67]. Correlations between the CMC of the non-ionic surfactants and the concentrations required for enzyme activation are seen in Table 10.9.

However, the nature of the activity–correlation plots is such that there is little difference between the behaviour of the three non-ionic surfactants. A direct comparison of deoxycholate, Triton X-100 and Tween 20, 60 and 80 on the

Table 10.9 Correlation of CMC with activation or inhibition of sarcosine dehydrogenase activity.

Surfactant	Concentration at midpoint of activation (g/100 ml)	Concentration at 50% inhibition	CMC (dye method) (g/100 ml)
Tween 80	0.0047		0.0036
Triton X-100	0.0095		0.0133
Brij 35	0.0127		0.0150
Deoxycholate		0.121	0.121
Sarkosyl NL 97		0.183	0.141

From [67].

activity of enzymes in the brush border of enterocytes has been made [68]. The surfactants have no effect on alkaline phosphatase activity but increase ATPase activity especially that of Na^+, K^+-ATPase at low surfactant concentrations. A progressive inhibition of activity is observed on increasing surfactant concentrations. Deoxycholate and Triton X-100 were active at concentrations around $0.2\,mg\,ml^{-1}$ whereas the Tween required much higher ($\times 100$) levels. Least active of the Tweens was Tween 60. The biphasic activity is shown best with Triton X-100 (Fig. 10.10) and has been referred to before [69]. Chan [70] has noted an increase in activity of Na^+, K^+-ATPase by 0.2 mM NaDS and complete inhibition by 1 mM detergent. Low concentrations probably cause a partial dissociation of lipoprotein complexes in the membrane, freeing certain active sites; high concentrations are the cause of progressive dissociation of lipoprotein. Interpretation is obviously complicated by the complex environment of the enzyme and the dependence of the enzyme on the maintenance of the integrity of the structure. Mammalian liver UDP glucuronyl transferase, for example, is firmly bound to microsomal membranes and its activity is strongly dependent on membrane structure and phospholipid composition and hence on the presence of membrane perturbants [70]. Triton X-100 has been found by some workers to activate this enzyme yet some have reported it to have no significant effect. A major factor in providing these discrepancies appears to be the species difference in response to detergents [71, 72] a factor that should be borne in mind in assessing the relevance of animal toxicity data. Rosenthal and Salton [73] have written 'membrane bound enzymes show a great deal of individuality in their responses to exposure to surface-active agents'. In their work on cardiolipin synthetase the biphasic action of both sodium deoxycholate and Nonidet P40 was clearly shown (see Fig. 10.11). The maximum effect for Nonidet P40 is at 0.05%, that of Triton X-100 at 0.25%. The former has a distinct advantage in biochemical work with this enzyme in that it maintains activity for a longer period than does Triton X-100, which brings about almost complete inactivation in 2 h.

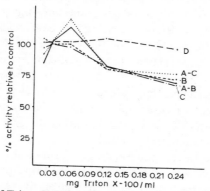

Figure 10.10 Effects of Triton X-100 on the activity of several enzymes in the intestinal brush border. B, Mg^{2+}-ATPase. C, Ouabain-insensitive ATPase. D, Alkaline phosphatase. (A–B), Total ATPase minus Mg^{2+}-ATPase. (A–C) Total ATPase minus ouabain-insensitive ATPase. From Mitjavila *et al.* [68] with permission.

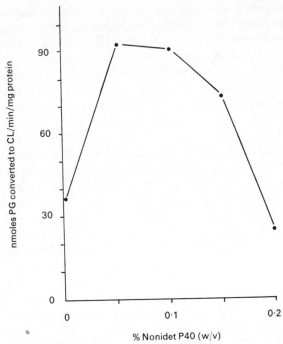

Figure 10.11 Stimulation of the cardiolipin synthetase activity of the shock-wash fraction by the nonionic surface active agent Nonidet P40. Conversion of phosphatidylglycerol to cardiolipin was determined after 20 min at 35° C with 10 μg protein, 0.2 mM ^{32}P-phosphatidylglycerol, 0.2 M Tris-HCl (pH 7.0) and various Nonidet P40 concentrations in a total volume of 0.1 ml. From Rosenthal and Salton [73] with permission.

The solubilizing ability of surfactant–protein complexes will be dependent on the nature of the complexes formed. Octylbenzene sulphonate, dodecylbenzene sulphonate and NaDS all bind to β-lactoglobulin in the same manner [74]. At low concentrations, two or three surfactant molecules interact with each protein molecule, which then undergoes a change in conformation resulting in an increased ability to bind detergent. A third type of binding appears to occur simultaneously between surfactant molecules already bound to the protein and those free in solution. The resulting binding has been called micellar in nature.

Hubbard [75] calculated that 180 to 200 molecules of digitonin and one molecule of rhodopsin formed 'complex micelles'. The molecular ratios quoted by Bridges [76] range from 250 for digitonin to 48 000 for hexylammonium chloride, and 1800 for decylammonium chloride. The mode of interaction is not clear. The suggestion is that the water-soluble aggregate forms in two stages: (i) monolayer formation by detergent on the peptide chains, and (ii) a second layer of cations amphipathically adsorbed on to the first layer. Blei's [77] conception of protein–detergent complexes is somewhat similar to this.

NaDS binds to linear peptides, forming what have been termed micelle-like clusters along the length of the polypeptide chain, able to solubilize lipophilic materials such as oil-soluble dyes 'Butter Yellow', Yellow OB and Orange OT [78, 79]. The solubilizing power of the NaDS-protein complexes towards these dyes was found to be comparable or slightly superior to that of the NaDS micelles and to be insensitive to the type of protein to which the surfactant was bound, in contrast to the findings of Steinhardt *et al.* [80] who showed that proteins differed in the extent to which they formed solubilizing complexes with NaDS. These authors used proteins which differed widely in their stability against denaturation by the surfactant. With Yellow OB as solubilizate Takagi *et al.* [78, 79] found uptake into a chemically modified BSA to increase by a factor of two over that solubilized in micellar solution, perhaps due to solubilization of the dye by the polypeptide itself.

It did not prove possible for a correlation to be obtained between dye

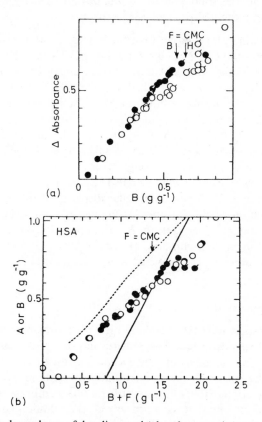

Figure 10.12 The dependence of dye dispersed (absorbance units) on amounts of NaDS bound by 0.1% solutions of HSA and BSA. Arrows indicate the points at which $F = CMC$ for the two proteins. Symbols: HSA (○); BSA (●). (b) The relation of total NaDS concentration $(B + F)$ of dye dispersed (A) and of Na bound (B) by 0.1% HSA. The solid diagonal represents the solubilization by micelles without protein. The broken line shows the free NaDS concentration (F). Symbols: (○) dye solubilization; (●) binding. From Steinhardt *et al.* [81] with permission.

solubilization by protein–surfactant systems and shape of the protein involved, its subunit structure, solubility, helicity, acid–base behaviour or amino acid composition [81]. A weak inverse relationship was, however, observed between solubilization and the size of the protein which is unremarkable if the uptake of dye occurs on the surfaces of molecules which are not fully unfolded. There are also strong indications of a dependence on the content of cationic groups. The dependence of dye dispersed (solubilized) in solutions of human serum albumin (HSA) and BSA on the amount of surfactant bound is clearly shown in Fig. 10.12a. It is obvious that solubilization occurs below the CMC of the NaDS and that the uptake of dye is by the complex. In Fig. 10.12b Steinhardt *et al.* demonstrate the relationship of total NaDS bound and free to dye solubilized and surfactant bound by 0.1 % HSA. The dotted line represents the free surfactant concentration and the solid line the solubilization capacity of NaDS in the absence of protein. Steinhardt *et al.* [81] explain:

'The free NaDS (F) is larger than the bound (B) at all values of ($B + F$) < 2.0. F reaches the CMC at ($B + F$) $\simeq$ 1.40 g l^{-1} where the absorbance and the binding are both about 0.60. Beyond this value of $B + F$, micelles must be present. They grow slowly as B (filled in circles) climbs to 0.85, increasing by about 0.38 (to 1.18) after exceeding the CMC. This increase should account for an increase in absorbance due to micellar solubilization of about 0.36 unit. An increment of only 0.27 absorbance unit is actually observed in the interval between ($B + F$) = 1.40 and 2.0. It appears therefore that the binding of NaDS to protein above the CMC, although it occurs, does not form additional complex capable of solubilizing DMAB. The absorbance and binding still run parallel above the CMC, due to micellar dispersal of the dye.

The fact that the binding by micellar complexes is parallel to the no-protein micellar curve but at *higher* values of $B + F$ supports the observation that some of the NaDS in protein complex, above the initial amounts bound, is *less* effective at solubilizing than normal (no-protein) micelles.'

10.3.5 Surfactant–viral particle interactions

The stepwise dissociation of the Semliki Forest (SF) Viral membrane with Triton X-100 [82] has been described. The SF virus has one of the simplest biological membranes; it has a spherical nucleocapsid consisting of RNA and one polypeptide. The nucleocapsid is surrounded by a lipid–protein membrane similar in composition to that of the host cell plasma membrane as it acquires this protective layer by budding as it leaves the cell. Binding of Triton to the membrane occurs below the CMC and increases with increasing surfactant concentration. Release of the nucleocapsids from the virus occurs when more than 0.2 to 0.4 mg Triton is bound per mg membrane. When higher concentrations of Triton were present (c 1.6 mg bound) the membranes dissociate to protein–lipid–surfactant complexes, followed at higher surfactant concentrations by delipidation of membrane protein (see Fig. 10.13). Lipophilic proteins isolated

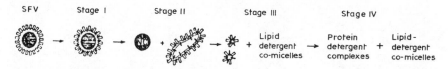

Figure 10.13 A schematic drawing of the breakdown of the SF virus membrane into soluble protein and lipid complexes caused by increasing concentrations of Triton X-100. NC, nucleocapsid. ●—, Triton X-100 SFV, Semliki Forest virus. Binding of Triton X-100 begins below the CMC (stage I). In stage II the SF virus membrane ruptures, releasing the nucleocapsid when about 9000 moles of surfactant are bound for each mole of virus (which corresponds to approximately 16 000 moles of cholesterol–phospholipid pairs and 590 moles of membrane polypeptides). Stage III is reached when about 40 000 molecules of Triton X-100 are bound, the result being lipoprotein–surfactant micelles and lipid–surfactant mixed micelles. In stage IV two types of surfactant–protein complex are obtained, differing in molecular weight. Delipidated proteins bind considerable amounts of the non-ionic surfactant (as do many lipophilic proteins). The final stage of the dissociation of the virus seems to involve exchange of lipid for surfactant. From Simons *et al.* [24] with permission.

from membrane and lipoproteins also bind considerable amounts of surfactants.

Sodium dodecyl sulphate readily inactivates rotavirus, an enteric pathogen belonging to the *Reoviridae*; it also causes loss of poliovirus infectivity – by disruption of virion proteins [83, 84]. Non-infective rotaviral particles lack the outer protein shell associated with infectivity. The initial step in infection of a cell by a virus is the adsorption of the virus to receptors on the surface of the cell. The outer layer of the viral particle must be obviously involved in this process and alterations to the capsid, say by detergent, may lead to a loss of adsorptive capacity and thus a loss in activity. NaDS causes such a loss in the ability of the rotavirus to adhere to CV-1 cells. However, most of the proteins of the outer shell seem to remain associated with the virions and the decreased adsorption may be an electrostatic effect due to adsorption of NaDS molecules on the virus surface.

A large range of surfactants has been considered for their effects on the infectivity of rotavirus (Table 10.10) [85].

Non-ionic detergents appear to stabilize the virus against the effects of NaDS. At a concentration of 0.01 % NaDS, inactivation occurred but the presence of 0.1 % Igepal Co-630 effectively protects the virus [85]; this effect is not evident at higher levels of NaDS.

Larin and Gallimore [86] have provided evidence to show that if viral protein is solubilized by surfactants this may result in loss or enhancement of specific viral antigenicity. Treatment of influenza virus with Triton X-100 significantly enhances the antigenicity of viral protein as judged by virus neutralization and haemagglutination inhibition tests; similar treatment by Tween 80, NaDS and deoxycholate led to a partial or complete loss of immunogenicity.

Viruses can induce fusion of cells. Detergents can modify this action and a report on the effects of detergents on virus- or chemically-induced fusion of erythrocytes has been published. The extent of activation or inhibition by surfactants of Sandai virus-induced fusion was determined primarily by the

Table 10.10 Effects of detergents and detergent-like compounds on the infectivity of rotavirus

Compound	Structural formula	% Recovery of virus after 60 min at 21°C*
Control [tris(hydroxymethyl)-aminomethane buffer]		100
Sodium octyl sulphate	$CH_3(CH_2)_7OSO_3^-Na^+$	75 (100)
Sodium decyl sulphate	$CH_3(CH_2)_9OSO_3^-Na^+$	85 (59)
NaDS	$CH_3(CH_2)_{11}OSO_3^-Na^+$	<0.025 (0.003)
Sodium dodecyl benzene sulphonate	$CH_3C_6H_4SO_3^-Na^+$	<0.025 (<0.001)
p-Toluenesulphonic acid	$CH_3C_6H_4SO_3^-$	100 (100)
Lauroyl sarcosine	$CH_3(CH_2)_{10}CON(CH_3)CH_2COO^-$	290 (0.003)
Sarkosyl O	$CH_3(CH_2)_5CH=CH(CH_2)_5CON(CH_3)CH_2COO^-$	260 (0.002)
Standapol ES-40	$CH_3(CH_2)_{13}OCH_2CH_2OSO_3^-$	270 (0.4)
Alipal Co-436	$CH_3(CH_2)_8C_6H_4O(CH_2CH_2O)_4SO_3^-$	230 (0.01)
Dodecyltrimethylammonium chloride	$CH_3(CH_2)_{11}N^+(CH_3)_3Cl^-$	5.0 (<0.001)
Nonyltrimethylammonium bromide	$CH_3(CH_2)_8N^+(CH_3)_3Br^-$	90 (97)
BTC-824 P-100	$CH_3(CH_2)_{13}N^+(CH_3)_2CH_2C_6H_5Cl^-$	<0.025 (<0.001)
Ethosperse LA-4	$CH_3(CH_2)_{11}O(CH_2CH_2O)_5H$	170 (100)
Igepal Co-630	$CH_3(CH_2)_8C_6H_4O(CH_2CH_2O)_9H$	210 (100)

* Rotavirus was diluted ten-fold into a 0.1% solution of each compound in 0.1 M tris(hydroxymethyl)aminomethane (pH 7.0), left for 60 min at 21°C, and assayed for recoverable plaque-forming units. Numbers in parentheses are values for reovirus recoveries after 20 min at 45°C. From [85].

hydrophobic residue of the surfactant molecule. Triton X-45, X-100, X-114 and Nonidet P40 (50 μg ml^{-1}) all enhanced virus-induced cell fusion and were not fusogenic in the absence of virus. Cell fusion was increased by a factor of 15 to 20 times. Less effective were the Span 20 and Span 80 surfactants (at 200 μg ml^{-1} and 1 mg ml^{-1}, respectively); Span 40 and 60 were inhibiting. Span 20 and 80 are fusogenic in the presence of Dextran 60 C [42, 81], Span 40 and 60 are not [87, 88].

Following some evidence that commercial emulsifiers used in crop spraying enhanced the sensitivity of a variety of cultured mammalian cells to some but not all viral infections, Lee *et al.* [89] studied the phenomenon in more detail. Enhanced sensitivity is not due to increased adsorption of virus; it is specifically related to single-stranded RNA viruses and not to double-stranded viruses. The optimum concentration of surfactant is just 'sub-toxic'. The mechanism eluded Lee and his co-workers who conclude 'it appears probable that the enhancing property is due to some intracellular activity of the emulsifier which either makes successful infection more likely . . . or allows a virus replicative mechanism some additional selective advantage in treated cells as compared to untreated'.

The effects of surface-active compounds on the thermal denaturation of DNA have been studied [90]. Anionic surfactants interact only weakly with DNA with the exception of *N*-lauroyl prolylprolylglycine and deoxycholic acid. Myristyl trimethylammonium chloride causes precipitation of DNA at the 1 mmol l^{-1} level probably by interaction with the phosphate residues of the DNA. The conclusion of this preliminary study was that the structure of the surfactant rather than its surface activity was the determining factor for interaction and that typical surfactants were unlikely significantly to affect DNA structure.

10.3.6 Artefacts due to solubilization of membrane-bound enzymes

Membrane-bound enzymes, as several authors have emphasized, are generally bound to lipid and resistant to solubilization unless the lipid is removed or lipid–protein interactions reduced. The forces between the lipid and protein elements are unlikely to be the same for all enzymes nor the same for the same enzymes from different species. Lipid removal by surfactants results in the exchange of bound lipid for bound detergent molecules [44]; whether or not this occurs without change in the conformation of the protein is open to question. Siekewitz [91] has pointed out other problems in interpreting results of solubilized proteins:

'What are the criteria for solubility? When are enzymes soluble? Certainly not when they resist centrifugation in a specified gravitational field, for perhaps a higher field will cause them to sediment. I think that we can only say that a protein is soluble when it is completely surrounded by water molecules. And in too many cases when proteins have been termed 'soluble' there is really no good reason to think this might be so . . . . On the other hand, when we extract and really purify a protein, and measure certain parameters, like pH optima, reaction rates, equilibrium constants, substrate binding constants, *we have no idea whether this same protein does have these characteristics when it is embedded within the cell.'*

Hence the difficulty in extrapolating the results of all these studies into points of toxicological significance. They are so surfactant-concentration dependent that it would be impossible to speculate on their toxicological significance in so far as ingestion of surfactant and its subsequent biological effects are concerned, for we nearly always neglect surfactant metabolism, and few have studied the interactions of surfactant metabolites with cells and cell components. Nonetheless it is plainly evident that surfactants are not inert species. One has simply to interpret *in vitro* biochemical results with caution.

As several papers show, however, even the *in vitro* experiments are sometimes complicated by the behaviour of the surfactants used. Dipeptidyl peptidase IV (dipeptidyl peptide hydrolase) solubilized from rat intestinal mucosal brush border with Triton X-100 was subjected to gel electrophoresis [92]. At low surfactant concentrations (0.5%) two peaks of enzyme activity were seen suggesting two electrophoretically distinct enzymes. With increasing concentration of Triton X-100, however, the slower species was converted to a faster form and only one major band of activity was observed at 10% Triton as shown in Fig. 10.14. The most probable explanation of this is that the slow moving peak

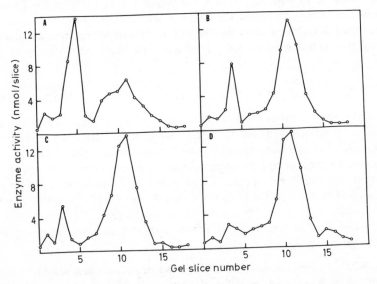

Figure 10.14 Effect of Triton X-100 concentration on solubilized dipeptidyl peptidase IV electrophoretic patterns. The enzyme was solubilized from rat intestinal mucosa and electrophoresed as described. The following concentration of Triton X-100 was used: A, 0.5%; B, 2%; C, 10%; D, an aliquot of the 0.5% supernatant was taken and the detergent concentration increased to 10%. The zero position on the gel slice axis represents the stacking gel. From Erickson and Kim [92] with permission.

represents an aggregate of proteins, lipids and detergent which is deaggregated at higher surfactant concentration, in spite of the fact that at 0.5% Triton the surfactant/protein ratio was 140 mg mg^{-1}. Obviously insufficient 'disruption' of the elements removed from the membrane obscures the unique size and charge properties of the protein of interest [93]. The molecular radius of partially purified adenylate cyclase (from rat liver plasma membrane) has been shown to be 4.9 nm, compared with the value of 3.9 nm for the enzyme before gel filtration [93]. This represents an approximate doubling of molecular volume implying aggregation of the enzyme with itself or other proteins during purification. Aggregation was not reversed by high concentrations (up to 1%) of Lubrol PX but electrophoresis in the presence of 0.1% Lubrol + 0.03 sodium deoxycholate decreased the molecular radius to 4 nm with greater than 90% recovery of activity: the net charge of the enzyme was, however, increased which indicated that the deoxycholate was adsorbed on to the protein surface. Ionic surfactants alone failed to deaggregate the enzyme at the low concentrations which were used to maintain activity; but higher concentrations may well deaggregate the system but would leave the enzyme in an inactivated form. The presence of the non-ionic Lubrol maintained activity, somehow protecting the enzyme from the ionic detergent.

The interpretation of chromatographic and electrophoretic behaviour in the presence of surfactants is not without its problems. If surfactant binds to the elements of the column then their charge characteristics might well be altered and thus the progress of protein and lipid through the column could be altered. The partition coefficient of NaDS monomer to Sephadex gels has been found to be abnormally high [94] and this has been attributed to the anomalous nature of the internal water of the dextran gel. Undoubtedly this has a bearing on the fractionation mechanism in gel chromatography as NaDS monomer can be concentrated above its CMC in the internal phase of Sephadex G-10. In Dextran T500 the effect of the external dextran concentration and surfactant concentration in the external phase on the concentration in the gel phase are shown in Fig. 10.15.

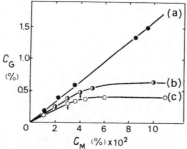

Figure 10.15 Effect of the external dextran (Dextran T 500) concentration on the CMC of NaDS. The data were obtained with Sephadex G-10 at 25° C. C_G = concentration of NaDS in the gel phase; C_M = concentration in the external solution. (a) In 40% dextran solution; (b) in 20% dextran solution; (c) in the absence of dextran (0.1 M NaCl). From Janado *et al.* [94] with permission.

Equilibrium and kinetic studies of the binding of proteins to N-(3-carboxy propionyl)aminodecyl Sepharose, an amphiphilic ampholytic adsorbant have shown that the introduction of 3.5 mM NaDS causes dramatic increases in the amounts of bound serum albumin and haemoglobin [95]. At concentrations of NaDS greater than 10 mM there is a fall in binding of all proteins, (Fig. 10.16) most likely due to competition between surfactant and protein molecules for binding sites. 3.5 mM NaDS also causes pronounced changes in the kinetics of adsorption of these two proteins. Thus the presence of surfactant may (by exposing additional hydrophobic or ionic sites on the protein) increase adsorption to the Sepharose, or by itself adsorbing on to the adsorbent alter the nature of that surface. By concentrating in the internal structure of some adsorbant gels it may influence the partitioning and solubility of small hydrophobic solutes.

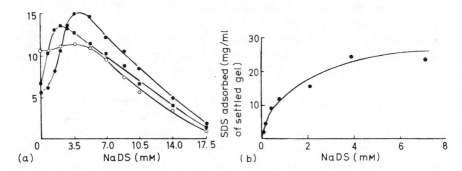

Figure 10.16(a) Effects of different concentrations of NaDS on the equilibrium adsorption of proteins at a constant total concentration. Binding experiments were performed, starting with 5 ml of a solution containing the protein at 1.5 mg ml^{-1} and various amounts of NaDS to which 0.5 ml (settled volume) of the adsorbent were added. The proteins were bovine serum albumin (●), haemoglobin (■) and soybean trypsin inhibitor (○). (b) Adsorption of NaDS by CPAD-Sepharose. Binding experiments were performed as above, but with omission of proteins and by using various starting concentrations of NaDS. From Yon and Simmonds [95] with permission.

The removal of surfactant used in solubilization studies is also a pre-requisite to the study of the intrinsic properties of enzymes. The low dialysis rate of non-ionic detergents is the result of their low critical micelle concentrations and thus low monomer concentrations. Bile salts with substantially higher CMCs than non-ionic surfactants are easier to remove; the removal of detergent from protein complexes will sometimes lead to the formation of amorphous precipitate or aggregation. In 1975 octyl-β-D glucoside was put forward as a new non-ionic surfactant for use in this type of work – its great advantage over commonly used substances such as Triton X-100 being its high CMC (0.7%) allowing more rapid removal by dialysis [96, 97]. Octyl glucoside and Triton X-100 can be removed by binding to Biobead SM2, a hydrophobic matrix which has a capacity (per g dry material) of 170 mg octyl glucoside.

10.3.7 Interactions with receptors: the influence of non-ionic surfactants

Although the cellular and molecular mechanism of action of Δ^9 tetrahydrocannabinol (THC) has not been fully studied, it has been suggested that THC combines with a specific receptor or selective membrane components, while others have suggested that it interacts non-specifically with neutral membranes. THC binds strongly to synaptosomal membranes–the membrane partition coefficient being of the order of 12 000 [98]. Membrane binding of THC is drastically reduced by Cremophor EL at concentrations of solubilizer as low as $8 \mu g\,ml^{-1}$ (Fig. 10.17). At concentrations of $0.4\,mg\,ml^{-1}$ the value of the partition coefficient P is decreased to 0.005 of its original value. The effect of Tween 80 is very similar. Thus the use of solubilizing agents to increase the solubility of THC, so that it can be readily used in pharmacological experiments, does not increase the membrane concentration of the active agent. Roth and Williams [98] conclude that the only advantage in using surfactant is to minimize the adsorption of the THC to glassware! Surfactants, however, have various effects on receptors. The cholinergic receptor of the electroplax and of skeletal muscle is resistant to the action of Triton X-100 and deoxycholate; the binding of D-tubocurarine to synaptosomal membrane is resistant to Triton X-100 treatment

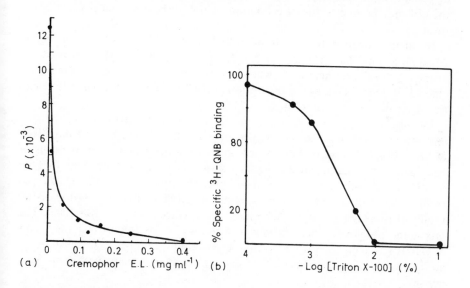

Figure 10.17(a) The partition coefficient (P) of THC in the presence of varying concentrations of Cremophor EL. Each point is the mean of at least 15 determinations. S.E.M. values (not shown) are on average 5.4% of the mean. From Roth and Williams [98] with permission. (b) Effect of Triton X-100 on the binding of [^{3}H]-QNB to synaptosomal membranes. The binding was carried out with 0.9 nM [^{3}H]-QNB in the presence of various concentrations of Triton X-100. The 100% represents the specific radioactivity of controls without detergent (i.e., 1.73 pmol/mg protein). The points are the means of triplicate determinations. From Aguilar *et al.* [100] with permission.

[99]. The binding of quinuclidinyl benzylate ($[^3H] - QNB$) to central muscarine receptors is, however, inhibited by Triton X-100, although the effect of the detergent can be prevented to some extent by protection of the receptor with atropine. In the absence of atropine the binding of $[^3H]$-QNB to the membranes is as shown in Fig. 10.17b, a function of surfactant concentration.

10.3.8 Miscellaneous interactions of non-ionic surfactants with cells

We have already discussed (Chapter 7) the effect of non-ionic surfactants on the permeability of cell membranes. Polysorbate 80 increases the permeability of ascites tumour cells [101, 102]. Reversible alterations to the properties of the surface of ascites cells have been induced by polysorbate 60 [103]. Polysorbate can cause swelling of the cells and it decreases the electrophoretic mobility as would be anticipated following adsorption of a non-ionic surfactant to a negatively charged membrane; it decreases the percentage of cells resistant to the action of trypsin. The cell surface area doubled after incubation with the polysorbate but the electrophoretic mobility fell by only 10 %; hence new charged groups must appear on the surface. Kay [101] also observed that internal changes occurred in that the rate of protein and DNA synthesis decreased while synthesis of phospholipids and nuclear RNA were stimulated. Metabolic changes within the cell may be due to the direct action of polysorbate on cytoplasmic structures.

In an investigation of the antimicrobial action of Triton X-45 and chlorhexidine against *Bacillus megaterium*, Nadir and Gilbert [104] measured the absorption of the surfactants by whole cell and cell wall preparation. Potassium chloride (0.35 M) increases the absorption of the Triton (Fig. 10.18) but decreases the absorption of chlorhexidine.

Increased salt enhances non-ionic surfactant surface activity by 'salting out' and the resultant increases in adsorption may assist in the bactericidal effect. Many cationic detergents have a non-specific disrupting effect on the cells of bacteria and tissue, thus precluding their systemic use. There is evidence that the non-ionic polysorbate 80 disrupts membrane structure but that there is a rapid reconstitution of cell membrane material after treatment [101]. Triton X-100 totally disrupts lysosomes, mitochondria, and erythrocytes [105, 106, 106a]. The minimum inhibitory concentrations of a series of non-ionic surfactants versus *S. aureus* obtained by Allwood [107] are given in Table 10.11. These show no straightforward trend, although the most hydrophilic of the two series are least active.

Theories of anaesthesia which depend on the assumption that the anaesthetic enters lipid membranes [108] may explain why the non-ionic $C_{12}E_9$ is a useful endo-anaesthetic. Polyoxyethylene ethers have, according to Bucher [109], a selective affinity for the myelin surrounding the afferent pathways of stretch and tactile receptors. In concentrations sufficient to cause disperse bacterial growth, polysorbate 80 and Triton A20 are completely innocuous to virulent tubercle bacilli; however, the latter detergent has a marked toxic effect on avirulent strains

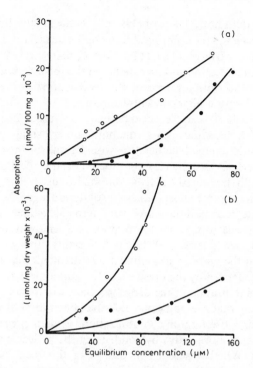

Figure 10.18 Absorption of Triton X-45 by whole cell (b) and cell wall (a) preparations of *B. megaterium* in water (●) and in 0.35 M potassium chloride (○). From Nadir and Gilbert [104].

Table 10.11 Growth-inhibition of *S. aureus* by non-ionic surfactants

Surfactant	Hydrophobe	HLB (calculated)	MIC (% v/v) after incubation for:	
			24 h at 37° C	48 h at 37° C
Triton X-35	Octyl phenol	7.8	0.03	0.03
Triton X-45	Octyl phenol	10.4	0.01	>1.0
Triton X-100	Octyl phenol	12.4	0.10	>1.0
Antarox CA 630	Octyl phenol	13.0	0.10	>1.0
Triton X-102	Octyl phenol	13.5	>1.0	>1.0
Triton X-114	Octyl phenol	14.6	0.03	>1.0
Triton X-305	Octyl phenol	17.3	>1.0	>1.0
Triton N-57	Nonyl phenol	10.0	0.01	0.01
Tergitol NP 14	Nonyl phenol	10.0	0.01	>1.0
Antarox CO 530	Nonyl phenol	11.3	0.01	>1.0
Tergitol NP 27	Nonyl phenol	12.3	0.01	>1.0
Tergitol TP 9	Nonyl phenol	13.0	0.01	>1.0
Antarox CO 630	Nonyl phenol	13.0	0.03	>1.0
Tergitol NPX	Nonyl phenol	14.0	0.02	>1.0
Tergitol NP 35	Nonyl phenol	15.0	>1.0	>1.0
Tergitol NP 40	Nonyl phenol	16.0	>1.0	>1.0

From [107].

and polysorbate does not. This probably reflects the greater hydrophobicity of the 'virulent' surface and the differing abilities of the surfactants to be adsorbed or to penetrate these surfaces [110, 111]. Dubos first showed the remarkable growth-stimulating effect of polysorbate 80 on the tubercle bacillus, thought to be due [112] to the concentration of the hydrophobe in the cell membrane supplying the cell with a high concentration of proved metabolite (in this case oleate, as polysorbate 80 is an oleic acid derivative). This led Eisman [113] to use surface-active tuberculostatic agents – diaminodiphenyl sulphones with ethylene oxide chains – to determine whether this would result in the concentration of sulphone in the cell surface. Increase in tuberculostatic effect of over 1000-fold was obtained with the surface-active compounds used. The effect of added polysorbate 80 on the tuberculostatic activity of these drugs was, however, erratic. 0.05 % polysorbate decreased the activity of 'Sterox 3500 times. This is possibly a case of mixed micelle formation, the polysorbate removing the surface-active drug from the tubercle surface. What is interesting, and what supports this possibility, is that the non-surface-active drugs, diaminodiphenyl sulphone and promine, are not affected by the presence of the non-ionic detergent. One might have expected solubilization of the drugs to occur, with resultant diminution of activity. It is likely that the non-ionic aids the penetration of the non-surface-active compounds, offsetting any solubilization. Eisman suggests that streptomycin may be made more active by condensing the molecule with hydrophilic chains, but says 'Whether actual penetration of this (bacterial) lipid layer is attained is only problematical' [113]. The experiments of Lovelock and Rees [114] suggest that not only do treated monocytes contain a high proportion of detergent but also that it is uniformly distributed and is freely available for the solution of lipoidal materials ingested by the monocytes. Certain macromolecular detergents are antituberculous agents *in vivo*, not *in vitro*, and it has been asserted that this property resides in those detergents having a high lipophile/hydrophile ratio; as the ratio is reduced, therapeutic activity is reduced until a so-called 'protuberculous effect' is obtained [107]. Their mode of action normally is not a direct one on the bacillus, but rather depends on the detergent rendering the organism vulnerable to the action of the phagocytes by modification of its outer layer. This normally takes place in the monocytes. The lack of activity *in vitro* is likely to be due to the concentration of detergent in bulk being insufficient to solubilize cell material; solubilization in the monocyte has been demonstrated. Lovelock and Rees [114] found that p-*tert*-octylphenyl formaldehyde polymeric detergents (I) with 10 to 12 ethylene oxide units are antituberculous, those with 30 units are inactive, and those with 60 or 90 units are protuberculous. This is consistent with the view that the more hydrophobic detergents displace cholesterol from the lipid layer and the hydrophilic compounds displace phospholipid. The amount of cholesterol removed from the cell membrane does not vary with the size of the ionic head of surface-active electrolytes, but a marked difference between the amount of lecithin and cephalin released by anionic and cationic detergents has been noted [115]. Penetration of the detergent is thought to be the result of the interaction of the alkyl chain with the cholesterol molecule;

release of phospholipid by short-chain cationic electrolytes is suggested to be caused by electrostatic attraction.

The differentiation between *in vitro* and *in vivo* activity may help to explain how the strongly active tuberculostatic Triton WR-1339 can be used as a substitute for polysorbate 80 in promoting the dispersed growth of tubercle bacilli in liquid culture media [116]. Fulton [117] has obtained favourable results in the

$$O(CH_2CH_2O)_xH \quad \left[\quad O(CH_2CH_2O)_xH \right] \quad O(CH_2CH_2O)_xH$$

(I) Triton WR-1339, $x = 17$-20

treatment of *Leishmania donovani* infections, in which the infecting agent also develops in the white cells of the host, with the non-ionic 'Macrocyclon' (a polyoxyethylene glycol ether with 12 to 13 ethylene oxide units) and Triton WR-1339.

Triton WR-1339 and related products are not directly antibacterial and most likely act via the host, probably accumulating in the macrophage lysosomes which are somehow induced to produce an antituberculous lipid or lipid not present or present at much lower levels in lysosomes of normal cells [118, 119].

10.4 Toxicology of surfactants

10.4.1 Non-ionic surfactants

Cremophor EL, widely used as a drug solvent, has been shown to have an antidiuretic effect in rats at doses of as low as 2.5 ml kg^{-1} *per os* [120]. Results suggest that this action is related to the laxative action of the surfactant noted during the study (see Table 10.12).

Cremophor EL has been implicated in several cases of adverse reactions to the compound present as a solubilizer in intravenous anaesthetic preparations. Adverse reactions may be due to direct pharmacological release of histamine, to immune-mediated mechanisms, or to activation of C3 complement leading to release of histamine [121]. It has been variously estimated that the incidence of reactions to Althesin which contains 20% Cremophor EL is 1 in 14 000 to 19 000, 1 in 900 or 1 in 1900 [122].

Cremophor has been shown to produce an anaphylactic response in man [123–125]. Wirth and Hoffmeister [126] induced in dogs a histamine-like response accompanied by a marked hypotension, but dogs seem to have a hypersensitivity to the surfactant given by the i.v. route. Death following Althesin

Table 10.12 Effect of Cremophor EL on diuresis

Treatment	Aqua fontis p.o. (ml kg^{-1})	NaCl 0.9% s.c. (ml kg^{-1})	Cremophor EL p.o. (ml kg^{-1})	Urine excretion as % of the water load
Aqua fontis	50	—	—	76.80
Aqua fontis + Cremophor EL	50	—	2.5	66.60
Aqua fontis + Cremophor EL	50	—	5.0	50.40
Aqua fontis + Cremophor EL	50	—	10.0	34.20
Aqua fontis + Cremophor EL	50	—	25.0	14.80
Aqua fontis	50	—	—	69.10
Aqueous solution of Cremophor EL	47.5	—	2.5	63.80
Aqueous solution of Cremophor EL	45	—	5.0	54.60
Aqueous solution of Cremophor EL	40	—	10.00	31.80
NaCl 0.9%	—	50	—	46.80
NaCl 0.9% + Cremophor EL	—	50	2.5	34.20
NaCl 0.9% + Cremophor EL	—	50	5.0	25.80
NaCl 0.9% + Cremophor EL	—	50	10.0	25.30
NaCl 0.9% + Cremophor EL	—	50	25.0	20.10

From [120].

reactions involve first exposure to the formulation and result from complement activation, according to Watkins [127]. Complement-mediated reactions to diazepam formulations employing Cremophor EL as a solvent have also been reported [128] with a frequency of 1 in 1000 patients. One patient developed a diffuse rash on face, neck and chest, rapidly followed by generalized rash immediately after i.v. injection of 10 mg diazepam in Stesolid MR. Respiratory distress and bronchospasm followed, the radial pulse became weak and heart rate increased to 150 beats min^{-1}.

The amount of histamine released in man following i.v. thiopentone, propanidid (Epontol) or Althesin is normally clinically insignificant. The pig resembles man in being relatively insensitive to the instant histamine releasing properties of Cremophor EL and has been used as an animal model with which to investigate the adverse reactions to these agents [129]. Following single injections of Althesin no significant increase in plasma histamine were detected [130]; second injections of Cremophor EL, Althesin or Epontol given 7 days after the first dose produced a high frequency of adverse responses. All injections were given over 30 s. Table 10.13 gives these results. The steroids are not without activity and there is a curious interaction between them and the Cremophor in eliciting responses [131]. Sometimes the results are contradictory as shown in Table 10.14. Cremophor RH40 has also produced adverse reactions in sensitive patients and cannot be considered as a replacement solubilizer.

Watkins' review [127] of anaphylactoid reactions to i.v. preparations should be consulted for further details of this complex problem.

Polysorbate 80 also induces histamine release on i.v. administration to dogs and this is accompanied by gastric secretion [132]. Normally antihistamines have no effect on gastric secretion induced by histamine [133]. However antihistamines *do* block this action of polysorbate 20 perhaps because the perme-

Table 10.13 Responses obtained in mini-pigs on the second administration of anaesthetic agents and solvents, 7 days after an initial exposure.

Anaesthetic agent	Solvent	No. of pigs	Abnormal clinical response	Hypertensive response > 50 mm Hg	Decrease in polymorph count > 50%	Increase in plasma hist-amine > 50%
Thiopentone	Water	6	0	0	0	2
Althesin*	20% Cremophor EL	4	3	3	3	2
Epontol[†]	20% Micellophor	6	5	5	6	3
Alphaxalone/ alphadolone	10% ethyl alcohol/ 25% propylene glycol	4	2	3	1	1
Propanidid	10% ethyl alcohol/ 25% propylene glycol	4	0	0	0	0
	20% Cremophor EL	5	4	5	5	4
	10% ethyl alcohol/ 25% propylene glycol	3	0	0	0	0

* Althesin (Glaxo) contains alphaxalone and alphadolone.   [†] Epontol (Bayer) contains propanidid. From [129].

Table 10.14 Skin and leukocyte tests with the various components of the intravenous steroid anaesthetic, Althesin, in a patient who developed an 'anaphylactic' reaction to this preparation

Test material	Skin test diameter (mm)		Leukocyte histamine release (% total cell content)
	Weal	Erythema	
Control			14
Vehicle (Cremophor) 1/1000	7	13	30
Althesin 1/1000	7	27	62
Alphadolone $10\,\mu g\,ml^{-1}$	8	29	< 4
Alphaxalone $10\,\mu g\,ml^{-1}$	8	32	< 4
Vehicle and Alphadolone $10\,\mu g\,ml^{-1}$	5		< 4
Vehicle and Alphaxalone $10\,\mu g\,ml^{-1}$	5		< 4
Alphadolone $50\,\mu g\,ml^{-1}$			< 4
Alphaxalone $50\,\mu g\,ml^{-1}$			< 4
Propanidid in Cremophor (Epontol) diluted 1/1000	7	17	23
Anti-IgE			61

(From Kessell & Assem, 1974) through [131].
Histamine release was induced by Cremophor but with neither alphaxalone nor alphadolone separately. Each of these steroids when added to Cremophor led to a reduction in the Cremophor-induced histamine release, even to levels well below the control value (spontaneous histamine release in the presence of the medium used, Tyrode solution). When both alphaxalone and alphadolone were added to the vehicle they produced a paradoxical increase in histamine release.

ability of the parietal cells is increased by contact with the surfactant [134] enabling the drug to reach the cell receptors.

A decrease in the viscosity of plasma following induction of anaesthesia with Cremophor-containing injectables has been reported [135] the effect persisting for 50 min (Fig. 10.19). This is not reckoned to be due to the sedative or

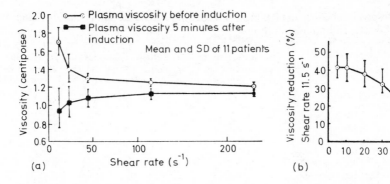

Figure 10.19(a) Plasma viscosity before and after induction of anaesthesia (11 patients) with intravenous anaesthetic formulations containing Cremophor EL; (b) the duration of this effect – a plot showing the viscosity of plasma at a shear rate of $11.5 \, s^{-1}$ as a function of time in minutes. Taken from Gramstad and Stovner [135] with permission.

sympatholytic effects of the induction as thiopentone, for example, causes smaller changes in viscosity. Triton X-100 produced a very similar effect on plasma viscosity [136]. Sodium oleate has previously been found to affect plasma viscosity initially increasing the charge on erythrocytes by adsorption. Low concentrations of sodium oleate decrease the viscosity and concentrations higher than 60 mg/100 ml increase the viscosity [137] perhaps due to the altered shape of the erythrocytes. Sodium oleate deaggregates erythrocytes and consequently would cause a reduction in viscosity by dispersing the asymmetric rouleaux. Poloxamer 188 has been suggested as an additive to perfusion systems due to its antisludging action and viscosity-lowering effects.

Local reactions of Cremophor-containing formulations are discussed. A significant decrease in the frequency of local vascular side effects from i.v. diazepam formulations can be achieved by replacing the propylene–alcohol solvent with Cremophor–water [139, 140]. The surfactant will modify the nature of the precipitation reaction as the solvent is diluted *in vivo*.

Repeated use of Cremophor EL-containing injections in animals has given rise to abnormal plasma lipid patterns [141]. Triton WR-1339 and linear non-ionic surface-active agents produce lipaemia associated with an increase in blood cholesterol and 'stripping' of lipids from the adrenal cortex [142].

Kellner *et al.* [143] reported that the administration of the non-ionics Triton A-25 and polysorbate 80 to rabbits and guinea-pigs resulted in sustained hyperlipaemia. Triton forms a complex with lipoprotein 'micelles' which has been observed by analytical ultracentrifugation [144]. Thus a consequence of Triton administration would be the trapping of lipids in the plasma and their resultant inability to exchange with tissue lipids [145]. Protein-detergent complexes have solubilizing powers, just as polymer-detergent complexes can solubilize oil-soluble materials [146, 147].

Kellner *et al.* [143] have also shown that Triton may initially protect cholesterol-fed rabbits from atherosclerosis, yet Scanu and Page [148] found this non-ionic detergent to be atherogenic. Van den Bosch and Billiau [149] have abolished convulsions in rabbits caused by high levels of circulating fatty acids by administration of Triton. The drop in plasma fatty acid was attributed to accelerated plasma removal. However, Scanu [145] states that 250 mg Triton per kg body weight given every 4th day results in a progressive increase of all plasma lipids – including free fatty acids. Despite the increased cholesterol levels in experimental animals on intravenous administration of detergents, there was significantly less atherosclerosis than in the controls fed on cholesterol [150]. The experiments suggest that the severity of atherosclerosis is reduced if the blood phospholipid levels are raised along with that of the cholesterol; it seems that cholesterol/phospholipid ratios are of vital importance. 'Increased blood phospholipid may modify or prevent the development of atherosclerosis.'

The foregoing illustrates the divergence of opinion and experimental evidence that abounds in the medical literature; it is apparent that precise physical measurements on well-defined systems is required in order to unravel the complex of processes contributing to atherosclerosis and the effect of surfactants, natural and synthetic, on them. Poloxamer 108 (Pluronic F. 38) has been used as a protein-precipitating agent for the isolation of the non-aggregated 7S IgG factor of immune serum globulin for intravenous administration. It is likely that residual poloxamer remains in the gamma globulin intended for intravenous use. It has been estimated [151] that the maximal concentration of the surfactant would be 10 mg ml^{-1} and that a patient would receive about 172 mg kg^{-1} for 3 to 4 weeks. An investigation of the possible toxicological consequences has been carried out. Light microscopy showed that administration to Sprague-Dawley rats of 0.15 to 4 g kg^{-1} poloxamer five times a week for 2 weeks, produced dose-dependent vaccualization of epithelial cells in the proximal renal tubules and of hepatocytes. These alterations in normal appearance were slowly reversible. The near absence of additional cellular changes suggested that the poloxamer was rapidly phagocytosed and well tolerated intravenously [151].

A problem in elucidating the effects of surfactant *in vivo* is the possibility of metabolism of the surfactant, since the metabolic products might have quite different physico-chemical properties and might well be devoid of surface activity. A specific example may be quoted here. Spolter and Rice [152] found that Triton X-100 increased the incorporation of [^{35}S] sulphate into lipid-soluble substances when it was used as an additive in experiments with supernatant enzymes of bovine or sheep adrenal cortex or rat liver, kidney and brain. TLC indicated that sulphate derivatives of the non-ionic detergents were formed by liver and adrenal enzymes, no such reaction occurring with kidney or brain preparations.

Two possible explanations for the effect of Triton on sulphate incorporation have been put forward [153]: (i) the detergent may stimulate sulphation by activating the sulphotransferases or substrates; or (ii) the surfactant itself may be enzymatically sulphated. Sulphation of the Triton X-100, of course, complicates efforts to determine its effect on the sulphation of other compounds present in the system. However, this could be ascertained by using enzyme preparations which did not affect the surfactant; it was then found that Triton X-100 stimulated the sulphation of cholesterol, pregnenolone, dehydroepiandrosterone and cholic and lithocholic acids. Brij 96 also increases $[^{35}S]$ sulphate uptake into the lipid fraction of rat liver and bovine and sheep adrenal cortex supernatants, probably due to sulphation of the detergent itself. In contrast polysorbate 80 reduces the sulphate incorporation in several systems; Spolter and Rice [152] suggest that its sulphation is prevented by the bulky nature of the polysorbate molecule which might block its access to the site of the sulphotransferase active sites.

It might be that this difference between Triton X-100 and polysorbate 80 may be the reason why the former is more toxic than polysorbate. Sulphate conjugation of non-ionic detergents yields anionic derivatives which are probably intrinsically more toxic. Different tissues produce different sulphate products [152]; in the liver (*in vitro*) the products of Triton X-100 sulphation have 5 to 11 oxyethylene units while in the adrenal cortex the products have a shorter chain length, of 2 to 5 ethylene oxide units.

Short- and long-term animal toxicity studies of a series of Spans have been conducted by workers at the British Industrial Biological Research Association (BIBRA) [153–155]. The toxicology of the Span and Tween range of emulsifiers was reviewed in detail by Elworthy and Treon in 1967 [1]. As with many toxicology studies, observed effects must be considered in the light of the dosages of substance administered. Rats have tolerated single oral doses of 20 g Span kg^{-1} without apparent harm during a 2 day observation period [156]. Rats exposed to Span 20 at levels of 25 % in diet for up to 60 days fared less well which is hardly surprising, only one rat surviving the test period and that was found on autopsy to have fatty changes in the liver [157]. At lower levels of intake (5 % diet) hamsters showed as the only sign of abnormality, a lower rate of body weight gain [158]. At the 15 % level diarrhoea was initially observed. No difference in blood density or change in the relative weight of organs was evident. Some histological evidence was adduced for irritation of the gastro-intestinal tract and mild degeneration of the kidney tubules. In the BIBRA study [153] rats were fed up to 10 % Span 20 for 13 weeks. Dose-related reductions in rate of body weight gain were associated with reduced intake of diet containing Span 20 suggesting that the animals disliked the food! Details of the other results obtained are not quoted here: several of the findings are in conflict with those of Harris *et al.* [157, 158] and the interpretation of the findings is not always clear. Several of the findings are possibly not due to the Span but to reduced food intake although an increase in relative kidney weight could have been a response to the need to excrete the surfactant or its metabolites. No evidence was seen of gastro-intestinal irritation, again in disagreement with Harris *et al.* who used slightly higher levels of surfactant.

As a result of a long-term toxicity study of Span 60 in mice, the 'no-untoward-effect' level was considered to be 2.0 % of the diet or approximately 2.6 mg kg^{-1} body weight per day [154]. Mice receiving 40 % Span 60 did display evidence of enlargement of kidneys and a higher incidence of nephrosis compared to the control animals. Span 80 administered to rats for 16 weeks [155] also produced kidney enlargement associated in females with tubular changes 'of uncertain pathological significance'. It was not possible to establish a 'no-untoward-effect' level for Span 80 as a result of the study.

Of particular pharmaceutical relevance are the studies conducted on the effect of emulsifiers on fat absorption, some aspects of which are discussed in Chapter 8. Oral administration of 6 or 20% Span 60 in olive oil to rats produced no changes in the rate of fat absorption, degree of hydrolysis or intestinal mobility 2 to 3 h after incubation [159]. Elworthy and Treon in their review [1] concluded that the ester bond of sorbitan fatty esters is hydrolysed in the gastro-intestinal tract, the sorbitan moiety being absorbed and excreted largely unchanged in the urine, while the fatty acid residue is metabolized in the normal way. Wick and Joseph [160] have found that at least 90% of sorbitan monostearate fed to rats as an oily solution was hydrolysed to stearic acid and anhydrides of sorbitol. Others (e.g. [161]) have found that single doses of 20 g sorbitan monostearate did not alter gastric mobility or acidity in man, important information when attempting to assess surfactant effects on drug absorption.

Pluronic L-101 is a potent inhibitor *in vitro* of human pancreatic lipase. Fed as 1 to 3% of diet to rats, a dose-dependent decrease in body weight has been observed [162] without a change in food consumption. Excretion of dietary fat was increased in a dose-dependent manner suggesting some 'antiobesity' function for this poloxamer. A related compound Pluronic F-68 – more hydrophobic than L-101 – a poor inhibitor of lipase, did not produce either decrease in body weight or increase in faecal fat excretion.

One recent paper has referred to the effect of surfactants on immune responses in mice [163]. Both stimulation and inhibition of the immune responses was revealed depending on the substance studied and the duration of the exposure. Triton X-100 causes stimulation of immunological activity. According to Ahkong *et al.* [88] surfactant adjuvants may facilitate the secretion of mediators by affecting fusion of membranes, e.g. of lymphocytes and macrophages. But no clear indication of mechanism of action was adduced.

Non-ionic surfactants (PEG esters of C_2-C_{18} fatty acids) have been found to modulate polymorphonuclear leukocyte locomotion [164]. Esters with shorter aliphatic chains had negligible effect whereas those with longer hydrocarbon chains (C_{16}, C_{18}) reduced locomotion, an effect perhaps mediated by alteration of membrane structure, for instance inducing lateral lipid phase separation which might alter the anchoring of the cytoskeleton and the position of membrane constituents responsible for adhesion of the cells to other cells and to various substrates [165].

Adverse reactions to ingested or topically applied (injection into the peri-ovarian space or sub-capsular injection in rats) surfactants of the Triton series of non-ionic surfactants include the formation of cysts on rodent ovaries. Goldhammer and McManus [166] found that the dose of Triton X-100 that killed by intraperitoneal injection was 50 to 100-fold higher than that required intravenously – the higher toxicity of the latter route arising from haemolysis. The Triton series of surfactants (X-35, X-45, X-165, N-100) are irritant when injected into tissues and lead to damage to the peritoneal lining with increase in permeability leakage of fibrin and some necrosis [167]. Increases in vascular permeability have been noted at the site of subcutaneous injection of Triton even at high dilutions. Injection of undiluted Triton leads to necrosis. Cyst formation in rodents was observed after oral injection of the surfactant and the mechanism for this is obscure. Most of the Triton X-100 given orally is recovered from the

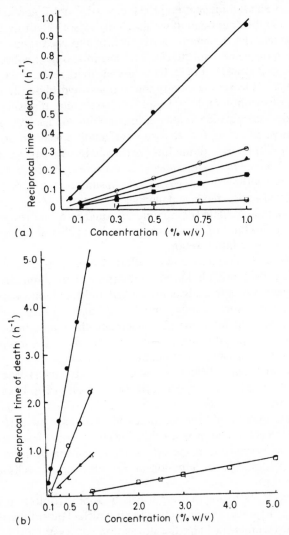

Figure 10.20(a) Reciprocal time of death of planarian as a function of surfactant concentration at 20° C. ●, polyoxyethylene 20 stearate; ○, polyoxyethylene 30 stearate; ▲, polyoxyl 40 stearate; ■, polyoxyethylene 50 stearate; and □, polyoxyethylene 100 stearate. (b) Reciprocal time of death of planarian as a function of surfactant concentration at 20 °C. ●, polysorbate 20; ○, polysorbate 40; △, polysorbate 60; and □, polysorbate 80. From Isomaa and Ahlroth [192] with permission.

urine in 24 h. It is possible that the absorbed surfactant exerts its action on serous surfaces [167].

Saski *et al.* [168] studied the effect of three homologous series of non-ionic surfactants on the death time (T) of the planarian *Dugesia lugubris*. In plots of $1/T$, the reciprocal death time against concentration of surfactant (Fig. 10.20), the slope of the line is related to the rate of absorption of the surfactant and its intrinsic toxicity. Hence in Fig. 10.20b the toxicity of the surfactants of the polysorbate series is in the order polysorbate 20 > 40 > 60 > 80, this is clearly dependent on the ester chain length (lauric, palmitic, stearic and oleic, respectively). No clear cut pattern was found in the alkyl esters but the toxicity of the polyoxyethylene esters (Fig. 10.20(a)) decreased with increasing polyoxyethylene chain length. Alkyl ethers have been studied for their effect on the death time of goldfish *Carassius auratus* [169]. Of the compounds studied in the Brij series, Brij 36T and Brij 30 were most active as can be seen in Fig. 10.21. At 0.1% surfactant concentrations, the effect of alkyl chain length is striking, peaking at C_{12} (Fig. 10.21). Wildish [170] found that polyoxyethylene (4) lauryl ether was fifteen times more toxic than polyoxyethylene (23) lauryl ether to salmon par. The bluegill sunfish (*Lepomis macrochirus*) is more susceptible to the toxic effects of non-ionic surfactants with shorter ethylene oxide chain length [171]. Vital to an understanding of toxic effects of surfactants on fish is the mode of penetration and uptake, metabolism and elimination of the components concerned. The main site of penetration is the gill tissues [172] and blood is the chief vehicle for transportation of the surfactant, the nonylphenyl E_{10} being found in cod blood 5 min after exposure to 5 ppm solutions. The liver rapidly takes up the surfactant and later highest concentrations are found in the gall bladder. The gill membrane is a likely site for toxic activity–swelling of the gill lamellae and changes in membrane permeability bring on asphyxiation as a main cause of acute poisoning [172]. Detergents have more than one activity. Hara and Thompson [173] speculate that in fish these might be removal of mucus, denaturation of proteins and alteration in membrane permeability and transport characteristics.

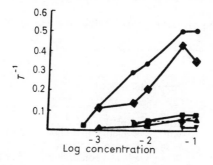

Figure 10.21 Reciprocal overturn time (min^{-1}) of goldfish immersed in non-ionic surfactant solutions. HLB values in parentheses.
(●) $C_{12}E_{10}$(12); (◆) $C_{12}E_4$(9.7); (■) $C_{12}E_{23}$(16.9); (▲) $C_{16}E_{20}$(15.7); (-▼--▼--) $C_{16}E_{10}$(12.9); (▼) $C_{18}E_{10}$(12.4). From [169].

Olfactory responses in whitefish (*Corogenus clupeaformis*) were reduced in the presence of concentrations of NaDS above $0.1\,\mathrm{mg}\,l^{-1}$. The initial sensory response in olfaction probably takes place at receptor membranes which interact with the surfactant. These receptor sites are believed to possess one cationic and one anionic subsite which are capable of interacting with stimulant amino acid molecules.

10.4.2 Anionic and cationic surfactants

The possession of charge confers an added dimension to the toxicity of ionic surfactants. Differences in the interaction of non-ionic surfactants and ionic agents with proteins have already been referred to. This, in itself, must lead to differences in biological behaviour. Surface activity *per se* is difficult to relate to biological activity but, nevertheless, the affinity of surfactants for membranes and macromolecules must confer upon all surface active agents a special toxicological dimension. Charge–charge interactions which can occur between ionic surfactants and biological molecules must also play their part in the complex spectrum of biological effects which ionic surfactants can exert.

The adsorption of linear alkylate sulphonates on human skin is assumed to result in some of the longer term adverse effects of these anionic surfactants. Different post-immersion treatments lead to differing levels of retention (see Table 10.15) [174]. Surfactants of all classes may exert their effects through increasing the absorption of other agents. In discussing the toxicity of the surface-active laxative dioctyl sulphosuccinate (DOSS), Dobbs *et al.* [175] refer to the possibility of DOSS increasing the assimilation of other laxative ingredients from formulated products. Attention has been drawn to the fact that the toxicity of oxyphenisatin in rats might be increased by DOSS [176]; higher systemic levels of oxyphenisatin might have been a factor in causing deaths in the experiments conducted, but Dobbs *et al.* conclude from their own experiments [175] (see Table 10.16) that DOSS may be the intrinsically toxic factor. Dujovne and Shoeman [177] have concluded that DOSS might be hepatotoxic.

Table 10.15 Adsorption of linear alkylate sulphonate (LAS) on the human skin after various treatments

Treatment (dipping)		Amounts of adsorbed LAS ($\mu\mathrm{g\,cm}^{-2}$)
Pretreatment*	Post-treatment†	
With LAS solution (pH 5.0)	None	19.54
With LAS (pH 5.0)	With water (pH 7.0)	18.87
With LAS (pH 5.0)	With alkali solution (pH 11.0)	9.64
With LAS (pH 5.0)	With toilet bar soap solution	8.67
With LAS (pH 5.0)	With glycerin and potash solution (pH 12.8)	6.29

* Dipped for 15 min.
† Dipped twice for each 1 min.
From [174].

Table 10.16 The effects of oxyphenisatin and DOSS on mice. Each group in the table refers to results from groups of 10 mice which were treated with oxyphenisatin alone, DOSS alone, or a mixture of the two at the dose levels shown, in a volume of 20 ml kg^{-1} 4% gum acacia, twice daily for a period of 5 days. It is clear from the table that the difference in acute toxicity between oxyphenisatin and DOSS is maintained when multiple doses are used.

Substance	Dose level (g kg^{-1})	Results in males	Results in females
Oxyphenisatin	2	No deaths	One dead by the sixth dose
Oxyphenisatin	4	One dead by the eighth dose	One dead by the sixth dose, three dead by tenth dose
Dioctyl sodium sulphosuccinate	0.5	Four dead by second dose	Five dead by seventh dose
Dioctyl sodium sulphosuccinate	1	All dead by eighth dose	All dead by tenth dose
Dioctyl sodium sulphosuccinate	2	All dead by fifth dose	All dead by fourth dose
Dioctyl sodium sulphosuccinate plus oxyphenisatin	0.5 2	Five dead by eighth dose	Six dead by seventh dose

From Dobbs *et al.* [175].

The difficulty of elucidating the mechanism of one toxic action, namely irritancy of surfactants on the skin, is illustrated in the various approaches that have been made to the problem. Ferguson and Prottey [178], in the introduction to their paper on the effects of surfactants on mammalian cells *in vitro*, list the laboratory studies with a physiochemical and a biochemical bias that have been reported in the literature. The effect of surfactants on water-binding capacity of skin, skin permeability and denaturation of proteins of the stratum corneus have been studied. The effects of surfactants on DNA and lipid metabolism, epidermal phospholipid metabolism and acid phosphatase activity have also been investigated. Ferguson and Prottey question whether any of these approaches is fully representative of the totality of interactions of surfactants with skin such as may be observed clinically. This is simply a fact of the relevance of *in vitro* studies to clinical situations. Compounds are applied directly to enzymes, cells or tissue samples in culture without the necessity of passing through lipid and other barriers which might prevent them reaching the site of action in the whole animal. Nevertheless, observational studies are useless without an understanding of the molecular processes involved in the interactions and their likely consequences.

The high incidence of sarcomas induced by polysorbate 80, Sodium Patent Blue V and by Sodium Blue VRS is in accord with observations made on the relation between surface activity and the incidence of local sarcomas in long-term injection tests [179, 180]. Surface-active food colourings and other additives had been found to produce sarcomas when administered over long periods in concentrated solutions, when the surface tension was below the 'critical lytic index' below which surfactants produce irreversible damage to cells. The physicochemical nature of the toxic action of such surfactants is illustrated by experiments with Sodium Patent Blue V [181] (Fig. 10.22). At 2% and 3% levels there was a high incidence of sarcoma induction but at 1% levels no sarcomas were produced. At 1% levels the surface tension lowering produces no cell

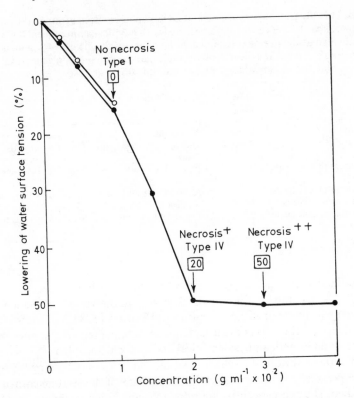

Figure 10.22 Relationship of surface activity, tissue damage, type of reaction and sarcoma production [n%] for salts of Patent Blue V (● sodium salt; ○ calcium salt). From Grasso [182] with permission.
Type I and II lesions are quiescent connective tissue lesions of self-limiting type. Types III and IV are progressive lesions characterized by persistent fibroblast proliferation and deposition of collagen.

damage. Grasso *et al.* [181] conclude from their contemplation of the physico-chemical factors determining sarcoma production that induction of sarcomas in the rat 'does not constitute an index of chemical carcinogenity'. Table 10.17 shows some of the physical properties of colouring agents in two categories, sarcoma inducers and non-inducers. Those compounds which have a low surface activity yet which cause sarcomas lead to high partition coefficients towards lipid (Eosine G and Fluorescein).

The macrophage response which was part of the type III responses at the site of injection of polysorbate 80 was found to be comparable to that observed on injection of food dyes, or iron-dextran or carboxymethylcellulose. It has been suggested [181] that micelle formation and association by surfactants in high concentration contribute to the macrophage response, but it is difficult to

Table 10.17 Physical properties in relation to sarcoma induction (colourings used as sodium salts)

Food colouring		Chemical category	Surface activity*	Partitioning tricaprylin†	Protein binding‡
No sarcoma induced					
Patent Blue V (Ca)	}	Triphenyl-	16.5	4.0	2.25
Green S	}	methane	1.0	0.1	12.5
Erythrosine	}	Xanthene	1.1	84.8	55.8
Violamine R	}		27.0	0.1	24.0
Amaranth	}	Azo	3.0	0.1	11.2
Sunset Yellow	}		2.0	0.1	1.0
Sarcoma induced					
Blue VRS	}		38.5	6.0	0.75
Patent Blue V (Na)	}	Triphenyl-	50.0	3.0	0.83
Fast Green FCF	}	methane	31.1	2.8	31.10
Brilliant Blue FCF	}		35.0	2.5	20.3
Light Green SF	}		32.0	4.0	25.6
Eosine G	}	Xanthene	6.4	68.0	29.4
Fluorescein	}		3.8	97.7	10.8

* Depression of surface tension of water (%). † Colouring partition in tricaprylin (%).
‡ g colouring/100 g rat serum protein.
From [182].

envisage why this should be so unless the size of the transient micellar species produces the response to a larger foreign body. This work adds a further element of caution (or complexity) to the design and interpretation of toxicity tests for surface-active compounds.

A rapid method for assessing the cytotoxicity of compounds has been suggested by Ferguson and Prottey [183] by measurement of inhibition of DNA synthesis. NaDS was one of the compounds studied, [³H] thymidine uptake being measured. At 0.01 mM there was observed a marked stimulation of thymidine uptake (perhaps due to increased fibroblast permeability or enzymic stimulation). At higher levels, as shown in Fig. 10.23, DNA synthesis is totally inhibited, lysis being observed concurrently. This technique was used to investigate the effect of a range of surfactants, results for which are shown in Table 10.18.

The concentration of surfactant required to inhibit DNA synthesis was always lower than that required to release histamine from most cells which was lower in turn than the concentration required for fibroblast lysis, measured by release of ^{51}Cr. In many studies of homologous series of surfactants, compounds with a C_{12} carbon chain appear to be optimally active. The same trend is shown here in the sodium caprylate–sodium palmitate series. Similar reports of the efficacy of the laurate ion in inducing skin irritancy have appeared, but Ferguson and Prottey do not claim that their test systems are indicators of irritancy. The molecular basis for the activity of the C_{12} hydrocarbon-chain surfactants is of considerable interest, however. Table 10.19 from a review of the molecular basis

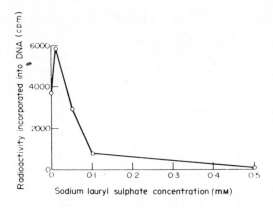

Figure 10.23 The effect of NaDS concentration on the incorporation of radioactivity from ^{3}HTdR (tritiated thymidine) after incubation of $10\,\mu$Ci of ^{3}HTdR with 168 000 fibroblasts for 17 h at 37° C in absence of serum. From Ferguson and Prottey [183].

of skin irritation [184] considers the activity of sodium alkyl carboxylates and sodium alkyl sulphates in several test systems. Human skin irritation, protein denaturation, extraction of stratum corneum, cell lysis *in vitro* and toxicity to mice are parameters observed.

Schott has suggested that the particular effectiveness of C_{12} compounds is due to the balance of two properties. As the homologous series is ascended the lipophilicity of the compounds increases but the critical micelle concentration falls thus limiting the concentration of monomers which can exist in the aqueous phase for higher members of the series. From C_8 to C_{12} with an increasing lipophilicity there is an increased opportunity for the surfactants to enter the biophase whereas from C_{12} to C_{18} while the thermodynamic tendency to partition persists, with decreasing concentration of monomers the surfactants can produce only a lower response [185]. It would seem coincidental that the C_{12} member of diverse series would always appear to be maximally active. Some workers have suggested that the C_{12} chain is of intrinsic biological relevance in relation to its ability to disrupt lipid bilayers, but this seems to have less relevance in skin irritation. It is an area obviously worthy of further study.

The irritancy of anionic surfactants such as sodium lauryl sulphate has been modified by combining them with a non-ionic detergent [186] it being suggested that complexation or mixed micelle formation occurs to lower the activity of the anion [187]. Irritation is a complex process involving many ill-defined phenomena, but a theoretical approach to the induction of irritancy by surfactants and the involvement of keratinic proteins has been proposed by Dominguez *et al.* [188] and a sequence of events (detergency, adsorption, interaction, denaturation, permeability, irritation) defined. In most of the experiments performed by this group with sodium alkyl sulphates, the C_{12} analogue has a maximal effect on the amount of cysteine formed during the treatment of human hair with

Table 10.18 Effect of surfactants *in vitro* on DNA synthesis in guinea-pig fibroblasts, on histamine release from rat peritoneal mast cells and on lysis of guinea-pig fibroblasts

Surfactant		Structure	Concentration (mM) of surfactant required to		
			Inhibit DNA synthesis by 50%*	Release all available histamine†	Release 50% of available ^{51}Cr*
Sodium	caprylate	C8:0	0.6	>5.0	>10.0
	caprate	C10:0	0.4	1.0	>10.0
	laurate	C12:0	0.1	0.4	7.1
	myristate	C14:0	0.2	0.5	7.7
	palmitate	C16:0	0.3	1.0	>10.0
Sodium	lauryl sulphate	$CH_3 \cdot [CH_2]_{10} \cdot CH_2 \cdot SO_4Na$	0.06	0.03	0.22
	lauroyl isethionate	$CH_3 \cdot [CH_2]_{10} \cdot COO \cdot [CH_2]_2 \cdot SO_3Na$	0.08	0.15	0.25
Lauryl	monoethoxylate	$CH_3 \cdot [CH_2]_{10} \cdot CH_2 \cdot OCH_2 \cdot CH_2OH$	0.06	0.46	0.21
	triethoxylate	$CH_3 \cdot [CH_2]_{10} \cdot CH_2 \cdot [OCH_2 \cdot CH_2]_3 \cdot OH$	0.01	0.03	0.30
	hexaethoxylate	$CH_3 \cdot [CH_2]_{10} \cdot CH_2 \cdot [OCH_2 \cdot CH_2]_6 \cdot OH$	0.01	0.03	0.02

* In guinea-pig fibroblasts.
† In rat peritoneal mast cells.

In each experiment, these surfactants were examined in batches. Sodium lauryl sulphate was included in each batch and the means of five estimations were 0.06 ± 0.003, 0.03 ± 0.004 and 0.22 ± 0.005 mM for the studies involving inhibition of DNA synthesis, release of histamine and release of ^{51}Cr, respectively.
From [178].

Table 10.19 Correlations of the structure of an homologous series of sodium alkyl carboxylates and sodium alkyl sulphates and their properties in various laboratory test methods, with observed irritant action on human and rat skin (from Prottey [184]).

Chain length of sodium alkyl carboxylate	Human skin irritancy	Rat skin irritancy	Collagen disc swelling	Extraction of stratum corneum		Ion conductance of stratum corneum
	Positive reactions to 22.5 mM solutions (%)	Assessed score for irritancy (macroscopic and microscopic). 0.25 M solutions for 15 min, once daily for 3 days	Change in thickness, critical micelle concentration (%)	Increase relative to water (%)	Proteins:Amino acids	Rate of change of conductance (Mohm/cm^2/min $\times 10^{-3}$)
C_8	2	2.5	62	126	0	0
C_{10}	19	13.5	94	197	43	1
C_{12}	70*	Reaction of skin* too strong for accurate assessment	136*	187	235*	45*
C_{14}	41	2.0	121	793*	148	40
C_{16}	4	1.0	19	148	161	2
C_{18}	0	2.0	0	ND	ND	ND

Chain length of sodium alkyl carboxylate

	Human skin penetration	Rat skin penetration	Inhibition of DNA synthesis in vitro	Mast cell lysis in vitro	Fibroblast lysis in vitro
	Permeability constants (μcm min^{-1}), 6 mM solution for 6 h in vitro	Amount penetrating 7.5 cm^2 of skin, applied in 0.1 ml of 6 mM solution for 15 min	Conc. (mM) required to inhibit uptake of ^{3}HTdR into fibroblasts in vitro	Conc. (mM) to release histamine from mast cells in vitro	Conc. (mM) to release ^{51}Cr-labelled cytoplasm from fibroblasts in vitro
C$_8$	ND	ND	0.6	> 5.0	> 10.0
C$_{10}$	5.4	1.8	0.4	1.0	> 10.0
C$_{12}$	18.2*	5.1*	0.1*	0.4*	7.1*
C$_{14}$	1.6	2.0	0.2	0.5	7.7
C$_{16}$	0.1	0.5	0.3	1.0	> 10.0
C$_{18}$	0.1	0.5	ND	ND	ND

Chain length of sodium alkyl sulphate

	Human skin irritancy	Denaturation of protein	Extraction of stratum corneum		Ion conductance of stratum corneum	Cell lysis in vitro	Toxicity to mice
	Positive reactions at 22.5 mM solutions (%)	Proportion of total sulphydryl groups liberated from ovalbumin (%)	Increase relative to water proteins : amino acids (%)		Rate of change of conductance (Mohm/cm^2/min $\times 10^{-3}$)	Conc. (mM) to release histamine from mast cells	Intraperitoneal injection mg/kg
C$_8$	5	0	ND	ND	0	ND	396
C$_{10}$	5	32	84	166	1	0.5	284
C$_{12}$	43*	93*	195*	238*	50*	0.03	250*
C$_{14}$	24	73	110	164	20	0.025*	342
C$_{16}$	5	21	ND	ND	4	0.05	356
C$_{18}$	0	ND	ND	ND	ND	ND	477

ND = Not determined.
* Denotes chain length with most potency in that test.

thioglycollic acid at pH 3.25. Adsorption experiments on human callus and virgin wool showed that the C_{12} alkyl sulphate is the most effectively adsorbed. Dominguez *et al.* have considered Schott's approach to the dominance of C_{12} compounds but have postulated that the unique properties are probably related to the conformation of the C_{12} alkyl chain, especially when adsorbed or interacting with protein. If the C_{12} chain is in an 'open cyclohexane' type structure a minimum length will be found in the C_6–C_{16} series at C_{12} (see Fig. 10.24) by adopting a compact conformation; Dominguez claims that the C_{12} chain has the dimensions to allow it, for example, to migrate deeper into the skin structure and thereby exert greater effects through its presence at higher concentrations than its more lipophilic analogues. This is very speculative, and a compact conformation would hardly explain a maximum in the adsorption of C_{12} surfactant from aqueous solution on to a hydrophobic surface. In pointing out that non-ionic surfactants decrease the irritancy of anionics the authors [188] suggest that the former alter the conformation of the ionic surfactant. The explanation of co-micellization is to be preferred as this would lead to a reduction in thermodynamic activity. That this is not a universal solution might be adduced from corneal irritancy tests of a cationic detergent and its mixture with a non-ionic detergent [189] which in rabbits was more toxic than the components of the mixture tested separately.

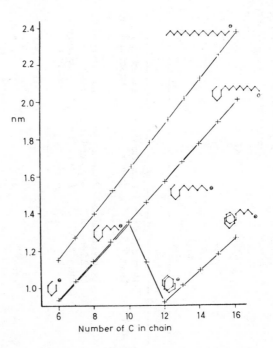

Figure 10.24 Length (A) for different conformations of alkyl sulphates: an attempt to explain the unique properties of the C_{12} alkyl chain in general surfactant series (see text). From Dominguez *et al.* [188].

The effect of alkyl chain length on the toxicity and pharmacology of a series of C_{10}–C_{20} has been studied in the female rat [190]. Short (C_1–C_{10}) alkyl chain alkyl trimethylammonium halides possess nicotinic, muscarinic and curare-like effects but the germicidal members of the series ($> C_{10}$) have an unknown toxicological profile. A report of the accidental intravenous infusion of cetrimide noted cardiac arrest, muscle paralysis and haemolysis following the injection, symptoms which were reversible. The acute intravenous toxicity of the series from C_{10} to C_{20} is recorded in Table 10.20. Toxicity decreases with increasing length of the alkyl chain up to C_{16}, but the depressor response of sublethal doses – an immediate and transient fall in arterial blood pressure – increases with increasing alkyl chain length (see Fig. 10.25). The acute toxicity leading to death is apparently due to respiratory failure; the depressor response is due to vaso-dilation [190]. Other experiments have shown that these compounds decrease axonal conduction [191] as well as decrease muscle contraction [192] in isolated preparations, an activity apparently related to the ability of these surfactants to alter the permeability of the cell membrane. However, the fact that the ability of the surfactants to depress the twitch response of the gastrocnemius muscle decreases with increasing alkyl chain length does not seem to be compatible with this simple explanation. On the other hand, experiments in whole animals and in isolated tissues cannot be directly compared because of the transport and distribution problems of the former.

Other examples of the cytological effects of surfactants have been reported in papers dealing with cell proliferation. Tween 60 brings about a rapid and sustained intensification of cell proliferation following application as a 10% solution to the epidermis [193]; there is rapid and sustained epidermal thickening accompanied by an increase in the thymidine-labelling index. Three surfactants have been studied as modulators of transformed cell proliferation [194]. Some surface active compounds have both *in vitro* and *in vivo* antitumour activity, and others inhibit the development of tumour metastases *in vivo*. Lack of *in vivo* effect

Table 10.20 Intravenous LD 50 of surface-active alkyltrimethylammonium bromides.

Surfactant	Mouse (♀)		Rat (♀)	
	μ mol kg^{-1}	mg kg^{-1}	μ mol kg^{-1}	mg kg^{-1}
C_{10}	10* (9–11)	2.8	19 (18–20)	5.5
C_{12}	17 (15–19)	5.2	22 (18–26)	6.8
C_{14}	40 (36–44)	12.0	45 (42–48)	15.0
C_{16}	87 (81–93)	32.0	120 (112–128)	44.0
C_{20}	48 (45–50)	20.0	—	—

* 19/20 confidence limits in parentheses.
From [190].

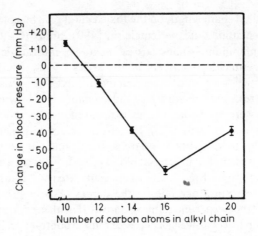

Figure 10.25 Effect of surface-active alkyltrimethylammonium bromides on arterial blood pressure in anaesthetized (sodium pentobarbitone $50\,mg\,kg^{-1}$, intraperitoneally) female rats. The surfactants were injected in the tail vein in a dose of $3\,\mu mol\,kg^{-1}$. Each point represents the mean of 3 to 6 separate experiments. Vertical bars indicate s. e. m. C_{10}, decyltrimethylammonium bromide; C_{12}, dodecyltrimethylammonium bromide; C_{14}, tetradecyltrimethylammonium bromide; C_{16}, cetyltrimethylammonium bromide; C_{20}, eicosyltrimethylammonium bromide. From Isomaa and Bjondahl [190] with permission.

is sometimes attributable to the poor bioavailability of the surfactant species [194]. A number of fatty acids, monoglycerides and fatty acid esters are active against Ehrlich ascites tumours in mice [195]. Using soluble fatty acid sucrose esters, Kato *et al.* [195] demonstrated their effect on the viability and transplantability of Ehrlich ascites tumour cells. The sucrose esters of lauric acid showed the highest antitumour activity *in vivo* and also exhibited a strong *in vitro* activity, but Tween 20 which had no antitumour activity had a strong haemolytic activity and cell lytic activity *in vitro*. Fig. 10.26 shows some of their results for haemolytic action, explaining the fact that compounds with and without antitumour activity may be haemolytic, reinforcing the view expressed above that transport of surfactant to the appropriate tumour site *in vivo*, as well as the power to interact with cell membranes, might well be a determinant of activity. Inoue, Sunakawa and Takayama [196] have examined the capacity of ten surfactants to induce morphological transformations in hamster embryo cells and *Salmonella typhimurium* cells. None of the surfactants were found to produce transformation at any of the doses tested, although a linear alkyl benzene sulphonate was cytotoxic at $50\,\mu g\,ml^{-1}$ and cetyltrimethyl ammonium chloride was very cytotoxic at $5\,\mu g\,ml^{-1}$; both Tween 60 and Span 60 were very cytotoxic at $200\,\mu g\,ml^{-1}$. Of the surfactants tested three (LAS, sodium alkyl poly(oxyethylene) sulphate and cetyl dimethylbenzylammonium chloride) have been found to be non-carcinogenic in animal experiments (see [196]). CTAB has also been found to be non-carcinogenic in rats [197].

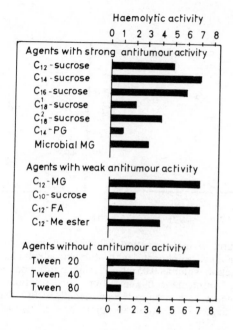

Haemolytic activity

Figure 10.26 Haemolytic activity of fatty acid esters. Samples were dissolved or suspended in 0.86 % NaCl solution at serial dilutions starting with 1 mg ml^{-1}, and an equal volume of the sheep red blood cell suspension in the NaCl solution was added. After the incubation at 37° C for 4 h, the haemolytic activity was determined. The activity was expressed by the highest dilution in which haemolysis was observed, PG, propylene glycol; MG, monoglyceride; FA, fatty acid; Me, methyl. Sheep red blood cells were used. From Kata *et al.* [195] with permission.

The influence of both alkyl chain length (of *n*-alkyl sulphates) and polyoxyethylene chain length of lauryl polyglycol ethers on the degree of tissue reaction following intravenous injection of solutions of surfactant at two levels, 0.9 % and 2.7 %, is illustrated in Fig. 10.27. Two ethoxylated decyl alcohols have been tested for their irritant action on the ears of female CF/L mice following four daily applications of 0.01 ml solution. Solutions of up to about 10 % were tested (Fig. 10.28). The irritant effects were much greater on animal skin than in the eye [198]. A single instillation of 0.1 ml 12.6 % aqueous solution of the surfactants produced a slight reddening of the conjunctiva at 24 to 48 h and at 1 to 2 h a transient corneal irritation or pitting was observed. Van Abbé suggests that the same surfactant present in an O/W emulsion of mineral oil was not irritant to the eye [199]. Cationic surfactants are much more toxic to the mucosa of the eye than most non-ionic surfactants. Table 10.21 compares the effect of non-ionic and cationic surfactants on rabbit eye mucosa [200]. Topical application of benzalkonium chloride results in marked cytotoxic effects which have been ascribed to breakdown of the outer layers of the corneal epithelium [201–203]. Clinical evidence seemed to support the view that topical administration of eye

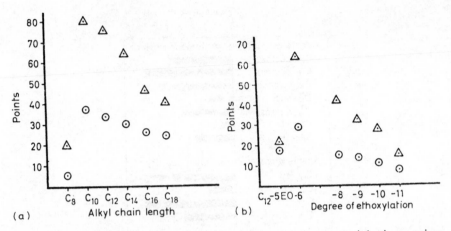

Figure 10.27(a) Degree of tissue reaction following intracutaneous injection – various concentrations of *n*-alkyl sulphonate with different alkyl chain lengths (C_8–C_{18}) (intracutaneous injection volume 0.05 ml). Maximum reaction: 80 points. (b) Tissue reaction following intracutaneous injection of solutions of dodecyl-poly-glycolethers – dependence on concentration and length of EO chain (injection volume 0.05 ml). Maximum degree of reaction: 80 points. $\triangle$ = 2.7%, $\odot$ = 0.9%. From Barail [198].

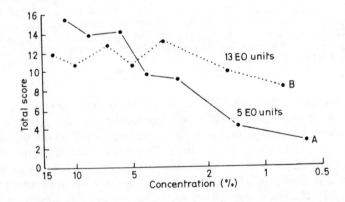

Figure 10.28 Irritancy to ears of female CF/1 mice due to four daily applications of 0.01 ml of solutions of oleyl alcohol ethoxylates. Curve A = aqueous dilutions of oleyl alcohol ethoxylate (5 EtO units per mole, nominal). Curve B = aqueous dilutions of oleyl alcohol ethoxylate (13 EtO units per mole, nominal). From von Abbé [199] with permission.

drops and preparations containing this surfactant would not produce toxic concentrations in man, but evidence has been gathered which proves the cytotoxic effect of benzalkonium chloride used accidentally to soak soft contact lenses [204], although endothelial damage was not observed.

Table 10.21 Effects of non-ionic and cationic surfactants on rabbit eye mucosa. In all cases the test concentration was 1 %. The figures indicate the degree of irritation

Product	Photo-phobia	Purulent exudate	Inflam-mation	Oedema	Corneal Opacity	Σ
Non-ionic compounds						
Sorbitanpolyoxyethylene-monolaureate	0	0	0	0	0	0
Sorbitanpolyoxyethylene-monooleate	0	0	0	0	0	0
Polyethyleneglycololeate	0	0	0	0	0	0
Polyglycollaureate	0	0	1	0	0	1
Cationic compounds						
Cetyldimethylethylene-ammoniumbromide	0	2	8	9	12	31
Alkyldimethylbenzyl-ammoniumchloride	2	4	8	6	16	36
Alkyldimethyl-3,4-dichloro-benzylammonium chloride	2	4	8	0	0	14

From [200].

10.4.3 Effect of surfactants on immunological activity of vaccines

Some surfactants possess an adjuvant activity. Hexadecylamine acts as an adjuvant for diphtheria and tetanus toxoid, and influenza and poliovirus in the guinea-pig [205] although there is species variation in its effectiveness. It is not, for example, active as an adjuvant for influenza virus in the rabbit. Asherson and Allwood [205] suggest that the activity of cationic agents may be related to their ability to adhere to and hence alter the negative charge on the surface of cell membranes. A few neutral surfactants, however, are active, e.g. Span 80 and Tween 80, although anionic agents were inactive. As non-ionic surfactants lower the negative charge in cells by adsorption, this tentative hypothesis could still hold.

Living *Mycobacterium bovis* strain Bacillus Calmette-Guérrin has been tested extensively in trials of cancer immunotherapy. Surfactant used to prepare emulsified mycobacteria fractions can affect antitumour activity. The nature of the formulation is also of importance, in particular the quantity of oil phase used in the emulsion [206] and the mode of preparation – which probably affects particle size. Ultrasonically prepared emulsions of the cell walls of the mycobacteria and trehalose-6-6'-dimycolate (TDM) required Tween 80 in an optimal range (around 0.18 %) above which tumour regressive action was diminished. Mechanically prepared systems retained this antitumour activity over a wider range of surfactant concentrations [207]. The mechanism of these effects is not understood. The conclusion is an empirical one: that in order to prepare emulsions of *Mycobacterium boris* or TDM which are immunologically active against murine fibrosarcoma, the Tween 80 concentration should be about 0.2 % and the oil concentration about 9 %. The toxicity of emulsified TDM had previously been shown [208] to be dependent on the size distribution of the mineral oil droplets.

Other factors that affect the acceptability of vaccines for clinical use can be ascribed to the chemical nature of the surfactant used in water-in-oil emulsion. Frequently W/O emulsions have proved to be unsatisfactory clinically due to local abscess at the site of injection. It has been suggested that this can be due to the oleic acid which occurs in some batches of Arlacel used as an emulsifer [205]. Hydrolysis of Arlacel A has been shown to occur in vaccine preparations with the liberation of free oleic acid [209]. Asherson and Allwood [205] have considered the various possible modes of action of adjuvant substances. These include delaying absorption of the antigen, causing aggregation of the various cell types involved in the immune responses, an action on membranes, leading to the release of chemotactic factors. One can deduce that surfactant adjuvants may participate in some of these effects but little direct evidence has been adduced.

The effect of surfactants *in vivo* may thus not be directly discernable especially in *in vitro* experiments on isolated tissues. The possible toxicological significance could lie in their ability to alter the absorption, distribution and metabolism of other substances. Increases in the absorption of phenolsulphonphthalein and pralidoxime caused by 17 mm NaDS have been ascribed to increased levels of cyclic AMP in the intestinal mucosa [210]. (The relationship between cyclic AMP and intestinal membrane on fluxes has been the object of many studies.)

10.4.4 Solubilization of carcinogens

The solubilization of toxic substances may result in their enhanced absorption. Carcinogens of the polycyclic hydrocarbon type are insoluble in body fluids and in river and tap water, where they may eventually appear as waste. The increasing use of surface-active agents in industry and in the home has increased the possibility of the solubilization of these materials, with the consequence of increased carcinogenic activity.

It is permissible to rule out the significance of solubilization by hydrotropy as substances likely to have this action would not be present in sufficient quantity in body tissues or in rivers. However, even in dilute solution, surfactants can solubilize large amounts of carcinogenic materials. A 2% Triton N solution solubilizes about 120 mg benzpyrene per litre [211]. Already present in the body are the natural association colloids, the bile acids and phosphatides, which have a similar effect. It is possible that the pulmonary surfactants which play such a large role in the dynamics of the lung are also responsible for the susceptibility of this organ to the development of carcinomas. Table 10.22 gives the solubilities of a number of typical carcinogens in potassium dodecanoate solutions at 25° C [212]. There is a general trend of reduction in solubility with increased bulk of the hydrocarbon molecule. Carcinogenic hydrocarbons maintain their ability to induce tumours when solubilized [138]. Ekwall [213, 214] points out that minute quantities of carcinogens can induce tumours when solubilized in association with colloid solutions; fore-stomach carcinomas were obtained with a total dose

Table 10.22 Solubility of carcinogenic hydrocarbons in 0.5 M potassium dodecanoate solutions at 25° C

Compound	Solubility $(g\,l^{-1})$
Benzene	30.6
Naphthalene	4.26
Acenaphthene	1.00
Fluorene, $C_{13}H_{10}$	0.728
Phenanthrene	1.21
Anthracene	0.155
Pyrene	0.453
1,2-Benzanthracene	0.145
Triphenylene	0.077
Chrysene	0.143
Naphthacene	0.023
1,2,5,6-Dibenzanthracene	0.024

From [212].

of only 0.6 to 4.0 mg 9,10-dimethyl-1,2-benzanthracene in a number of solubilized systems. For very low carcinogen concentrations, the average response to a fixed dose of dimethylbenzanthracene decreases with increasing colloid concentration [208].

Carcinogenic hydrocarbons which have entered the stomach with the ingested foods may be solubilized by the bile acids and fatty acids and brought into solution. Other substances present in the stomach which can solubilize hydrocarbons are the butyric and lactic acids produced by bacterial fermentation processes *in vivo* (see Table 10.23). The presence of these third components may

Table 10.23 Maximum solubilizing power of surfactant solutions for polycyclic hydrocarbons

Surfactant	Mol solubilizate/mol micellar substance				
	DMBA	BA	BP	MC	DBA
Potassium myristate	2.25	2.47	1.59	0.68	0.11[8]
Sodium dodecyl sulphate	3.25	4.58	1.38	1.12	
Triton-N	32.26	25.64	16.39	6.13	2.27
Sodium cholate	15.87	8.13	7.14	2.59	
Sodium deoxycholate	27.78				

1. DMBA 9,10-dimethyl-1,2-benzanthracene

2. BA Benzanthracene

3. DBA 1,2,5,6-dibenzanthracene

4. MC 20-methylcholanthrene

5. BP 3,4-benzpyrene

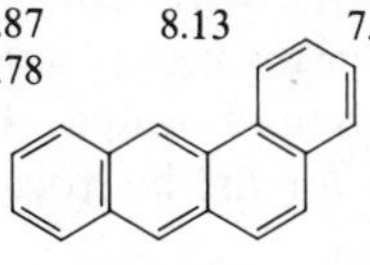

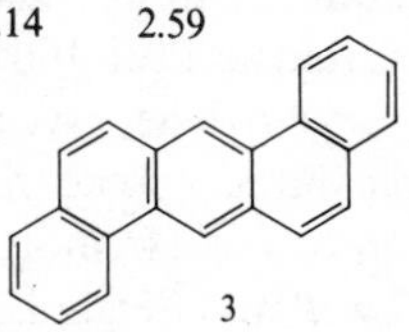

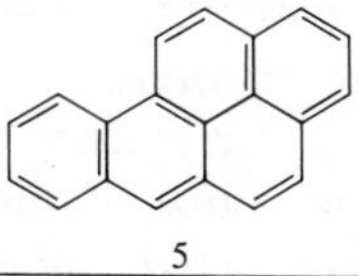

From [215].

result alternatively in a decrease in the CMC of the association colloid. It is interesting to note that the carcinogenic activity of four of the hydrocarbons listed in Table 10.23 increases in the same order as their solubility in aqueous colloid solutions. The compounds are: 1,2,5,6-dibenzanthracene 20-methylcholanthrene 3,4-benzpyrene 9,10-dimethyl-1,2-benzanthracene.

Two important aspects of the solubilization process should be noted when considering biological environments; the effect of salt concentration and the effect of time. Fig. 10. 29 shows the effect of salt on the uptake of 3,4-benzpyrene by cetyltrimethylammonium bromide [216]. About 90 h were required to reach equilibrium in the CTAB-3,4-benzpyrene-water system, although after 24 h 92% saturation is reached. Equilibration times must always be compared with the biological time-span in the ingestion, digestion, and excretion cycle. The effect of salt and polyphosphates on the solubilization of the same compound by alkyl benzene sulphonates in distilled water and tap water has been studied by Bohm-Gossl and Kruger [217]. Critical micellar concentrations were considerably lower in tap water than in distilled water (CMC of Marlon A in distilled water 1.8×10^{-3} mol l^{-1}; in tap water 1.2×10^{-4} mol l^{-1} – a figure comparable to its CMC in 0.2 M aqueous NaCl). This might suggest that studies on pure systems cannot always be extended to the biological situations.

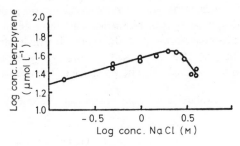

Figure 10.29 The effect of salt concentration on the solubilization of 3,4-benzpyrene by cetyltrimethylammonium bromide. From Guerritore *et al.* [216].

Investigators have employed solubilizers to attain the concentrations required to produce carcinomas in experimental animals. 1,3,7,9-tetramethyluric acid has been used as a solubilizer for benzpyrene [218]. Setala [219] used water-soluble polyoxyethylene glycols and found that solid Carbowax derivatives, when employed as spreaders for the hydrocarbons, apparently promoted the penetration of the carcinogen.

Ekwall and Setala [220] reported on the production of cutaneous tumours in mice treated with 9,10-dimethyl-1-2-benzanthracene solubilized in aqueous sodium cholate solutions. Beck [221] has studied the effect of the incorporation of solubilized 20-methylcholanthrene and 3,4-benzpyrene in the drinking water of rats. Feeding studies of this kind can be complicated if the solubilizer has an unpleasant taste, as the animals will avoid the 'laced' mixture. Molecular compounds containing 2.4 mol bile acid have been prepared with certain polynuclear aromatic hydrocarbons such as 1,2,5,6-dibenzanthracene [222]. These complex choleic acids have been suggested to be useful in the study of insoluble carcinogens.

10.4.5 Hazards of exposure to household surfactants

Estimates of human exposure to surfactants from the most common source of residues on dishes and utensils have ranged from 0.3 to 1 mg day^{-1}. This and the intake of surfactant from toothpastes and contaminated water supplies, Swisher estimates [224] to amount to a possible daily oral intake of up to 3 mg day^{-1}. This has been thought of as a conservative estimate, Moncrieff [225] estimating that as much as 15 to 20 mg might be ingested daily. Vastly discrepant views exist on this point. Oral intake on an *annual* basis of about 100 mg has been claimed [225], a level confirmed by studies using radiolabelled surfactants [227]. It was suggested that the greatest hazard arose from the contact of the skin with surfactants and their subsequent percutaneous absorption. Charlesworth has reviewed some toxicological studies on the alkyl benzene sulphonates (ABS) which are widely used in household surfactants [228], and has concluded that 'it seems unlikely that ABS products in normal use represent a hazard to the housewife'.

10.4.6 Cutaneous toxicity of surfactants

The irritant effects of surfactants have been discussed earlier. Cutaneous toxicity has recently been dealt with in some detail in a symposium volume edited by Drill and Lazar [229]. The damage surfactants can cause to the barrier layer of the skin can potentiate the toxicity of other substances. 5% NADS doubles the permeation rate of water after 1 h [230] and markedly reduced the amount of bound water in the tissues [231]. Increased penetration induced by surfactants is largely due, however, to keratin denaturation accompanied by tissue swelling [232].

A surfactant has been found to increase the topical toxicity of bupivacaine [223] following intratracheal and intravesicular administration in rabbits (Table 10.24). The lethal dose of bupivacaine when infused intravenously to rabbits

Table 10.24 Topical toxicity in rabbits after administration to the trachea or bladder

Compound	$LD_{50} \pm$ s.d.	
	Intratracheal route (mg kg^{-1})	Urinary bladder (mg kg^{-1})
DL-Bupivacaine	12.5 ± 2.8 ($n = 30$)	64.0 ± 6.7 ($n = 20$)
Surfactant (S)*	15 ($n = 10$)	75 ($n = 10$)
DL-Bupivacaine $+0.25\%$ S	8.2 ± 1.3 ($n = 30$)	44.5 ± 6.2 ($n = 30$)
DL-Bupivacaine $+1.0\%$ S	7.3 ± 1.0 ($n = 30$)	41.0 ± 6.7 ($n = 30$)

From [223].
*Surfactant = a dodecyl polyoxyethylene ether.

($1 \, mg \, kg^{-1} \, min^{-1}$, $0.32 \, ml \, min^{-1}$) was $6.9 \pm 0.7 \, mg \, kg^{-1}$. When the solution contained $0.25 \, mg \, kg^{-1}$ of the non-ionic surfactant used in this work the lethal dose was almost unchanged. The mechanism of the increased toxicity of the drug administered to the trachea or to the bladder of rabbits is probably an increased absorption of bupivacaine caused by the surfactant.

The fate of cutaneously applied surfactants is thus of obvious interest both from the point of view of elucidating their primary toxicology and how they influence the toxicity of other substances. The metabolism and distribution of a range of surfactant types following topical application is described by Drotman [233]. 88% of the applied cationic surfactant, dioctadecyl dimethyl $[1-^{14}C]$ ammonium chloride (II) remained at the test site of rabbits. Total recovery of the alkyl benzene sulphonate (III), from rat, rabbit and guinea-pig skin sites

$$C_{17}H_{35}{}^{14}CH_2$$
$$\diagdown$$
$$N^+(CH_3)_2 \quad Cl^-$$
$$\diagup$$
$$C_{18}H_{37}$$

$$C_{12}H_{25} \!-\! \langle \ \rangle \!-\! {}^{35}SO_3{}^- \ Na^+$$

(II) Dioctadecyl dimethyl $[1-^{14}C]$ ammonium chloride

(III) Dodecyl benzene sulphonate

$72 \, h$ after application, determined by complete digestion and analysis of the treated skin sample, was at a similar level. This method of analysis is, of course, not possible in human studies. In one study in man a skin site wash revealed 50% of the applied dose, the total recovered from urine being 0.3% and faeces 0.3%, a total of 51%. The non-ionic surfactant dodecyl dimethyl amine oxide (DDAO) (IV) was absorbed, a comparison of the oral and cutaneous dosing results being shown in Table 10.25.

$$^{14}CH_3$$
$$|$$
$$C_{12}H_{25}N \!\rightarrow\! O$$
$$|$$
$$CH_3$$

(IV)

Of perhaps greatest interest to pharmaceutical scientists must be Drotman's data on the radiolabelled poloxyethylene alkyl ethers using $*C_{12}E_6$, $*C_{13}E_6$ and $*C_{15}E_7$ labelled on the first carbon of the hydrocarbon chain nearest the ether linkage (e.g. $C_{11}E_{23}*CH_2(OCH_2CH_2)_6OH, *C_{12}E_6$) and $C_{12}*E_6$, $C_{13}*E_6$ and $C_{14}*E_7$ (labelled in the first carbon of the polyoxyethylene chain e.g. $C_{12}H_{25}(OCH_2CH_2)_5OCH_2*CH_2OH$, $C_{12}*E_6$). There are some metabolic differences between these closely related homologues, as shown in Table 10.26 which shows also a comparison of the distribution following oral dosing of the animals. Each of the compounds penetrate the skin, evidenced by the fact that over 20% of the ^{14}C appeared in the excretory products. The urine was the chief route of elimination of $C_{12}E_6$ and $C_{13}E_6$ but with $C_{15}E_7$ the ^{14}C appeared in the expired CO_2, an unexplained result.

Table 10.25 Distribution of radioactivity after oral and cutaneous dosing of [^{14}C] DDAO to rats

Sample	% of administered radioactivity*	
	Cutaneous dosing	Oral dosing
CO_2	2.5 ± 0.3	18 ± 1.8
Urine	14.2 ± 1.9	37.8 ± 7.1[†]
Faeces	1.8 ± 0.2	21.2 ± 8.7
Bile	—	3.6 ± 0.4
Total excreted	18.5 ± 2.4	80.8 ± 9.0
Carcass	15.6 ± 2.7	14.8 ± 5.8
Test-bile skin	48.0 ± 3.2	—
Cage wash	6.1 ± 1.3	
Total recovered	88.8 ± 1.6	95.6 ± 5.4

* Mean $\pm$ S.E.M. $n = 4$ † Includes cage wash.
From [233].

Table 10.26 Distribution of radioactivity 72 h after cutaneous application of radioactive alkyl ethoxylate to rats

Sample	% of administered radioactivity†					
	*$C_{12}E_6$	*$C_{13}E_6$	*$C_{15}E_7$	C_{12}*E_6	C_{13}*E_6	C_{14}*E_7
CO_2	4.2 ± 0.4	9.1 ± 1.2	21.7 ± 3.4	3.4 ± 0.2	2.3 ± 0.2	2.4 ± 0.0
Urine	23.4 ± 4.6	21.1 ± 4.3	6.4 ± 1.5	24.8 ± 8.2	15.1 ± 2.8	16.1 ± 2.0
Faeces	6.2 ± 0.5	6.4 ± 0.9	1.8 ± 0.6	$3.4 + 2.3$	4.9 ± 0.7	5.4 ± 0.5
Total carcass	10.2 ± 0.6	9.3 ± 1.0	11.7 ± 0.8	7.2 ± 1.3	16.6 ± 0.7	14.2 ± 1.2
Test-site skin	41.7 ± 3.2	42.4 ± 4.6	47.1 ± 8.9	59.7 ± 13.1	43.6 ± 14.1	56.9 ± 5.7
Cage wash	3.9 ± 0.7	3.1 ± 0.0	6.2 ± 4.0	2.5 ± 1.3	$17.5 + 12.6$	6.6 ± 2.8
Total recovered	89.6 ± 2.2	91.4 ± 1.7	94.9 ± 1.7	101.0 ± 5.0	100 ± 5.8	101.6 ± 2.6

Distribution of radioactivity 72 h after oral dosing of radioactive alkyl ethoxylate to rats

Sample	% of administered radioactivity†					
	*$C_{12}E_6$	*$C_{13}E_6$	*$C_{15}E_7$	C_{12}*E_6	C_{13}*E_6	C_{14}*E_7
CO_2	3.6 ± 0.4	20.4 ± 1.8	53.7 ± 5.4	3.2 ± 0.3	2.4 ± 0.2	1.9 ± 0.1
Urine	48.7 ± 6.4	45.1 ± 1.2	12.3 ± 3.4	52.1 ± 8.1	53.9 ± 5.9	54.9 ± 8.4
Faeces	25.4 ± 5.1	13.3 ± 3.0	7.9 ± 1.0	27.0 ± 4.8	25.6 ± 2.6	22.7 ± 2.1
Total carcass	9.2 ± 0.7	4.7 ± 0.5	8.9 ± 2.0	1.7 ± 0.2	1.9 ± 0.5	2.7 ± 1.1
G.I. wash	0.8 ± 0.2	1.1 ± 0.5	0.2 ± 0.1	0.8 ± 0.2	0.6 ± 0.3	0.4 ± 0.2
Cage wash	2.6 ± 1.0	2.2 ± 0.7	3.0 ± 0.9	2.4 ± 1.4	1.9 ± 0.5	2.9 ± 1.1
Total recovered	90.3 ± 2.2	86.8 ± 4.8	86 ± 8.2	87.2 ± 3.8	86.3 ± 2.5	85.5 ± 7.4

† Mean $\pm$ S.E.M. $n = 4$.
From [233].

10.5 Surfactants and plant systems

10.5.1 Phytotoxicity of surfactants

Surfactants find widespread use as components of herbicidal formulations. In the absence of herbicide many surfactants show an inherent phytotoxicity at high concentrations (around 1%). Some surfactants at low concentrations stimulate growth. Surfactants have been reported to influence a variety of plant processes, photophosphorylation [234], protoplasmic streaming [235], mitosis [236], elongation of root hairs [237], and permeability of cell walls [238].

As with animal and human studies, isolated plant cell data cannot always be extrapolated to the whole plant or organ. The data of St John *et al.* [239] on isolated leaf cell permeability changes following surfactant contact cannot necessarily be interpreted for the whole system as the significance of the effects depend on the ability of the surfactant to penetrate the leaf surface to affect cells inside the leaf. Some of the results on isolated cells are shown in Table 10.27.

Table 10.27 Release of intracellular ^{14}C-material from cells in the presence of surfactants

Surfactant and concentration		Total radioactivity (cpm/100 μl of solution + cells)	Radioactivity leaked in 45 min (cpm/100 μl of solution − cells)	(% of total)
Wild onion cells				
Untreated control		8 672	1 784	21
AHCO DD 50*	0.01%	14 169	12 675	93
AHCO DD 50	0.005%	10 076	9 629	96
AHCO DD 50	0.001%	8 772	3 997	46
Sterox SK†	0.01%	10 160	6 454	64
Sterox SK	0.005%	9 320	2 762	30
Sterox SK	0.001%	9 281	1 667	18
Daxad 21‡	0.01%	8 781	1 257	14
Tween-20	0.01%	9 855	1 989	20
Soybean cells				
Untreated control		40 840	1 547	4
AHCO DD 50*	0.1%	40 603	37 355	92
AHCO DD 50	0.01%	45 061	37 085	82
AHCO DD 50	0.005%	39 319	26 752	68
AHCO DD 50	0.001%	41 544	2 958	7
Sterox SK†	0.1%	42 172	39 642	94
Sterox SK	0.01%	41 759	29 903	72
Sterox SK	0.005%	40 986	1 649	4
Sterox SK	0.001%	41 035	1 472	4
Daxad 21‡	0.1%	41 690	2 919	7
Daxad 21	0.01%	38 221	4 696	12
Tween-20	0.1%	40 200	2 814	7
Tween-20	0.01%	39 325	4 287	11

* Alkyl benzene quaternary ammonium halide.
† Polyoxyethylenethioether.
‡ Mono-calcium salt of polymerized aryl alkyl sulphonic acids.
From [239].

Surfactants that increased cell permeability also inhibited photosynthetic $^{14}CO_2$ fixation, indicating that their effect was not restricted to the outer cell membrane. Reduced photosynthetic CO_2 fixation is possibly the result of disruption of thylakoid membrane sites of ATP and $NaDPH_2$ formation, or inhibition of one or more of the CO_2 fixing enzymes [240]. Disorganization of all membranes of cotton cells was caused by the cationic surfactant 'WSCP' [polyoxyethylene(dimethylimino ethylene)ethylene(diethylimino)ethylene dichloride] whereas Sterox SK and Renex 36 disrupted only the chloroplast grana-intergrana thylakoids causing the formation of abnormal grana, but to a lesser extent than the cationic surfactant in the study [240]. The integrity of cell structure was maintained in these experiments. Jansen previously reported [239] the following order of phytotoxicity for the surfactants described in Table 10.27: AH CO DD50 > Sterox SK ≫ Daxad 21 = Tween 80. The data in tabular form reveals the same order of effect on cell permeability. Other results (e.g. on the permeability of red beetroot tissue [238]) support these correlations. In an electrochemical study of transport of biocides in plant systems [242] it was found that in all instances when an emulsifier was introduced on the cuticle side of the membrane an increase in transport was observed due to partial or complete removal of the lipid from the leaf surface. In this paper [242] Allen refers to one of the surfactants Dobanol 25-3 (a C_{12}–C_{15} ethoxylated alcohol) as a 'phytotoxic biocide'. Dobanol 25-3 reduced transport in the soybean leaf epidermis-cuticle preparation and it has been suggested that it affects the 'gateing' mechanism of the pores causing a partial or complete stasis of the gates which would result in restricted flow through the epidermis.

Aqueous solutions with a surface tension of around $70 \, mN \, m^{-1}$ do not enter the stomatal pore, but penetration of the pore occurs when the surface tension is lowered by addition of surfactant [243]. The mechanism of penetration is by no means simple. The contact angle between the solution and the surface of the pore is important, but there is a suggestion that the morphology of the pore wall is equally important [244]. However, in the pear *Pyrus communis* L. cr. Bartlett, stomatal penetration of aqueous solution into leaves was promoted by surfactants according to their surface activity [245]. Dioctyl sodium sulphosuccinate was the most effective surfactant used and Tween 20 least effective. However, only 0.5 to 4.5 % of the stomata were penetrated suggesting that stomatal penetration is, indeed, a relatively unimportant pathway of entry to the leaf.

Wallihan *et al.* [246] have produced results with spray solutions on citrus trees which suggest that spray solutions were at least partly entering the leaves by way of the stomata; solutions with surface tensions below $30 \, mN \, m^{-1}$ have since been shown to be able to penetrate into the sub-stomatal regions of leaves [244], the cuticle in this cavity being generally thinner than the external cuticle and thus more readily penetrated by chemicals.

Vieitez *et al.* [247] interpreting the disparate effects of Tween 20, 40 and 80 on the elongation of *Avena* coleoptile sections concluded that the fatty acid moiety of the surfactant accounted for any intrinsic biological activity. Tween 20 has a growth inhibiting effect, Tween 40 has little effect, whereas Tween 80 enhances the elongation of the section. Non-ionic, amphoteric, anionic and cationic surfactants have been studied for their effect on intrinsic inhibition of the primary pea root meristem. Of the 22 compounds studied at a concentration of 0.1 %, 16 of the compounds caused significant inhibition [248, 249] there being no clear indication of the importance of the class of surfactant. Vernon and Shaw [250, 251] have shown that Triton X-100 causes uncoupling of photophosphorylation in spinach chloroplasts at concentrations as low as 7×10^{-3} %. Tween 20 at very low concentrations also stimulates root respiration and succinoxidase activity of root mitochondrial fractions from tobacco plants [252].

10.5.2 Influence of surfactants on penetration of solutes into plants

Considerable evidence has been gathered to show that the addition of a surfactant to solutions of phytotoxic substances may either enhance, diminish or not affect the biological responses [231, 253, 254] – the behaviour is thus similar to that of drug–surfactant systems. Concentration-dependent effects are clear, too, as with drug–surfactant systems and many surfactants, as we have seen above, enhance foliar absorption. Corns and Dai [255] found that the saplings of *Populus tremuloides* and *P. balsamifera* could be killed with 2000 ppm picloram, but in the presence of 1 % of a non-ionic mixture 500 ppm had the same effect. The surfactant system was less effective in enhancing the effectiveness of 2,4D and 2,4,5T, achieving only a doubling of activity.

Jansen [256] has reviewed the effect of surfactants on herbicide entry into plants. Polyoxyethylene sorbitan monolaurate and polyoxyethylene trimethyl-nonyl alcohol ether were compared. The former has little influence on the growth and appearance of corn, whereas the latter is phytotoxic at a concentration of 1 %. The first is a straight-chain compound (mol wt 1200); the latter is a branched-chain alcohol derivative, of mol wt 500. The trimethylnonyl ether is known as an efficient solubilizer used in disrupting cell membranes *in vitro*. Jansen [256] suggests that it might 'disrupt the integrity of the cuticle by partial solubilization of cuticular components and solubilization of the cytoplasmic membranes of underlying cells'. The degree of phytotoxicity is a function of concentration. No relation was found between surface activity, interfacial tension, and the activity of DNBP (4,6-dinitro-*o-sec*-butylphenol)–non-ionic mixtures at 0.1 % and 1.0 % concentrations, the surface characteristics being relatively constant above the CMC. As activity varied with concentration of detergent, this suggests the importance of some solubilizing action. Other workers [257] have found that surface tension effects *are* important, but also conclude that herbicide–surfactant interaction was at least as important. Dybing and Currier [258], using a fluorescent dye technique for measuring foliar penetration, also found that surfactants enhanced the rate of entry of herbicides. The effect of surfactants on translocation is unclear.

There are a number of possible modes of action of solubilizers on activity which have been discussed by Parr and Norman [259] but we must limit our discussion of solubilizers to their action as co-solvents and their reaction with the herbicide. As Staniforth and Loomis [260] noted that the major effect on herbicidal activity was found in the surfactant concentration range above the CMC, surface-tension reduction cannot play a large role. Furmidge [261] observed that the micelles of the detergents can, by solubilization, remove large areas of the wax from leaf surfaces, thus enabling penetration of the cuticle. Two routes of entry are possible: a polar (aqueous) route and a non-polar (lipoidal route). This perhaps explains why any one surfactant may both promote and retard the penetration of compounds which enter by different routes. Barrier and Loomis [262] noted increased absorption of 2,4-D, but not of ^{32}P through soybean leaves in the presence of Triton B-1956. Tween 80 reduced and Triton X-100 increased the absorption of ^{32}P.

In a study of the influence of surfactants on the toxicity of glyphosphate [N-(phosphonomethyl) glycine] to common milk weed (*Asclepias syriaca L.*), and Remp dogbane (*Apocynum cannabinum L.*) Wyrill and Burnside [263] found that solution contact angle was not related to surfactant effectiveness. The two plants are known to have quite different cuticle structures yet both responded similarly to a wide range of surfactants. The authors suggest that if the surfactants exerted a major influence on glyphosphate movement through the cuticle, greater differences might be expected. Diffusion of the active substance was, in fact, unaffected by the presence of one of the effective surfactants and it is likely, therefore, that the primary effect of the surfactant is on the plasmalemma, it having been found that surfactants increase the permeability of this membrane [264].

Temple and Hilton [253] determined the solubilities of ametryne, diuron, and atrazine in a wide range of 0.5% surfactant solutions. Weight for weight the solubility is greatest in the cationic Nalquat and Aliquat series. The Carbowaxes and dimethylformamide have no effect on the solubility of these herbicides (Table 10.28). These authors point out that when no surfactant is present, saturated solutions applied to the foliage will precipitate as the spray evaporates. When a solubilized preparation is used, two effects should be considered: firstly the increased amount of surfactant in the initial spray, and secondly, the solubility of the herbicide in the remaining surfactant after the water has evaporated. It has been found that the solubility (4 to 15%) of ametryne, diuron, and atrazine in some pure surfactants is sufficient to keep all of the herbicide present in a saturated spray solution in solution on the leaf surface but it is possible that solubilized preparations are more readily lost from leaf surfaces in rainwater.

0.25 to 0.5% of a polyoxyethylene thioether increased the activity of sodium 2,2-dichloro-propionate (dalapon) in the control of Johnson grass [266]. Jansen [267] has made a comparison of non-ionic surfactants at equimolar concentrations. His results are shown in an unusual diagram reproduced as Fig. 10.30. Neither dalapon nor DNBP is toxic to corn. With non-toxic surfactants the activity of dalapon is enhanced as solubilizer concentration rises. The combination of DNBP and tridecyl-E_6 suggests some sort of synergism even at the 10 mmol level. The monoethylene glycol ether of nonylphenol was almost insoluble and provided no enhancement of either herbicide's activity.

Table 10.28 Increase in solubility of ametryne, diuron, and atrazine in 0.5% aqueous surfactants at $23 \pm 1°C$
Solubility in water: ametryne 185 ppm, diuron 33 ppm, atrazine 56 ppm (= 1.00)

| Commercial name | Surfactant type | Increase in solubility over aqueous solubility | | |
		Ametryne	Diuron	Atrazine
O Samul 05	Amine salt of an alkyl sulphonic acid	4.03	3.33	2.67
HD9	Sodium alkylaryl sulphonate	3.68	3.00	2.35
Vatsol OT	Disodium N-octadecyl sulphosuccinate	3.65	2.52	—
Ultra vel K	Alkyl benzene sodium sulphonate	3.51	3.03	—
Santomerse 1	Sodium dodecyl benzene sulphonate	3.49	3.33	1.32
Sterox SK	Polyoxy-thioether	3.59	4.81	1.75
Triton X-100	Iso-octyl phenyl PEG	2.84	3.39	1.34
Tergitol NPX	Nonyl phenyl PEG	2.78	4.24	1.52
Nalquat G8.12	1-(2-hydroxyethyl)-2-methyl-1-benzyl imidazolinium chloride	8.43	9.03	2.15
Aliquat 204	Dilauryldimethylammonium chloride	4.39	15.76	2.75
Aliquat 4	Lauryltrimethylammonium chloride	2.64	2.57	1.54
—	1% PEG 200	1.04	0.91	1.00
—	600	1.08	0.91	0.84
—	6000	1.04	0.97	1.00

From [253].

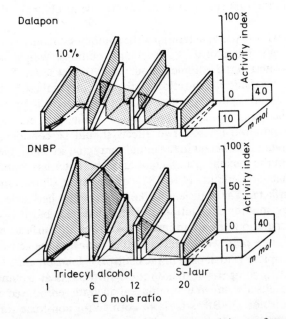

Figure 10.30 Toxicity responses of corn to sprays containing surfactants alone (clear blocks) and in combination with DNBP and dalapon (cross-hatched blocks) and their relationship to surfactant concentration and structure. Shaded planes show extrapolated relation of 1% standard concentration to concentrations on a molar basis. S. laur. = sorbitan monolaurate. From Jansen [267] with permission.

The activity of anionic surfactants is intimately related to the configuration of their alkyl chains [267]. At concentrations at which surfactants differently affected herbicide activity, differences in surface activities and conductivities could be detected, but could not be correlated with the biological action; Jansen concludes that surface activity does not affect the biological activity *per se*, but rather it is the solution properties of the surfactants which are influential. Solubility of the herbicides in surfactant solutions was found to be a useful guide to activity, but was not sufficiently related directly to it to make this one characteristic a foolproof pointer [268]. It is essential that precise physical data are obtained on herbicide–surfactant–water systems and that their ability to interact with lipoidal membranes is determined.

Smith *et al.* [269] also have studied the effect of surfactant structure of non-ionic detergents on the activity of water-soluble herbicides. Fig. 10.31 illustrates the influence of ethylene oxide chain length (at equimolar surfactant concentrations) on the toxicity of paraquat, dalapon, and amitrole. The reduction in the toxicity index with longer chain-length surfactants may be explained as follows. The authors [269] suggest that non-ionic detergents align themselves at the surfaces of cracks or fissures in the leaf surface with the polyoxyethylene chains oriented away from the surfaces (shown diagrammatically in Fig. 10.32). Water-soluble herbicides such as those studied will thus be able to diffuse down by this route. Longer polyoxyethylene chains hinder the passage of the herbicide. The results of Babiker and Duncan [270] on amitrole penetration into bracken fronds under the influence of several non-ionic surfactants would tend to support this view.

On leaf surfaces the loss of the insecticidal properties of DDT solutions may result from the solution of the DDT in the cuticle. DDT sprays should contain no solubilizers which might be capable of promoting this effect.

Buchanan and Staniforth's [271] results on the toxicity of surfactants towards *Elodea* and *Glycine max* generally showed a decreased toxicity with increase in ethylene oxide chain length.

Although non-ionic surfactants do not always maximally promote foliar penetration, their presence in formulations is often preferred because of their greater spectrum of compatibilities especially when concentrates are diluted with water of low quality, hard water and water of high salt content.

The relationship between surfactant HLB and pesticidal activity has been referred to as 'variable and yet intricate' [272]. This is perhaps an understatement, as the same lack of success in relating HLB to activity has been found with herbicides as with drugs. Sirois [273], for example, found a direct relationship between the lipophilicity of a series of non-ionic polyoxyethylene ethers, their inherent phytotoxicity and their ability to enhance the activity of 2,4D versus *Lemna minor*, but Evans and Eckert [274] found an *inverse* correlation in the paraquat–*Bromus tectorum* system. Morton and Coombs [275] suggest that surfactants of HLB 13.3–15.4 have the greatest effect on picloram-2,4,5-T triethylamine salts on seedlings of three woody species. Surfactants with ether linkages have been found to be more effective than those with ester linkages when

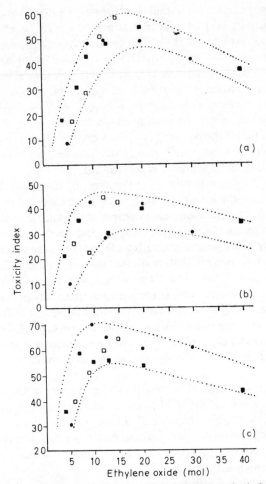

Figure 10.31 The relationship between the number of moles of ethylene oxide in octyl phenyl (●), nonyl phenyl (■), or lauryl phenyl (□) polyoxyethylene glycol ether surfactant molecules and the toxicity index in mixtures with (a) paraquat, (b) dalapon, and (c) amitrole on corn plants. Surfactant concentration 0.005 M. The toxicity index is calculated by expressing fresh weight for each treatment as a percentage of untreated control and subtracting this value from 100. From Smith *et al.* [269] with permission.

Figure 10.32 Diagrammatic representation of the 'hydrophilic channels' described by Smith *et al.* [269], and the effect of: (a) a short ethylene oxide chain non-ionic detergent, and (b) a long-chain non-ionic detergent on the passage of a water soluble herbicide (●) into the leaf cuticle.

applied to *Prosopis juliflora* but to be equally effective on *Quercus virginiana* and *Smilax bona-nox*.

10.5.3 Effect of surfactants on absorption of mineral nutrients

Surfactants are often used to increase the foliar uptake of minerals by plants. Nelson and Garlich [276] have examined the effects of over seventy surfactants and found several anionic, cationic and non-ionic surfactants which increased Fe uptake without burning the leaf tissue. There is some conflict in the literature; individual authors have found that one surfactant will enhance the absorption of one inorganic species but prevent another's absorption. For example, Triton X-100 greatly enhances the absorption of P by apple leaves but reduces Mg^{2+} uptake [277]. To clarify some of these problems, Beauchamp [278] has made a detailed study of manganese and zinc absorption with octylphenyl non-ionic surfactants in the Triton X series using soybean leaf tissues (Fig. 10.33). Absorption of both was increased in tissues floating on solutions of the ions and surfactants. Neither surface activity or HLB could be correlated with the effect; in most experiments Triton X-114 (with 7 or 8 ethylene oxide units) was generally the most effective of the series, although Triton X-100 and X-102 have similar effects on Zn^{2+} uptake. These results should be compared with those shown earlier in Fig. 10.31 on dalapon and amitrole uptake.

There seems to be little more evidence in the literature to allow a more coherent discussion of the effects of surfactants on herbicide activity and explanation of their intrinsic phytotoxicity. It may be that with such a diversity of plant structures the task of rationalizing the activity of surfactants (with the complex pattern of wetting effects, evaporation retardation, solubilization, particle size alteration of precipitated active ingredient, membrane permeability effects and intrinsic biological effects on enzyme systems) will be more daunting than with the relatively simple problem that must be faced with surfactant effects on the human organism. It is likely, however, that study of surfactant effects on plant cells will give useful information to help in the elimination of effects on mammalian cells.

10.5.4 Other determinants of surfactant effects

It is not only the physical effect of surfactants on membrane permeability or on enzyme systems that will determine toxicity of the surfactant itself or of formulations containing an active ingredient. Because the mode of application of solutions to plant surfaces is generally by spraying, the physical properties of the spray solution, particle size, dynamic surface tension, evaporation, bouncing from the leaf surface and retention of droplets are all factors which have to be considered in evaluating formulations [279, 280]. In addition the removal of active material by rain is a factor that could be altered by the presence of solubilizers. These factors are discussed in detail by Hull [272]. Tadros [281] has taken a physico-chemical approach to the investigation of the spreading and

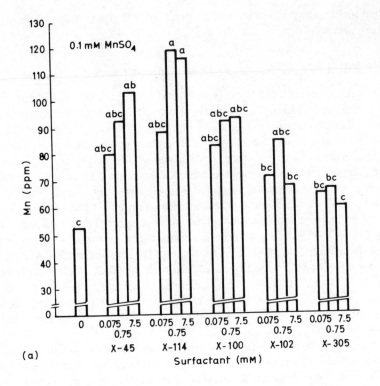

(a)

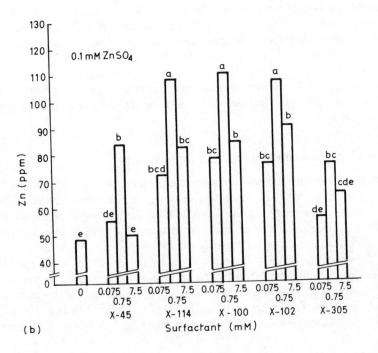

(b)

retention of surfactant solutions on wheat leaves. The spreading coefficient S may be defined as (see Chapter 1)

$$S = \gamma_{L/A} (\cos \theta - 1) \quad (10.1)$$

where $\gamma_{L/A}$ is the surface tension of the solution and θ the contact angle of the drop on the leaf surface. If the advancing (θ_a) and receding (θ_r) contact angles are measured when the leaf surface is tilted at an angle of 40° from the horizontal, the retention factor, F, as defined by Furmidge [282],

$$F = \theta_M \left[\gamma_{L/A} (\cos \theta_a - \cos \theta_r)/\rho \right]^{1/2} \quad (10.2)$$

where θ_M is the arithmetic mean of θ_r and θ_a and ρ is the density of the solution, can be calculated. Fig. 10.34 shows values of F for different values of advancing

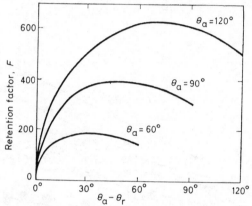

Figure 10.34 Variation of retention factor F with contact angle hysteresis calculated from Equation 10.2 (assuming surface tension of $50 \, \text{mN m}^{-1}$ and density of $1 \, \text{g cm}^{-1}$. From Ford *et al.* [287] with permission.

Figure 10.33(a) Influence of concentration of octylphenoxyethanol (Triton) surfactants on the Mn concentration in soybean leaf tissue discs in 0.1 m M MnSO$_4$ solution after a 1-h absorption period. From Beauchamp [278] with permission.
(b) Influence of concentration of octylphenoxyethanol (Triton) surfactants on the Zn concentration in soybean leaf tissue discs in 0.1 mM ZnSO$_4$ solution after a 1-h absorption period.

Surfactant	Molecular weight	Moles ethylene oxide per mole surfactant	% active ingredient	Calculated HLB
Triton X-45	446	5	100	10.4
Triton X-114	566*	7–8	100	12.4
Triton X-100	662*	9–10	100	13.5
Triton X-102	806*	12–13	100	14.6
Triton X-305	1852	30	70	17.3

* Average molecular weight.

contact angle and $(\theta_a - \theta_r)$. Fluorocarbon surfactants were required in much lower concentrations than the hydrocarbon chain surfactants to produce the same spreading coefficient, but the retention factor was lower for the fluorinated surfactants. To maintain reasonable wetting as well as retention of the spray the best compromise suggested by Tadros is to use a mixture of hydrocarbon and fluorocarbon surfactants, with the latter at low concentrations. Fig. 10.35 shows results for S and F of non-ionic surfactants $C_{16}E_{17}$ and Monflor 51 $C_{10}F_{19}O(CH_2CH_2O)_{23}C_{10}F_{19}$ and their mixtures.

The consequences of improved spreading have been demonstrated, for example, in the use of $FeSO_4$ sprays with and without surfactant [283]. Sprays without surfactant were not effective in greening the chloratic citrus trees that were being treated; the spray formed into large droplets on the leaf surfaces and

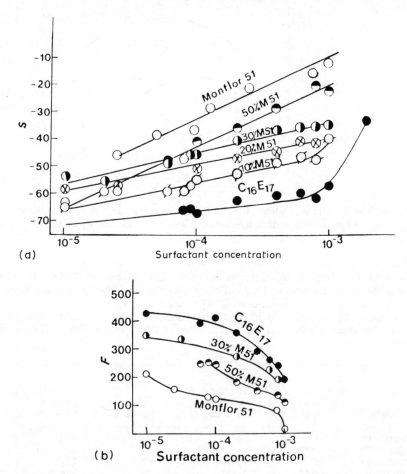

Figure 10.35(a) S–$\log C$ curves for 'Monflor' 51, $C_{16}E_{17}$ and their mixtures. (b) F–$\log C$ curves for 'Monflor' 51, $C_{16}E_{17}$ and their mixtures. Substrate: wheatleaves. From Tadros [281] with permission.

the iron penetrated in localized areas giving rise to numerous small green spots where chlorophyll had been synthesized. Leaves containing a non-ionic silicone block co-polymer L77 (Union Carbide) were uniformly wetted by a thin film and large patches of leaf and sometimes whole leaves were turned green by the treatment. The iron content of the leaf had fallen but the surfactant-containing solutions were more effective in colour production.

Deviations from expected behaviour may occur if the surfactants used in formulations interact with natural substances present in the leaf. Cationic surfactants may interact with surface-active ions and long-chain acids present in the leaf waxes, anionic surfactants may interact with long-chain alcohols but non-ionic surfactants should show little evidence of interaction. The importance of such interactions can readily be envisaged if changes in surface tension and critical micelle concentration occur. Simple adsorption of surfactant on to large areas from dilute solutions can result in increases in surface tension. Results on the change in surface tension of solutions of a series of cationic surfactants following immersion of a known surface area of leaves are shown in Fig. 10.36. The results have been explained [287] by the formation of surface-active anion–cation complexes leading to so-called 'activation' of the surfactant. The lowering of surface tension corresponds to an *apparent* increase in surfactant concentration.

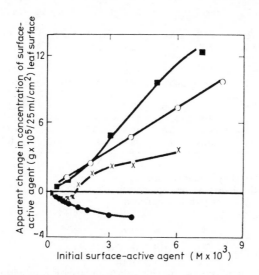

Figure 10.36 Activation of a range of cationic surface-active agents after immersion of leaves.

■ – ■ – ■ Dodecylpyridinium chloride
○ – ○ – ○ Tetradecylpyridinium chloride
× – × – × Cetylpyridinium chloride
● – ● – ● Octadecylpyridinium chloride

From Ford *et al.* [287] who note that 'the octadecyl complex is insoluble, and the overall effect is to decrease the surface activity of the solution; as the cation chain-length decreases, the complex becomes more soluble so that the cetyl, tetradecyl and dodecyl complexes all increase the surface activity of the solution.'

Surfactant effects on adsorption of herbicides on to soil have been investigated and suggested to be a factor to be considered in the overall effect of surfactant on toxicity towards the plant. The degradation, mobility and uptake of one such compound, picloram [4-amino-3,5,6-trichloropicolinic acid] ($pK_a = 3.4$) is affected by adsorption–desorption processes in solids. Picloram adsorption on to soils at pH 5 was reduced by 1% anionic surfactant [284]. The mechanism involved in picloram adsorption included protonation of the molecule, metal-ion bridging and interaction with metal ions. Picloram adsorption was enhanced by cationic surfactants, suggesting that hydrophobic adsorption of the cationic monomers on to the soil provides a cationic surface for interaction of the anionic picloram. Different soils with different pH values resulted in some variations in these effects which are presented in Table 10.29.

Table 10.29 Picloram adsorption on Oregon soils from aqueous and cationic, anionic and non-ionic surfactant solutions containing 21 μM picloram

| | | | Surfactant types | | | | | | |
| | | | Cationic | | | Non-ionic | | Anionic | |
Soil series	Organic matter (%)	0 (μg/5g)	0.1% (μg/5g)	1% (μg/5g)	10% (μg/5g)	1% (μg/5g)	10% (μg/5g)	1% (μg/5g)	10% (μg/5g)
Minam loam	6.6	8.7	16.6	41.0	16.0	6.3	3.5	5.8	12.4
Minam loam	3.8	5.4	12.0	42.4	17.6	3.4	2.4	2.7	7.8
Minam loam	0.7	0	1.8	41.0	35.6	0	0	0	0.3
Woodcock loam	7.7	5.8	17.4	47.6	7.2	7.5	5.9	4.9	11.0
Woodcock loam	4.3	9.9	40.9	nd*	nd	nd	nd	nd	nd
Woodcock sandy loam	1.6	7.1	40.6	32.9	7.1	14.4	15.2	4.1	7.5
Kinney clay loam	7.4	34.6	43.9	49.6	16.9	34.0	33.0	20.6	31.0
Kinney clay loam	2.5	25.4	40.6	49.5	23.7	36.1	35.0	11.9	17.2
Kinney clay loam	0.8	20.5	39.7	48.8	35.4	35.3	36.0	10.1	12.6

* Not determined.
Cationic surfactant: cetyl pyridinium chloride.
Anionic: sodium olefin dodecylbenzene sulphonate.
Non-ionic: octylphenyl polyoxyethylate surfactant.
From [284].

10.5.5 Surfactants in fungicidal formulations

Backman [285], in a review of the relationship of fungicide formulation to biological activity, has suggested that plant pathologists are generally unaware of the nature or 'multivalent activities' of formulations. One explanation may be that formulation ingredients are usually considered to be inert and the 'nature of these 'inert ingredients' is protected under the veil of trade secrets'. He bemoans the fact that in spite of the demonstrable effects of formulation ingredients especially surfactants, this promising area of research (the manipulation of activity by proper formulation) is frequently overlooked as a possible means of improving performance in the field. Many of the factors that are important in herbicide formulations apply equally to fungicides; the following list relates to fungicides [285]. The presence of surfactants

(1) increases spray retention where plant surfaces are of low wettability,
(2) decreases spray retention through runoff where surfaces are readily wettable,
(3) increases penetration by increasing the area of contact with the leaf,
(4) increases the period for penetration by acting as a humectant, keeping spray droplets moist indefinitely,
(5) increases cuticular penetration by acting as a co-solvent or as a solubilizing agent, or by affecting permeability,
(6) improves stomatal penetration by lowering surface tension of the spray solution,
(7) facilitates movement along cell walls after entry into the foliage by lowering interfacial tensions.

The intrinsic fungicidal activity of many surfactants will also contribute to the activity of the formulation. As a general rule it has been found (in experiments with *Monolinia*, *Alternaria* and *Puccinia* [286] that the fungicidal activities follow the order cationics > anionics > non-ionics, which parallels the general rule for phytotoxicity. It is considered that the surfactants increase membrane permeability, cause leakage of cell contents and subsequently death, but in the case of Triton X-100 and Triton X-114 which were fungicidal only to *Puccinia*, cell death was caused without increase in permeability. One of the few commercially available surfactant fungicides is a cationic compound, dodine which is *n*-dodecyl guanidine acetate. This has systemic as well as contact fungicidal activity.

References

1. P. H. ELWORTHY and J. F. TREON (1967) in *Nonionic Surfactants* (ed. M. J. Schick), Marcel Dekker, New York, pp. 923–970.
2. C. GLOXHUBER (1974) *Arch. Toxicol.* **32**, 245.
3. G. E. PETROWSKI (1976) *Adv. Food. Sci.* **22**, 309.
4. G. E. PETROWSKI (1975) *Food Technol.* **29**, 52.
5. N. R. ARTMAN (1975) *J. Am. Oil Chemists Soc.* **52**, 49.
6. P. J. CULVER, C. S. WILCOX, C. M. JONES and R. S. ROSE (1951) *J. Pharmacol. Exp. Therap.* **103**, 377.
7. J. F. TREON, L. E. GONGIVER, M. F. NELSON and J. C. KIRSCHMAN (1967) *Proc. IVth Int. Cong. Surface Activity* **3**, 381.
8. R. B. DROTMAN (1980) *Toxicol. Appl. Pharmacol.* **52**, 38.
9. Z-Y. J. WANG and I. J. STERN (1975) *Drug Metabolism Disposition* **3**, 536.
10. B. ISOMAA (1975) *Fd. Cosmet. Toxicol.* **13**, 231.
11. R. M. LEVINE, M. R. BLAIR and B. B. CLARK (1955) *J. Pharmacol. Exp. Therap.* **114**, 78.
12. J. K. FINNEGAN, P. S. LARSON, R. B. SMITH, H. B. HAAG, J. D. REID and M. L. DREYFUSS (1953) *J. Pharmacol. Exp. Therap.* **109**, 422.
13. R. A. CUTLER and H. P. DROBECK (1970) in *Cationic Surfactants*, (ed. E. Jungermann), Marcel Dekker, New York, pp. 527, *et seq.*
14. G. M. POWELL and A. H. OLAVSEN, University of Wales, Cardiff, private communication.
15. W. R. MICHAEL (1968) *Toxicol. Appl. Pharmacol.* **12**, 473.

16. F. YANAGISAWA and M. WATANABE *et al.* (1962) *Tokyo-to Ritsu Eisei Kenkyusho Nenpo*, **14**, 51.
17. J. B. KNAAK, J. S. KOZBELT, L. J. SULLIVAN (1966) *Toxicol. Appl. Pharmacol.* **8**, 369.
18. T. L. STECK (1974) *J. Cell. Biol.* **62**, 1.
19. W. H. D. DENNER, A. H. OLAVESEN, G. M. POWELL and K. S. DODGSON (1969) *Biochem. J.* **111**, 43.
20. B. BURKE, A. H. OLAVESEN, C. G. CURTIS and G. M. POWELL (1975) *Xenobiotica*, **5**, 573.
21. B. BURKE, A. H. OLAVESEN, C. G. CURTIS and G. M. POWELL (1978) *Xenobiotica*, **8**, 145.
22. B. BURKE, A. H. OLAVESEN, C. G. CURTIS and G. M. POWELL (1976) *Xenobiotica*, **6**, 667.
23. R. S. PRATT and G. M. W. COOK (1979) *Biochem. J.* **179**, 299.
24. K. SIMONS, A. HELENIUS and H. GAROFF (1973) *J. Mol. Biol.* **80**, 119.
25. T. GULIK-KRZYWICKI (1975) *Biochem. Biophys. Acta* **415**, 1.
26. B. LOIZAGA, I. G. GURTUBAY, J. M. MARCULLA, F. M. GONI and J. C. GOMEZ (1979) *Biochem. Soc. Trans.* **7**, 148.
27. I. G. GURTUBAY, E. AZAGRA, A. GUTIÉRREZ-ARRANZ, J. C. G. MILICUA and F. M. GÒNI (1979) *Biochem. Soc. Trans.* 150.
28. D. M. FOSTER, C. F. HAWKINS, D. FIFE and J. A. JACQUEZ (1976) *Chem. Biol. Interactions*, **14**, 265.
29. F. H. FITZPATRICK, S. E. GORDESKY and G. V. MARINETTI (1974) *Biochim. Biophys. Acta.*, **345**, 154.
30. D. A. WHITMORE, L. G. BROOKES and K. P. WHEELER (1979) *J. Pharm. Pharmacol.* **31**, 277.
31. S. FELDMAN and M. REINHARD (1976) *J. Pharm. Sci.* **65**, 1460.
32. K. WALTERS, P. H. DUGARD and A. T. FLORENCE (1981) *J. Pharm. Pharmacol.* **33**, 207.
33. M. YONEZAWA (1977) *Nihon. Univ. J. Med.* **19**, 125.
34. T. NADAI, R. KONDO, A. TATEMATSU and H. SEZAKI (1972) *Chem. Pharm. Bull.*, **20**, 1139.
35. P. M. LISH (1961) *Gastroenterology* **41**, 580.
36. H. NECHELES and J. SPORN (1966) *Am. J. Gastroent.* **46**, 481.
37. B. ISOMAA and G. SJÖBLÖM (1975) *Fd. Cosmet. Toxicol.* **13**, 517.
38. S. FELDMAN, R. J. WYNN and M. GIBALDI (1968) *J. Pharm. Sci.* **57**, 1493.
39. D. A. WHITMORE, L. G. BROOKES and K. P. WHEELER (1980) *J. Pharm. Pharmacol.* **32**, 62.
40. J. A. LUCY (1970) *Nature* **227**, 815.
41. G. R. A. HUNT (1980) *FEBS Lett.* **119**, 132.
42. Q. F. AHKONG, J. I. HOWELL, W. TAMPION and J. A. LUCY (1974) *FEBS Lett.* **41**, 206.
43. J. G. DOMINQUEZ, J. L. PARRA, M. R. INFANTE, C. M. PELEJERO, F. BALAGUER and T. SASTRE (1977) *J. Soc. Cosmet. Chem.* **28**, 165.
44. A. HELENIUS and K. SIMONS (1972) *J. Biol. Chem.* **247**, 3656.
45. S. LAPANJE (1978) *Physicochemical Aspects of Protein Denaturation*, chapter 3, John Wiley, New York, pp. 156 *et. seq.*
46. R. R. GENNIS and A. JONAS (1977) *Ann. Rev. Biophys. Bioeng.* **6**, 195.
47. A. J. REYNOLDS, S. HERBERT. H. POLET and J. STEINHARDT (1967) *Biochemistry* **6**, 937.
48. J. STEINHARDT and J. A. REYNOLDS (1969) *Multiple Equilibria in Proteins*, Academic Press, New York.
49. Y. NOZAKI *et al.* (1974) *J. Biol. Chem.* **249**, 4452.
50. S. MAKINO *et al.* (1973) *J. Biol. Chem.* **248**, 4926.
51. A. WISHNIA and T. W. PINDER (1966) *Biochemistry*, **5**, 1534.
52. T. S. SEIBLES (1969) *Biochemistry*, **8**, 2949.
53. A. WISHNIA (1969) *Biochemistry* **8**, 5070.
54. J. A. REYNOLDS and R. H. SIMON (1974) *J. Biol. Chem.* **249**, 3937.
55. S. MAKINO *et al.* (1974) *J. Biol. Chem.* **249**, 7379.
56. B. BORGSTRÖM and C. ERLANSON (1973) *Eur. J. Biochem.* **37**, 60.
57. A. HELENIUS and K. SIMONS (1975) *Biochim. Biophys. Acta.* **415**, 29.

58. Z. WASYLEWSKI and K. KOZIK (1979) *Eur. J. Biochem.* **95**, 121.
59. M. N. JONES, H. A. SKINNER, B. TIPPING and A. WILKINSON (1973) *Biochem. J.* **135**, 231.
60. K. NAKAYA, A. USHIWATA and Y. NAKAMURA (1976) *Biochim. Biophys. Acta.* **439**, 116.
61. E. FELDBAU and C. SCHWABE (1971) *Biochemistry* **10**, 2131.
62. P. CANIONI, R. JULIEN, J. RATHELOT and L. SARDA (1980) *Lipids* **15**, 6.
63. H. SARI, S. GRANON and M. SEMERIVA (1978) *FEBS Lett.* **95**, 229.
64. D. H. KEMPNER and B. J. JOHNSON (1976) *J. Pharm. Sci.* **65**, 1799.
65. M. D. BURKE, J. W. BRIDGES and D. V. PARKE (1975) *Xenobiotica* **5**, 261.
66. H. DENK, J. B. SCHENKMAN, B. BACCHIN, F. HUTTERER, F. SCHAFFNER and H. POPPER (1971) *Exp. Molec. Path.* **14**, 263.
67. J. T. PINTO and W. R. FRISELL (1975) *Proc. Soc. Exp. Biol. Med.* **148**, 981.
68. M. T. MITJAVILA, S. MITJAVILA, N. GAS and R. DERACHE (1975) *Toxicol. Appl. Pharmacol.* **34**, 72.
69. A. T. FLORENCE and J. M. N. GILLAN (1975) *Pesticide Sci.* **6**, 429.
70. P. C. CHAN (1967) *Biochem. Biophys. Acta* **135**, 53.
71. A. B. GRAHAM and G. C. WOOD (1972) *Biochim. Biophys. Acta* **276**, 392.
72. A. B. GRAHAM and G. C. WOOD (1973) *Biochim. Biophys. Acta* **311**, 45.
73. S. L. ROSENTHAL and M. J. SALTON (1974) *Microbios* **11**, 159.
74. R. M. HILL and D. R. BRIGGS (1956) *J. Amer. Chem. Soc.* **78**, 1590.
75. R. HUBBARD (1954) *J. Gen. Physiol.* **37**, 38.
76. C. D. B. BRIDGES (1957) *Biochem. J.* **66**, 375.
77. I. BLEI (1960) *J. Colloid Sci.* **15**, 370.
78. T. TAKAGI, K. TSUJII and K. SHIRAHAMA (1975) *J. Biochem.* **78**, 939.
79. T. TAKAGI, K. KUBO and T. ISEMURA (1980) *Biochim. Biophys. Acta.* **623**, 271.
80. J. STEINHARDT, N. STOCKER, D. CARROLL and K. S. BIRDI (1974) *Biochemistry* **13**, 446.
81. J. STEINHARDT, J. R. SCOTT and K. S. BIRDI (1977) *Biochemistry*, **16**, 718.
82. A. HELENIUS and H. SÖDERLUND (1973) *Biochim. Biophys. Acta.* **307**, 287.
83. R. L. WARD and C. S. ASHLEY (1979) *Appl. Environ. Microbiol.* **38**, 314.
84. R. L. WARD and C. S. ASHLEY (1980) *Appl. Environ. Microbiol.* **39**, 1148.
85. R. L. WARD and C. S. ASHLEY (1980) *Appl. Environ. Microbiol.* **39**, 1154.
86. N. M. LARIN and P. H. GALLIMORE (1971) *J. Hyg.* **69**, 35.
87. C. A. HART, Q. F. AHKONG, D. FISHER, A. H. GOODALL, T. HALLINAN and J. A. LUCY (1975) *Biochem. Soc. Trans.* **3**, 733.
88. Q. F. AHKONG, D. FISHER, Q. TAMPION and J. A. LUCY (1973) *Biochem. J.* **136**, 147.
89. S. H. S. LEE, K. R. ROZEE, S. H. SAFE and J. F. S. CROCKER (1978) *Chemosphere* **7**, 573.
90. K. TSUJII and F. TOKIWA (1978) *J. Amer. Oil Chem. Soc.* **54**, 585.
91. P. SIEKEWITZ (1972) *Ann. Rev. Physiol.* **34**, 117.
92. R. H. ERICKSON and Y. S. KIM (1980) *Biochim. Biophys. Acta.* **614**, 210.
93. A. C. NEWBY and A. CHRAMBACH (1979) *Biochem. J.* **177**, 623.
94. M. JANADO, Y. YANO, H. NAKAMORI and T. NISHIDA (1980) *J. Chromatog.* **193**, 345.
95. R. J. YON and R. J. SIMMONDS (1979) *Biochem. J.* **177**, 417.
96. C. BARON and T. E. THOMPSON (1975) *Biochem. Biophys. Acta.* **352**, 276.
97. J. T. LIN, S. RIEDEL and R. KINNE (1975) *Biochem. Biophys. Acta.* **557**, 179.
98. S. H. ROTH and R. J. WILLIAMS (1979) *J. Pharm. Pharmacol.* **31**, 224.
99. E. DE ROBERTIS, J. M. AZCURRA and S. FISZER (1967) *Brain Res.* **5**, 45.
100. J. S. AGUILAR, M. CRIADO and E. DE ROBERTIS (1980) *Europ. J. Pharmacol.* **63**, 251.
101. E. R. M. KAY (1965) *Cancer Res.* **25**, 764.
102. E. A. MODJAMOVA and J. M. VASILIEV (1965) *Tsitologiya* **7**, 410.
103. A. G. MALENKOV, S. A. BOGATYREVA, V. P. BOZHKOVA, E. A. MODJANOVA and J. M. VASILIEV (1967) *Exp. Cell Res.* **48**, 307.
104. M. T. NADIR and P. GILBERT (1979) *Microbios* **26**, 51.
105. C. DE DUVE, R. WATTIAUX and M. WIBO (1962) *Biochem. Pharmacol.* **9**, 97.
106. G. WEISSMANN (1965) *Biochem. Pharmacol.* **14**, 525.
106a G. WEISSMANN and H. KEISER (1965) *Biochem. Pharmacol.* **14**, 537.

107. M. C. ALLWOOD (1973) *Microbios* **7**, 209.
108. A. M. SHANES and N. L. GERSHFELD (1960) *J. Gen. Physiol.* **44**, 345.
109. K. BUCHER (1956) *Schweiz. Med. Wochschr.* **86**, 94.
110. R. J. DUBOS (1948) *J. Exptl. Med.* **88**, 81.
111. R. J. DUBOS (1945) *Proc. Soc. Exptl. Biol. Med.* **58**, 361.
112. R. J. DUBOS (1947) *J. Exptl. Med.* **85**, 1.
113. B. EISMAN (1948) *J. Exptl. Med.* **88**, 189.
114. J. E. LOVELOCK and R. J. W. REES (1955) *Nature*, **175**, 161.
115. I. KONDO and M. TOMIZAWA (1966) *J. Colloid Sci.* **21**, 224.
116. G. B. MACKANESS (1954) *Amer. Rev. Tubercul.* **69**, 690.
117. J. D. FULTON (1960) *Nature*, **187**, 1129.
118. P. D'ARCY HART (1968) *Science*, **162**, 686.
119. P. D'ARCY HART, A. H. GORDON and P. J. JACQUEZ (1969) *Nature* **223**, 672.
120. G. COPPI, G. BONARDI and S. CASIDIO (1971) *Toxicol. Appl. Pharmacol.* **19**, 721.
121. J. WATKINS, A. CLARK, T. N. APPLEYARD and A. PADFIELD (1976) *Br. J. Anaesth.* **48**, 881.
122. ANON (1978) *Br. Med. J.* **3**, 648.
123. C. M. CONWAY and D. B. ELLIS (1970) *Br. J. Anaesth.* **42**, 249.
124. H. L. THORNTON (1971) *Br. J. Anaesth.* **26**, 490.
125. W. LORENZ, A. DOENICKE, R. MEYER, H. J. REIMANN, J. KUSCHE, J. BARTH, others (1972) *Br. J. Anaesth.* **44**, 355.
126. W. WIRTH and F. HOFFMEISTER, quoted by P. VANEZIS (1979) *Practitioner*, **222**, 249.
127. J. WATKINS (1979) *Br. J. Anaesth.* **51**, 51.
128. M. S. HÜTTEL, A. S. OLESEN and E. STOFFERSEN (1980) *Br. J. Anaesth.* **52**, 77.
129. J. B. GLEN, G. E. DAVIES, D. S. THOMPSON, S. C. SCARTH and A. V. THOMPSON (1979) *Br. J. Anaesth.* **51**, 819.
130. J. B. GLEN, G. E. DAVIES, D. S. THOMPSON, S. C. SCARTH and A. V. THOMPSON (1978) in *Adverse Response to Intravenous Drugs* (ed. J. Watkins and A. M. Ward), Academic Press, London, p. 129.
131. E. S. K. ASSEM (1977) in *Drug Design and Adverse Reactions to Drugs* Alfred Benzon Symposium, Munksgaard, Copenhagen.
132. A. C. IVY *et al.* (1948) quoted in M. I. GROSSMAN *et al.*, *Proc. Soc. Exptl. Biol. Med.* **68**, 550.
133. S. LINDE (1959) *Acta Physiol. Scand.* **21**, Suppl. 74.
134. J. D. BLUM and M. VALERIO (1946) *Schweiz. Med. Wochr.* **76**, 737.
135. L. GRAMSTAD and J. STOVNER (1979) *Br. J. Anaesth.* **51**, 1175.
136. L. GRAMSTAD and J. STOVNER (1980) *Br. J. Anaesth.* **52**, 108P.
137. A. M. EHRLY (1968) *Biorheology*, **5**, 209.
138. F. L. GROVER, R. S. KAHN, M. W. HERON and B. C. PATON (1973) *Arch. Surg.* **106**, 307.
139. A. S. OLESEN and M. S. HÜTTEL (1980) *Br. J. Anaesth.* **52**, 609.
140. M. A. K. MATTILA, M. RUOPPI, M. KORHONEN, H. M. LARNI, L. VALTONEN and H. HEIKKINEN (1979) *Br. J. Anaesth.* **51**, 891.
141. H. B. NIELL (1977) *New Engl. J. Med.* **296**, 1479.
142. J. W. CORNFORTH, P. D'ARCY HART, G. A. NICHOLLS, R. J. W. REES and J. A. STOCK (1955) *Br. J. Pharmacol.* **10**, 73.
143. A. KELLNER, J. W. CORRELL and A. T. LADD (1951) *J. Exptl. Med.* **93**, 373.
144. A. SCANU and P. ORIENTE (1961) *J. Exptl. Med.* **113**, 735.
145. A. SCANU (1965) *Adv. Lipid Res.* **3**, 63.
146. S. SAITO (1957) *Kolloid-Z.* **154**, 19.
147. S. SAITO (1959) *Kolloid-Z.* **165**, 162.
148. A. SCANU and I. H. PAGE (1961) *J. Lipid Res.* **2**, 161.
149. J. VAN DEN BOSCH and A. BILLIAU (1963) *J. Exptl. Med.* **118**, 515.
150. A. T. LADD, A. KELLNER and J. W. CORRELL (1949) *Fed. Proc.* **8**, 360.
151. C. D. PORT, P. J. GARVIN and C. E. GANOTE (1978) *Toxicol. Appl. Pharmacol.* **44**, 401.

152. L. SPOLTER and L. I. RICE (1980) *Biochim. Biophys. Acta.* **612**, 268.
153. B. R. CATER, K. R. BUTTERWORTH, I. F. GAUNT, J. HOOSON, P. GRASSO and S. D. GANGOLLI (1978) *Fd. Cosmet. Toxicol.* **16**, 519.
154. R. J. HENDY, K. R. BUTTERWORTH, I. F. GRANT, I. S. KISS and P. GRASSO (1978) *Fd. Cosmet. Toxicol.* **16**, 527.
155. A. J. INGRAM, K. R. BUTTERWORTH, I. F. GAUNT, P. GRASSO and S. D. GANGOLLI (1978) *Fd. Cosmet. Toxicol.* **16**, 535.
156. Summaries of Toxicological Data (1970) *Fd. Cosmet. Toxicol.* **8**, 339.
157. R. S. HARRIS, H. SHERMAN and W. W. JETTER (1951) *Arch. Biochem. Biophys.* **34**, 243.
158. R. S. HARRIS, H. SHERMAN and W. W. JETTER (1951) *Arch. Biochem. Biophys.* **34**, 259.
159. H. TIDWELL and M. E. NAGLER (1952) *Proc. Soc. Exp. Biol. Med.* **81**, 12.
160. A. N. WICK and L. JOSEPH (1953) *Fd. Res.* **18**, 79.
161. F. STEIGMANN, E. M. GOLDBERG and H. M. SCHOOLMAN (1953) *Am. J. dig. Dis.*, **20**, 380.
162. K. COMAI and A. C. SULLIVAN (1980) *Int. J. Obes.* **4**, 33.
163. J. SZYMANIEC, M. ZIMECKI and Z. WIECZOREK (1980) *Int. Archs. Allergy Appl. Immun.*, **63**, 88.
164. C. DAHLGREN, I. RUNDQVIST, O. STENDAHL and K. E. MAGNUSSON (1980) *Cell. Biophysics.*, **2**, 253.
165. J. M. OLIVER, J. A. KRAWIEC and E. L. BECKER (1978) *J. Reticuloendothel Soc.* **24**, 697.
166. H. GOLDHAMMER and W. R. MCMANUS (1960) *Nature* **186**, 317.
167. H. GOLDHAMMER, W. R. MCMANUS and R. A. OSBORN (1970) *J. Pharm. Pharmacol.* **22**, 668.
168. W. SASKI, M. MANNELLI, M. F. SAETTONE and F. BOTTARI (1971) *J. Pharm. Sci.* **60**, 854.
169. A. T. FLORENCE, K. A. WALTERS and P. H. DUGARD (1978) *J. Pharm. Pharmacol. Suppl.* **30**, 29P.
170. D. J. WILDISH (1972) *Water Res.* **6**, 759.
171. K. J. MACEK and S. F. KRZEMINSKI (1975) *Bull. Environ. Contam. Toxicol.* **13**, 377.
172. A. GRANMO and S. KOLLBERG (1976) *Water Res.* **10**, 189.
173. T. J. HARA and B. E. THOMPSON (1978) *Water Res.* **12**, 893.
174. S-I. TOMIYAMA (1975) *J. Amer. Oil Chem. Soc.* **52**, 135.
175. H. E. DOBBS, R. L. F. DAWES and B. A. WHITTLE (1972) *N. Z. J. Med.* **76**, 213.
176. H. GODFREY (1971) *J. Amer. Med. Assoc.* **215**, 643.
177. C. A. DUJOVNE and D. SHOEMAN (1971) *Pharmacologist* **13**, 288.
178. T. F. M. FERGUSON and C. PROTTEY (1967) *Fd. Cosmet. Toxicol.* **5**, 601.
179. S. D. GANGOLLI, P. GRASSO and L. GOLDBERG (1967) *Fd. Cosmet. Toxicol.* **5**, 601.
180. P. GRASSO and L. GOLDBERG (1966) *Fd. Cosmet. Toxicol.* **4**, 269.
181. P. GRASSO, S. D. GANGOLLI, L. GOLDBERG and J. HOOSON (1971) *Fd. Cosmet. Toxicol.* **9**, 463.
182. P. GRASSO (1970) *Chemistry in Britain* **6**, 17.
183. T. F. M. FERGUSON and C. PROTTEY (1974) *Fd. Cosmet. Toxicol.* **12**, 359.
184. C. PROTTEY (1978) in *Cosmetic Science*, Vol. 1, (ed. M. M. Breuer), Academic Press, London, pp. 275 *et. seq.*
185. H. SCHOTT (1973) *J. Pharm. Sci.* **62**, 341.
186. J. A. FAUCHER and E. D. GODDARD (1978) *J. Soc. Cosmet. Chem.* **29**, 323.
187. J. A. FAUCHER, E. D. GODDARD and R. D. KULKARNI (1979) *J. Amer. Oil Chemists Soc.*, **56**, 776.
188. J. C. DOMINGUEZ, J. L. PARRA, M. R. INFANTE, C. M. PELEJERO, F. DALAGUER and T. SASTRE (1977) *J. Soc. Cosmetic Chem.* **28**, 165.
189. L. L. GERSCHBEIN and J. E. MCDONALD (1977) *Fd. Cosmet. Toxicol.* **15**, 131.
190. B. ISOMAA and K. BJONDAHL (1980) *Acta. Pharmacol. et Toxicol.* **47**, 17.
191. U. KISHIMOTO and W. J. ADELMAN (1964) *J. Gen. Physiol.* **47**, 975.
192. B. ISOMAA and T. AHLROTH (1979) *Acta. Pharmacol. et Toxicol.* **45**, 387.
193. H. WIRTH, M. GLOOR, E. WEILAND and U. W. SCHNYDER (1979) *Fette Seifen Anstrichm.* **81**, 290.

194. J. TOBLER, M. T. WATTS and J. L. FU (1980) *Cancer Res.* **40**, 1173.
195. A. KATO, K. ANDO, G. TAMURA and K. ARINA (1971) *Cancer Res.* **31**, 501.
196. K. INOUE, T. SUNAKAWA and S. TAKAYAMA (1980) *Fd. Cosmet. Toxicol.* **18**, 289.
197. B. ISOMAA, J. REUTER and B. M. DJUPSUND (1976) *Arch. Tox.* **35**, 91.
198. L. C. BARAIL (1960) *J. Soc. Cosmet. Chem.* **11**, 241.
199. N. J. VAN ABBÉ (1973) *J. Soc. Cosmet. Chem.* **24**, 685.
200. S. S. HOPPER, H. R. HULPIEU and V. V. COLE (1949) *J. Amer. Pharm. Assoc.* **38**, 428.
201. A. R. GASSET, Y. ISHII, H. E. KAUFMANN and T. MILLER (1974) *Am. J. Ophthalmol.*, **78**, 38.
202. A. TONJUM (1975) *Acta. Ophthalmol.* **53**, 335.
203. K. GREEN and A. TONJUM (1973) *Acta Ophthalmol.* **53**, 348.
204. A. R. GASSET (1977) *Amer. J. Ophthalmol.* **84**, 169.
205. G. L. ASHERSON and G. G. ALLWOOD (1969) in *Biological Basis of Medicine* (ed. E. Bittar and N. Bittar), vol. 4, Academic Press, New York, pp. 327 *et. seq.*
206. E. YARKONI, M. S. MELTZER and H. J. RAPP (1977) *Int. J. Cancer* **19**, 818.
207. E. YARKONI and H. J. RAPP (1980) *Cancer Res.* **40**, 975.
208. E. YARKONI and H. J. RAPP (1978) *Infect. Immun.* **20**, 856.
209. M. C. HARDEGRE and M. PITTMAN (1966) *Proc. Soc. Exp. Biol. Med.* **123**, 179.
210. G. BRISEID, K. BRISEID and K. KIRKEVOLD (1976) *N. S. Arch. Pharmacol.* **292**, 137.
211. P. EKWALL (1954) *Extrait de Acta. Int. Cancer* **10** (3) 44.
212. H. KLEVENS (1950) *Chem. Rev.* **47**, 1.
213. P. EKWALL (1952) *Finska Kemists Medd.* **61**, 6.
214. P. EKWALL and L. SJÖBLÖM (1952) *Acta Chem. Scand.* **6**, 96.
215. P. EKWALL, K. SETÄLÄ and L. SJÖBLÖM (1951) *Acta Chem. Scand.* **6**, 96.
216. D. GUERRITORE, L. BELLELLI and M. L. BONACCI (1964) *Ital J. Biochem.* **13**, 222.
217. TH. BÖHM-GOSSL and R. KRUGER (1965) *Kolloid-Z.* **206**, 65.
218. H. WEIL-MALHERBE (1946) *Biochem. J.* **40**, 351.
218a H. WEIL-MALHERBE (1946) *Cancer Res.* **6**, 171.
219. K. SETÄLÄ (1949) *Acta Path. Microbiol. Scand.* **26**, 280.
220. P. EKWALL and K. SETÄLÄ (1949) *Acta Path. Microbiol. Scand.* **26**, 795.
221. S. BECK (1943) *Nature* **152**, 537.
222. L. F. FIESER and M. S. NEWMAN (1935) *J. Amer. Chem. Soc.* **57**, 1602.
223. G. ÅBERG and G. ADLER (1976) *Arzneim Forsch* **26**, 78.
224. R. D. SWISHER (1968) *Archs. Envir. Health* **17**, 232.
225. R. W. MONCRIEFF (1969) *Soap Perfum. Cosmet.* **42**, 447.
226. H. WEDEL (1964) *Proc. IVth Int. Cong. Surface Activity*, Vol. 4, (ed. C. Paquot), Gordon and Breach, New York.
227. J. SCHMITZ (1973) *Tenside* **10**, 11.
228. F. A. CHARLESWORTH (1976) *Fd. Cosmet. Toxicol.* **14**, 152.
229. V. A. DRILL and P. LAZAR (Eds) (1977) *Cutaneous Toxicity*, Academic Press, New York.
230. H. J. BAKER and A. M. KLIGMAN (1967) *Arch. Dermatol.* **96**, 441.
231. I. H. BLANK and E. B. SHAPIRO (1955) *J. Invest. Dermatol.* **25**, 391.
232. R. J. SCHEUPLEIN and L. ROSS (1970) *J. Soc. Cosmet. Chem.* **21**, 853.
233. R. B. DROTMAN (1977) in *Cutaneous Toxicity*, (ed. V. A. Drill and P. Lazar), Academic Press, New York, pp. 95 *et seq.*
234. J. A. NEUMANN and A. JAGENDORF (1965) *Biochim. Biophys. Acta* **109**, 382.
235. E. HAAPALA (1970) *Physiol. Plant.* **23**, 187.
236. A. A. NETHERY (1967) *Cytologia* **32**, 321.
237. W. T. JACKSON (1962) *Plant Physiol.* **37**, 513.
238. D. L. SUTTON and C. L. FOY (1971) *Bot. Gaz.* **132**, 299.
239. J. B. ST. JOHN, P. G. BARTELS and J. L. HILTON (1974) *Weed Sci.* **22**, 233.
240. C. A. TOWNE, P. G. BARTELS and J. L. HILTON (1978) *Weed Sci.* **26**, 182.
241. L. L. JANSEN, W. A. GENTNER and W. C. SHAW (1961) *Weeds* **9**, 381.

242. M. J. ALLEN (1979) *Bioelectrochem. Bioenergetics.* **6**, 197.
243. C. D. DYBING and H. B. CURRIER (1961) *Plant Physiol.* **36**, 169.
244. J. SCHÖNHERR and M. J. BUKOVAC (1972) *Plant Physiol.* **49**, 813.
245. D. W. GREENE and M. J. BUKOVAC (1974) *Amer. J. Bot.* **61**, 100.
246. E. F. WALLIHAN, T. W. EMBLETON and R. G. SHARPLES (1964) *Proc. Am. Soc. Hort. Sci.*, **85**, 210.
247. E. VIEITEZ, J. MENDEZ, C. MATO and A. VAZQUEZ (1965) *Physiol. Plant* **18**, 1143.
248. A. A. NETHERY (1967) *Amer. J. Bot.* **54**, 646.
249. A. A. NETHERY (1967) *Cytol.* **32**, 321.
250. L. P. VERNON and E. SHAW (1965) *Plant Physiol.* **40**, 1269.
251. L. P. VERNON and E. SHAW (1964) *Biochemistry* **4**, 132.
252. F. D. H. MACDOWALL (1963) *Canad. J. Bot.* **41**, 1281.
253. R. E. TEMPLE and H. W. HILTON (1963) *Weeds* **11**, 297.
254. C. L. FOY (1962) *Weeds* **10**, 35.
255. W. G. CORNS and T. DAI (1967) *Canad. J. Plant Sci.* **47**, 711.
256. L. L. JANSEN (1964) *Weeds* **12**, 251.
257. V. H. FREED and M. MONTGOMERY (1958) *Weeds* **6**, 386.
258. C. D. DYBING and H. B. CURRIER (1959) *Weeds* **7**, 214.
259. J. F. PARR and A. G. NORMAN (1965) *Bot. Gaz.* **126**, 86.
260. D. W. STANIFORTH and W. E. LOOMIS (1949) *Science* **109**, 628.
261. C. G. L. FURMIDGE (1959) *J. Agr. Fd. Chem.* **10**, 267, 274.
262. G. E. BARRIER and W. E. LOOMIS (1957) *Plant Physiol.* **32**, 225.
263. J. B. WYRILL and O. C. BURNSIDE (1977) *Weed Sci.* **25**, 275.
264. D. L. SUTTON and C. L. FOY (1971) *Bot. Gaz.* **132**, 229.
265. C. G. MCWHORTER (1963) *Weeds* **11**, 83, 265.
266. L. L. JANSEN (1964) *J. Agric. Fd. Chem.* **12**, 223.
267. L. L. JANSEN (1965) *Weeds* **13**, 117.
268. L. L. JANSEN (1965) *Weeds* **13**, 123.
269. L. W. SMITH, C. L. FOY and D. E. BAYER (1966) *Weed Res.* **6**, 233.
270. A. G. T. BABIKER and H. J. DUNCAN (1975) *Weed Res.* **25**, 123.
271. G. A. BUCHANAN and D. W. STANIFORTH (1966) *Weed Soc. Amer.* Abstr. p. 44.
272. H. M. HULL (1970) *Residue Reviews* **31**, 1.
273. D. L. SIROIS (1967) Ph.D. dissertation, Iowa State University.
274. R. A. EVANS and R. E. ECKERT (1965) *Weeds* **13**, 150.
275. H. L. MORTON and J. A. COOMBS (1969) *Weed Sci. Soc. Amer.* Abstr. p. 65.
276. P. V. NELSON and H. A. GARLICH (1969) *J. Agric. Fd. Chem.* **17**, 148.
277. E. G. FISHER and D. R. WALKER (1955) *Proc. Am. Soc. Hortic. Sci.* **65**, 17.
278. E. G. BEAUCHAMP (1973) *Can. J. Bot.* **51**, 613.
279. G. S. HARTLEY and I. J. GRAHAM-BRYCE (1980) *Physical Principles of Pesticide Behaviour* Vols 1 and 2, Academic Press, London.
280. J. W. VAN VALKENBURG (1967) in *Solvent Properties of Surfactant Solutions* (ed. K. Shinoda), Marcel Dekker, New York, pp. 263 *et seq.*
280a J. W. VAN VALKENBURG (ed.) (1969) *Pesticidal Formulations Research.* Advances in Chemistry Series **86**, American Chemical Society, Washington.
281. TH. F. TADROS (1978) in Wetting, Spreading and Adhesion (ed. J. F. Padday), Academic Press, London, p. 423 *et. seq.*
282. C. G. L. FURMIDGE (1962) *J. Colloid Sci.* **17**, 309.
283. P. M. NEUMANN and R. PRINZ (1974) *J. Sc. Fd. Agri.* **25**, 221.
284. J. D. GAYNOR and V. V. VOLK (1976) *Weed Sci.* **24**, 549.
285. P. A. BACKMAN (1978) *Ann. Rev. Phytopathol.* **16**, 211.
286. F. R. FORSYTH (1964) *Canad. J. Bot.* **42**, 1335.
287. R. E. FORD and C. G. L. FURMIDGE (1967) *Monograph 25* Society of Chemical Industry, London, 417.

11 *Reactivity in surfactant systems*

11.1 Introduction

The transfer of a reactive solute into a micelle changes several features of its existence that can alter its inherent stability. In the micelle the molecular environment of the solute molecules will have changed drastically, from an aqueous to a relatively non-polar milieu, depending on the depth of insertion into the micelle. The simple fact that the solute may be protected from attacking species such as H^+ or OH^- ions will give rise to stabilization of labile molecules such as esters; in some charged micelles, the surface characteristics will be such that there is a concentration of ions which would result in a more rapid breakdown than in simple aqueous solution.

The micellar environment is sufficiently different from the simple aqueous environment that reaction rates may sometimes be dramatically changed; because of this there has built up a school of chemistry in which micellar systems are used deliberately to alter the rates and directions of chemical reactions. In pharmaceutical formulations the influence of surfactant on the stability of the pharmaceutical is generally secondary to its main purpose, but surfactants may be used to stabilize labile pharmaceuticals, and an understanding of the mode of action and interactions can avoid the problem of destabilization which might unwittingly occur.

In this chapter we wish to explore not only the influence of micelles on reaction rates and the course of reactions, both chemical and photochemical, but also the stability of surfactants themselves and how aggregation can affect their stability. The chemical modification of surface-active agents and attempts to polymerize surfactant micelles will also be covered. The literature on reactivity in micellar systems has grown enormously since 1968 when an account of the pharmaceutical aspects was given in the first edition of this book [1], to the extent that a book has been devoted to the subject reviewing and collating the data in the literature prior to mid-1974 [2]. Here we can probably only hope to extract some of the salient features of the subject, and could certainly not claim to be comprehensive. The reference list, however, contains several reviews which should be consulted for more detailed treatments. The analytical consequences of solubilization of chromophoric species and change in the apparent dissociation constants of compounds in the presence of surfactants is also discussed at the end of the chapter.

11.2 Chemistry at interfaces

As a prelude to the discussion of reactions at micellar interfaces and in the interstices of micelles we should examine what is special about the nature of reactions at liquid–liquid interfaces. Adsorption of substances at interfaces can lead to an ordering of molecules that is not encountered in bulk solution. The topic has been reviewed by Menger [3, 4] who has illustrated the possibilities of reactions at interfaces by investigation of the reaction of the water-insoluble ester *p*-nitrophenyl laurate in heptane with imidazole in an adjacent aqueous phase. Because of the insolubility of the ester in water any reaction between the species must occur at the interface. Migration of the reactants to the interface was partially rate-determining; small amounts of laurate ion which adsorb at the interface retarded the interfacial hydrolysis of the ester. The adsorbed laurate ion probably impedes the transport of one or more of the reactants to the interface, suggesting a means of control of reactions not available in normal bulk reactions.

Where water-insoluble agents have to be reacted with water-soluble reagents, addition of an emulsifier can assist the reaction by increasing the surface area available for reaction [5]. Brij 35 (0.006 M) reduces the reaction time for the hydrolysis of α,α,α-trichlorotoluene to benzoic acid in 20 % NaOH from 60 h to 11 h at 80° C. Polymerization at interfaces is discussed later in the chapter.

Adsorption of solutes on to micellar surfaces will also affect their reactivity; *p*-nitrophenyl esters bound to the surface of anionic micelles of NaLS do not contribute to the overall reaction rate, probably due to the repulsion of OH^- ions; in contrast ester adsorbed on to cationic micelles is more rapidly broken down as both the ester and the OH^- ion will accumulate at the micellar surface [6].

11.3 Micellar reactions

It seems [7] that most reported cases of catalysis by micelles in aqueous solution concern reactions having a charged transition state and occurring in an aggregate of opposite charges. The ability of cationic micelles to promote the hydrolysis of *p*-nitrophenyl esters where there is electrostatic stabilization of a reactive intermediate is one example. But in non-ionic detergents the factors affecting stability must be more concerned with the micellar environment, its relative state of dehydration and the possibility of the ordering in space of the reactant molecules.

Micelles not only can alter rate constants of reactions but they can alter the conformation of molecules and thus affect the outcome of a reaction [8]. Self-association of alkyl tyrosines enhances the population of one conformer (Fig. 11.1) (a) over conformer (b) and (c) (Table 11.1). The special aspect of conformer (a) is that it is the only one which has a *trans* carboxylate co-planar with the hydrocarbon and could expose its ionic group to the water phase while the alkyl group points to the centre of the aggregate, thus it will be preferred. Self-associating solutes are a special class of compounds (so called 'functional micelles') which are dealt with separately. It is more common to experience solutes which are more simply

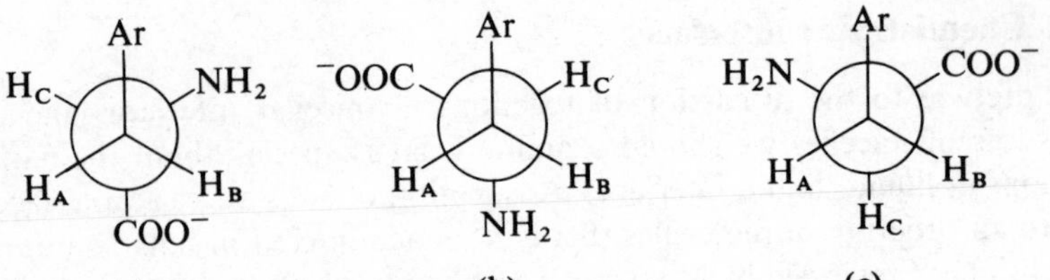

Figure 11.1 Representation of conformation of substituted tyrosines conformer (a) has a *trans* carboxylate which is coplanar with the hydrocarbon. A more conventional representation of the molecule is shown.

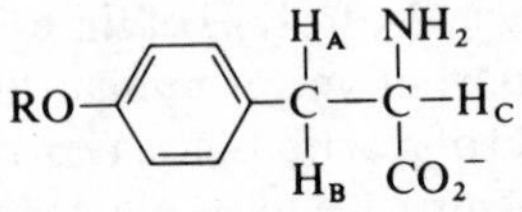

When R = C_6H_{13} or C_8H_{17} (but not H or CH_3), the amino acids form micelles. The relative populations of the following three staggered rotamers in the monomeric and aggregated states are shown in Table 11.1. From [12].

Table 11.1 Fractional rotamer populations for *o*-alkylated tyrosines in basic aqueous solutions at 33°C

Compound	Aggregation State	*a*	*b*	*c*
L-Tyrosine	Monomeric	0.48	0.23	0.29
o-Methyl-L-tyrosine	Monomeric	0.52	0.24	0.24
o-Hexyl-D,L-tyrosine	Micellar	0.67	0.12	0.21
o-Octyl-D,L-tyrosine	Micellar	0.65	0.05	0.30

From [12].

associated with micelles and solubilized at various sites within or on the micelle, although with some amphipathic compounds co-micellization with a surfactant occurs to alter its reactivity. Co-micellization of (I) with sodium lauryl sulphate changes the stereochemical course of the reaction, from complete inversion in water to 56% inversion with an approximately 1:1 ratio with NaLS [9]; strong

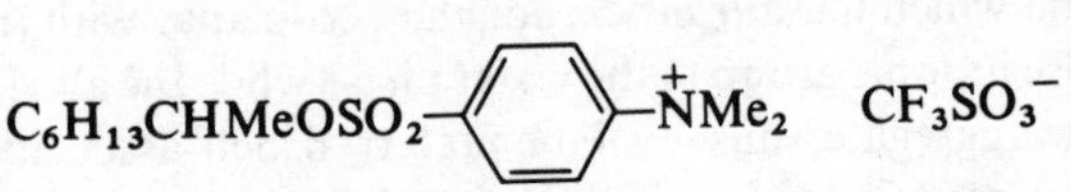

(I)

head group interactions between (I) and NaLS probably force water molecules away from the reaction centre. CTAB has little effect on reaction rate or stereochemistry, head group repulsion allowing access of water. Ihara *et al.* [10, 11] have emphasized the importance of three factors in enantioselective reactions: (a) the asymmetric centre and active site must exist close together in the reaction system; (b) strong interactions must exist among the reagents; and (c) the catalysed reaction must occur in a hydrophobic environment in order to avoid reaction with non-selective hydroxyl ions. CTAB micelles seem to satisfy these criteria when enantioselective ester hydrolysis of optically active amines (L- or D-II) by hydroxamic acids is considered. Optically active amines with amino groups at position α to the asymmetric carbon atom showed effective enantioselectivity [10] while none was shown with amino acids with the amino group distant from this centre.

$$C_6H_5CH_2OCONH\overset{*}{C}HCOOC_6H_4NO_2$$
$$CH_2C_6H_5$$

(II)

The subject of micellar catalysis and inhibition of reactions can be divided into the types of reaction occurring, e.g. base-catalysed and acid-catalysed hydrolyses, oxidation, etc., or in terms of mechanisms, e.g. juxtaposition of reactive groups in micelles, attraction of counterions to an oppositely charged micellar surface, protection by solubilization within non-ionic micelles, etc. It is not possible to adhere rigidly to either scheme but we will attempt here to consider, in turn, hydrolysis, oxidation in aqueous micelles, reactions in inverse micelles, reactions involving drugs and miscellaneous reactions of interest. Bunton's summary of the topic in his recent review of the subject is worth repeating here [12]:

'Micelles of non-functional surfactants (detergents) can catalyse bimolecular reactions by bringing reactants together in an environment conducive to reaction and they inhibit reactions by keeping reactants apart but they affect rates of unimolecular reactions by providing a submicroscopic medium. The relation between rate and surfactant concentration can be explained in terms of the distribution of reactants between the aqueous and micellar pseudophases which can also be perturbed by added solutes. Catalysis depends upon the charge type of the reaction and reactant hydrophobicity.'

Hartley [13] formulated a sign rule to explain shifts of indicator equilibria in the presence of anionic or cationic surfactant micelles, to which a corollary has been added for kinetic systems which says 'reactions involving anions are catalysed by cationic micelles and inhibited by anionic ones . . . . Reactions involving cations are catalysed by anionic micelles and inhibited by cationic ones . . .' [14]. These concepts have been illustrated on many occasions, as we will see below. The acid-catalysed hydrolysis of benzylideneaniline to benzaldehyde is inhibited by CTAB micelles [15]. Cationic micelles increase the rate of addition of CN$^-$ ion to *N*-alkylpyridinium ions [16] and the effects increase markedly on increase in the alkyl chain length of the pyridinium ion, illustrating that charge effects are only one aspect of the problem. Other examples where Hartley's rule is obeyed would include the inhibition by both cationic and anionic micelles of the coupling of *p*-nitrobenzenediazonium ion (III) and 2-naphthol-6-sulphonate ion (IVa). Micellar CTAB will solubilize anionic substrate (IVa) but exclude (III) and anionic NaLS will solubilise (III) but exclude (IVa) [17]. What is called 'this

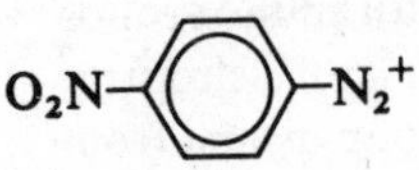

$$O_2N-\langle\ \rangle-N_2^+$$

(III)

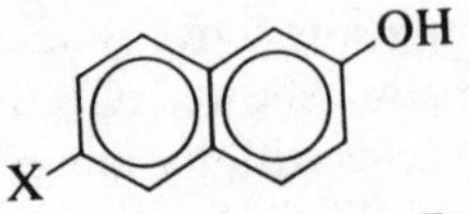

(IV) (a, X = SO$_3^-$;
b, X = H)

$$n\text{-}C_{16}H_{33}\overset{+}{N}(CH_3)_2CH_2-\langle\ \rangle-X, Br^-$$

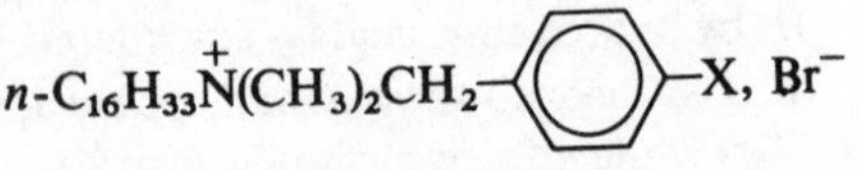

(V) (a, X = NH$_2$;
b, X = N$_2^+$, Br$^-$)

electronic impasse' [18] can be circumvented by adding the aryldiazonium moiety to a cationic surfactant molecule; anionic substrates would still be solubilized but the diazonium ion would not be excluded. Azo-coupling with the micellar diazonium ion surfactant (V) was faster than azo-coupling with the diazonium ion by factors of 122 for 2-naphthol and 244 for 2-naphthol-6-sulphonate demonstrating, Moss and Rav-Acha suggest, a circumvention of Hartley's rules.

One of the most comprehensive studies has been carried out by Bruice *et al.* [19] who studied the rate of solvolysis of neutral, positively and negatively charged esters when incorporated into non-functional and functional micelles of neutral, positive and negative charges. The second-order rate constants for alkaline hydrolysis, $k_{OH}[OH^-]$ were found to *decrease* with increasing concentration of surfactant for all cases studied. The association of the esters with non-nucleophilic micelles must either decrease the availability of the esters to OH$^-$ attack or provide a less favourable medium for the hydrolysis reaction to occur. This is another circumvention of the simple electrostatic rules as the kinetic effect seems to have nothing to do with the concentration or restriction of access of the hydroxyl ions in the Stern layer of the micelles. Presumably the labile ester bond is not positioned near the surface of these micelles, but the molecules are oriented as shown in Fig. 11.2.

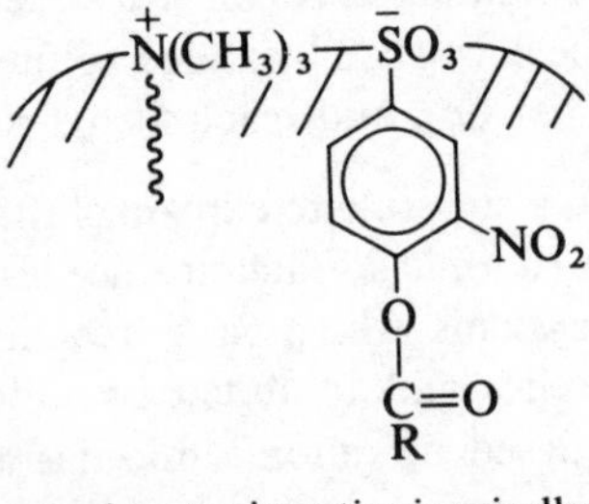

Figure 11.2 Orientation of anionic ester in cationic micelle as envisaged by Bruice *et al.* [19].

Interactions between organic anions and cations in the bulk phase may also affect the outcome of the experiment. Using acylcholinesters, Nogami and co-workers [20–22] found that in anionic surfactants a good correlation existed between the apparent distribution of the acylcholinester into the micellar phase and the magnitude of the observed stabilization. However, as Tomlinson *et al.* [23] point out, where the degrading solute is ionized and where the surfactant is ionic and of opposite charge, ion-pair formation or complex coacervation reactions may take place. The latter re-examined Nogami's results and suggest that complexation may be primarily responsible for the protection of the esters from hydrolysis. If micellar solubilization was the mechanism of protection against degradation, it would be expected that addition of salt, which would lower the CMC of the surfactant, would result in greater protection. The reverse is, however, the case. Tomlinson *et al.* produced the following scheme to explain the results in an alternative manner, explaining how complex formation reduces the rate of reaction by resulting in a reduction in the available free base in solution and so reducing the amount available for hydrolysis.

They consider the interaction between a monovalent alkylsulphate ion, A^-, and a monovalent cation B^+. This will be governed by the solubility product (K_s) between the two ions, thus:

$$A^- + B^+ \rightleftharpoons A.B \qquad (11.1)$$

such that for the solubility product value we may write in concentration terms:

$$K_s = [A^-][B^+]. \qquad (11.2)$$

If it is assumed that X is the amount of complexing ion (A^- and B^+) removed from solution upon complexation, then for the solubility product we may write:

$$K_s = [A_o^- - X][B_o^+ - X] \qquad (11.3)$$

where the subscript o refers to the original concentration of each ion. This treatment assumes that the K_s value is constant and that the concentration of the electrolytes does not vary considerably from their activities in these solutions.

If A^- is in excess of B^+ then the formed complex (which is hydrophobic) may be solubilized by the micelles of the alkylsulphate. As B^+ remaining in solution hydrolyses, then some of this (solubilized) complex will be mobilized. This serves to maintain the B^+ monomer concentration in solution.

Above the solubility product, if B_s is the amount of B^+ monomer remaining in solution, and if this is degrading by a first-order process, then the change of total base with time may be written as:

$$\frac{dB}{dt} = B_o - B_s K_1 t, \qquad (11.4)$$

where K_1 is the first-order rate constant.

With no precipitate the total base will degrade in a first-order manner, thus its half-life ($t_{1/2}$) will be given by:

$$t_{1/2} = \frac{0.693}{K_1}. \tag{11.5}$$

With surfactant,

$$\frac{B_o}{2} = B_o - \frac{K_s K_1 t_{1/2}}{A_s}, \tag{11.6}$$

where A_s is the amount of alkylsulphate monomer remaining in solution. Thus

$$\frac{B_o A_s}{2} = B_o A_s - K_s K_1 t_{1/2} \tag{11.7}$$

$$t_{1/2} = B_o A_s / 2 K_s K_1. \tag{11.8}$$

Assuming that the first-order rate constant for hydrolysis in the absence of surfactant is independent of alkyl chain-length of base, then in the presence of the surfactant,

$$t_{1/2} = \frac{k}{K_s}, \tag{11.9}$$

where $k = B_o A_s / 2 K_1$.

For interacting systems of large organic ions of opposite electrical charge it has been shown that the stoichiometric solubility product changes with alkyl chain-length of either ion, such that:

$$\log K_s = a - bn \tag{11.10}$$

where n is the carbon number of the alkyl chain and a and b are constants. By substituting this expression into Equation 11.9 and taking logarithms,

$$\log t_{1/2} = \log k - a + bn \tag{11.11}$$

or

$$\log t_{1/2} = k' + bn, \tag{11.12}$$

Equations 11.11 and 11.12 signify that stability of the acylcholinester should increase linearly with increased alkyl chain length of either ion, and Equations 11.3 and 11.9 that stability should increase with increasing surfactant concentration. Micellar interaction and ion complexation are different and we are not here simply indulging in semantic problems; above the saturation limit the complex may precipitate from solution, although excess surfactant can solubilize such precipitates. Normally, however, there is no doubt that micellar solubilization is the main cause of changes in reaction rate, certainly when there is catalysis which might be difficult to explain on the basis of complexation theory.

Observed rate constants are often the result of several processes: reaction in the bulk phase, reaction in the micelle and perhaps reaction of adsorbed material. Analysis of the rate constants is necessary to determine the contribution of each. If the micelles are considered to be a pseudophase, the concentration of micelle can be approximated by subtracting the CMC from the total concentration in solution. Using this approach Menger and Portnoy [6] considered the following equilibrium between surfactant, M, and substrate, S,

$$M_n + S_w \overset{K}{\rightleftharpoons} M_nS_w$$

$$\downarrow k_0 \qquad \downarrow k_m$$

$$P \qquad P$$

Scheme 11.1

where P is the product of the reaction both below and above the CMC of the surfactant. k_0 and k_m are, respectively, the first order rate constant of the substrate S_w in water, and in the micelle. The rate equation for this equilibrium is given by

$$\frac{-d([S]+[MS])}{dt} = -\frac{d[S]_t}{dt} = \frac{d[P]}{dt} \tag{11.13}$$

and

$$\frac{d[P]}{dt} = k_0[S] + k_m[MS] \tag{11.14}$$

where $[S]_t$ is the concentration of the substrate at time t. If f_0 and f_m are the fractions of the free and solubilized substrate, the observed rate constant, k_ψ, may be written

$$k_\psi = k_0 f_0 + k_m f_m = \frac{-d[S]_t}{dt} \bigg/ [S]_t. \tag{11.15}$$

The equilibrium constant, K, can be expressed either in terms of f_0 and f_m or in terms of concentration thus, where the micellar concentration is $[M]$,

$$K = \frac{[MS]}{([S]_t - [MS])[M]} \frac{f_m}{[M](1-f_m)}. \tag{11.16}$$

Combining Equations 11.15 and 11.16 one gets

$$k_\psi = \frac{k_0 + k_m K[M]}{1 + K[M]}. \tag{11.17}$$

The assumptions made in the derivation of this equation are:

(a) the substrate does not complex with the surfactant monomer;
(b) substrate does not perturb micellization;
(c) substrate associates with the micelles in a $1:1$ stoichiometry;
(d) the concentration of micelles is given by {concentration of surfactant $-$ CMC}/n, where n is the aggregation number of the surfactant.

Although this equation holds for a number of systems, the assumption that there is a $1:1$ stoichiometry between micelle and substrate is not always valid. The distribution of solubilizate into the micellar phase, especially when reactant concentrations are comparable with or higher than micellar concentrations, has been deemed crucial [24]. Moroi [24] has considered the distribution of molecules per micelle to be a random, Poisson or Gaussian distribution and considered the kinetic consequences. Equation 11.17 may be written,

$$\frac{k_\psi - k_0}{k_m - k_\psi} = K[M].$$ (11.18)

The equation derived assuming a Poisson distribution of reactant takes the form,

$$\frac{k_\psi - k_0}{k_{m_1} - k_\psi} = K[M] \exp(R)$$ (11.19)

where R is the average number of reactant molecules solubilized in each micelle and k_{m_1} is the rate constant for a micelle containing one solubilizate molecule. On the assumption of a random distribution of molecules the relevant equation is found to be

$$\frac{k_\psi - k_0}{k_{m_1} - k_\psi} = K[M](1 + R)$$ (11.20)

It is charged surfactant systems which have given rise to most detailed analyses of mechanisms. In particular the effects of added salts on micelle-modified reactions have been the subject of detailed experimental and theoretical study. The question of exchangeable ionic species in such reactions has been dealt with recently [25]. Five cases were treated:

(i) the binding of reactive ions to charged micelles in the presence or absence of salt or buffer;
(ii) the first-order reaction rate of an ionic substrate in the micellar phase;
(iii) the second-order reaction of an ionic nucleophile with a neutral substrate solubilized in the micellar phase;
(iv) the effect of micelles on the dissociation of weak acids; and
(v) the second-order reaction of the corresponding conjugate base.

Ion–ion and ion–head group reactions are non-co-operative, allowing the formulation of equations which deal only with the number of ions in an aggregate and with the number of ions in free solution. Where X_f and Y_f are the concentrations of free X^- and Y^- ions and MY_iX_j represents a micelle with i bound Y^- and j bound X^- ions one can write [25]

$$X_f + MY_{m-n+1}X_{n-1} \underset{nk_{-1}}{\overset{(m-n+1)k_1}{\rightleftharpoons}} MY_{m-n}X_n + Y_f.$$ (11.21)

$$n = 1, 2, \ldots, m.$$

A series of equations containing only accessible parameters have been developed by these authors to deal with the situations (i) to (v) above. The subscript b in later equations refers to bound ions. If the charged micellar phase were in reality a distinct phase it might be possible to describe ion exchange between micellar and aqueous phases by an equilibrium equation of the type:

$$X_f + Y_b \overset{k_{x/y}}{\rightleftharpoons} X_b + Y_f \tag{11.22}$$

and to define a selectivity coefficient,

$$K_{x/y} = \frac{X_b Y_f}{X_f Y_b}. \tag{11.23}$$

Quina and Chaimovich [25] find that for a unimolecular reaction of an ionic substrate S^-, whose charge is opposite to that of a micelle $D^+ Y^-$, the observed rate constant should be given by,

$$k_{\psi m} = \frac{k_m K_{s/y}(Y_b/Y_f) + k_0}{1 + K_{s/y}(Y_b/Y_f)}. \tag{11.24}$$

Computed curves for activation ($k_m > k_0$) are shown in Fig. 11.3(a). The upper diagram shows the calculated effect of the values of $K_{s/y}$ and S_T (the total analytical concentration of substrate). The lower diagram illustrates the effect of added salt for given values of $K_{s/y}$ and S_T. $[B^+ Y^-]$ is the concentration of added salt. Calculated rate enhancements for a bimolecular reaction between an uncharged substrate and a reactive added univalent ion whose charge is opposite to that of the micelle are shown in Fig. 11.3(b); the key factor in determining the outcome is the value of X_b in this case. Quina and Chaimovich's text [25] should be consulted for the full details of the analysis.

11.3.1 Surfactant effects on hydrolysis

Hydrolysis of esters can occur by acid-catalysed or base-catalysed reactions or the process may be pH-independent. Base-catalysed hydrolysis seems to be best understood; it proceeds by bimolecular attack of the OH^- ion on the carbonyl group forming a tetrahedral intermediate followed by elimination of ROH. As the intermediate is negatively charged it has been postulated that cationic micelles enhance base-catalysed hydrolysis not only by increasing the attraction of OH^- but also by stabilizing the intermediate.

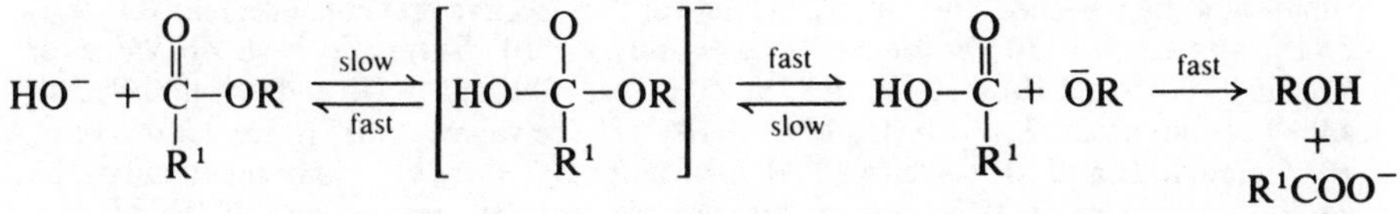

$$HO^- + \underset{\underset{R^1}{|}}{\overset{\overset{O}{\|}}{C}}{-}OR \underset{fast}{\overset{slow}{\rightleftharpoons}} \left[HO{-}\underset{\underset{R^1}{|}}{\overset{\overset{O}{|}}{C}}{-}OR \right]^- \underset{slow}{\overset{fast}{\rightleftharpoons}} HO{-}\underset{\underset{R^1}{|}}{\overset{\overset{O}{\|}}{C}} + \overset{-}{O}R \overset{fast}{\longrightarrow} \begin{matrix} ROH \\ + \\ R^1COO^- \end{matrix}$$

Simple approaches do not always hold. It has been reported that while CTAB at concentrations about twice its CMC increased the rate of hydrolysis of ethyl *p*-aminobenzoate, higher concentrations decreased the rate [26]; also Mitchell [27] has reported decreased base-catalysed hydrolysis of several esters by Cetrimide BP above its CMC. Meakin *et al.* [28] obtain the results shown in Fig. 11.4 an

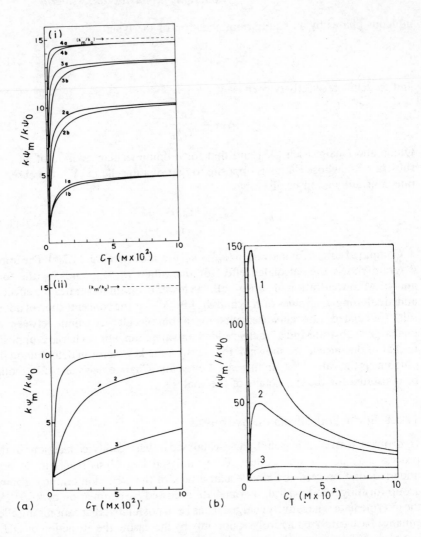

Figure 11.3(a) Model calculations of activation. $k_m/k_0 = 15$. The true catalytic ratio k_m/k_0 is indicated by a dashed line. (i) Dependence of the effective rate enhancement (k_{ψ_m}/k_{ψ_0}) on $K_{s/y}$ and S_T. $S_T = 10^{-6}$M for curves 1a–4a and $S_T = 10^{-3}$M for curves 1b–4b. Values of $K_{s/y}$ are 0.10 (1a and 1b), 0.50 (2a and 2b), 2.0 (3a and 3b), and 10 (4a and 4b). (ii) Effect of added common salt. $K_{s/y} = 0.50$ and $S_T = 10^{-6}$M. The values of $|BY|_T$ are 0M (curve 1), 10^{-2}M (curve 2), and 10^{-1}M (curve 3). If, for example, $k_m = k_0$, k_{ψ_m} will be unaltered by the addition of detergent. If $k_m > k_0$ a rate increase is to be expected and, if $k_m < k_0$, an inhibition will be observed. Since, where α is the degree of ionization,

$$\lim_{C_T \to \infty} K_{s/y} \frac{Y_b}{Y_f} = K_{s/y} \frac{(1-\alpha)}{\alpha}$$

a limiting value of k_{ψ_m}/k_{ψ_0}, independent of S_T, is approached as C_T increases. The larger the value of $K_{s/y}$ and the smaller the values of S_T and $|BY|_T$, the lower the detergent

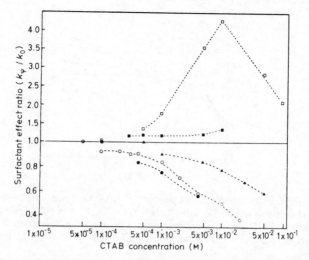

Figure 11.4 The effect of CTAB on the hydrolysis of four esters expressed as a ratio (k_ψ/k_0) of the first order rate constants obtained in the presence (k_ψ) and absence (k_0) of the surfactant. ●, EPAB; ▲, PAPA; □, PNPA; ■, EPNB; all in Delory and King's buffer; ○, EPAB in Sørensen's glycine buffer. The esters used were ethyl *p*-aminobenzoate (EPAB), *p*-nitrophenyl acetate (PNPA), ethyl *p*-nitrobenzoate (EPNB) and *p*-aminophenyl acetate (PAPA). From Meakin *et al.* [28].

increase in rate being observed only with PNPA (*p*-nitrophenyl acetate.)

Examination of these results shows that the nature of the ester grouping on the aromatic ring does not determine whether CTAB increases or decreases the observed rate of hydrolysis, although it might alter the magnitude of the observed effect. The group on the para position appears to be more influential, probably by determining the nature or site of the interaction with the micelle. Meakin *et al.* conclude that where CTAB increases the rate of hydrolysis it is likely that the site of interaction is the surface, where the ester linkage would be in a region of high hydroxyl ion concentration. The NO_2 groups on PNPA and EPNB (see Fig. 11.4) are likely to be attracted to the cationic head groups. When the solubilizate molecules are drawn deeper into the micelle, physical protection occurs, and in addition the different 'solvent' conditions result in lowered affinities of the ions for each other.

If hydroxide concentrates at the micellar surface the relationship of the micellar surface concentration, $[OH]_m$ to the bulk concentration, $[OH]_b$ may be obtained from [29]

$$[OH]_m = P_{OH}[OH]_b \exp(F\psi_0/RT) \tag{11.25}$$

concentration necessary to attain the limiting value of k_{ψ_m}/k_{ψ_0}. Furthermore, the larger $K_{s/y}$, the closer the limiting value of k_{ψ_m}/k_{ψ_0} approximates the true catalytic ratio k_m/k_0. (b) Dependence of k_{ψ_m}/k_{ψ_0} on added $|BY|_T$ in the absence of buffer: $K_{x/y} = 0.5$, $K_s = 3 \times 10^2 M^{-1}$, $X_T = 10^{-3} M$. The values of $|BY|_T$ are $0 M$ (curve 1), $10^{-2} M$ (curve 2), and $0.1 M$ (curve 3). From Quina and Chaimovich [25].

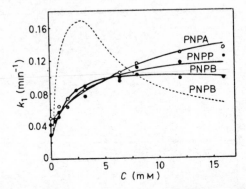

Figure 11.5 First-order rate constants plotted against CTAB concentrations for the three *p*-nitrophenyl esters in the AMP buffer (pH 9.59) at 25° C. The solid lines are calculated from Equation 11.26. The dashed line represents the rate–CTAB concentration profile for *p*-nitrophenyl butyrate in a 0.01 M carbonate buffer (pH 9.87) at 25° C. From Funasaki [29].

where P_{OH} is the distribution coefficient of hydroxide ions at a surface potential, ψ_0, of zero.

Martinek and co-workers [29a] derived a rate equation which yields the equation,

$$k_\psi = \frac{k_a + k_m [OH]_b \, P_{OH} \cdot K[M] \quad K_a/K_{a_i}}{1 + K[M]}. \tag{11.26}$$

Here K_a is the acid dissociation constant in the presence or absence of a surfactant, and K_{a_i} is the acid dissociation constant at zero surface potential.

The effect of CTAB concentration on the first-order rate constants for three *p*-nitrophenyl esters in 2-amino-2-methyl-1,3-propanediol (AMP) buffer is shown in Fig. 11.5. Results for PNPA are included, results for PNPB, the *p*-nitrophenyl butyrate, possess a quite different profile. The solid lines are calculated using Equation 11.26. The ratio of rate constants in micellar and non-surfactant systems did not depend on the chain length of the acyl group to any great extent as can be seen, but the nature of the buffer ions has a significant effect on the profiles obtained as can be discerned in Fig. 11.5. Carbonate buffer components and triethylamine are probably incorporated into the micelle at its surface, whereas AMP is not. Values for the pK_a of the groups at the surface of the micelle were estimated from the pK_a values of indicator (Thymol Blue) adsorbed on to the surface. A complication of analysis is that it is known that basic hydrolysis of substances such as PNPA is catalysed by general bases as well as hydroxide ion. This general reaction would, therefore, proceed in bulk and in micellar solutions, it would only occur at the micellar surface if the buffer components interacted with the micelle surface.*

---

* Fendler and Fendler [2] have tabulated the many results of studies published on surfactant effects in ester hydrolyses where cationic, anionic and non-ionic surfactant species are involved; values of k_ψ/k_0 are generally quoted.

Funasaki [30] has found that as the concentration of CTAB is increased above the CMC, the surface hydroxide ion concentration decreased and bromide counterions were bound to the surface of the micelle in place of carbonate and hydrogen carbonate ions. The rate constant at the micellar surface, which could be estimated by analysis of the results, was found to be proportional to surface OH^- concentration and to decrease with increasing Br^- concentration at the surface, as in the bulk phase. This, he suggests, explains the profile for PNPB hydrolysis in CTAB–carbonate buffer systems, shown in Fig. 11.5. Counterion binding was estimated by measurement of bromide ion activities, from which the degree of dissociation of the bromide can be calculated.

Micellar catalysis of amides has been less widely studied [31]. Gani and co-workers [32–34] have investigated the effects of CTAB and related compounds on the basic hydrolysis of acetanilides and found both acceleration and retardation of reaction rate. Others [35, 36] have found the rate of hydrolysis of several anilides to be increased by CTAB and inhibited by NaLS. The catalysis of the hydrolysis of four nitroacetanilides has been studied by O'Connor and Tan [31], the reactions following the simple treatment embodied in Scheme 11.1. This catalysis was inhibited by salts, in particular those of the arene sulphonate class and also by large carboxylate ions (see Table 11.2). Inhibition was analysed using the scheme adopted by Bunton [37] assuming that inhibition was competitive. In Scheme 11.2 below, I is the inhibiting anion and MI is the inhibitor–micelle complex, K_1 being the inhibitor binding constant.

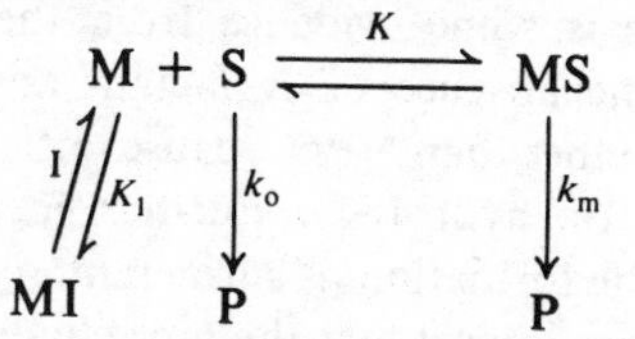

$$
\begin{array}{ccc}
\text{M + S} & \xrightleftharpoons{\;K\;} & \text{MS} \\
I\Big\downarrow K_1 \quad \Big\downarrow k_o & & \Big\downarrow k_m \\
\text{MI} \qquad \text{P} & & \text{P}
\end{array}
$$

Scheme 11.2

Table 11.2 Inhibition of the micellar catalysis of the hydrolysis of 4-nitroacetanilide

Inhibitor	$10^3 C_1 (\mathrm{mol\,l^{-1}})$	$10^3 C_{CTAB} (\mathrm{mol\,l^{-1}})$	$10^5 k_\psi (\mathrm{s^{-1}})$
CH_3COONa	—	15	14.1
CH_3COONa	5	15	13.1
CH_3COONa	10	15	11.2
CH_3COONa	15	15	11.0
CH_3COONa	20	15	10.7
CH_3COONa	25	15	9.38
$p\text{-}CH_3C_6H_4SO_3Na$	2	15	9.95
$p\text{-}CH_3C_6H_4SO_3Na$	5	15	9.21
$p\text{-}CH_3C_6H_4SO_3Na$	7	15	7.68
$p\text{-}CH_3C_6H_4SO_3Na$	10	15	6.14
C_6H_5COONa	2	17	13.2
C_6H_5COONa	7	17	8.44
C_6H_5COONa	10	17	7.31
C_6H_5COONa	20	17	4.80
C_6H_5COONa	40	17	3.33

From [31].

This analysis gives rise to the equation,

$$\frac{k_m - k_0}{k_\psi - k_0} = 1 + \frac{N}{K[M]} + \frac{K_1 C_1 N}{K[M]} \qquad (11.27)$$

Use of Equation 11.27 to estimate values of K_1 gave the following results: sodium acetate 92 litre mol^{-1}, sodium benzoate 707 litre mol^{-1}, and sodium p-toluene sulphonate 950 litre mol^{-1}. The results of the study suggest that a negatively charged compound capable of hydrophobic interactions is required to prevent effective interaction of the CTAB micelles with the substrate. With amphipathic compounds such as the p-toluene sulphonates there is the possibility of the type of interactions that were discussed by Tomlinson *et al.* [23]. Independent measurement of binding or interaction constants thus would be required before this mechanism of inhibition was elucidated. Charge reduction at the surface is likely to inhibit the access of OH^- ions to the substrate incorporated into mixed 'micelles'.

In much of this book we have emphasized the application of non-ionic surfactants. Non-ionic surfactants generally decrease the rates of hydrolysis of solubilized materials. An example would be the reduction in the rate of hydrolysis of a monochlorotriazine dye (VI) by a series of nonylphenyl polyoxyethylene condensates (NPE_{10}–NPE_{30}) [38]. At 80 and 100°C the longer the ethylene oxide chain length, the more effective the protection afforded to the dye molecule. Given the size of the molecule it is unlikely that solubilization occurs as with smaller molecules; there is some evidence from the results that at 60° dye aggregation occurs in the absence of surfactant and that addition of small quantities of surfactants other than NPE_{10} cause initial deaggregation of the dye and an increased rate of hydrolysis. Further increases in the surfactant concentration result in the formation of surfactant–dye aggregates in which the dye is protected, it seeming logical that the more hydrophilic the surfactant the greater the effect. This should explain why the rate constants in the presence of NPE_{15}–NPE_{30} do not extrapolate at zero surfactant to the expected value of k_0, as can be seen in Fig. 11.6.

(VI)

Mitchell [39, 41], in his analysis of the reduction of rate of base-catalysed hydrolysis of propyl benzoate in cetomacrogol solutions, found that incorporation of the ester in the micelle protected the compound against degradation, the

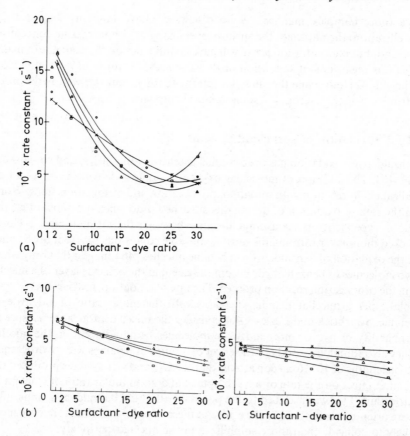

Figure 11.6(a) Hydrolysis of reactive dye (VI) in presence of nonylphenol–poly(oxyethylene) at 60° C and pH 10.98. × NP10; ○ NP15; △ NP20; □ NP30. (b) Hydrolysis of dye (VI) in presence of nonylphenol–poly(oxyethylene) at 80° C and pH 10.60. × NP10; ○ NP15; △ NP20; □ NP30. (c) Hydrolysis of dye (VI) in presence of nonylphenol–poly(oxyethylene) at 100° C and pH 10.35. × NP10; ○ NP15; △ NP20; □ NP30. From Craven *et al.* [38] with permission.

rate of hydrolysis depending on the degree of saturation of the system. This degree of saturation he defined as R, where

$$R = C/C_s,$$

C being the concentration of ester and C_s its solubility in the surfactant solution. As with the oxidation of aldehydes in surfactant systems [40], the rate of hydrolysis was found to be independent of the concentrations of ester and surfactant except in so far as these controlled R. The relationship between R and reaction rate holds for non-ionic surfactants but not for cationic and anionic systems for the oxidation of aldehydes [41], although it holds to some extent for propyl benzoate hydrolysis [42]. It is not surprising that ionic surfactants behave in a more complex manner. As we discussed above (and in Chapter 3 on micellization) the nature of the surface layer changes with increasing concentration and thus concentration *per se* will have an influence on the outcome of many reactions regardless of saturation ratio. At the same saturation ratio, however, Mitchell [42] found that the rates of reaction of propyl benzoate were in the order cetrimide > cetromacrogol > sodium lauryl sulphate.

11.3.2 Oxidation of surfactant systems

Benzaldehyde oxidation has been studied extensively by Carless and co-workers [43–46]. The addition of molecular oxygen to liquid aldehydes involves a free radical chain reaction with initiation, propagation and termination steps. Work on the rate of oxidation of drying oils adsorbed onto silica has shown that the rate is dependent on the average distance between the oil molecules, a close-packed monolayer forming the most favourable conditions for reaction. Studies of the oxidation of benzaldehyde in betaine micelles [46] suggest that only when two molecules of benzaldehyde become adjacent in the palisade layer of a micelle will the propagation reaction proceed. The rate of oxidation, measured as oxygen uptake per g micellar betaine, increases with the molar ratio of aldehyde to betaine. Swarbrick and Carless [46] consider the micellar ratio 'as a measure of probability of two or more aldehyde molecules being adjacent' in the micelle. They also found, in a study of the effect of a series of betaines with hydrocarbon chain lengths of 10–16 carbon atoms, that the C_{14} and C_{16} systems produced the greatest effect on the rate of oxygen uptake at a given molar ratio of reactant to surfactant. This was ascribed to the possibility that the benzaldehyde molecules were oriented more towards the centre of these micelles and therefore were more likely to coincide than those solubilized in the micellar periphery.

The main conclusion of the work was that oxidation of the aldehyde in betaine micelles did not depend simply on the saturation ratio R as defined earlier and which was shown to hold for aldehyde–cetomacrogol systems [44]. Detailed studies of betaine–benzaldehyde–water systems were carried out by Swarbrick and Carless [47] in order to elucidate the impact of the complex nature of these systems and the role of the different phases that could occur. Their results showed that, in general, it was wrong to assume that all material in excess of that taken up into micelles of the L_1 phase formed liquid dispersions in water. This situation obtained only with the C_8 and C_{10} betaines; in the others the presence of liquid crystalline phases as an alternative phase for solubilization was identified. Oxidation in the various phases of these systems was studied in some detail [45]. The maximum rate of oxidation of the aldehyde was related to the concentration of aldehyde present in any one phase, the conclusion being that the use of the saturation ratio which expresses concentrations in one phase in terms of another phase is an unnecessary consideration when attempting to establish explanations for the behaviour of *t*-aldehydes in these systems.

The autoxidation of linoleic acid has been followed in non-ionic surfactant systems [48] consisting of L_1 phases or L_1 plus emulsions. The maximum rate

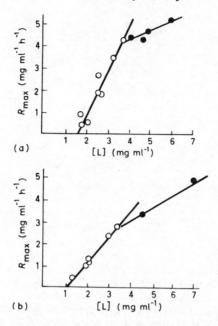

Figure 11.7(a) Variation of rate of hydroperoxide formation with linoleic acid concentration: systems containing $C_{16}E_{24}$, A24 (○, L_1 phase; ●, emulsions). (b) Variation of rate of hydroperoxide formation with linoleic acid concentration: systems containing $C_{16}E_{30}$, A30 (○, L_1 phase; ●, emulsions). From Rhodes [48] with permission.

was found in L_1 phases to be linearly related to linoleic acid concentration but the maximum rate in emulsions was less than in similar systems containing only L_1 phase. Some results are shown in Fig. 11.7.

The rate of oxidation of β-lactoglobulin in NaLS solutions is decreased when the oxidizing agent is ferricyanide but little affected when the oxidizing agents are *o*-iodosobenzoate or 2,6-dichlorophenol indophenol [49]. To explain the different effects, it was suggested that the steric effect of the bound surfactant molecules hinders the formation of a transition state in the former case, this transition state involving the sulphydryl group of the protein, a heavy metal ion and the ferricyanide ion. The most striking difference between the reaction involving ferricyanide and the other agents mentioned is the strong requirement for the presence of heavy metal cations. These may well be bound to the surfactant and, therefore, be unavailable for the transition state.

11.3.3 Reduction in surfactant systems

Relative rates of reduction of the carbonyl group of several steroids solubilized by hexadecyltrimethylammonium borohydride micelles have been related to the structure of the steroids [50] shown in Fig. 11.8. Relative rates are shown underneath the structures in the diagram and show that micelles cause an increase

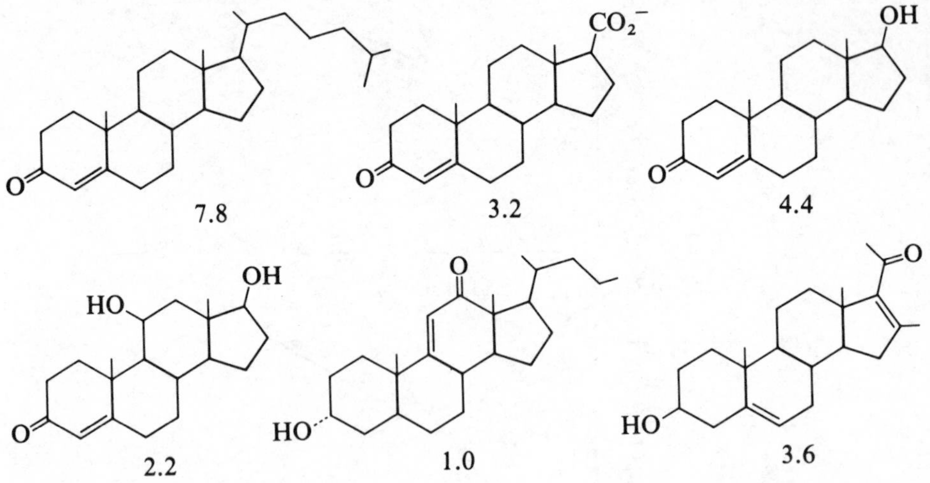

7.8   3.2   4.4

2.2   1.0   3.6

Figure 11.8 Relative rates of carbonyl reduction of several steroids adsorbed in hexadecyltrimethylammonium borohydride micelles. The second-order rate constant for reduction of 3-methyl-2-cyclohexen-1-one by borohydride in bulk water at 25.0° C was assigned a relative value of 1.0. The results demonstrate the insensitivity of the rates to substituents in the C and D rings, and to the position of the carbonyl group. From Bonicamp [50].

in rate of reduction ascribable to the favourable accumulation of the carbonyl groups and BH_4^- ions at the interface. Reduction rates depend only slightly on the nature of the substituent on ring D which varies from non-polar to ionics. Results suggest that the carbonyls of all the steroids are located in the interface; the implication for the steroid with both carbonyl and carboxylic acid grouping is that it lies flat in the interfacial region.

11.3.4 Hydrated electrons in micellar systems

The generation of hydrated electrons in radiolysis experiments and their subsequent interaction with micellar species or with species solubilized within micelles has been investigated by Grätzel and co-workers [51, 52]. NaLS and CTAB, as examples of surfactants which might be expected to react with e_{aq}^-, have been found to be quite unreactive [53], but can alter the reactivity of a variety of substrates towards the hydrated electron. Anionic micelles reduce the rate constant and cationic micelles enhance the rate [54–56]. The fate of e_{aq}^- in CTAB solutions has been studied by labelling the micelles of CTAB with highly reactive cetylpyridinium chloride (CPyCl). Three processes occur as shown in Fig. 11.9, namely (a) trapping of the hydrated electron by the potential field of the micelles, (b) intramicellar scavenging by a reactive probe on the micelle surface, and (c) electron migration from micelle to micelle in solution. The migration of e_{aq}^- through micellar systems is thus altered and subsequent reactions of the hydrated electron with other species will be affected.

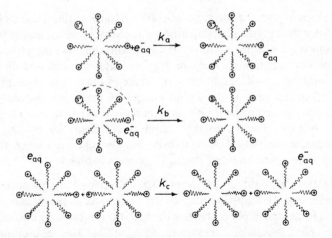

Figure 11.9 Schematic diagram for the behaviour of the hydrated electron in a system of positively charged micelles. The e_{aq}^- trapping step is illustrated by k_a, intramicellar electron scavenging by k_b, and electron migration by k_c. The S^+ notation indicates a reactive probe. From Patterson and Grätzel [51].

An electron in water which is to react with a substance dissolved in the lipid interior of a micelle, must transfer through the interphase between micelle and bulk. An electron cannot move spontaneously from water into a lipid environment but it may 'tunnel rapidly through the electrical double layer' [52] if an acceptor in the micelle has unoccupied levels with appropriate energies*. Using 9-nitroanthracene and pyrene as acceptors, interactions in CTAB and NaLS have been compared. The anthracene reacts much faster with e_{aq}^- in CTAB than it does in NaLS, the respective rate constants being 9×10^{10} and 1.5×10^9 $(mol\,l^{-1})^{-1}\,s^{-1}$. On addition of sodium sulphate to the latter the surface charge is reduced and the rate constant is increased to 4×10^9 $(mol\,l^{-1})^{-1}\,s^{-1}$. Reaction of the electron with the acceptor occurs at practically every encounter of the hydrated electron with a micelle containing the acceptor. Thus the net positive charge of the CTAB micelles assures a large surface area for contacts. Grätzel *et al.* [52] using Smoluchowski's equation ($k = 4\pi V 10^{-3} Dr$) have related the rate constant k for reactions between e_{aq}^- and the acceptor to D, the diffusion coefficient of e_{aq}^- and micellar radius, r. The expression gives a value for the rate constant of 6.8×10^{10} $(mol\,l^{-1})^{-1}\,s^{-1}$ using a value of D of $4.5 \times 10^{-5}\,cm^2\,s^{-1}$. Grätzel *et al.* say this for reactions in the sodium lauryl sulphate:

'In the case of the NaLS-micelles, a repulsive force acts on the approaching hydrated electron. It seems difficult to understand why the hydrated electron

* If the overlap of the occupied levels of the donor with the unoccupied levels of the acceptor is not good, transfer is slow, irrespective of how thermodynamically favourable it may be. These levels are shifted, relative to one another, by the charge on the micellar head group, which determines the direction of the electrical double layer at the interface. Reaction rates are, therefore, influenced both by the charge on the micelle and the concentration of electrolyte in the aqueous phase.

can as efficiently react as observed, i.e. diffuse to the surface, lose its hydration shell, enter the lipoidic phase against the negative potential across the double layer and react with the dissolved 9-nitroanthracene. We therefore postulate that the electron tunnels from its trap in water through the double layer directly into the 9-nitroanthracene molecule in the micelle. This process can occur efficiently only if the Fermi energy of the redox system A/A^- (A: acceptor molecule) in the micelle is essentially lower than that of the system aq/e_{aq}^-. A high rate of reaction requires sufficient density D_{unocc}^A (ε) of unoccupied states of the A/A^- system at electronic energies ε where the density D_{occ}^e (ε) of occupied levels of the aq/e_{aq}^- system is high.'

The ionization by light at 347.1 nm (3.57 eV energy) of phenothiazine incorporated in NaLS micelles in water has been attributed to the rapid tunnelling of an electron from excited phenothiazine through the double layer into unoccupied electronic redox levels of the system aq/e_{aq}^- [57]. This photoionization is promoted by co-solubilization of duroquinone which prevents ejection of electrons from the phenothiazine into the water phase. It is suggested [57] that the phenothiazine/water/quinone/micelle system offers a simple model for electron transfer in photosynthetic systems and for the heterogeneous catalysis of the photodecomposition of water via the freed electrons. A schematic representation of the processes when the surfactant is anionic [58] is shown in Fig. 11.10.

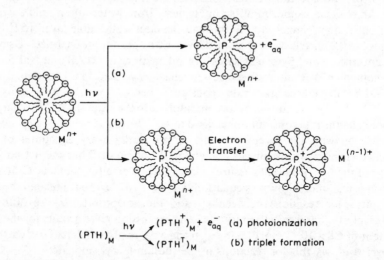

Figure 11.10 Schematic representation of photoionization and electron transfer processes in solutions of surfactant micelles containing a solubilized photoactive probe P. The electron acceptor is M^{n+} located in the Stern layer of the micelle and the electron is transferred through the Stern layer from the triplet (P^T). Hydrated electrons produced by the photoionization process (a) cannot re-enter the micelle and recombine with parent cations. The most likely fate of e_{aq}^- in micellar solutions is conversion into H_2 via the bimolecular reaction:

$$e_{aq}^- + e_{aq}^- \rightarrow 2OH^- + H_2$$

From Alkaitis *et al.* [58].

Ionization was also achieved in cationic micelles in the presence of the electron scavenger naphthoquinone sulphonate adsorbed at the periphery of the micelle. In the absence of scavenger the probability of geminate ion recombination is high and therefore the photoionization efficiency is low.

The photophysics of molecules and molecular excited states in micellar systems have been reviewed by Kalyanasundaram [59] and Lindig and Rodgers [60], the latter quoting some 56 references from the 1979 literature. Turro *et al.* [61] have also reviewed the field of photophysical and photochemical processes in micelles.

11.3.5 Luminescence and fluorescence in micellar systems

'Proximity, favourable orientation and microscopic environment' are factors which contribute to the optimization of various biological processes [62]. While the study of energy transfer in micellar systems might have little direct biological relevance, the information gained helps in our understanding of more complex biological systems. Efficient energy transport has been demonstrated from solubilized naphthalene in NaLS micelles to terbium chloride. In the absence of micelles in this system there is no energy transfer because of the low efficiency of the theoretically feasible process [62]. The role of the micelle is to allow the compartmentalization of no more than one donor into each micelle interior while concentrating a large number of acceptor molecules at the micellar surface. There are two requirements to be satisfied before energy transfer can be achieved: (i) the energy level of the donor has to be similar to that which is required to raise the acceptor to its excited state; and (ii) the solubilized naphthalene triplet must be able to diffuse to the terbium chloride (in this specific case) within its lifetime. The size of the micelle thus will determine whether or not this is possible. An estimate of the time required for this can be estimated from

$$t = \frac{X^2 3\pi\eta}{kT} \tag{11.28}$$

where X is the mean path length, taken to be the radius of the NaLS micelle (2 nm), η is the microscopic viscosity, taken to be 92 cP, k is the Boltzmann constant and T the temperature. Substituting the values into Equation 11.28 leads to a value of t of 262 ns. The triplet is sufficiently long lived to diffuse from the micelle interior to the Stern layer and transfer its energy to the acceptor there, as the fluorescence lifetime of the naphthalene triplet has been found to be about 1×10^{-5} s. Compartmentalization separates the naphthalene molecules and prevents the triplet annihilation ($^3N^* + {}^3N^* \rightarrow$ quenching) which would otherwise occur.

The role of micellar dimensions and what has been called the 'spatial extent' of species in intramicellar kinetic processes has been considered [63]. Three qualitatively different types of reaction were studied: (i) the diffusion of a confined excited species to a reactive surface; (ii) energy transfer between two separated reactants; and (iii) chemical reaction between species which are restricted in their diffusion to the surface of the micelle. Table 11.3 summarizes the main findings when r and D are fixed for these three cases.

Table 11.3 Comparison of effective rate constants for the three cases studied for the common choice of parameters: $r_0 = 2\,\text{nm}$ and $D = 5.0 \times 10^{-7}\,\text{cm}^2\text{s}^{-1}$

	Theory (s^{-1})
Diffusion to a surface	1.2×10^8
Compartmentalized reactants	1.1×10^7
Diffusion on the surface	3.4×10^6

Table 11.4 Effective rate constants k_{eff} and lifetimes τ (ns) for representative values of the diffusion constant D^e ($10^7\,\text{cm s}^{-1}$) and micellar radius r_0 (nm)

r_0	D			
	1.0	3.0	5.0	10.0
2.0	2.64	7.41	12.34	24.67
	37.92	13.49	8.10	4.05
1.8	3.17	9.14	15.23	30.46
	31.56	10.94	6.56	3.28
1.5	4.44	13.15	21.93	43.86
	22.56	7.60	4.56	2.28
1.3	5.86	17.52	29.20	58.40
	17.06	5.70	3.42	1.71

Upper number of each entry is k_{eff} in units of $10^7\,\text{s}^{-1}$ and the lower number is τ in ns. From [63].

Effective rate constants and lifetimes for reactions in which diffusion to a reactive surface must occur are shown in Table 11.4 for a range of values of r and D. The latter are a quantification of expected trends which show k_{eff} to increase with increasing diffusion coefficient and to decrease with increasing micellar radius. In spite of good correspondence between experiment and theory there is some caution expressed by the authors in their paper in view of the uncertainty that macroscopic equations for 'normal' chemical kinetics apply in the reactions explored by them. The problem, they say, is that micellar kinetics is a non-equilibrium phenomenon which can only be treated by taking the geometry of the system explicitly into account in any formulation of the process.

Singhal *et al.* [64] have found that energy migration from thionine to Methylene Blue in NaLS micelles is most efficient just above the CMC of the surfactant when the mean distance between donor and acceptor is estimated to be 1.6 nm. These two dye cations are believed to be adsorbed on to the micelle surface, and so correspond to one of the reactions explored by Hatlee *et al.* [63]. Energy transfer from 1,3-dioctylalloxazine (DOA) (VII)) to proflavine (PF^+) (VIII) and acriflavine (AF^+) (IX) in NaLS is mainly determined by the probability of finding the donor and acceptor pairs at the same time in the same micelle [65].

(VII) DOA

½ (SO₄)²⁻

(VIII) PF⁺

CH₃ Cl⁻

(IX) AF⁺

Energy transfer efficiency falls with increasing concentration of surfactant above the CMC. The probability of finding n molecules in a given micelle can be estimated from the equation used by Dorrance and Hunter [66],

$$P_n = \frac{N^n M}{(N+M)^{n+1}} \tag{11.29}$$

when N and M are the total numbers of molecules which can be incorporated into the micelles and the number of micelles respectively. If N is taken to be the total number of donor and acceptor molecules in the system, the partitioning of fluorophores in the micelles as given by a Poisson distribution as shown in Fig. 11.11, remembering that n represents two molecular species such that, for example, P_5 is the sum of six terms,

$$P_5 = P_{DDDDD} + 5P_{DDDDA} + 10P_{DDDAA} + 10P_{DDAAA} + 5P_{DAAAA} + P_{AAAAA} \tag{11.30}$$

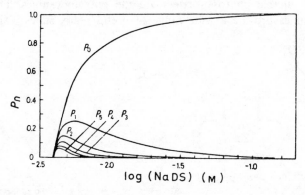

Figure 11.11 Poisson distribution of fluorophore molecules among micelles. P_n represents the probability that n fluorophore molecules are incorporated into a micelle under the following conditions: [donor] $= 4 \times 10^{-6}$ M and [acceptor⁺] $= 2 \times 10^{-5}$ M. From Matsuo *et al.* [65].

The first and last terms of the r.h.s. of the equation are not expected to make any contribution to the reaction since they represent micelles with only one trapped species. The energy transfer efficiency is related to P_1-P_5 in a complex fashion (see Equation 9 in [65]).

Electron transfer rates between adrenaline and related benzene diols and complexes of iron(III) with some substituted 1,10-phenanthrolines have been reported [67] in surfactant systems. In cationic systems the reactions take place in the aqueous phase and reaction rates are lower than they are in simple aqueous systems, but in anionic surfactant systems the reaction rates are enhanced, reactions probably taking place at the micellar interface. The rates of exit and entrance of aromatic compounds from and into micelles have recently been studied using phosphorescence decay measurements [68]; exit rate constants of aromatic hydrocarbons are of the order of 10^3 to $10^4 \, s^{-1}$, whereas values of 10^7 to $10^8 \, (mol \, l^{-1})^{-1} s^{-1}$ have been reported for intramicellar energy transfer processes. Release of aromatic phosphorescence probes from micelles followed by their deactivation in the aqueous phase is hence expected to be an important mode of deactivation of the triplet state [69]. Kinetic schemes for triplets that are partitioned between aqueous and micellar phases are considered for the cases of single occupancy and double occupancy of the micellar units.

The photochemistry of solubilized chlorophyll is similar to that of chlorophyll dissolved in organic solvents, but is not the same as the photochemistry of chloroplasts or their fragments which provide greater specificity and higher efficiency [70]. At low light intensities photoreduction rates were some 70 times higher in chloroplast fragments. Micellar systems are being investigated as means of increasing photochemical reactivity (e.g. [71]). Photoproduction of H_2 from water can be achieved when three component systems are used comprising an electron donor (e.g. cysteine), an electron acceptor (methyl viologen [MV^{2+}]) and a photosensitizer with a suitable catalyst. The use of photosensitizers which absorb in the near infra-red would allow a large fraction of the solar spectrum to be utilized; suitable photosensitizers are the phthalocyanines. The metallo-phthalocyanines provide a wide range of useful redox states. Darwent [71]

(X) Zn(pcts)

(XI) MV^{2+}

reports that zinc(II) tetrasulphophthalocyanine (X) is an inefficient photo-sensitizer in homogeneous solution since dimers are formed and since aggregation occurs with methyl viologen (MV^{2+}) (XI). Such complexes can be prevented by solubilizing the zinc phthalocyanine into cationic micelles; anionic micelles of NaLS are ineffective. Concentrations of methyl viologen up to $0.1 \, mol \, l^{-1}$ lead to only 15% fluorescence quenching whereas, in simple aqueous solutions low concentrations of the viologen efficiently quench the fluorescence.

The photoreduction of viologens in surfactant systems has also been described by Massini and Voorn [70]. An amphipathic derivative of methyl viologen ($C_{14}MV^{2+}$) (XII) has been employed as an electron acceptor in reactions involving the charge-transfer excited state of the ruthenium complex $Ru(bpy)_3^{2+}$; the reaction products MV^+ and $Ru(bpy)_3^{3+}$ of the reaction between MV^{2+} and this ruthenium complex, are capable of producing hydrogen and oxygen from water, but to improve the quantum efficiency of the process the back reaction between the products must be retarded. The amphipathic derivative shown here,

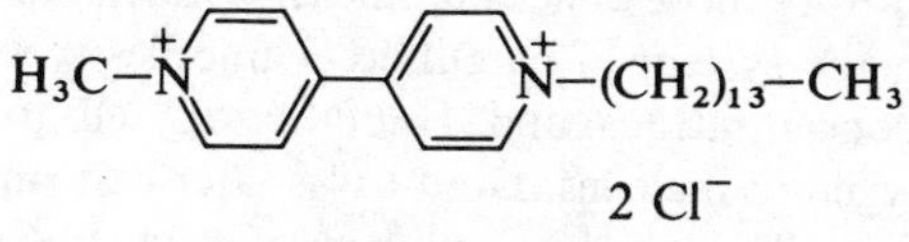

$$H_3C-\overset{+}{N}\diagdown\diagup-\diagup\diagdown\overset{+}{N}-(CH_2)_{13}-CH_3$$

$$2 \, Cl^-$$

(XII) $C_{14}MV^{2+}$

forms micelles above about 7×10^{-3} and the rate constant for the back-electron transfer is at least 100-fold smaller in mixed micellar solution with CTAC than in water. In the absence of CTAC the $C_{14}MV^+$ compound disappears rapidly; this species is much more hydrophobic than the parent $C_{14}MV^{2+}$ species and hence will show a greater affinity for the micellar phase. Thus during the course of the reaction its position in the system will change. It will be trapped in the CTAC micelles immediately after electron transfer has occurred and this results in blocking of the subsequent back reaction by repulsion of $Ru(bpy)_3^{3+}$ by the cationic head groups [72].

Proflavine (PF) has also attracted attention as a possible sensitizer in the photoproduction of hydrogen from water [73]. Anionic micelles stabilize cationic radicals, preventing recombination of PFH^{2+} and hydrated electrons by the repulsive potential of the head groups.

Some studies have also been carried out in non-ionic micellar systems. Costa and Macanita [74] for example, have further probed the influence of micellar structure on quenching of fluorescence in non-ionic (Triton X-100) systems, calculating reaction distances and effective diffusion coefficients in these systems. Although the use of an average diffusion coefficient seems to be in error (although used in papers discussed earlier) because of the existence of a mobility gradient in the micelle interior, the view has been taken that given the high viscosities of the micellar phase (an effective viscosity of 12 cP was estimated for Triton X-100) and the fact that under these conditions the excited probe is only affected by processes occurring in a small volume of the whole micelle it is possible to define an average value of D.

The rapid decay of the excited state of pyrene P* in micelles of hexadecyl pyridinium chloride, which is a known quencher of fluorescence, is in contrast to the lifetimes of P* in nonquencher micelles such as Brij 35 and NaLS where lifetimes are of the order of 300 to 350 ns. In mixed Brij 35–CPC micelles, even when there are only a few molecules of CPC per micelle the net lifetime of pyrene falls to around 80 ns [75]. Preliminary experiments on quenching of photochemical reactions using menadione (vitamin K_3) as solute show that on photolysis in aerated methanol this compound produces a fluorescent product. Photolysis at about 350 nm of menadione solubilized in NaLS or hexadecylpyridinium chloride micelles produces different results; anionic surfactant solutions containing menadione exhibit fluorescence after photolysis, whereas the cationic solutions do not. Decomposition of menadione in NaLS was estimated to be 3.5 times higher than in hexadecylpyridinium chloride solutions.

The effects of oxygen on the photocyclization of N-methyldiphenyldiamine to N-methylcarbazole have been investigated in hexane, water and in aqueous surfactant systems [76]. Fluorescence lifetimes in surfactant systems are sensitive to oxygen levels in the system [77]. Different micelles seem to have different effective oxygen concentrations and Hautala was led to suggest [78] the sequestering of oxygen in micelles. Doubt was later cast on this interpretation by Turro and his colleagues [79] who preferred the view that micelles offer different degrees of protection from the effects of oxygen. In aerated micellar solutions the quantum yield of N-methylcarbazole is significantly higher than in hexane but the rate constants of the unimolecular reaction show no solvent dependence [76]. The dependence of the quantum yield on the solvent has been ascribed to differences in the overall oxygen concentration in the systems. Oxygen does not interact strongly with micelles and there is no evidence for binding of oxygen which may be possible in some biological systems. Although the solubility of oxygen in hydrocarbons is high (the solubility of oxygen in micelles has been found to be twice that in water [80]) the extra time the molecules spend in a micelle compared with an equivalent volume of water is not considered to be of importance in micellar systems due to the small total volume occupied by micelles. There seems to be no evidence that the micellar surface (of NaLS and CTAB at least) offers a barrier to oxygen diffusion.

The lifetime of the excited state of oxygen (O_2^*) in solutions of NaLS and CTAB studied by laser photolysis, has been found to be 53 ± 5 μs, longer than values previously measured in D_2O [80]. Lifetimes measured in nonionic surfactant solutions (e.g. Brij 35, Igepal CO 630 and Igepal CO 660) were found to be considerably shorter (21 to 26 μs) probably due to the loss of electronic excitation and to vibrational modes of the terminal hydroxyl groups of the polyoxyethylene chains of these surfactants [81].

11.3.6 Some reactions involving metal ions

Monoesters of some dicarboxylic acids are subject to metal ion hydrolysis. The effect of micellization on cupric ion-promoted hydrolysis of dicarboxylic acid hemiesters has been the subject of an investigation by Ong and Kostenbauder [82].

The interest in these hemiesters arises from the use of succinate or glutarate hemiesters of 21-hydroxysteroid as water-soluble derivatives in formulation. There is some evidence too that corticosteroid 21-phosphate esters associate in solution [83] and it is likely that they form mixed micelles with surface-active agents. Sodium *n*-decyl oxalate was selected as a model compound likely to mimic the behaviour of these steroidal derivatives. It was found that the rate of cupric ion-catalysed hydrolysis of this agent when solubilized in micelles of NaLS was about 50 times faster than the reaction in bulk solution. Although the formation constant of the chelate complex intermediate was reduced in the micellar phase this was more than compensated for by the increased rate of attack of the OH⁻ ion on the chelate complex. Undoubtedly copper ions bind to the surface of the lauryl sulphate micelle, but the observed micellar rate enhancement for attack of the hydroxide ion on the positively charged intermediate chelate complex is a complete contrast to the inhibition of hydroxide ion attack on such solubilized compounds as benzocaine [84] or on lauryl sulphate ions in micellar array [85]. The various reactions occurring in the copper–decyl oxalate surfactant system are represented below, where M represents the NaLS micelles, K_b and K are the substrate–micelle and Cu^{2+}–micelle binding constants respectively; K_f and K_f' are the formation constants of the chelate complex $CuDOx^+$ in non-micellar (subscript o) and micellar (subscript m) phases, respectively.

$$(DOx^-)_o + M \overset{K_b}{\rightleftharpoons} (DOx^-)_m$$

$$Cu^{2+} + M \overset{K}{\rightleftharpoons} (Cu^{2+})_m$$

$$(DOx^-)_o + Cu^{2+} \overset{K_f}{\rightleftharpoons} (CuDOx^+)_o$$

$$(CuDOx^+)_o + OH^- \underset{slow}{\overset{k_o}{\rightleftharpoons}} products$$

$$(DOx^-)_m + Cu^{2+} \overset{K_f'}{\rightleftharpoons} (CuDOx^+)_m$$

$$(CuDOx^+)_m + OH^- \underset{slow}{\overset{k_m}{\longrightarrow}} products$$

Scheme 11.3

Ong and Kostenbauder's analysis of the data from this system is given in Table 11.5 where values of k_o, k_m, K_f and K_f' are quoted. Sodium lauryl sulphate catalyses the interaction between Ni^{2+} and the ligand pyridine-2-azo-*p*-dimethylaniline (PADA), the effect residing in the ability of the micelle to concentrate the reactants at the micellar surface, since the rate of water loss from the inner co-ordination sphere of Ni^{2+}, which is the rate-determining step in the reaction in water, is little changed in surfactant micelles [86]. Triton X-100 fails to catalyse the reaction and, in fact, results in a small decrease in rate probably due to the partitioning of one reagent, PADA, into the micelle while Ni^{2+} is excluded.

Table 11.5 Second-order rate constants for hydroxyl attack on the chelate complex (k_o and k_m) and formation constants (K_f and $K_{f'}$) of the complex in the absence and presence of sodium lauryl sulphate micelles at pH 5.00 and ionic strength 0.1 M

Temperature (°C)	Nonmicellar		Micellar	
	$10^{-5}k_o$ $((\text{mol}\,1^{-1})^{-1}\text{s}^{-1})$	$10^{-3}K_f$ $((\text{mol}\,1^{-1})^{-1}\text{s}^{-1})$	$10^{-7}k_m$ $((\text{mol}\,1^{-1})^{-1}\text{s}^{-1})$	$10^{-3}K_{f'}$ $((\text{mol}\,1^{-1})^{-1}\text{s}^{-1})$
40	0.67 ± 0.09	1.09 ± 0.25	0.32 ± 0.06	0.54 ± 0.13
45	0.93 ± 0.09	0.94 ± 0.13	0.47 ± 0.07	0.43 ± 0.07
50	1.45 ± 0.39	0.74 ± 0.29	0.68 ± 0.22	0.32 ± 0.05

From [82].

Fig. 11.12 shows the concentration dependence of the observed relaxation times for the interaction between Ni^{2+} and PADA.

The neutral bidentate ligand pyridine-2-azo-*p*-dimethylaniline (PADA) interacting with Ni^{2+} ions

The maximum in the curve in Fig. 11.12 corresponds with the CMC in the system. The decrease in rate has been attributed by Holzwarth *et al.* [87] to the dilution of the reactants over a greater available surface as the concentration of NaLS is increased; the decrease would confirm the notion that the reaction only took place at the micelle surface. Some experimental difficulties centred around batch variation in the surfactant. Sodium lauryl sulphate is notoriously difficult to purify to an extent to guarantee pure surfaces [88]; while it will be readily purified sufficiently well for the measurement of bulk properties, reactions at the micellar surface would be sensitive to impurities.

These effects are noted especially in region D of the plot in Fig. 11.12a; the results for three samples of NaDS are shown in Fig. 11.12b. In this work the nickel concentration at the micellar surface was estimated by titration with murexide, a chromophoric anion. This interaction has been studied in more detail; the reaction between Ni^{2+} and murexide (below) results in a pronounced colour change making the reaction easy to follow. Murexide being hydrophilic is not solubilized by the micellar species and thus can be used as an indicator of the appearance of 'micellar surface'.

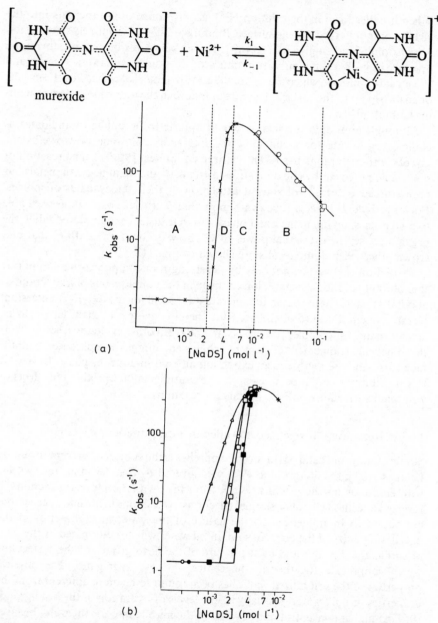

Figure 11.12(a) Reciprocal relaxation time versus concentration of NaDS for the reaction of Ni^{2+} with PADA; $[Ni^{2+}] = 10^{-3}$ mol l^{-1}, $[PADA] = 5 \times 10^{-6}$ mol l^{-1}, pH = 8.0, $T = 25\,^\circ C$. Measurements performed using continuous-flow ($\times$), stopped-flow ($\square$), and dye-laser ($\circ$) methods. (b) Same plot and conditions as in (a), continuous flow measurements. NaDS supplied by BDH ($\bullet\circ$), Merck ($\blacktriangle\triangle$) and Fluka ($\blacksquare\square$). Open symbols: both solutions mixed together contain NaDS; closed symbols: only one solution contains NaDS. From Holzwarth *et al.* [87].

When micelles form in the solution Ni^{2+} is partitioned out of the bulk into the vicinity of the charged micelle surface. Binding of $NiMu^+$ is unlikely since the ion is hydrophilic and singly charged [89]. The nickel murexide systems, therefore, operate somewhat differently from indicator dyes such as Acridine Orange and Pinacyanol Chloride which are lipophilic and are either solubilized by the micelle or adsorbed on to the surface. The chromophore in the murexide system remains in the bulk phase.

Although non-ionic surfactants would appear to be unlikely candidates as complexing agents for metal ions, the interaction of some polyoxyethylene glycols with metal ions has recently attracted interest [90, 91]. The reaction of non-cyclic polyoxyethylene derivatives with alkali and alkaline earth metals has been studied by means of solvent extraction of their thiocyanates or iodides. Polyoxyethylene dodecyl ethers with more than 7 ethylene oxide units were able to bind potassium ion in the water phase and to transfer the complexed salt to the organic phase; the extracting power of $C_{12}E_8$ was about one sixth of that of a crown ether [92]. Some results are shown in Fig. 11.13.

Extraction is governed not only by the strength of complex formation but by the solubility of the surfactant–ion complex in the non-aqueous phase. A ratio of about 0.8 : 1 surfactant : potassium picrate was found for the $(C_{12}E_{25})$-potassium picrate system. Because of the need for the ethylene oxide chain length to be greater than 7 for complexation to occur, and from other evidence, it seems likely that a helical arrangement of the PEG chains encourages complexation with the metal ions and their subsequent extraction into a non-aqueous phase. In view of this possibility it might be necessary to be cautious in interpreting the effects of non-ionic surfactants on metal-catalysed reactions.

11.3.7 Reactions in reversed micelles in non-aqueous solvents

Solubilization and catalysis in reversed micelles is the subject of a recent review by Kitahara [93]; the literature to 1976 was covered by Fendler in his review [94] with emphasis on the extensive work from his own laboratories. Reactions in reversed micelles will not be simple reflections of reactions in normal micelles, but are bound to be influenced by the nature of the water in the interior of the micelles. The size of the pools of solubilized water will be determined by the ratio of surfactant to water and by the nature of the head groups of the surfactants which congregate together in the centre of these aggregates. The physical properties of the solubilized water has been found to be quite different from the properties of bulk water especially at low levels of hydration of the head groups [95]. At higher concentrations of water in the micelle interior the water behaves more like bulk water. Fluorescence probe analysis of the micellar core has indicated a very rigid interior state with a viscosity of over 40 cP [96–99].

A surfactant widely studied in non-aqueous systems is di-2-ethylhexyl sodium sulphosuccinate (Aerosol OT) which, despite its ionic character, dissolves freely in hydrocarbon solvents. A 0.1 M solution of Aerosol OT in octane can solu-bilize nearly 10% water. Imidazole inside water pools is capable of

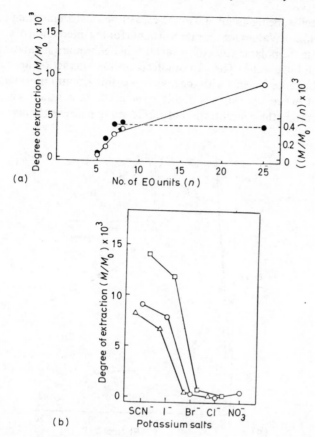

Figure 11.13(a) Extraction of potassium thiocyanate into dichloromethane by homogeneous poly(oxyethylene) monododecyl ethers. Thiocyanate concentration $[M_0] = 0.5 \, \text{mol} \, \text{l}^{-1}$. Polyether concentration: $0.01 \, \text{mol} \, \text{l}^{-1}$. $\bigcirc$, homogeneous poly(oxyethylene) monododecyl ethers (The ether having 25 EO units is not homogeneous.); $\bullet$, the degree of extraction per one oxyethylene unit. ———: Average value; (b) Extraction of potassium salts into dichloromethane by poly(oxyethylene) derivatives. Concentration of each salt $[M_0] = 0.5 \, \text{mol} \, \text{l}^{-1}$. POE concentration: $0.01 \, \text{mol} \, \text{l}^{-1}$. $\square$, 18-Crown-6: $\bigcirc$, $C_{12}H_{25}EO_{25}-H$; $\triangle$, PEG 1000. From Yanagida *et al.* [92].

catalysing the hydrolysis of esters added initially to the external octane phase [4, 100]. Partitioning of the substrate into the pools from the octane is important but is not rate limiting. Enzymes can retain their activity when dissolved in water pools [101]. Solubilized chymotrypsin denatures over several days in Aerosol OT–hexane systems but the sigmoidal pH-rate profile for the catalysis by the enzyme of a specific substrate such as *N*-acetyl-L-tryptophan methyl ester, shifts to a higher pH by 1.5 units in the water pool [4]. Surfactant aggregates in benzene have been found to be capable of solubilizing a large molecule such as Vitamin B_{12a} [102] although the number of surfactant molecules required to solubilize the vitamin is considerably greater than the aggregation numbers generally found for reversed micelles, being about 300 in the case of dodecyl ammonium propionate (DAP) in benzene. Values for the rate constants for the interaction of vitamin B_{12a} with glycine in the polar cavities of surfactants in benzene are significantly greater than those in bulk water. The largest effects on the interaction are observed in surfactant–benzene systems with the lowest possible amount of water m, as seen in Fig. 11.14, when the water is highly structured. At constant water concentration, increasing the concentration of surfactant in benzene results in increased

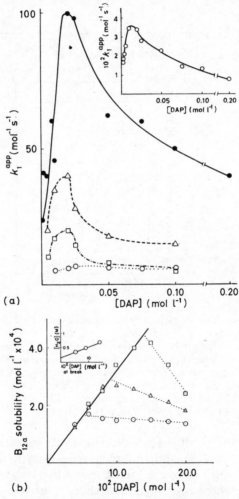

(a)

(b)

Figure 11.14(a) Plot of k_1^{app} versus DAP concentration for the interaction of vitamin B_{12a} with glycine in benzene in the presence of (●) 0.010M, (△) 0.031M, (□) 0.50M, and (○)0.10M solubilized water at 25.0° C. Insert shows a plot of k_1^{app} versus stoichiometric DAP concentration for the same reaction in benzene in the presene of 0.010M solubilized water. (b) Plot of solubility of vitamin B_{12a} in benzene in the presence of DAP versus stoichiometric DAP concentration: (○)0.35M, (△)0.48M, and (□)0.69M water. From Fendler *et al.* [102] with permission.

values of apparent rate constant up to a maximum, followed by a decrease with further increase in concentration. It is speculated that the rate enhancements are due to the tightening of the solvation shell around the reactants; the results parallel the solubility measurements for vitamin B_{12a} shown in Fig. 11.14b where increasing the surfactant concentration can be seen to increase the solubility of the vitamin up to a maximum, after which the number of water molecules per surfactant decreases with resultant decrease in solubility and rate enhancement. Imidazole, on the other hand, does not bind strongly to DAP as it is soluble in benzene and thus its interaction with Vitamin B_{12a} in water solubilized by 0.2 M DAP or Aerosol OT is not significantly different from that in pure water: this is in conditions of relatively high water content and high surfactant levels, where the water is normal. At low water and imidazole levels there is a substantial rate enhancement with respect to water, but the observed rate constant decreases with increasing imidazole concentration [102]. Favourable orientation of the substrate in the polar core and enhanced hydrogen bonding have been invoked to explain a variety of rate enhancements such as the increased rate of mutarotation of 2,3,4,6-tetramethyl-α-D-glucose [103], the decomposition of a Mesenheimer complex [104], the aquation of *tris* (oxalato) chromate (III) and cobaltate (III) anions [105] and the *trans-cis* isomerization of diaquabis (oxalato) chromate (III) anion [106]. In a series of recent papers, O'Connor and Ramage have investigated the reactivity of *p*-nitrophenyl esters in non-aqueous surfactant solutions [107–110]. For the decomposition of *p*-nitrophenyl acetate (PNPA) in dodecyl-ammonium propionate (DAP) in dry benzene and in benzene containing water, the rate of reaction is given by

$$\text{rate} = k_1[\text{PNPA}] + k_2[\text{PNPA}][\text{DAP}] \tag{11.31}$$

but in toluene and cyclohexane the rate equation changes to [108]

$$\text{rate} = k_1[\text{PNPA}] + k_3[\text{PNPA}][\text{DAP}]^2. \tag{11.32}$$

The observed pseudo first-order rate constant, k_ψ, for PNPA in DAP in benzene is shown in Fig. 11.15, and it agrees with previous findings of El Seoud *et al.* [115].

Possible sites of orientation of the PNPA in the micelles are shown in Fig. 11.16. Either the PNPA is located where specific interactions between the nitro group and the surfactant head groups can occur or alternatively, the aromatic nucleus may interact with the hydrocarbon chains of the surfactant, not far from the polar core while the acyl portion interacts with the head groups.

Rapid proton transfer between the dodecylammonium head group and the ester in the close environment of the micellar core is probably followed by the rate-determining attack by the carboxylate ion to form the mixed anhydride [107] (Scheme 11.4)

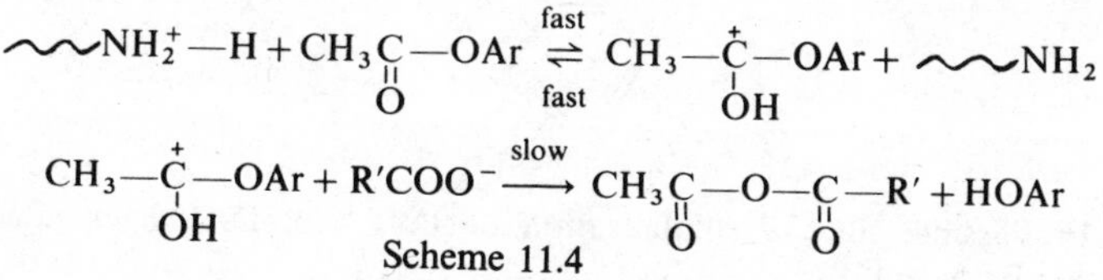

Scheme 11.4

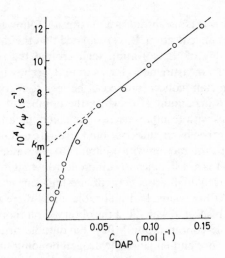

Figure 11.15 Rate profile for the reaction of PNPA in benzene in the presence of DAP at 298 K. From O'Connor and Ramage [107].

(a)

(b)

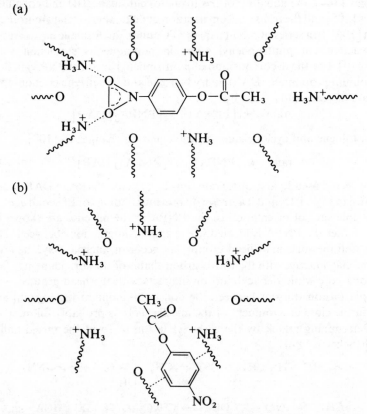

Figure 11.16 Possible sites of solubilization of PNPA in DAP aggregates. From O'Connor and Ramage [107].

The propionate part of the surfactant may also play its part by nucleophilic attack a process, depicted in Scheme 11.5, which could occur both in the micelle or in bulk solution.

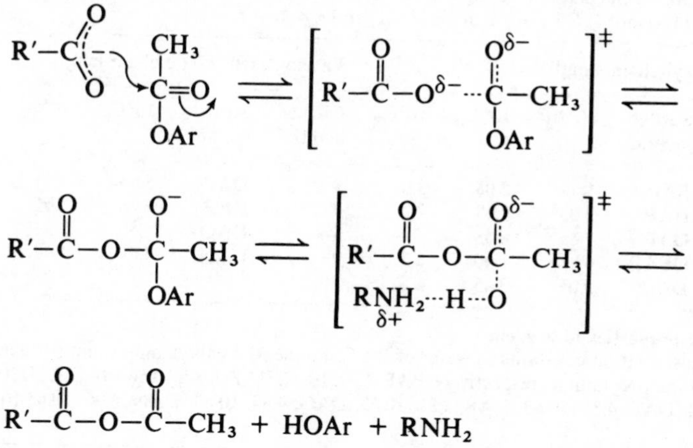

Scheme 11.5

The rate of decomposition of PNPA in cyclohexane solutions of DAP is $44.6 \times 10^{-4} s^{-1}$ and in toluene solutions of DAP k_ψ is $11.0 \times 10^{-4} s^{-1}$ which compares with the value of $9.63 \times 10^{-4} s^{-1}$ in benzene [108]. PNPA is more soluble in benzene than in cyclohexane so this may explain the trend of results. The decrease in rate constant on the addition of water to benzene solutions of DAP may be attributed to inhibition of proton transfer from surfactant head groups and to the competition of water molecules with the ester for occupancy of the micellar core.

PNPA has also been studied in benzene solutions of a series of alkylammonium propionates with alkyl chains from C_4 to C_{12} and a series of dodecylammonium carboxylates, $CH_3(CH_2)_{11}NH_3^+ {}^- O_2C(CH_2)_m CH_3$ where $m = 1, 2, 6$ or 7. Results are shown in Tables 11.6 and 11.7. k increases with increase in

Table 11.6 Rate constants for the reaction of PNPA in benzene solutions of alkylammonium carboxylates containing water at 298 K

H_2O ($\mu l/10$ ml solvent)	$10^4 k_\psi$ (s^{-1}) for surfactants (concn in mol l^{-1} in parentheses)					
	BAP (0.106)	HAP (0.120)	OAP (0.101)	DEAP (0.100)	DAB (0.100)	DAN (0.100)
0	4.56	4.96	5.55	8.63	8.36	11.0
5	4.09	4.79	5.03	8.53	7.87	9.90
10	3.59	4.39	4.47	7.58	7.29	9.48
15		3.95	4.46	7.30	5.80	9.26
20		3.47	3.54	6.39	6.55	8.98

BAP, HAP, OAP and DEAP are the butyl, hexyl, octyl and decyl ammonium propionates; DAB and DAN are dodecylammonium butanoate and nonanoate. From [109].

Table 11.7 Rate constants for the reaction of PNPA in alkylammonium carboxylates in benzene at 298 K

k_M, micellar rate constant in the presence of 0.056 mol l^{-1} water. k'_M, micellar rate constant in the absence of water, calculated from $k_\psi - k_2 \times 0.100$ mol l^{-1}, where k_ψ is the observed rate constant at [DAP] = 0.100 mol l^{-1}. k_2 in l mol^{-1}s^{-1}; k_M and k'_M in s^{-1}

Varying alkyl chain length					Varying carboxylate chain length				
Chain length	Sur-factant	$10^3 k_2$	$10^4 k_M$	$10^4 k'_M$	Chain length	Sur-factant	$10^3 k_2$	$10^4 k_M$	$10^4 k'_M$
4	BAP	1.49	2.05	3.0	3	DAP	5.05	3.45	4.6
6	HAP	2.03	1.69	2.9	4	DAB	6.08	1.61	2.3
8	OAP	1.85	2.48	3.7	8	DAO	5.59	1.32	2.1
10	DEAP	5.80	1.40	3.3	9	DAN	7.88	1.49	3.0
12	DAP	5.05	3.45	4.6					

Surfactant properties in benzene

The acid dissociation constants in water of the components of alkylammonium surfactants are for pK_a (acid) and pK_a (amine) respectively; BAP, 4.87, 10.77; HAP, 4.87, 10.56; OAP, 4.87, 10.65; DEAP, 4.87, 10.64; DAP, 4.87, 10.63; DAB, 4.81, 10.63; DAO, 4.89, 10.63; DAN, 4.96, 10.63 (from [111]).

Surfactant	CMC (mol l^{-1})	N	T(K)	Method
BAP	$(4.5-5.5) \times 10^{-2}$	4 ± 1	306–308	^{1}H n.m.r. [112]
	$(4.0-5.0) \times 10^{-2}$		298	u.v. dye absorption [2]
HAP	$(2.2-3.2) \times 10^{-2}$	7 ± 1	303	^{1}H n.m.r. [112]
OAP	$(1.5-1.7) \times 10^{-2}$	5 ± 1	306–308	^{1}H n.m.r. [112]
DEAP	$(8-10) \times 10^{-3}$		306–308	^{1}H n.m.r. [112]
DAP	2.0×10^{-3}		299	water solubilization [113]
	$(3-7) \times 10^{-3}$		303	^{1}H n.m.r. [112]
	4.5×10^{-3}	3.4	303	vapour pressure osmometry
	8.7×10^{-3}		295	positron annihilation [114]
DAB	1.8×10^{-2}		299	water solubilization [113]
	4.5×10^{-3}	3.2	303	vapour pressure osmometry
DAO†	2.5×10^{-2}		299	water solubilization [113]
	4.8×10^{-3}	2.9	303	vapour pressure osmometry
DAN	*			

* No reported values. † DAO = dodecylammonium octanoate

alkylammonium chain length, but there is no simple relationship between chain length and the calculated values of k_m in the alkylammonium propionate series, nor is there a clear trend of k_m with carboxylate chain length [108]. Overall the highest values of k_m were observed for DAP, but these results are to be compared with those of O'Connor *et al.* [116] on the hydrolysis of 2,4-dinitrophenyl sulphate when k_m was found to increase with both alkylammonium and carboxylate chain length in a series of alkylammonium propionates and octylammonium carboxylates, respectively. Although the dissociation constants of the head groups in the central water pool will differ from the values in bulk water, the pK_a values determined in aqueous solution should be a reasonable guide to the change in dissociation in a surfactant series when the micellar dimensions are similar. O'Connor and Ramage attempt to rationalize the results for PNPA by drawing attention to the increase in k_2 with decreasing pK_a of the amine (with increasing acidity), i.e. $k_2(\text{BAP}) < k_2(\text{OAP}) < k_2(\text{HAP})$, results consistent with the idea that the alkylammonium ion is involved in facilitating the *p*-nitrophenoxide leaving group. It is, however, apparent that other factors are influencing the outcome as the pK_a values for OAP, DEAP and DAP are apparently identical, yet k_2 for OAP is significantly lower than for the two other longer chain surfactants. The small differences in the pK_a values for the carboxylate series do not really allow any interpretation along these lines.

The value of k_2 for PNPA hydrolysis in a series of dodecylammonium phenoxides in the absence of water increases with increasing pK_a of the phenolic group [110]. This may be explained by the fact that increase in the basicity of the phenol may shift the equilibrium,

$$\text{RNH}_2 \ldots \text{H—OAr} \rightleftharpoons \text{RNH}_3^+ \ldots {}^-\text{OAr}$$

to the left thus increasing the concentration of the attacking nucleophile, the unprotonated amine or causing a significant contribution to the overall rate constant if the phenoxide ion removes a proton from the tetrahedral intermediate. Assuming that the position of the equilibrium shown above favours the unprotonated amine it is concluded that the reaction being observed is aminolysis of the ester by the amine derived from the surfactant.

It has been found that the interiors of DAP micelles have a strong buffering action [117]. The pK_a values of fluorescein decreased in DAP micelles in water–methanol mixtures. The buffering action is greater in systems containing 0.2M H_2O than in systems with 0.5M H_2O. The results in Fig. 11.17 are obtained for the number of protons necessary for the neutralization of the propionate groups, using data on aggregation number and water content per micelle obtained by Correll *et al.* [118], and assuming that the protons were supplied by the acid added to the water [117]. The number of monomers per micelle is 8 at 0.2M H_2O and 54 at 0.5M H_2O; solubilized water molecules below about 0.37M water are

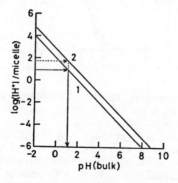

Figure 11.17 The log proton number per micelle, consumed by the protonation of the propionate, as a function of the bulk pH. Curves 1 and 2, 0.2 and 0.5 M H_2O, respectively; concentration of DAP, 8.0×10^{-2} M; number of propionate groups/micelle, 8 at 0.2 M H_2O and 54 at 0.5 M H_2O. From Miyoshi and Tomita [117].

thought to be bound to the head groups while the remainder are free. In constructing Fig. 11.17, Miyoshi and Tomita assumed that at a bulk pH of 1, all the propionate groups would be protonated thus log (number of protons required per micelle) is about 1 and 1.7 for aggregation numbers of 8 and 54, respectively. The initial rate of oxidation of 1,3-diphenylisobenzofuran (DF) at various bulk pH values in DAP reversed micelles is shown to be dependent on pH. DF is oxidized by the singlet oxygen generated by the photosensitization of fluorescein (^{3}F) (see Fig. 11.18). The buffering action within the micelles can thus, in a qualitative fashion, be said to result in changes in the nature of reactions occurring within the micelles.

Another aspect of the properties of the inverse micellar water pool is the ability referred to by Balny *et al.* [120] to circumvent freezing at zero degrees. This supercooling is a feature observed also in water-in-oil emulsions, and opens up

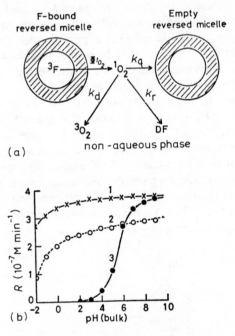

Figure 11.18(a) Simplified diagram for the photochemical reaction in DAP reversed micellar solutions containing F and DF. ^{3}F, triplet state of F; Φ^1O_2, quantum yield for singlet oxygen production in cyclohexanic phase; k_r, rate constant for DF oxidation by singlet oxygen; k_d, rate constant for physical decay of singlet oxygen in cyclohexane; k_q, rate constant for the quenching of singlet oxygen by empty DAP reversed micelles. DF is hydrophobic and thus will be found largely in the external phase. From Miyoshi and Tomita [119]. (b) The initial rate (R) of DF oxidation at various bulk pHs. Concentrations of F, DAP and DF, 3.6×10^{-6}, 8.0×10^{-2} and 4.0×10^{-5} M, respectively; Curves 1 and 2, DAP reversed micellar solutions containing 0.2 and 0.5 M H_2O, respectively; Curve 3, H_2O–MeOH (1:1, v/v) mixed solutions; temperature, 40° C. From Miyoshi and Tomita [117].

the possibility of the study of enzyme-catalysed reactions at subzero temperatures. It has been proposed that the water in inverted micelles has properties akin to water structures in cellular structures [120]. In probing Schiff base formation between pyridoxal and amino acids in reversed micelles, Kondo et al.[121] found that both forms of Schiff base (SB 1 and SB 2 in Scheme 11.6 below) were present in the micelle interior, indicating that the polarity of the environment is intermediate between that of water and that of an organic solvent. There is a 10 to 100-fold enhancement of rate of formation of the pyridoxal Schiff base compared to that in normal aqueous media when Aerosol OT is used as a surfactant. The less polar environment and its restricted nature lead to a more intimate encounter between the reacting species and an enhanced reactivity. The SB 1 species is predominant for alanine and arginine while methionine results in the production of more SB 2 species.

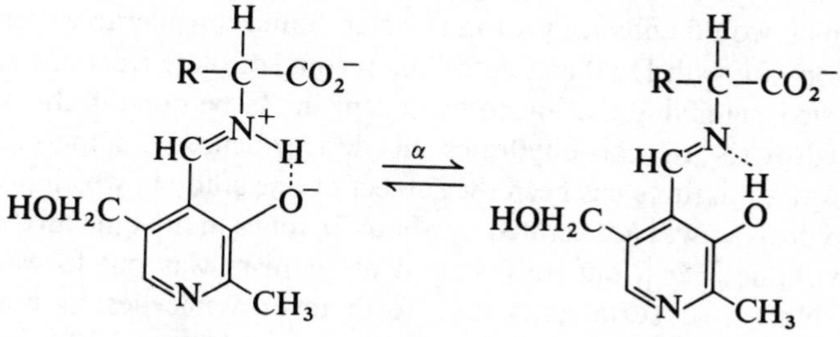

Scheme 11.6

Murexide, earlier mentioned in relation to investigations of reactions with ions in aqueous micellar systems, has also been used in studies on ion–ligand reactions in reversed micellar systems [122] of Aerosol OT in hexane, the size of the micelles being varied over a range by altering the surfactant/water ratio. The reaction between murexide and Ni^{2+} is seen to comprise two steps, the first involving the transfer of the reagents into the same droplet (a communication step) and the second involving the complexation of the reagents within the water pool. The rate constants of these reactions are designated k_1 and k_2. From the results in Table 11.8 it is seen that k_{-2} is independent of the size of the droplet.

Table 11.8 Rate of complexation of Ni^{2+} and murexide in Aerosol OT solutions in hexane from [122]

$R^{-1} = [AOT]/[H_2O]$	Radius of aqueous core $(10^{-10}m)$	No. of water mols in each droplet	$k_2(s^{-1})$	$k_{-2}(s^{-1})$
5.8	10.0	280	60	2.8
8.6	14.0	600	36	2.8
11.4	17.0	1060	23	2.8
16.9	22.0	2240	9.7	2.8

k_{-2} is a measure of the lifetime of the metal–ligand complex and has a value close to that obtained in bulk water, namely $2.4\,s^{-1}$. k_2, on the other hand, varies with droplet size, which is to be expected. As the volume of the aqueous core increases the chances of finding the reactants in close contact are reduced.

This system has been evaluated in more detail by Robinson *et al.* [123]. When chromium (III) or cobalt (III) complexes are solubilized by alkylammonium carboxylate surfactants in benzene in the presence of water, they may undergo rapid aquation, the rate of which depends on the surfactant and the amount of water present [124]. Rate constants for the aquation of $[Cr(C_2O_4)_3]^{3-}$, $[Co(C_2O_4)_3]^{3-}$ and *cis*-$[Co(en)_2(N_3)_2]^+$ (en = ethylene-diamine) in benzene solutions of DAP were 10^6, 1500 and 11 times greater, respectively, than the corresponding reactions in bulk water. These, sometimes massive, increases in rate have been attributed to hydrogen-bond formation between the oxygen atoms of the substrate and the ammonium surfactant head groups which would enhance proton transfer. Some considerable complexity in the variation of k with DAP concentration of some of these reactions has recently been reported, signifying that much work remains to be done in this field [125].

The hydrolysis of carbohydrates in dodecylbenzene sulphonic acid in dioxane–water mixtures has been the subject of one study in which it was found that the hydrolysis was accelerated by about 21 times in dioxane mixtures above 60% by volume [126], but no coherent mechanism was put forward for the catalysis. Non-ionic surfactants may form inverse micelles in non-aqueous solvents in the presence of small amounts of water. Triton X-100, for example, micellizes in carbon tetrachloride on addition of water. This system, which obviously does not suffer the problems which result from the dissociation of the head groups of ionic surfactants in the water pool, has been used to investigate the hydration reaction of acetaldehyde [127]. This acid-catalysed reaction is increased by a factor of 10 000 over that in water (Table 11.9). In spite of the non-ionic nature of the peripheral head groups surrounding and penetrating the aqueous core, the nature of the water is such that ionization of solubilized species is changed.

The apparent pK_a of dyes such as Malachite Green increases with increase in the surfactant–water ratio in Igepal CO-530 inverse micelles in benzene [128]. If the water is largely bound to the PEG chains in the micellar interior it is likely that the effective dielectric constant will be quite different from that of bulk water and thus the apparent simplicity of the non-ionic system over an ionic reversed system is disproved. Non-ionic surfactants have also been found to increase the rate of the

Table 11.9 Second-order rate constants (k_{obs}) of catalysed hydration of acetaldehyde [128]

Catalyst	Solvent	Temperature (°C)	k_{obs} (s^{-1})
H_2O	H_2O	36	3.6×10^{-4}
$HClO_4$	H_2O	36	625 ± 30
H_2O + Triton X-100	CCl_4	36	5.4 ± 0.3
$H^+ + H_2O$ + Triton X-100	CCl_4	36	1500 ± 70

reaction haemin $+ CN^-$ in benzene containing 0.2% methanol [129]. The dimer of haemin predominates in water, whereas in the micelle the monomer is the observed species, a transition favoured by the decrease in the dielectric constant of the water pool.

11.4 Stability of drugs in surfactant systems

Many of the investigations carried out on mechanisms of catalysis and inhibition in surfactant systems have employed model compounds. Such is the observed specificity of reactions that whatever understanding is gained from the work there can be little confidence in extrapolation of the results to complicated drug structures. This section of the chapter surveys some of the work which has been carried out with pharmaceuticals. A later section will consider the stability of surfactant molecules themselves.

The aim in pharmaceutical formulations will be to reduce rates of degradation by appropriate choice of surfactant system, and to avoid surfactants which catalyse reactions. Cetomacrogol 1000 reduces the rate of hydrolysis of aspirin in its unionized state but does not stabilize the ionized form of the drug. Above a pH of about 3 it seems that only drug in the aqueous phase undergoes hydrolysis, so considering the micellar solution as a two-phase system [131] one can write

$$k_\psi = \frac{1-v}{P_m \cdot v + (1-v)} k_w \tag{11.33}$$

where k_w is the rate constant in water, P_m is the apparent partition coefficient of the solute to the micelle, and v is the volume fraction of the micellar phase. It is obvious that k_ψ is always less than k_w if the reaction takes place in the aqueous phase and is not affected by the presence of surfactant monomers. Equation 11.33 is derived as follows [130]. If C is the initial reactant concentration (mol 1^{-1} total system), C_w the initial reactant concentration in the aqueous phase (mol 1^{-1} aqueous phase), and x the concentration reacting in time t (mol 1^{-1} total system), then, on the assumption that hydrolysis in a solubilized system takes place only in the aqueous phase, the concentration of reactant remaining after time t, is $C_w(C-x)/C$ mol 1^{-1} aqueous phase or $C_w(1-v)(C-x)/C$ mol 1^{-1} total system. The differential rate equation for the concentration change with respect to the total system is:

$$\frac{dx}{dt} = k_w C_w (1-v) \frac{(C-x)}{C} \tag{11.34}$$

which on integration and rearrangement gives:

$$k_w = \frac{C}{C_w(1-v)} \frac{1}{t} \ln \frac{C}{C-x} \tag{11.35}$$

The distribution coefficient is given by

$$P_m = \frac{C - C_w(1-v)}{C_w v} \tag{11.36}$$

which on rearrangement becomes:

$$\frac{C}{C_w} = P_m v + (1 - v).$$ (11.37)

Substitution into Equation 11.35 gives:

$$k_w = \frac{P_m v + (1 - v)}{1 - v} \cdot \frac{1}{t} \ln \frac{C}{C - x}$$ (11.38)

where $1/t \ln C/(C - x)$ is the observed rate constant k_ψ, assuming that hydrolysis takes place only in the aqueous phase. Substitution of k_{obs} into Equation 11.38 and rearrangement leads to Equation 11.39:

$$k_\psi = \frac{1 - v}{P_m v + (1 - v)} k_w.$$ (11.39)

If the reaction partially takes place in the micellar phase an additional term must be added as the total rate constant is now the sum of k_w and the rate constant in micellar phase, k_m. This leads to equations of the form,

$$k_\psi = k_m + \frac{(k_w - k_\psi)}{P_m v},$$ (11.40)

when $v = 0.01$ or less. The kinetics of alkaline hydrolysis of indomethacin in non-ionic surfactant systems have been investigated [132, 133]. Protection is observed, in contrast to the effect of a cationic surfactant as can be seen in Fig. 11.19. Increasing the length of the hydrophobic chain of a series of polysorbates had little effect on the rate of the reaction [133]. It is suggested [132] that ionized

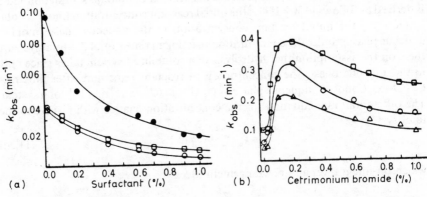

Figure 11.19(a) Observed rate constant, k_{obs}, versus varying concentrations of surfactants in alkaline aqueous solutions at 30.3° C. $\bigcirc$, ethoxylated lanolin in 0.002 M hydroxide-ion concentration; $\bullet$, ethoxylated lanolin in 0.005 M hydroxide-ion concentration; and $\square$, polysorbate 80 in 0.002 M hydroxide-ion concentration.
(b) Observed rate constant, k_{obs}, versus varying concentrations of cetrimonium bromide in alkaline aqueous solutions at 30.3° C. (Hydroxide-ion concentration): $\triangle$, 0.002 M; $\bigcirc$, 0.003 M; and $\square$, 0.005 M. From Dawson *et al.* [132].

indomethacin will confer a charge on the micelles of non-ionic surfactant and that this might explain deviations from expected 'model' behaviour.

Relatively small increases in stability of chloramphenicol in polysorbate 80, Myrj 59, polysorbate 20 and Brij 35 have been observed on autoclaving [134]. Selection of a suitable concentration of polysorbate 80 and adjustment of solutions to pH 4.6 reduced to half the autoxidative degradation of methyl-prednisolone even in the presence of oxygen [135]. Contrary to electrostatic theories of stabilization the base-catalysed hydrolysis of procaine was inhibited by non-ionic, anionic and cationic micelles as shown in Table 11.10. The order of inhibition of hydrolysis was NaLS > CTAB > PLE > NDB; the order of partition coefficients (P_m) was found to be NaLS > CTAB > NDB > PLE (see table for abbreviations).

Table 11.10 The observed rate constants for hydrolysis of procaine in various surfactant solutions

PLE (10^{-2} mol l^{-1})	$k_{obs} \times 10^3$ (min^{-1}, pH = 11.8)	NaLS (10^{-2} M)	$k_{obs} \times 10^3$ (min^{-1}, pH = 11.8)	$k_{obs} \times 10^6$ (min^{-1}, pH = 7.0)	CTAB (10^{-2} mol l^{-1})	$k_{obs} \times 10^3$ (min^{-1}, pH = 11.8)	NDB (10^{-2} mol l^{-1})	$k_{obs} \times 10^3$ (min^{-1}, pH = 11.8)
0.00	9.58	0.00	9.58	44.5	0.40	5.83	0.20	7.83
0.83	4.74	1.00	3.32	12.5	0.70	4.40	0.67	5.53
1.67	3.02	1.50	2.35	9.25	1.00	3.55	1.33	3.98
2.50	2.25	2.00	2.00	7.48	2.00	2.24	2.00	3.02
3.33	1.68	2.50	1.62	5.45	2.50	1.94	2.66	2.35
		3.00	1.30		3.33	1.52	3.33	2.03
		3.33	1.08	3.15				

From [136].
PLE = Brij 35; NDB = *N*-dodecyl betaine.

Meakin and others have reported that CTAB decreases the rate of hydrolysis of ethyl *p*-aminobenzoate and *p*-aminophenylacetate [28] (see above) contrary to expectation, a phenomenon ascribed to the orientation of these molecules in the CTAB micelles which facilitates degradation. However, although procaine was found to be located in the outer layers of CTAB and NDB micelles stabilization was still observed [136].

Nicotinic acid esters which are located both in the interior and exterior portions of non-ionic micelles are stabilized by interaction with the micellar phase, the higher the binding factor the greater the stabilizing effect. This implies that lipophilic esters are stabilized to a greater extent than more hydrophilic esters. Lippold *et al.* [137] plot *Q*, defined as the stabilization quotient, against the percentage of ester bound to the surfactant (Fig. 11.20). The results for a series of surfactants with the same ethylene oxide chain length (approximately 22 units) and increasing alkyl chain length are superimposed, suggesting that the effect of alkyl chain length is related only to the degree of binding.

The longer the ethylene oxide chain in a series of stearyl ethers the greater the degree of protection afforded to the ester at 20° C, a feature most prominent with

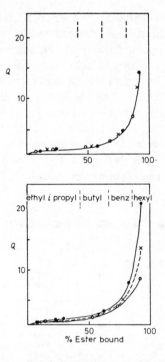

Figure 11.20 Stabilization quotient Q of nicotinic acid esters (ethyl, isopropyl, butyl, benzyl and hexyl) as a function of percentage binding of ester, (a) to three non-ionic surfactants based on PEG 1000: ○ PEG 1000 – lauryl ether, X PEG 1000 – myristyl ether, ● PEG 1000 palmityl ether, and (b) to three stearyl surfactant derivatives, ○ PEG 900 stearyl ether, X PEG 1400 stearyl ether and ● PEG 2000 stearyl ether. In both (a) and (b) ionic strength is 0.083, temperature $20.0 \pm 0.1°$ C and borate buffer pH 10.00 ± 0.02 used. Redrawn from Lippold *et al.* [137], by permission.

the hexyl ester (Fig. 11.20b). At 40° C this effect is not detectable and suggests that the dehydration of the ethylene oxide chains causes them to lose some of their ability to protect. The hydrolysis of benzocaine has been widely studied (e.g. [139]); a reciprocal relationship has been found between the pseudo-first order rate constant k and the apparent solubility of the benzocaine in solutions of POE (26), lauryl ether and polysorbate 80. For example, at 70° C this relation may be expressed as,

$$\log k_{70°} = -0.025C - 0.5 \tag{11.41}$$

where C is the concentration of the lauryl ether surfactant; for polysorbate 80 the relation is

$$\log k_{70°} = -0.017C - 0.5. \tag{11.42}$$

On a percentage basis the lauryl ether is more effective than polysorbate 80 in retarding the hydrolytic degradation of benzocaine. In the same paper the authors [139] investigate homatropine stability at elevated temperatures; as before the ability of the non-ionic surfactants to protect the drug at 70° C is poor, and about 10 % polysorbate 80 is required to make any significant impact on $t_{1/2}$ values. Investigation of the base-catalysed hydrolysis of benzocaine (ethyl *p*-amino benzoate) and two more liposoluble homologues, *n*-butyl and ethyl *p*-(*n*-butyl amino benzoate) butyl aminobenzoates in the presence of a non-ionic surfactant $C_{16}E_{24}$ [138] showed that greater protection was afforded to the latter two compounds. The overall rate of hydrolysis can be expressed by the following equation,

$$-(V_a + V_m)\frac{d\overline{C}}{dt} = k_a V_a C_a + k_m V_m C_m, \tag{11.43}$$

where V_m is the volume of micelles, V_a is the volume of the alkaline aqueous phase, C_a is the concentration of ester in the aqueous phase, and C_m the concentration in the micellar phase, k_a and k_m being the corresponding pseudo-first order rate constants; $\overline{C}$ = the average concentration of ester i.e. $\overline{C} = [(V_a C_a + V_m C_m)/(V_a + V_m)]$. The partition coefficient of the ester is P_m as before. Substituting P_m leads to the following equation where k is the measured rate constant:

$$k = -\frac{d\ln\overline{C}}{dt} = \frac{k_a - k_m}{1 + P_m (V_m/V_a)} + k_m. \tag{11.44}$$

Good agreement was obtained between experimental points and theoretical lines drawn according to this equation. The rate constants reported by Smith *et al.* [138] are a linear function of the reciprocal solubility $(1/S)$ of a given ester in the surfactant solutions, as suggested by Hamid and Parrott [139]. The results for each ester fall on a different line so we must conclude that the reactions proceed at quite different rates within the micelles (Fig. 11.21). The emphasis on model drug compounds such as benzocaine has perhaps obscured the need for a wider study of drug stability in surfactant solutions. It is thus worth citing studies on penicillin

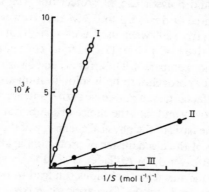

Figure 11.21 A plot of the reaction rate constant k of three compounds as a function of their reciprocal solubility $(1/S)$ in surfactant solutions, drawn from data calculated from Table 1 of Smith *et al.* [138]. I: benzocaine, II: *n* butyl- and III ethyl *p* *n*-butyl amino benzoates. The relationship between k and $1/S$ holds only for individual compounds and is obviously not a general trend.

and cephalosporin stability. A catalytic effect of CTAB and benzalkonium chloride on the stability of cephalexin has been reported [140]. At neutral pH the degradation involves intramolecular nucleophilic attack on the side chain α-amino group of the β-lactam carbonyl. The pseudo-first-order rate constant for degradation in the absence of surfactant was $0.045\,h^{-1}$ at pH 6.5. In the presence of CTAB and benzalkonium chlorides above the respective CMCs, the degradation of cephalexin increased by a factor of about 9 to 14. Positively charged tetraethylammonium chloride which does not micellize did not affect the rate. However, this catalytic effect could be overcome to a great extent by adjusting the ionic strength (μ) of the medium. In the presence of 20 mM CTAB the reaction rates were suppressed with increasing salt concentration so that, for example at $\mu = 2.0$, the rate constant for degradation was reduced to $0.076\,h^{-1}$, less than double that observed in the absence of cationic surfactant.

In acidic media the picture is different. Cationic (CTAB) and a non-ionic surfactant ($C_{12}E_{23}$) reduced the degradation of several penicillins by a factor of 4 to 12, while an anionic surfactant (NaLS) increased the rate [141]. In all cases the rate constants first increased or decreased rapidly and then approached a constant value above the solubilizer CMC. In contrast to the penicillins, the acid degradation of cefazolin, a relatively acid-unstable cephalosporin, was not influenced by the presence of any of the surfactants, suggesting that this antibiotic is not sufficiently bound to any of the surfactant micelles. The log P values for the affected penicillins (at pH 2.1) are in the range 2.7 (propicillin) to 1.70 (penicillin G) while that of cefazolin is 0.39. It is fairly clear that stabilization of the pencillins is the result of the decreased hydrogen ion activity in the vicinity of the cationic head groups of the micelles. Results for propicillin are shown in Fig. 11.22.

Current interest in radiation sterilization of pharmaceuticals led Fletcher and Davies [142] to investigate the sensitivity of benzocaine to irradiation in aqueous solution. Cetrimide and polysorbate 80 protect the drug from the deleterious effects of a ^{60}Co source at doses up to 0.3 M rad. However, ethanol, glycerol and polyoxyethylene glycol 200 when used as co-solvents at concentrations up to 40% were more effective [142] (Fig. 11.23) (equal concentrations of surfactant and co-solvent were not compared). These results emphasize the need to compare critically the different approaches to both solubilization and stabilization. High concentrations of surfactant, even non-ionic surfactant, are usually impracticable on the grounds of toxicity and thus the choice of solubilization technique cannot always be made on the basis of the physical chemistry of the competing methods.

The interpretation of the mechanisms of protection against gamma rays may not always be straightforward, for Al-Saden *et al.* [143] have found that gamma-irradiation of non-ionic surfactant solutions can lead to polyoxyethylene chain scission. This in turn leads to the formation of mixed micelles between the surfactant and the more hydrophobic, degraded species, with the result that the micelles grow and the cloud point is lowered. Cross-linking may also occur in polyoxyethylene glycol solutions with subsequent gel formation [144].

On exposure to γ-irradiation from a ^{60}Co source in the absence of oxygen,

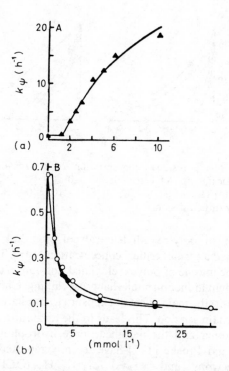

(a)

(b)

Figure 11.22 Plots of the observed first-order rate constant, k_ψ (h^{-1}) (ordinate) against (a) the NaLS concentration (▲); and (b) the POE (○) and CTAB (●) concentrations, C_D (mM) (abscissa) for the acid degradation of propicillin at pH 1.6, 35° and ionic strengths A 0.03 M and B 0.5 M. The curves were calculated from Equation 11.45 and the parameters listed below. From [141] with permission.

$$k_\psi = \frac{k_o + k_m K[M]}{1 + K[M]} \tag{11.45}$$

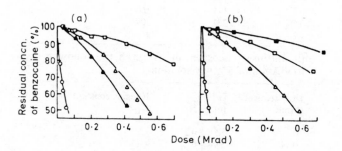

Figure 11.23 (a) and (b) Plots of % residual concentration of benzocaine against dose of radiation for the water–cosolvent mixtures. 1.25×10^{-4} M, benzocaine in (a) ethanol (○—0% ▲—10% △—20% □—40% w/v) and in (b) PEG (○—0% △—5% □—10% ■—40% v/v). (continued overleaf)

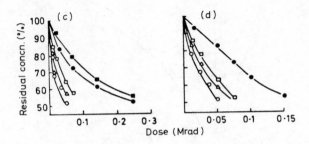

(c) and (d) Plots of % residual concentration of benzocaine against radiation. 1.25×10^{-4} M benzocaine in (c) cetrimide ($\bigcirc$—0; $\triangle$—10^{-5}; $\square$—10^{-4}; $\bullet$—10^{-3}; $\blacksquare$—10^{-2} M) and in (d) Tween 80 ($\bigcirc$—0; $\triangle$ 10^{-2}; $\square$—10^{-2}; $\bullet$—10^{-1} %). From Fletcher and Davies [142] with permission.

aqueous solutions of poly(oxyethylene) glycol cross-link provided that the polymer has reached a certain critical concentration [143]. The proximity of the glycol chains in the micelle of polyoxyethylated non-ionic surfactants suggested that irradiation might induce intramicellar cross-linking. Under the conditions so far studied, however, the main effect of exposure to irradiation has been to cause polyoxyethylene chain scission. This leads to the formation of mixed micelles of the starting monomer and the modified, more hydrophobic, surfactant molecules. Henglein and Proske [145] have recently reported on the irradiation (high-energy electrons and γ-rays) of $C_{16}H_{33}(OCH_2CH_2)_{21}OH$ and $C_{14}H_{24}(OCH_2CH_2)_3SO_3Na$: γ-irradiation of the former at concentrations

Table 11.11(a) Effect of radiation dose on the intrinsic viscosity of polyoxyethylene, molecular weight 600 000, at 0.5% and 1% concentration

Dose received (Mrad)	Intrinsic viscosity, $[\eta]$ (ml g^{-1})	
	1%	0.5%
0.0	178	
0.3	171	160
0.6	231	209
1.2	gel	gel

(b) Intrinsic viscosities of Triton X-100 following γ-irradiation of 5% w/v solutions

Dose received (Mrad)	$[\eta]$ (ml g^{-1})	Huggins constant K_H
0	5.55 ± 0.2	1.59
0.44	5.98 ± 0.2	1.25
0.74	6.43 ± 0.09	0.97
1.95	5.51	2.44
10	7.01	3.65

From [143].

above 0.005 mol l^{-1} (0.58 %) resulted in increases in viscosity and the solutions became turbid at a certain critical dose, e.g. ~ 0.8 Mrad for an 0.93 % solution. These effects were explained in terms of cross-linking between the surfactant molecules, but reduction in effective hydrophilicity following chain scission results in an increase in viscosity and apparent micellar weight.

Some tabulated results are given below comparing the effects of γ-irradiation on the viscosity of PEG 6000 and Triton X-100 (Table 11.11) and Triton X-405, cetomacrogol 1000 and a Brij 92–96 mixture (Table 11.12). Surface tension concentration plots for Triton X-100 before and after irradiation are shown in Fig. 11.24.

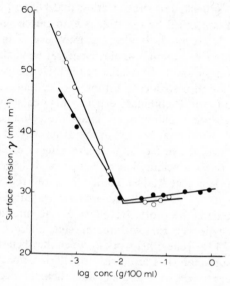

Figure 11.24 Surface tension versus log (concentration plots) of Triton X-100 (●) and Triton X-100 irradiated as a 5 % aqueous solution for 24 h (○). From Al-Saden *et al.* [143] with permission.

Table 11.12 Effect of γ-irradiation on the intrinsic viscosity of non-ionic surfactants

Surfactant system	Concentration (%)	Dose received (Mrad)	$[\eta]$ (ml g^{-1})	Huggins constant K_H
Triton X-405	6.62	10	~ 9.5	~ 4
Triton X-405	6.62	0	8.46	2.09
Cetomacrogol 1000	3	0.44	8.41	0.81
	3	1.95	8.57	1.58
	3	10	9.5	5
	3	0	8.0	1.09
Brij 92-96 mixture	0.775	0.74	13.32	43.6
	0.775	0	22.66	27.8

From [143].

Henglein and Proske [145] also describe the effects of exposure of $C_{16}E_{21}$ and $C_{14}H_{29}(OCH_2CH_2)_3SO_3Na$ to short pulses of high-energy radiation. This produces in the water OH radicals as well as hydrated electrons, the latter being rapidly converted (within 10^{-8} s) to OH˙ radicals by reacting with dissolved nitrous oxide $[N_2O + e_{aq}^- + H_2O = N_2 + OH^- + OH˙]$. The OH˙ radicals interact with the micelles producing macroradicals such as —O—CH—CH₂—O—.

11.5 Stability of surfactant solutions

The stability of surfactants themselves must thus be considered in experiments when the system is subjected to stress. Carless and Nixon [146] suspected that half of the oxygen consumed by benzaldehyde in cetomacrogol solutions was taken up by the detergent molecules themselves, and in the study of the autoxidation of emulsified methyl linoleate several emulsifiers including non-ionic surfactants were found to be oxidized [147]. Hamburger et al. [148] have studied the formation of peroxides in solutions of cetomacrogol using oxidized cetomacrogol as initiator. Preliminary experiments had shown that benzocaine hydrochloride developed a yellow colour in aqueous cetomacrogol after about a week, the reaction being retarded by antioxidants; it was suspected that the breakdown products of the surfactant were attacking or catalysing attack on the aromatic group of the benzocaine. The rate of peroxide formation was fastest at the lowest level of detergent studied, namely 3%, and reduced with increasing concentration up to 20%. As all the material is in micellar form at all concentrations studied, the protective effect of the micelles must be due to interaction of the hydroperoxide with the micellar phase.

Hamburger et al. [148] cite other work in which they believe that autoxidation of the non-ionic surfactant has given rise to deterioration, giving examples of discoloration of certified dyes, and chloramphenicol decomposition in the presence of polyglycols. Thus reaction kinetics in the presence of non-ionic surfactants liable to decomposition must be interpreted with care.

Degradation of surfactants on storage can lead to unexplained behaviour in use, and perhaps lead to false interpretation of results. Aqueous solutions of polysorbate 20 autoxidize on storage, the pH falling and levelling off, as does the surface tension of the solutions. The cloud point decreases until turbidity is noted at room temperature [149] as illustrated in Fig. 11.25. The main disadvantage in the use of polysorbate 20, apart from the rapid appearance of clouding which indicates phase separation, is the potential for hydrolysis to form lauric acid. Sample variation can be a cause of confusion. Some of the data assembled by Donbrow et al. is given in Table 11.13a and shows the variability which is possible. Several samples contain considerable quantities of peroxide which is known to accelerate the decomposition of drugs such as benzocaine [150], corticosteroids [151] and vitamin A acetate [152]. Methods for the removal of peroxides from non-ionic surfactants have been reported recently [153, 154].

The pitfalls in the use of commercial non-ionic surfactants in biochemical work

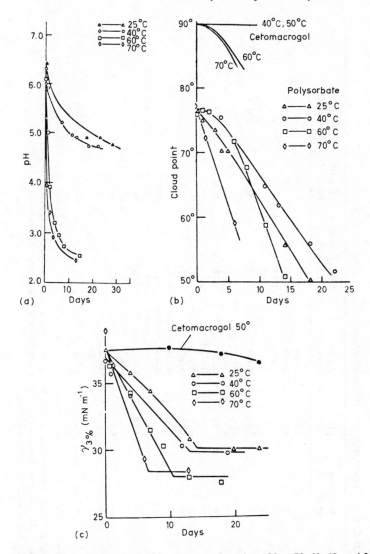

Figure 11.25(a) The pH change of 3 % aqueous polysorbate 20 at 70, 60, 40, and 25° C in daylight with no catalyst, with time. (b) Rate of change of cloud point of 3 % aqueous polysorbate at 70, 60, 40, and 25° C in daylight with no catalyst, measured after addition of 1 M NaCl. (Cetomacrogol is shown for comparison.) (c) Rate of change of surface tension, γ, of 3 % aqueous polysorbate 20 at 70, 60, 40, and 25° C in daylight with no catalyst. (Cetomacrogol, 3 % aqueous solution at 50° C, is shown for comparison.) From Donbrow *et al.* [149] with permission.

have been highlighted by Chang and Bock [154] who found many surfactant samples to contain significant amounts of oxidizing agents, some obviously peroxide. Some detergents, however, including widely used materials such as Brij 35, have added to them deliberately oxidizing agents to bleach the yellow colour

Table 11.13(a) Physicochemical data for surfactants from different sources

Surfactant	Sample*	Solubility† at 20% (w/v)	pH	Acidity (mEq g^{-1})	Peroxide number (mEq kg^{-1})	Cloud‡ Point	γ_1% (mN m^{-1})
Polysorbate 20	A	+	4.6	0.040	14.6	76°	—
	B	+	3.8	0.050	4.5	—	40.9
	C	+	4.2	0.016	0.13	77°	41.2
	D	+	4.5	0.027	0.07	76°	40.0
	E	+	5.2	0.033	0.05		—
Polysorbate 40 (polyoxyethylene 20 sorbitan monopalmitate)	A	—	6.6	0.016			—
	B	—	4.9	0.023	0.21	72.8°	42.8
	Ca	+		0.022 d§			
	Cb	—	—	0.033 d			
	Da	—	3.6	0.065 f§	0.04 f	74.3°f	42.3 f
	Db	—	4.2 d, 4.1 f	0.02 d, 0.014 f	0.36 f	80.3°f	
Polysorbate 60 (polyoxyethylene 20 sorbitan monostearate)	A	—		0.013 d, 0.0095 f	0.1 f	69.5°f	
	B	—	5.1 f	0.012 d, 0.006 f	0.5 f	76.8°f	45.9 d, 46.0 f
	C	—	4.3 d, 4.0 f	0.33 d, 0.017 f	0.06 f	80.7°f	
	D	—		0.021	3.7	65°	
Polysorbate 80 (polyoxyethylene 20 sorbitan mono-oleate)	A	+	6.1	0.022	0.17		
	B	+	6.5	0.034			
	D	+	5.5				
Cetomacrogol	F	+	7.2	0.025	0	90.2°	38.0
1000 BPC	G	+	7.4	0.0	0	89.2°	89.2

* A, B and C. I. C. I. Atlas, Wilmington, Del.; D. Sigma, St. Louis, Mo.; E, NBC, Cleveland, Ohio; and F and G, Glovers, Liverpool, England. Sample b is about 1 month older than Sample a. † All surfactants stated by the manufacturers to be soluble at this concentration (+, soluble; and −, turbid). ‡ In 3% solution containing 1 N NaCl. § d = dispersed, and f = filtered.
From [149].

(b) Concentration of sulphydryl oxidizing agents in various 1 % non-ionic detergent solutions*

1% detergent solutions	Concentration of oxidizing agents	
	Fresh solution (μM)	Solution after 8 weeks[†] (μM)
Brij 35[‡]	140	—
Brij 35[§]	26	$\geqslant$ 152
Brij 56[¶]	< 1	< 1
Brij 96	70	$\geqslant$ 152
Emulphogene BC-720	8	$\geqslant$ 152
Lubrol PX	139	$\geqslant$ 152
Lubrol PX (purified)[‖]	< 1	—
Lubrol WX[¶]	< 1	16
Renex 30	24	$\geqslant$ 152
Triton X-100	< 1	136
Triton X-100[**]	< 1	35
Tween 20	< 1	$\geqslant$ 152
Tween 80	3	$\geqslant$ 152

* The lower and upper limits of sensitivity of these assays are 1 μM to 152 μM.
[†] 1 % detergent solution in buffer was prepared, just before assay, from 10% aqueous stock detergent solutions kept at room temperature for 8 weeks in closed bottles.
[‡] From Fisher Scientific Co.
[§] From Sigma Chemical Co.
[¶] These products contain the antioxidant t-butylated hydroxytoluene (0.01 %) and 0.005 % citric acid.
[‖] Purified as described in [154].
[**] Prepared similarly to solution described in footnote [¶] except that the 10% stock detergent solution contained 1 mM EDTA and was stored about 1 year at room temperature.
From [154].

of the product. When such contaminated samples are used in biochemical work, oxidation of sulphydryl groups might occur on prolonged contact, with consequent intramolecular or intermolecular disulphide bond fomation. Some of Chang and Bock's findings are given in Table 11.13b.

The stability of the non-ionic surfactant, sucrose monolaurate, below and above its CMC has been studied. Below the CMC first-order kinetics are observed for its hydrolysis to lauric acid, but this is not so above the CMC. In buffered systems at a pH when the lauric acid is ionized, the product of hydrolysis appears to form mixed micelles with the sucrose monolaurate, producing negatively charged micelles which appear to protect the ester from attack and thus reduce rates of hydrolysis [155]. Sodium dodecyl sulphate is more stable to hydrolysis in alkaline solutions above its CMC than in acid solutions where the rate of hydrolysis is proportional not only to the hydrogen ion concentration, but also to the concentration of the detergent itself [156]. This agrees with the observation of Kurz [157] that the proton-catalysed rates of hydrolysis of monoalkyl sulphates were accelerated and the hydroxyl ion-catalysed rates of reaction were retarded by micelle formation. The kinetics at $60°$ C of the acid-catalysed hydrolysis of NaLS in the presence of hexadecanol has been found to be first order above the CMC and to depend on the ratio of alcohol to alkyl sulphate [158]. Up to a molar ratio of 0.75 the rate constant increases; attempts to correlate the maximum in the rate constant with the phase changes that occur in this system were not successful. It was, however, qualitatively explained by discussion of packing of alcohol molecules and surfactant head groups in micelles and liquid crystalline phases. Addition of dodecanol and Pluronic F-38 to solutions of NaLS above its CMC causes an increase in its hydrogen ion-catalysed hydrolysis [159]. This was explained by the reduction in charge density of the sulphate groups at the mixed micellar surface which resulted in the sulphate groups becoming weaker bases. The non-ionic molecule thus lowers the effective dielectric constant of the surface which should strengthen electrostatic inter-actions and increase rates of hydrolysis. However, as Barry and Shotton point out [158], it seems more likely that an alternative explanation is required, as it has been shown that only about 20% of the head groups on a NaLS micelle are ionized. Separation of the head groups by dilution with non-ionic molecules would allow greater ionization, and thus greater attraction of the attacking hydrogen ions.

11.5.1 Biodegradation of surfactants

Studies on the stability of surfactants in aqueous solution have some relevance in the understanding of the biodegradability of surfactants in the environment, although this is often complicated by the additional action of enzymes. Degradation of Tween and Span non-ionic surfactants has been discussed by Delgado *et al.* [160]. The modes of biological breakdown are shown in Scheme 11.7 below. Studies on the degradation of a non-ionic polyoxyethylene derivative, Dobanol 45–7 [161], have shown that in activated sludge the surfactant molecules initially lose the alkyl moiety leaving polyoxyethylene glycols which are subsequently degraded via acidic intermediates; mechanisms compatible with the work of others [162, 163]. Tobin *et al.* [164] have shown that the ethylene oxide moiety of Dobanol 25–9 (a commercial $C_{12}/C_{15} E_9$ non-ionic surfactant) is slow to degrade in the laboratory cultures studied, the species remaining after the initial rapid breakdown being unlikely to be degraded by micro-organisms likely to be encountered in the environment. It has been suggested that ozone could be used for the conversion of refractory organic matter to biodegradable com-pounds; ozonization of a branched chain non-ionic surfactant (a nonyl phenyl ethoxylate) significantly increased biodegradation, by initial conversion of the molecule to more readily degradable species [165]. High molecular weight polyoxyethylene glycols have to be degraded to compounds with molecular weights of about 300 before they can be utilized by bacteria; the oxidized alkyl phenol moiety with hydroxyl and carboxylic acid groups will be more bio-degradable. Gamma irradiation of water has been attempted as a means of

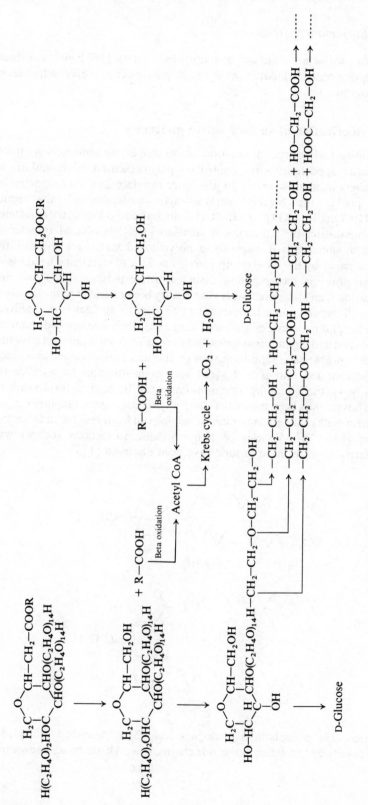

Scheme 11.7

From [160].

reducing the surfactant load of contaminated waters [166] and has been successfully demonstrated to reduce the concentration of linear alkyl sulphates in aqueous solution.

11.6 Polymerization of surface-active molecules

The proximity and regular orientation of surface-active molecules at interfaces presents opportunities for controlled polymerization. Cross-linking of diacrylic esters in monolayers at an air–water interface has been reported by Dubault [167]; the weakly surface-active molecules of the series $CH_2 = CH.CO.(CH_2)_n.O.OC.CH = CH_2$ can stabilize oil-in-water emulsions after polymerization by ultraviolet irradiation [168]. Interfacial transfer of paramagnetic spin labels through these polymerized surfactant layers is the subject of a paper by the same group [169]. The idea of stabilizing unstable or labile dispersions by incorporating a surface-active monomer and inducing polymerization is an attractive one. It has recently been extended to the liposome field and will assume some importance in the search for new drug delivery systems. Phospholipid vesicles are now being extensively investigated as model membranes and as drug carrier systems, but they suffer from fragility of structure which causes problems in long-term storage and use. Surface-active monomers can be incorporated into phospholipid vesicles in attempts to stabilize the structures or synthetic amphipathic monomers can be used to form synthetic vesicles. It has recently been shown that simple synthetic organic amphiphiles, e.g. of long-chain dialkyldimethylammonium halides, can form vesicles under certain conditions [170–174]. Vesicles of two synthetic surfactants (below) with 'micropolarity' reporter head groups have been observed [175].

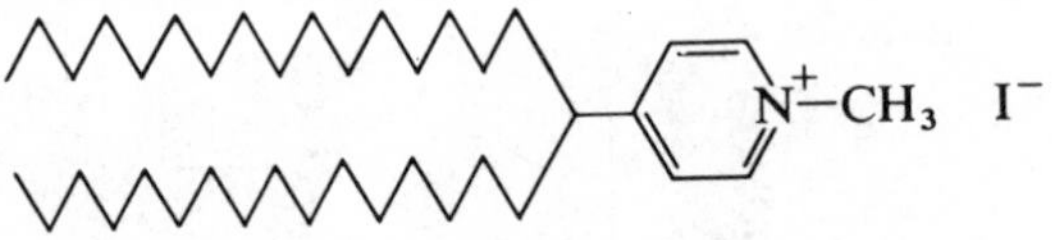

(XIII)

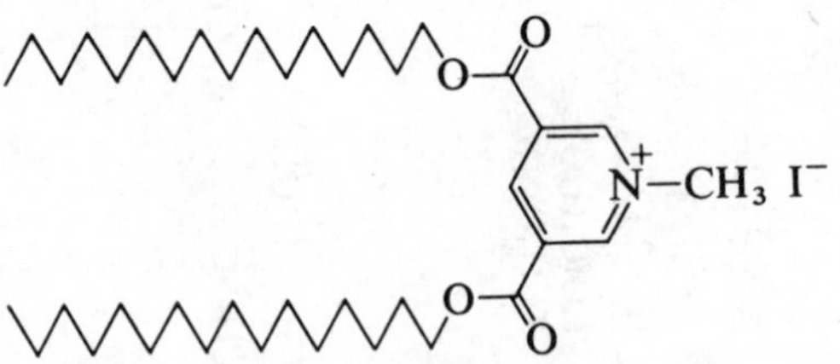

(XIV)

Polymerizable phospholipid analogues have been described [176] which contain diacetylene moieties in the side chains; diacetylenes being known to be polymerizable by ultraviolet irradiation in the solid state and in oriented monolayers and multilayers:

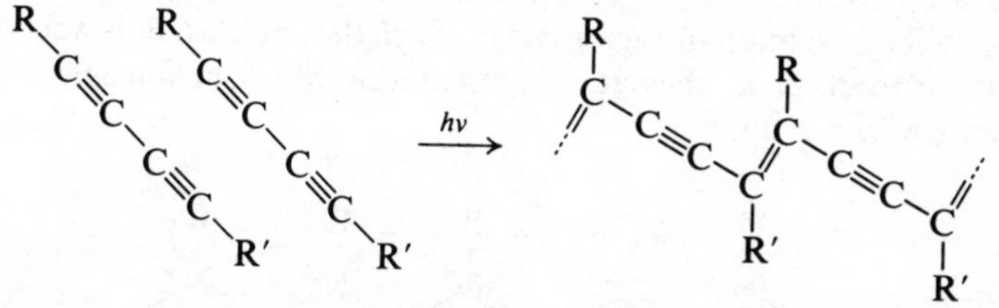

As the resulting polymers have a red colour due to the conjugated enyne structure the reaction can readily be followed. The shape of the monomer vesicles formed from these compounds by the action of ultrasound is retained during the polymerization (see Scheme 11.8); their dispersions are stable, dilution with 50 % ethanol, for example, does not lead to precipitation.

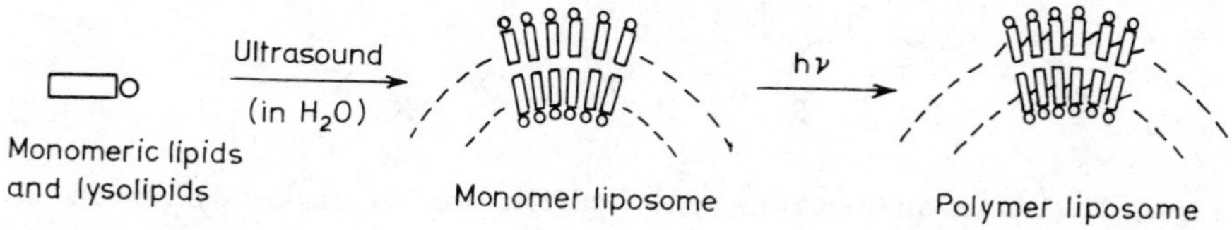

Scheme 11.8 Formation of polymer liposomes from polymerizable lipid analogues.

From [176].

The retention of the original shape and structure of the vesicles indicates that a topochemical reaction has occurred. These conclusions are supported by the results obtained on other synthetic vesicles [177]. Aggregation of micelle-forming polymerizable monomers tends not to result in 'polymerized micelles' as, on induction of cross-linking, the surface-active properties of the monomers change, and because of the lability of micellar systems the micelle tends to grow. The formed units are thus many times larger than the original micelle, although Elias originally intended to polymerize surfactant micelles to give an indication of their size and shape in the original solution [178]. Elias and co-workers [179] synthesized the monomer

$$CH_2 = C(CH_3)\text{—}CO.N(R)(CH_2)_{10}CO.X$$

where $R = CH_3, C_2H_5, n\text{-}C_3H_7, i\text{-}C_3H_7$ and $X = ONa$ or $(OCH_2CH_2)_5OH)$. The monomers used polymerized to measureable conversion only at high initiator concentrations and long reaction times and the products are most likely coagulates of polymers. Florence *et al.* [180] have obtained similar results with the acryloyl derivative of Triton X-100 and related non-ionic surfactants. Several examples of polymerization of micelle-forming monomers are discussed by Martin *et al.* [181]. These authors draw attention to the work of Kabanov

[182,183] who found an increase in the spontaneous polymerization of N-methylated 2-, 3- and 4-vinylpyridinium salts in water above a monomer concentration of 3 mol l^{-1}. This increase in polymerization rate has been ascribed to the formation of aggregates in which the double bonds were optimally oriented for reaction, a schematic representation of the arrangement proposed being shown in Fig. 11.26.

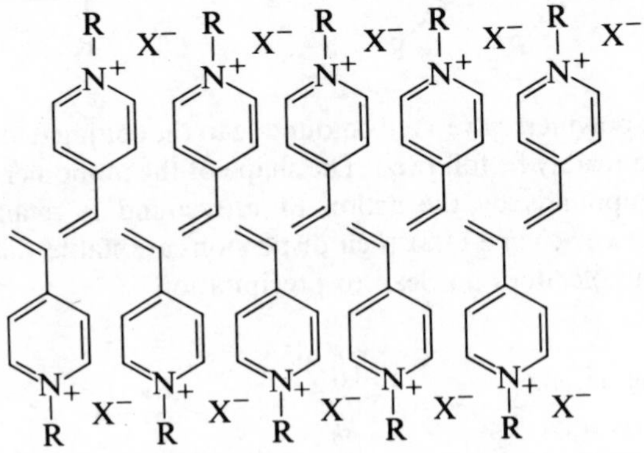

Figure 11.26 Schematic representation of micellar oriented vinylpyridinium salts in water.

Similar results have been obtained by Mielke and Ringsdorf [184] when they were studying the spontaneous polymerization of 4-vinylpyridinium perchlorate. Below the CMC a low molecular weight ionene was formed, but above the CMC larger polymers were formed (see Scheme 11.9). Ideally, cross-linked polymerized micellar systems while theoretically capable of solubilizing poorly soluble drugs, would not be subject (as uncross-linked micelles are) to disruption by dilution or addition of solvents which do not favour micelle formation. While attempts at cross-linking micelles of acryloyl derivatives of non-ionic surfactants or poly-

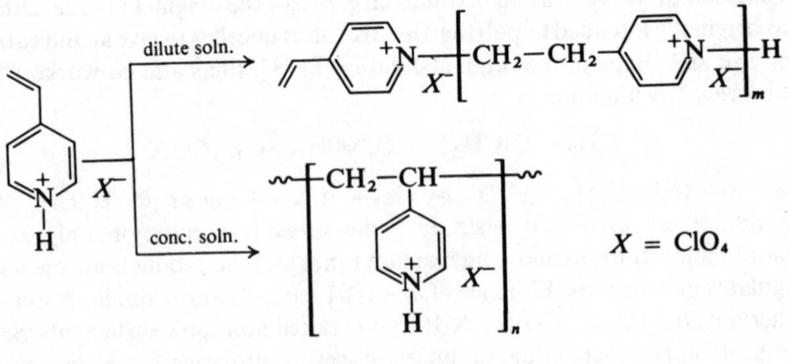

Scheme 11.9

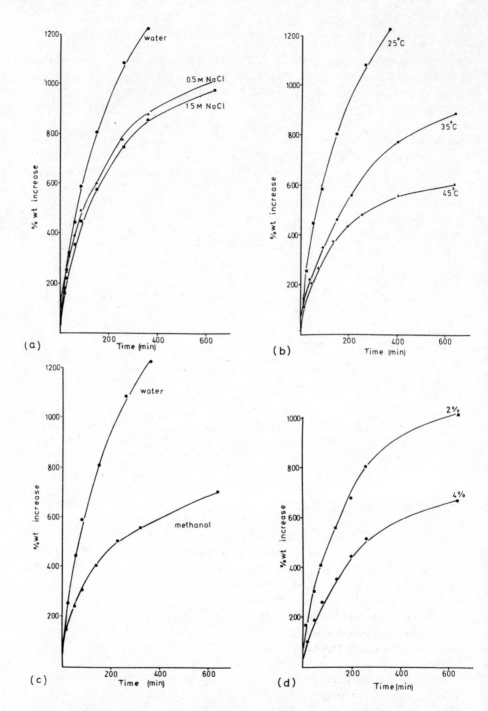

Figure 11.28 Swelling capacity of the gels expressed as % weight increase vs time under different conditions. (a) gel of 2% Pluronic F68 in different salt solutions, (b) gel of 2% Pluronic F68 at different temperatures, (c) gel of 2% Pluronic F68 in different solvents, (d) gel of 2% and 4% Pluronic F88. From Al-Saden *et al.* [185] with permission.

acetanilide was degraded by 6 and 12 h irradiation, respectively, whilst in the poloxamer solution (1 % L64), 5 % and 20 % of the drug was degraded after the same exposure time. This indicates the protection afforded by micellar solubilization.

11.6.1 Nano-encapsulation

When polymerizable monomers are solubilized in surfactant micelles it should be possible to produce polymerized micelles containing co-solubilized drug. The process has been successfully achieved by Speiser [186]. Fig. 11.29 represents an inverse micelle containing a water-soluble drug in which the monomeric species is located in the outer part of the micelle and is cross-linked to form a 'nanocapsule'

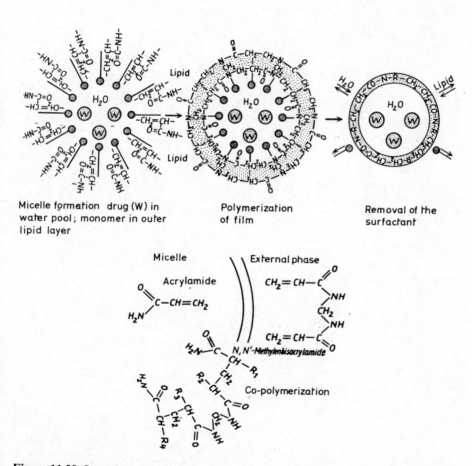

Micelle formation drug (W) in water pool; monomer in outer lipid layer

Polymerization of film

Removal of the surfactant

Figure 11.29 Secondary solubilization and polymerization of acrylamide in schematic form. From Speiser [186].

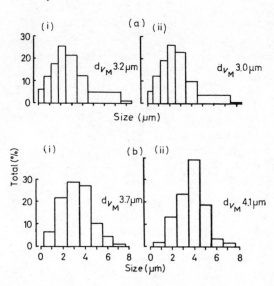

Figure 11.31 Size distributions of one sample of microcapsules (a) determined by light microscopy; (i) before and (ii) after *in vitro* release experiments (321 and 308 microcapsules, respectively were sized). (b) determined by measurement of the SEM photographs. Same sample as in (a) (i) before and (ii) after *in vitro* release experiments, for which were sized 187 and 114 microcapsules, respectively. dv_m = mean volume diameter. From Jenkins and Florence [198].

surgery. The methyl ester in solid form degrades most rapidly yielding formaldehyde which is believed to be responsible for its histotoxicity. The butyl ester, with a slower degradation rate, is well-tolerated *in vivo* [200].

Florence *et al.* [201] showed that methyl 2-cyanoacrylate and butyl 2-cyanoacrylate dissolved in an oil phase-formed polymeric film in contact with an aqueous phase and suggested that these monomers could be used in the preparation of microcapsules which would degrade *in vivo*. Florence *et al.* [202] describe the preparation of microcapsules formed from butyl 2-cyanoacrylate monomer. The reaction involves a base-catalysed anionic mechanism with the monomer dissolved in the oil phase of a water-in-oil emulsion and *in situ* polymerization occurring at the interface of the aqueous disperse phase. Aqueous protein solutions have been used as the microcapsule core materials; the protein is believed to act as an initiator and therefore to become a cross-linking agent on the inner wall surface. When aqueous solutions of ^{125}I-labelled albumin are encapsulated, 30 to 40% of the albumin becomes incorporated into the microcapsule membrane [202]. This method of micro-encapsulation does not require the presence of reactive monomers in the core material, a disadvantage of micro-encapsulation using interfacial polycondensation.

Investigation of micelles and solubilized systems as sites for polymerization should lead us to a consideration of the potential of microemulsions as reaction media, a topic explored by Mackay *et al.* [203] and Barden and Holt [204]. It has been claimed [205] that microemulsions provide the possibility of studying the effect of reactant separation more readily than micellar systems as they can accommodate larger molecules without distortion; when large molecules are solubilized by normal micellar systems, gross distortion of the original micelle often occurs, as we have seen. Microemulsions, micelles and vesicles have been compared as media for photochemistry [206]. Relatively few chemical reactions have been studied in microemulsions, and the kinetics of reactions has not yet been treated quantitatively as it has for micellar systems. The important factor appears to be the concentration of reactants at appropriate surfaces. Reactivities in microemulsions have sometimes been observed to differ from those in similar micellar systems. The rate of incorporation of cupric ions into protoporphyrin dimethyl ester is about 20 000 times faster in aqueous NaLS than in CTAB, but in microemulsions, cupric ion incorporation into tetraphenylporphyrin is faster in cationic-stabilized microemulsions than in anionic or non-ionic systems [207]. On the other hand, chlorophyll photochemistry is similar in microemulsions, micelles and organic media [208].

11.7 Some analytical consequences of surfactant presence

The most obvious consequence of solubilization on the properties of a chromophore will be shifts in ultraviolet and visible absorption spectra, which if ignored can give rise to erroneous results. However, there are other effects which must not be ignored. We will deal with some of these in this section.

Many examples are available of the effect of solubilizers on the spectroscopic properties of chromophoric substances. A clear example is the effect of the non-ionic surfactant $C_{18}E_{20}$ on the absorption spectrum of Bromophenol Blue [209]; Fig. 11.32 indicates the extent of change of absorbance at 591 and 603 nm.

The interaction of dyes with surfactants has received wide attention in view of the importance of this phenomenon in the textile industry. There has been some suggestion that dyes interact with some non-ionic monomers rather than the micelles; Craven and Datyner [210] have observed that complexation between dye and surfactant increases with the increasing hydrophobicity of the dye, is inhibited by branching of the alkyl chain of the surfactant [211], but is increased by a lengthening of the ethylene oxide chain of the surfactant. There is an apparent contradiction between the idea of hydrophobic interaction between the species and increases in complexation with increasing hydrophilicity of the surfactant [212] but it is argued that this is removed if dye interacts with the monomer rather than with the micellar species. Spectroscopic measurements and gel chromatography have convinced others that anthraquinoid dyes are adsorbed on to the surface of non-ionic micelles [213]. The dye (XV) shown below is certainly a large molecule and might not be expected to be located in the centre of spherical micelles.

In NaLS solutions above the CMC, one molecule of this dye appears to be associated with one molecule of surfactant [214]. Interaction between other acid dyes, e.g. CI Acid Blue 120 (XVI) has been observed above and below the CMC of non-ionic nonylphenyl and octylphenyl ethers. At low concentrations the dye:surfactant ratio is around 1:2 to 1:3 and at higher concentrations larger complexes may be formed with dye:surfactant ratios of between 1:10 and 1:30 [215]. Similar conclusions have been reached on the interaction of three non-ionic dyes with NaLS [216]. Detailed kinetic studies of dye–surfactant interactions have been published [217] in which the process of surface adsorption and subsequent incorporation of the dye in the interior of the micelle has been investigated, the latter process being referred to as absorption. Using anionic micelles, James *et al.* [217] found that neutral dyes were absorbed more rapidly than positively charged dyes; increased counterion binding, however, increases the rate of absorption of the latter.

Some of the observed spectral changes which occur on solubilization will be due to change in the dissociation of the chromophoric species. The pK_a of three dyes in water and 0.025 M NaLS is compared in Table 11.14. An increase in pK_a is observed in the micellar state. The ionization of *p*-nitrophenol in the same surfactant calculated from the absorbance of the phenolate ion at 400 nm has been assessed. Fig. 11.33 shows the fractional ionization, α, as a function of pH in a range of surfactant concentrations. Herries *et al.* [218] calculate theoretical ΔpK_a as follows. Let C_{NPH} be the concentration of unionized *p*-nitrophenol in

Table 11.14 pK_a of dyes in aqueous solution and when absorbed into NaLS micelles [217]

Dye	pK_a (water)	pK' (0.025 M NaLS)
1. AO⁺	10.4	12.4
2. NB⁺	9.9*	12.2*
3. AB²⁺	7.95	10.6

* [Na⁺] maintained at 0.075 M by the addition of NaCl.

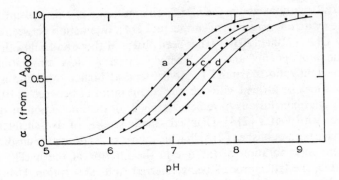

Figure 11.33 Ionization of *p*-nitrophenol in the presence of sodium dodecyl sulphate. The fractional ionization, α, was calculated from the absorbance of the *p*-nitrophenolate ion at 400 nm. The total *p*-nitrophenol concentration was $1 \times 10^{-4}\,\text{mol}\,l^{-1}$. The solvent was 0.004 M sodium phosphate buffer, or 0.004 M glycylglycine buffer for the higher pH values, adjusted to the pH indicated and to a constant ionic strength of 0.1 M with NaCl. The solid lines are standard titration curves for the dissociation of a monobasic acid. The NaLS concentrations in g ml^{-1} were: a, 0; b, 0.0144; c, 0.0288; d, 0.0576. From Herries *et al.* [218].

aqueous phase; C_{NP^-} the concentration of ionized *p*-nitrophenol in aqueous phase; C'_{NPH} and C'_{NP} the equivalent concentrations in micellar phase; P_{NPH} and P_{NP^-} the partition coefficients for the indicated forms; α the fraction of total *p*-nitrophenol in ionized form. Then

$$\frac{\alpha}{1-\alpha} = \frac{\beta C'_{\text{NP}^-} + (1-\beta)C_{\text{NP}^-}}{\beta C'_{\text{NPH}} + (1-\beta)C_{\text{NPH}}} = \frac{(\beta P_{\text{NP}^-} + 1 - \beta)C_{\text{NP}^-}}{(\beta P_{\text{NPH}} + 1 - \beta)C_{\text{NPH}}}. \tag{11.46}$$

Now,

$$\text{pH} = \text{p}K'_{\text{app}} + \log\frac{\alpha}{1-\alpha} \tag{11.47}$$

Therefore,

$$\text{pH} = \text{p}K' + \log\frac{C_{\text{NP}^-}}{C_{\text{NPH}}}, \tag{11.48}$$

where pK' refers to the ionization of *p*-nitrophenol in the absence of micelles.

$$\text{p}K'_{\text{app}} - \text{p}K' = \Delta\text{p}K' = \log\frac{(\beta P_{\text{NPH}} + 1 - \beta)}{(\beta P_{\text{NP}^-} + 1 - \beta)} \tag{11.49}$$

P_{NP^-} is assumed to be zero. Substituting for β and rearranging one obtains

$$\Delta\text{p}K' = \log\frac{1 + \bar{v}(P_{\text{NPH}} - 1)C_{\text{m}}}{1 - \bar{v}C_{\text{m}}}, \tag{11.50}$$

$\bar{v}(P_{\text{NPH}} - 1) = 94.3$ is obtained directly from experiment.

Table 11.15 The influence of sodium dodecyl sulphate on the ionization of *p*-nitrophenol*

NaDS concn. (g ml^{-1})	$\Delta pK'$ Calcd.*	Obsd.
0.0144	0.38	0.32 ± 0.04
0.0288	0.58	0.52 ± 0.04
0.0576	0.83	0.80 ± 0.04

* The total ionic strength was 0.1 M and the temperature 25° C.

If $\bar{v}$ is assumed to be 0.9 ml g^{-1}, $\Delta pK'$ can be calculated for various values of C_m. The results are shown in Table 11.15.

The nature of the interactions between dye and surfactant is perhaps of little practical relevance; the immediate problem faced by the analyst is the change in the absorption spectrum which results from these interactions. Other problems arise in the spectroscopic determination of solutes. The effect of surfactants on the determination of phosphate by a spectrophotometric determination – the heteropoly blue method using either ascorbic acid or stannous chloride as a reducing agent – has been described. Cationic surfactants interfered, while biodegradable linear alkyl sulphate (LAS) surfactants did not [219]. Pakalns and Farrar have conducted a series of experiments on the effect of surfactants on the determination of soluble iron [220], sulphate [221] and aluminium [222] in water. Recommended methods for the determination of iron are the phenanthroline and tripyridine methods. Using the former method and a standard solution of iron (20 μg Fe in a 25 ml sample aliquot) the recoveries depicted in Fig. 11.34 were obtained, showing the gross error that can be introduced. The tripyridine and another method involving biquinoline, were not affected until much higher concentrations of surfactant were achieved. A turbidimetric method can be applied to the determination of sulphate [221]. Cationic and non-ionic surfactants were found not to interfere, but anionic detergents had to be kept below 0.5 mg l^{-1}. Mechanisms of the effect have not been discussed. In the case of aluminium assay by complexation with chromophores there is clear evidence of marked changes in absorbance of the complex at 370 nm [222].

It has been suggested that micellar catalysis could be exploited in analytical chemistry to increase the rate of derivative formation prior to spectroscopic measurement of the product [223, 224]. This has been attempted in the assay of amino acids and peptides following reaction with 1-fluoro-2,4-dinitrobenzene (this undergoes aromatic nucleophilic substitution by amines to give arylated amines) [225]. For this reaction some amines require up to 20 min. In the presence of cetrimonium bromide, catalysis was achieved, although absorbances some 10% higher were obtained in the presence of surfactant.

Polarographic determinations can be affected by low levels of surfactant, because of adsorption of surfactant at the dropping mercury electrode surface.

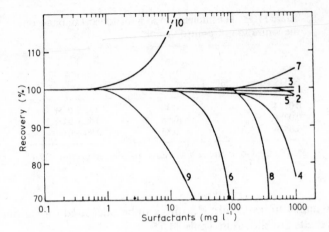

Figure 11.34 The effect of surfactants on the determination of iron. Phenanthroline method; 20 μg Fe in 25 ml. Curve (1) cationic detergent (100 % active material); (2) anionic detergent (100 % active material); (3) non-ionic detergent (100 % active material); (4) industrial LAS-type detergent (74 % LAS and 6 % sodium tripolyphosphate); (5) formulated detergent (13 % active material); (6) washing powder (15 % LAS and 25 % sodium tripolyphosphate); (7) mixed detergent (7 % LAS and 3 % non-ionic detergent); (8) sodium pyrophosphate (anhydrous); (9) sodium tripolyphosphate; (10) soap. Broken line indicates turbidity. From Pakalns and Farrar [220].

The effect of Triton X-100 on the current–voltage curve of the iron (III)-diethylenetriaminepentaacetic acid complex has been explained as follows [226]: at low levels of surfactant the mercury is only partly covered by surfactant and the reversible reduction proceeds at the uncovered portions of the undisturbed surface and at the covered areas a penetration-controlled reaction occurs. As the Triton concentration is increased so the penetration-controlled reaction contribution increases. When the coverage is complete only the second wave is observed on the polarogram. The shift in the half-wave potential with increasing surfactant concentration is probably due to increased adsorbed layer thickness and is shown along with the polarograms and electrocapillary capillary curves in the absence and presence of Triton X-100 in Fig. 11.35.

The effect of adsorption of surfactants at the mercury electrode on electrocapillary curves, drop times, maximum suppression have been discussed by Malik *et al.* [227] and by Barradas and Kimmerle [228], the former attempting to measure the CMCs of non-ionic surfactants by changes in their effects on electrocapillary curves. By tagging micelles with a cation which is reducible at the dropping mercury electrode, Novodoff *et al.* [229] have developed a method of measuring the diffusion coefficient of the surfactant system using the Ilkovic equation,

$$i_d = 607\, nD^{\frac{1}{2}} C m^{2/3} t^{1/6}, \tag{11.51}$$

i_d being the average diffusion current related to the concentration c of the

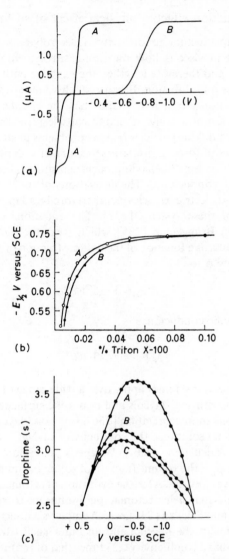

Figure 11.35 Polarograms of mixtures of 5×10^{-4} M iron(II) and 5×10^{-4} M iron(III) in 10^{-2} DTPA at pH 4.6. Concentration of Triton X-100: 0 % (curve A) and 0.1 % (curve B). (b) Variation of the half-wave potential of the second wave with the concentration of Triton X-100. Supporting electrolytes: ammonia buffer pH 9.6 (curve A) and acetate buffer pH 4.6 (curve B). (c) Electrocapillary curves of 10^{-3} M iron(III) and 10^{-2} M DTPA in ammonia buffer pH 9.4. Concentration of Triton X-100: 0 % (curve A), 0.01 % (curve B) and 0.1 % (curve C). From Kalland and Jacobsen [226].

reducible species, m being the mass of drop and t the drop time, n the valency of the reduced species. A review on the use of surfactant and micellar systems in analytical chemistry by Hinze with over 300 references is to be found in [230].

11.7.1 The obstruction effect: a physical effect of surfactant micelles

Non-electrolytes can affect the conductivity of electrolyte solutions. Wang [231] has suggested that in viscous flow the non-electrolyte molecules distort the streamlines of flow and therefore lengthen the effective path lengths of moving ions in conductance and diffusion. Because of this hindrance to flow, which is termed the obstruction effect, it is found that the specific conductivity of a suspension or solution of a non-conducting particle (κ) is less than that of the medium which surround the particles (κ_0). As problems of steady current flow in conductors and lines of force in insulators are formally identical, the mathematical treatment of the dielectric behaviour of suspensions and solutions is identical with that of specific conductance. The derivations of the dielectric constant of suspensions of low-dielectric particles can therefore be adapted for discussion of the conductivity of these systems [232]. The equations of Rayleigh [233], Böttcher [234] and Bruggeman [235] which treat the dielectric constant of inhomogeneous media as a function of the volume fraction of the dispersed phase (ϕ) all reduce at low ϕ to

$$\varepsilon/\varepsilon_0 = 1 - 1.5\phi \tag{11.52}$$

or, in terms of specific conductance

$$\kappa/\kappa_0 = 1 - 1.5\phi. \tag{11.53}$$

These limiting equations do not apply over a wide range of ϕ values; hydrated units such as polyoxyethylene glycols and non-ionic surfactant micelles present further complications to interpretation [236]. For poly(oxyethylene glycol) and poly(vinylpyrrolidone) solutions, the conductivity of NaCl and KCl at infinite dilution can be calculated using Wagner's equation: $\kappa/\kappa_0 = (1 - \phi_h)/[1 + (\phi_h/2)]$, where ϕ_h is the volume fraction of solute including hydrating water, but this equation gives low values for the hydration of cetomacrogol micelles. The movement of the salts in micellar cetomacrogol solutions is explicable in terms of the model of Pauly and Schwan [237], which employs a nonconducting core (the hydrocarbon interior of the micelle) and an outer shell (the poly(oxyethylene glycol) layer) which has a conductivity 0.23 times that of the medium. The original derivation was obtained by Pauly and Schwan in terms of dielectric constants, as shown in Fig. 11.36. Their equation can be written

$$\frac{\varepsilon_m - \varepsilon}{2\varepsilon_m + \varepsilon} = \phi_h \frac{(\varepsilon_m - \varepsilon_s)(2\varepsilon_s + \varepsilon_p) + (\varepsilon_m + 2\varepsilon_s)(\varepsilon_s - \varepsilon_p)f}{(2\varepsilon_m + \varepsilon_s)(2\varepsilon_s + \varepsilon_p) + 2(\varepsilon_m - \varepsilon_s)(\varepsilon_s - \varepsilon_p)f} \tag{11.54}$$

In the analogous form for conductivities, this becomes

$$\frac{\kappa_0 - \kappa}{2\kappa_0 + \kappa} = \phi_h \frac{(\kappa_0 - \kappa_s)(2\kappa_s + \kappa_p) + (\kappa_0 + 2\kappa_s)(\kappa_s - \kappa_p)f}{(2\kappa_0 + \kappa_s)(2\kappa_s + \kappa_p) + 2(\kappa_0 - \kappa_s)(\kappa_s - \kappa_p)f}, \tag{11.55}$$

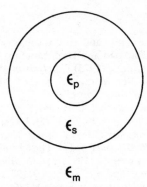

Figure 11.36 Model of Pauly and Schwan for particle with shell. ε is the dielectric constant of the suspension, ε_m is the dielectric constant of the medium, ε_p is the dielectric constant of the core, and ε_s is the dielectric constant of the shell. From Pauly and Schwan [237].

where f is the fraction of inner-phase volume to the total volume of the dispersed particle. For cetomacrogol, f can be calculated from published data on core and micellar volumes to be 0.08. κ_p can be considered to be negligible. Hence, the equation becomes

$$\frac{1 - \dfrac{\kappa}{\kappa_0}}{2 + \dfrac{\kappa}{\kappa_0}} = \frac{\phi_h}{2} \frac{(2+f) - \dfrac{\kappa_s}{\kappa_0}(2-f)}{2(2+f) - \dfrac{\kappa_s}{\kappa_0}(f-1)}. \tag{11.56}$$

Table 11.16 compares some experimental and calculated data for the ratio κ/κ_0 for cetomacrogol systems containing NaCl or KCl, indicating that the model chosen gives reasonable, if not unique, results.

Table 11.16 Experimental and calculated* conductivity ratios at infinite dilution in cetomacrogol

Salt	[Cetomacrogol] (%)	k/k_0(exptl)	k/k_0(calcd)	Deviation, ±%
NaCl	1.0	0.980	0.972	−0.82
	2.5	0.937	0.930	−0.75
	5.0	0.870	0.864	−0.69
	7.5	0.801	0.800	−0.12
KCl	1.0	0.982	0.972	−1.02
	2.5	0.940	0.930	−1.06
	5.0	0.869	0.864	−0.57
	7.5	0.806	0.800	−0.74

* Pauly and Schwan equation, ω (the hydration) = 1.9 g g^{-1}, $k_s/k_0 = 0.23$.
From [236].

References

1. P. H. ELWORTHY, A. T. FLORENCE and C. B. MACFARLANE (1968) *Solubilization by Surface Active Agents*, Chapman and Hall, London.
2. J. A. FENDLER and E. J. FENDLER (1975) *Catalysis in Micellar and Macromolecular Systems*, Academic Press, New York.
3. F. M. MENGER (1972) *Chem. Soc. Rev.* **1**, 229.
4. F. M. MENGER (1979) *Pure and Appl. Chem.* **51**, 999.
5. F. M. MENGER, J. U. RHEE and H. K. RHEE (1975) *J. Org. Chem.* **40**, 3803.
6. F. M. MENGER and C. E. PORTNOY (1967) *J. Amer. Chem. Soc.* **89**, 4698.
7. J. M. BROWN (1979) in *Colloid Science* Vol. 3, The Chemical Society, London, p. 253.
8. F. M. MENGER and J. M. JERKUNICA (1977) *Tetrahedron Lett.* 4569.
9. C. N. SUKENIK, B. A. WEISSMAN and R. G. BERGMAN (1975) *J. Amer. Chem. Soc.* **97**, 445.
10. H. IHARA, S. ONO, H. SHOSENJI and K. YAMADA (1980) *J. Org. Chem.* **45**, 1623.
11. K. YAMADA, H. SHOSENJI and H. IHARA (1979) *Chem. Lett.* 491.
12. C. A. BUNTON (1977) *Pure and Appl. Chem.* **49**, 969.
13. G. S. HARTLEY (1934) *Trans. Faraday Soc.* **30**, 444.
14. E. DUYNSTEE and E. GRUNWALD (1959) *J. Amer. Chem. Soc.* **81**, 4540.
15. K. G. VAN SENDEN and C. KONINGSBERGER (1966) *Tetrahedron* **22**, 1301.
16. J. BAUMRUCKER, M. CALZADILLA, M. CENTENO et al. (1972) *J. Amer. Chem. Soc.* **44**, 8164.
17. M. POINDEXTER and B. MCKAY (1972) *J. Org. Chem.* **37**, 1674.
18. R. A. MOSS and C. RAV-ACHA (1980) *J. Amer. Chem. Soc.* **102**, 5045.
19. T. C. BRUICE, J. KATZHENDLER and L. R. FEDOR (1968) *J. Amer. Chem. Soc.* **90**, 1333.
20. H. NOGAMI, J. HASEGAWA and M. IWATSURU (1970) *Chem. Pharm. Bull.* **18**, 2297.
21. H. NOGAMI, S. AWAZU, K. WATANABE and K. SATO (1960) *Chem. Pharm. Bull.* **8**, 1136.
22. H. NOGAMI, S. AWAZU and M. IWATSURU (1963) *Chem. Pharm. Bull.* **11**, 1251.
23. E. TOMLINSON, S. S. DAVIS and J. E. BROWN (1979) in *Solution Chemistry of Surfactants*, Vol. 1 (ed. K. L. Mittal), Plenum, New York, pp. 867 *et seq.*
24. Y. MOROI (1980) *J. Phys. Chem.* **84**, 2186.
25. F. H. QUINA and H. CHAIMOVICH (1979) *J. Phys. Chem.* **83**, 1844.
26. S. RIEGELMAN (1960) *J. Amer. Pharm. Ass (Sci. Ed.)* **49**, 339.
27. A. G. MITCHELL (1962) *J. Pharm. Pharmacol.* **14**, 172.
27a. A. G. MITCHELL (1964) *J. Pharm. Pharmacol.* **16**, 43.
28. B. J. MEAKIN, K. K. WINTERBORN and D. J. G. DAVIES (1971) *J. Pharm. Pharmacol.* **23**, Suppl. 25S.
29. N. FUNASAKI (1979) *J. Phys. Chem.* **83**, 237.
29a. K. MARTINEK, A. K. YATSIMIRSKI, A. V. LEVASHOV and I. V. BEREZIN (1977) in *Micellization, Solubilization and Microemulsions* (Ed. K. L. Mittal) Vol. 2, Plenum Press, New York, pp. 489 *et seq.*
30. N. FUNASAKI (1978) *J. Colloid Interfac. Sci.* **64**, 461.
31. C. J. O'CONNOR, A-L TAN (1980) *Aust. J. Chem.* **33**, 747.
32. V. GANI and C. LAPINTE (1973) *Tetrahedron Lett.* 2775.
33. V. GANI, C. LAPINTE and P. VIOUT (1973) *Tetrahedron Lett.* 4435.
34. V. GANI and P. VIOUT (1978) *Tetrahedron* **34**, 1337.
35. T. J. BROXTON, L. W. DEADY and N. W. DUDDY (1978) *Aust. J. Chem.* **31**, 1525.
36. T. J. BROXTON and N. W. DUDDY (1979) *Aust. J. Chem.* **32**, 1717.
37. C. A. BUNTON, E. J. FENDLER, L. SEPULVEDA and K. U. YANG (1968) *J. Amer. Chem. Soc.*, **90**, 5512.
38. B. R. CRAVEN, A. DATYNER and T. P. DOYLE (1968) *Aust. J. Chem.* **21**, 1007.
39. A. G. MITCHELL (1963) *J. Pharm. Pharmacol.* **15**, 761.
40. J. E. CARLESS and A. G. MITCHELL (1962) *J. Pharm. Pharmacol.* **14**, 46.
41. A. G. MITCHELL (1960) Ph.D. Thesis, London University.
42. A. G. MITCHELL (1964) *J. Pharm. Pharmacol.* **16**, 43.

43. J. E. CARLESS and J. SWARBRICK (1962) *J. Pharm. Pharmacol.* **14**, 97T.
44. J. E. CARLESS and A. G. MITCHELL (1962) *J. Pharm. Pharmacol.* **14**, 46.
45. J. SWARBRICK and J. E. CARLESS (1964) *J. Pharm. Pharmacol.* **16**, 670.
46. J. SWARBRICK and J. E. CARLESS (1964) *J. Pharm. Pharmacol.* **16**, 596.
47. J. SWARBRICK and J. E. CARLESS (1963) *J. Pharm. Pharmacol.* **15**, 507.
48. C. T. RHODES (1967) *Can. J. Pharm. Sci.* **3**, 16.
49. J. LESLIE and F. VARRICCHIO (1968) *Can. J. Biochem.* **46**, 625.
50. J. BONICAMP, reported by F. Menger (1979) *Accounts of Chem. Research*, **12**, 111.
51. L. K. PATTERSON and M. GRÄTZEL (1975) *J. Phys. Chem.* **79**, 956.
52. M. GRÄTZEL, A. HENGLEIN and E. JANATA (1975) *Ber Bunsen Gesell. fur Phys. Chem.* **79**, 475.
53. K. M. BANSAL, L. K. PATTERSON, E. J. FENDLER and J. H. FENDLER (1971) *Int. J. Radiat. Phys. Chem.* **3**, 321.
54. S. C. WALLACE and J. K. THOMAS (1973) *Radiat. Res.* **54**, 49.
55. J. H. FENDLER, N. V. KLASSEN and H. A. GILLIS (1974) *JCS Faraday Trans I*, **70**, 149.
56. P. P. INFELTA, M. GRÄTZEL and J. L. THOMAS (1974) *J. Phys. Chem.* **78**, 180.
57. S. A. ALKAITIS, M. GRÄTZEL and A. HENGLEIN (1975) *Ber Bunsen Gesell Phys. Chem.*, **79**, 541.
58. S. A. ALKAITIS, G. BECK and M. GRÄTZEL (1975) *J. Amer. Chem. Soc.* **97**, 5723.
59. K. KALYANASUNDARAM (1978) *Chem. Soc. Rev.* **7**, 453.
60. B. A. LINDIG and M. A. J. RODGERS (1980) *Photochem. Photobiol.* **31**, 617.
61. N. J. TURRO, M. GRÄTZEL and A. M. BRAUN (1980) *Angew. Chem. Int. Ed.*, **19**, 675.
62. J. R. ESCABI-PEREZ, F. NOME and J. H. FENDLER (1979) *J. Amer. Chem. Soc.* **99**, 7749.
63. M. D. HATLEE, J. J. KOZAK, G. ROTHENBERGER, P. P. INFELTA and M. GRATZEL (1980) *J. Phys. Chem.* **84**, 1508.
64. G. S. SINGHAL, E. RABINOWITCH, J. HEVESI and V. SRINIVASAN (1970) *Photochem. Photobiol.* **11**, 531.
65. T. MATSUO, Y. ASO and K. KANO (1980) *Ber Bunsenges. Phys. Chem.* **84**, 146.
66. R. C. DORRANCE and T. F. HUNTER (1972) *JCS Faraday Trans. I* **68**, 1312.
67. E. PELIZZETTI, E. PRAMARRO and D. CROCE (1980) *Ber. Bunsenges. Phys. Chem.* **84**, 265.
68. M. ALMGREN, F. GRIESER and J. K. THOMAS (1979) *J. Amer. Chem. Soc.* **101**, 279.
69. N. J. TURRO. and M. AIKAWA (1980) *J. Amer. Chem. Soc.* **102**, 4866.
70. P. MASSINI and G. VOORN (1968) *Biochim. Biophys. Acta.* **153**, 589.
71. J. R. DARWENT (1980) *JCS Chem. Comm.* 805.
72. P. A. BRUGGER and M. GRÄTZEL (1980) *J. Amer. Chem. Soc.* **102**, 2461.
73. M-P PILENL and M. GRÄTZEL (1980) *J. Phys. Chem.* **84**, 2402.
74. S. M. B. COSTA and A. L. MACANITA (1980) *J. Phys. Chem.* **84**, 2408.
75. A. V. SAPRE, K. V. S. RAMA RAO and K. N. RAO (1980) *J. Phys. Chem.* **84**, 2281.
76. N. ROESSLER and T. WOLFF (1980) *Photochem. Photobiol.* **31**, 543.
77. M. W. GEIGER and N. TURRO (1975) *Photochem. Photobiol.* **22**, 273.
78. R. R. HAUTALA, N. E. SHORE and N. J. TURRO (1973) *J. Am. Chem. Soc.* **95**, 5508.
79. N. J. TURRO, K-C LIU, M. F. CHOW and P. LEE (1978) *Photochem. Photobiol.* **27**, 523.
80. I. B. C. MATHESON and A. D. KING (1978) *J. Colloid Interfac. Sci.* **66**, 464.
81. B. A. LINDIG and M. A. J. RODGERS (1979) *J. Phys. Chem.* **83**, 1683.
82. J. T. H. ONG and H. B. KOSTENBAUDER (1976) *J. Pharm. Sci.* **65**, 1713.
83. G. L. FLYNN and D. J. LAMB (1970) *J. Pharm. Sci.* **59**, 1433.
84. S. RIEGELMAN (1960) *J. Amer. Pharm. Assoc. (Sci. Ed)* **49**, 339.
85. J. L. KURZ (1962) *J. Phys. Chem.* **66**, 2239.
86. A. D. JAMES and B. H. ROBINSON (1978) *JCS Faraday Trans. I.* **74**, 10.
87. J. HOLZWARTH, W. KNOCHE and B. H. ROBINSON (1978) *Ber. Bunsenges Phys. Chem.* **82**, 1001.
88. K. J. MYSELS and A. T. FLORENCE (1968) in *Clean Surfaces* (ed. G. Goldfinger), Marcel Dekker, New York.

89. M. FISCHER, W. KNOCHE, B. H. ROBINSON and J. H. M. WEDDERBURN (1979) *JCS Faraday Trans. I.* **75**, 119.
90. T. SOTOBAYASHI, T. SUZUKI and K. YAMADA (1976) *Chem. Lett.* 77.
91. I. M. PANAYOTOV, D. T. PETROVA and C. B. TSVETANOV (1975) *Makromol. Chem.* **176**, 815.
92. S. YANAGIDA, K. TAKAHASHI and M. OKAHARA (1977) *Bull. Chem. Soc. Jap.* **50**, 1386.
93. A. KITAHARA (1980) *Adv. Colloid Interfac. Sci.* **12**, 109.
94. J. H. FENDLER (1976) *Acc. Chem. Res.* **9**, 153.
95. M. WONG, J. K. THOMAS and T. NOVAK (1977) *J. Amer. Chem. Soc.* **99**, 4730.
96. M. WONG, M. GRÄTZEL and J. K. THOMAS (1975) *Chem. Phys. Letters* **30**, 329.
97. M. WONG, J. K. THOMAS and M. GRÄTZEL (1976) *J. Amer. Chem. Soc.* **98**, 2391.
98. D. J. MILLER, U. K. A. KLEIN and M. HAUSER (1977) *JCS Faraday Trans I* **73**, 1654.
99. U. K. A. KLEIN, D. J. MILLER and M. HAUSER (1976) *Spectrochim Acta* **324**, 379.
100. F. M. MENGER, J. A. DONOHUE and R. F. WILLIAMS (1973) *J. Amer. Chem. Soc.* **95**, 286.
101. K. MARTINEK, A. V. LEVASHOV, N. L. KLYACHKO and I. V. BEREZIN (1977) *Dokl. Akad. Nauk SSSR* **236**, 920.
102. J. H. FENDLER, F. NOME and H. C. VAN WOERT (1974) *J. Amer. Chem. Soc.* **96**, 6745.
103. J. H. FENDLER, E. J. FENDLER, R. T. MEDARY and V. A. WOODS (1972) *J. Amer. Chem. Soc.*, **94**, 7288.
104. J. H. FENDLER, E. J. FENDLER and S. A. CHANG (1973) *J. Amer. Chem. Soc.* **95**, 3273.
105. C. J. O'CONNOR, E. J. FENDLER and J. h. FENDLER (1973) *J. Amer. Chem. Soc.* **95**, 600.
106. C. J. O'CONNOR, E. J. FENDLER and J. H. FENDLER (1974) *J. Amer. Chem. Soc.* **96**, 370.
107. C. J. O'CONNOR and R. E. RAMAGE (1980) *Aust. J. Chem.* **33**, 757.
108. C. J. O'CONNOR and R. E. RAMAGE (1980) *Aust. J. Chem.* **33**, 771.
109. C. J. O'CONNOR and R. E. RAMAGE (1980) *Aust. J. Chem.* **33**, 779.
110. C. J. O'CONNOR and R. E. RAMAGE (1980) *Aust. J. Chem.* **33**, 1301.
111. R. C. WEAST (1974) *Handbook of Chemistry and Physics* 55th edn, Chemical Rubber Co., Cleveland, Ohio.
112. J. H. FENDLER, E. J. FENDLER, R. T. MEDARY and O. A. EL SEOUD (1973) *JCS Faraday Trans. I* **169**, 280.
113. A. KITAHARA (1970) in *Cationic Surfactants* (ed. E. Jungermann) Marcel Dekker, New York, p. 289.
114. Y.-C. JEAN and H. J. ACHE (1978) *J. Amer. Chem. Soc.* **100**, 6320.
115. O. A. EL SEOUD, A. MARTINS, L. P. BARBUR, M. J. DA SILVA and V. ALDRIGUE (1977) *J. Chem. Soc., Perkin Trans* 2 1674.
116. C. J. O'CONNOR, E. J. FENDLER and J. H. FENDLER (1973) *J. Org. Chem.* **38**, 3371.
117. N. MIYOSHI and G. TOMITA (1980) *Z. Naturförsch* **35b**, 736.
118. G. D. CORRELL, R. N. CHESER, F. NOME and J. H. FENDLER (1972) *J. Amer. Chem. Soc.* **94**, 7288.
119. N. MIYOSHI and G. TOMITA (1980) *Z. Naturförsch* **35b**, 731.
120. C. BALNY, E. KEH and P. DOUZOU (1978) *Biochem. Soc. Trans.* **6**, 1277.
121. H. KONDO, H. YOSHINAGA and J. SUNAMOTO (1980) *Chem. Lett.* 973.
122. B. H. ROBINSON, A. D. JAMES and D. C. STEYTLER (1978) *Protons and Ions Involved in Fast Dynamic Phenomena*, Elsevier, Amsterdam pp. 287 *et. seq.*
123. B. H. ROBINSON, D. C. STEYTLER and R. D. TACK (1978) *JCS Faraday Trans I.* 41.
124. C. J. O'CONNOR, E. J. FENDLER and J. H. FENDLER (1974) *JCS Dalton Trans.* 625.
125. C. J. O'CONNOR and R. E. RAMAGE (1980) *Aust. J. Chem.* **33**, 695.
126. K. ARAI, Y. OGIWARA and K. EBE (1976) *Bull. Chem. Soc. Japan* **49**, 1059.
127. O. A. EL-SEOUD (1976) *JCS Perkin Trans II* 149.
128. F. NOME, S. A. CHANG and J. H. FENDLER (1976) *JCS Faraday Trans I* 296.
129. W. HINZE and J. H. FENDLER (1975) *JCS Dalton Trans.* 238.
130. A. G. MITCHELL and J. F. BROADHEAD (1967) *J. Pharm. Sci.* **56**, 1261.
131. H. YAMADA and R. YAMAMOTO (1965) *Chem. Pharm. Bull.* **13**, 1279.
132. J. E. DAWSON, B. R. HAJRATWALA and H. TAYLOR (1977) *J. Pharm. Sci.* **66**, 1259.

133. H. KRASOWOKA (1979) *Int. J. Pharmaceutics* **4**, 89.
134. M. A. KASSEM, A. A. KASSEM and A. E. M. EL-NIMR (1971) *Die Pharmazie* **33**, 359.
135. M. I. AMIN and J. T. BRYAN (1973) *J. Pharm. Sci.* **62**, 1768.
136. H. TOMIDA, T. YOTSUYANAGI and K. IKEDA (1978) *Chem. Pharm. Bull.* **26**, 148.
137. B. C. LIPPOLD, K. THOMA and E. ULLMAN (1972) *Arch. Pharm.* **11**, 803.
138. G. G. SMITH, D. R. KENNEDY and J. G. NAIRN (1974) *J. Pharm. Sci.* **63**, 712.
139. I. A. HAMID and E. L. PARROTT (1971) *J. Pharm. Sci.* **60**, 901.
140. M. YASUHARA, F. SATO, T. KIMURA, S. MURANISHI and H. SEZAKI (1977) *J. Pharm. Pharmacol.* **29**, 638.
141. A. TSUJI, M. MATSUDA, E. MIYAMOTO and T. YAMANA (1978) *J. Pharm. Pharmacol.* **30**, 442.
142. G. FLETCHER and D. J. G. DAVIES (1974) *J. Pharm. Pharmacol.* **26**, *Suppl.* 82P and 83P.
143. A. A. R. AL-SADEN, A. T. FLORENCE and T. L. WHATELEY (1981) *Colloids and Surfaces* **2**, 49.
144. J. STAFFORD (1970) *Makromol. Chem.* **135**, 57; 71; 86; 99; 113.
145. A. HENGLEIN and T. L. PROSKE (1978) *Makromol. Chem.* **179**, 2279.
146. J. E. CARLESS and J. R. NIXON (1957) *J. Pharm. Pharmacol.* **9**, 963.
147. B. TADROS and K. LEUPIN (1965) *Pharm. Acta Helv.* **40**, 407.
148. R. HAMBURGER, E. AZAZ and M. DONBROW (1975) *Pharm. Acta Helv.* **50**, 10.
149. M. DONBROW, E. AZAZ and A. PILLERSDORF (1978) *J. Pharm. Sci.* **67**, 1676.
150. E. AZAZ, M. DONBROW and R. HAMBURGER (1973) *Pharm. J.* **211**, 15.
151. J. W. MCGINITY, J. A. HILL and A. L. LAVIA (1975) *J. Pharm. Sci.* **64**, 356.
152. E. AZAZ and R. SEGAL (1977) *J. Pharm. Pharmacol.* **29**, 322.
153. R. SEGAL, E. AZAZ and M. DONBROW (1979) *J. Pharm. Pharmacol.* **31**, 39.
154. H. W. CHANG and E. BOCK (1980) *Analyt. Biochem.* **104**, 112.
155. R. A. ANDERSON and A. E. POLACK (1968) *J. Pharm. Pharmacol.* **20**, 249.
156. H. NOGAMI, S. AWAZU and Y. KANAKUBO (1963) *Chem. Pharm. Bull.* **11**, 13.
157. J. L. KURZ (1962) *J. Phys. Chem.* **66**, 2239.
158. B. W. BARRY and E. SHOTTON (1967) *J. Pharm. Pharmacol.* **19**, 785.
159. V. A. MOTSAVAGE and H. B. KOSTENBAUDER (1967) *J. Pharm. Pharmacol.* **19**, 785.
160. C. DELGADO, J. R. BERGUEIRO and M. BAO (1979) *Quim. Ind. (Madrid)* **25**, 275.
161. K. A. COOK (1979) *Water Research* **13**, 259.
162. J. R. NOOI, M. C. TESTA and S. WILLEMSE (1970) *Tenside* **7**, 61.
163. S. J. PATTERSON, C. C. SCOTT and K. B. E. TUCKER (1970) *J. Amer. Chem. Soc.* **47**, 37.
164. R. S. TOBIN, F. I. ONUSKA, B. G. BROWNLEE, D. H. J. ANTHONY and M. E. COMBA (1976) *Water Research* **10**, 529.
165. N. NARKIS and M. SCHNEIDER-ROTEL (1980) *Water Research* **14**, 1225.
166. D. M. ROHRER and D. D. WOODBRIDGE (1975) *Bull. Environ. Contam. Toxicol.* **13**, 31.
167. A. C. DUBAULT, C. CASAGRANDE and M. VEYSSIE (1975) *J. Phys. Chem.* **79**, 2254.
168. M. BREDIMAS, M. VEYSSIE, L. STRZELECKI and L. LIEBERS (1977) *Colloid Polymer Sci.* **255**, 975.
169. M. BREDIMAS, C. SAUTEREY, C. TAUPIN and M. VEYSSIE (1978) *Colloid Polymer Sci.* **256**, 459.
170. T. KUNITAKE and Y. OKAHATA (1977) *J. Amer. Chem. Soc.* **99**, 3860.
171. K. DEGUCHI and J. MINO (1978) *J. Colloid Interfac. Sci.* **65**, 155.
172. C. D. TRAN, P. L. KLAHN, A. ROMERO and J. H. FENDLER (1978) *J. Amer. Chem. Soc.* **100**, 1622.
173. T. OKAHATA and T. KUNITAKE (1979) *J. Amer. Chem. Soc.* **101**, 5231.
174. A. ROMERO, C. D. TRAN, P. L. KLAHN and J. H. FENDLER (1978) *Life Sci.* **22**, 1447.
175. E. J. R. SUDHOLTER, J. B. F. N. ENGBERTS and D. HOEKSTRA (1980) *J. Amer. Chem. Soc.* **102**, 2467.
176. H. H. HUB, B. HUPFER, H. KOCH and H. RINGSDORF (1980) *Angew. Chem. Int. Ed.* **19**, 938.
177. S. L. REGEN, B. CZECH and A. SINGH (1980) *J. Amer. Chem. Soc.* **102**, 6640.

178. V. KAMMER and H-G ELIAS (1972) *Kolloid Z. Z. Polymere* **250**, 344.
179. H-G. ELIAS (1973) *J. Macromol. Sci. (Chem).* **A7**, 601.
180. A. T. FLORENCE, A. A. AL-SADEN and T. L. WHATELEY, unpublished results.
181. V. MARTIN, H. RINGSDORF and D. THUNIG (1977) in *Polymerisation of Organised Systems* (ed. H-G Elias) Gordon & Breach, London, pp. 175ff.
182. V. A. KABANOV (1967) *Pure Appl. Chem.* **15**, 391.
183. V. A. KABANOV, T. I. PATRIKEEVA and V. A. KARGIN (1967) *J. Polymer Sci.* **C16**, 1079.
184. I. MIELKE and H. RINGSDORF (1972) *Makromol. Chem.* **153**, 307.
185. A. A. AL-SADEN, A. T. FLORENCE and T. L. WHATELEY (1980) *Int. J. Pharmaceutics* **5**, 317.
186. P. P. SPEISER (1976) *Prog. Colloid and Polymer Sci.* **59**, 48.
187. P. P. SPEISER (1979) in *Lysosomes* (eds J. T. Dingle, P. J. Jacques and I. H. Shaw) North-Holland, Amsterdam, pp. 653–68.
188. G. BIRRENBACH and P. P. SPEISER (1976) *J. Pharm. Sci.* **65**, 1763.
189. J. KREUTER (1978) *Pharm. Acta Helv.* **53**, 33.
190. P. COUVREUR, P. TOLKENS, M. ROLAND, A. TROUET and P. P. SPEISER (1977) *FEBS Letters*, **84**, 323.
191. A. J. GLAZKO, R. A. OKERHOLM and F. E. PETERSON, unpublished data given by Speiser in [187].
192. J. J. MARTY, R. C. OPPENHEIM and P. P. SPEISER (1978) *Pharm. Acta Helv.* **53**, 17.
193. J. J. MARTY and R. C. OPPENHEIM (1977) *Austral. J. Pharm. Sci.* **6**, 65.
194. H. W. MACKINNEY (1963) *S. P. E. Trans.* **3**, 71.
195. T. M. S. CHANG (1964) *Science* **146**, 524.
196. T. M. S. CHANG, F. C. MACINTOSH and S. G. MASON (1966) *Canad. J. Physiol. Pharmacol.* **44**, 115.
197. A. T. FLORENCE and A. W. JENKINS (1976) in *Microencapsulation* (ed. J. R. Nixon) Marcel Dekker, New York, pp. 39–55.
198. A. W. JENKINS and A. T. FLORENCE (1973) *J. Pharm. Pharmacol.*, **25** Suppl. 57P.
199. A. T. FLORENCE, A. W. JENKINS and A. H. LOVELESS (1974) *J. Pharm. Sci.* **63**, .
200. F. LEONARD (1970) *Adhesion in Biological Systems* Academic Press, New York, pp. 185–199.
201. A. T. FLORENCE, M. E. HAQ and J. R. JOHNSON (1976) *J. Pharm. Pharmacol.* **28**, 539.
202. A. T. FLORENCE, T. L. WHATELEY and D. A. WOOD (1979) *J. Pharm. Pharmacol.* **31**, 422.
203. R. A. MACKAY, K. LETTS and C. JONES (1977) in *Micellization, Solubilization and Microemulsions* (ed. K. L. Mittal) Vol. 2, Plenum Press, New York, 801ff.
204. R. E. BARDEN and S. L. HOLT (1979) in *Solution Chemistry of Surfactants* (ed. K. L. Mittal) Vol. 2, Plenum Press, New York, pp. 707ff.
205. S. J. GREGORITCH and J. K. THOMAS (1980) *J. Phys. Chem.* **84**, 1491.
206. J. H. FENDLER (1980) *J. Phys. Chem.* **84**, 1485.
207. S. HOLT, unpublished, through [206].
208. P. MASSINI and G. VOORN (1968) *Biochim. Biophys. Acta.* **153**, 589.
209. W. LUCK (1960) *Angewandte Chemie* **72**, 57.
209a W. LUCK (1958) *J. Soc. Dyers and Colourists* **74**, 211.
210. B. R. CRAVEN and A. DATYNER (1967) *J. Soc. Dyers Colourists* **83**, 41.
211. A. DATYNER and M. J. DELANEY (1971) *J. Soc. Dyers Colourists* **87**, 263.
212. I. D. RATTEE (1974) in *The Physical Chemistry of Dye Adsorption* (ed. M. M. Breuer) Academic Press, London, pp. 133ff.
213. T. D. TUONG and S. HAYANO (1977) *Chem. Lett.* 1323.
214. T. D. TUONG, K. OTSUKA and S. HAYANO (1977) *Chem. Lett.* 1319.
215. Y. NEMOTO and H. FUNAHASHI (1977) *J. Colloid Interfac. Sci.* **62**, 95.
216. A. DATYNER (1978) *J. Colloid Interfac. Sci.* **65**, 527.
217. A. D. JAMES, B. H. ROBINSON and N. C. WHITE (1977) *J. Colloid Interfac. Sci.* **59**, 328.
218. D. G. HERRIES, W. BISHOP and F. M. RICHARDS (1964) *J. Phys. Chem.* **68**, 1842.
219. P. PAKALNS and H. STEMAN (1976) *Water Research* **10**, 437.

220. P. PAKALNS and Y. J. FARRAR (1979) *Water Research* **13**, 987.
221. P. PAKALNS and Y. J. FARRAR (1979) *Water Research* **13**, 991.
222. P. PAKALNS and Y. J. FARRAR (1977) *Water Research* **11**, 387.
223. R. DIXON and T. CREWS (1978) *J. Anal. Toxicol.* **2**, 210.
224. R. E. WEINFELD, H. N. POSMANTER, K. C. KHOO and C. V. PUGLISI (1977) *J. Chromatogr.* **143**, 581.
225. K. A. CONNORS and M. P. WONG (1979) *J. Pharm. Sci.* **68**, 1470.
226. G. KALLAND and E. JACOBSEN (1963) *Acta Chem. Scand.* **17**, 2385.
227. W. U. MALIK, P. CHAND and S. M. SALEEM (1968) *Talanta* **15**, 133.
228. R. G. BARRADAS and F. M. KIMMERLE (1966) *J. Electroanal. Chem.* **11**, 1128.
229. J. NOVODOFF, H. L. ROSANO and H. W. HOYER (1972) *J. Colloid Interfac. Sci.* **38**, 424.
230. W. L. HINZE (1979) in *Solution Chemistry of Surfactants* (ed. K. C. Mittal) Vol. 1, Plenum Press, New York, pp. 79ff.
231. J. H. WANG (1954) *J. Amer. Chem. Soc.* **76**, 4755.
232. R. H. STOKES and R. A. ROBINSON (1965) *Electrolyte Solutions* 2nd Ed., Butterworths, London, p. 311.
233. LORD RAYLEIGH (1892) *Phil. Mag.* **34**, 481.
234. C. J. F. BÖTTCHER (1952) *Theory of Electric Polarization* Elsevier, New York, p. 419.
235. D. A. G. BRUGGEMAN (1935) *Ann. Phys.* **24**, 636.
236. P. H. ELWORTHY, A. T. FLORENCE and A. RAHMAN (1972) *J. Phys. Chem.* **76**, 1763.
237. H. PAULY and H. P. SCHWAN (1959) *Z. Naturforsch.* **146**, 125.

Index

779